The fact that bodies decay after death has concerned humans throughout the ages. Many cultures have attempted to arrest this decay, so that bodies are preserved (or mummified) in a state as near to life as possible, but spontaneously mummified bodies are also found. Mummies are being studied increasingly to answer questions about the health, social standing and beliefs of the population from whence they came, and the lessons that they have for modern populations. This authoritative reference work explores the reasons why people mummify bodies and the mechanisms by which they are preserved, details study methods and surveys the myriad examples that can be found world-wide, evaluates the use and abuse of mummified bodies throughout the ages, and how mummified remains can be conserved for the future. Lavishly illustrated, *The Scientific Study of Mummies* will become the classic work for all those interested in paleopathology, archaeology and anthropology.

ARTHUR C. AUFDERHEIDE is Professor in the Department of Pathology at the University of Minnesota Duluth. His research interests focus on the patterns of disease in ancient populations and the cultural anthropology of arctic populations. He has also written *The Cambridge Encyclopedia of Human Paleopathology* with Conrado Rodríguez-Martín (ISBN 0 521 55203 6).

THE SCIENTIFIC STUDY OF MUMMIES

Arthur C. Aufderheide, M.D.
University of Minnesota, Duluth

CAMBRIDGE
UNIVERSITY PRESS

PUBLISHED BY THE PRESS SYNDICATE OF THE UNIVERSITY OF CAMBRIDGE
The Pitt Building, Trumpington Street, Cambridge, United Kingdom

CAMBRIDGE UNIVERSITY PRESS
The Edinburgh Building, Cambridge CB2 2RU, UK
40 West 20th Street, New York, NY 10011–4211, USA
477 Williamstown Road, Port Melbourne, VIC 3207, Australia
Ruiz de Alarcón 13, 28014 Madrid, Spain
Dock House, The Waterfront, Cape Town 8001, South Africa

http://www.cambridge.org

First published 2003
Reprinted 2004

Printed in the United Kingdom at the University Press, Cambridge

Typeface Ehrhardt 9.5/13 pt *System* QuarkXPress™ [SE]

A catalogue record for this book is available from the British Library

Library of Congress Cataloguing in Publication data

Aufderheide, Arthur C.
The scientific study of mummies / Arthur C. Aufderheide.
p. cm.
Includes bibliographical references and index.
ISBN 0 521 81826 5 (hb)
1. Mummies. 2. Paleopathology. I. Title.

R134.8.A934 2002
616.07′093–dc21 2002024698

ISBN 0 521 81826 5 hardback

This book is lovingly dedicated to
Mary L. Aufderheide
who made it all possible

Contents

Contents

Contents

The plate section is between pp. 174 and 175.

Preface

As this volume goes to press, in the USA the scientific study of mummies is still an orphan discipline. It has no committed academic training programs, standing university faculty positions, students, dedicated funding sources or journals. In Europe the situation is better, but only by a small margin. Yet, in spite of this lack of structure, the last two decades have witnessed a rapidly increasing number of reports by scientists from a broad array of disciplines who have focused their expertise on specimens from ancient mummified human remains. This upwelling of interest reflects the realization that such specimens house valuable information, much of which can be extracted by appropriate application of modern laboratory instrumentation. While many of the features unique to the study of mummified tissues remain to be elucidated, enough have been defined to establish an expanding database on which hypotheses can be constructed and tested.

The field, however, still suffers serious problems that constrain its progress. The absence of consensus for a theoretical basis and mission tends to limit the coherence and directionality of its activities. Lack of a stated purpose restricts even the definition of its content and its parameters. Many authors of its articles disappear after their first or second contribution.

In brief, mummified human and animal tissues have been shown to harbor a vast body of unique biomedical and bioanthropological information that laboratory workers can recover. Interest in (indeed, enthusiasm for) working in this area of investigation is already manifest. Were the infrastructure that characterizes most modern scientific disciplines available to it, research on mummified tissues could be expected to flourish.

This publication is a very modest response to that defined need. It was conceived as an attempt to provide in a single volume much of what we now know about mummies. Space limitations soon dictated that each chapter could be viewed more realistically as an introduction to an individual aspect of mummy studies, with guidance to the sources of further information on the topic. The book highlights areas that have already given promise of research rewards, but also accents the field's "black holes" of ignorance. Hopefully the assembly of such data will fulfill a novitiate's need to gain overall perspective of the nature and current status of this field of study. As well, those seeking specific information about a feature of mummies may be guided by the textual organization buttressed by an extensive index. For the professional seeking verification or the student desiring expanded discussion of a topic herein offered, I have inserted what some may even regard as an excessive level of citations. Thus this book may be used as a basic text around which to build a curriculum, or it can serve as a research reference source.

Anthropogenic mummification is so inseparably linked to a population's ethos that, whenever the information was available, I have devoted a substantial fraction of textual space to the cultural features characterizing a group practicing mummification. The connections may not yet always be evident, but their presentation may invite future study to establish such links. In addition, all of Chapter 2 is devoted to the question of mummification's general purpose. To do less would reduce this volume to little more than a manual or gazetteer.

Like most single-author texts that attempt coverage of an entire field of study, this volume shares some limitations. The depth of coverage varies among the topics discussed, reflecting differences in my personal experience and knowledge. Some topics, e.g. Egyptian mummies, are so broad that their discussion merits an entire dedicated volume, as Ikram & Dodson (1998) have carried out so admirably. Such sections in my volume needed arbitrary selection decisions, with citations directing interested readers to alternative sources for further details.

Inevitably, individual or groups of mummies are discussed more than once, emphasizing certain of their features in one chapter and others in different chapters. In such cases the general characteristics of a group are elaborated in the chapter where it appears most appropriate (usually Chapter 4: The geography of mummies). Discussion of postmortem changes in Chapter 5 (Soft tissue taphonomy) left me with no realistic option other than to assume that the reader has at least a minimal background in chemistry.

Because this field of research employs laboratory methods, many of which are not part of all readers' experience, introduction to some of these seems appropriate. These are presented in Chapter 6 (Mummy study methodology), most of which is devoted to morphological methods including a step-by-step guide to the performance of a mummy autopsy. Again, selection for inclusion may not meet with universal reader approval. Because this field's methods frequently have been borrowed from other disciplines where they were designed for modern specimens, the potential hazards of their application to partly decayed, ancient soft tissues are emphasized.

The text is presented in the form of first person singular. The reasons for this choice are (1) much is derived from my personal experience in the dissection of more than 500 mummies, and (2) in many areas incomplete available information permits more than one possible interpretation. I have tried to be certain that the reader understands when I indulge in speculation or the topic is controversial, though it is doubtful all would agree that this effort always succeeds. A single-author book has the potential for presentation of a logical, progressive flow of information delivered with a consistent, overall perspective. I can only hope these virtues will mitigate the noted shortcomings.

The short-term future of the study of mummified remains is not at all clear. Large caches of mummified bodies are available at only a few geographical locations. A growing movement to discourage their study through legislative restrictions is approaching global distribution. Nevertheless, mummified tissues are now known to house a treasure of knowledge that can not be acquired from other stores. Once a source of knowledge has been identified, the historical record documents the ultimate futility of efforts to suppress its recovery. The rebound of knowledge acquisition that began with the Renaissance grossly exceeds what was lost during the medieval Dark Ages. I am confident that the value of information resident in mummified tissues will be recognized and its retrieval will be encouraged. I hope this small step to embody much of its present knowledge base in one volume and share it with interested workers will enhance that process.

Finally, attention needs to be called to the importance of illustrations. Much of the scientific study of mummies is morphologically based. The detailed presentation of guidance for a mummy autopsy performance clearly calls for illustrative support. Many of the soft tissue changes in desiccated mummies have no precedence in modern hospital autopsies and require illustration. The literature reveals an almost irresistible impulse for authors to select for illustrations those mummies having the most life-like appearance. It is important for the serious student to become aware of the range of departure from such an ideal that is normally encountered in mummy studies. With the demonstration that unique, recoverable information is resident in mummified tissue and that it can be integrated with the general biomedical and bioanthropoloical database, the scientific study of mummies appears to have come of age. I would feel rewarded most generously if, subsequently, the formal structure of a scientific discipline would be constructed upon the compilation of the field's present knowledge base presented in this book.

Acknowledgments

Principal among the many to whom I am indebted for the contents of this book is my life's partner, Mary L. Aufderheide. Uncomplainingly she shared the grime of the excavation, the scent of death, the desert's plague of flies, all the while employing her personal warmth and language skills to open the doors for access to mummies around the globe. I also owe a particular obligation to Marvin J. Allison, Ph.D., of the Department of Pathology at the Medical College of Virginia, who not only expended extraordinary effort to provide me with opportunities for study and experience with mummies, but also taught me the inestimable value of integrating my findings with the databases in archaeology, ethnology, anthropology and history. Calogero Santoro and the other members of the San Miguel Museum staff in Arica, Chile, enthusiastically responded to my interest in mummy studies there. Without the enthusiastic support of George (Rip) Rapp and Stanley Aschenbrenner, my Paleobiology Laboratory may never have been initiated. It was Michael Zimmerman's invitation to participate in a mummy dissection that lured me into the field of mummy studies. Patrick Horne's familiarity with the literature and his eagerness to share it with me enriched the coverage of innumerable topics in this book to the point almost deserving coauthorship in some sections. It was Gino Fornaciari who organized our initial "mummy hunt" in Europe in 1983 that proved to be such a profoundly fruitful mummy experience. Our Colombian friends Professor Gonzalo Correal and Felipe Cárdenas-Arroyo provided us with invaluable knowledge and access to that country's mummies. The opportunity to participate in the massive excavation of the Chiribaya site in southern Peru directed by Jane Buikstra, Ph.D., was a special privilege. The generous sharing of his assistance and knowledge by Bernardo Arriaza, Ph.D., was and still is priceless. Bob Brier provided many time-saving suggestions and eclectic mummy data sources. Paul Sledzik of the National Museum of Health and Medicine catalyzed access to many of the museum's specimens and photos. The opportunity of a Guanche mummy study arranged by Rafael González and Conrado Rodríguez-Martín in Tenerife, Canary Islands, was a momentous experience and led to lifelong Spanish friendships and even coauthorships. Doctors Tony Mills, J.L. Molto and Peter Sheldrick made possible a field study of Egypt's Dakhleh Oasis mummies, while Mario Rivera and Cora Moragas enabled examination of mummies of Iquique and Pisagua in Chile. Professor Wang Bing Hua (Director, Xinjiang Archeological Research Institute) introduced us to the Xinjiang mummies. Other helpers included Julie Moller, M.D. (translation), Johan Reinhard, Constanza Ceruti, Mario Castro, and Juan Schobinger (mountain-top mummies), Jon Kramer (mammoth), Ezio Fulcheri and Francisco Etxeberria (mummified saints), Sonja Jerkic (Beothuk mummy), Juan Rofes (guinea pig mummies), Jane Wheeler (llama mummies), Karen Stothert (Inca mummy), Ronald Wade and Pat Remler ("sugar mummies"), Gabrielle Weiss and Ildikó Cazan (overmodelled skull), Egypt Exploration Society (photos from *Journal of Egyptian Archaeology*), Bernd Herrmann and Marie Gaida (Chavín mummy), Chauncey Loomis (Charles Francis Hall mummy), I. Morimoto (Japanese mummies), Nguyen Lan Cuong (Vietnam mummies), Cristina Ara de Mackey (Eva Perón mummy), John Tygart (Guanajuato mummies), James Ramsey (Hamrick mummies), Michael Zimmerman and T.A. Reyman (photomicrographs), Ekkehard Kleiss (Vienna mummies), Zhang Yang (Chinese mummies), the Greenland Museum (Greenland mummies), Swedish Society for Anthropology and Geography (Andrée mummies), Wellcome Institute Library (St Blaise caricature), National Museum of Denmark, National Museum of Ireland, Drents Museum, British Museum, Mosegård Museum (bog body photos), Instituto Nacional de Antropología y Historia de Mexico (jade suits), Luigi Capasso, Konrad Spindler (Iceman), Royal Collection Enterprises (Leonardo drawing), Victor Mair (Xinjiang mummy), Arie Nissenbaum (Dead Sea asphalt), William Pestle and Nina Cummings (Field Museum Egyptian photos), Donna Anstey, Yale University Press (divining tablet), Miriam Mandelbaum, New York Academy of Medicine (Vesalius drawing), Mauritshuis den Haag (Rembrandt painting), Rosalie David (Manchester Museum), Neale W. Watson (Estes diagram), Owen Beattie (Franklin Expedition), Derek Notman (X-ray unit), Rev. Padre Superiore della Confraternita de Cappuccini (Palermo), American Museum of Natural History (Jivaro heads and Copper Man), Corky Ra (Summum), State Historical Society of Wisconsin (embalmed twins), Library of Congress (Civil War embalming), E.C. Johnson (embalming history), Barbara Knowles and Christine Jeans (collagen drawing), US Naval Historical Center (John Paul Jones mummy), Anya Lukasewycz and Saras Aturaliya (taphonomy), Otto Appenzeller (Neurology), G.J.R. Maat (sickle cell scanning electron micrograph), University of Kansas Museum of Natural History (Comanche photos), École Nationale Vétérinaire d'Alfort (Fragonard mummy), Tony K. Meunier (seal mummy), Rose Tyson (Museum of Man mummy collection), Rosa Boano (photomicrographs), Shin Maekawa (Getty Conservation Institute), Larry Cartmell (Martindale mummy), Phyllis Janik (poem), Royal Pharmaceutical Society of Great Britain (mummy jar), Glen Doran (mummy brain), Sonja Guillén (Chinchorros), Sylvia Horwitz (Pompeii mummies), John Varano (Peruvian mummies), Carter Lupton (Egyptian mummies), JoAnn Wallgren (laboratory analyses), Conrado Rodríguez-Martín (almost everything), Sandi Irvine (copy-editor) and many others. Acquisition of the illustrations was made possible by the generosity of Rondi Erickson and Guilford Lewis. The quality of the illustrations is the product of dedicated application of the photographic and computer graphics genius of Dan Schlies and Mark Summers. To all of these I am deeply indebted. It was, however, Sara Hammer whose total and selfless dedication to this and my other mummy-related projects that made possible not only the secretarial services, but also the implementation of logistics and my many ideas that required contact with often foreign personnel and agencies.

Photographic credits are acknowledged in the illustration legends. ACA means photo by Arthur C. Aufderheide; PBL means Paleobiology Laboratory, University of Minnesota, Duluth School of Medicine, Duluth, MN; NMHM, means National Museum of Health and Medicine, Washington, DC. Also in the legends, SM means spontaneously mummified and AM anthropogenically mummified.

Chapter 1. History of mummy studies

Introduction

A book about mummies needs to begin by defining the term "mummy". Historically the term is a misnomer. The English word "mummy" has taken a circuitous course to reach its present meaning. Pursuit of this serpentine path exposes us to several fascinating eras of ancient history. One of these is detailed in Chapter 10 (see Mummy as a drug) but can be summarized here. As early as the first century AD the Roman historian Pliny the Elder (AD 23–78) was lauding the medical virtues of a black, tarry material oozing spontaneously from earth fissures in several locations of what we now call the Middle East. Especially popular was a site in modern Iraq (ancient Persia) near Babylon in the territory of Darábgerd. This material, which we would call asphalt or bitumen, was called *múmiyá* by the Persians (perhaps because its consistency resembled that of wax, called *mûm* by the Arabs) (Pettigrew, 1834:1). When its medical popularity exceeded its supply in about the thirteenth century, the black, crystalline resin found in ancient, mummified Egyptian bodies was substituted for the bitumen, since its gross appearance suggested chemical identity. The term *múmiyá* was then transferred to such resins (Dawson, 1927f). Still later, when the resins' apparent superior clinical efficacy was attributed by European physicians to the desiccated, light-brown muscle fragments often accidentally included within the mummies' resin, the term was again transferred – this time to the preserved nonskeletal (soft) tissues of the human body that were thereafter employed as medical therapeutic agents. When that practice subsided in the eighteenth century the term finally evolved into its present application, referring to any form of mummified human remains.

In adjectival form the term "mummified human" can be applied conveniently to even the smallest nonskeletal fragment of a human body surviving a postmortem interval long enough normally to anticipate complete decay. Few, however, would use the noun "mummy" to designate an isolated fingernail, lock of hair or dry skin fragment. This indicates that, for most people "mummy" induces a mental image of a corpse with soft tissues sufficiently preserved to resemble a once-living person. This is far from the type of precision we normally accept within scientific terminology, but if language is to be an effective communication medium it must conform to popular usage. Hence, this book will employ the term "mummy" in this general sense, modifying it with descriptive adjectives where necessary.

Justification of studies

But why study mummies at all? The tombs of the world's largest and most examined group of mummies in Egypt have contributed much information useful to recovery of Egyptian views of life and the afterworld through study of inscriptions on the tomb walls, on the mummy's container (cartonnage, coffin or sarcophagus) or on linen wrappings and included papyrus scrolls. The principal focus of interest on the body itself, however, has been the documentation of specific features of the mummification process. These would then be used to establish a chronological sequence of mummification features that can be used for dating purposes in much the same way that the form, composition, and decoration of ceramic features are commonly employed in archaeology. Even the evidence of disease occurring in Egyptian mummies has been derived largely on the basis of simple gross observations of skeletal and some soft tissue malformations, commonly published as isolated cases or simply as addenda to archaeological reports.

The above only hint at the vast residue of latent information that is resident in both the soft and skeletal tissues in these bodies. But how available is this information, and what needs to be done to obtain it? Some of the answers to these questions were provided at several pioneering seminars by the University of Manchester, England (David, 1979, 1986), the mummy dissections by members of the Paleopathology Club and the Paleopathology Association, and especially in the 151 manuscripts read at the First World Congress on Mummy Studies held in Puerto de la Cruz on Tenerife in the Canary Islands on February 3–6, 1992 (Archaeological and Ethnographical Museum of Tenerife, 1995). The more than 300 scientists gathered at that Congress described the application of a broad range of biochemical, biophysical, radiographical, radiometric, photo-imaging, molecular biological and other techniques to spontaneously ("naturally") and anthropogenically ("artificially") preserved human remains from sites on nearly every continent. The wealth of new data presented at that meeting represents a quantum leap forward in the study of ancient populations. Above all, it brought into focus the existence of the enormous legacy of biomedical and sociocultural information present in mummified human remains. Even extant technology is capable of extracting a significant fraction of it. Such recovered information can be integrated with the existing biomedical and bioarchaeological database; these data are unique because they can not be acquired by other means.

This is the very definition of a new branch of science. Because science is the pursuit of truth, it is its own justification. The term *science* is used here to include all forms of scholarly endeavor including, but not limited to, history, archaeology, anthropology, ethnology, art, biology, museology and medicine. The purpose of this book is to document the field's current database as well as identification of the nature and potential of methods employable to expand it.

Ethical aspects of mummy studies

A growing world-wide movement emphasizing ethnic identity is affecting the study of human remains. In some areas ethnoconsciousness is so powerful that it overrides nationalistic concerns. The roots of these forces are so deep and complex they may be unknowable in their entirety. They surely, however, include such factors as disappointment over the failure of both science and many experimental forms of population governance systems to solve basic social problems. They also involve the perception of anonymity of population subgroups in increasingly densely populated areas, and the extension of communication systems that permit the people of even the remotest part of our planet to be aware of conditions elsewhere. The effect of these and other contributing factors is a search for identity and socioeconomic solutions among population subgroups sharing common cultural heritage practices and world views as well as language, and, in some cases, religious hallmarks of ethnic identity. Ironically, this process of shrinking one's sociocultural space coincides with the increasing awareness by large nation states of their global interactions and interdependence.

This process has had a profound impact on the study of antiquity through examination of human remains. One aspect of the several, different responses to these changes has been increasing interest in one's ancestral origins, and is reflected not only in a flurry of related publications (Haley, 1976) but also in grass roots efforts to recover the features of the life of aboriginal populations (Tenerife's 1990 Cronos Project, others). Simultaneously such growing identification with ancestors has also extended to their physical remains in many areas, leading to restriction of their exhumation and study by dissection or other methods that are viewed as offensive and irreverent. Australia, USA, Israel and other countries have regulated such practices, and increasing minority militancy is making similar demands in parts of South America, the Canadian Arctic and elsewhere. Understanding of such ethical concerns is imperative for the success of a scientist involved in mummy studies.

Because it is within this global climate that we must carry out our work, ethical concerns need to be dominant in the minds of scientists involved in the dissection or sampling of mummified human remains. Investigators of mummy studies will be most comfortable in subsequent work if they first evaluate their own views by any criteria or features perceived to be relevant and arrive at a comfortable conviction. It may be necessary to participate in one or more dissections to achieve the experience necessary to make this informed decision. The next, equally important step is the recognition that all others concerned with a possible mummy dissection have an equal right to reach other, but equally valid, viewpoints. In the perfect world those influencing the decisions regarding the extent of dissection (museum board, institutional administrators, legislators, etc.) would make the same effort to be as informed as possible when making such judgments. Since this seldom occurs spontaneously, it is in the investigator's interest to prepare such a report or oral presentation. In such circumstances, a highly *persuasive* stance by the investigator is rarely successful. Usually more useful is a presentation identifying the nature of the proposed study, the types and value of information expected to be gained and possible alternative limitations of the extent or modifications of the dissection. Risks, inconveniences or costs as well as possible museum applications of the data generated need to be pointed out. Optimally, all this can be carried out in a manner transmitting the investigator's recognition of normal concerns about dissection and respect for the ultimate decision. Once the decision has been made, all concerned with the examination must honor whatever restrictions are imposed without manifesting judgment.

The degree of survivors' identification with the physical bodies of deceased relatives varies enormously both between and within cultural groups. I have encountered the full range of concerns about the extent of dissection, varying from total indifference to complete prohibition. Furthermore, if the decision-makers can be fully informed, the degree of dissection limitation is not necessarily correlated with emotional concerns for the remains. In one such study in which I participated, the controlling group's members' sincere and deepest reverence for the remains was obvious, yet after becoming fully informed they imposed as their only restriction the prohibition of public display and the prompt return of the remains after study completion.

This situation is not very apt to change in the near future. Hopefully the information in this book will help both scientists and those controlling access to mummified human remains to reach a level of understanding that will permit both to be comfortable with such studies. Human behavior is best understood when it is studied in the context in which it occurred. It seems appropriate, therefore, that we begin our process of understanding this new scientific discipline by reviewing its history.

The Archaic Period

The study of mummified human remains is an outgrowth of the study of the bodies of recently deceased humans. Not surprisingly, then, the history of human mummy dissections is an inseparable part of the pursuit of human anatomical knowledge. We can begin our review, therefore, with Preceramic hunter–gatherer groups. Certainly, early hunting groups dissected their prey both as part of the butchering process

Fig. 1.1. Clay model of sheep liver.
Cuneiform inscriptions on this clay model of a sheep liver suggest its employment for divination purposes. (Courtesy of author Albrecht Goetze, 1947: Plate CXXXIV, and Yale University Press (copyright).)

and also for magicoreligious purposes such as sacrifice or divination. One might anticipate that, by opening animal or human bodies, these early populations would have gained anatomical knowledge from such rituals. Archaeological findings at Mesopotamian sites have uncovered clay models of animal livers that presumably were employed as instructional devices to teach neophyte clerics the interpretation of changes useful for the translation of divine messages (Goetze, 1947; Mujais, 1999) (Fig. 1.1). A comparable bronze model of a sheep's liver with inscribed subdivisions found in Italy apparently served a similar divination purpose for Etruscans (Lyons & Petricelli, 1987:67, 232). Both ethnographical and archaeological data indicate that examination of intestines extracted from human sacrifice probably was based on a similar divination principle (haruspicy) among Pre-Hispanic Peruvian populations (Allison *et al.*, 1974c). Guides to the Hebrew practice of examination of animals and meat in the Babylonian Talmud of the second century AD include reference to several specific visceral structures (Garrison, 1929). However, as Ackerknecht (1943) points out, there is no evidence that any such dissections led to an understanding of detailed human anatomy, the functional interaction of various body tissues or any other useful anatomical information beyond that needed to satisfy the requirements for the ritual purposes of such dissections. Benedict (1934) has established a theoretical basis for support of a concept that "only in the context of a culture pattern oriented towards a kind of 'science' do dissections furnish anatomical knowledge . . .".

Classical Greek and Roman periods

Flickers of the beginning of such a scientific movement about 500 BC can be read into the animal dissections carried out by the southern Italian physician Alcmaeon of Crotona. In his book *Concerning Nature* detailing his dissections, Alcmaeon made it clear that his goal was a better understanding of human anatomy that would be helpful in his medical practice. Lyons & Petrucelli (1987:192, 399) consider him to be the first medical scientist, and note also that a subsequent Hippocratic publication about shoulder anatomy most likely was the product of a human dissection. The intellectualism of classical Greece, however, did not include human dissection. Thus it is not surprising that the commitment to a search for truth that formed the basis for the existence of the Alexandrian museum and library would lead men such as Herophilus of Chalcedon and Eristratus of Chios to base their learning of human anatomy on human dissections there during the third century BC. Tragically, the prohibition of human dissection by Rome in 150 BC arrested this progress and few of their findings survived. By the second century AD, Galen, the Greek physician who practiced principally in Rome, popularized the humoral theory of health and disease with the aid of only two human skeletons and animal dissections but without laboratory-based opportunities for human soft (nonskeletal) tissue dissection to guide him (Long, 1965:9–11; Lyons & Petrucelli, 1987:399; Porter, 1996:60).

The Dark Ages

Rome's fall initiated the Dark Ages, during which Galenic concepts were perpetuated but not tested (Fig. 1.2). While during this period the Church did not forbid human dissections in general, certain edicts were directed at specific practices. These included the *Ecclesia Abhorret a Sanguine* in 1163 by the Council of Tours and Pope Boniface VIII's command to terminate the practice of dismemberment of slain crusaders' bodies and boiling the parts to enable defleshing for return of their bones. Such proclamations were commonly misinterpreted as a ban on all dissection of either living persons or cadavers (Rogers & Waldron, 1986), and progress in anatomical knowledge by human dissection did not thrive in that intellectual climate.

Fig. 1.2. The medieval teaching of Galenic concepts. This fourteenth century painting pictures Galen as a feudal lord surrounded by his courtiers, reflecting the dominance of his anatomical and medical ideas during the Dark Ages. (Courtesy of Masters and Fellows of Trinity College Cambridge (Trinity MS 0.2.48, f.65a). Porter, 1996: 64.)

The Renaissance

The sweeping changes of the Renaissance involved not only the arts but also the sciences, including the construction of European universities. The revival of

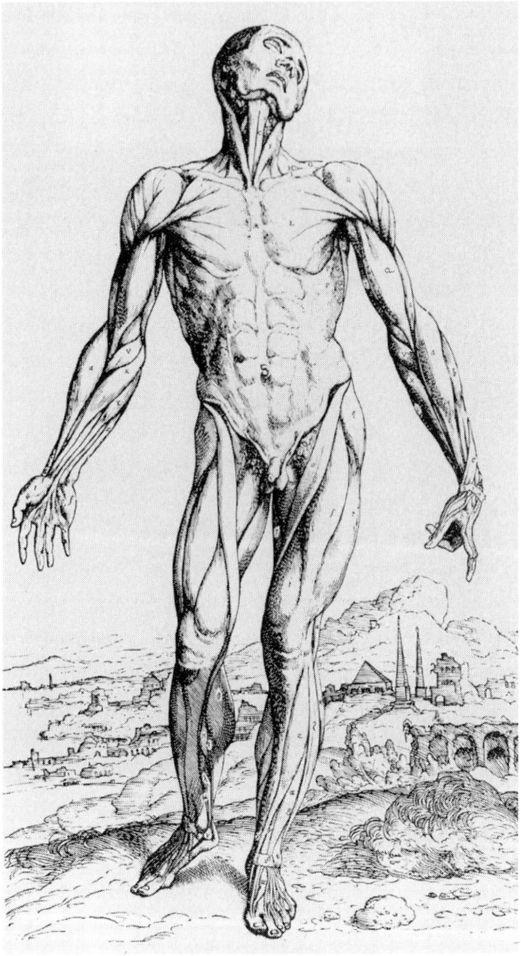

Fig. 1.3. Reproduction of an illustration from the sixteenth century anatomy text published by Andreas Vesalius.
Vesalius' 1543 text dethroned Galen by establishing that the ultimate authority of anatomical knowledge was the anatomical features of the human body itself as revealed by dissection. (From illustration Primae Muscolorum Tabulae, p. 170 of his 1543 book titled *De Humani Corporis Fabrica Libri Septem.* Published by Johannes Oporinus, Basel. Courtesy of New York Academy of Medicine Library.)

human dissection became part of this movement. Soon, however, dissections were performed for reasons much more proximate to those characterizing modern mummy dissections. The aim of this practice gradually shifted from expanding knowledge of normal anatomy to the understanding of disease through study of pathological anatomy. One of the earliest postmortem dissections carried out for the specific purpose of identifying the cause and mechanism of death was the autopsy of a plague victim by Fra Salimbene di Adam at an Italian monastery in Cremona in 1286 (Rogers & Waldron, 1986). It was, however, the newly established universities at Bologna and Padua that pioneered the routine performance of autopsies. In 1302 Bartolome da Varignana carried out medicolegal autopsies at Bologna, while Gentile da Foligno held regular public human dissections at Padua University by 1341. This practice proved so popular that it was imitated and institutionalized by most major universities during the subsequent century (Garrison, 1929). It was the Padua graduate Andreas Vesalius (1514–1564) (Fig. 1.3) who maximized this trend, organized his dissections in a focused manner and challenged the many anatomical errors of Galenic anatomy when he published his richly illustrated book, *De Humani Corporis Fabrica* in 1543 (Lyons & Petrucelli, 1987:416). Thereafter it became clear that the final arbiter of an anatomical controversy would not be the opinion of a well-known physician but the body itself as revealed by dissection. Many hours of observing human dissections carried out by anatomists enabled Michelangelo to paint the muscled physiognomy of the Sistine Chapel's figures, while Leonardo da Vinci's numerous personally conducted dissections of human bodies helped him to execute his dramatic anatomical illustrations (Yesko, 1940) (Fig. 1.4). All of these dissections, coupled to close observations in living humans and animals led to the revolutionary definition of the circulation of blood in William Harvey's 1628 *Exercitatio anatomica de Motu Cordis et Sanguinis in Animalibus* (Porter, 1996:159).

Scientific rustlings in the eighteenth century

With human anatomy now firmly rooted in human dissections, interest became increasingly focused on pathological anatomy: the morphological changes

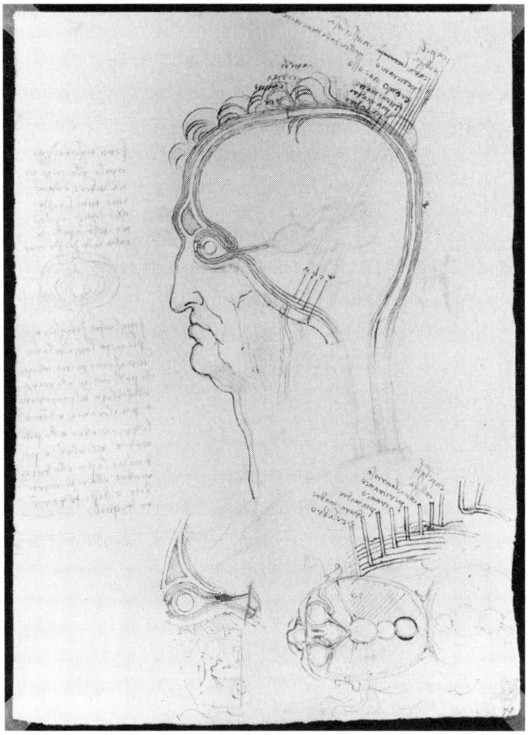

Fig. 1.4. Da Vinci anatomical sketch.
This pencil drawing by Leonardo da Vinci was clearly a sketch of a human anatomical dissection. (Photo courtesy of The Royal Collection © 2001, Her Majesty Queen Elizabeth II.)

caused by various diseases. Men such as Nicolaes Tulp, a practicing Amsterdam physician, taught anatomy by human cadaver dissection in the Dutch medical schools and both described and illustrated many pathological entities that he encountered in such dissections in a 1641 text entitled *Observationes Medicae* (Fig. 1.5). Compilations of autopsy reports for teaching purposes began to appear in the latter part of the seventeenth century and early eighteenth century, culminating eventually in the 1761 publication of the five-volume *De Sedibus et Causis Morborum* [Seats and Causes of Disease] by Battista Morgagni (1682–1771). As great as that contribution was, it was soon eclipsed by events in Vienna. There the Austrian emperor Joseph II, son of Maria Teresa, in 1784 rebuilt the Allgemeine Krankenhaus [General Hospital] into a 1600 bed hospital to serve the health of the poor (Porter, 1996:212) (Fig. 1.6). The eventual decision to autopsy all deaths at that hospital had an enormous impact on human dissection everywhere in the Western world, together with stimulating an immense increase in knowledge about both normal and pathological human anatomy (Long, 1965:102–105).

Surely there were occasional examinations carried out on mummified human bodies during the eighteenth century, most of which probably went unrecorded. Several that we know about include the opening of several Egyptian mummy bundles by Blumenbach (1794), some of which proved to be frauds (see Chapter 10: Mummy as deceptions). Granville (1825) lists a variety of sporadic studies dating back as far as 1662. Few of these, however, were significantly informative. One such study of the mummification method performed on an Egyptian mummy was documented by John Hadley (Dr William Hunter was a participant as well) as early as 1763. In a letter to a friend, Hadley lists his findings but notes that they probably differ little from those of others who had done similar examinations (Hadley, 1764).

The autopsy in the nineteenth century

It was Karl Rokitansky who took over leadership of the pathology department at Vienna's Allgemeine Krankenhaus in 1832 and turned it into an autopsy machine. In addition to all the cases other members of his staff performed, Rokitansky personally carried out or immediately supervised about 30000 autopsies during the subsequent 30 years, and created what is still the world's largest museum of pathological lesions (Long, 1965:102–106; Malkin, 1993:114). This emphasis on the value of autopsies was still in practice in 1947 when I had the privilege of studying pathology there briefly under Dr Hermann Chiari. The information generated by these autopsies (much of it included in Rokitansky's five volume *Handbook of Pathological Anatomy*) catapulted Vienna into the position of being

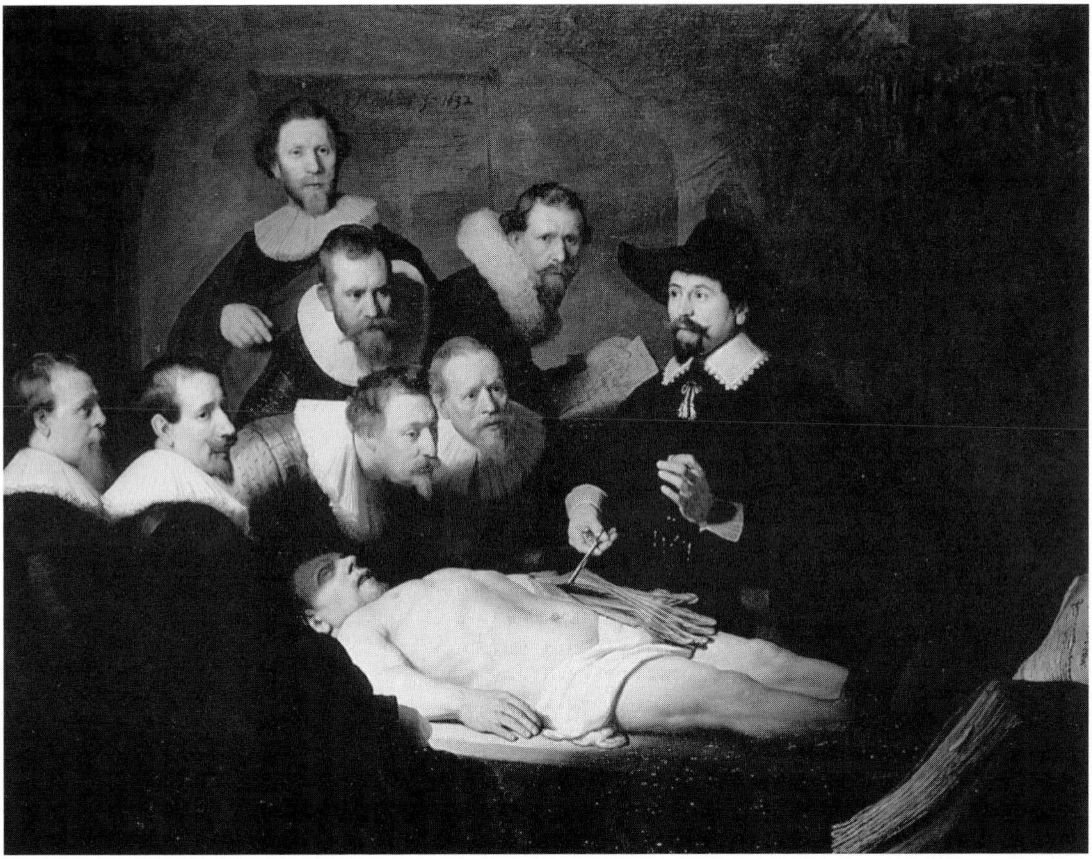

Fig. 1.5. ***A Lesson in Anatomy.***
Rembrandt painted Nicolaes Tulp (1593–1674) of Amsterdam, demonstrating anatomical features by actual dissection of a human cadaver in 1632. (Courtesy of Stichting Vrienden van Het Mauritshuis, den Haag [Royal Cabinet of Paintings, Mauritshuis, The Hague, NL].)

the major medical teaching center of the Western world. Within a few decades, however, medical leadership passed into German hands. Berlin's Rudolf Virchow, also using both the autopsy and surgical specimens, shattered Rokitansky's adherence to remnants of Galen's humoral physiology by eventually placing pathological anatomy and the understanding of human disease on the firm basis of cellular pathology, where it remains today.

The serpentine path of the development of biomedical knowledge that had been initiated in the darkness of human sacrifice was subsequently enlightened by anatomical dissections and finally burst into efflorescence with the autopsy. These established methods, so effective in generating knowledge of contemporary disease, were now ready to be applied to ancient mummified remains in an effort to understand modern disease even better by linking it to its prehistory.

Mummy dissections in the nineteenth century

Oddly, it was a military commander, Napoleon Bonaparte, who indirectly created the milieu that led to mummy studies. In a precedent-setting action Napoleon carried about a hundred scientists (from

Fig. 1.6: Vienna's Allgemeine Krankenhaus (General Hospital).
Rebuilt by Joseph II in 1784, the pathological anatomical knowledge acquired by this hospital's autopsy program propelled Vienna into world medical leadership. (Courtesy of Historiches Museum der Stadt Wien.)

what formerly had been called the French Academy of Sciences and was so renamed a few years later) with his army that invaded Egypt in 1798. These scientists were charged with drawing and recording all aspects of Egyptian society, technology and natural history. During the next two decades they published their findings in nineteen volumes (*Description de l'Égypte*), filled with fascinating hieroglyph-covered monuments, tombs, portraits and even mummies (though no dissections were performed) (Denon, 1821:219–220). These enchanting folios, together with the cracking of the hieroglyphic code by Young (Peacock, 1855) and Champollion (Davies, 1987:47–68) in 1819–1822, are principally responsible for the wave of Egyptomania that swept the Western world and generated intense interest in all aspects of Egyptian culture, including their mummies.

In such an atmosphere it was almost inevitable that hucksters would exploit the opportunity. Extracting the bodies of Egyptian mummies from European pharmacies (leftovers from the earlier days of the production of medicinal mummy powder – see Chapter 10: Mummy as a drug) or importing them directly from Egypt, these promoters carried out pseudoscientific mummy "unrollings" for paying public admissions. These were little more than entrepreneurial spectacles. Eventually, however, they evolved into serious, educational studies. As early as 1833 the Egyptologist John Davidson (cited in Sluglett, 1980) formally reported his study and findings of *Embalming and Unrolling of a Mummy*. Among others who rescued mummy studies from the frauds was Dr Thomas Pettigrew F.R.S., professor of anatomy at Charing Cross Hospital in London (Fig. 1.7). His first

Fig. 1.7. Thomas Pettigrew.
A British pioneer of mummy investigations, this physician's 1834 book (A *History of Egyptian Mummies*) was among the first to emphasize the bioanthropological information that was resident in ancient mummified remains. (From Dawson, 1934. Drawing by Charles Baugniet. Courtesy of the Egypt Exploration Society.)

unrolling of an Egyptian mummy at that hospital in 1833 was attended by many physicians as well as a few nonmedical dignitaries. During the next 18 years he examined 13 more. He established public courses of six meetings' length, in which the first five were lectures of two to three hours each on a wide variety of Egyptological aspects, the final one devoted to the unrolling of a mummy. Pettigrew did charge fees, but at that time it was standard practice for professors to charge students fees for each lecture as this was their only remuneration; they were not paid by the institution. These courses were obviously of a high quality and designed for educational purposes (Dawson, 1934). His 1834 book (A History of Egyptian Mummies) includes much of the information incorporated into such lectures and, in spite of its title, does include consideration of non-Egyptian mummies as well, such as the Canary Islands Guanches, Palermo's Capuchin monks, etc.

Another popular nineteenth century academic passion was craniometric analysis designed to differentiate the human races. In an attempt to identify the native Egyptian population's founders, such studies were also carried out early on, notably by the American anthropologist Samuel George Morton in 1844 (he is also author of *Crania Americana* published in 1839). Influenced by the reigning concept that all the world's populations were Noah's descendants, he concluded that Egyptians were Caucasian, of the branch of Ham. Among the more medically oriented studies was the dissection of an Egyptian mummy that had been purchased in Egypt, and investigated by the well-known British gynecological surgeon Augustus Bozzi Granville in 1825. His report includes a drawing (Fig. 1.8) of a large, cystic ovarian tumor he found in the mummy's abdomen (Granville, 1825). A Viennese laryngologist, Johann Czermak, carried out what probably represents the first published microscopic examination of Egyptian mummy tissues in 1852. At about this same time Samuel Birch (1850) reported on his observations of a mummy from Egypt's Late Period. In Canada, Daniel Wilson (1865) compared prehistoric American mortuary practices with those in Europe in a treatise designed to emphasize that human progress was not biologically defined for any given population but depended, instead, on environment and social learning. He described his findings when he unwrapped (but did not dissect) several spontaneously mummified bodies from the Atacama Desert near the modern city of Arica that is now in Chile but was at that time in Peru (Berger, 1990). In 1892 Germany's R.W. Schufeldt described the scientific value of studying ancient human remains and introduced the term

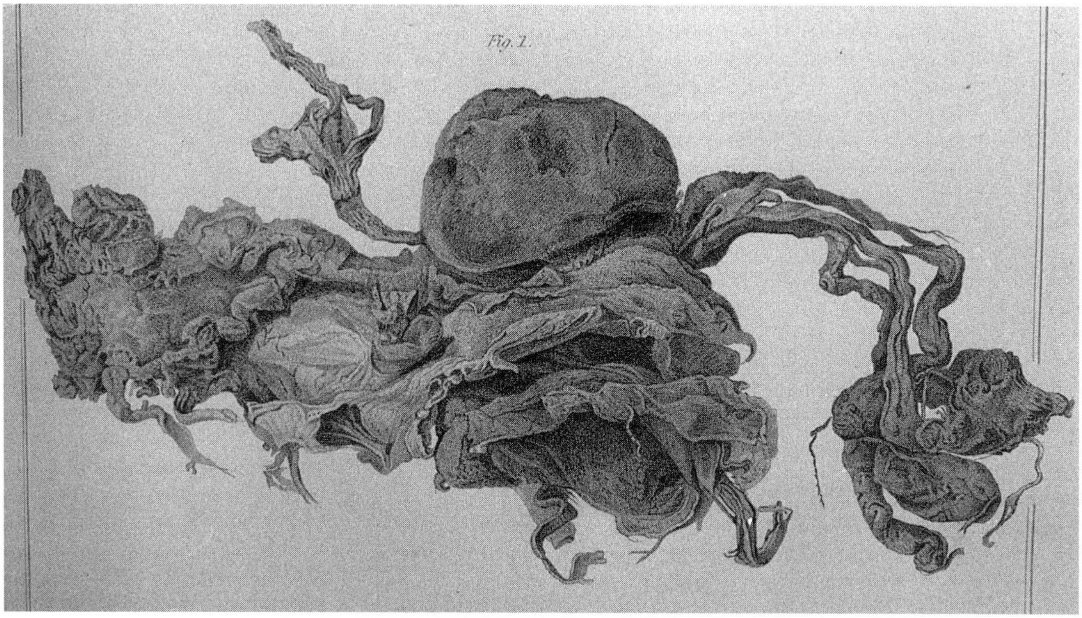

Fig. 1.8. Ovarian cyst in a mummy; AM, 1549–1069 BC.
Granville (1825) found this desiccated, large ovarian cyst in an Egyptian mummy. (Reproduced from *Philosophical Transactions of the Royal Society of London*.)

paleopathology to define the field. This period also was marked by the founding of the United States Army Medical Museum (Fig. 1.9), stimulated by the desire to understand better and improve results of limb amputation for gunshot wound injuries and other military health concerns during the American Civil War. The later collections of this institution (now the National Museum of Health and Medicine) have made major contributions to the study of mummified tissues of globally distributed provenience (Sledzik & Ousley, 1991).

In an obscure pamphlet, W. Koenig (1896) published an X-ray of an Egyptian mummy bundle of a cat about a year after Roentgen announced his discovery of X-rays (Fig. 1.10).

These scattered, uncoordinated and largely unfocused efforts reflect the beginnings of studies on mummified human remains, at least some of which, like Granville's dissection, represent application of the autopsy approach developed in the previous era as a method of obtaining medically useful information.

The twentieth century

Egyptomania of the first quarter

This is a remarkably productive period for study of mummified remains. The period began with a dermatohistological study on mummy skin (Wilder, 1904), followed by a pioneer effort by Grafton Elliot Smith applying to mummies the newly discovered technique using X-rays (see Chapter 6: Physical methods: radiographical methods), and a survey on brain mummification (Lamb, 1901). In addition, in Germany, Schmidt (1908) tried to find hemoglobin by using chemical tests on Egyptian mummy tissue. Lucas (1908) initiated the chemical studies that later became his principal contribution to Egyptology by identifying and quantifying the various salts that compose natron, the Egyptian embalmers' dehydrating agent. Shattock (1909) reported finding calcified atherosclerosis in Pharaoh Menephtah's aorta. In Munich, Meyer extracted human mummified muscle tissue, injected it into

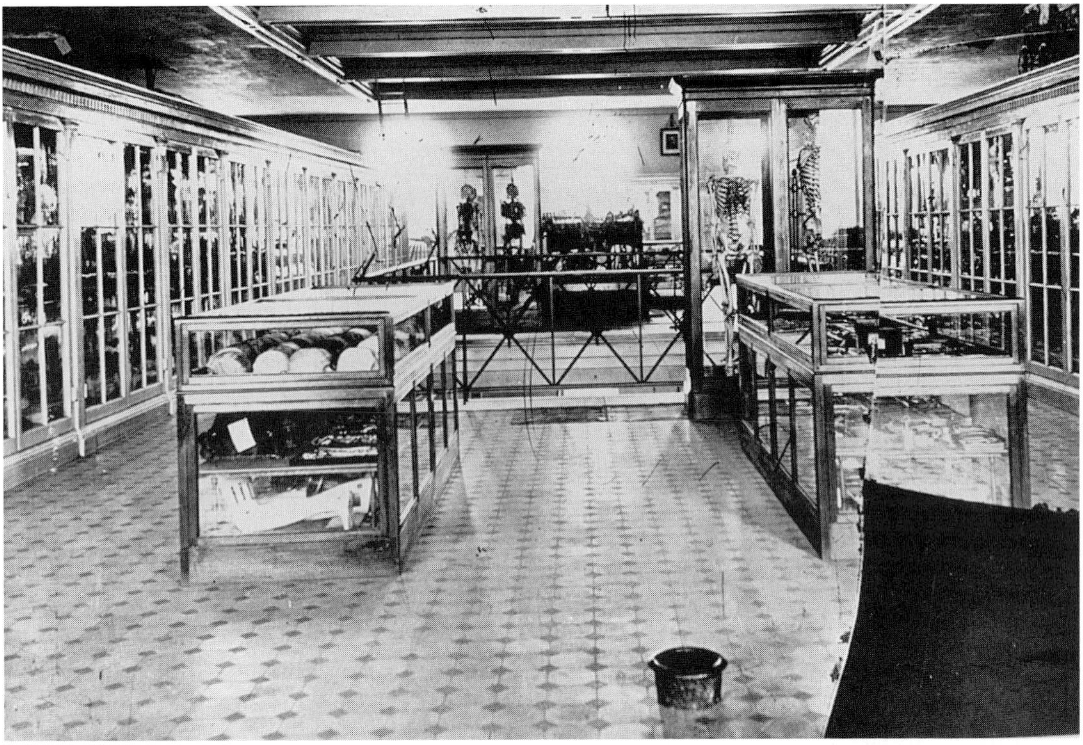

Fig. 1.9. US Army Medical Museum.
This photo of the museum reveals its appearance when it was located for eight years beginning 1967 in what is now Ford's Theatre in downtown Washington, DC. It was later transferred to the campus of the Armed Forces Institute of Pathology and is now called the National Museum of Health and Medicine. (Courtesy of Paul Sledzik.)

rabbits and used an immunological precipitin technique in an effort to develop a test for human blood identification (Meyer, 1904). But it was Marc Armand Ruffer (1859–1917) who provided the major impetus. Born in Lyons, France, of aristocratic parents he was educated at Oxford, gained a medical degree at University College London and became a clinical bacteriologist at the Pasteur Institute in Paris under its founder. While there, Ruffer came under the influence of the French physician Daniel Fouquet, who had become interested in the medical potential of studying diseases in mummies. Fouquet had traveled to Egypt and carried out an autopsy on a 1000-year-old mummy in the Cairo Museum in 1891 (Fouquet, 1897). Suffering from paralytic sequelae of diphtheria (acquired during study of a vaccine's effectiveness) Ruffer moved to Egypt to recuperate. There he became professor of bacteriology at the Cairo Medical School in 1896. Extensive excavations in Nubia (salvage archaeology in anticipation of flooding resulting from raising the Aswan Dam in 1907) by the American archaeologist George Andrew Reisner made prodigious numbers of mummies available for study. Ruffer joined Grafton Elliot Smith (Australian anatomist), F. Wood Jones, Douglas Derry and others in the investigations carried out on these bodies. Although Ruffer (1911b) employed both gross and histological methods, he is best remembered for his microscopic contributions. A flow of reports (some published posthumously) in scientific journals between 1910 and 1921 record his findings in these Nubian and Egyptian mummies. Many of the diseases that he identified in

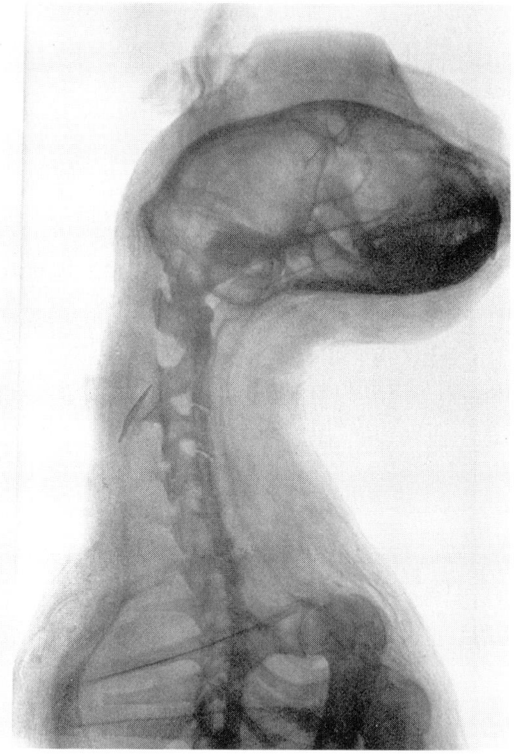

Fig. 1.10. Early X-ray of Egyptian cat mummy bundle. W. Koenig (1896) published this photo of an X-ray of an unwrapped small Egyptian mummy bundle revealing a mummified cat only about a year after Roentgen had discovered the existence of X-rays.

these mummies reflect his training in infectious diseases: schistosomiasis (Ruffer, 1920), abscesses, tuberculosis, pyorrhea (Ruffer, 1921a) and splenomegaly (malaria?), but he also recognized atherosclerosis (Ruffer, 1911b), achondroplasia and rectal prolapse, as well as a host of skeletal and dental pathological conditions (Ruffer, 1913). Ruffer's career was terminated tragically when the ship on which he was returning from a Red Cross consultation in Greece was torpedoed and sank in 1917 (Garrison, 1917). His American friend and colleague Roy Moodie collected his scattered reports and published them under Ruffer's name in book form in 1921: *Studies in the Paleopathology of Egypt*.

Grafton Elliot Smith (1887–1937) was an Australian neuroanatomist with an appointment at Cambridge University who accepted a faculty position at the Cairo Medical School 1900–1909. His skills were quickly recruited to evaluate the many skeletons and mummies that the archaeologists were excavating. Beginning with the examination of a series of mummies ranging from the Old Kingdom to the Coptic Period (*The Ancient Egyptians*, 1911a) he responded to an invitation to study all the recovered mummified bodies of the pharaohs (*The Royal Mummies*, 1912b). During those investigations he used Roentgen's newly discovered X-rays to estimate the individual's age at death by the degree of epiphyseal fusion in the mummy of Tuthmosis IV. The Egyptologist W.M.F. Petrie was among the first to use X-rays to study a human mummy in 1897 (but published in 1898), within two years after Roentgen's scientific presentation of his discovery in December, 1895. Perhaps his largest project involved the examination of Reisner's Nubian mummies (Reisner, 1910), a study carried out with F. Wood Jones (Smith, 1910).

Smith probably dissected more Egyptian mummies than any other worker had done before (and perhaps even since) his time. It is unfortunate, therefore, that we do not have the detailed record of each dissection. Together with mummies excavated by other archaeologists, Dawson (1938) estimates that Grafton Elliot Smith was responsible for the examination of 30000 mummies (obviously a number prohibitive of detailed dissection and recording of findings (Dawson, 1938). This number is rivalled only by the efforts of Rokitansky, who achieved it with hospital autopsies at Vienna's Allgemeine Krankenhaus over a longer period (30 years), with the help of his staff and with few conflicting other duties (Malkin, 1993:114). Instead, we are left primarily with Smith's summary statements as recorded in the books noted above, as well as a variety of individual articles. Sadly, these do not permit re-examination nor restudy with the synthesis of new conclusions. We need, however, to consider the enormous volume of mummies Smith examined while he carried out his other responsibilities and activities simultaneously (medical teaching, continuing his personal neuroanatomical research, buttressing his anthropological knowledge with self-study, lecturing unceasingly at international meetings and publishing a continual flow of articles and books). Hence we probably should be more grateful for, than critical of, what he has left us. Unfortunately he evolved a highly improbable concept of the global diffusion of mummification practice that originated in Egypt. He proposed that such practices spread throughout the world via marine travel by ancient populations indulging in mummification rituals (*The Migrations of Early Culture*, 1915). Defense of this concept, that was supported by only the most fragile evidence, occupied much of his later years and eroded some of his scientific credibility among his contemporaries.

The scholar pursuing the history of mummy studies inevitably will encounter Warren R. Dawson (Fig. 1.11). Born in Ealing (London) in 1888, the death of his father when Dawson was 15 years old terminated his education. Minor posts in the insurance industry eventually led him to create his own agency. The acquisition of a partner resulted in time available for self-education. Wallis Budge, Keeper of Egyptian antiquities of the British Museum, responded to Dawson's thirst for knowledge. In 1914 (at age 30) Dawson studied hieroglyph interpretation. Much of the remainder of his subsequent life became devoted to translations of papyri. These brought him into contact with renowned scientists. By cataloging their collections and publications he received fellowship appointments in the Medical Society of London, the Linnaean Society, Imperial College London and the Royal Society of Edinburgh. He became particularly interested and knowledgeable in methods of Egyptian mummification and medical practice in Egypt. He persuaded Grafton Elliot Smith to write *Egyptian Mummies* (Smith & Dawson, 1924). Though he is listed as a co-author, Dawson himself actually wrote that entire book except for the last two chapters. He made several other major contributions, and when he died in 1968 this unlettered, self-taught, amateur anthropologist left a profoundly impressive quantum of scholarly publications behind him for which the field of Egyptology and mummy studies has been exceedingly grateful (James, 1969).

Fig. 1.11: Warren Royal Dawson.
This British insurance agent-turned scholar translated many papyri, coauthored articles with Grafton Elliot Smith and recorded biographies of many early egyptologists. (Biography by James, 1969. Photo courtesy of Egypt Exploration Society.)

Another major figure in the history of mummy studies during this period was Alfred Lucas (1867–1945). Among the physical sciences, chemistry was developing as rapidly as was bacteriology in the biological sciences. Lucas was a British chemist in Egypt's Department of Survey and, after 1923, served in the Department of Antiquities there. His friend and colleague John Wilson (1964:224) describes him as highly informed in forensic chemistry including poisons and even ballistics and handwriting. Lucas' 1926 tome (*Ancient Egyptian Materials and Industries*)

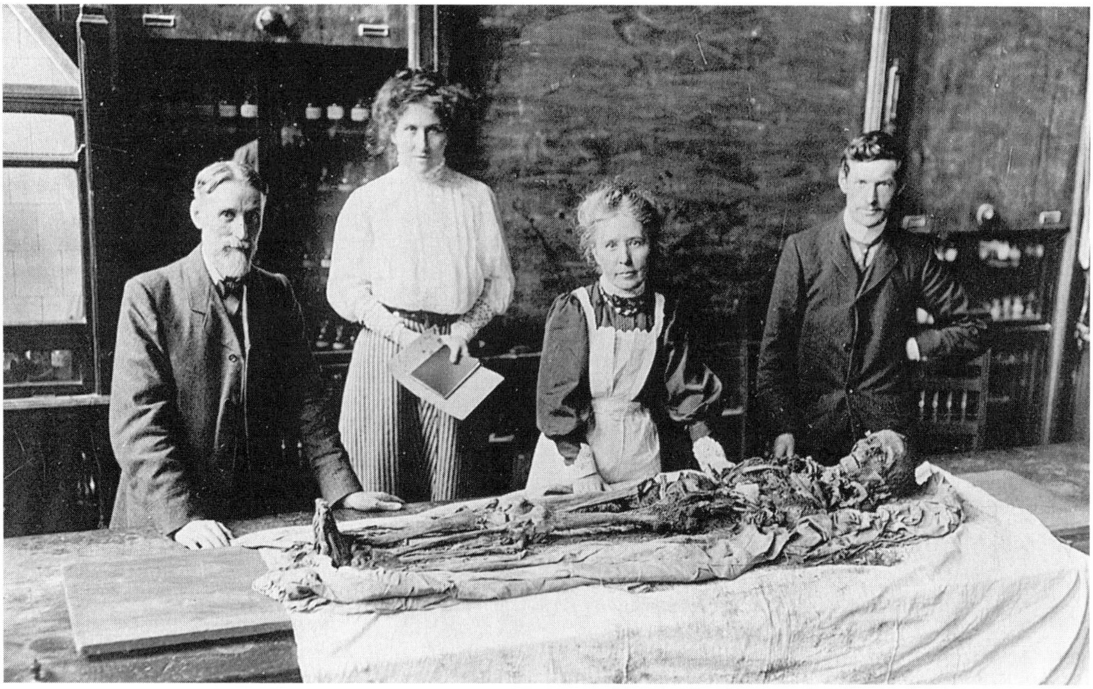

Fig. 1.12. Margaret Murray and associates.
Murray and six colleagues studied two of the Manchester Museum's Egyptian mummy collection in 1910. (Courtesy of Dr Rosalie David and the Manchester Museum, Manchester, England.)

is filled with his chemical analytical applications to materials encountered in Egyptian mummies, with an entire chapter devoted to substances employed by ancient Egyptian embalmers, including resins, Dead Sea bitumen, natron, beeswax, spices, oils, gum, henna, ointments, onions, palm wine, salt, sawdust, fats, minerals and a host of packing materials. The meticulous care he employed in his analyses and his conservative interpretations of analytical results, coupled to his knowledge of the ancient literature dealing with these substances, are constant features of his discussions. They are particularly prominent in his review of analyses relating to the possible use of Dead Sea bitumen as a substitute for resins in mummies. He was preparing for retirement when Carter found Tutankhamun's tomb in 1922, whereupon he offered his services and made valuable contributions to studies of that tomb's materials. His 1926 text was so thorough that it was reprinted in 1962 with additions and deletions by J.R. Harris. This edition remains highly useful today.

Contemporary with the early work in Egypt by Ruffer was the initiation of what would eventually evolve as an eminent research program at the Manchester Museum. This museum curates a major collection of Egyptian mummies and Margaret Murray at that institution began a study of two Egyptian mummies from a single tomb. It was unique in that the examination was interdisciplinary, involving the curator (Murray), a physician (John Cameron), three chemists (Paul Haas, H.B. Dixon, E. Linder) and two textile specialists (Thomas Fox, Julius Huebner) (Fig. 1.12). While soft tissue preservation proved to be less than desirable when the wrappings were removed, a remarkable amount of information was generated by the study (Murray, 1910). The next significant step in the evolution of that program took place in the 1970s (see below).

Finally, the discovery of the now well-known mummies of the Chinchorro people on the coast of northern Chile by the German archaeologist Max Uhle (1917) needs to be noted, even though the bulk of the studies on this group was carried out many years later (see Chapter 4: Mummies from Chile).

The irresolution of the second quarter

Oddly, the auspicious momentum generated in the field of mummy studies in the previous period by such pioneers as Ruffer, Smith and other colleagues was not maintained during the second quarter of the twentieth century. Smith died in 1937, devoting much of his final decade to defending his theory of the spread of mummification out of Egypt. Flinders Petrie transferred the British School of Archaeology out of Egypt to Palestine. The failing eyesight of Harvard's George Reisner (Fig. 1.13) restricted his field activities long before he died in 1942, and Herbert Winlock of the New York Metropolitan Museum of Art left the field to become its director in 1932 (Wilson, 1964). Termination of field work in Egypt by these giants of Egyptian archaeology was paralleled by a decrease in the scientific study of mummies. Although Ehrenberg (1927) published a thoughtful article justifying the place of paleobiology among the general biological sciences, his optimism was not realized during the two and a half decades following 1925. The exuberant 1920s were soon eclipsed by the economic depression of the 1930s, and these were succeeded immediately by World War II, which dominated the 1940s. The social, economic and military instability of this period was not conducive to expansion of the academic field of mummy studies.

Nevertheless, a few, substantive contributions from this period merit citation. Warren Dawson published "Making a mummy" in 1927, and his valuable "Bibliography of works relating to mummification in Egypt", appeared in 1929. The American pathologist H.U.Williams (1927) dissected several Peruvian mummies early in this period, identifying, among other conditions, a calcified thrombus in a leg artery. Boyd & Boyd initiated their serological studies for blood typing on mummy tissues in 1934. In Germany, Graf (1949) extracted a bioactive substance from Egyptian mummified muscle whose action suggested it was probably histamine. While the indefatigable Roy Moodie poured many dozens of publications into the literature before 1931, most of them deal with skeletal pathology (Moodie, 1923). A notable exception is his 1931 landmark radiographical study of the large mummy collection at Chicago's Field Museum (Moodie, 1931). He found a number of them were frauds whose external, wrapped, anthropoid shapes suggested that they contained children's bodies, but whose X-rays indicated that their content consisted only of an adult human leg, animal bones or sticks.

Fig. 1.13. Egyptologist George Reisner. One of the well-known excavators of the late nineteenth and early twentieth centuries, he established collections for Harvard and the Boston Museum of Fine Arts. (Photo courtesy of John Larson and the University of Chicago's Oriental Institute. Photo by Leslie Thomson, Nov. 1935. Negative no. P30077/N. 16329.)

Shaw (1938) reported a detailed microscopic study of the mummified tissues of the Egyptian mummy Harmose. Julio Tello, a Peruvian archaeologist, excavated many spontaneously mummified bodies during this period, but was especially interested in the *Treponema*-like skeletal changes (Tello, 1943). It was also during the late 1920s and 1930s that Ales Hrdlička of the Smithsonian Institution in Washington, DC, acquired a group of Aleut mummies from Kagamil Island, one of Alaska's Aleutian Islands. However, his studies of them suggested that he was more interested in their skeletal than their soft tissues (Hrdlička, 1941). It can truly be said that, with a few other isolated exceptions such as Zuki & Iskander's (1943) study of Amentefnekht's mummy, progress of the scientific study of mummies was moderated during this century's second quarter prior to its subsequent advance after 1950.

Reawakening of the third quarter

The extraordinary impetus to development of the technological sciences generated during World War II created a proliferation of new technical applications in the nonmilitary disciplines. Following the first postwar decade, a surge of articles and books gives evidence that mummy investigators were participating in this technological transfer. By 1959 electron microscopical techniques were applied to mummified tissues (Leeson, 1959). Beginning in mid 1950s, Sandison reported his modification of Ruffer's technique for rehydration of desiccated mummified tissues prior to preparation of histological sections (Sandison, 1955a). Attempting to establish relationships between Tutankhamun and his other family members, Connolly & Harrison (1969) employed newer methods of blood antigen serology.

A spate of books and monographs from this period also reflect renewed interest in the study of ancient human remains. Most of these focus on paleopathology, primarily the study of "dry bones". Nevertheless, many include at least certain soft tissue changes. Recognizing the value of understanding postmortem changes in mummified human remains by those who study such tissues, Evans' (1963) treatise *The Chemistry of Death* represents a pioneering effort in soft tissue taphonomy. A published symposium chaired by Jarcho (1966) includes his lament concerning the apathy of paleopathologists for the past several generations. While skeletal pathology is clearly the primary area of interest in this publication, Jarcho (1966) himself notes the potential contributions of trace element analysis, paleoserology, immunodiffusion methodology, X-ray diffraction and serological techniques when applied to mummified tissues. Brothwell *et al.* (1963) produced an edited text (*Science in Archaeology*) that addresses the value of a broad range of both physical and chemical methods of value in studying ancient remains. They included some biological areas such as hair studies and paleoserology methods in a 14 page chapter on the study of mummified human remains written by Sandison. Four years later Brothwell & Sandison (*Diseases in Antiquity*, 1967) published what may still represent the single most comprehensive review of the state of knowledge about diseases in ancient populations. It includes descriptions of parasitic diseases and various types of biological calculi. Janssens' 1970 text *Human Paleopathology* contains only a three page chapter on diseases of the soft tissues.

These major publications, together with a scattering of individual articles, permit us to characterize this interval as a period of paleopathology renaissance. Following several generations of relative inactivity, the general field of paleopathology seemed to require an inventory of existing knowledge in this field before it could resume its progress. Though limited in degree, we also can detect a willingness to apply new laboratory technology to the study of mummified ancient remains. These preliminary proceedings set the stage for the explosion in both numbers and breadth in the range of mummy study methods that began in the 1970s.

Dynamism of the fourth quarter

The symposia, seminars and books about general paleopathology that took place in the previous period

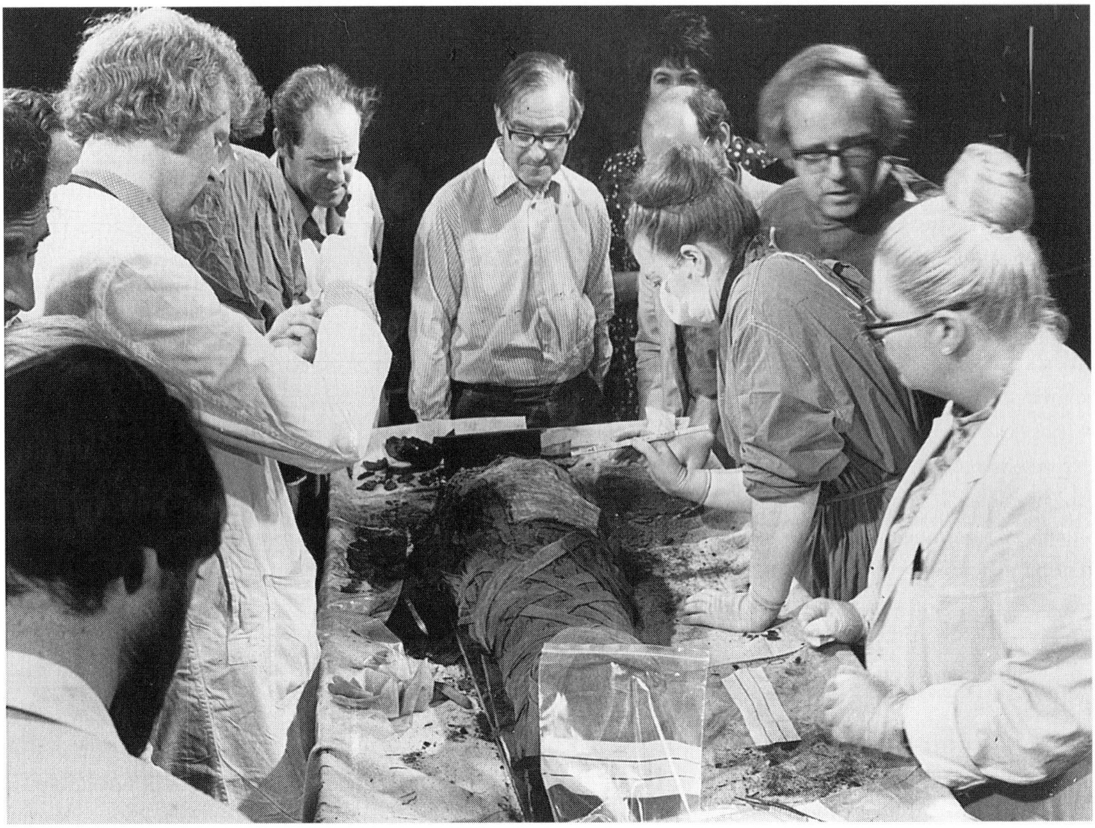

Fig. 1.14. A. Rosalie David and colleagues examining an Egyptian mummy.
Interdisciplinary studies on mummies of the Manchester Museum collection were reported in two seminars in 1979 and 1984. (Courtesy of A. Rosalie David and Manchester Museum, Manchester, England.)

rekindled interest in research on ancient human remains and set the stage for the emerging field of mummy studies. Three events in the early 1970s proved to be epochal for this field of study. The first was the report of Zimmerman's doctoral studies in which he compared the microscopic appearance of synthetically mummified modern tissue specimens with those of Egyptian mummies (Zimmerman, 1972). This became the model for subsequent workers who applied contemporary laboratory techniques within the structure of scientific methodology. The second involved Marvin Allison, Ph.D., of the Medical College of Virginia in Richmond, who initiated a series of summer research visits to Ica, Peru, for the anatomical dissections of Peruvian mummies. Joined soon by the Argentinian physician Enrique Gerszten, M.D., from the same institution, they supplemented their anatomical investigations first with studies of coprolites (dried feces), then radiography and subsequently a host of other, morphological and nonanatomical laboratory methods. By the middle of the decade they had formed a study group, The Paleopathology Club, that still meets concurrently with the annual meetings of the International Academy of Pathology. It also distributes a quarterly newsletter, each issue of which includes a color transparency of a paleopathological lesion. Almost simultaneously Aidan Cockburn, a British epidemiologist working in Detroit, Michigan, and his wife, Eve, organized the dissection of several Egyptian mummies by a multidisciplinary team that eventually included more than 75 scientists (Cockburn *et al.*, 1975; Cockburn, 1978). This, too, led to the formation of an organization, The Paleopathology Association, that also meets annually (coordinated with the meeting of the American Association of Physical Anthropologists) and issues a quarterly newsletter. These newsletters communicated results of their mummy studies widely and quickly, and their organizations' conferences provided face-to-face meetings of interested workers. The consequence of these activities was a rapid increase in interest in mummy studies, the number of studies carried out, and the new instrumental applications to those studies. These activities soon expanded to have international and even eventually global scope.

Several spectacular mummies were studied during this period. They include the spontaneously frozen-dried, sacrificed Inca bodies on Andean mountain tops (Reinhard, 1998), the more than 5000-year-old alpine hunter preserved in Tyrolean glacier ice (Spindler, 1994), the cache of hundreds of Chinchorro mummies in Arica, Chile (Allison, 1984b; Arriaza, 1995a), and Acha Man (Aufderheide *et al.*, 1993) who at 9000 years old may be not only the earliest Chinchorro mummy but perhaps one of the oldest mummies discovered anywhere.

During the decade of the 1970s the program of scientific study of the Manchester Museum's mummy collection in England, originated by Margaret Murray in 1910 (see above), was reactivated under the direction of A. Rosalie David (Fig. 1.14). Its initial results were reported by Dr David at an international seminar in 1979, and further studies at a second one in 1984. In addition, an extensive interdisciplinary study (the Cronos Project) of Tenerife's Canary Islands aboriginal population, the Guanche mummies (Fig. 1.15), was carried out in the late 1980s (Rodríguez-Martín, 1996). The results of this project were reported at the First World Congress on Mummy Studies held at Puerto de la Cruz in Tenerife during February, 1992 (Fig. 1.16). Among its goals was the opportunity for the widely scattered mummy investigators to meet each other and develop collaborative research projects. The manuscripts presented there were published three years later by the Archaeological and Ethnographical Museum of Tenerife (1995). The more than 300 attendees from 17 countries were so responsive that the Congress was institutionalized, now meeting regularly every 3 years.

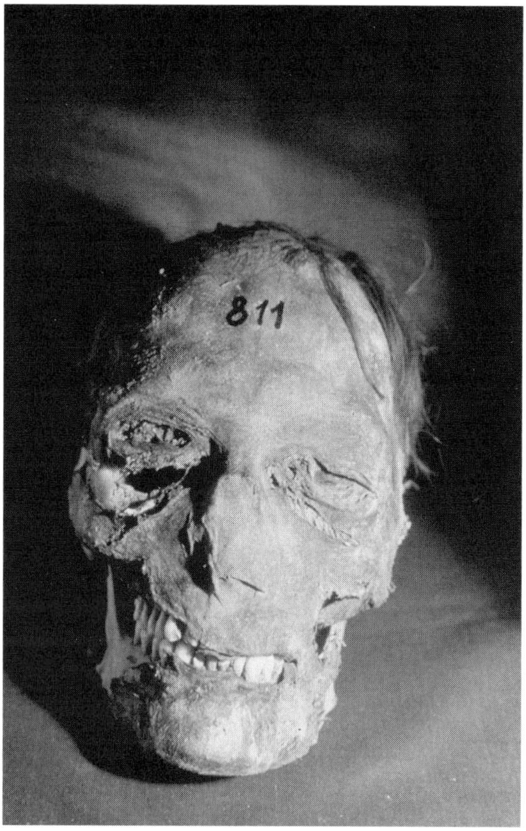

Fig. 1.15. Head of a Guanche mummy, ca. twelfth century.
Chiefs of the indigenous population of Tenerife, Canary Islands, were mummified anthropogenically and placed in caves. (Photo by ACA.)

These organizations' activities have continued to recruit an ever-expanding range of investigators from other disciplines. Zimmerman's many studies (1977, 1979, 1980, 1985) have demonstrated the potential contributions that both gross and microscopic anatomical studies can make. The elegant measurements of lead and lead isotope ratios in the tissues of frozen mummified bodies of the arctic Franklin expedition by

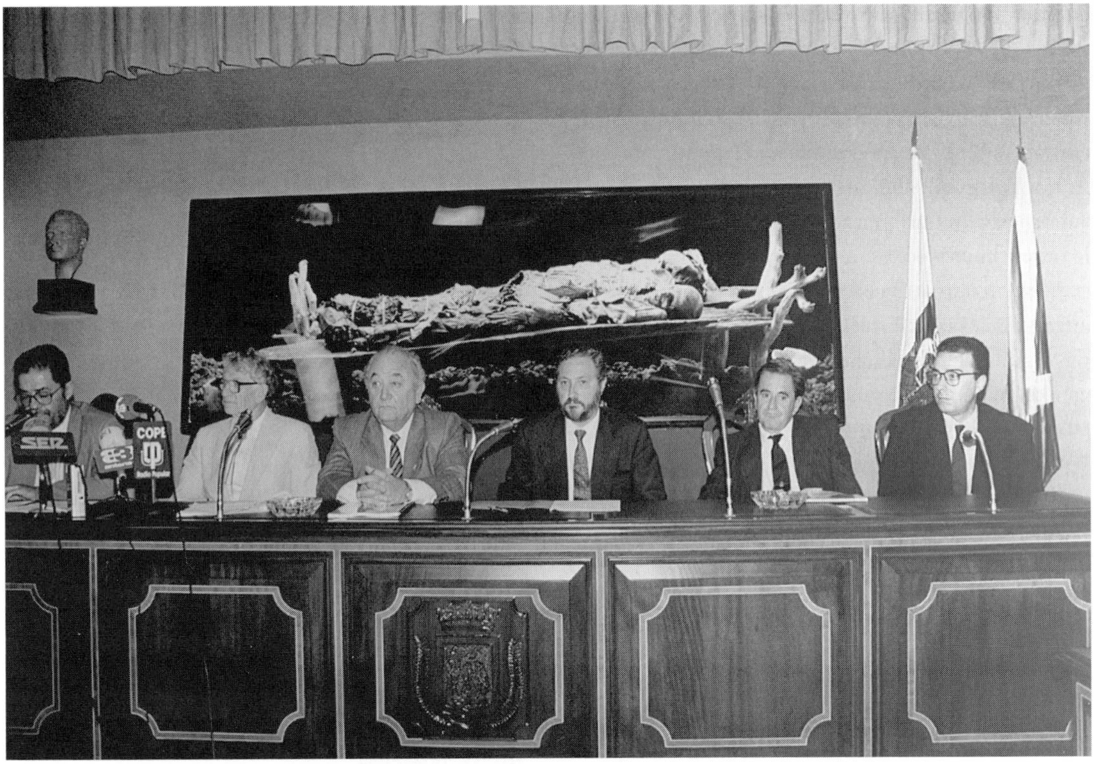

Fig. 1.16. Officials of the First World Congress on Mummy Studies, 1992.
From left to right: Rafael González, Director of Archaeological Museum of Tenerife; Arthur C. Aufderheide, President of Scientific Committee; Victoriano Rios, President of the Canarian Parliament; Adan Martín, President of Tenerife Cabildo (government); Manuel Hermoso, Vice-President of the Canarian government; and Miguel Zerolo, President of Museums and Centers of Tenerife's Cabildo. (Photo courtesy of Miguel Zerolo.)

archaeologist O.B. Beattie and colleagues (Kowal *et al.*, 1991) clearly demonstrate how mummy studies sometimes can resolve otherwise unfathomable archaeological problems. Although Flinders Petrie and Grafton Elliot Smith had X-rayed Egyptian mummies shortly after Roentgen's great discovery of this form of radiation, it was not until the popularization of computed tomography (CT scan) in the 1970s with its three-dimensional and other software applications that it became commonly employed in the study of mummified human remains. The ability to gain information without unwrapping a mummy or even removing it from its container was profoundly appealing to museum curators. Some even became so enamored of this procedure that they declared physical dissection was no longer necessary, until it became obvious that the application of this medical instrument to mummies was generating far more information of anthropological interest than biomedical data. While postmortem epithelial cell changes have limited some of electron microscopy's (EM) traditional types of information, when coupled with an electron probe capable of energy dispersive X-ray analysis (EDXA), it is able to identify the elemental nature of crystals. In a study of a diffusely fibrotic destructive lung disease in a mummy, EDXA proved the condition to be due to silicate pneumoconiosis, predictive of dusty occupational (farming) exposure (el-Najjar *et al.*, 1985). Similarly by demonstrating the characteristically deformed red blood cells in mummified tissue, EM enabled diagnosis of sickle cell disease in a prehistoric body (Maat & Baig, 1990).

When Reinhard (1990) found the ova of fish tapeworm in coprolites from the intestines of South America's Chinchorro mummies, he not only confirmed the archaeological suggestion of their maritime subsistence, but also that the Chinchorro people had ingested uncooked fish. Coprolite analysis can also suggest ingested dietary items but skeletal trace element analysis and stable isotope ratio studies in soft tissues can refine prediction of the diet's principal food classes (Aufderheide & Allison, 1995a,b; Tieszen *et al.*, 1995a, b; Aufderheide, 1996). While the ultimate position of immunological studies has not yet been defined, optimism has been created by certain specific achievements such as the immunohistological demonstration of smallpox virus and the trypanosome parasite of Chagas' disease in human mummified tissue (Fornaciari & Marchetti, 1986; Fornaciari *et al.*, 1992a). The identification of cocaine in the hair of South American mummies using techniques of radioimmunoassay and gas chromatography/mass spectrometry permitted Cartmell *et al.* (1991) to supplement archaeological findings by defining the antiquity, geography and demography of prehistoric coca-leaf chewing practices in western South America. Chromatographic techniques also made it possible for Nissenbaum (1992) to demonstrate that bitumen from the Dead Sea instead of the usual resins was employed by some late Egyptian embalmers. Another unique method used in mummy studies involved application of nuclear magnetic resonance spectroscopy to black, radio-opaque material from a mummy's intervertebral disc space. This procedure identified it as resinous embalming material rather than the previously suspected biological deposition of homogentisic acid that would be present in an individual suffering from alkaptonuria (Wallgren *et al.*, 1986). The methods of molecular biology proved the presence of recoverable DNA in human mummies (Pääbo, 1985), and enabled detection of an infectious agent (tubercle bacillus) in a Pre-Columbian mummy by identifying the presence of that bacterium's unique DNA structure (Salo *et al.*, 1994). Other infectious agents identified by this technique in mummies include leprosy bacillus (Rafi *et al.*, 1994); *Trypanosoma cruzi*, the cause of Chagas' disease (Guhl *et al.*, 1999); a leukemia–lymphoma-associated virus HTLV-1 (Li *et al.*, 1999); *Clostridium* spp. in mummy coprolites (Ubaldi *et al.*, 1998); *Bacillus* species from amber-embedded insects (Cano & Borucki, 1995) and others. Mitochondrial DNA patterns in mummified human remains have also helped to define trans-Beringeal human migrations into the New World (Monsalve *et al.*, 1994).

Presently all the limitations for applications of most of these methods to mummified tissues have not yet been defined. Nevertheless, successes to date promise the generation of a new biomedical and biophysical database unimaginable even a generation ago. The relevance and value of such a new body of data will, however, need to be made unmistakably evident. The expense of many of these procedures will place an unprecedented strain on the budget that traditional research funding agencies have made available in the past for investigations of human mummified remains.

This review clearly discloses that the tap root of the scientific discipline of mummy studies is inseparable from the study of both normal and pathological human anatomy. It is equally evident that the pursuit of understanding the human body's structure and function caused the initial, purely morphological methodology to be supplemented by physical, biological, physiological and behavioral techniques. The secondary rootlets of mummy studies became entwined among these methodologies as well. Nourishment from all these sources is not only responsible for the current efflorescence of the scientific study of mummies but has also shaped its present interdisciplinary nature. The future of this discipline lies in the hands of training program directors who recognize how it came to be what it is today, and that its future – indeed its survival – will depend on the nurturing of its interdisciplinary soul.

Chapter 2. The purpose of anthropogenic mummification

Introduction

It is virtually impossible to view an anthropogenically mummified human body without questioning the purpose of such a mortuary practice. Nor is that an idle question for those who are serious about using mummy findings to enhance their understanding of an ancient culture. An activity so exotic must reflect extraordinarily powerful motivations of either a secular or a religious nature. If such purposes varied among ancient populations, then individual motives may be associated with certain archaeological mortuary expressions with sufficient regularity to be predictive. The lure, then, for the pursuit of the answer to the question of mummification's purpose is that appropriate interpretation of archaeological findings related to mummification might be predictive of its purpose. This, in turn, could provide a level of cultural understanding of an archaeological population not always achievable by traditional interpretations of archaeological assemblages.

But where to begin? Ideally, to achieve this we need tested and calibrated archaeological models. Perhaps one way to start is to consider the anthropogenically altered cadaver itself as an artifact. One could then expect to identify the purpose of its alteration by subjecting it to the same archaeological principles governing interpretation of other material items included in a tomb. Unfortunately, there is little current agreement about the nature of these principles.

Prior to 1960 "traditional" interpretation was based on assumptions (viewed by later critics as untested) that human mortuary behavior was sufficiently consistent that similar patterns of archaeological features reflected not only similar purposes but also similar cultural concepts and structure. With this approach, the degree of archaeological similarity between two sites becomes a measure of the populations' cultural distance.

Conveniently for us, mummification has been tested by these criteria already, perhaps somewhat indirectly but usefully. Early in the twentieth century an Australian anatomist working at Cambridge University, Grafton Elliot Smith, applied this concept to excavated Egyptian and other tombs throughout the Old and New Worlds. He assembled a selected group of anthropological and archaeological features, all of which he felt shared a "heliolithic" focus, and of which mummification was of central importance. On the basis of similarity of these features, he then attempted to define the route of diffusion of such practices that were assumed to have originated in Egypt (Smith, 1915). Although parts of his study relating to African areas immediately adjacent to or near Egypt have been found to conform to this pattern (Hoffman, 1979:259), subsequent information has demonstrated that the bulk of his reconstructions are erroneous. Indeed, recent radiocarbon dating of northern Chile's elaborately mummified bodies has documented that these were prepared millennia before the time of the first pharaoh (Allison *et al.*, 1984; Aufderheide *et al.*, 1993). Admittedly the *purpose* of mummification was not

Smith's central goal, and mummification was only one (though the principal) of the items in his "heliolithic" features assemblage, but the failure of mummification to conform to this concept in that study discourages pursuit of such an approach.

The decades of the 1960s and 1970s initiated a period of challenge (so-called "New Archaeology") to traditional archaeological interpretation. This challenge was characterized by concern about process, by insistence that artifacts be interpreted in a documented cultural context (situational adaptation) and by expectations that such interpretations conform to the testable methodology of the natural sciences (Binford, 1965; Saxe, 1970:9; O'Shea, 1984:3). It is not our purpose to review the complex history of this movement; sufficient for our purposes is that, not content with the limitations of the contributions that the New Archaeology approach has made to the development of general theory for mortuary analysis, the post-New Archaeology (postprocessual) period has been characterized by a plethora of variations and alternatives much of which are directed at attempts to derive the symbolism and structure of culture (Hodder, 1986; Gibbon, 1989). These have not arrived at a degree of concordance sufficient to be employed comfortably as a tested, interpretive model for treating a mummy as an artifact.

Their legacy, however, does include several findings relevant to our interest. Of these, probably the most important is the intensive ethnological study of certain cultural practices among modern populations to predict how they would be reflected archaeologically. Models based on such studies successfully predicted the social status of the deceased, for example, and also reflected the social complexity of the population of which the interred individual was a member (Saxe, 1970:9; O'Shea, 1984:3). Equally important, however, is that such studies also revealed that differing purposes could generate similar mortuary practices. This indicates that caution is necessary when one attempts to derive cultural concepts from archaeological assemblages unless ethnological studies of a group's practices relating to the archaeological features of interest have been carried out.

Most of these archaeological approaches have one common feature: they focus on the deceased. It is not, however, facetious to recall that the deceased does not mummify her/his own body. The effort and cost of anthropogenic mummification can be enormous. It seems inappropriate to expect survivors to carry it out unless those responsible for the mummification anticipate benefit from it (Hoffman, 1979:299). In anthropological analysis, altruism enjoys a low level of credibility as an explanation for human behavior.

The approach to the "purpose question" detailed below has been shaped by the above observations. It involves examination of mummification practices of a variety of populations about whom sufficient ethnological or ethnohistorical data are available to evaluate their possible motivations for mummification. This involves identification of conditions or circumstances that require a human response, followed by a search for tangible evidence for the nature of that response. The quest for understanding the motive of the response focuses on the survivors responsible for the mummification. Interpretation is based on the assumption that mummification serves a perceived, often pragmatic need directly related to the mummifiers, and only secondarily to the interests of the deceased.

Paleopsychology surely ranks among the most treacherous of intellectual occupations, especially when it involves the assignment of motives for the actions of preliterate individuals or societies. Proof of the validity of such reconstructions is rarely absolute. Readers will judge the degree of logic and skill employed in the pursuit and therefore the plausibility of the solutions offered to the question in the following dissertation.

Enhancement of royal authority in a theocracy

The Inca empire

The abbreviated duration of the Inca empire, coupled with the absence of a native-authored historical record, provides us with a much more restricted database for

Inca cultural features than that from Egypt. Inca people become recognizable as a cultural group about AD 1200. The more major expansion into an empire was largely launched by Inca Pachacutec (ca. 1438–1471) (Rowe, 1945, cited by Morris, 1988:236). It is not possible to extract from either ethnohistorical accounts or archaeologically derived evidence how much anthropogenic mummification was carried out on the bodies of nonroyal or at least nonelite members of their population. It is clear from historical and ethnohistorical records (de la Vega, 1609/1987:323–324; Cobo, 1653/1990:251–252), however, that the deceased rulers of the Inca people were mummified and their preserved bodies were stored in family homes and in the Temple of the Sun at Cuzco. Garcilaso de la Vega (1609/1987:323–324) tells us only that they were eviscerated and dried, that their mummified bodies were so light they were easily carried, and that their facial features were "perfectly" preserved. He had an opportunity to view five of them before their permanent disappearance and presumed destruction. Father Cobo (1653/1990) refers to the use of "bitumen and concoction" in respect of their preservation, and Prescott (1936) attributes desiccation to the "action of exceedingly dry, cold, highly rarefied atmosphere of the mountains". Not one of the bodies of the Inca emperors has been identified and described since the time of the conquest, denying us archaeological supplementation to the ethnohistorical accounts. The climate at Cuzco's altitude is not amenable to predictable spontaneous mummification.

But if the chroniclers provide little information relevant to procedural details of mummification, they are positively loquacious about the context in which this practice was executed. Their records do contain some inconsistencies regarding the period of initiation of royal Inca mummification practices. This is probably due to the later Inca kings' practice of publicly attributing to the revered Manco Capac (legendary founder of the Inca dynasty) much of the body of laws, religious concepts and symbolism that these later kings actually themselves initiated and that subsequently came to form their material and spiritual dogma (Gheerbrant in de la Vega, 1609/1961:82, footnote 2). All agree that Inca Viracocha, listed by Garcilaso de la Vega as the eighth Inca king, was "embalmed" (Garcilaso actually claims to have seen his body), as were all subsequent deceased Inca kings until the last – Atahualpa – who was executed by the Spanish general Pizarro. It is probably no accident that it was also Viracocha who militarily initiated a significant increase of Inca territory, and whose son, Pachacutec, expanded it greatly to the level of an empire. Tradition also attributes the origin of the Inca kings' claim to divine status to the probably mythical Manco Capac. However, most of the royal decrees that constitute later Inca law are now believed by historians to have been initiated by Inca Viracocha and his son Pachacutec (Gheerbrant in de la Vega 1609/1961:82, footnote 2). Since these decrees were claimed to be divine utterances, an argument can be made that Viracocha and Pachacutec were the authors (or at least strong popularizers) of the concept that Inca kings were divine, the children of the sun god, and that those who broke their law were also committing sacrilege. Emphasis on the divine continuity of Inca kings by requiring them to marry their sisters was stated by Garcilaso de la Vega (1609/1961:53, footnote 5) to have been instituted by Tupac Inca Yupanqui, listed by him (de la Vega, 1609/1961:270) as Pachacutec's son but now often regarded as a synonym for Pachacutec himself (Gheerbrant in de la Vega, 1609/1987:291, footnote 5).

Thus these observations define a modern historical trend to view many of the features characterizing Inca royalty (including their divinity and the routine mummification of their corpses) as being at least crystallized and canonized, if not actually initiated, by those Inca rulers responsible for the active military expansion and organization of the empire. Furthermore, practices were established that constantly reminded the Inca public of the mummies' presence. The mummified Inca kings' bodies were not buried, but kept in special locations in Cuzco's Temple of the Sun or in their family's home. On state occasions these mummies were dressed in royal clothing (proclaiming their divinity) and displayed in parades (Fig. 2.1). Their continuing

Fig. 2.1. Display of Inca ruler's mummy, ca. sixteenth century.
Drawing by the chronicler Poma de Ayala, showing an Inca ruler's mummy being transported, probably participating in an Inca ritual. (Image was hand sketched from Poma de Ayala (1613/1956) *Nueva Coronica y Buen Gobierno*, Biblioteca Ayacucho. Franklin Pease, transl.)

existence was emphasized by ritual preparation of food for them. They were even incorporated physically into some rituals such as the graduation ceremony for the Inca military academy students. On that occasion the mummy occupied a chair among the reviewing officials. At times their advice was even sought by official diviners. Such roles for the mummified bodies must have carried a clear message to the viewing public that the spirit of the deceased, divine king was still playing an active role in the lives of the populace through his continuing contacts with his divine son, the currently reigning monarch, and interceding with the immortal gods for the benefit of the Inca nation. In this manner mummification practices acted to enhance central authority. Since they appear to have been initiated by the kings themselves and at a time of political instability related to rapid military expansion, it seems appropriate to postulate that enhancement of power centralization was the kings' motive for the establishment of royal mummification.

Early Egypt

We are most apt to understand Egyptian mummification if we focus on the forces impinging on the initiators of this mortuary practice. Traditionally the unification of Egypt has been conceived as the product of military action. More recent evidence has been used to suggest that the process was driven primarily by sociological forces. For the purpose of the search for the innovating stimulus for anthropogenic mummification, it is immaterial which of these two mechanisms brought about the structure we call kingship about 3050 BC. They both would present the designated leader with the same problem: how to acquire and maintain the power that was perceived to be necessary to govern such a widespread population and one that had previously functioned as semi-independent small communities (Kemp, 1989). I have chosen to present the situation using the traditional, military explanation, knowing that the logic presented will be equally effective if unification was the result of a sociological process.

The earliest detectable effort of Egyptian mummification is the isolated, linen-wrapped arm found by Petrie in the tomb of Pharaoh Djer, the third ruler after the unification of Upper and Lower Egypt. Following the conquest of the Nile's northern people by Upper Egypt's legendary ruler Menes about 3050 BC, the first rulers of this combined empire must have struggled with the problem of how to gain the cooperation of the defeated, foreign subjects. One of the mechanisms they chose, divine kingship, succeeded so brilliantly that it shaped the nature and form of the world's first unified national state for the subsequent 3000 years (Hoffman, 1979:335). These rulers proclaimed that the living pharaoh was not human, but instead, was Horus – Upper Egypt's principal deity. Not his representative, but the actual, physical presence of Egypt's living, almighty god. Furthermore, the ruler's death was explained simply as a translocation to the realm of the gods. After Dynasty 5, the deceased pharaoh was alleged to continue his divine status as Osiris, ruler of the netherworld (David & David, 1992:97). Archaeological suggestions of the conceptual principles of a divine status for the regional leader are detectable already in the late Predynastic Period (Hoffman, 1979:349). However, evidence of proliferation and elaboration of monumental structure is abundant early in Dynasty 1. This suggests that the overt claim to central authority by means of divine kingship was at least publicly embraced by the rulers to an unprecedented degree at about the time of Egypt's unification (Frankfort, 1948:32). The image of regality and omnipotence of the early kings was intensified by the addition of several names reflecting their status. One of these was a "Horus name", and later records identify such a one (Narmer) for Menes, the founder of Dynasty 1. While most scholars regard Narmer as a specific, historical individual, the nature of Menes is less well established and might refer to Narmer, a series of military leaders or even a legendary figure.

To be accepted, however, so dramatic a claim as divinity required considerable and repetitive public amplification. Archaeological evidence records a rapid increase in size and complexity of tomb-related, sacred temples, together with the creation of compulsory daily religious rituals. An entire cadre of professional (and perpetual) clergy was created to administer these public rites, focused on the divinity of the current and former rulers. The image of the inseparability of church and state was also enhanced by localizing certain civic functions (e.g., payment of taxes) into the temple milieu (Hoffman, 1979:335). The effects of these acts undoubtedly served as a constant reminder of the pharaoh's divine status.

Furthermore, if the king was divine, then everything related intimately to him also was, including his corpse. Clearly, its perpetuation would help to establish the continuity between the world of the living ruler and that of the underworld. The deceased king was identified as Osiris, lord of the underworld. During the Old Kingdom only kings could achieve such immortality. Later, during the Middle Kingdom, Osiris was elevated to the position of principal deity and immortality was offered also to nonroyal subjects. However, as Osiris, the deceased pharaoh did not merely continue to exist in perpetuity. Importantly, as the underworld god he was the gatekeeper: he judged and determined the fate of subsequently deceased commoners, thus remaining forever in a position to influence the behavior of the living. It is not surprising, then, that the most primitive postmortem efforts to preserve the appearance of the living body (the aforementioned arm of Djer or his queen that had been merely wrapped tightly in linen bands) can be identified already in the tomb of the third ruler of Dynasty 1, probably only about a century after Egyptian unification. Because liquid resin hardened upon air exposure, the wrapping of the corpse in resin-soaked linen bandages during the first two dynasties resulted in retention of the body shape even if the encased body itself skeletonized. The finding of Hetepheres' (Dynasty 4, Cheops' mother) mummified viscera in a wooden chest is evidence of progression of the evolution of the mummification procedure about 700 years later to include the essentials for successful soft tissue preservation: evisceration and desiccation (David, 1982:65). While only a few mummies from the Old Kingdom have survived the effects of time, archaeologists have documented a parallel celerity in the evolving size and complexity of the royal tombs. This reflects emphasis on the importance of the mortal remains during the Archaic Period and early part of the Old Kingdom, culminating in the vast pyramidal superstructures of Dynasty 4 (Fig. 2.2). The spectacular ceremonies related to the interment of a deceased pharaoh within a pyramid or other mortuary superstructure must have had a profound effect on the observing general populace. Subsequently, sight of that monument would act to recall the image of the royal mummy that was still housed within that structure.

It has been possible, then, for us to demonstrate that

Fig. 2.2. The Great Pyramid at Giza.
The tombs of Dynasty 4 rulers in a Cairo suburb are pyramids of enormous size and must have acted to remind the viewing populace of the continuing presence of these rulers' mummified bodies, and their divine status. (Photo by ACA.)

military conquest of Lower Egypt, claim to divine kingship by the victors, dramatic mortuary architectural expansion and, soon thereafter, mummification practices all evolved almost concurrently. These rulers incorporated into a single, sanctified structure (1) the awe-inspiring, monumental tomb, (2) its temple for the supreme god, (3) the deceased pharaoh's forever-mummified body and (4) daily public religious rituals to serve its immortal spirit. This sacred complex functioned as a daily reminder of the past ruler's divine status and therefore also that of the current pharaoh. Within that context mummification clearly contributed to the royal divinity claim. This buttressed the central authority of the Egyptian crown precisely at a time when political stability appeared to be essential for the conquerors to exploit the potential of the now militarily unified Egypt.

Of course, it would be naive to imply that the early dynastic rulers deliberately crafted a novel, self-serving religious dogma based on a mortuary cult that was totally foreign to the population, specifically designed to enhance their political power, and that they then foisted this on their unreceptive subjects. Such a simplistic suggestion would not accommodate the enormous range of complex influences and variables that normally shape human behavior. When dealing with an equivalent enigma in his attempt to understand the medieval British view of a king whose touch was believed to cure tuberculous scrofula, Marc Bloch (1973:48) points out that

> for all religious phenomena, there are two traditional explanations. One . . . prefers to see the fact under study as a conscious work of an individual thought, very sure of what it is doing. The other, on the contrary, looks rather for the expression of social forces of an obscure and profound nature . . . these two kinds of interpretation are only apparently in contradiction. If an institution marked out for particular ends, chosen by an individual will, is to take hold upon an entire nation, it must also be borne along by the deeper currents of the collective consciousness.

A search for such "deeper currents" within the collective consciousness of the Egyptian public reveals that Hoffman (1979:257, 349) and others have called attention to abundant evidence that the germs of the Old Kingdom royal tenets very likely preceded the dynastic period. They were far from being imposed arbitrarily. Many different ideas preexisted within the population, including some relatively dormant or poorly focused concepts. Of the latter, those that were perceived by the rulers to be favorable to their interests were crystallized, coupled with a cluster of homologous features and nourished into efflorescence by royal actions. In this way the resulting structure could be viewed by the general public not as an alien importation, but rather as one consistent with a past tradition and one that could be accommodated by the popular belief system without disruption.

Preservation of the royal bodies in this context can be seen as only a part, though an important one, of a complex of actions all related to the establishment in the public mind of the ultimate in status and authority of the king's office: divinity itself.

As social conditions in Egypt changed, so did the reasons for mummification. Motivations for mummification during the Middle Kingdom and subsequent periods are discussed later in this chapter.

Summary of mummification practices related to political power

Mummification practices in early Egypt and among the Incas shared several distinctive features:

1. Mummification was initiated at a time of great population stress. In both cases this occurred at, or closely followed, a period of great military expansion.
2. In both cases mummification was coincident with the establishment of a theocracy.
3. Initially, at least, mummification was a royal monopoly.
4. The mummies themselves were used to enhance the claim for royal divinity of the reigning monarchy, an action that can be expected to have increased the authority and power of the throne as well as homogenized the population. Employment of the mummies for this purpose assumed several forms:
   (a) Continuing public display of the deceased ruler's mummified body (Incas) or only a single, pompous display at the time of the funeral, but continuing display of an ostentatious tomb monument (e.g., pyramid) that housed the mummy (Egyptians).
   (b) Establishment of rituals designed to remind the general populace that the deceased divine ruler continues to exist and influences the lives of the living through contact with his divine son, the living and reigning ruler. This includes such activities as incorporation of the mummy as a ritual participant (Incas), the institution of a clerical profession to carry out public religious rituals relating to the mummy (both Egyptians and Incas), and regular provision of food for the mummy (Egyptians; irregularly for Incas).

Archaeological features reflecting these cultural behavior patterns can be expected to include the following:

1. Evidence of social stratification in the society.
2. Initiation of mummification practices coincident with major population stress requiring a strong central authority, especially a large military expansion.
3. Mummification restricted (at least initially) to a population's elite subgroup.
4. Distinctive, publicly evident tomb differentiation.

5. Evidence of repeated mummy use, directly (Incas) or indirectly (mummy-related rituals; Egyptians).

Historical records also identify some examples of behavior that almost parallel that of early Egypt and the Incas, but in which a surrogate for the body is employed. Herodian's description of a practice in the Roman Imperial Age is clear and cryptic (Herodian of Antioch ca. AD 250/1961:110–112; Workman, 1964:191–192; Carey, 1987:21). Deification of a deceased emperor was carried out posthumously. It is important to note that Herodian emphasizes that this was performed only if the emperor left an heir, and that the process was initiated by that heir, the then ruling emperor (it would be detrimental to the self-interest of a usurper to deify the emperor he deposed). After disposal of the body, usually by cremation, a wax image of the emperor was created and displayed in a public area, lying on a bed and attended by living representatives of the Roman senate and family members. Living doctors attended the "sick emperor's" image daily and after several weeks they declared that the ailing emperor had died. The effigy was then removed to a huge, wooden, prepared bier (five, successively smaller stories in height) and placed within it. Several weeks of dramatic public performances followed, during which many distinguished visitors filled the bier with expensive gifts. Finally the bier was set alight. As the flames consumed the structure an eagle, imprisoned in a container at the top, was released and flew out and upward, believed to be bearing the emperor's soul "from earth to heaven". The Roman senate then declared the deceased to be divine and permanent rituals to worship him or her were established. Herodian, an historian, is, of course, not always completely objective and we know the described event was not always carried out. Even if it was not always performed, however, the analogy remains applicable. Augustus, Julius Caesar's adopted son, established deification of Caesar, validating his act by noting the appearance of a comet shortly after his death, and the later emperor Decius even deified himself and instituted regular worship rituals (Moorhouse, 1997:139).

Clearly, during his politically vulnerable period of power transfer at the time of his throne assumption, the same principles of enhancement of power and authority of the surviving and ruling son were incorporated into this practice, as was the case among the Inca and early Egyptian rulers – only the mummified body was replaced with a wax effigy. However, the examples from early Egypt and the Peruvian Incas, using anthropogenically mummified bodies, are themselves sufficient evidence for the perception of mummification's purpose in such circumstances so that no further pursuit employing examples of effigies rather than mummified cadavers appears necessary.

Other cultural groups in which the rulers claimed at least some form of divine kingship include the following. Mesopotamia's Sumerian kings are sometimes accredited divinity in the literature. However, while they had a close relation to the deity, they themselves were viewed principally as the people's representative to their god. Admittedly, since he was the only representative they had, the king's role was vital to their existence. Nevertheless, though chosen, or at least approved by the deity, the king was not seen as the incarnation of a divine being, but rather as a very specially selected human intermediary (Saggs, 1962:359ff). Since that role ended with the king's death, it is not surprising that mummification did not become a feature of the belief system.

Ross & Robins (1989:133) maintain that Celtic Druids considered themselves divine. While they make a plausible argument that Lindow Man, the British bog body, was a Druid, it would require demonstration of a pattern involving many examples to conclude that preservation of that body was an intentional practice.

Others, such as Greece's Mycenaean kings also declared themselves divine (the word for king and god are identical: wanax), but were not associated with mummification (LeRoy, 1972:257). Perhaps the climate in such areas led them to seek alternative methods of enhancing centralized power.

Personal or population status and/or security

Mechanisms that fueled the later, widespread practice of Egyptian mummification

Just as biological evolution alters the form and function of an animal to adapt best to a changing environment, so too do the changing needs of a population generate adaptational evolution of its social practices. Just as the motivations for the development of the mummification practices in Egypt can be reconstructed most easily by scrutiny of the events associated with its initiation, subsequent evolution of mummification practices can be understood only in connection with later social innovations.

Unfortunately, archaeological interests in Egypt have focused primarily on royal or elite class tombs. We have only a small amount of detailed data from excavation of sites representing the tradesman class and even less of peasant (fellah) burial sites. Until these other classes are well-documented in the archaeological record, it will be difficult to derive motivational factors for the social diffusion of mummification practices to nonroyal classes. Nevertheless, we can anticipate some of such evidence by scrutinizing the public response to divine kingship as documented in the following known events.

Up through Dynasty 4 (for dynastic chronology see Chapter 4: Mummies from Africa and Egypt) the king had successfully claimed he was the incarnation of a god, with the resulting absolute earthly power that implied. Creation of a specialized clergy to perform the daily rituals served as public reminders of the continued existence of the deceased pharaoh's spirit. This, however, eventually produced a group with sufficient political strength to influence governmental policy decisions. The power of this class is detectable as early as Dynasty 5 when a late pyramid text describes how the deceased pharaoh joins the sun god in Re's boat during his daily journey across the sky (Spencer, 1982:140). By the First Intermediate Period the sun god Re, considered the god supreme to all others by the clergy at Heliopolis, had become elevated to that status throughout all Egypt, and Horus (the living king) was declared to be only his son (Frankfort, 1948:45). Royal cartouches as early as Dynasty 5 include both Horus and Re. By the beginning of the Middle Kingdom, following the upheavals of the First Intermediate Period, two changes of significance to our discussion are apparent:

1. Re was replaced as head of the Egyptian pantheon by Osiris. The living pharaoh in this new light is still Horus, the son of Osiris, and remains capable of functioning as a vitally effective intermediary between the populace and their gods. The demotion of Re had a suppressive effect on the power of Re's priesthood. The deceased pharaoh's role as Osiris never changed throughout the rest of ancient Egyptian history. However, beginning with the Middle Kingdom, Osiris became available also to nonroyal individuals who now could share an afterlife that had formerly been a royal monopoly.
2. The clergy gained personal mummification privileges. Expansion of the state bureaucracy and clergy, whose memberships earlier had been drawn largely from among the pharaoh's relatives, now diluted their composition with increasing numbers of nonroyal persons. The previous exclusively royal privilege of mummification was extended to the nonroyal clergy and bureaucratic elite. This enhanced the status of their role in maintaining the ideology of divine kingship and hence their political power, which was important to the perpetuation of their profession. In brief, the clergy benefited from participation in mummification by the same mechanism previously operational for only royalty.

Following the tumult of the Second Intermediate Period, the onset of the New Kingdom ushered in an interval of stability for the Egyptian public. By now, more than a millennium of Egyptian mortuary practices had developed a belief system powerful enough to generate its own driving forces. The Theban princes who had driven out the foreign invaders (Hyksos) by military force reunited Egypt. Predictably they elevated their local god (Amun) to supremacy by fusing him with Re, producing the new god Amen-Re. The king was viewed as born of a godly father (Amen-Re) and an earthly mother of royal blood. The power gained by local (Theban) priests eventually threatened royal authority. An effort by Akhenaten to strip the priesthood of power did not survive him and thereafter centralization of power gradually eroded. With the firm entrenchment of Egyptian religious dogma, mummification to achieve participation in the pleasures of a spiritual eternity was demanded by members of all socioeconomic levels. Unfortunately, we simply do not yet have enough archaeological or historical information about the lower socioeconomic classes to identify with certainty all of the benefits deriving to their survivors, motivating them to carry out this potentially fiscally crippling procedure. Initially the general public had accepted the divine kingship dogma with its related mummification practice, even though that credo was devoid of obvious, immediate reward to the populace as long as it was restricted to royalty. Such tolerance may well have been motivated by a perception of their desperate need for the re-establishment of order (Frankfort, 1948).

Those who saw to it that a body was mummified were provided an opportunity to gain status within the living population by indulging in an especially ostentatious mortuary ritual, including expensive forms of mummification. Some suggest that additional benefits enjoyed by the living who mummified the deceased commoners also included participation in "community continuity", i.e., those who help the deceased to achieve immortality through mummification hope their acts will serve as a model to their own heirs (Frankfort, 1948). Manipulation of the deceased's spirit for gain by the living (a motive described in more detail below) is evidenced by a letter from a surviving husband to his deceased wife, inscribed on a Dynasty 10 stele, in which he requests her to intercede for him (Parkinson, 1991:142). Thus the later periods of Egyptian history reflect a general trend of erosion of central authority as first the nonroyal elite and later the general populace demanded participation in what the kings had established as a belief in a desirable afterlife. Since this could be achieved only by the mummifying activity of survivors, social pressure to approve such behavior rose to a level sufficient to make mummification the norm in later Egyptian history. The mummifiers gain personal status and the social group gains collective religious security.

War trophies

The Jivaros

The Jivaros of the northeastern Ecuadorian jungles continued their "head-hunting" practices into historical times and considerable data are available about them from anthropologically trained observers. Traditionally the Jivaros used the "vengeance system" to help to establish their security from attack by hostile neighbors. Their societal values required murders of enemy tribal members to avenge living or dead relatives, and social status accrued to the avengers. Claims to such status were verified by public display of the preserved head of the avenger's victim at designated public rituals (Fig. 2.3). During such an avenging raid on an enemy group, the slain individual was decapitated, and the skin was meticulously dissected away from the skull. During subsequent boiling of the skin it shrank to about one-fourth of its original size, leaving the hair unaffected. Hot stones and ashes were then packed into the boneless skin to dry it. A hole at the top of the scalp admitted a rope by which the head was suspended and displayed. The prestige value to the avenger is self-evident. In addition, such preserved heads (*tsantsas*) were believed to bring good fortune to the preparer. While the earlier *tsantsas* in modern museum collections appear to have been looted from tombs, ethnological observations suggest that *tsantsas* lose their magical value following the prescribed public ceremonies and may then be discarded (although the hair was often removed and worn proudly around the waist) (Stirling, 1938; Harner, 1972). By establishing societal values strong enough to motivate avenging acts among their members, the tribal group gained security through the deterring effects of such acts of vengeance on hostile neighbors.

Fig. 2.3. Jivaro native (Ecuador) with shrunken head, ca. twentieth century
The acquisition of a "trophy" head of an enemy tribe is steeped in ritual. Public display of the shrunken head brings esteem to the warrior and the tribal members gain collective security by that tradition. (Photo courtesy of Felipe Cárdenas and the journal *Bioantropología,* Universidad de las Andes, Bogotá, Colombia.)

The Sausas

The Sausas were members of the Huancas nation from the Jauja Valley in Peru, encountered by the Incas during early expansion of their empire. Garcilaso de la Vega (1609/1961:199) described them as an aggressive group that skinned their military captives before they killed them. Furthermore, they sometimes filled up the separated skin with ashes, sewing up the skin and erecting "skin-mummies" in their temple as war trophies. The Popoyan population from southern Colombia are reported to have created mummies in a similar way, displaying them on the walls of their dwellings (Stirling, 1938). No motivation for such acts other than prestige deriving to the slayers of enemies is obvious, except the security value, again, deriving to the tribal group that established such status symbols.

Such an ethnologically documented behavior pattern may be archaeologically recognizable in the presence of isolated, mummified heads without bodies accompanying a skeletonized body in a burial within a cemetery that is devoid of evidence for mummification of its own group's members. It should be noted that not all isolated, preserved heads are war trophies; for example, the Mundurucús of Brazil mummified the heads of respected relatives, though the "mummifier" even in these cases probably also acquired status, and may have been motivated to exploit the spiritual force thought to reside in such a body or body part (vide infra).

Regulation of spiritual force of the deceased

A common belief in nontechnological cultures is that the deceased's vital force or spirit remains in the vicinity of the body for a variable but usually finite period of time. Prioreschi (1990:22, 203) suggests this idea is widespread because the deceased commonly appears in the survivors' dreams. Many mortuary practices incorporate taboos designed to pacify and/or hasten the departure of such spirits. Some groups, however, believe that these vital forces have magical power which the survivor may exploit if appropriate rituals are employed. The "stone heads" (*moais*) of Easter Island (Rapa Nui) could be considered such an example (Fig. 2.4). Often they are erected above an ossuary containing the bones of family members, and are oriented to overlook the family property. The vital force of a distinguished, powerful ancestor was felt to be entrapped within this lithic structure by a ritual that also enabled their gods' entry into these stone effigies. This permitted interaction between the living humans and the gods, facilitated by their ancestors' spirits (van Tilburg, 1994:129). Mummification of the body, although only symbolized as lithic heads by Easter Islanders, has evolved in several other groups embracing this type of belief as a method of entrapping such

Fig. 2.4. *Moais* on Easter Island (Rapa Nui).
While not composed of mummified human remains, these lithic surrogates serve as intermediaries between the living and their deceased ancestors whose bones often are located beneath the moais. (Photo by ACA.)

"power" and subsequently regulating it, or dealing with the transiently "homeless" spirit in a nonthreatening manner (see Aleuts, below).

Aleuts

The aboriginal population of the Aleutian Islands ("Aleut") was practicing selective mummification at the time of Russian contact about the mid eighteenth century. This island chain, extending from the Alaska Peninsula to within about 1200 kilometers of Russia's Kamchatka Peninsula has been occupied by Aleuts for about 4000 years. The far eastern and central islands' populations were engaging in mummification rituals when Russian explorers arrived there about AD 1760. Mummies have been found on Umnak, Shugamin, Anarak, Ship Rock and others of these islands, but the largest number has been retrieved from Kagamil Island (Island of the Four Mountains), where a hot volcanic cave enhanced desiccation.

While both ethnohistorical (Veniaminov, 1840) and archaeological evidence from recovered mummies suggest some inter-island variations, selective mummification practices based on social status are clearly evident. Deceased local rulers were eviscerated, placed in cold running water allegedly to remove fat, dried and had their abdomens stuffed with dry grass. Wrappings included waterproof seal intestine and sea otter hides. Such mummified bodies were then placed in rock caves (when available) or rock overhangs, often separated from the shelter's floor by interposed wood planks. Accompanying the bodies were many grave goods most of which were utilitarian items related to subsistence activities. Societal members not part of the elite subgroup usually were not eviscerated, but simply wrapped and commonly stored under a rock overhang. Infant mummies were often kept in a cradle in the home.

Salamatov, a Russian Orthodox priest, described how the living would consult with the mummified body to gain advice as well as prophecies for hunting species (Laughlin, 1980a). On Kodiak Island, mummy parts, especially those of a whaler, were carried on fishing trips because the entrapped vital force of the deceased was expected to enhance success. Such power needed to be used sparingly and appropriately to avoid undesirable consequences. Children, therefore, were warned not to approach such burial sites.

The concept of spiritual power entrapment was expressed culturally among the Aleuts in another form: the practice of dismemberment of a slain enemy. This was based on the assumption that fragmentation of the enemy's body would release the vital force, depriving the victim's family of the opportunity for entrapping that power by mummification and using it against the slayer to avenge the victim.

It is noteworthy that in these traditions and belief systems no tangible benefit to the deceased is apparent. The benefits of the perceived entrapped power accrue only to the living. Even the inclusion of lavish quantities of grave goods (whose value to survivors would be great) tend to enhance the status of those mummifying and burying the body, by the action of the potlatch "wealth-destruction" principle.

The concept of deriving power from a deceased's spirit by possession of the deceased's mummified remains also is firmly embedded in mythology. Northern Teutonic sagas tell how Mimir, a water demon, was the guardian of a fountain in which all wisdom and knowledge were hidden. Odin, the principal deity, coveted the fountain's secrets. Following Mimir's death, Odin mummified his head and pronounced magic formulas over it, permitting Mimir's head to retain the power to answer Odin's questions and reveal to him the fountain's knowledge (Guirand, 1978:257).

In bodies mummified for the reasons of personal status, security or spiritual power control, the archaeological identification of social status would pose few interpretive problems. However, recognition of motivation for regulating the spiritual force of the deceased could present formidable interpretational obstacles, and probably would require such a conclusion to be reached by exclusion.

Temporary mummification

The concept that the spirit of a deceased remains in or near the body for some period of time is global in distribution and ancient in time. Humans deal with their relationship to such spirits in a myriad ways. Some groups view the spirit as necessarily neither malevolent nor benevolent, accepting its presence and occasional perceived manifestations with an almost bemused demeanor, sensing few special obligations toward it. More commonly, however, the spirit is perceived as capable of actions either harmful or beneficial to the survivors, depending on the degree of the survivors' conformity to traditional expectations during that postmortem interval. Practices reflecting such beliefs include temporary postmortem "storage" of the body, during which time prescribed rituals are often performed, followed by later, permanent disposal of the corpse. Examples include such Amerindian rituals as the temporary placement of a corpse on a tree platform with periodic ritual offerings for about a year, followed by permanent interment ("secondary burial") accompanied by a major burial ceremony among the Sioux. A variation of this pattern can be observed in the Huron practice in southern Ontario of burying the body in a shallow pit quite promptly after death, exhuming such accumulated bodies every decade or so for common deposition in an ossuary within a larger excavation ("Feast of the Dead").

In these examples the body is usually allowed to skeletonize during the initial period. Among other groups, the specifics of the belief system or the nature of the rituals required preservation of the body's soft tissues. The Egyptians felt that a finite, though unmeasured, interval after death passed before the

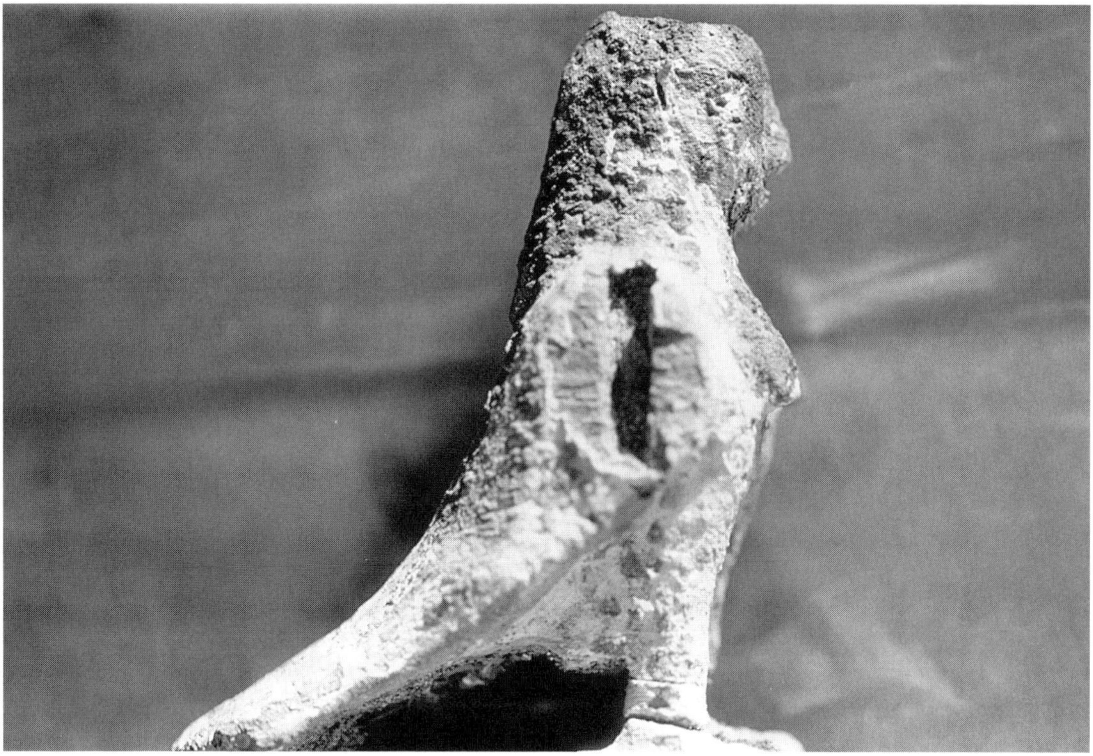

Fig. 2.5. The ba.
The ba is believed to represent the living person's spirit that frequently revisits the deceased's home and sites after death, returning to the tomb and mummified body frequently. The ba is pictured as a human-headed bird. Pictured here is a wood carving of a ba found in a tomb at the Roman Period Kellis site in Egypt's Dakhleh Oasis. (Photo by ACA.)

spirit (ka) was presented to Osiris for judgement and possible permanent admission to the underworld. During this period both the ba (Fig. 2.5) and ka spirits frequently left the body but returned to it regularly and therefore needed to be left food for nourishment at regular intervals (David, 1982:79). More specifically, Smith (1915:57) describes how the Bangala from the Upper Congo displayed the body of the deceased in a hut for some time before burial, enabling this practice by first eviscerating and then smoking and drying the body to retard decay long enough to accommodate the public exposure. He also describes a similar practice among the Khasisas of the Himalayas, who prepared the body for such display by evisceration, filling the abdomen with honey and spices, covering the body with resin and wax overlaid with gold leaf; this preserved the body until its later cremation. Perhaps the best-studied practice of this type is that employed by native residents of certain scattered islands in the Torres Strait between Australia and Papua New Guinea. Upon death the corpse was eviscerated. The body was then desiccated by lashing it to a wooden trellis for air-drying. A major ceremony was carried out much later (often months). Final disposal of the body does not appear to have been rigidly prescribed, and varied from decay by exposure to burial. The principal, apparent function of the rather poorly effective mummification was its preservation until the extended family and friends could gather for the ceremony and carry out the necessary mortuary rituals.

The nature of the rituals carried out during the initial postmortem interval in many of these examples make it clear that an important goal of the survivors is pacification of the deceased's spirit to prevent undesirable actions against them. However, almost invariably the final body disposal event is accompanied by a major ritual attended by relatives and friends as well as members of neighboring groups. Acquiring the accouterments for such a ceremony, food preparation as well as notification of, and travel by, the attendees all require substantial time. Using temporary mummification to delay disposal of the body provides an obvious benefit to the survivors. It makes time available to plan and prepare the final ritual. Indeed, the now almost routine, modern North American practice of embalming the body immediately after death received the major impetus leading to its expansion during the American Civil War in the mid nineteenth century. Initially the bodies of several high-ranking officers were embalmed so they could lie in state for a major public funeral. This was followed by a rapidly spreading desire for similar services from families of slain soldiers of all ranks to permit the return of a battle casualty's body. So great became the demand that the Yankee army contracted with private operators (initially surgeons) to carry out the embalming necessary to retard decay during the slow journey that was often made by ox-drawn carts. The flexibility for funeral planning made available by such temporary "mummification" was a feature so well appreciated that embalming was incorporated as a common – even expected – mortuary practice after the Civil War ended (Iserson, 1994:172). In the past few decades cremation has become increasingly popular, perhaps because modern communication and transportation technology has minimized the need for prolonged postmortem intervals, and wakes are generally no longer carried out in the home.

Thus temporary mummification was and is employed by many societies embracing a belief system that incorporates the presence of a deceased's spirit in the corpse's vicinity for a defined period prior to its departure for final residence in some other world. The survivors' benefit deriving from the mummification effort is avoidance of vengeful spiritual actions by satisfying ritual requirements or the opportunity to exploit the spirit for personal benefit. An additional, pragmatic benefit is the time gained for ceremonial planning and preparation.

The less thorough mummification techniques employed for temporary mummification are recognizable archaeologically. Secondary burials of skeletal tissue are well known to be highly selective, often involving principally only the skull, mandible and long bones (Aufderheide *et al.*, 1994). This may well reflect merely the fact that smaller bones are lost or decayed during the initial postmortem period of storage. The regularity of the pattern, however, suggests the practice may be dictated by a belief system. If so, it is conceivable that this principle might also apply to a temporarily mummified body. It is probable that the finding of some isolated heads or headless bodies in tombs may be reflecting this type of mortuary activity.

It is probably not possible to use archaeological findings in order to separate temporary mummification carried out for spiritual reasons from that performed simply to gain time for ritual preparation, though the latter may well have been only a secondary consideration in antiquity.

Summary of ethnohistorical observations

Modern scholars' quest for the reasons ancient people mummified their dead has led to the study of ancient mummification practices as well as rituals observed at the time of contact by technologically developed countries. The findings among these indicate that more than a single purpose stimulated anthropogenic mummification. In some groups such as the Incas, Egyptians and (via effigies) the Romans, mummification acted to enhance the power of a central, governing authority. In others the population gained physical security by granting status, reflected in mummification. Still others employed mummification to exploit the power of, or to appease, the deceased's spirit. Most modern or recent mummification practices are recognizable variants of these principles. We have considered many of the mummifying groups about whom enough is known to permit evaluation. Some motivation factors may well exist that have not been included in my consideration. For example, it could be argued that the modern North American practice of embalming could reflect an attitude of death denial by the American public. If so, however, the rapidly growing practice of cremation would be incompatible with such an interpretation or signal a dramatic reversal of viewpoint. Nevertheless, no matter how complete our review of current or recently past practices might be, the possibility exists that ancient populations embraced special needs unique to those times that would not be included in the above review of recent or modern peoples.

Some of the purposes discussed above may be reflected directly in the archaeological assemblage. In some circumstances the tomb contents themselves may suggest several possibilities, some of which may be given more weight than others if the findings are integrated with other archaeological, historical or similar information. In still others, available information may not permit diagnostic differentiation. In spite of these limitations it is felt that diligent examination of the evidence presented, integrated with all other available data sources, will predict the purpose of mummification by a specific, studied group at a useful level of certainty often enough to reward the effort.

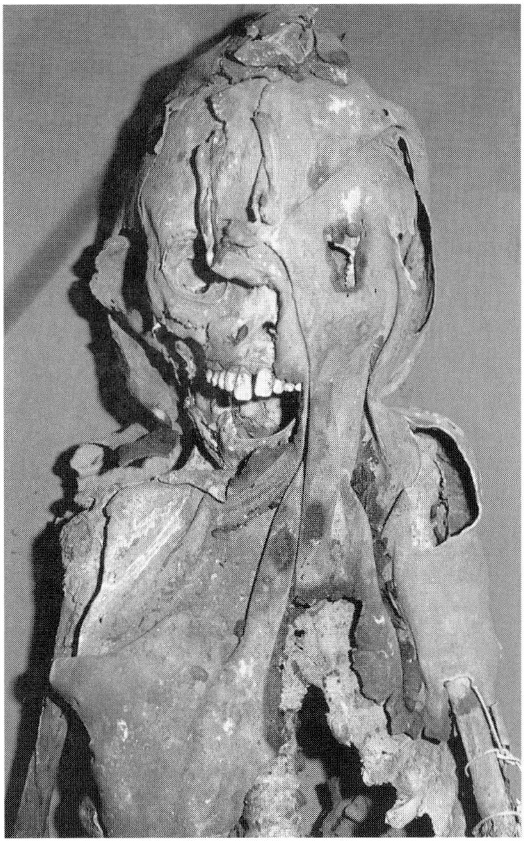

Fig. 2.6. A Guanche mummy; AM, ca. AD 1000. This anthropogenically mummified body of an indigenous resident of Tenerife, Canary Islands, was wrapped in goat hides. In this socially stratified population, it was principally the elite whose bodies were prepared in this manner. Mummy no. Ten-M1. (Photo courtesy of Rafael González, Director of the Archaeological and Ethonographical Museum of Tenerife.)

Application of defined mummification purposes to ancient populations

To test the applicability of the above-derived motivations to ancient populations, we can apply them to two prehistoric groups that I have studied.

The Guanches of Tenerife

The Guanches were the aboriginal inhabitants of Tenerife, one of the volcanic Canary Islands off the northwestern coast of Africa. The Phoenicians, and shortly thereafter the Romans (200 BC), found the islands uninhabited, although the first settlers arrived about a century later. In the fourteenth century, several nations attempted to dominate this cluster of seven inhabited islands. Spain conquered the aboriginal Guanche population of Tenerife in 1409. They found an agropastoral society with language and cultural traits suggestive of Berber origin. A few chroniclers (conquistadors and clerics) wrote some rudimentary and often conflicting descriptions of mortuary practices. These described anthropogenic mummification by evisceration, body cavity replacement with soil and vegetals, restoration of the dried extremities' form by subcutaneous insertion of gravel, and burial in caves (Fig. 2.6). Burial of chiefs in coffins was also mentioned.

During and immediately following conquest, the island suffered rapid depletion of the aboriginal population secondary to deaths in battle and from imported epidemic disease augmented by extensive exportation in the slave trade. The burial caves were looted and the bodies sold. Today fewer than 50 mummies in widely scattered museums are available for study (Rodríguez-Martín, 1995).

These remaining mummies reveal variable degrees of soft tissue preservation. Most are near-skeletons, probably because the better preserved were looted, sold and exported. Nevertheless, the data available include the following:

1. Only a small fraction of the total population was mummified.
2. Evidence of anthropogenic mummification is unmistakable.
3. Only adults were mummified.
4. Both sexes were mummified.
5. Insufficient archaeological data are available to judge social stratification but ethnohistorical documents describe a definitely stratified society of only moderate complexity with chosen leaders. The island's population functioned as subgroups distributed among steep-walled, lava-formed valleys.
6. Chemical dietary reconstruction data identify a predominance of terrestrial meat and dairy products for most of the population, but a distinct and statistically significantly higher meat diet for the mummified group. Consistent with postulated origin from the Berber mountain people, marine resources formed only a minimal part of their diet.
7. Radiocarbon dates indicate mummification did not begin until about AD 400.
8. Archaeological and linguistic criteria suggest the island was probably occupied about 100 BC.

Now, applying to the Guanches the expected archaeological findings for societies in which mummification enhanced central political authority through the claim to royal divine status, we find some parallels:

1. Both ethnohistory and dietary reconstruction support the presence of social stratification among the Guanches.
2. We cannot document population stress at the initiation of mummification, but, if our estimate of the date of peopling on Tenerife is correct, mummification began abruptly about 500 years after occupation of the island. Even if there was no threat from external groups, this is a period long enough for the entire island to have been populated, raising the possibility of internal stress from land competition.
3. Both ethnohistorical and dietary studies confirm that mummification was limited to an elite (royal?) subgroup of the population.
4. The criterion of publicly evident tomb differentiation is weakly supported. Provenience data are lacking for most of the mummies, but ethnohistorical records suggest that only the chiefs were buried in coffins, sometimes under a small pyramid of stones.
5. No archaeological evidence of repeated mummy use can be documented, although an occasional item of folklore suggests that mummies were consulted in times of impending catastrophe.

While not every one of these items conforms with great certainty to the model of the political power-enhancing role of mummification, the findings appear to resemble this model to an extent greater than the others discussed. Accordingly, it appears useful to employ this model in future Guanche research, testing it continually by the findings from investigations structured on this hypothesis.

The Chinchorros of north Chile

The Chinchorros were a maritime population dated to about 7000–2150 BC. They occupied about an 400 kilometer coastal strip of what is now northern Chile in South America between latitudes 23° and 27° S, extending southward from about the present southern Peruvian border. Their origin is unknown. Their artifacts are almost entirely marine related. Their chemically reconstructed diet throughout their period of archaeological recognition (5000 years) uniformly reveals that about half of it was acquired from the sea. The remainder was split between terrestrially hunted

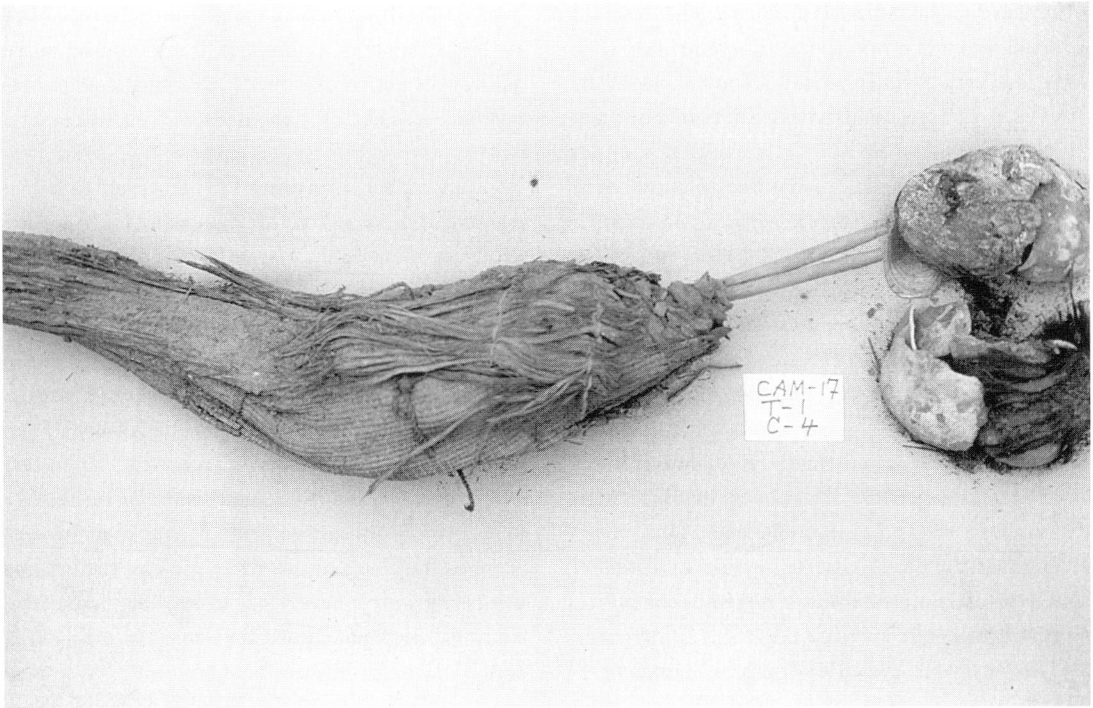

Fig. 2.7. A Chinchorro mummy; AM, ca. 4830, BC.
The body of this Chinchorro child was mummified anthropogenically. Note the clay mask that replaces facial features. Clay, vegetals and wood splints have been employed to refashion the skeletonized body. Mummy no. Cam-14, T-1, C-4. (Aufderheide *et al.*, 1993. Photo by ACA.)

meat and gathered vegetals. Few habitation sites have been found; bodies are buried at beach sites, usually simply placed in sand without tomb structures. Associated artifacts are limited to those found within the woven split reed wrappings. About half of the bodies are anthropogenically mummified; the remainder show spontaneous soft tissue preservation secondary to the arid climate of the Atacama Desert. Anthropogenic mummification efforts are major, often involving complete defleshing followed by extensive reconstruction of body form using clay, vegetals and hide. Clay face masks are common, some of which reveal damage with repair and repainting (Fig. 2.7). Both sexes were mummified, as were bodies of all ages; indeed, at least one-half of the bodies are those of subadults. They appear to have lived in small groups, probably at the band level. These operated out of the mouths of the occasional river, eating primarily shellfish, sea lions, marine fish and shore birds. Paleopathological findings revealed a relatively healthy population without evidence of battle trauma. Bioanthropological studies revealed minimal evidence of contact with their highland neighbors from whom they were separated by the Atacama Desert (Guillén, 1992; Aufderheide & Allison, 1994; Aufderheide, 1996).

Testing this group by the same criteria used on the previously detailed Guanche mummy group we find far fewer parallels with the power-enhancing mummification model.

1. There is no detectable social stratification among these bodies.
2. Population stress at the onset of mummification can neither be affirmed nor negated. The oldest

available body is radiocarbon dated at 9000 years BP (7000 BC) and is spontaneously mummified. The next oldest was anthropogenically mummified 7800 years ago. Because both types of mummified bodies have been found in many periods, this does not permit precise identification of the mummification practice's initiation. The group loses its archaeological identification at 2000 BC. However, paleopathology reveals no evidence of war injuries, so the group's occupation appears uncontested throughout the 5000 years of its nearly unchanging culture.

3. While not all bodies were anthropogenically mummified, those that were certainly do not represent an elite subgroup. Not only was no elite status identifiable for the Chinchorros, but the mummified group included about half of the population.
4. Distinctive, publicly evident tomb differentiation was not demonstrable because no tombs of any type of structure were built.
5. There was some evidence of repeated mummy use: several facial masks had been repaired and repainted several times.

Of the five items of the "political power" model tested, only the last was a feature of Chinchorro mortuary practices (and this is probably the weakest of the features – even the Egyptians did not reuse their mummies). The small, scattered bands do not lend themselves to a monarchial model, nor is there evidence of a threat to the population by a competing group.

However, their ecological niche is a narrow one, with marked dependence on marine resources. The most obvious natural threats to this group of mariners must have been (1) massive tidal waves (tsunamis) secondary to the earthquakes in this area, common even today, and (2) the periodically recurring climatic phenomenon known as El Niño. The latter is characterized by a sudden warming of the coastal waters off Peru, dramatically reducing their productivity. Further north, flooding of the desert coastal areas with otherwise rare, drenching downpours of rain occurs. These violent and uncontrollable natural catastrophes probably were perceived to be of divine origin. Conceivably, intercession with the deity would appear to be appealing to people threatened by such formidable events. Of the models suggested by our ethnohistorical review, one dealing with use of a mummified body to communicate with its spirit (either to exploit its power or, more likely in this case, to appease it or employ it as an intermediary with the deity) seems the most likely. It would be consistent with repeated use of the mummies. The excavated areas may not even represent true burial sites, but might, instead, be convenient dry storage areas from which they could be retrieved readily when needed.

In concluding this topic we note that the ultimate answer to the question central to this chapter is simply that the purpose of mummification is to satisfy some special need of the survivors that mummify the body. At different times and in different places these needs vary from consolidation of political power to the need for communication with supernatural forces. Archaeological patterns, based on ethnohistorical analogies, sometimes (but not invariably) can be helpful in predicting the motivation for mummification in specific, ancient societies.

Chapter 3. Mechanisms of mummification

Introduction and definition

Mummification involves the transformation of a once-living body or tissue into a state of arrested decay. The first chapter of this volume defined "mummy" as a physically preserved corpse or tissue that resembles its living morphology but resists further decay for a prolonged postmortem interval. That duration may be as long as 20000 years or more in the case of some mammoths (Zimmerman & Tedford, 1976), 9000 years for some of northern Chile's Chinchorro people (Aufderheide *et al.*, 1993) or a matter of months for the aborigines of Darnley Island in the Torres Straits; in the last instance, mummification is designed to preserve the cadaver only until completion of certain mortuary rituals (Hamlyn-Harris, 1912a).

Because the mummification methods effecting such transformations are multiple, it is desirable to separate them into individual categories to enhance understanding and memory. In the absence of approved standards, I will employ the following taxonomy throughout this book:

1. Anthropogenic ("artificial") mummification.
2. Spontaneous ("natural") mummification.
3. Spontaneous-enhanced mummification.
4. Indeterminate type of mummification.

Each of these mummification forms can be achieved by multiple mechanisms. The next portion of this chapter is devoted to the elucidation of these mechanisms.

Principles of mummification

The postmortem decay process

The usual decay process following death of a living body is dominated by enzymatic action. Enzymes are chemicals designed to enhance specific chemical reactions. In the usual decay process the chemical reactions of interest are those that break down the large protein, fat and carbohydrate molecules that compose the body's various structures. The final products of this progressive splitting of large molecules are ever smaller fragments primarily (though not exclusively) produced by successive, enzyme-enhanced and enzyme-directed actions. These end products are small molecules, most of which are soluble in water or evolved as gas. These molecules then join the general biochemical pool in water, soil and atmosphere from which all new living tissue ultimately is synthesized. The physical effect is seen initially as tissue softening and finally liquefaction or gasification of the involved tissues. These simple end products may react with environmental substances (e.g., air, soil compounds, etc.) or be ingested by living agents (bacteria, other microbiota, insects, predators) that use other enzymes to transform them into new compounds.

The enzymatic process of autolysis (self-destruction) begins immediately after death. It is initiated by intracellular enzymes normally present within structures called lysosomes. During life these are highly

Fig. 3.1. Insect contributions to soft tissue decay.
Dermestid beetles were found to have fed on brain tissue in the cranial cavity of this spontaneously mummified Egyptian body (30 BC to AD 395) excavated from cemetery 1 at the Kellis site in the Dakhleh Oasis (Roman Period). The live fly in the picture's center landed there just as I was pressing the shutter, after which it departed precipitously. Mummy no. 102. (Photo by ACA.)

regulated, and activated only when their effects are useful to the cell's welfare. After death such regulation ceases and the released enzymes soon produce autodigestion and liquefaction of the cells. Arrival of bacteria from either endogenous sources within the body (oral and fecal flora) or from the environment external to the body then initiates the second stage. These bacteria will secrete more enzymes resulting in further tissue liquefaction in order to generate additional molecules that the bacteria can absorb for their own nutrition. Flies deposit eggs whose larvae (maggots) consume tissue. These are followed by beetles who feed primarily on fly eggs and larvae (Fig. 3.1). Some, especially dermestids, secrete enzymes that permit them to feed directly on corpse tissues (Haskell *et al.*, 1997). Finally, the odor of gases released by these decay reactions attract scavengers that ingest some of the liquefied and the remainder of decaying tissue voraciously. The process eventually terminates with environmental effects involving dissolution of skeletal tissue by interaction of bone mineral with ions in the groundwater (Evans, 1963:13–21).

It becomes obvious, then, that an understanding of mummification mechanisms requires knowledge of factors that will influence the action of the chemicals in this enzyme-dominated decay process.

Factors influencing enzyme action

Aqueous medium. Most enzymes are designed to operate in a watery environment; hence removal of water from dead tissue can retard its decay.

Both spontaneous and anthropogenic desiccation can be very effective mummification mechanisms. Water can be removed from a corpse by heat, osmosis and evaporation. This is the principle employed in dry food processing.

Acidity (pH). Most enzymes operate best at a specific degree of acidity called optimum pH. The greater the departure from an enzyme's optimum pH in its environment, the less will be the enzyme's activity. Strong acids and alkalis have a chemical caustic effect on tissues, but mild forms are tolerated better. Nevertheless, while pH alteration can make a supplemental contribution, it is seldom sufficient as the sole agent to achieve mummification.

Temperature. A given enzyme is commonly exquisitely sensitive to its optimal temperature. Enzyme inhibition is usually more effective at lower temperatures than at higher levels (e.g., meat keeps longer in a refrigerator than in an incubator).

Substrate specificity. Most enzymes are designed to react only with very specifically structured molecules. A significant change in molecular structure can render a molecule immune to the action of a specific enzyme. Alcohol and formaldehyde denature (modify protein structure) so effectively that few naturally occurring enzymes can deal with such altered tissue, making them ideal tissue-preserving agents.

Inhibitors. The action of enzymes can be affected by the presence of chemicals that retard or prevent enzyme activation. Such inhibitors may react directly with the enzyme or retard its action indirectly through several of the above features. Several heavy metal ions sometimes present in groundwater can paralyze most decay-enzyme systems (see below).

Any or several of these enzyme features can be operative in mummification methods. In addition to enzymatic action, tissue breakdown or interaction can also occur by nonenzymatic chemical changes such as oxidation, hydrolysis, esterification and others. We will now turn our attention to various mummification mechanisms in which these features are operational.

Mummification mechanisms

The mechanisms by which mummification is achieved include the following:

1. Desiccation (dehydration; drying)
2. Thermal effects
3. Chemical effects
4. Anaerobiasis
5. Excarnation
6. Miscellaneous
7. Indeterminate

These are discussed individually below.

Desiccation

Drying of body tissues is undoubtedly the most common form of both spontaneous and anthropogenic mummification. Not surprisingly, spontaneous mummification occurs most commonly in the world's hyperarid deserts, including those of Egypt (Sahara), China (Gobi, Tamir Basin, etc.), Peru and northern Chile (Atacama) and even the southwestern USA. However, microclimates leading to desiccation are well documented in otherwise humid climates in which spontaneous desiccation would not be expected (see Chapter 4: Mummies from Peru). Kagamil Island in the foggy, rainy zone of the Aleutian Islands contains a volcanically heated cave that has maintained or even enhanced preservation of anthropogenically-mummified bodies (Dall, 1878; Hrdlička, 1941). Bodies from the southwestern North American Anasazi people were sometimes placed in stone-lined cists designed for grain storage. The summer sun, shining directly on the south-facing cists, could convert such storage units into a crude form of bake-oven, transforming the contained corpses into spontaneously mummified bodies (el-Najjar *et al.*, 1985). Predynastic Egyptian bodies, as well as some of those of the Roman Period, shallowly buried in Egyptian desert soil, commonly became mummified spontaneously (Brier, 1994:19; Aufderheide *et al.*, 1999b). Experimental studies detailed in Chapter 5 have demonstrated that conditions which enhanced the removal of water from the skin surface also accelerated a buried body's desiccation

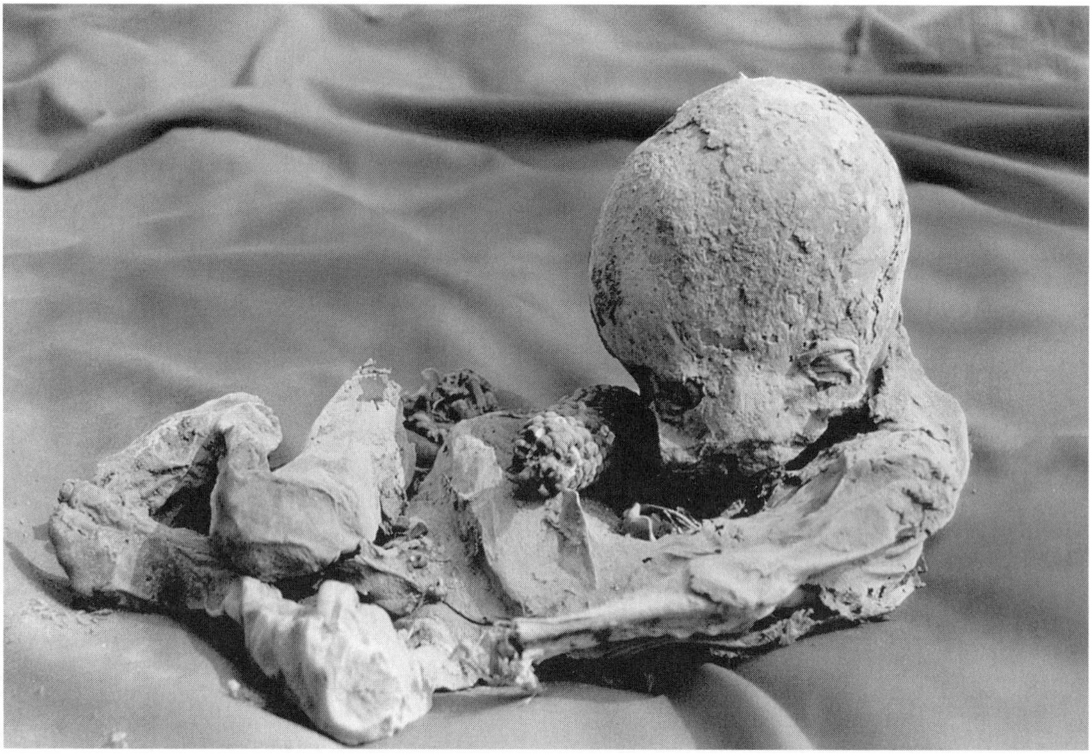

Fig. 3.2. Spontaneously mummified Andean body; ca. AD 200.
The soil and hyperarid climate of the Atacama Desert resulted in spontaneous mummification of this child's body. ca. AD 100–300. Mummy no. AZ-75,T-1,C-1. (Photo by ACA.)

rate, promoting spontaneous mummification. Porous clothing as well as porous soil (sand) were found to function in this manner (Aturaliya & Lukasewycz, 1999).

In the most arid portion of the Atacama Desert (one of the world's driest areas) in northern Chile, soil interment is commonly followed by spontaneous mummification via desiccation (Fig. 3.2) (Aufderheide, 1996). As in Predynastic Egyptian burials, capillary action secondary to the porosity of sand has been assumed to be the process that accelerates removal of enzyme-laden body fluid from the skin surface of these bodies, enhancing spontaneous mummification. It should be noted, however, that the much-cited mummification by burial in Egypt's hot, dry sand during predynastic times implies that heat hastened desiccation. However, while heat applied by other means certainly does accelerate the removal of water from a corpse, sand's contribution to mummification is most probably by its "wicking" action secondary to its porosity. The sun's heat is limited to only the most superficial few centimeters. I measured the temperature of the Atacama Desert's sand at a depth of only a few centimeters and found it to be 47 °C on a sunny day, but at a depth of only 14 cm it was merely 24 °C.

To my knowledge we do not know the precise degree of tissue desiccation necessary to achieve long-term human soft tissue preservation. The dehydrated-food industry seeks a state approaching absolute dry weight for such preservation (Barbosa-Cánovas & Vega-Mercado, 1996:2). Nevertheless, experimental studies in my laboratory on modern, fresh animal tissues indicated that about 80% of their weight was removable water (Table 3.1). Similar tissues obtained from some

Table 3.1. *Water content of living tissues*

Tissue[a]	% dry wt of living tissue	% water content of living tissue	N[b]
Lung	20.9 (0.85)[c]	79.1 (0.85)[c]	3
Liver	27.3 (0.68)	77.7 (0.68)	3
Muscle	24.9 (1.90)	75.2 (1.90)	2
Heart	20.9	79.1	1
Kidney	21.2	78.8	1
Spleen	20.0	80.0	1
Thymus	20.6	77.2	1

Notes:

[a] Tissues from Wistar rats.

[b] N = number of specimens examined.

[c] One standard deviation in parentheses.

Source: From Aturaliya & Lukasewycz, 1999.

recently excavated, spontaneously mummified Andean bodies from the Atacama Desert in northern Chile had reached this state of absolute desiccation from which no further water could be extracted (Aturaliya & Lukasewycz, 1999).

Pharaonic Egypt, however, employed the tissue preservation action via desiccation most dramatically. Exploiting the hydrophilic (water-loving) feature of a local ore (natron), the mummifiers enveloped the eviscerated deceased's corpse in a bed of natron for a period of about six weeks, during which interval sufficient water transferred from the body to the natron to achieve an anthropogenic form of mummification by desiccation. Recently, this result was confirmed experimentally in a human body using Egyptian natron ore (Brier, 1994:64) (Fig. 3.3). Several other methods to achieve desiccation of a corpse are somewhat more eclectic. Sublimation is the term used to describe the transformation of a solid substance into a gas without passing through a liquid phase. The secular term for this process is "freeze-drying" when it is applied to water. Classic examples include the frozen, desiccated human bodies of Inca sacrifices found at Andean mountain peaks (Schobinger, 1991) (Fig. 3.4). A similar effect occurs spontaneously when water in the frozen tissues of human bodies entrapped within glacial ice sublimates into the air bubbles commonly present within such ice, slowly desiccating such bodies. The mummified, frozen, desiccated body of the Iceman trapped in a Tyrolean glacier for more than 5000 years had undergone this change (Spindler, 1994:155).

Even more rare is desiccation by osmosis. When two solutions are separated by a semipermeable membrane, water will flow from the more dilute to the more concentrated until both solutions are at equal concentrations. Thus a body immersed in honey has a lower sugar concentration than that of the honey. Consequently water will move out of the body into the honey until the concentrations are equal. If the volume of honey is much greater than that of the body, the body theoretically can become desiccated. It has been alleged (Crane, 1983:240), but unsatisfactorily documented, that the body of Alexander the Great was immersed in honey at least initially when he died in Babylon in 323 BC (Budge, 1889). Other bodies said to have been preserved in honey include those of Democritus, that of Agesipolis, king of Sparta in Libya in 360 BC (Ransome, 1937:81) and those of the sixteenth century earls of Southampton (Wriothesleys) in a church crypt in Hampshire, England (Crane, 1983:240).

The matter of mummification by honey, however, is more complex than the simplistic rendering in the previous paragraph. First, experiments by Majno (1975:139) demonstrated that preservation of small tissue fragments was effectively enhanced by immersion in honey, but that fragments large enough that honey could not gain immediate access to all parts of it resulted in bacterial putrefaction of the tissue. I know of no equivalent reported test of honey's ability to preserve a large organic mass such as a human body, but an extension of Majno's experience would seem to invite skepticism.

In addition, honey has been found to have an antimicrobial effect other than its osmolar mechanism. In the 1960s an enzyme (glucose oxidase), secreted by the bee's pharyngeal gland, was described that oxidizes glucose to gluconolactone and hydrogen peroxide

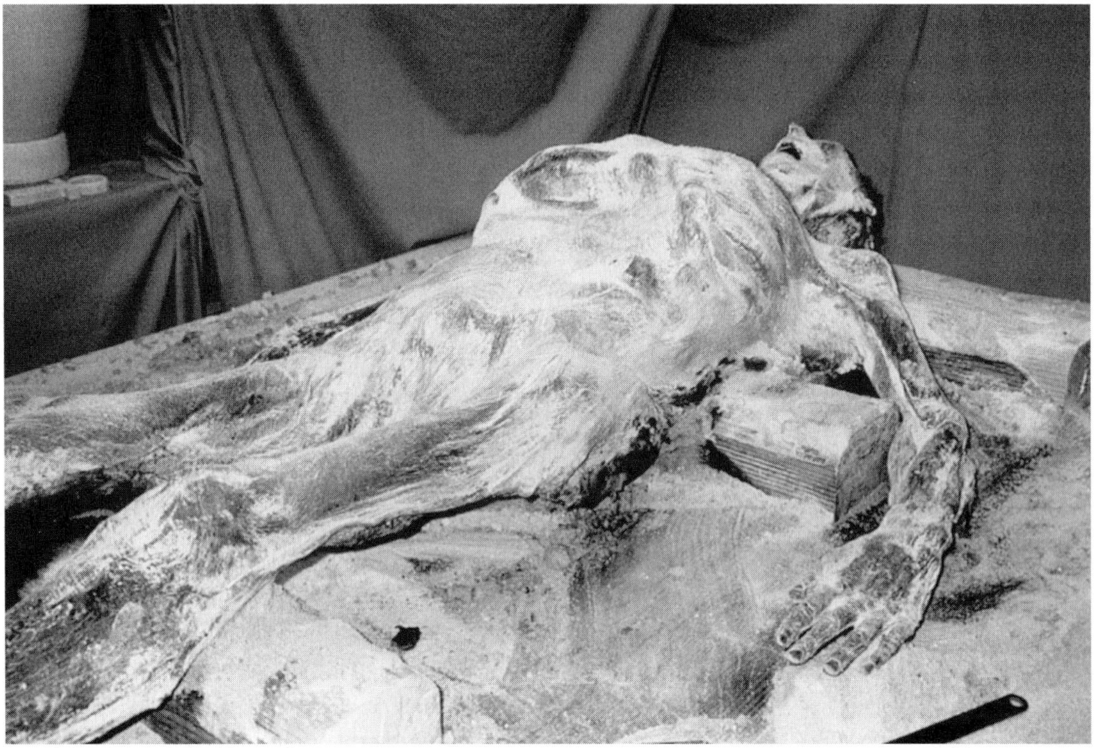

Fig. 3.3. Experimental mummification with natron.
Using powdered natron ore from Egypt, Brier (1997) desiccated a human body successfully. (Photo by Pat Remler, courtesy of Bob Brier and Ronald Wade.)

(White, 1963; Shepartz & Subers, 1964). Nascent oxygen released by the peroxide is responsible for the enzyme's additional antibacterial effect. Because this enzyme (called inhibine) is sensitive to heat and light, its effect could be measured separately. Finally in the late 1960s, an additional honey component (propolis) was identified that had a similar effect but was resistant to heat and light (Lindenfelser, 1967). When applied to local wounds, honey quickly becomes diluted, losing much of its osmolar effect when diluted to less than about 25% of its original concentration (Fig. 3.5). However the antimicrobial effect of inhibine and propolis is retained down to a concentration of less than 11% (Cooper *et al.*, 1999).

It is clear, then, that honey has two different effects that contribute to postmortem tissue preservation: an osmolar effect that extracts water from tissue and also from bacterial bodies sufficiently to kill them, and several enzymes that produce hydrogen peroxide in amounts lethal to bacteria and fungi. Reliable evidence indicates that these effects are sufficient for long-term preservation of small pieces of tissue, but it remains to be demonstrated that these mechanisms are sufficiently effective to preserve whole bodies.

The oft-cited references to examples of whole-body preservation in honey, however, enjoy distinctly lesser support. Original descriptions of what happened at the time of Alexander the Great's death have not survived. What are available to us today are fragmentary copies of copies, most of which are obviously biased by the chaos of political and religious events that swept the then-known world after the demise of its unquestioned leader. Typical is the English translation by Budge (1889) of a Syriac translation by a Christian priest from

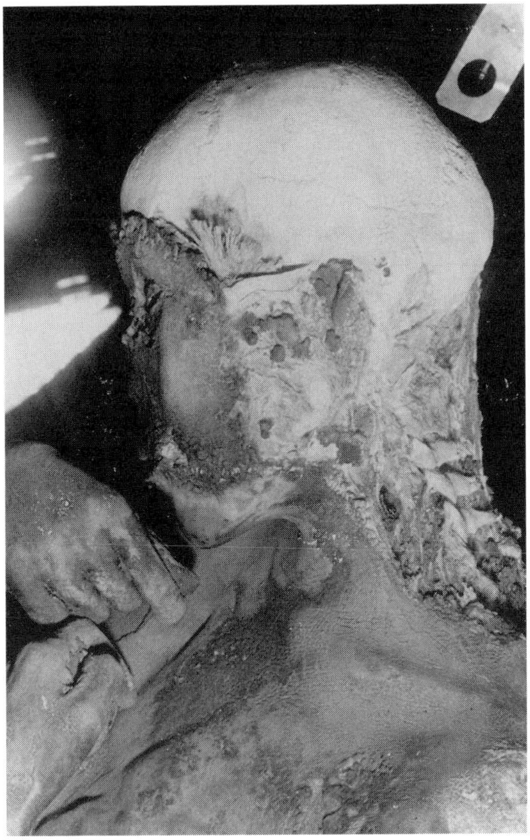

Fig. 3.4. Mummy el Torro from Andean peak.; SM, AD 1400
The frozen and desiccated body of this mummy was found atop the Andean mountain Cerro el Torro at the site of an Inca sacrificial platform. (Photo by and courtesy of Juan Schobinger.)

an Arabic translation of an alleged Greek original document sometime between the seventh and ninth centuries (Budge, 1889). This "History of Alexander" maintains that Alexander's will, dictated just before he died, requested "a coffin of fine gold . . . and let them fill it with white honey which has not been melted . . .". However, whether that was carried out as allegedly requested is not established. The only other item of interest on this topic is a comment by Herodotus (ca. 450 BC?1970, I:198) while describing Babylonians that their "dead are embalmed in honey for burial, and their dirges are like to the dirges of Egypt . . .". While Strabo and others also make comments on this issue, there appears no reason to believe that any account of this issue is completely reliable.

We do have evidence that a long interval, perhaps approaching 2 years, passed before Alexander's body was moved from Babylon. We also know that he was taken to Egypt and that an elaborate sepulcher, designed to enhance his prestige, was constructed to house his corpse in Alexandria. Furthermore, the last record of his appearance was that by Octavian (Caesar Augustus) nearly three centuries later, who touched the body and inadvertently broke off Alexander's nose (Dio Cassius, ca. AD 200/1970, 51:16). This certainly sounds like a desiccated body consistent with an Egyptian type of preparation (Graves, 1957:63). If so, it is still not evident whether that mummification was carried out after Alexander's body arrived in Egypt or whether it was so prepared in Babylon, since it is conceivable Egyptian embalmers may have been traveling with Alexander's army.

In summary, therefore, the historical literature available to us today provides no plausible evidence that Alexander's body was preserved in honey. While most of the other allegations of human bodies preserved in honey fall within the category of anecdotal evidence, one deserves additional scrutiny. The Jewish historian, Flavius Josephus (AD 37–101), writing a history of his people, records a minor incident of interest. The period described dealt with that when Julius Caesar, having taken control of Rome, had interests that clashed with those of Pompey who was campaigning in the eastern part of the realm. To expand his influence in the area of Judea, Caesar freed the imprisoned Maccabean (Hasmonean) leader Aristobolus and gave him two legions to put down unrest in Syria. However, Josephus describes how Pompey's party poisoned Aristobolus and how "his dead body also lay, for a good while, embalmed in honey, till Antony afterwards sent it to Judea, and caused him to be buried in the royal sepulcher . . ." (Josephus, ca. AD 80/1970, XIV:7:4). While Josephus can certainly be criticized for disingenuous recording of some of his motives relating to occasional political items, he clearly prides himself on

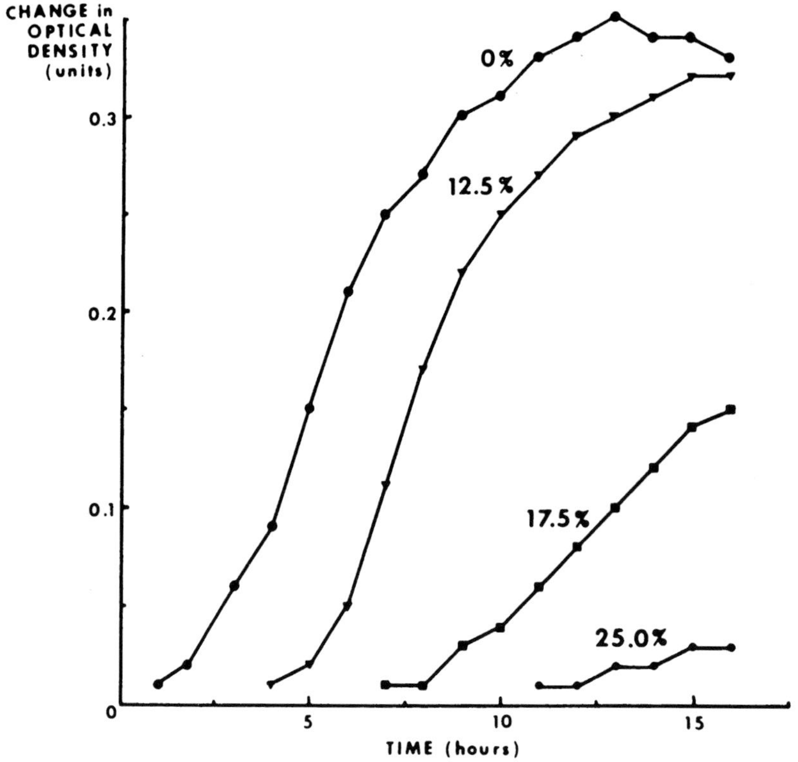

Fig. 3.5. Bacteriostatic effect of honey.
The growth of *Staphylococcus aureus* in broth culture was determined by optical density measurements of the broth. Complete growth inhibition was achieved with broth containing 25% honey, but lesser concentrations of honey had lesser degrees of inhibition. (From Estes, 1993: 70 (revised edition) with permission of publishers Watson Publishing International/Science History Publications.)

historical accuracy in the description of events. The above incident happened more than a century before his recording of it, so we can not search for eye witness evidence. However, information provided by an historian normally viewed with considerable respect does not allow us the liberty of simple rejection of that account in cavalier fashion. Only a formal test by submersion of a large animal carcass in honey can now provide credible evidence for or against the hypothesis of mummification by immersion in honey.

Thermal effects

It is well known that cooling a body results in slower decay of cadaver tissues. From what has been stated above about the optimum temperatures for enzymatic action, this is what can be expected of endogenous enzymes as well as those of external and internal bacteria. While some bacteria, called psychrophiles, have been found whose optimum temperature is near 0 °C and others (thermophiles) thrive at temperatures between 50 and 100 °C, the time required for multiplication by most bacteria that colonize the mammalian gut rises to almost infinity below 4 °C and, of course, ceases completely at 0 °C (Micozzi, 1991:41). Preservation of meat for consumption in the home employs temperatures between +4 and −10 °C, depending on the desired interval of preservation.

Many examples of ancient bodies spontaneously preserved by freezing have been reported. These

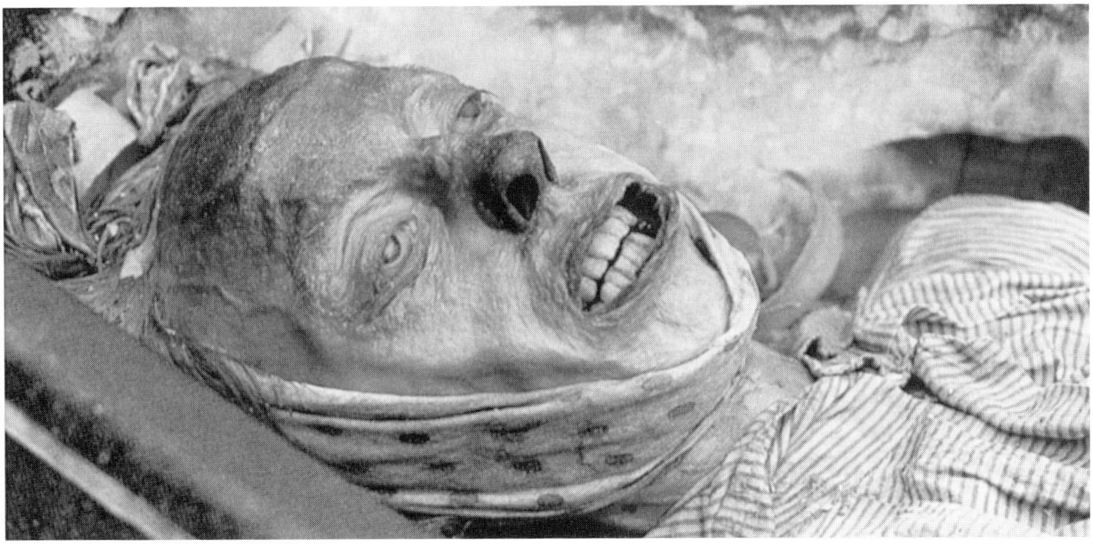

Fig. 3.6. Preservation by freezing; SM, AD 1846.
Buried in permafrost soil in the Canadian Arctic (Beechey Island) in the mid nineteenth century, this body of the crew member (John Torrington) of the Franklin expedition demonstrated superb soft tissue preservation. (Photo courtesy of archaeologist Owen Beattie.)

include several from Alaska in which catastrophes such as mud slide immersion (Zimmerman & Smith, 1975) and Arctic Ocean pack ice submersion occurred (Zimmerman & Aufderheide, 1984). In these cases the bodies were flattened against permanently frozen ground (permafrost) and held thus in position by overlying soil or ice until re-exposed or excavated centuries later. When the British arctic explorer John Franklin buried several sailors in Canadian arctic soil in the mid nineteenth century, he placed them within permafrost. Excavated a century later the tissues of the still-frozen bodies were almost ideally preserved (Beattie & Geiger, 1987) (Fig. 3.6). The frozen bodies of Scythian chiefs at Pazyryk, Russia, are similar examples of preservation by freezing in permafrost burials (Rudenko, 1970). Siberian mammoths recovered from glacial ice as long as 44 000 years after death maintained very good gross morphology, though chemical evidence of considerable protein breakdown was attributed to postdiscovery thawing of the specimens (Goodman *et al.*, 1979). Frozen Inuit (Eskimo) bodies in Greenland were preserved spontaneously both by freezing and freeze-drying (Hart Hansen *et al.*, 1991:139). In all of these, low temperatures essentially suspended enzyme action and in some cases the freeze-drying sublimating action contributed to desiccation as well.

Deliberate mummification by freezing has become a reality. Suggested as early as 1966 (Ettinger), commercial mortuary services are now available to provide this option. Customers for this procedure are attracted by the concept that, if freezing occurs rapidly, the usual destructive effects of ice crystal formation can be avoided. If an appropriate rapid thawing method can be devised, the possibility of revivification at a later time is enticing. Rapid freezing by immersion in liquid nitrogen is relatively simple and can achieve that initial criterion. Thawing with maintenance of life after such freezing has been achieved for individual cells (ova, sperm) and simple, multicellular living creatures but not yet for intact large animal or human bodies.

Elevated temperatures can also contribute to mummification. Experimental evidence has demonstrated that most increases above body temperature of humans

Fig. 3.7. Heat-dried mummy; AM, ca. AD 1300. Bodies of the Muisca tribe on the savanna of Bogotá, Colombia, were desiccated by heat from a fire. (Photo courtesy of Felipe Cárdenas-Arroyo.)

can decrease the time required for bacterial multiplication, increasing their numbers more rapidly (Micozzi, 1991:41). In an experimental study of rat carcasses at controlled temperatures elevation of environmental temperature to 43 °C demonstrated more rapid desiccation and mummification but at the expense of more extensive tissue destruction because of more rapid bacterial growth (Aturaliya *et al.*, 1995). Most bacteria present in normal mammalian bodies demonstrate an ability to grow in laboratory cultures at temperatures substantially above normal body temperatures, but I have been unable to find studies on intact animal bodies focusing on the critical threshold levels at which growth ceases. We do know that seventeenth and eighteenth century European physicians created such a demand for desiccated, powdered mummy tissues to be used as a medicinal agent (Chapter 10: Mummy as a drug) that illegal production of "mummies" from recently deceased corpses by Egyptian hucksters was carried out by baking the bodies in makeshift ovens. Such bodies could be kept in a state of good preservation for years (Dawson, 1927c). Presumably such oven temperatures substantially exceeded the limits of bacterial growth or even survival. The Muisca and other aboriginals of Pre-Hispanic Colombia mummified their deceaseds' bodies by suspending them above or between fires (Cárdenas-Arroyo, 1990) (Fig. 3.7). The eventual desiccation of these bodies was the feature that provided long-term soft tissue preservation, but the fires probably also raised body temperatures high enough to retard bacterial growth while desiccation took place. It has been suggested that deliberate exposure of corpses to the hot summer sun contributed to the mummification of the Canary Islands Guanche bodies (Rodríguez-Martín, 1995), Incas and even Egyptians (Majno, 1975:134), but ethnographic accounts are vague and such sun exposure would imprint no recognizable diagnostic tissue change on the body. Hence these alleged spontaneous-enhanced mummification methods have remained unconfirmed.

Chemical effects

Heavy metals

Mercury (Hg), arsenic (As), copper (Cu), lead (Pb) and other heavy metals have long been known to be toxic to living humans. Small amounts of some of these (e.g., Cu) are essential for certain body metabolic needs, but in excess they can be toxic. Clinical symptoms of such poisonings frequently are brought about by inactivation of enzyme systems necessary for normal metabolism and health. Exposure of cadavers to high concentrations of these metals can also inactivate enzymes participating in the chemical reactions of the decay process. For example, the local action of copper

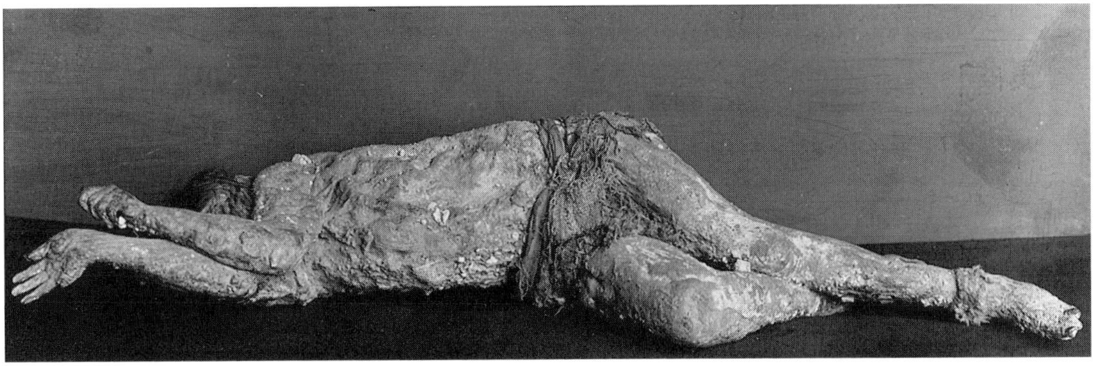

Fig. 3.8. The "Copper Man" mummy; AM, ca. AD 1000.
While copper ions in the copper mine that trapped this ancient miner have been suggested as responsible for the body's soft tissue preservation, tissue tests could identify copper only in the superficial tissues. (Photo courtesy of the Department of Library Services of New York's American Museum of Natural History, Negative no. 45503. Photo by R.E. Dahlgren.)

ions has been seen in the form of soft tissue preservation restricted to the immediate vicinity of copper artifacts such as bracelets, eye covers and others (Morris, 1981). The binding of copper ions to polypeptides can also have a "tanning" (protein-denaturing) effect. More extensive exposure to copper ions can produce a diffuse effect, resulting in preservation of much of the body's skin as occurred in an aboriginal infant from British Columbia (Canada) (Schulting, 1995) and an ancient copper miner who was trapped in a cave in northern Chile (Fig. 3.8) (Bird, 1979). Currently it is not yet clear whether enough copper can permeate the entire body quickly enough after death to preserve the viscera spontaneously, though Schulting believes that this is what occurred in the infant he examined. However, tests on "Copper Man" from Chile reveal that copper was limited to the skin, suggesting his mummification was the product of the Atacama Desert's hyperarid environment (Preston, 1980).

Arsenic, however, has a long record of deliberate use for purposes of soft tissue preservation. It was the principal ingredient in many American and European solutions used for intra-arterial injection by nineteenth century embalmers. Probably the most dramatic forensic anthropological example is the case of Elmer McCurdy, a young adult, nineteenth century outlaw cowboy of the western USA. This case is detailed in Chapter 10 (Mummy as display), but it can be pointed out here that apparently arsenic-containing embalming fluid had retained gross body structure sufficiently to permit use of his body for display at carnival side shows for more than a century after his death. The histological appearance of his soft tissues revealed surprising detail (Snow & Reyman, 1977) (Fig. 3.9). A group of mummies from northern Chile (Cam-9), buried near the Camarones River, whose water contained high levels of arsenic, was found to demonstrate high arsenic content in the soft tissues and had skin lesions compatible with those seen in chronic arsenic poisoning. However, bodies buried in the soil of that region (the Atacama Desert) often undergo spontaneous mummification, so that we cannot be certain what role the arsenic played in the soft tissue preservation (Allison *et al.*, 1995–1996). Similarly, the soft tissue preservation found in the body of the arctic explorer Charles Francis Hall in Denmark (discussed in Chapter 4: Mummies from northern Scandinavia) can be explained by thermal effects even though the body had been exposed to arsenic before death and to arsenic in the groundwater after burial. We still have no clearly defined examples of postmortem arsenic exposure capable of reaching tissue arsenic concentrations high enough to

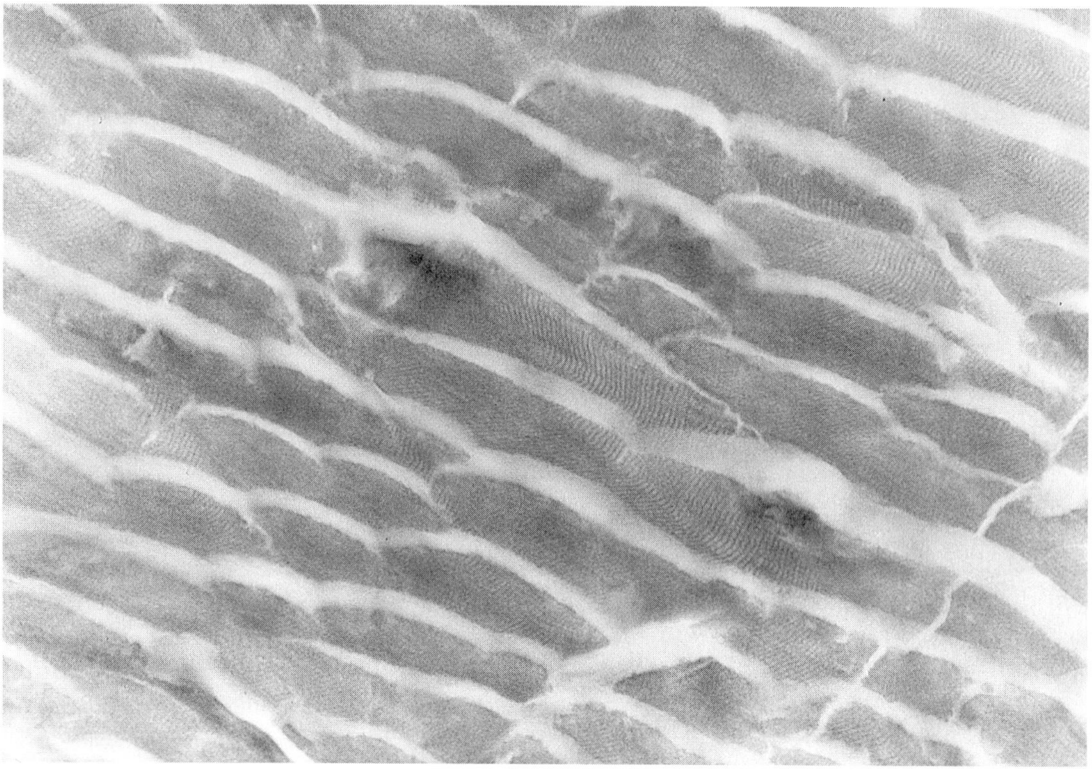

Fig. 3.9. Retention of cytological detail by arsenic ion effect.
The individual muscle fibers have retained their shape, together with preservation of their characteristic cross-striations more than 90 years after death. Mummy of Elmer McCurdy (AM, ca. AD 1911). (Photo by and courtesy of T.A. Reyman.)

result in significant body-wide soft tissue preservation effect except by intra-arterial injection. A superbly preserved nineteenth century body of a child in Palermo's Capuchin monastery is suspected to be the product of arsenic embalming, though no chemical studies have yet been performed on it (Fig. 3.10).

While lesser accessibility of mercury compounds may be responsible for the metal's less frequent employment for anthropogenic mummification, we do have examples of its use. Best known is the application of mercury-containing paste to a Renaissance Italian juvenile's body for purposes of soft tissue preservation (Fornaciari, 1985) (Fig. 3.11). A similar use of mercury in the form of its ore (cinnabar) was found on the mummified body of a Swiss Franciscan monk from the sixteenth century (Kaufmann, 1996a). However, no chemical assays of tissue mercury were performed, so we do not know what role, if any, mercury played in the soft tissue preservation. Yamada *et al.* (1990) also detected mercury in soft tissue fragments of 329-year-old remains of a human from Japan. The striking soft tissue preservation noted in the 2000-year-old body of a Chinese female was attributed at least in part to the presence of mercury ions in the coffin fluid. However, the small amounts present (not quantitatively reported) probably required contributions to the body's mummification by currently unidentified other factors as well (Peng, 1995).

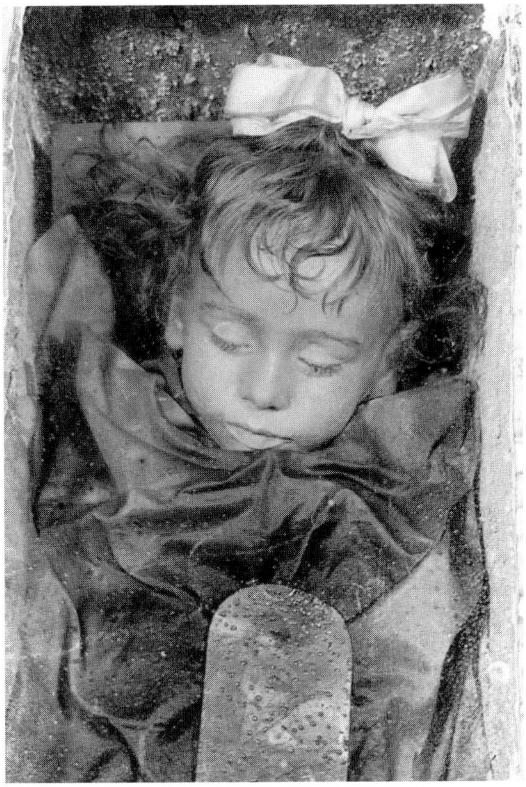

Fig. 3.10. Mummified child in Sicilian Capuchin monastery (AM, ca. AD 1900).
The intra-arterial embalming solution used in this body may have had a high arsenic content. (Photo courtesy of Padre Superiore della Confraternita dei Cappuccine in Palermo, Sicily.)

Chelation

Some substances called chelating agents tend to combine with heavy metals. In peat bogs containing sphagnum moss a compound called sphagnan chelates calcium. Calcium ions are essential for normal enzyme action within most bacteria. By chelating these ions, sphagnan makes calcium ions in peat unavailable to bacteria. This action limits bacterial multiplication and that effect is a major contribution sphagnan makes to the preservation of soft tissue in peat bog bodies such as those of Lindow Man (Painter, 1991). This chelation of calcium is also responsible for the decalcified state of bog mummies' bones. In addition, sphagnan reacts with ammonia, amino acids and polypeptide amino groups to form a highly stable chemical product (Maillard reaction); this nitrogen-trapping effect profoundly reduces the availability of nitrogenous nutrients required for bacterial growth, further limiting microbial multiplication. The much-cited tanning effect of peat bogs by tannin, while not absent, probably makes only an additional contribution.

Alteration of chemical composition of tissue matrix

Enzymes are extremely specific in their selection of chemical sites at which to carry out their action. Some postmortem changes involve such an extensive alteration of chemical structure that few, if any enzymes exist that can react with the transformed chemical. The effect of this will act to preserve such altered soft tissue. Two common examples encountered in mummification are provided below.

Adipocere formation

Under appropriate conditions, often in a watery environment, body fat can undergo chemical changes that result in a crumbly, wax-like product (adipocere) that resists subsequent chemical change and thus tends to preserve the tissue's gross morphology (Fig. 3.12). The precise chemical chronology is presently still incompletely understood. Studies by Takatori *et al.* (Takatori & Yamaoka, 1979; Takatori *et al.*, 1987) indicate that the process probably involves breakdown of neutral fat into fatty acids. Subsequent oxidation occurs involving enzymes contributed by certain bacteria (*Clostridium* spp.). These oxidized fatty acids assume forms not commonly present in nature. Consequently these are not reactive with most of nature's enzyme pools. The process may act focally in bodies, but a common effect is the production of a surface shell that shields enclosed soft tissues from

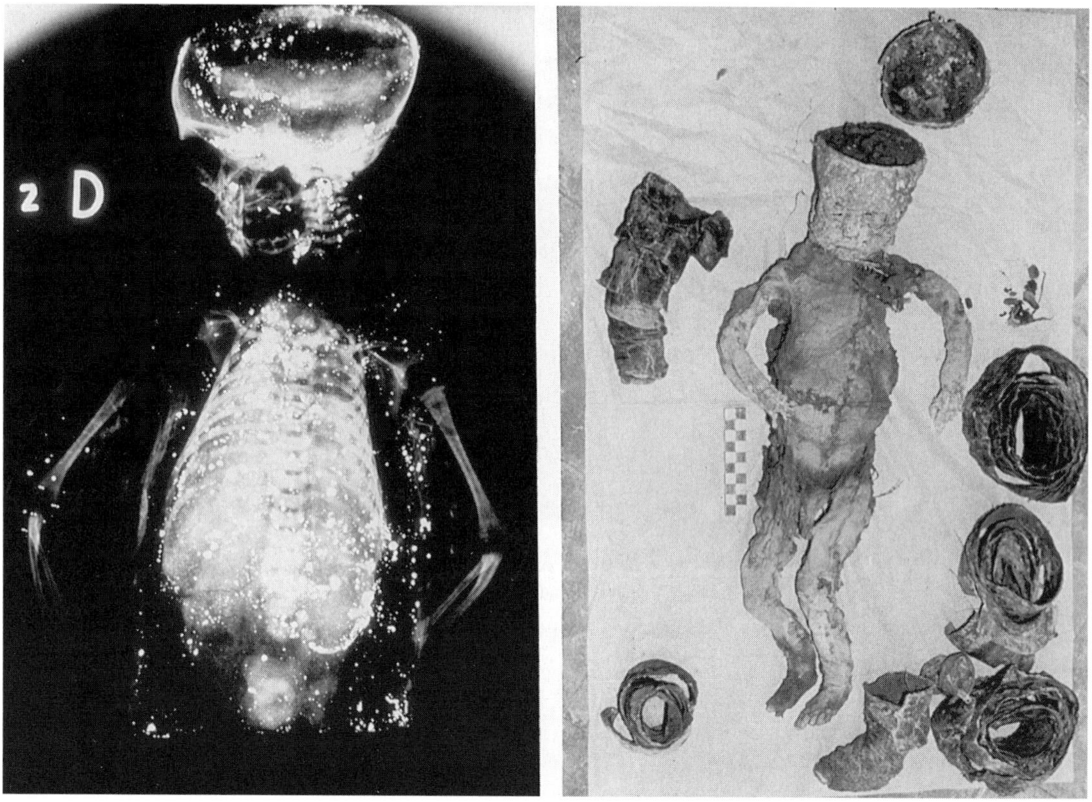

Fig. 3.11. Use of mercury for mummification; AM, ca. AD 1530.
Right: The body of the sixteenth century Italian child was eviscerated and the skin surface covered with a mercury-containing paste. *Left*: X-ray radioopacities represent mercury in the paste. (Photo by and courtesy of G. Fornaciari.)

further decay (Cotton *et al.*, 1987). In some corpses adipocere formation can involve the entire body, generating mummified human remains (Fig. 3.12) and it is commonly a component of glacier mummification (Bereuter *et al.*, 1996a). The high fat content of brain tissue can also result in brain preservation by this process (Tkocz *et al.*, 1979).

Tanning

The Maillard reaction produced by sphagnan in peat bogs also causes cross-linking of collagen fibers that expels water and resists rehydration – a process that stabilizes and preserves tissue collagen ("tanning"). This type of change can also be produced by the immersion of soft tissue in formaldehyde. Modern soft tissue specimens excised at surgery or autopsy dissection are commonly preserved for further study by this technique.

Resins and spices

Resins are tree saps or bark distillate products. Liquid when heated, they harden to crystalline density at room temperature. Most popular among those employed for mummification were pine resins from juniper trees (biblically often called "cedars of Lebanon") as well as myrrh and frankincense (from

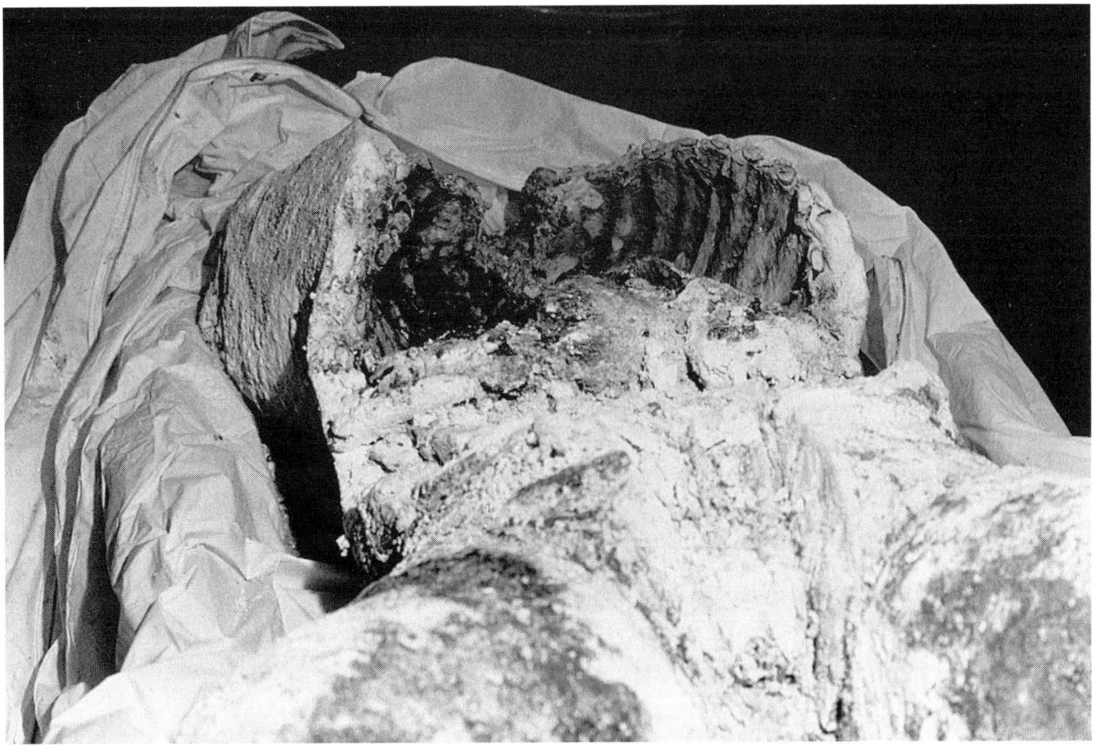

Fig. 3.12. Adipocere formation.
The body of this young adult male freshwater (Lake Superior, USA) drowning victim demonstrates massive adipocere formation in the trunk and proximal extremities. (Photo by Ned Austin, PBL.)

trees of the Burseraceae family). The last two are particularly aromatic. Frankincense was obtained from the horn of Africa (modern Somalia). Myrrh was present there also as well as at the southern tip of the Arabian peninsula (modern Yemen). Juniper resin was procured primarily from the area of modern Lebanon. Egyptian embalmers specialized in their use of these materials. Following evisceration and body desiccation with powdered natron, the chest and abdominal cavities were painted with resins. They were also applied to the skin surface. Dabs of resin were used to hold bandages and wrappings in place. It is not clear whether embalmers used them primarily as agents to prevent rehydration or as antiseptics. Majno (1975:217–218 and his Fig. 5.11) confirmed the antibacterial effect of myrrh, using modern hospital bacterial antibiotic sensitivity techniques (Fig. 3.13). Spices such as cinnamon (*Cinnamomum cassia*) were also employed, though little formal study of their tissue preservation potential has been reported.

Other chemical effects

Lime

Lime is calcium oxide (CaO), derived by burning limestone or ground seashells (calcium carbonate: $CaCO_3$) to drive off the carbon dioxide. This residual lime has a very high affinity for water, forming a crumbly, white mass (calcium hydroxide: $Ca(OH)_2$) when exposed to

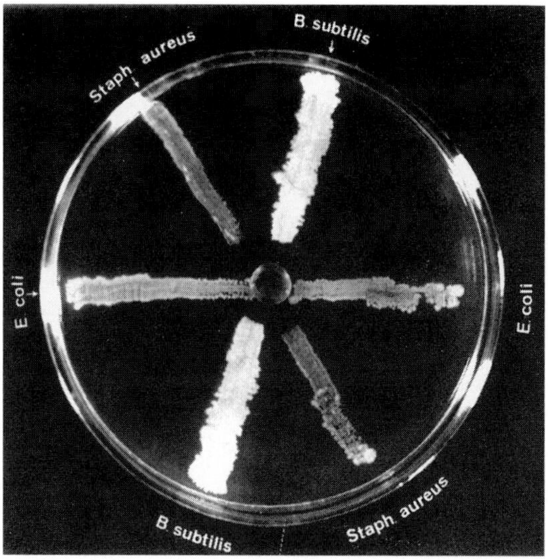

Fig. 3.13. Antibacterial effect of myrrh.
The myrrh that diffused from the central well into the agar prevented the growth of *Staphylococcus aureus* and *Bacillus subtilis* near the well, but had no effect on *Escherichia coli*, demonstrating myrrh's antibacterial potential. (From Majno, 1975: 217. Photo by and courtesy of G. Majno, E. Shorer and Sylvia Dersi.)

water, with the evolution of a great amount of heat. Encasing a body with lime, then, would have multiple mummifying effects including heat desiccation and enzyme inactivation by raising the pH substantially. Surface application of lime to the body of Francisco Pizarro, the Spanish conquistador (Peru), contributed to its mummification (McGee, 1894) as it did also to some of the bodies in the initial interment site of the members of the Russian Czar Nicholas' family (Quinn-Judge, 1998) (see Chapter 4: Mummies from Russia and Ukraine). It is also believed to have been used on some of the Capuchin monks' bodies in Palermo, Sicily (Matranga, 1983). Other examples include medieval Italian mummies from Urbania (Fornaciari, 1982) and 300-year-old mummies from Japan (Yamada *et al.*, 1996). In several spontaneously mummified bodies from Ireland desiccation was attributed to the hygroscopic effect of the limestone composing their mausolea (McKinley, 1977). Pure lime, however, can have a tissue-corrosive effect that mitigates its value as a preservative, and its combination with water can produce a brittle crust.

Lye

This consists of potassium or sodium hydroxide, commonly obtained by leaching wood ashes. It has a highly caustic action and a profoundly alkalinizing effect. While the dramatic change in pH would effect mummification, the caustic effect is too tissue destructive to be attractive as a deliberate, tissue-preserving agent. Nevertheless, occasionally at least partly preserved bodies have been found that were preserved in this manner (Kaufmann, 1996a). The gentler effect of whole wood ashes on tissue pH was used more commonly.

Anaerobiasis and plaster mummies

In the past, anaerobic conditions were quite commonly evoked as an explanation for mummification. Anaerobiasis was cited as a major factor contributing to the preservation of a more than 2000-year-old body of a female from China's Han Dynasty (Hunan Medical University, 1980). I have been unable to identify scientific reports that support this allegation. Because the human colon commonly harbors some anaerobes (or at least their spores) one can expect that they would flourish in an anaerobic environment and effect tissue destruction through putrefactive processes. The peculiar properties of peat bogs commonly produce an anaerobic zone in their deeper layers to which the mummifying effect of such bogs has been attributed, but the effects of sphagnan, described above, are now believed to be the principal factors (Painter, 1991). In fact, shortly after blood circulation ceases, most body tissues promptly become oxygen depleted.

Plaster mummies

It is probably safe to assume that efforts to encase parts of, and sometimes whole, bodies in plaster of some type

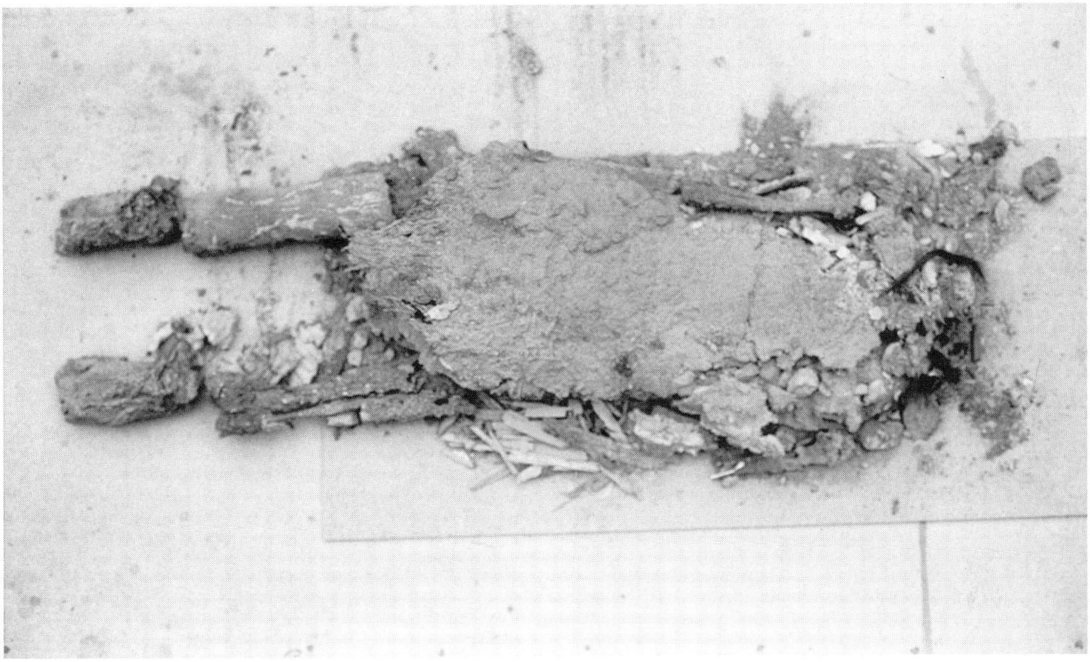

Fig. 3.14. Chinchorro "plaster" mummy; AM, 4120 BC.
This mummified body from the Chinchorro population of northern Chile was encased in a slurry of sand and mud containing a hardening agent. Little, if any, soft tissue survived this treatment. (Photo by ACA.)

were based on the assumption that exclusion of air would enhance preservation. Admittedly, prevention of access to the body by insects or scavengers may have played a role by suppressing soft tissue loss in some of these, but plaster mummies did not live up to such expectations.

Probably the oldest effort to preserve human tissue by isolating it from the environment occurred during Egypt's first dynasty, where the simple wrapping of a body's arm was noted in Pharaoh Djer's tomb (Brier, 1994:81). Predictably, it failed: the contents were skeletonized. By the sixth dynasty (at Giza), some heads and some entire bodies were enveloped in plaster (Junker, 1914). At the latter end (Ptolemaic) of the Pharaonic Period, linen-wrapped baboon bodies were placed in coffins that were filled with plaster-like material (Emery, 1970). Human bodies so treated, however, usually skeletonized (Delattre, 1888, 1905, 1929). In Peru and northern Chile the earlier, highly complex mummification methods carried out by the Chinchorro people gradually became more simplified until, at about 2000 BC, this procedure was characterized by simply encasing the entire body with a slurry of sand and mud combined with an as yet undetermined proteinaceous hardening agent, which, when dried, effected a plaster-like consistency (Fig. 3.14). The earliest such body had been eviscerated but thereafter no other obvious method seems to have been made to preserve tissue other than use of plaster (though Arriaza (1995b):114–115) suggests that the body may have been smoked). Soft tissue preservation in these bodies is not good, though complete spontaneous skeletonization occurred only occasionally. It must be noted that spontaneous mummification of a buried Chinchorro (or other) body is common in the Atacama Desert, so that preservation of some soft tissue secondary to spontaneous desiccation there is to be expected. Plaster-enveloped bodies in Algeria and Tunisia from

Fig. 3.15. One of Pompeii's plaster mummies; SM, AD 79.
When hot ash covered the bodies of those trapped in Pompeii, Italy, as Mt Vesuvius erupted in the first century, subsequent rain compacted the ash. The bodies decayed, but when the "mummies" were encountered during excavation, archaeologists filled the space (originally occupied by the body) with plaster, then excavated away the surrounding ash to reveal the original body form. (Photo courtesy of Sylvia Horwitz.)

the Roman Period are not uncommon in Christian cemeteries. Green (1977) has presented an excellent survey of plaster burials from this period not only in North Africa but also in Europe, including Italy, Germany, France and Britain. In some of these, laboratory tests have identified the plaster as gypsum. In England, forty such plaster burials were found in York, and nearly as many at Dorchester. The total of plaster burials that Green describes must number in the hundreds. While he rarely describes the bodies' degree of preservation, he appears to regard the second century soft tissue-preserved body of St Cecilia, a plaster burial in Rome's Catacombs of Priscilla, as sufficiently remarkable to suggest that she may have been embalmed. This, together with his occasional comment of "skeletal remains" within the plaster, suggests that the latter was the common condition.

Perhaps the most dramatic evidence for the soft tissue preservation ineffectiveness by isolation of the body from its environment are the "plaster people of Pompeii" (Place, 1995b:28). As hot ash rained down upon the Pompeii residents fleeing the AD 79 eruption of Mount Vesuvius in Italy, many fell dead. The hot ash covered them, and subsequent rain compacted and then hardened it. At excavation nearly 2000 years later, archaeologists found nearly all soft tissue traces of the bodies themselves had disappeared. Only the hollow spaces conforming to the bodies' shape (reconstructed by filling these spaces with plaster before excavating the hardened ash around them) remained to identify their origin (Fig. 3.15). In summary, plaster burials offer little support for the suggestion that anaerobiasis contributes significantly to soft tissue preservation.

Miscellaneous mummification methods

Tree coffins

About 1000 BC, contemporary with the end of Egypt's New Kingdom, an agropastoral population in what is now Denmark buried the corpses of their elite members in oak coffins. They cut down oak trees of a size not seen in modern Danish lands; some of these are up to 2 meters in diameter. A 3 meter-long segment of such a trunk was sawed longitudinally, the two halves hollowed out and, following placement of the body in the lower half, the opposing half was placed on it as a lid. The body was wrapped in textile and cow

Fig. 3.16. Danish tree coffin; ca. 1000 BC.
The coffin was created by bisecting longitudinally a two meter segment of an oak tree trunk, hollowing each half and placing the body in the lower half while using the other half as a lid. Preservation is assumed to have been the effect of tannin leached from the trunk by groundwater. (Photo courtesy of Nationalmuseet. Danske Afdeling, Danmarks, Oldtid, Kobenhavn.)

hides, accompanied by grave goods (Fig. 3.16). A huge mound of soil covered the buried coffin.

Several such mounds (barrows) have been excavated, one near Aarhus and another near modern Egtved close to Silkeborg in Denmark. These bodies are almost skeletonized but hair and patches of skin are commonly preserved. It is assumed that groundwater seeping through the oak coffin absorbed enough tannin from the wood to exert a tanning effect on these tissues. Viscera and muscle are usually absent though brain tissue sometimes survives, probably as adipocere (Kaner, 1996:96).

The central African baobab tree (*Adansonia digitata*) has a unique structure. Its trunk is huge – some are more than 30 meters in circumference. It thrives in areas of low rainfall by virtue of a depression at the top of its blunted trunk that catches and stores rainfall. The wood has a porous fiber structure that enhances desiccation of stored material. Natives are said to achieve a degree of spontaneous mummification via desiccation by suspending human bodies in such structures during dry seasons (Wright & Kerfoot, 1966).

While the Scythian rulers in Russia's Altai (Chapter 4: Mummies from Russia and Ukraine) were also buried in hollowed-out logs, their excellent preservation is not the result of a tanning effect, but rather that of freezing in permafrost.

Salt as a preserving agent

When the missionary-explorer David Livingstone died at a village on Lake Bangweulu in what is now modern Zambia in 1873, several of his party "embalmed him" so his corpse would not decay during the six months required to carry it through the central African jungle to Zanzibar on the Tanzanian coast 2400 kilometers away. Actually it was a process of mummification. They first eviscerated his body, placing these structures in a tin box and burying this ceremoniously at the base of a tree into which they carved an appropriate inscription (now preserved in a London museum). The abdominal and chest cavities were then filled with salt. Thereafter the body was dried, principally by exposing it to the sun for two to three weeks, turning it regularly so all aspects would be desiccated. After a reasonable degree of dehydration had been achieved the deeper areas of larger muscles were excarnated, the legs incised at the knees, and the body folded into the length of the head and trunk. It was then wrapped in canvas and bark, after which the bundle was waterproofed with tar. Arriving at port six months later, the bundle was put aboard a ship in a wood coffin. When that coffin was opened at arrival in London, his friends were astonished to recognize Livingstone's face well preserved and partly covered by his long hair, while the trunk revealed the shrunken, wrinkled skin of a desiccated body. His remains rest today in Westminster Abbey (Monro, 1933:98–99; Ransford, 1978:304–310).

Commercial modern mummification

The intra-arterial embalming carried out on most modern American deceased individuals can be viewed as a form of mummification. It is designed to maintain the body in a viewable condition until mortuary rituals ("funeral services") have been completed. Since the chemical basis of such preservation is formaldehyde, this eventually evaporates, following which the tissue slowly decays.

A novel form of more prolonged (permanent?) mummification is offered by an organization located in Salt Lake City, Utah. Called Summum, this nonprofit organization functions as a religious group that provides mummification services. After customary funeral services, the body is injected with a variety of chemicals, the nature of which is not disclosed. The entire body then is submerged in more of this liquid in a purpose-designed tank. At least 30 to 80 days of such preparation is required. The outer container ("Mummiform"©), chosen by the individual before death or by representatives of the deceased is then prepared, individualized by an appropriate artist. The container may be composed of bronze, stainless steel or gold. The prepared body is then placed within the container and all residual space within the container is obliterated by the introduction of a liquid resin that subsequently hardens.

The organization claims to have treated successfully a number of cadavers from anatomy departments in this manner. Its clients to date have been pets (birds, dogs, cats), though in 1996 the organization stated that 137 living human clients had contracted for future services at time of death (London Observer Service, 1996). During my visit in 1998 the procedure was courteously demonstrated to us, and we were informed that insurance arrangements for payment of these services to be provided after death were available.

Bat guano as preservative

Lovelock Cave is located about 24 kilometers south-southwest of Lovelock, Nevada at about 39° 50′ N and 118° 32′ W. Its floor was covered with bat guano to a depth of about 2 meters. Buried within the midst of this guano were numerous mummified birds as well as the remains of 29 adult humans and 16 children. These were excavated in 1912, though some of them had been found and discarded into a waste heap by a crew that dug about five carloads of guano from the cave in 1887 and sold it as fertilizer. Most of the human remains were only skeletons, though two adult males and a child were found to be mummified by the guano crew. The remains were transferred to the Nevada Historical Society. In addition a newborn infant mummy with attached umbilical cord and placenta were excavated in 1912 (Loud & Harrington, 1929:29–32, 167–169). Preservation of these tissues is obviously the consequence of spontaneous desiccation, as is evident in Loud's photos (Loud & Harrington, plate 11). Factors contributing to this desiccation are speculative, but the most eligible candidates would seem to be the arid climate, the time of year, the porosity of guano as well as its alkalinity and, in the case of the newborn, its nearly sterile body at death.

Preservation by taxidermy

Born in 1721 as an African prince, Angelo Soliman was kidnapped and sold into slavery at age 7 years. A succession of masters led to his being educated and eventually achieving a high level of social service to a Lichtenstein prince until his death in Vienna in 1796. At the prince's request, the body of Soliman was flayed and his preserved skin fitted onto a specially prepared wood model that displayed him in a dramatic, heroic presentation in the prince's quarters. Here he remained on view to the prince's guests until consumed by a fire 48 years later (Seipel, 1996). The records of the Austrian Academy of Sciences do not detail how the skin was preserved but possibilities would seem to include tanning or desiccation.

In the storage rooms of the National Museum of Health and Medicine in Washington, DC, is another example of preservation by taxidermy. A plastic model of only the trunk of a human body is covered with a surface of preserved human skin (Plate 1). The museum's records provide no provenience or other data about this acquisition, but it is most probably of Japanese origin because every square millimeter of the skin surface is covered with brilliant, polychrome tattoos of artistically displayed figures resembling Japanese motifs. Again, the most likely method of skin preservation is tanning or desiccation. Chapter 7 (Non-Egyptian mummified animals) also describes several famous horses preserved by taxidermy tanning methods.

Mummification by ventilated air drying

Desiccation mummification secondary to exposure of a corpse to a flow of air appears to be the key mechanism in only a few reported cases. In a discussion of medieval mummification, Kaufmann (1996a) describes the custom in the Baltic area of mummification by hanging a recently deceased body in a church belfry that was open enough to admit a constant exchange of air. He comments on its effectiveness, noting that the only disadvantage was a strong odor. An additional example Kaufmann cites is the body of Johann Philipp of Hohensax that revealed skin changes suggestive of such a process; his viscera were well preserved.

The ventilation factor probably also played a role in the spontaneous mummification presented by the bodies of the Peruvian Chachapoya mummies described in Chapter 4 (Mummies from Peru). These were placed in specially constructed mausolea in the Andean mountains. These "mummy warehouses" were in an elevated position with prominent window-like openings that achieved excellent (and probably frequent) air movement, resulting in desiccation of the bodies in spite of the humid environment (Kauffmann-Doig, 1998a) (Fig. 3.17).

Fig. 3.17. Chachapoya mummy (SM, ca. AD 1350) placed in window of mausoleum.
Placement of mummified bodies in a well-ventilated location may have played an important role in survival of the Chachapoya bodies' soft tissues. (Photo by and courtesy of Federico Kauffmann-Doig, Instituto de Arqueología Amazonica (Lima).)

The spontaneously mummified bodies in the churchyard in Guanajuato may provide confirmatory

Fig. 3.18. Overmodeled skull (AM, ca. AD 1800) from Sepik River, Papua, New Guinea.
After removing all soft tissues from the head after death, a vegetal paste was applied to the anterior half of the bare cranium, facial features modelled, real hair or a wig implanted and designs painted after drying. (Photo by ACA, courtesy of Gabriele Weiss and Ildiko Cazan, Museum für Voelkerkunde, Vienna.)

evidence of the role of ventilation (see Chapter 4: Mummies from Mexico). They have been exhibited for many years within a building near the graveyard, many in an erect position. When several of these bodies were transferred into a closed, glass-fronted container, the skin quickly became overgrown with a green fungus.

Finally, the mummies on Darnley and others of the Torres Strait's eastern islands were prepared by evisceration and some heat-assisted desiccation. During part of the latter stage the body was exposed to wind by tying it to a wood lattice erected to maintain the body in a vertical position (Chapter 4: Mummies from Oceania).

Experimental mummification has confirmed the desiccating effect of ventilation (Aturaliya & Lukasewycz, 1999, Fig. 1). While the variables of the rate of air exchange, the humidity and other factors would probably operate to make prediction of the mummification quality difficult, when conditions are ideal it is conceivable that ventilation could occasionally produce successful mummification. In most cases, however, it was probably only one of the contributing mummification mechanisms.

Excarnation (defleshing)

Defleshing the body's skeletal structure could be viewed as self-defeating or at least not falling within the range of tissue preservation associated with the term "mummification". However, the vagaries of mummification's definition as well as the fact that excarnation is rarely complete led me to include this method. Those forms discussed hereunder have been selected because, following soft tissue removal, they involve efforts at reconstruction of body form.

The simplest form of excarnation has been and still is carried out by the residents of the island group Vanuatu (formerly New Hebrides, westnorthwest of Fiji) where they simply allow the usual postmortem decay process to skeletonize the body. The skull is then recovered and a vegetal fiber paste is applied to the face, shaped to resemble the individual's living appearance. The surface is smoothed with a clay slip and subsequently painted, often with totemic or other designs reflecting the status achieved during life. After drying, the "overmodeled" head is often attached to a wood-and-paste body effigy and placed in a sacred building to be used not only as a family memorial but also as a spiritual intermediary. Prolonged, detailed rituals accompany every step of the process (Huffman, 1976). A similar overmodeling of the skull is carried out by the people along segments of the Sepik (Fig. 3.18), Fly and other rivers in Papua New Guinea, where group members' skulls may be so prepared in order to use them for magic purposes (Meyer, 1995:220, 221 and fig. 230). More ancient examples include prepottery Neolithic groups at Jericho and other Middle East sites between 7300 and 6300 BC (Bienert, 1991).

Northern Chile's coastal residents, the Chinchorros (about 5000–6000 BC), carried out a very extensive process of excarnation, including disarticulation of most of the skeleton. The absence of cut-marks suggests removal of soft tissue may have been allowed to proceed by the usual postmortem decay process in some of these bodies (Arriaza, 1995c:105). The skin was commonly removed (and sometimes later replaced) before excarnation was carried out. After defleshing was completed, the skeleton was rearticulated with the help of splints, the body shape restored with vegetals and clay, and the skin (original or animal) replaced. The face, however, was only rarely sculptured in detail; instead, a simple, flat clay mask with a few token depressions or clay balls to represent facial features was applied to the skull or its clay replacement (Fig. 3.19) (Allison *et al.*, 1984).

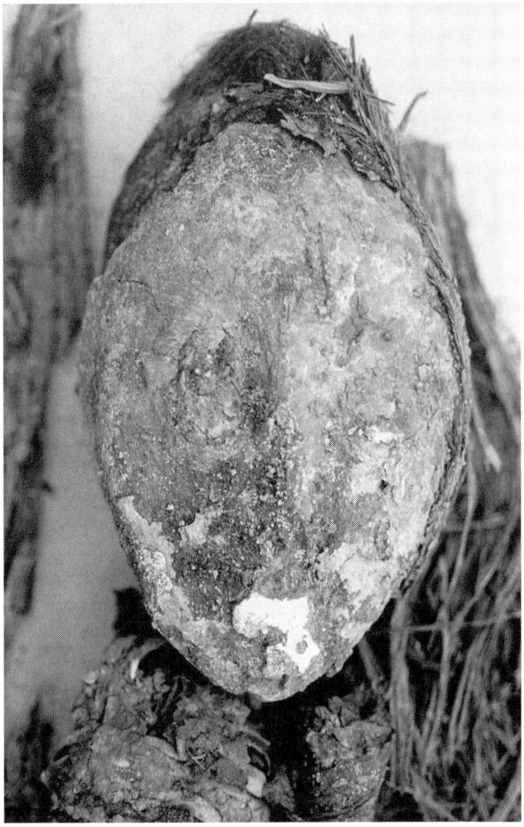

Fig. 3.19. Chinchorro mummy clay face mask; AM, ca. 4830 BC.
In contrast to facial reconstruction in overmodelled skulls (Fig. 3.18), the Chinchorro population made no effort to model realistic facial features on the clay mask with which they covered the facial area. Mummy no. Cam-17,T-1,C-3 from Camarones Valley near Arica, Chile. (Photo by ACA.)

Similarly, body parts of individual saints were and often are retained in containers as relics and placed in elaborate reliquaries for veneration. Here, too, the methods of their preparation and preservation are only rarely known. Not only was such preservation considered miraculous, but it was one of the requirements for sainthood. It is probable that such usually small body fragments (finger, heart, etc.) provided sufficient surface area to result in spontaneous drying before the remains were placed in a reliquary (Fig. 3.20).

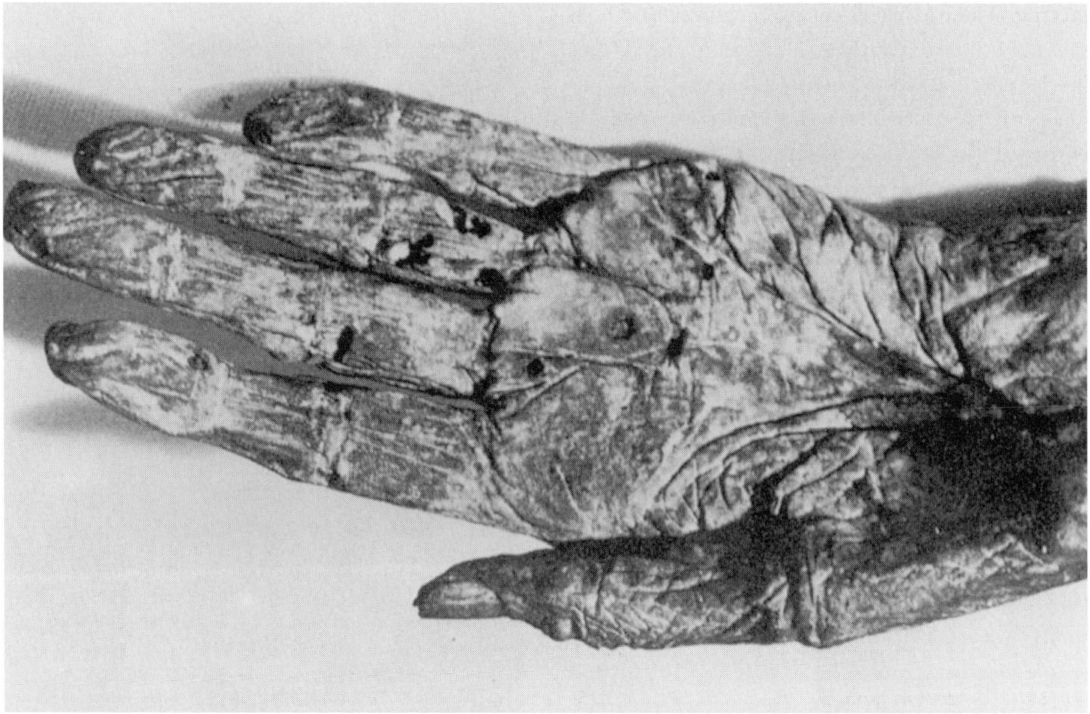

Fig. 3.20. Relic of Santa Margherita of Cortona, Italy; SM, AD 1464.
This mummified hand is characteristic of the type of a saint's body part kept as a relic. (Photo courtesy of E. Fulcheri.)

Occasionally, mummified bodies are found under other circumstances, the mummification mechanism of which is not obvious. Most of these bodies appear to be desiccated. In Chapter 4 these mechanisms are discussed in detail in the section Mummies from Italy.

Readers with a special interest in mummification of saints will enjoy the publications by Fulcheri (1996), Camporesi (1988), Marin & Cappalletti (1997) and Pinto (2000).

Indeterminate mummification mechanisms

"Catacomb" mummies

This is a term I use for bodies that undergo spontaneous mummification by desiccation when placed in underground, rock-lined chambers. During the medieval and Renaissance periods clerics and later elite laic individuals were frequently buried in tombs beneath churches, monasteries or other religious buildings. In some cases at least some of such bodies became mummified (Fig. 3.21). The detailed circumstances of burial routines of most of these are only occasionally available, though in many it appears highly probable that the mummification was spontaneous.

In Venzone, a village of northern Italy's mountainous Frioli region, residents have entombed both deceased clergy and laity in rock chambers beneath the local church. A small fraction of these bodies underwent spontaneous mummification by desiccation (Fig. 3.22). Local lore attributes this effect to a white fungus they call *Hypha bombycina*. This organism is identified in Vol. 1 of an 1822 issue of *Mycologia Europaea* by Persoon as producing a white, flocculent appearance that would fit literally dozens of different organisms in modern taxonomy. The concept that fungi can effect

Fig. 3.21. "Catacomb mummy" in Vienna; SM, AD 1750).
Bodies placed in underground stone chambers beneath Vienna's St Francis church sometimes mummified spontaneously. This adult female's soft tissues are quite well preserved. (Photo by and courtesy of Ekkhard Kleiss, Merida, Venezuela.)

soft tissue preservation by extracting water from the skin surface is still prevalent, though actual experience suggests that skin surface postmortem fungal growth can be destructive to the epidermis. It is also difficult to understand how surface fungi would desiccate deeper-lying tissues (Aufderheide & Aufderheide, 1991:79–86), though the lung conceivably could function as a portal of entry (Horne *et al.*, 1996). I suspect it is the low humidity of the catacomb chambers, produced by the porosity of the chambers' limestone walls, and protection from rain by the overlying cathedral that are the principal factors operating to desiccate the occasional Venzone body in those chambers. Almost all other "catacomb mummies" of which I am aware are in chambers with limestone walls, usually protected from rain and groundwater by overlying buildings.

Most of the bodies in the catacombs of Rome are skeletonized. It is, however, in those at the Capuchin monastery in Palermo, Sicily, that the phenomenon of catacomb mummification occurs with the greatest regularity. Here more than 2000 bodies are stored. A large fraction of these is alleged to have undergone at least partial mummification. The mummification methods employed are described in detail in Chapter 4 (Mummies from Italy). They vary from spontaneous to anthropogenic methods.

Another example of mummification whose precise mechanism is not obvious occurred at a mineral spring within an old peat bog located near Titusville, Florida, at a site called Windover Farms. Multiple skulls of paleo-Indians were found there containing brains in varying states of preservation. Some had recognizable gross cerebral morphology and histology. Radiocarbon dates cluster near 6000 BC. No adipocere formation is evident, as is usual when brain tissue survives in peat

Fig. 3.22. "Catacomb mummies" in Venzone, Italy; SM, ca. AD 1700.
Spontaneous mummification occurs in occasional bodies buried in stone chambers cut into limestone rock beneath St Andrew Cathedral in the community of Venzone in northeastern Italy, where they are displayed proudly. (Photo by Mary Aufderheide.)

bogs. DNA was cloned from several of the brain samples. The spring has a very high mineral content and is of neutral pH (Doran *et al.*, 1986). A similar degree of preservation was found in the skull of a nineteenth century explorer who died in Alaska. The body's return to Chicago occupied several months, after which he was buried there. Exhumed more than a century later the soft tissues of the entire body, including the brain, had been transformed into adipocere (Fig. 3.23).

Ross (1967) describes Celtic traditions relating to a "cult of the head", citing Diodorus of Sicily (ca. 50 BC/1947, V:29) and Strabo (ca. 10 AD/1977, IV:4, 5), in that ancient Celts preserved heads of their most distinguished enemies in cedar oil. If the heads were first desiccated and then painted with resin from an appropriately selected species of cedar, that described practice seems plausible, though it lacks archaeological confirmation.

History of modern embalming practices

Up to this point we have reviewed the various methods of mummification employed by ancient populations. While at least some of the Neanderthals (30000–300000 BP) buried their dead, we have no evidence that anyone from this period attempted postmortem body preservation. The Chinchorros' mortuary practices described above can qualify as one of the oldest groups to engage in anthropogenic mummification, albeit many would dispute that their product represents embalming. Though the *Oxford English Dictionary* describes embalming as the impregnation of a dead body with spices to preserve it from decay, or the anointing of it with aromatic spices, oil, etc., most modern readers visualize embalmment as the internal introduction of some substance having a soft tissue-preserving effect. This would exclude the earlier Hebrew and later Christian customs of external application of aromatic plants and enshrouding the body (some with cerecloth). Sporadic alleged use of honey immersion (insufficiently documented), resin (Ethiopian) or wax envelopment (allegedly by Persians but also inadequately evidenced) fall outside that definition, as would the medieval practice of desiccating small fragments of soft tissue for preservation as relics.

Exceptions to the rule are few. The body of a young girl found in Grottarossa, Italy, had been mummified by an Egyptian technique (see Chapter 4: Mummies from Italy; Ascenzi *et al.*, 1996). Charlemagne's body (AD 618) was allegedly embalmed, though we do not know the mechanism. Though the corpse (or at least its separated head) of Charles I (executed 1649) appears to have retained its soft tissue for about 165 years, Pettigrew (1834:251–253) comments only on the use of cerecloth and its burial in a sealed lead coffin. We know even less about the mortuary preparation of the cadaver of El Cid (died 1099), the Spanish hero whose corpse folklore affirms to have been mounted erect on his horse and led into battle to frighten the enemy. A few convent nuns were mummified in Italy between AD 1300 and 1450. The principal feature contributing to soft tissue preservation in these appears to have been evisceration. Even that procedure, in some cases,

Fig. 3.23. Brain preservation in a skull; SM, AD 1856.
This skull is part of a body in which almost all soft tissues had been converted into adipocere, including the brain. (Photo by and courtesy of Chip Clark, Smithsonian Institution.)

appears to have been motivated by desire to obtain relic tissues (Fulcheri, 1996).

Thus our review of modern embalming practices can begin in the seventeenth century. In Europe several social factors operated to spur interest in soft tissue preservation. Not the least of these was a widening interest in anatomical dissection, initiated in the Italian universities at Padua and Bologna, as described in Chapter 1. Sculptors such as Michelangelo viewed these dissections as aids to achieve artistic goals while others such as Leonardo da Vinci simply sought anatomical knowledge (Leonardo actually attempted specimen preservation by injecting into veins). Bodies with preserved soft tissues permitted anatomy laboratory sessions to be scheduled. By 1628 William Harvey had described the circulation of blood and later that century the Dutch physician Frederick Ruysch had employed that concept by injecting solutions into arteries for the purpose of preparing teaching aids for anatomical dissection (Mendelsohn, 1944a).

The eighteenth century brought about increasing need for at least temporary body preservation. The practice of public display of politically important persons' bodies before burial, or simply the need for time to permit the family to gather for the burial all required a form of preservation that would postpone decay for at least several weeks (Habenstein & Lamers, 1981:198).

By 1750 evisceration was being practiced quite frequently but efforts to rid the body of fluid were commonly focused on making multiple deep incisions of the skin, down through underlying muscles and expressing fluid from them by compression and gravity. Aromatics rubbed into these, however, contributed little other than to mask offensive odors. Pettigrew (1834:157) cites the practitioners of this approach to include Paré, Sue, Penicher, Guybert and Charas, denigrating their skills. It was the Hunter brothers, John and (especially) William, however, who formalized the embalming procedure by incorporating the various individual features that had been found useful, and who shaped it into a predictably effective process. William Hunter used intra-arterial injection of turpentine and oils of chamomile and lavender, adding vermilion pigment to generate a more life-like skin color. He then eviscerated the body, treating these organs separately after squeezing much fluid out of them and filling the body cavity space with camphor powder and resin, plugging body orifices with this same powder. Aromatics were applied to the skin and then the entire body was laid on a bed of plaster of Paris to withdraw the remaining fluid. The plaster was to be renewed every 4 years (Pettigrew, 1834:251–260). William Hunter's fame grew when he embalmed a local character's wife whose last will and testament gave her surviving husband control of her estate only as long as her body remained uninterred. Following Hunter's preparation, her husband put her embalmed body in a glass case in her home and established visiting hours for her public viewing. Later Hunter also embalmed the mistress of his assistant, John Sheldon. His assistant was so pleased that he kept her in his bedroom until he yielded to his subsequent wife's complaints (Broaddus & Harmony, 1984).

The eighteenth century also initiated the practice of pouring preserving fluids (through a defect created in the chest wall) into the thoracic and abdominal cavities without subsequent evisceration. Gabriel Clauderus (Germany) used ammonium chloride and tartar salts for this purpose (Mendelsohn, 1944b; Pettigrew, 1834:255–256). In Colonial North America the common practice consisted of simply keeping the body in the home until offensive odors forced burial. However, relatives of the deceased sometimes gained time to gather friends and family by evisceration of the corpse or immersion in alcohol as well as by the use of alum and a ceresheet shroud (Habenstein & Lamers, 1981:199). Benjamin Franklin even considered that total immersion in wine might be effective, though there is no evidence that he ever tried it. It is clear, however, that solutions containing sufficiently high concentrations of alcohol can preserve human bodies (see below and Chapter 4: Mummies from North America, and Mummies from France).

The nineteenth century opened to a mixture of methods for delaying the onset of soft tissue decay. Airtight and waterproof coffins were still in high demand, based on the assumption that the decay process required oxygen. This myth was probably capable of perpetuation because, once sealed, the coffin was ultimately buried without being opened again. If it had been opened, two constraints to body preservation would have been exposed – the biological myth about the essential presence of oxygen for decay, and also the fact that many nineteenth century coffins leaked.

This century also saw the employment of novel, emergency methods of body preservation. When Britain's navy under Lord Horatio Nelson defeated that of the French in 1805 at Trafalgar, the price of victory included Nelson's death. Reluctant to return to England without the hero's body, his shipmates placed it in a keg of rum. Similarly, a father preserved his daughter's body in a cask of alcohol when she died during a transatlantic voyage (Habenstein & Lamers, 1981:199). In still another response to a similar need when a female invalid died aboard ship in transit from Naples to London, her escort eviscerated the body, then painted both the insides of the body cavities and the outside of the whole corpse with conventional ship's tar, finally enveloping the treated corpse in a tarred sheet (Pettigrew, 1834:259–260).

It was, however, with chemicals that the battle of the embalmers was waged. Heavy metal ions (arsenic, zinc and mercury) formed the initial "ammunition" for this battle. Of these, arsenic was among the first (and the last) to be used. Early during this century Tranchina of Naples used intra-arterial injection of arsenicals for embalming. Arsenic was usually used as arsenic trioxide ("white arsenic") or arsenous acid. Tranchina dissolved the arsenic in five times its weight of wine. These compounds, however, are not very soluble in cold water and were usually prepared in boiling water

– a hazardous occupation because of arsenic's toxicity. Typical of such preparations are the following, listed in a nineteenth century embalmers' formulary (Hatfield, 1886:318–319):

Morrell's antiseptic liquid

Arsenous acid	14 parts
Caustic soda	7 parts
Water	20 parts

Carbolic acid [phenol] added plus water to make 100 parts

This solution approximates a 7% (weight/weight) concentration of arsenic.

Another formula (below) had about a 5% concentration of arsenic (Barnes, 1896:340).

Creosote	6 ounces
Arsenous acid	4 ounces
Alumina sulphate	6 pounds
Water	1 gallon

Arsenic solutions were widely used both in Europe and in North America during this century (see Chapter 4, Miscellaneous mummies of the USA). In France, however, it did not even survive the first half of the century, being outlawed there in 1846 for forensic reasons. Allegedly, a woman was being tried in court for poisoning her lover when chemical tests on the body revealed high arsenic content. She was saved from the gallows when a chemist (J.N. Gannal, see below) testified that the arsenic was the result of the embalming process. The legal profession declared that such an embalming practice was akin to legalizing murder (poison the victim; then have the body embalmed, making the fatal dose undetectable). Whether or not the details of this anecdote are accurate, the fact remains that the French government recognized the validity of the complaint and banned the use of arsenic in embalming fluids in 1846. It would be not until 49 years later in 1895 that the first of the American states (Michigan) prohibited its use. Just 5 years before it was banned in France, the first US patent (no. 15972) granted for an arsenical embalming fluid was awarded to J.A. Gaussardia, October 28, 1856, for a 7% (wt/wt) solution of arsenic in pyroligneous acid injected intra-arterially. The body was then charged with electricity (one pole in the mouth, the other in the anus). After anointing the body with aromatics and arsenicals, it was sealed in a metal coffin filled with an alcoholic arsenical solution (US Patent Office, 1856; Johnson *et al.*, 1990).

The search for a substance to fix (i.e., preserve) tissues and even whole bodies for anatomical dissection also led to the use of another heavy metal – mercury, popularized in France in the early part of this century by Francois Chaussier. He recommended immersing an eviscerated cadaver in a solution of mercuric chloride. By the second quarter of the century this substance was also being injected intra-arterially by men like Louis Jacques Thénard. This agent, however, was commonly employed in lower concentration with other agents because of its tendency to coagulate, desiccate and deform specimens. By the mid 1800s J.P. Souquet had also added zinc chloride to embalming solutions and in 1846 the Jefferson Medical College was using a 2.5% solution of zinc chloride for anatomical tissue specimen preservation. Other agents used zinc either alone or, more commonly, mixed with agents noted above, including alum (Wickersheimer, 1879), 4% chloral hydrate by W.W. Keen (Mendelsohn, 1944b) and especially the acetate and sulfate forms of aluminium salts used so extensively by J.N. Gannal (1838/1840). While Hofmann introduced formaldehyde as early as 1867, the Berlin physician Aronsohn was the first to inject it intra-arterially as an embalming fluid in 1891. Formaldehyde remains the basic ingredient of embalming fluids today.

Interestingly, while both evisceration and intra-arterial injection had been introduced during the eighteenth century, the unpredictable and often unsatisfactory result of intra-arterial embalming without evisceration using contemporary compounds had remained an unsolved problem. The solution came during the nineteenth century when Samuel Rogers in 1878 patented the use of a trocar. This was a hollow tube with a beveled cutting tip that could be connected to the embalming fluid pump. It was originally

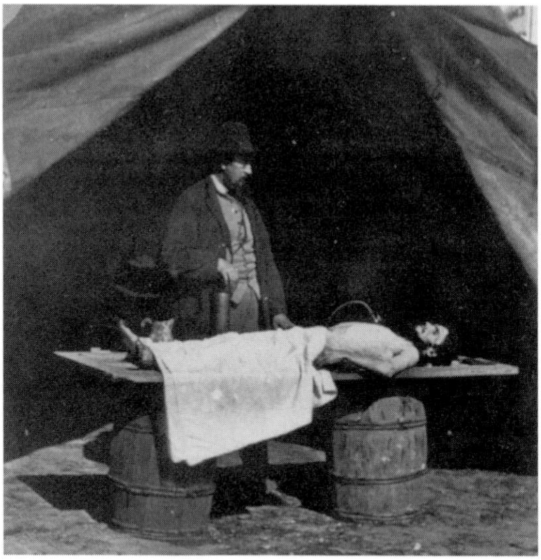

Fig. 3.24. Embalming an American Civil War casualty in the field; AM, AD 1856.
Richard Burr is shown embalming the body of a soldier near the battlefield during America's Civil War. (Reproduced courtesy of Library of Congress.)

designed for aspirating fluids from body cavities and for infiltrating leg muscles when the arteries were blocked by a clot. This permitted the embalmer first to inject the entire body intra-arterially. He could then attach the trocar, push it into the abdominal cavity and, with the pump in operation, penetrate the bowel and individual viscera, sterilizing the gut and fixing the various abdominal and thoracic organs without evisceration (Johnson *et al.*, 1990:454). Thereafter suction applied to the trocar could remove surplus fluid.

Initially in Europe the various surgical manipulations required for embalming tended to place this procedure in the hands of physicians and surgeons. As the search for the appropriate selection and method of preparing chemicals widened, apothecaries and chemists such as J.N. Gannal in France entered the fray. At the beginning of the nineteenth century all of these were in competition with each other. However, as embalming practices in Europe declined in popularity, the struggle shifted to North America. By mid century the search for a safe but antiseptic embalming solution had become an item of health concern. Richard Harlan, M.D., a member of Philadelphia's Health Council and responsible for the anatomical dissection room where anatomy was taught, traveled to Europe to study disease control. He was impressed with embalming as a valuable contribution to public hygiene. After his return he translated J.N. Gannal's (1838/1840) treatise on the history of embalming into English, making it available to American embalmers. Gannal had evolved a fluid for anatomical dissection preparation consisting of a mixture of aluminum sulfate and aluminum acetate suspended in a very low concentration (<1%) arsenical solution. His compound for whole-body embalming use is not revealed in his book but presumably was similar. This 1840 translation quickly became the standard American embalming reference (Johnson *et al.*, 1990).

In addition to health concerns, undoubtedly the principal stimulus for embalming's popularity in North America was its Civil War (1861–1865). Early in this war an aide to President Lincoln, Col. Elmer Ellsworth, was killed while defending Lincoln against a would-be assassin. To allow his body to lie in state in Washington, DC, it was embalmed. The family was so impressed that demands for similar services became common in the north, so that corpses of battle casualties could be returned for burial in their home community. This eventually led to army contracts with civilian embalmers to satisfy these expectations (Fig. 3.24). When Lincoln was killed shortly after the war ended, his body was also embalmed for public viewing.

The man whose activities were central to these events was Thomas Holmes. While the details of his earlier career are not generously documented, it is certain he was at one point registered in Columbia University's medical school in 1844–1845. Holmes developed a method of preparation that included evisceration and intra-arterial injection of an embalming fluid containing arsenic, mercury and zinc salts (Fig. 3.25). With experience he apparently modified this solution several times so the exact final mixture is uncertain. He failed to patent it before the war because the basic ingredients were already known. Instead, he prepared it himself and sold it for three dollars per gallon. He did patent a fluid pump. Other inventions

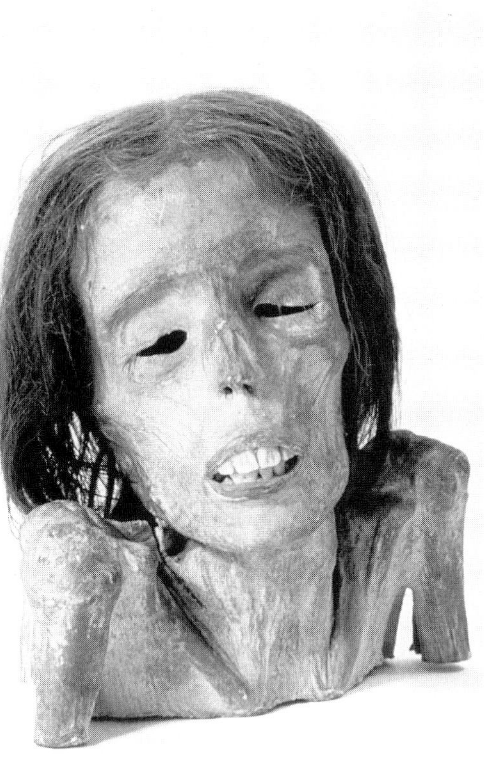

Fig. 3.25. Current appearance of body embalmed by Thomas Holmes; AM, ca. 1860.
The 12-year-old girl's body shown here was probably injected intra-arterially. This body today is housed at the National Museum of Health and Medicine in Washington, DC. (Photo courtesy of Paul Sledzik.)

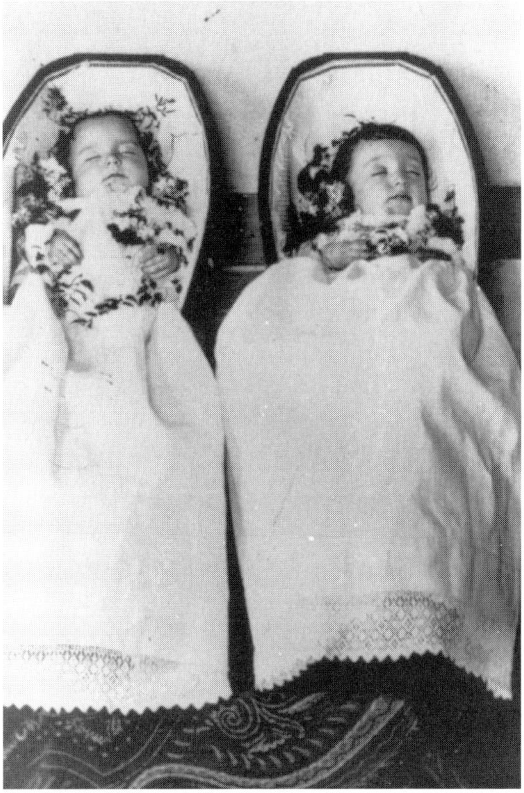

Fig. 3.26. Example of modern American embalming methods; AM, ca. AD 1935.
The American children in this photo were prepared by intra-arterial embalming using formaldehyde-based fluids during the 1930s. (Photo courtesy of the State Historical Society of Wisconsin, Negative (V2)724.)

by Holmes included the "body pouch", that soldiers sometimes employed as a surrogate sleeping bag. During the American Civil War, Holmes, together with another embalmer Benjamin Wheatley (taught by Holmes), probably embalmed about 4000 bodies (Habenstein & Lamers, 1981:205–211).

Following that war, Holmes appears to have had a thriving business with the sale of his embalming fluid. Initially, sales were slow, until he taught the necessary intra-arterial injection techniques to untrained undertakers. Physician interest declined and nonphysician undertakers took over the embalming practice. The economy of the use of ice for corpse preservation during transportation competed successfully for a short interval. The first embalming school was initiated in Cincinnati, Ohio, in 1882, stimulated by a coffin salesman, Joseph H. Clarke, and a second later that year was established in Rochester, New York, by Dr August Renouard (Harmer, 1984). The Funeral Directors Association of the USA and Canada was formed that same year, resulting subsequently in standardization of embalming practices and organization of the profession and industry much as it remains today, with formaldehyde still forming the principal effective agent in most formulae (Johnson *et al.*, 1990:448) (Fig. 3.26).

Chapter 4. The geography of mummies

This chapter is devoted to the geographical distribution of mummies and the correlation of mummification and its types with environmental, geophysical, climatological and cultural features.

MUMMIES OF NORTH AMERICA

Mummies from Canada

Mummies from the Canadian Arctic

Of the limited number of mummies that have been found in Canada probably the mummified remains of three sailors on an arctic expedition are of greatest interest. In May of 1845 Sir John Franklin left England with two ships – the *Erebus* and the *Terror* to find the postulated Northwest Passage through the North American arctic islands from the Atlantic to the Pacific. None of the 134 crew members survived after the ships became frozen into the arctic ice off the northwest coast of King William Island in the Canadian Arctic. However, the grave sites of three crewmen were identified on a different (Beechey) island. During the summers of 1981 and 1982 the Canadian archaeologist Owen Beattie found and sampled some skeletal tissue of Franklin's men on King William Island. In 1984 and 1986 he excavated the graves of the three on Beechey Island (Beattie & Geiger, 1987).

These three had died before the ships became ice locked, and had been buried in the usual European manner: in wood coffins in a pit dug 1.5 meters down into the permanently frozen soil (permafrost). When rain trickled down through the gravel used to fill the hole above the coffins, it froze promptly. Like the permafrost around it, this did not thaw again during the subsequent 138 summers. At excavation, the bodies appeared remarkably well preserved, with evidence of only mild desiccation during their frozen state of more than a century. Facial details, including half-open eyes were clearly recognizable (Beattie, 1995) (Figs. 4.1 and 4.2). A field autopsy on one body examined in 1984 revealed all viscera intact and identified the presence of pleural adhesions (healed pneumonia episode) and carbon-laden lungs (anthracosis) with mild emphysema. Loss of histological detail suggested that a significant period of time intervened between death and burial. Microscopic evidence of exudate in lung tissue suggested that acute pneumonia may have been the cause of death (Amy *et al.*, 1986). Autopsies on the others also revealed that one of the two had been subjected to postmortem dissection. Field radiological studies also demonstrated evidence of tuberculosis in the spine (Pott's disease) (Notman & Beattie, 1995).

Chemical studies, however, detected what probably incapacitated Franklin's crew and led ultimately to their demise. Beattie's bone specimens of the men that died on King William Island as well as those on Beechey Island were found to bear a lead content of

Fig. 4.1. The frozen mummy of John Hartnell; SM, AD 1846.
Melting ice reveals the emerging face of John Hartnell after more than a century in the frozen state. (Photo courtesy of Owen Beattie.)

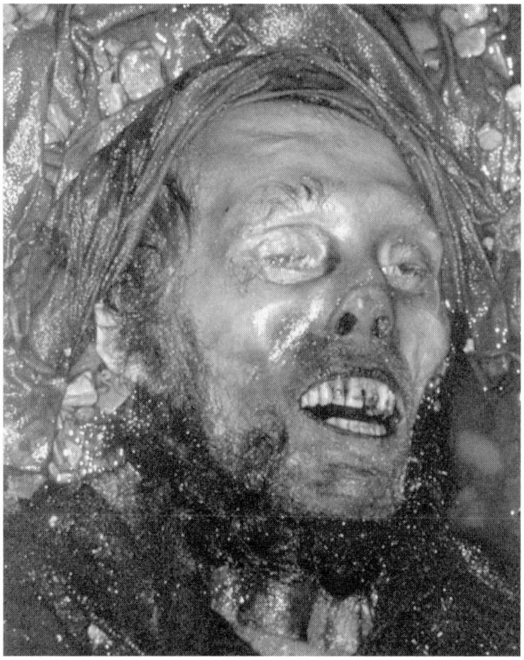

Fig. 4.2. Thermal preservation; SM, AD 1846.
Encased in the ice of arctic permafrost for nearly 140 years, William Braine's soft tissues demonstrate superb preservation. (Photo courtesy of Owen Beattie.)

about 133 parts per million (ppm). Hair samples of the Beechey burials had two- to fourfold greater quantities (Kowal *et al.*, 1989). Values of that order of magnitude can be expected to be associated with serious complications of lead poisoning (Handler *et al.*, 1986). Samples of Inuit (Eskimo) bones and those of caribou had only 2–5 ppm of lead and those of modern Vancouver (Canada) had 10–30 ppm. In an elegant analytical study, the source of the lead proved to be the lead-tin solder used to seal the canned food tins prepared for the expedition. This relationship was established by demonstrating virtually identical ratios of lead isotopes ($^{206}Pb/^{204}Pb$) in bones of the Franklin crew, in solder from their food tins and in solder from other, contemporary, British tins but different ratios in lead of Inuit bones (Kowal *et al.*, 1991). Clearly the lead solder in the food tins had seeped into the food within those cans and seriously poisoned Franklin's crew as an inadvertent contaminant of their daily rations.

Mummies from Western Canada

The enzyme-inhibiting effects of copper ions from the burial container and pectoral artifacts are believed to have been responsible for preservation of much of the soft tissue of an infant's body buried about AD 1800 in a copper urn near Yale, British Columbia, territory of the Salish tribe. Although the viscera had degenerated, the skin and muscles of the chest, extremities and head were quite intact (Schulting, 1995). Green staining of even the interior of the bones reflected the penetration and permeation of the copper ions throughout body tissues, and this distribution was also demonstrated by X-ray fluorescence. Portions of several other bodies from that region had also been preserved where the soft tissue directly underlay copper ornaments. As explained in Chapter 5, copper ions are capable of inhibiting the actions of decay enzymes. These are all examples of spontaneous mummification under conditions that, in the absence of copper, probably would have resulted in skeletonizing decay of the soft tissues.

Two spontaneously mummified bodies of Native Americans have also been reported from the west coast of Vancouver Island (Melbye & Stuart-Macadam, 1995). These bodies were wrapped in cedar bark and placed into cedar boxes that were decorated with elaborate carvings. One box had been stored in a cave and the other beneath a cedar board shelter. The bark probably enhanced water transfer from the bodies; cedar is also very resistant to decay. The spontaneous mummification of these bodies may owe their effectiveness to these factors. Radiological study demonstrated multiple lytic lesions in various bones of the male, characteristic of metastatic carcinoma. The female had advanced dental disease. Smith (1950:15) notes the frequent presence of cedar box burials from the middle Columbia and middle Fraser rivers in British Columbia as well as among the Coast Salish of the straits of Georgia and northward.

Canadian Iceman

Eight years after the discovery of the more than 5000-year-old body of the Iceman in a Tyrolean glacier had made world news, the body of another man was found by hunters, this time on a glacier in Canada's province of British Columbia. The thawing ice had apparently exposed the body shortly prior to its discovery, since it was still very well preserved. It was accompanied by a woven hat, fur cloak, leather pouch filled with dried fish, a spear thrower and a carved walking stick. However, radiocarbon dating revealed he had died only about 550 years before he was found in 1999 (Holden, 1999b; Pringle, 1999).

Mummies from Central and Eastern Canada

In 1991 the female body of a Caucasian (Irish) pioneer was exhumed from the Cheyne Pioneer Cemetery near Brampton, Ontario, for transfer to another site because of road construction. It had been interred there in 1879, more than a century earlier, under 2 meters of clay. Extensive, hard, whitish-yellow, friable material having the shape of a human trunk and thighs encased the skeletal tissue (Pfeiffer, 1991) (Fig. 4.3). This type of adipocere production is not always predictable. In this case, the body of her husband, buried immediately adjacent to his wife in a similar manner had completely skeletonized without adipocere formation. Microbiological cultures revealed only *Aeromonas hydrophilia*, a common soil organism. Conceivably the amount of body fat could have been a factor in the development of adipocere.

The body of a 5- to 6-year-old Native American child was found on Burnt Island in Notre Dame Bay in Canada's maritime province Newfoundland. This Native American was a member of the now extinct Beothuk tribe. Wrapped (except for the head) in an animal hide, the skull retained no soft tissues but the skin and muscles (though no viscera) were preserved over the trunk and extremities. No evidence of anthropogenic effort at mummification was evident. No Harris lines of bone growth arrest could be identified. A conventional radiocarbon date of AD 549 $\pm$ 62 years (1 standard deviation) suggests that death occurred prior to European contact, consistent with a diet (reconstructed from stable isotope measurements) about half of which was of marine food origin (Jerkic *et al.*, 1995).

Mummies from the USA

Mummies from Alaska

St Lawrence Island

About 1600 years ago an Inuit woman sat alone in her semisubterranean hut near the edge of a low beach cliff at the foot of a steep slope at Klegiak Point on St Lawrence Island in the Bering Sea. It had rained frequently and the water had saturated the surface soil and penetrated until it reached the permanently frozen soil just a few inches below the surface. With the next

Fig. 4.3. Soft tissue preservation by adipocere formation; SM, AD 1869.
Adipocere formation can maintain the gross morphology of basic body structures. (Photo courtesy of Susan Pfeiffer.)

rain the weight of the waterlogged surface sod overcame its frictional component with the frozen surface of the underlying permafrost. Abruptly the soil near the slope's upper level became detached. Sliding down the precipitous hill it accelerated rapidly, transforming itself into an avalanche of mud. Before the woman could escape, the mud enveloped the hut, incorporated her and her dwelling, and buried her, flattened against the permafrost. There she asphyxiated, after which the deeper soil's subzero temperature froze her body. Insulated now from warmer surface temperatures by the overlying carpet of muddy soil, her frozen body lay silent and immobile for more than 1½ millennia. This described scenario can be reconstructed with reasonable plausibility from archaeological and later autopsy findings (Zimmerman & Smith, 1975).

In 1972 storm-driven waves reached the beach of this site, washing away the mud covering the area and exposing the body. Three Inuit hunters the next day found the body, still frozen, and reinterred it at a more protected site. The following summer an official of the National Park Service transported the body to the Fairbanks, Alaska, Health Center, where it was kept frozen until autopsied.

The thawed body was that of a postmenopausal female. Soft tissues were moderately desiccated, a common feature of frozen bodies. Both arms and the back of her hands were tattooed with dots, lines, oval- and heart-shaped figures, though surprisingly, none were on her face (facial tattooing, especially on the chin, was a common Inuit practice until recently). Age-at-death was estimated at 50–60 years on the basis of dental and pelvic bones, later confirmed by osteon density of the microscopic structure of a cross-section of long bone (Smith *et al.*, 1978) and still later refined further to 53 ± 5 years by aspartic acid racemization technique (Masters & Zimmerman, 1978). Two radiocarbon tests clustered around AD 400. Dissection findings identified the moss-filled bronchi as the cause of death by asphyxia, a microfracture of the right temporal bone, coronary artery atherosclerosis, pulmonary anthracosis, pleural adhesions (reflecting an episode of pneumonia with recovery and healing), calcified mediastinal lymph node (healed tuberculosis or fungal infection?), arteriosclerosis of the aorta, and a fish parasite (*Cryptocotyle lingua*) ovum in a coprolite.

The Utqiagvik mummies of Barrow

A similar scenario can be reconstructed for the mummified bodies found in a crushed wooden hut at Barrow, Alaska's most northern point on the Arctic Ocean shore. Row upon row of ancient dwellings (the older ones built of whale bones) testify to the centuries-long occupation of this site. At this site in the fall of about AD 1500 a family of five Inuit were sleeping in their half-buried driftwood hut waiting out a violent northern autumn gale, when a phenomenon natives call *ivu* developed. The storm drove drifting ice before it, pushing it onto the shore. Subsequent arriving ice began to override the accumulating mass of tumbled ice fragments on the beach. As the pile of ice floes grew ever higher it became more and more unstable. Finally several huge pieces of rafted ice were pushed up almost vertically to the top of the pile, balanced briefly at the peak and then slid down the shore side of the heap, impacting with a thunderous crash – directly on top of the wooden hut. The main roof beam snapped and the roof collapsed upon its occupants. The three sleeping on the hide-covered wood platform were flattened by the falling roof. One of the other two women was struggling to get into her clothes when the broken roof timber felled her, crushed her chest and pinned her to the permafrost that formed the hut's floor. Roof and wall debris similarly compressed the other woman's body against the frozen soil of the hut's floor (Zimmerman & Aufderheide, 1984).

During the subsequent 500 years the above-freezing summer arctic temperature every year thawed the bodies of the three that had been trapped on the wood platform sufficiently to accommodate the normal decay processes, gradually skeletonizing them. The two women who had been flattened against the never-thawing soil of the hut's floor, however, remained frozen or at least cold enough that most of the soft tissue remained intact. Uncovered first by local, modern artifact hunters and then by archaeologists, the two frozen mummies were flown to Fairbanks for study. Thawing revealed superb preservation of the nondesiccated soft tissue (even the parathyroid glands were identifiable), but histological anatomical integrity was seriously compromised. Dissection revealed empty stomachs and full bladders, confirming that this probably occurred in the early morning hours. Both women were osteoporotic, had multiple rib fractures with pleural effusion and showed mild aortic atherosclerosis. In addition, the older woman (40–45 years) also had prominent, lactating breasts and a large resolving ovarian corpus luteum suggesting recent pregnancy. A calcified diaphragmatic cyst suggested possible trichinosis and a calcified lung nodule implied healed granuloma. Severe anthracosis was most prominent in the younger woman (24–31 years as estimated by aspartic acid racemization (Masters, 1984) and bone osteon density (Thompson & Cowen, 1984)). This produced sufficient fibrosis to impair lung function. Thus the bodies on the wood platform and not in contact with the permafrost gradually skeletonized during the 500 years after death, while the two in direct contact with permafrost thawed sufficiently to lose histological structure but retained gross soft tissue morphology. Autopsy findings in addition to the traumatic changes included osteoporosis, atherosclerosis, mitral valve calcification, trichinosis, lactation and pleural adhesions reflecting recovery from an episode of pneumonia.

In both of these cases, spontaneous mummification was achieved thermally. The Barrow bodies, however, were not sufficiently insulated. The three bodies on the wood platform were not in contact with the permafrost soil; the Arctic's summer thawing temperatures permitted incremental soft tissue loss during each season, which resulted ultimately in complete skeletonization.

Even the bodies in contact with underlying permafrost were not insulated sufficiently by the overburden to resist thawing completely. The result in these two was retention of gross organ structure, but extensive histological tissue destruction including fungal growth. Such small differences emphasize how narrow is the margin between long-term soft tissue destruction and preservation (Zimmerman & Aufderheide, 1984).

The Aleutian Islands

That this region of almost perpetual rain, sleet, fog and gale harbors a permanent human population is a testament to human adaptability; that such conditions tolerate long-term, postmortem soft tissue preservation is evidence that microclimates exist within any region's general climate. While few examples of spontaneous mummification are evident among the residents (Aleuts), anthropogenic mummification has emerged in some of the eastern Aleutian Islands. As could be expected, natives of this area evolved a largely marine hunting (sea mammals, fish) and collecting (shellfish) subsistence. Caucasian discovery of these islands occurred when Vitus Bering, leader of an exploratory fleet sponsored by Russia's Peter the Great (and of his second discovery expedition funded by Empress Anna) sailed from the Kamchatka Peninsula. After a long voyage his two ships, separated by a storm, made independent contact in 1741 with several of the Aleutian Islands including Umnak, Attu and Adak. The economic value of the Aleut's prey animals, especially the sea otter, was immediately apparent. By 1761 Russia had subdued the natives and organized what became a lucrative fur harvest through enforced Aleut labor (Jochelson, 1933).

Though the Aleuts had occupied the islands for more than 4000 years (Vasilievsky, 1975), we do not know precisely when mummification there began. Ethnographic accounts suggest it had not been in practice for more than a few generations prior to Russian contact and did not survive the Russian domination for more than two generations (about AD 1800). It is clear, though, that it was initiated in the eastern island cluster, and Laughlin (1980b) believes it was spreading westward at the time of the Russian arrival. We are indebted to the Russian priest Veniaminov (1840) for information about pre-Russian Aleut customs but Dall (1878), an officer of the US Coast Guard and Geodetic Survey Agency and active in the Aleutian area, summarize their method of anthropogenic mummification as follows:

1. Evisceration (usually via perineal, occasionally abdominal incision).
2. Abdomen filled with dry grass.
3. Body placed in a stream of cold, running water (allegedly to remove remaining abdominal fat).
4. Body tied in flexed position and dried (usually air dried, occasionally in hot, volcanic cave; perhaps sometimes by fire). Skin is frequently wiped.
5. Body wrapped (initially in fur, then in vegetal mat, then in waterproof bird skin held in position by vegetal net).
6. Body suspended on a wood rack in cave or rock shelter.

It is not clear to me how cold, running water would remove intact fat, nor do I know of precedence for this feature. It would seem improbable that it would be an enzymatic phenomenon, but perhaps those circumstances somehow enhance nonenzymatic hydrolysis of neutral fat, the running water constantly removing the fatty acid products whose accumulation might retard the process by excessive acidification. Almost all writers indicate that the described method was employed regularly for the Aleutian leaders and other elite members. The number and nature of grave goods accompanying elite burials allows easy recognition of social status, because "wealth destruction" was the ultimate symbol of status among Aleuts (Lantis, 1947). Commoners and children often were not eviscerated (Zimmerman *et al.*, 1971).

Once dried, the mummified bodies were placed in rock shelters. In some locations these were actual caves, while in others simple rock overhangs were exploited. For the latter, additional shelter was sometimes supplied by creating a wall of boulders and driftwood around them. The rock overhangs are often quite crude shelters with inadequate protection from the sea, rain or scavengers. The cave from which the largest number of best-preserved mummified remains have been removed is the one Hrdlička (1941) calls the "warm mountains". This is found on Kagamil, one of a small cluster of islands collectively referred to as islands of "the four mountains". That island is dominated by a 1000 meter high active volcano, and the "warm" cave is heated by a nearby fissure whose visible steam emanations make its location easily identifiable. The walls and floor of the cave are perpetually warm to the touch and the effort expended in recovering mummies that had been left there made the cave's temperature seem unpleasantly warm. The dry, warm microclimate of this cave undoubtedly was responsible for the long-term maintenance of its mummified bodies. Nearby is a cave of similar size but without a volcanic vent to warm it; this has been termed the "cold cave". It, too, contained mummies, though fewer of them were in excellent condition. Indeed, feral island foxes had ravaged many of them. Some caves are low enough to be flooded occasionally by high seas.

The practice of mummification was not uniformly distributed in the Aleutian Islands. Examples have been found in many of the eastern islands (even a few on Kodiak Island), smaller numbers in the central Aleutian islands of Amlia, Ilak, Kanaga and Tanaga, and none in the far western group (Laughlin, 1980a:99). This pattern has led Laughlin to suggest that the practice originated in the eastern segment, and was diffusing toward the west when it was interrupted by the Russian arrival. Use of the warm cave on Kagamil can be dated roughly by a local legend (Dall, 1875) in which a chief committed use of that cave as a mausoleum for his family when his son and daughter-in-law both died at almost the same time, as the result of two separate accidents. About 5 years later the chief himself died and was buried there; the following year the Russians arrived at Kagamil. Using that date, the earliest use of that cave for mummification was probably about AD 1756 (Dall, 1878). Other bodies may have been moved into that cave for their protection following the Russian arrival.

The purpose of mummification is seldom easily reconstructed for an extinct population, but in the case of the Aleuts we have ethnographic data that are of considerable value. The Aleuts' belief system included a concept of continuation of power in the dead, and that such power could be exploited. In the event of a slain enemy, the spirit's power could be dissipated by dismemberment of the body, thus frustrating revenge efforts. For a family member, however, that same concept of power could be exploited in one's favor through "sealing in" the spirit by mummifying the body. Subsequent to mummification, the body was then employed as a source for advice regarding fishing or hunting conditions, protection from an enemy, and similar requests (Laughlin, 1980a). In spite of the apparent conversions of Russian orthodox priests, such a mummy consultation was witnessed as late as 1862 (Laughlin, 1963:102). Possession of a small part of such a mummy (e.g., a portion of a finger) on a whale hunt was expected to guarantee not only safety but also hunting success (Dall, 1875).

The USA purchased Alaska from Russia in 1867 and promptly gave trading rights to a private concern: the Alaska Commercial Company (ACC). The government also permitted and encouraged the Coast Guard and Geodetic Survey (CGGS) to assist and cooperate with that company, instructing them to recover items of cultural and historical value. In 1874 Captain Hennig of the schooner William Sutton recovered 12 mummies and two skulls from Kagamil's warm cave. Two of these mummies were donated to the California Academy of Science's museum in San Francisco and nine sent to the Smithsonian Institution's Natural History Museum in Washington, DC. In 1928 New York's American Museum of Natural History launched its Stoll McCracken expedition, during which the archaeologists Edward Weyer Jr, and Julius Bird removed four mummies from Split Rock Island near Unalaska Island. The bodies were in a cave high on its steep-sided mountain and one was buried in a low log shelter on the island's flat top (May, 1951). Between 1936 and 1938 Ales Hrdlička (1941) of the Smithsonian's Physical Anthropology section became interested in these mummies and visited the Aleutians with the help of the ACC and the CGGS. On his first voyage he removed at least 50 mummies from the warm cave at Kagamil and from the nearby cold cave he removed one intact mummy and 24 sacks of mummy fragments and bones. Some of the mummies had been eviscerated via the perineum and others through the chest or abdomen (May, 1951). Subsequent removal of mummies from Ship Rock and other islands added many dozens more. Laughlin (1980a:96) estimates that 234 mummies were eventually removed from the Kagamil caves as well as additional ones from caves on Amlia, Ilak, Kanaga and Tanaga islands.

Few of all of these mummies have been studied formally. In his zeal to classify ancient populations by using skeletal criteria, Hrdlička may have stripped soft tissue from six of the nine mummies that Capt. Hennig (ACC) had originally removed from Kagamil's warm cave and sent to Washington. No detailed record of those soft tissue examinations remain, if indeed they were ever studied. Whether Hrdlička really studied the soft tissue of the remainder is not clear. Zimmerman *et al.* (1971) studied one uneviscerated member of that mummy group at the Smithsonian Institution and another in 1981 (Zimmerman *et al.*, 1981a) that had found its way to Harvard's Peabody Museum.

They found that the first of these died of lobar pneumonia, and that both had pulmonary anthracosis with fibrosis, as well as atherosclerosis of the aorta and some of its branches. The interdisciplinary teams also contributed useful information with other methods (Horne, 1979; Laughlin *et al.*, 1979). Lombardi (1997) examined an Aleutian mummy housed at Tulane University in New Orleans and found evidence of a violent death including a scalp wound and strangulation perhaps related to Aleut resistance to Russian domination.

Prince William Sound

A cluster of islands in Prince William Sound on Alaska's southern coast near the port city of Cordova at about 60° N and 148° W represents the unlikely location of about 50 spontaneously mummified bodies of Native American Indians. Unfortunately the overwhelming majority of these bodies were removed by modern adventurers and tomb robbers. De Laguna's (1956) exhaustive survey and excavations document the many island sites of both burials and cave deposits of these mummies. Most bodies she found were skeletonized but a very few had some remaining tissue and one from the principal cave on Mummy Island had intact, preserved soft tissue. We know of most of the others only through ethnographic reports of their removal. The mechanisms that led to their spontaneous mummification remain unknown. The unfortunate loss of all the potential information that these mummified remains could have contributed can only be lamented.

Alaska's panhandle area

Six anthropogenically mummified heads of Native American Indians from the Alaska "panhandle" near Petersburg at about 57° N were reported by de Laguna (1933). They were found by local residents in a cave. Each head was packaged in a separate red cedar box. The heads were enclosed in cedar shavings enveloped by a mat. The area is within Tlingit territory. Tlingit mortuary practice for their own people is cremation. However, de Laguna reports that Tlingits decapitated slain enemies, occasionally retaining the heads as trophies. The nonskeletal tissue of three of these heads had decayed. Preservation of soft tissue in the surviving three heads seems to be the result of anthropogenic desiccation, though by what method is not apparent. The brain had been removed in all three and sticks employed to hold the cheeks in position. Local lore cites decapitation for the purpose of acquiring trophy heads of massacred Wrangell natives by the Kagwantan (wolf) people of Sitka, Alaska, in the nineteenth century but the location of the heads reported by de Laguna seems to convince her they were a Tlingit product.

In summary, the bulk of Alaskan anthropogenically mummified bodies are those of Aleuts from several of the Aleutian Islands. The best-preserved ones were found in a volcano-heated cave. A spontaneously mummified but little studied group from Prince William Island on Alaska's southern coast was largely looted and disappeared before examination by professionals became possible. Most of the occasional, other individual bodies, both human and animal, from the northern Alaska regions became frozen or freeze-dried as a consequence of contact with permafrost.

Western USA

Four Corners region

Most of the United States mummies have been retrieved from the area commonly termed "Four Corners", a point around which the four states of Arizona, New Mexico, Colorado and Utah are clustered. This high plateau averages nearly 1000 meters' altitude, most of which is arid land with long, hot summers and little rainfall. Several deep canyons lacerate the landscape, the cave and rock overhangs of which have created microclimates conducive to spontaneous mummification. Canyon de Chelly (pronounced shay) and Canyon del Muerto are both quite large and most mummies from this area were found in caves of these canyons. The bulk of these mummies are members of the agricultural groups collectively termed the Anasazi Tradition (or Basketmakers) that occupied this area between approximately AD 100 and 1200. The bodies are usually flexed, bound into a sitting position, wrapped commonly in fur blankets and located within cave recesses or, quite commonly, in stone-lined cists originally created for grain storage. The degree of preservation under these conditions predictably has been found to be quite variable, but certainly some of North America's best-conserved ancient bodies are in this group. About 100 mummies scattered through this region have been reported, less than one-third of which found their way into museum collections, curated at state museums in Tempe and Tucson (Arizona), Boulder (Colorado) and Las Cruces (New Mexico) (el-Najjar *et al.*, 1998). Most of these either have been reburied or can be expected to be reinterred as the consequence of the Native American Graves Protection and Repatriation Act (NAGPRA), a law that is designed to return Native American human remains and artifacts to the tribes to whom those items are affiliated. Only a small number of these have been studied biomedically. El-Najjar *et al.* (1985) reported findings in six examined mummies, identifying a bladder calculus, silicate pneumoconiosis, abdominal stab wound and probable tooth aspiration. In another group of four, a 1-year-old infant had died of congenital heart disease (common atrioventricular canal) (Robertson, 1988) and an adult male suffered an epidural occipital hematoma due to a head injury. An intestinal parasite (*Enterobius vermicularis*) was also identified (el-Najjar *et al.*, 1980). Williams (1929a) dissected a single mummy from this area, given to him by the anthropologist E.A. Hooton, and found a caseous granuloma in the lung that probably represented a primary tuberculous infection. Birkby's study of the collection at Arizona University Museum in Tucson found that 44% were infected with headlice in the hair (pediculosis) (el-Najjar *et al.*, 1998). Although all of these studied bodies have been preserved by spontaneous mummification secondary to the region's dry, hot climate, Lombardi (1998) found a photograph of an Anasazi mummy in the files of Tulane University at New Orleans, Los Angeles, that demonstrates clearly a sutured transverse abdominal incision, suggesting probable evisceration. This photograph is probably one of many taken by Charles Lang during the early exploration of these sites by a local rancher, Richard Wetherill, which are stored at Tulane. This body probably is the one Wetherill called Cut-in-Two Man, though the incision in the photograph is limited to the abdomen (Siegel, 1998). An article about American mummies by Dawson (1928a) includes descriptions of two mummies from Utah curated at that time by a Salt Lake City museum, and states that they "bear evidence of having been suspended and smoke-dried . . .", though he does not identify the nature of that evidence. Yarrow (1881) quotes eighteenth and nineteenth century writers stating that evisceration and desiccation by fire-heat were the principal processes of North American embalmers, but our subsequent experience has confirmed use of that method commonly only in Central and South America, especially Colombia.

Mummies from other western states

Elsewhere in the western USA only a few mummified bodies have been found. Two flexed, poorly preserved male bodies wrapped in cowhide and buffalo robes were found in a small rock shelter near Pitchfork in northwestern Wyoming (Gill, 1976). These spontaneously mummified bodies were from the late prehistoric period, since one was wearing a British or Canadian military coat. Lice were present in the hair of one (Gill & Owsley, 1985).

An unusual form of mummification was employed by an early American colonial settler. Willie Keil, a child, wanted to accompany his father on the family trek from Missouri to Oregon in a covered wagon on the Oregon Trail. When he died in 1855 just before departure, his father placed his body in an alcohol-filled zinc container, strapped the container to the wagon and carried him along on the long and arduous journey to Oregon (Gibbons, 1986:154).

Recently the rock shelters of northwestern states of Washington and Nevada have produced some of the most ancient human remains found in North America. About ten bodies have been identified in Nevada whose radiocarbon dates exceed 8000 years before the present (BP); several are in excess of 9000 years old (Dansie, 1997). Most of these, like Nevada's Wizard's Beach body, are skeletonized. Among these bodies, however, is one early Holocene burial of a male whose lower body is skeletonized but the upper half retains much soft tissue, including hair. It was found in 1940 in a small rock shelter called Spirit Cave beneath another body, wrapped in a vegetal mat and covered with wind-blown sand. Believed at that time to be only about 2000 years old, it was placed in storage at the Nevada State Museum. More than 50 years later the box was opened and his hair was sampled as part of an experimental methodological study of radiocarbon dating. The reported value of 9415 ± 25 BP (mean of seven samples tested) astonished everyone and to date remains the world's oldest mummy, 400 years older than northern Chile's well-known Chinchorro Acha Man (Aufderheide *et al.*, 1993). Analysis of Spirit Cave Man's coprolites has identified fragments of small freshwater fish probably netted from shallow water (Eiselt, 1997; Napton, 1997). Small amounts of pine and other pollen also found in his coprolites probably represent only airfall, wind-blown pollen (Wigand, 1997). Initial examination revealed a skull and pelvis with male features, numerous congenital anomalies of vertebrae, spondylolysis of the fifth lumbar vertebra, dental abscesses secondary to pulp exposure by attrition and a healing skull fracture (Edgar, 1997). Skeletal measurements were most similar to those of ancient Japanese Ainu and Norse collections, but beyond limits of variation even of these groups; they bore little resemblance to those of modern Mongoloids or Amerindians (Jantz & Owsley, 1997). Facial reconstruction suggests Caucasoid features. As of the date of this review (2002) further analysis has been arrested while disputes over ownership of these remains (based on NAGPRA) are resolved.

Similarly, in August 1997 the partially mummified remains of an ancient body were recovered as it was eroding from a bank of the Columbia River near Kennewick, Washington (Lyke, 1997). The body was promptly dubbed "Kennewick Man". Radiographic computed tomography (CT) scans revealed a lithic spear point embedded in the pelvis. Radiocarbon dates initially indicated an age of about 9300 years, later refined to a 2 standard deviation range of calibrated dates of about 8340 to 9200 BP or 7250 to 6390 BC (Taylor *et al.*, 1998). Facial reconstruction also suggested Caucasoid features. An extract was prepared to study mitochondrial DNA (mtDNA) for comparison with various populations including Native Americans, but the procedure had to be suspended because of legal disputes regarding access to the remains for study or burial (McDonald, 1998). After prolonged court proceedings, during which the find site was buried under tonnes of gravel by a federal agency, the body was subjected to intensive attempts to study its origin including mtDNA testing (McManamon, 1999). Finally, in September 2000 when laboratory efforts to extract DNA from the bones failed, a court ruling assigned disposition of the body to Native American groups (Hughes, 1998; Holden, 1999a; Bonnichsen & Schneider, 2000). However, to date (2002) appeals have delayed implementation of that decision.

In 1962 a desiccated, spontaneously mummified body was excavated by Robert Edgard under the direction of Harold McCracken from a dry cave (aptly named Mummy Cave) located 55 kilometers west of Cody, Wyoming. The lower humidity in the cave is partly due to the semi-arid climate and is partly the product of its geomorphology and geology. The multiple, dated cultural layers of soil inside the cave indicate human presence approaching 10000 years ago, while that in which the body was found was dated to AD 734. The mummy was curated at the Cody Historical Center for 25 years, after which it was examined by a multidisciplinary team including Michael Zimmerman, M.D., at the Medical College of Ohio in Toledo, the project being under the direction of Frank Saul, Ph.D. Chemical dietary reconstruction indicated the man's diet had been composed of about 60% meat (principally mountain sheep as indicated by related animal bones in the cave) and 40% vegetals. Coprolite study by Patrick Horne (Toronto) found abundant spruce (*Picea*) pollen that is normally produced only in April and May, suggesting that interval as his season of death. The body has been suggested to be that of a pioneer fur trader (McCracken *et al.*, 1978).

Texas is another area from which early American mummies have been recovered. Mummies from the caves of the lower Pecos River area have demonstrated many Harris lines of growth arrest in bones of children from the Archaic Period, suggesting malnutrition and perhaps starvation. Phytoliths of succulent plants are believed to be responsible for the pathological degrees of dental attrition (el-Najjar *et al.*, 1998).

The cave mummies from the Southeast

The vast limestone outcroppings bordering the Mississippi River and extending into the adjacent states of Kentucky, Tennessee and even Arkansas contain a constellation of caves and rock shelters varying from those of diminutive size to the massive Mammoth Cave of Kentucky (the latter now being a national park). The constant temperature and low humidity of the larger caves has resulted in spontaneous mummification by desiccation of several bodies found there. These must be the product of additional, environmental, microclimatic factors, since most human remains in these caves have been purely skeletal. Unfortunately the caves of the mummies' origins were privately owned and the caverns' spectacular, geological beauty commercialized as a tourist attraction. The mummies were exhibited to enhance the visitors' interests. The subsequent chicanery could be a testimonial to the methods of that greatest of all showmen, P.T. Barnum. The discovery conditions are shrouded in secrecy, as their finders smuggled them to other caves in order to claim ownership by discovery. We do know they were all spontaneously desiccated, wrapped in animal hides and textile blankets (some with additional feather blankets), and that none of them were examined professionally by dissection.

A mummified infant was found in 1811 by saltpeter miners in Short Cave (near Mammoth) and destroyed by them, but in 1813 they encountered an adult female buried in a stone cist in a sitting position, wrapped as noted above. Subsequently called Fawn Hoof, she was initially exhibited in Mammoth Cave for 2 years, then went on tour for several years, ending her career in the Smithsonian Institution. There, after a quarter century of exhibition, she was stripped of her soft tissues and filed as a box of disarticulated bones. A second body from that cave, an adult male, was acquired quite promptly by a commercial museum in New York city (Scudder's Museum), and the third (adult female) by a Cincinnati museum; both of these were destroyed ultimately when the museums caught fire. All three were advertised as "Mammoth Museum mummies", though none actually had originated in that cave. In 1875 still another spontaneously mummified, desiccated body, soon called "Little Alice" was found in Salts Cave, transported to another privately owned cave (Wright's Cave) and displayed there. Little Alice became the most traveled of all mummies, exhibited at many conventions and meetings including the Chicago World's Fair. In the 1950s she was finally subjected to a professional evaluation by Dr Louise Robbins, who reported that Little Alice was actually a 9-year-old boy (thereafter known as Little Al) (Meloy & Watson, 1969).

Probably the most well known of the Kentucky cave mummies is Lost John. About 2200 years ago this 45-year-old man entered Mammoth Cave to collect some of the gypsum crystals that form on the cave's limestone roof. Carrying a brush torch to light his way, a hammer to remove the gypsum and a bag to collect and transport it, he climbed a ledge and crouched on it as he chipped away at the crystal. Accidental dislodgement of small supporting rocks caused a huge slab of the roof to collapse, crushing him beneath it. For more than 2000 years he lay there, his chest and left arm crushed and fragmented as his body gradually dried to leather-like consistency. Eventually, in 1935, two cave guides found him, his hammer, bag and partly burned torch at his side. The National Park Service (who now owned the cave) built 10 meter high towers, attached pulleys and cables and gently lifted the 6 tonne rock sufficiently to extract Lost John's spontaneously desiccated body. He was placed in a glass vitrine and respectfully viewed by thousands of cave visitors for many years (Pond, 1937; Neumann, 1938; Meloy, 1971).

The smaller caves and shelters in Tennessee and Arkansas seem to have been less effective for soft tissue preservation. Most human remains in these are skeletons. In 1810 two bodies were removed from a "copperas" (ferrous sulfate) cave in Warren County on the Cant Fork of the Cumberland River. One was a young adult male and the other an adolescent female. Wrapped in a linen shirt, deerhides and cane mat and placed in cane boxes their bodies had desiccated spontaneously. The male was described as intact "excepting the bowel . . .". No further biomedical study was carried out (Miller, 1812). The limestone bluffs near the Arkansas–Missouri border were also used for shelter as well as a burial site by Native Americans more than 1000 years ago. A small fraction of the bodies found there were similarly spontaneously desiccated, nine of which now form part of the collections at the University of Arkansas in Fayetteville. They have not yet been dissected, but a radiological study that included CT scanning indicated a large variation in the quality of soft tissue and visceral preservation resulting from their spontaneous desiccation (Riddick, 1995). Three of these had been acquired from a limestone rock shelter in southwestern Missouri. Similar small, dry caves (Echo Cave, Kettle Hill Cave) in south central Ohio have produced spontaneously mummified bodies found by Boy Scouts in 1925. One of these found its way into the Ohio State Archaeology Museum.

Kentucky is also known as the site of a very different type of mummy. When a local factory worker (Speedy Atkins) drowned in the Ohio River in 1925 at Paducah and no one claimed the body, the local funeral director injected it intra-arterially with an experimental solution. Except for turning the body's black skin a somewhat red color, the process was so successful that it was displayed (without fees) for 66 years until the undertaker's widow tired of caring for it and buried it in 1994. Unfortunately the embalmer never revealed the composition of the fluid he injected (Fisher, 1994).

What are probably the most unique mummified human remains in the USA are the mummified brains from a Florida sinkhole. A water-filled basin of a peat bog near Titusville in central Florida (Windover site) yielded a group of paleo-Indian skulls up to nearly 9000 years old (Kunerth, 1985; Stephenson, 1985). Several of these contained preserved brains that had shrunk to one-half or less of their original size but retained the normal convolutions of the brain surface.

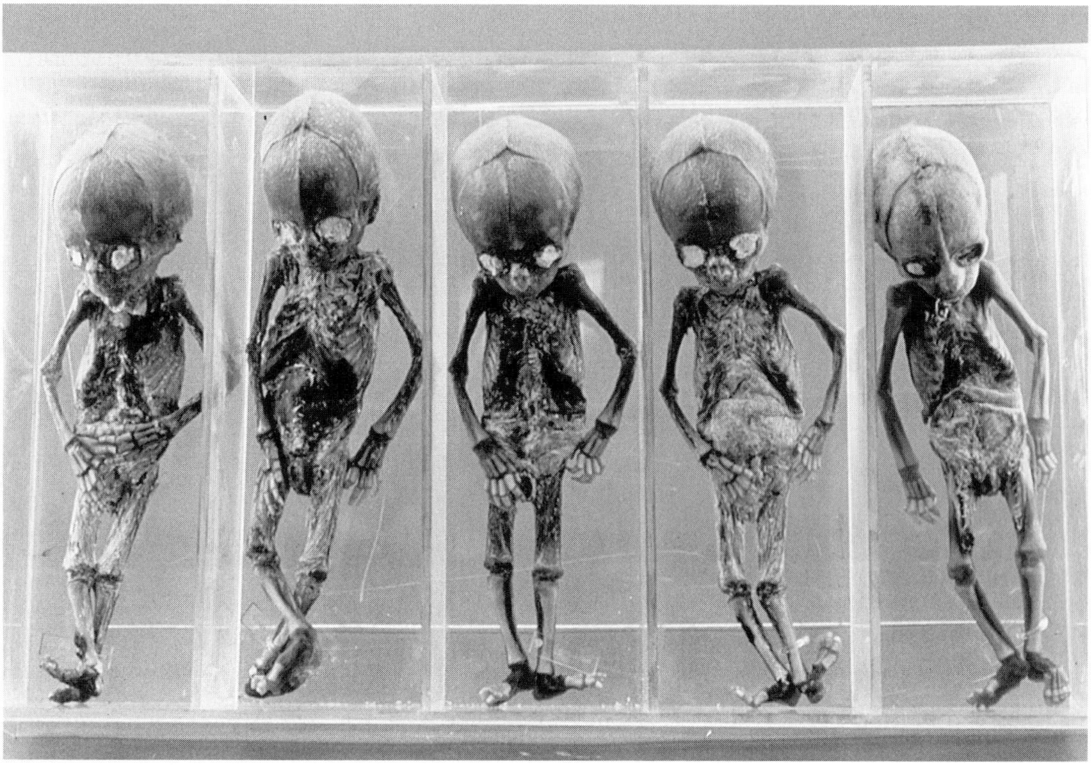

Fig. 4.4. Lyon quintuplets; AM, AD, 1896.
These bodies were originally embalmed by methods common in 1896, but subsequently became desiccated, probably spontaneously. (Photo courtesy of Paul Sledzik and the National Museum of Health and Medicine.)

In some, brain slices revealed many of the principal, grossly recognizable brain structures. Histologically neurons and some of their processes could be identified and cloning techniques were successful for the recovery of intact DNA segments (Doran *et al.*, 1986; Ballinger, 1988). A similar site at Little Salt Spring near Charlotte Harbor, Florida, contained preserved bones but no soft tissue (Clausen *et al.*, 1979).

Miscellaneous mummies

Lyon quintuplets

On April 29, 1896, residents of the community of Kevil, Kentucky, were astonished at the announcement of an unprecedented incident: the birth of live quintuplets to a local mother – Mrs Oscar Lyon. Without the availability of modern neonatal intensive care medical services, all five infants died within two weeks. The intense curiosity of the public caused the grieving mother to fear for the security of the deceased and embalmed bodies (Fig. 4.4), so she arranged for their transfer to the Army Medical Museum in Washington, DC, now called the National Museum of Health and Medicine. There they remain (specimen no. 43411) on exhibit in a five-chambered glass container (Quigley, 1998; P.S. Sledzik, 1998, personal communication).

Holmes mummy

As detailed in Chapter 3, Thomas Holmes played a central role in the development of intra-arterial embalming during America's Civil War. Thereafter he, probably more than any other individual of that period, was responsible for popularizing the practice of embalming much as we know it today. We still have at least part of one body he embalmed with an arsenic solution in 1860: a 12-year-old female (Fig. 3.25). How this specimen arrived at its present location is not recorded but it can be seen today (specimen no. 558844) in the National Museum of Health and Medicine in Washington, DC. This 110-year-old body's superb preservation is a testament to Holmes' skill.

Alleged Virginia mummies

Except for the Alaska Aleuts, the evidence for the occasional suggestion of anthropogenic mummification among Native North Americans is slender (see Western USA, above). This section would, however, be incomplete without mention of such an alleged practice in several European publications during the North American Colonial Period. The earliest of these is a book published in 1705 in London whose author, Robert Beverley, was a resident of the state of Virginia. On pages 214–216 he describes mortuary rites of local indigenes reserved for their tribal leaders. He indicates that they "first . . . flay off the Skin . . . pick all the Flesh off from their Bones . . . dry the Bones a little in the Sun and put them back in the Skin again . . . fill up the vacuites with a very fine white Sand . . . sew up the Skin again . . . they lay it upon a large Shelf . . . for the Corps to rest easie on . . .". He goes on to describe drying of the excarnated tissues that were preserved in a basket laid near the corpse's feet. The description is accompanied by a highly stylized drawing of a high-roofed building containing a platform upon which such bodies are laid side by side and a fire attended by a "priest" clad in a highly imaginative garment (Fig. 4.5). We know little of this author's qualifications relating to the history of such a practice. A subsequent publication (Willoughby, 1907) essentially repeats Beverley's description so closely that it suggests Beverley may have been Willoughby's source. Suffice it to say that no such preserved bodies have come to my attention in modern times, nor have I been able to find earlier reports by any one claiming eye-witness evidence. Hence, unless convincing evidence emerges, these descriptions will remain unconfirmed anecdotes.

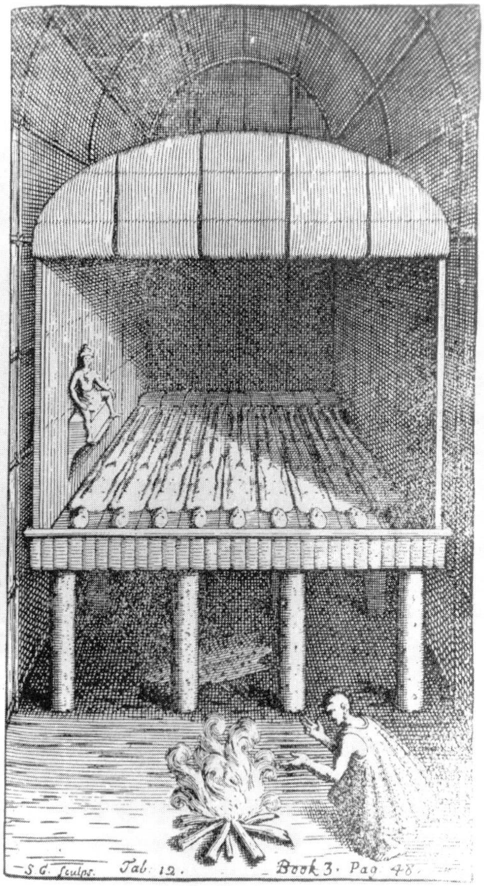

Fig. 4.5. Alleged ancient North American mummification practices.
This illustration from an early colonial publication by Beverley (1705/1947) is surely the product of his imagination. Archaeology has not documented any such practices anywhere in the continental area of what is now the USA.

We have, however, two anthropogenic mummies from West Virginia. In the village of Philippi of that state, a local embalmer (Graham Hamrick)

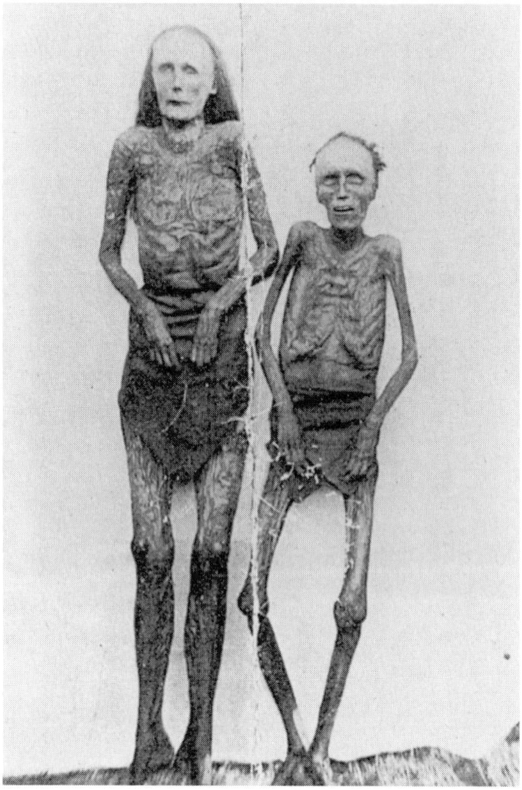

Fig. 4.6. North American experimental embalming; AM, AD 1888.
After the American Civil War, fierce competition among funeral directors led to embalming experiments such as these West Virginia bodies from a local institution for mental illness. (Photo courtesy of James Ramsey, curator of Barbour County Historical Museum.)

experimented with various embalming fluid ingredients. Gaining permission from a local judge, he injected one of these into two bodies he acquired from deaths at a local insane asylum. He patented this technique. The Barbour County Historical Museum today curates these two adult female bodies. Hamrick's published method involved injection (intra-arterial?) of a saltpeter solution. Their relatively well-preserved status more than a century later in spite of surviving two flooding episodes is a testimony to the method Hamrick originated. The bodies, however, have a des-

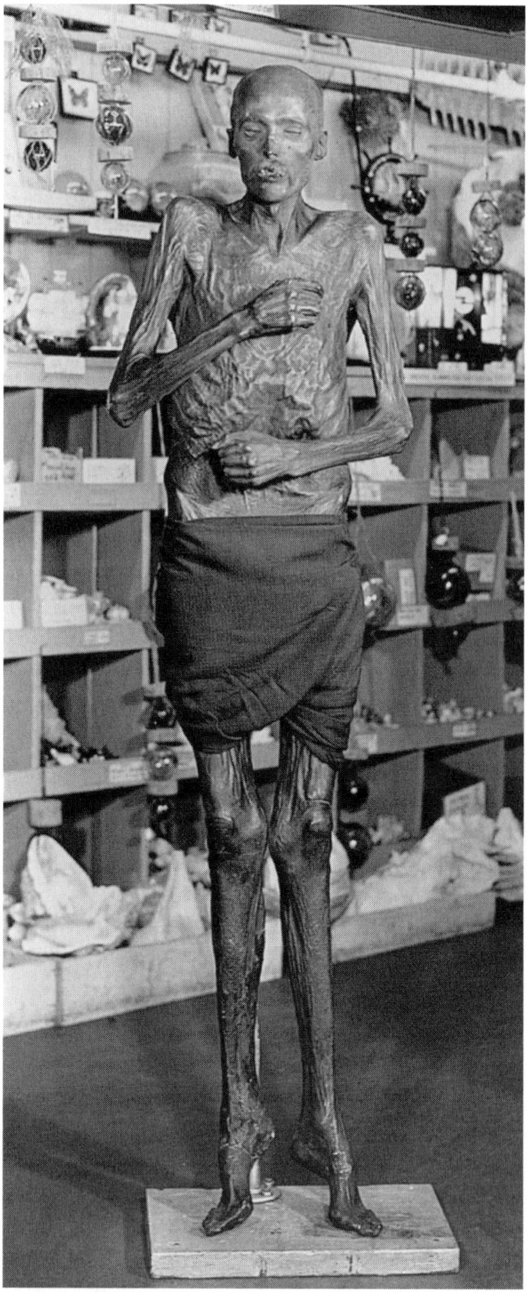

Fig. 4.7. The mummified body of Sylvester; ca. AD 1890.
This is the desiccated body of an adult male from a desert area in the western USA. Preservation method is uncertain. (Photo courtesy of Smith-Western Co., Tacoma, WA.)

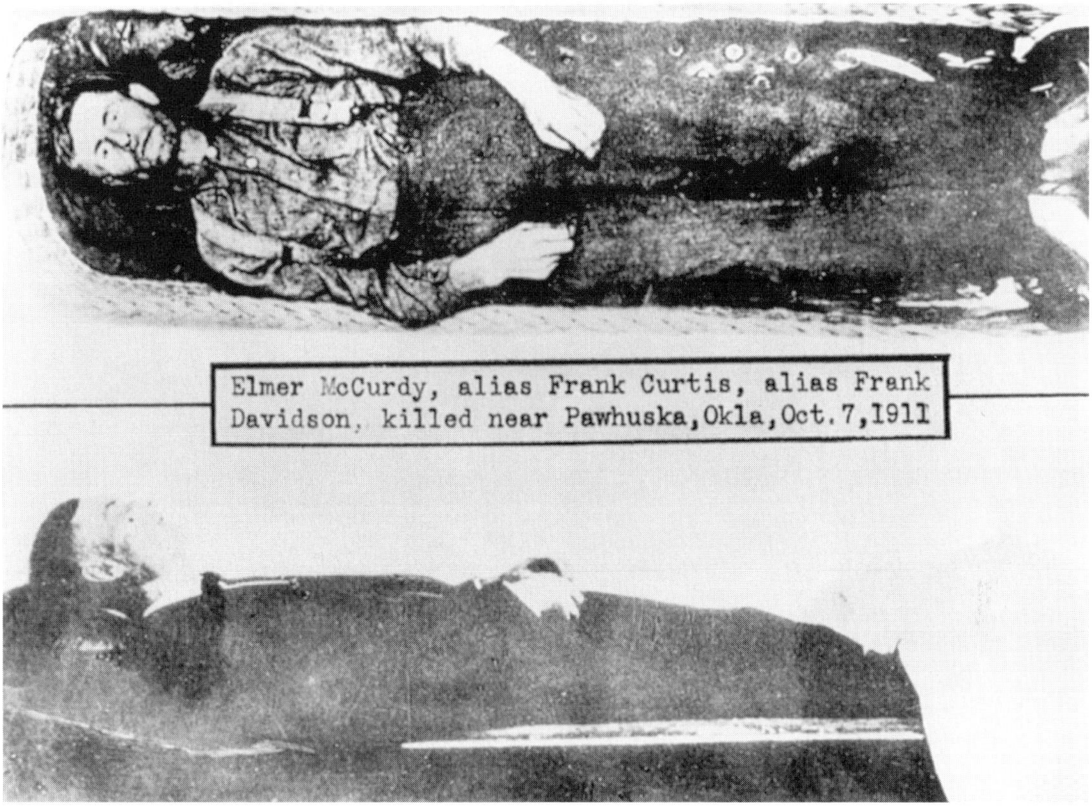
Elmer McCurdy, alias Frank Curtis, alias Frank
Davidson, killed near Pawhuska,Okla,Oct.7,1911

Fig. 4.8. Elmer McCurdy; AM, AD 1911.
The arsenic solution apparently used for intra-arterial injection of this late nineteenth century body preserved soft tissues superbly for more than 90 years. (Photo courtesy of Western History Collections. University of Oklahoma Libraries.)

iccated appearance (Fig. 4.6) so their preparation and subsequent preservation may well have included additional steps not identified in Hamrick's patent description (Quigley, 1998:124).

"Sylvester and Sylvia"

The World Famous Ye Olde Curiosity Shop in Seattle, Washington, houses several spontaneously mummified bodies. One of these, a male dubbed "Sylvester", was allegedly found in the Gila Bend Desert of central Arizona half-buried in sand. Recent radiological studies demonstrated intact visceral organs (Fig. 4.7). The other, called Sylvia, is a spontaneously mummified female, allegedly found in a mountainous area of Central America. She also has preserved viscera. Recent analysis showed high tissue content of arsenic in Sylvester.

Elmer McCurdy

Finally, the body of Elmer McCurdy, a late nineteenth century Oklahoma outlaw, is a superb example of mummification by intra-arterial injection of an arsenic solution (Fig. 4.8). The interesting account of this widely travelled body was summarized in Chapter 10 (Snow & Reyman, 1977; Quigley, 1998:60–70). The high quality of histological detail led to suspicion of

Fig. 4.9. Pulmonary tissue of Elmer McCurdy; AM, AD 1911.
These soft tissue samples are easily recognizable as lung nearly 100 years after embalming with arsenic. (Photo courtesy of T.E. Reyman.)

arsenic injection, subsequently confirmed by chemical analysis of tissue biopsies. Gross specimens of lung tissue are easily recognizable (Fig. 4.9) and intrapulmonary hemorrhage is demonstrable in the histological sections (Fig. 4.10).

Mummies from Mexico

The mummies from Guanajuato

Within the city of Guanajuato in Mexico's central valley, about 250 kilometers northwest of Mexico City, is a distinctly unique museum – the Museo de las Momias [the Museum of the Mummies]. Because the local San Sebastian cemetery became full, the community constructed a new municipal cemetery in 1861. Four years later the body of a French physician was exhumed. When the coffin lid was opened, the excavators were astonished to find themselves staring at a well-preserved, spontaneously desiccated mummy. This mummy was placed in a building and exhibited in 1870. Since that time more than a hundred similarly mummified bodies have been added to that museum, the most recent in 1979 (Anonymous, 1991). They remain on exhibit even today, many in an upright position and some at least partially clothed. Sagging of the mandible secondary to postmortem facial masseter muscle degeneration is common in spontaneously mummified bodies when the jaw is not supported with wrappings. In the Guanajuato mummies this has resulted in open-mouthed, grotesque facial features (Fig. 4.11).

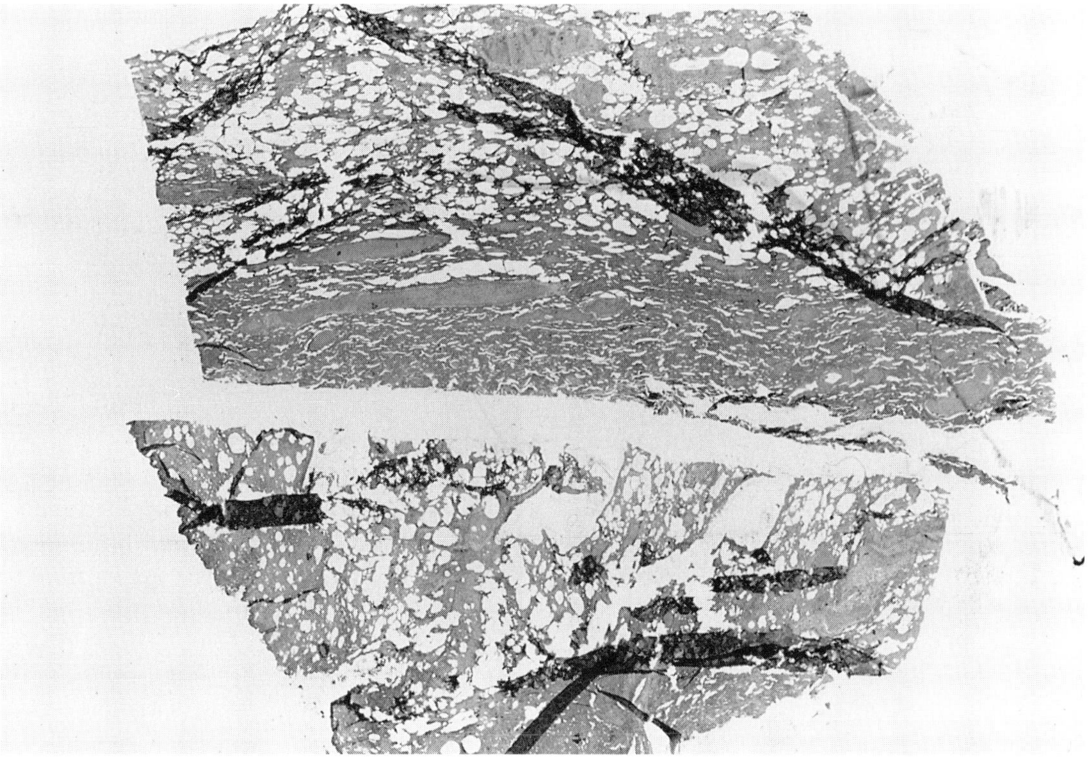

Fig. 4.10. Lung tissue of Elmer McCurdy; AM, AD 1911.
The histological structure of the lung is well-preserved, including evidence of intrapulmonary hemorrhage, probably from a bullet wound. (Photo courtesy of T.E. Reyman.)

The skin and muscles of the extremities seem to be very well preserved on many of the bodies, though visceral preservation is uncommon. The museum's catalogue attributes the spontaneous mummification effect to the dry, porous, nitrate-containing soil in that part of the cemetery in which mummification occurs most frequently, though it also notes that even there many interred bodies become skeletonized. The desert climate of this region most likely makes a significant contribution by minimizing groundwater and maintaining a low humidity in the ambient air of the museum (Fig. 4.12).

The rows of mummified, upright cadavers are reminiscent of the similar arrangement in the Capuchin monastery of Palermo in Sicily. However, the mummification in Guanajuato appears to be purely spontaneous, whereas in Palermo many of the bodies attained mummification as a result of at least some anthropogenic contribution (see Mummies of Italy, below). Probably the most dramatic of the publicly available photos of Guanajuato mummies can be found in the text by Lieberman & Bradbury (1978). The emotional response to the viewing of these bodies by a poet is movingly recorded in Chapter 10 (Mummy as language, literature and film).

Mummies from other Mexican sites

Two mummies were found in the bat guano of a cave at about 27° 40′ N and 108° 36′ W in the desert of the Mexican state Chihuahua between Loreto and Santa Ana in 1966. They were transported to San Diego in California where they eventually found their way to the Museum of Man in that city, and were studied by

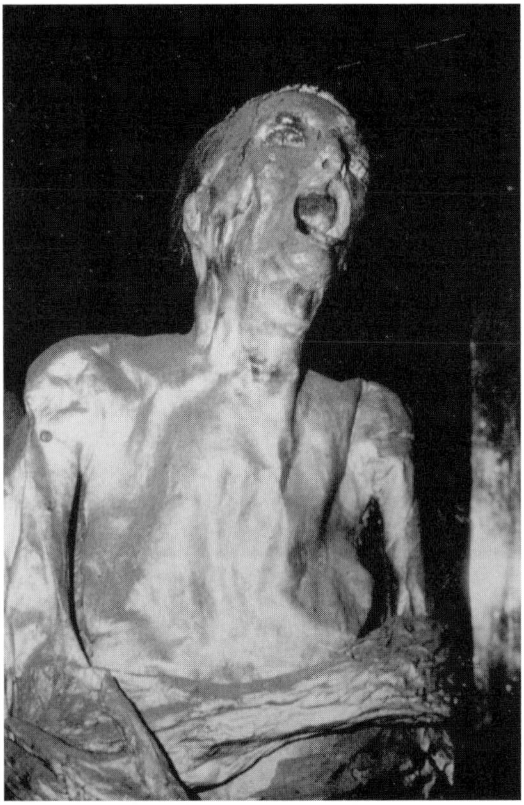

Fig. 4.11. Guanajuato mummy.
This is one of numerous spontaneously mummified bodies exhumed from a church cemetery between 1870 and 1979. (Photo courtesy of John Tygart.)

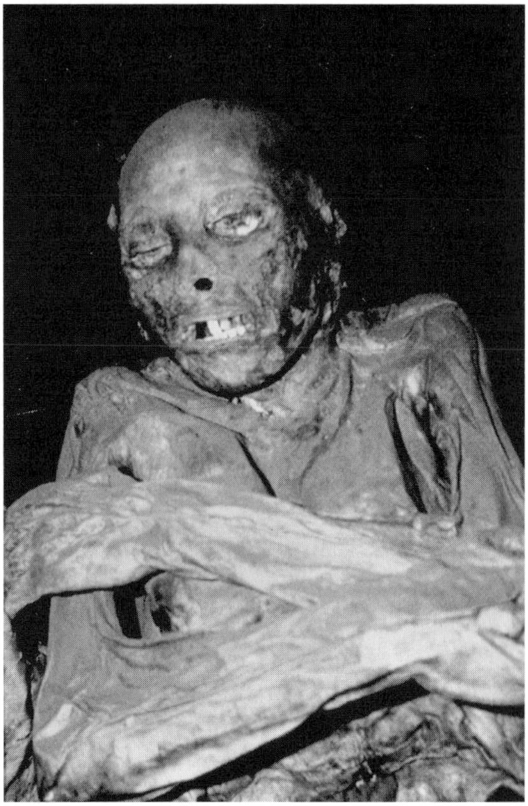

Fig. 4.12. Guanajuato mummy.
Another of the mummies described in Fig. 4.11. Preservation of soft tissue is attributed to the arid climate, porous soil and good drainage. (Photo courtesy of John Tygart.)

staff there. The infant was wrapped in bark and buried so near the surface it was partly destroyed by grazing cattle. The other was a 15-year-old female of about 155 cm stature. Calibrated radiocarbon date range was AD 690–910. That mummy bundle was 70 cm below the surface in bat guano. She was maximally flexed and wrapped in a mat whose sides were tucked in "like an egg roll . . .". Radiography identified a 28–32 week fetus in the abdomen. Cranial porotic hyperostosis was noted (Tyson & Elerick, 1985:1–28). Examiners suspect she may have died of placental hemorrhage following a traumatic injury, but this remains speculative.

A cluster of mummies found also in Mexico's Yucatán Peninsula are those from the Santa Elena *municipio*, approximately 21°11′N, 90° 00′W. A church in that community had been constructed in 1592 but expanded considerably in 1779. In 1980 reconstruction led to the discovery of several coffins beneath the floor of the church annex. These contained the spontaneously mummified bodies of four infants and children, ages 1–7 years. They were clothed, wrapped in cotton shrouds and wore flowered headdresses. Beneath them, within the coffin, were branches and twigs with leaves. Radiographs revealed multiple Harris lines of growth arrest and periostitis suggestive of generalized infection. Excavators sug-

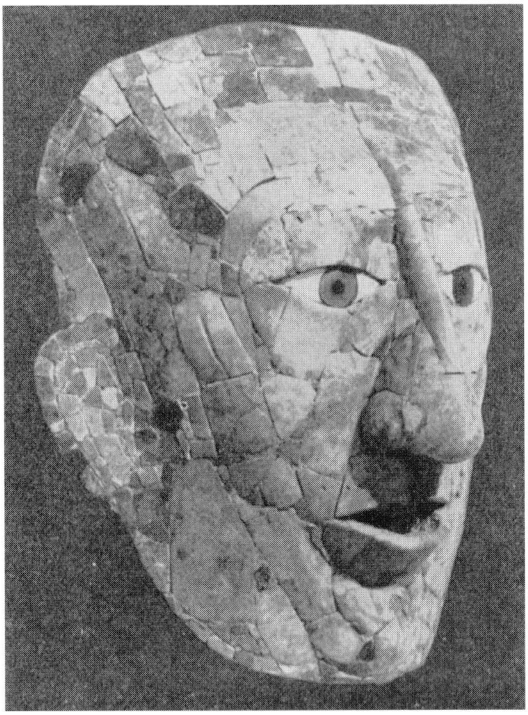

Fig. 4.13. Jade mask from Palenque.
The face and other parts of this Maya ruler's body (Pacal; SM, AD 600) at Palenque were covered with jade tiles, analagous to those used in ancient China (see Mummies from China). Mask restored by Alberto García Maldonato. Photo courtesy of CONACULTA.-INAH.-MEX. Instituto Nacional de Antropología y Historia).

gested that factors contributing to spontaneous mummification in these cases included the hot climate with a prolonged dry summer season, protection from rainfall and subsequent groundwater by the overlying church building and alternating layers of earth and clay that provided good drainage. They also felt that the branches, leaves and clothing contained tannins with antimicrobial effects (Márquez-Morfín & González-Crespo, 1985), though the quantities described seem too small to make a substantial contribution to preservation.

Another mummy preserved in the guano of a bat cave was recovered from a site southwest of the city of Durango in the desert climate of Mexico's state of Durango. The Chalchihuites' chiefdom occupied this site; the mummy was dated to about AD 950 ± 300. The body was that of a 3–3½ year old child of indeterminate sex. Xeroradiography revealed a large mass in the left lower chest whose nature could not be determined by histological study of samples. It was not further dissected (Heinemann, 1974).

A huge sarcophagus found beneath the Mayan Temple of Inscriptions at Palenque in Mexico's Yucatán province in 1952 housed the remains of a Mayan ruler who had ruled there for 80 years until his death in AD 683. While these were only skeletal remains it is conceivable that the jade facial mask (Fig. 4.13), necklace, rings, loincloth and jade pieces in both hands and mouth may have been intended to enhance soft tissue preservation as was once the custom in China and elsewhere (Lhuillier, 1953; Cortez, 1996).

Stirling (1938) indicates that contemporary sixteenth century sources suggest that Mexican Indians skinned the heads of their sacrifice victims and displayed them on poles or spears, but confirming archaeological evidence is meager indeed.

At the site of Tlayapacan, Morelos, near Cuernavaca, at about 18°53′ N and 98°55′ W two male children and a 12-year-old adolescent were discovered in 1982. They had been placed in wooden coffins that were still intact. These had been buried at a depth of 1.6 meters in soil beneath the nave of a church that was dedicated in AD 1574. The arid, hot climate combined with the protection from rain and groundwater by the church building had resulted in spontaneous, desiccating mummification of the bodies (Oliveros, 1990).

Two mat-wrapped, spontaneously mummified bodies were found in a mountainous cave of Mexico's Sierra Madre Occidental, buried within bat guano (Ascher & Clune, 1960). Three spontaneously mummified bodies also found in northern Mexico's Sierra Tarahumara included an adult male and two children. These were studied by CT scans and endoscopy. Findings included skull fracture with calcified epidural hematoma in the male as well as osteoporosis with compression fracture of the fourth lumbar vertebra (Chavez *et al.*, 1995).

Mummies from Central America

The tropical climate prevailing in most of Central America's land mass would not be expected to be conducive to spontaneous mummification nor to the maintenance of anthropogenically mummified bodies. To date I have found only historical references but no published examples of mummies from this region. Stirling (1938) cites a comment by an early missionary (Father Pozzi) that the Indians in the interior of the British Honduras indulged in head shrinking practices. However, Stirling states that this clearly applies to the Jivaro people of Ecuador. A 3000-bone ossuary in a cave of the Olancho Valley in northeastern Honduras contained skeletal remains with defleshing cut-marks but no mummies (Brady *et al.*, 1995). Nevertheless, Bradley *et al.* cite a seventeenth century missionary doctor's report that "in the Olancho area soldiers found many houses with vaulted underground compartments where they buried the dead and there the bodies stayed dried and whole with their meals . . .". I have not found references in scientific publications to modern recovery of any mummies from Central America.

MUMMIES FROM SOUTH AMERICA

Mummies from Venezuela

Most of the relatively few mummies found in Venezuela have been recovered from the Andean area of this country. Sixteenth century descriptions of funerary customs describe binding the body into a flexed position after death, wrapping it in a mat or blanket, suspending it from a tree branch and keeping a fire burning beneath it for a month. After the body dried it was placed in a mountain cave or subterranean rock-lined cist. Berrizbeitia (1991, 1992, 1995) has examined several such bodies curated in Caracas and Merida. She confirmed the flexed position and in several found soot on the skin and in the enveloping textiles, suggesting that they had been preserved by fire and smoke. She also identified a periostitis suggestive of treponematosis, including "sabre shin" deformity in several. Several children and a newborn curated in Caracas have a desiccated appearance suggestive of spontaneous mummification. She emphasized that the details of the method of desiccation have not been established with certainty. In 1989, exploring parties identified at least five funerary caves in the Andean area of that country. Human remains in some appeared abundant and included some with preserved soft tissues, though to date they have not been studied.

Mummies from Ecuador

The Jivaro (Shuar)

The Jivaro of Ecuador were head-hunters that reduced their enemies' heads to a doll-like size and mummified them. They occupied an area approximately between 02–04° S and 77–79° W (Fig. 4.14). Situated on the eastern slopes of the Andes at an altitude of 1200 meters in the west to 400 meters in the east, this area's climate is that of the hot, wet rain forest environment of the upper Amazon basin. Isolated by a steep western Andean escarpment, this group rebelled against Spanish occupation successfully and were relatively unmolested by Caucasians until the late nineteenth century. Hence our knowledge about them is more detailed than about those acculturated soon after the Spanish conquest. This group, therefore, will be used in this text as a "head-hunter" model for head mummification and the others noted only to the extent that we know they differed.

Harner (1972) states that the Jivaro's reputation as "an anarchistic population engaged in extensive internal hostilities and feuds . . ." is justified. Their purpose in head-hunting is more complex, however, than mere vengeance. The acquisition of spiritual power (*arutum*) is believed to render a Jivaro immune to death by violence and the power of an enemy's spirit (*muisak*) can be transferred to the slayer as well. These beliefs are incorporated into survival and vengeance concerns to produce an elaborate set of behavioral traditions and taboos relating to the taking of an enemy's head (Allen,

Fig. 4.14. Location of principal ancient populations with mummies in South America.
1, Chavín; 2, Paracas-Nasca; 3, Maranga; 4, Moche; 5, Huari; 6, Tiwananku; 7, Sican Chimu; 8, Inca; 9, Chinchorro etc.; 10, Muisca; 11, La Plata; 12, Yupka; 13, Jivaro.

1981:52–54). Protected from the west by the Andean terrain, the Jivaros regarded all the half-dozen tribes to the east as enemies, and it is upon these that they carried out their head-hunting raids.

The reduction and mummification of the slain enemy's head begins promptly at the time of the raid. The steps involved are as follows (as detailed in Stirling, 1938 and Harner, 1972).

1. Excise and reflect the skin of the shoulder at the level of the clavicle.
2. Transect the neck at the level of the clavicle, pass a head band through the mouth and out of the neck, attach a cord and flee the raided village, carrying the head over the shoulder.
3. Arriving at a prepared trail camp, make a posterior skin incision from the neck to the crown of the head.
4. Dissect out the skull including the face; discard the skull.
5. Boil the skin in water (without herbs) for less than a half-hour.
6. Dry the skin (which now has about a 50% size reduction).
7. Turn skin inside out and scrape off all flesh.
8. Turn skin right side out and suture the incision.
9. Sew a vine into the skin of the base of the neck to retain its shape.
10. Tie the lips together with a vine.
11. Insert a half-dozen stones into the "skin bag", each about 5 cm diameter that have been preheated in a fire. Repeat when stones cool.
12. When the "skin bag" has become too small to accept stones, continue with heated sand.
13. As the skin shrinks, tighten the neck vine to maintain proportional neck size.
14. During the several days required to return home, repeat the hot sand treatment several times.
15. Rub the facial skin regularly with hands and warm stones to flatten it and prevent wrinkles.
16. Rub the skin daily with finely powdered balsa wood charcoal to blacken it (preventing the *muisak* spirit from escaping).
17. Dry the lips with a heated machete blade.
18. Close the lips with three chonta palm pins and lash them together with a string made of kumai inner bark.
19. If eyes become sunken, insert a seed under the eyelids.

This process requires about one week, and the final size of the shrunken head is about that of a man's fist. The hair, of course, remains its original length (Fig. 4.15). When the slayer reaches the home village he creates a hole in the top of the head skin, inserts a kamai string, ties it to a stick that he places inside the skin and loops the string around his own neck, carrying the product (now called a *tsantsa*) prominently and proudly suspended on his chest.

Fig. 4.15. Jivaro shrunken heads; AM, ca. AD 1750. These heads were meticulously flayed and the soft tissues with attached hair preserved as described in the text. (Photo courtesy of Department of Library Services, American Museum of Natural History, Negative no. 118222.)

The entire village holds a celebration feast a few days following the return. The slayer then sponsors a second, five-day feast, often about a year later (to give him time to raise the necessary food) and, finally a third, six-day feast after the second. In addition to acquisition of prestige, an important function of these feasts is to maintain the capture of the enemy's *muisak* spirit, which otherwise could avenge the victim. The feasts, therefore are carried out with great respect for their many taboos. Following the third feast the *muisak* is expelled and exiled. At this point the lip's wood pins are removed and two kamai strings, each about 30 cm long are passed through each of the three holes and allowed to dangle from the lips; horizontal red stripes are often painted on them. After these feasts have been held, the spiritual purpose of the *tsantsa* has been met and its use thereafter is secular. It may be kept permanently as a house ornament and worn on festive occasions; it may be sold to a *mestizo* (half-caste), or it may be buried with the slayer when he dies (Harner, 1972:191).

In addition to this most common form, several alternative types of *tsantsas* exist. If the raid circumstances do not permit the slayer to recover the victim's head, a *tsantsa* may be borrowed from some other tribal member and used after a shaman has magically transferred the slain persons' spirits into it. Alternatively, a *tsantsa* made from a tree sloth may also be so employed. Preparation of a sloth *tsantsa* by an adolescent youth is also a "coming-of-age" requirement among the Jivaros (Harner, 1972:147–149).

During the nineteenth century an active international trade in *tsantsas* evolved. When the demand exceeded the supply, dealers began to create tsantsas – at first only from sloths and later from other animals. Zimmerman *et al.* (1993) indicates that even the heads of unclaimed human paupers' bodies were so employed. By 1938 Stirling suggested that perhaps the majority of *tsantsas* in private collections were frauds. This proved to be the case recently at the Smithsonian Institution where only 5 of 21 examined *tsantsas* were found to be genuine Jivaro products; of the remainder, nine were of non-Jivaro human origin and five had been prepared from other nonsloth animals including a dog, horse and monkey (Mann *et al.*, 1992). These authors also note that features almost invariably present in Jivaro *tsantsas* that can be used to separate them from frauds include:

1. Three mouth loops
2. Multiple cotton lip strings
3. Narrow heads
4. Brown skin color
5. "Pinched", distorted face
6. Upturned nose
7. Suspension cord
8. Posterior median sutured incision

Fig. 4.16. Mundurucú head; AM, ca. AD 1750.
The Mundurucú of the western Amazon jungle preserved heads but did not shrink them. (Photo courtesy of Gabriele Weiss and Ildikó Cazan and the Museum für Voelkerkunde Wien, Foto-Archiv, Foto Nr. 41.644.)

9. Small, bulbous chin
10. Size of human fist

Perhaps the most brazen hoax was the shrinking, not only of the heads but the entire bodies. One was the body of a Spanish military officer and the identity of the other is unknown, but fashioned to resemble a jungle native. A Lima physician, Dr Gustave Struve, who had lived with the Jivaro for several years has been suggested to have used Jivaro methods to prepare these bodies while he was a medical student in Guayaquil, Ecuador. He is alleged to have reduced their size dramatically (one was only 53 cm tall). Because he apparently did not bother to do the detailed dissection required to remove the hand and foot bones, these appeared disproportionately large. These bodies were acquired by New York's Museum of the American Indian in 1923 whose staff eventually buried them about 70 years later (Alexander, 1994).

The Jivaros, however, were not the only headhunters. In the western Amazon the Mundurucú Indians also hunted heads (Horton, 1948:271–282). No archaeological evidence exists to support the assertion that they shrank them. Instead, they removed the brain and filled the cranial cavity with cotton. They dried the head with the skull intact, covered with its soft tissues (Fig. 4.16). A cord was laced through the lips (Page, 1897). Vreeland (1998:177) also cites the notation by a chronicler (Estete, ca. 1535) that the coastal people at Pasao, Ecuador, shrank heads, but no archaeological or even eye-witness accounts are available to support this, although Stirling (1938:65) cites not only Estete but also another chronicler, Zárate, to the same effect. Other mummification practices Estete attributes to them is the flaying of the body, burning the remaining soft tissues off the body, with subsequent "dressing" of the skin, stuffing it with straw and hanging it from the temple roof (Fleming *et al.*, 1983). Other head-hunters in South America took heads from slain war victims and displayed them as suspended trophies; the Paracas and especially the Nasca from Peru are among the most well known indulging in these practices. While early Nasca art suggests that some of these may have been reduced, there is to date no archaeological evidence. The ones that have been found have not had the skulls removed; they are suspended via a hole in the frontal bone (Proulx, 1971) (see Mummies from Peru, below).

Mummies from Colombia

The geography of Colombia

The northward progression of the Andes mountains enters Colombia with peaks more than 5000 meters high at its southwestern corner and soon divides into three branches, creating a trident pattern. The central range is directed nearly straight north while the eastern and western branches of the trident progressively diverge, gradually dwindling in height as they approach the Caribbean coast until they blend with the Caribbean lowlands there. Between the three arms of the trident lie the 3000 meter high Colombian highlands – savannas of incomparably beautiful grasslands flanked by snowcapped mountains. These highlands are drained by the north-flowing Magdalena river that eventually traverses the lowlands to enter the Caribbean at Barranquilla. The other half of Colombia lies east of the Andean range, progressing through grassland zones to lowlands. The southern half of this drains into the Amazonian rain forest while the northern part is drained by the Oronoco River, which courses through Venezuelan lowland, entering the Caribbean near Trinidad. Most of Colombia's mummies have been recovered from the savannas. Only a few were found in the lowlands of the Pacific plains (that lie between the western arm of the trident and the Pacific coast) at an altitude of less than 1000 meters near the Colombia–Panama border. None of the published reports of mummies have come from the lowlands east of the Andes.

Pre-Hispanic mummifying Colombian populations

The highlands of Cundinamarca and Boyacá on the Eastern Cordillera were occupied by the Muisca people, including the area of Colombia's modern capital, Bogotá. This was a complex confederation of multiple chiefdoms controlled mainly by the four principal villages of Hunza, Duitama, Sogamoso and Bacatá people, none of which survived beyond the eighteenth century (Cárdenas-Arroyo, 1995). A host of small chiefdoms occupied the western highlands and the Cauca River basin that drains them. The Yuko, a remote group relating to a small, northward projection of the Andean trident's eastern arm (Sierra de Perijá) about 100 kilometers west of Venezuela's Lake Maracaibo, is one of very few tribes that survived the Spanish conquest and its aftermath to the present day (Metraux & Kirchhoff, 1948). The Cueva people from the Darien area near the Gulf of Uraba occupied that lowland area immediately north of the Colombia–Panama border (Cárdenas-Arroyo, 1994). All of these noted aboriginal societies contributed to the group of Colombian mummies that have been reported or are still in collections today and described below.

The original peopling of South America traditionally is viewed as merely a gradual extension of the trans-Beringean migration. Following a postulated "bottleneck" effect in Central America, the flow is believed to have progressed initially southward via the

Andean highlands with later deviations to the east and to the Pacific coast. Recent studies dealing with the distribution of an "Asian marker" in mitochondrial DNA extracted from Colombian mummies, however, suggests a much greater population diversity than the traditional model would suggest (Monsalve *et al.*, 1996).

Mummification methods of Colombia

Both spontaneous and anthropogenic mummification methods were employed by pre-Hispanic Colombian populations. Examples of the spontaneous form have been reported only at highland sites. The anthropogenic form was also employed by some highland groups while, as might be expected, the populations occupying the humid, tropical lowlands mummified the corpses exclusively by a variety of anthropogenic techniques. In addition, Cárdenas-Arroyo (1994) notes that the finding of only ashes, clay and charred seeds in 37 funerary wrappings suggestive of mummy bundles at a highland site (Samacá) suggests a failed mummification effort. Radiocarbon-dated Colombian mummies range from AD 200 to 1520.

Unfortunately modern tomb looters (*huaqueros*) preceded trained archaeologists into most of the caves and rock shelters used by pre-Hispanic Colombian populations for storage of their mummified bodies. Schottelius (1946), for example, describes how the body parts of broken and fragmented mummies were so numerous when Jose Antonio Bárcenas (the site's discoverer) first entered the highland burial cave called Cueva de los Indios that it was difficult not to step on them. Yet Schottelius found only one intact mummy and three other partial bodies of value for inclusion in permanent museum collections. Additionally, for reasons indicated below, Spanish colonial officials commonly burned mummified bodies when they became aware of them. As a consequence, of the thousands of mummies that must have been stored in Colombia's caves and rock shelters, only a few dozen remain in public collections today, most of which were also acquired directly or indirectly from tomb looters. Our efforts to reconstruct Colombian mummification practices, therefore, are heavily dependent upon this small number, amplified by not always trustworthy descriptions by the post-Conquest chroniclers.

Fig. 4.17. Muisca male mummy from Colombia; AM, date not determined.
The Muisca people desiccated the body with heat from a fire. Mummy no. 38-I-771. (Photo courtesy of Felipe Cárdenas-Arroyo.)

The bulk of Colombian mummies have been recovered from highland sites. While desiccation characterizes the preservation mechanism for all mummies from this country, this was not always achieved in an identical manner. Simple envelopment of a corpse in several layers of textile wrappings followed by storage in a highland cave resulted in sufficiently rapid dehydration to preserve soft tissues in a significant number of bodies of the Muisca people (Fig. 4.17). These reported mummies have been found principally in the states of Boyacá and Santander (Table 4.1), and several

Table 4.1. *Mummies curated at the Colombian Institute of Anthropology, Bogotá*

Mummy number	Identification Number	Age (yrs) or cohort	Sex	Origin	^{14}C yrs before present ± SD	Date AD	Culture	SM/AM	Curator	Comments
1	42-IX-3956	20–24	F	Cisas, Boyacá province	1480 ± 100	ca. 470	Lache	AM	ICANH	Gallstone. Possible traumatic injury, right tibia, midshaft
2	423-A-423	2(?)	I	Ubaté(?), Cundina marca province	480 ± 110	ca. 1470	Muisca	AM	ICANH	Textile makings on soft tissue. Arms and feet tied and flexed
3	38-I-776	15–19	M	Unknown	750 ± 60	ca. 1200	Muisca(?)	AM	ICANH	Charred right side, thorax
4	38-I-777	Adult[a]	I	Unknown	710 ± 60	ca. 1240	Muisca(?)	AM	ICANH	Gallstone
5	38-I-771	35–40	F	Socotá, Boyacá province	100 ± 1.2 (Unreliable)	—	Muisca	AM	ICANH	^{14}C sample may have been contaminated
6	00-OS-0001	2.5	I	Unknown	120 ± 100 (Unreliable)	—	Muisca(?)	AM	ICANH	^{14}C sample may have been contaminated. Calcinated skull bones
7	MO.1	30–34	I	Pisba, Boyacá province	430 ± 100	ca. 1520	Tunedo	SM	MO	Cotton ball inside anal canal
8	41-III-2535	Adult	I	Mesa de los Santos,[b] Santander province	No date available	—	Guane	SM	ICANH	
9	41-III-2536	4–5	M	Cueva de los Indios, Santander province	390 ± 100	ca. 1560	Guane	SM	ICANH	Body in extended position
10	41-III-2599	Adult	M	Mesa de los Santos, Santander province	No date available	—	Guane	SM	ICANH	Skull with soft tissue
11	41-III-2603	Adult	M	Mesa de los Santos, Santander province	No date available	—	Guane	SM	ICANH	Skull with soft tissue
12	42-IX-3957	5–6 month	I	Chiscas, Boyacá province	1750 ± 55[c]	ca. 200	Lache	AM	ICANH	Child inside textile bundle
13	42-IX-3958	—	—	Chiscas, Boyacá province	No date available	—	Lache	—	ICANH	—
14	00-OS-0002	—	I	Unknown	No date available	—	—	SM	ICANH	—
15	00-OS-0003	—	I	Unknown	No date available	—	—	SM	ICANH	—
16	1838-11-11-1	Adult	F	Leyva, Boyacá province	1100 ± 60	ca. 850	Muisca	AM	BM	Artificial cranial deformation
17	1842-11-12-1	Adult	I	Gachantivá, Boyacá province	790 ± 60	ca. 1160	Muisca	AM	BM	Abdominal evisceration
18	?	Child	F	Unknown	No date available	—	—	—	INHM	Hydrocephalus(?), artificial cranial deformation (?)

Notes:

F, female; I, intermediate; M, Male; SM, Spontaneously mummified; AM, anthropogenically mummified; ICANH, Columbian Institute of Anthropology and History; MO, Museo del Oro; BM, Bogotá Museum; INHM, Instituto Nacional de Historia Museo.

[a] Adult is defined as over 16 years of age.

[b] The Mesa de los Santos is a large flatland with two main caves containing human remains: Cueva de la Loma and Cueva de los Indios. This mummy was not assigned a particular cave of origin and, therefore, is listed here as coming from that general area of La Mesa de los Santos.

[c] This date was previously reported incorrectly as AD 1750. The correct date is 1750 radiocarbon years before the present.

Source: Table constructed from information provided by Felipe Cárdenas-Arroyo of the Department of Anthropology, Universidad de los Andes, Bogotá, Colombia.

Fig. 4.18. Muisca mummy from Colombia; AM, ca. AD 1300.
The body was desiccated with heat from a fire. Mummy no. M01. (Photo courtesy of Felipe Cárdenas-Arroyo and Museo del Oro, Bogotá.)

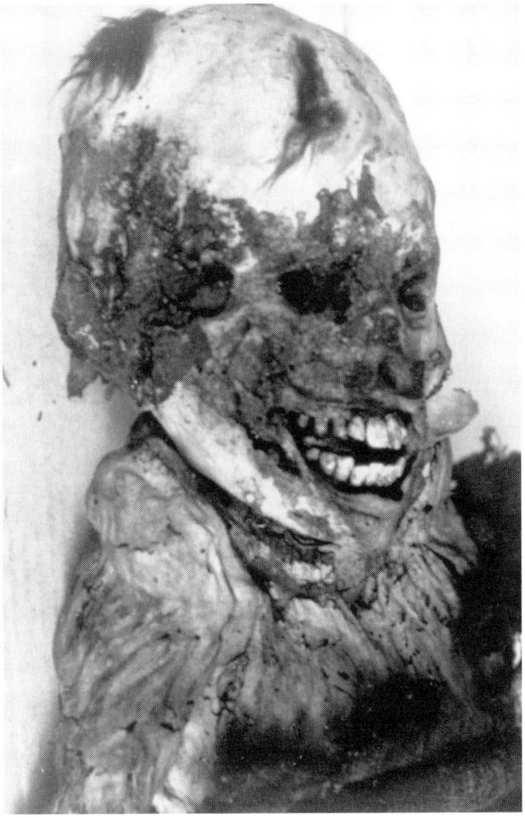

Fig. 4.19. Mummy from Guane culture, Colombia; AM, ca. AD 1400.
This spontaneously mummified child is from the Guane culture from the Bogotá savanna, late Pre-Hispanic Period. (Photo by and courtesy of Gonzalo Correal Urrego.)

are among the collection at the Colombian Institute of Anthropology and History (ICANH).

The general principle involved in Colombian anthropogenic mummification involved desiccation by heat from a fire. Details of heat application varied among the many tribes, as described below. Interestingly mummification by fire desiccation was generally achieved without evisceration, though among the tropical lowland groups visceral extirpation was more commonly performed than in the Andean highlands. In most groups the complexity of mummification and related funerary rituals were directly related to social status. Mummification variations among the chiefdoms follow.

Among the highland groups, the Muiscas fire-dried the corpse (Fig. 4.18), usually without evisceration, and deposited it in a sanctuary (most commonly a cave) (Cárdenas-Arroyo, 1998). Five such bodies are in the ICANH collection, as are two more in Berlin's Voelkerkunde Museum (Silva Celís, 1945; Herrmann & Meyer, 1993:126–129). The neighboring Guanes used methods similar to those of the Muisca (Fig. 4.19); they also practiced cranial deformation (Pérez-Martínez, 1960). Another Muisca neighbor, the Laches, employed similar methods (Fig. 4.20) but included some distinctive variations such as filling the

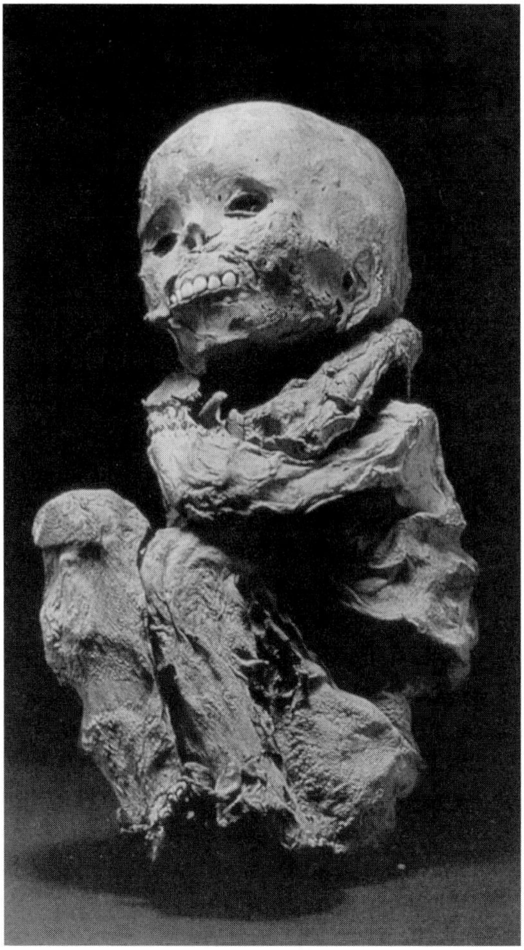

Fig. 4.20. Mummy from Colombia; AM, ca. AD 1300. The infant's body is desiccated, probably by heat from a fire. Part of a collection at the Instituto Colombiano de Antropología in Bogotá. Mummy no. 00–OS-0001. (Photo courtesy of Felipe Cárdenas-Arroyo.)

eviscerated abdominal cavity with ritual items and jewels. One of the two Muisca mummies removed by British officials to the British Museum is of this type, as described by Dawson (1928b), Holden (1989) and Cárdenas-Arroyo (1989). Silva Celís (1945) also notes that some bodies, enveloped in six or more textile blankets and buried in dry soil, mummified spontaneously. Areas of charred skin on some bodies testify to the exposure to fire that was used to induce desiccation. Among the horde of small tribes occupying the Cauca Valley were some that practiced endocannibalism (ingestion of tissue from the bodies of deceased individuals of their own group). This was a feature that some groups in the Amazonas lowlands incorporated into their culture in the form of drinking a suspension of powered ash of their deceased compatriots' bones (Linne, 1929). The people from the area of Popayán of Colombia's southern Andean area employed a distinctive form of body preservation by flaying the corpse, filling the "skin bag" with ashes and suspending the bundle in their huts (Stirling, 1938:66). The Yuko from the Sierra de Perijá near Venezuela's Lake Maracaibo placed the body on a platform or hut floor for a month, then sewed the collected bones into a cloth bag that was suspended from a hut roof for several years until final disposal by placement in a cave (Metraux & Kirchhoff, 1948). Metraux & Kirchhoff also describe how the Spanish found heads, arms and legs stuffed with straw in the houses of the neighboring Corbago tribe.

The humid, tropical climate of Colombia's lowlands accommodated fewer options for mummification. Desiccation by exposure of the body to the heat from one or more fires was practiced in the entire lowland areas from Costa Rica, through Panama to the Amazonas lowlands of southeastern Colombia. In the Darien area near the Panama–Colombia border, the Cuevas people suspended the body and placed one or more fires that were kept burning continually under it until drying of the corpse was complete, collecting the drippings in clay vases (Cárdenas-Arroyo, 1995). The body was then hung in a hut. In AD 1514 an early chronicler, Oviedo, described how a chief would be placed on a stool or rock in a sitting position and desiccated by surrounding him with fires in multiple braziers, usually without evisceration. His dehydrated body would be placed then adjacent to that of his father in a separate building maintained for that purpose while nonroyal bodies were kept in a hut hammock and buried 1 year after death (Linne, 1929). On the Isla de Piños of Panama the body was placed in a hammock hung in a grave and covered with leaves, after which the grave was filled with soil (Stirling, 1938:66). Similar methods were employed for groups in the interior of the Mosquito coast, though there clay masks sprinkled with gold dust were also placed on the face (Linne, 1929). The Cunas and Caimenes people on the lowlands of the Atrato River just south of Darien in northwestern Colombia wrapped their chief's fire-dried body in cotton cloth that was lavishly and expensively decorated, though they only preserved the heads of the common people (Stirling, 1938:66).

Human mummies, whether achieving their mummified status spontaneously or anthropogenically, appeared to have survived their storage environment (predominantly in caves) quite intact in Colombia's Andean highlands. All of the mummies in this country's public collections are from these areas. In contrast, lowland mummification (always anthropogenic) appears to have been sufficiently effective to satisfy the immediate needs of the various groups' mortuary rituals. However, none of these bodies have been recovered and our knowledge of those mummification practices is derived from literary sources. Undoubtedly the humid climate at these tropical sites is responsible, particularly since the paucity of caves there resulted in storage practices under less sheltered conditions. In some parts of the Darien the goal of the practice appears to have been temporary mummification, because the mummy was burned following the ceremony of the first anniversary of the deceased's death (Linne, 1929).

In addition to these mummies, attention must be called to the "San Bernardo phenomenon". This small community is in the highland savanna of Bogotá, 87 kilometers from the capital in the state of Cundinamarca. The community of San Bernardo met a local financial crisis by requiring a recurrent fee to maintain the burials of bodies interred in the local cemetery. Delinquent fee payments were followed by prompt exhumation of the buried body. To their surprise, the porous soil coupled with the arid highland climate had resulted in spontaneous mummification of some of the bodies (Fig. 4.21). Currently about 30

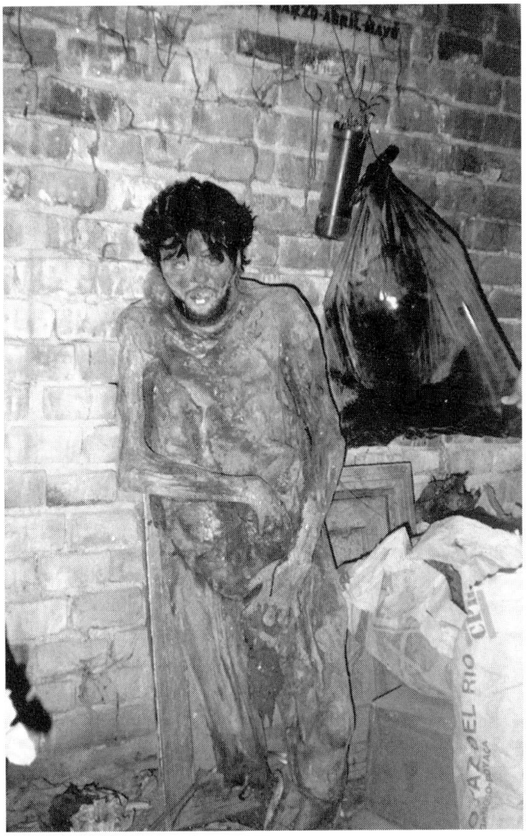

Fig. 4.21. Mummy of Girardot, Cundinamarca, Colombia; SM, ca. AD 1900.
The spontaneous mummification of the bodies from the cemetery in San Bernardo is attributed to the climate's aridity and the soil's porosity. (Photo courtesy of William Mauricio Romero.)

intact, mummified human remains of men, women and children are stored temporarily in an underground chamber (Esguerra, 1991).

Purpose of Colombian mummification

Cárdenas-Arroyo (1990, 1998) notes that Muisca mummies participated in rituals and were carried into the battlefields. In Chapter 2 we pointed out that these features are indicators for the use of mummies to reinforce the centralization of political power. This would be consistent with the observation that the individual chiefdoms that collectively were called Muisca were exceptionally skilled in military arts and in frequent military confrontation with their neighbors. The common practice of body orifice closure with a tampon is less specific, but the knowledge that some groups from the lowland Darien area consulted their mummies about religious matters (Cárdenas-Arroyo, 1998) strongly suggests that the motive for mummification among those groups was to imprison the spirit within the body (including the obstruction of potential escape routes via body orifices) not only to prevent an undesirable spirit's acts but also to manipulate its (predictive ?) power.

Curation and bioanthropological studies

Table 4.1 identifies the current location of Colombian mummies in public collections. While the number of bodies is not great, the mummies at ICANH are currently under intensive investigation as a part of the Colombian Mummy Project initiated in 1989 and directed by Felipe Cárdenas-Arroyo. A registry has been initiated, photography has been completed, X-rays and CT scans have been finished of the majority. Most have been radiocarbon-dated and their diets have been reconstructed chemically by both trace element and stable isotope ratio measurements as well as by coprolite examinations. Textiles are also under study.

Findings to date include a range of noncalibrated dates from AD 200 to 1520 (Cárdenas-Arroyo, 1995). Their diet (highlands only) included a major maize fraction (Holden, 1989; Cárdenas-Arroyo, 1993c). Pulmonary calcification consistent with tuberculosis was found in two mummies (Correal-Urrego & Florez, 1995). Antibodies have been extracted from these mummies (Guhl *et al.*, 1992) and these populations' practice of chewing coca leaves has been documented by chemically identifying cocaine in samples of their hair (Cárdenas-Arroyo, 1993a). In addition, the presence of gallstones was identified (Cárdenas-Arroyo, 1998) and such anatomical features as cranial deformation (Correal-Urrego & Florez, 1995) and hydrocephalus (Herrmann & Meyer, 1993) established. Many other studies are ongoing.

Mummies from Peru

Peruvian mummies are presented in this volume grouped according to the cultural time intervals commonly employed by archaeologists working primarily in that country. Lacking universal acceptance of exact dates limiting these intervals, the dates assigned in this presentation are approximations, guided particularly by Richardson (1994:27). Figure 4.22 has been constructed from multiple sources to provide cultural perspective.

Preceramic Period (prior to 1800 BC)

To date I have been able to identify scientific reports of only four human mummified Peruvian bodies reliably dated to the Preceramic periods. Two of these are radiocarbon dated to 4000 BC, the Middle Preceramic Period. Engel (1977) describes them as coming from a dry cave (Tres Ventanas I) in the highland section of the Chilca Valley of the central portion of the Peruvian coast. In this cave he found the mummified body of a baby wrapped in a vegetal mat, and adjacent to it was that of an adult enveloped in a painted vicuña fur hide. Although these bodies have not been studied formally yet the mummification process in these was probably spontaneous, because he makes no remarks regarding evidence of evisceration.

This same author also reports two other mummified human remains, these having been recovered in the dry cave Tres Ventanas III from the same area. Radiocarbon dates for both of these are about 2000 BC, placing them in the Late Preceramic Period. One is an infant (length 0.95 meters) from an early agricultural group, wrapped in fur surrounded by a mat enveloped in a textile robe, adorned with a shell pectoral and a bone needle. The other is also an infant (length 0.90 meters) slightly flexed, lying on its right side, with a leather sling binding the skull. Again, the absence of any comment regarding evidence of evisceration

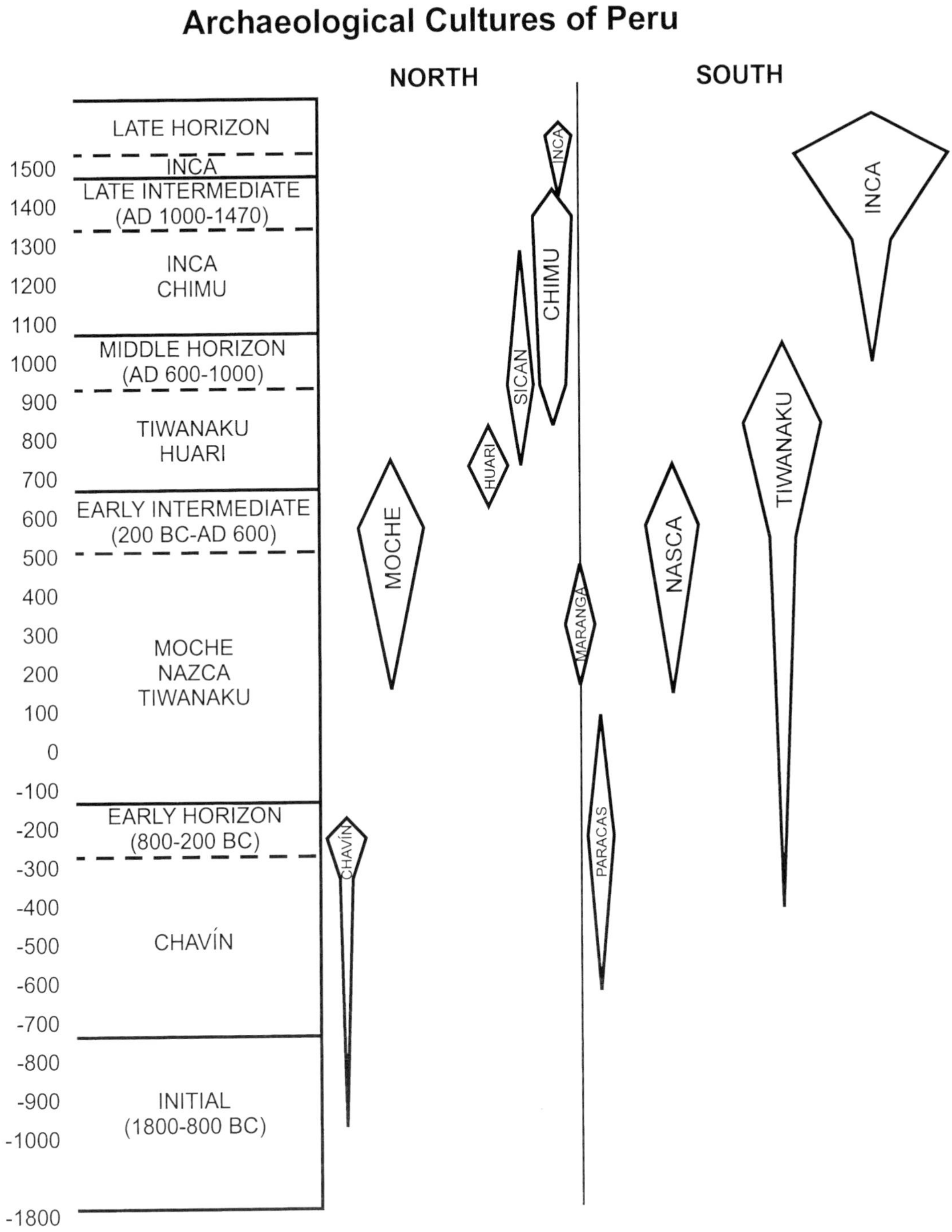

Fig. 4.22. Sketch of Peru's cultural chronology.
Maximum width of each cultural group's symbol indicates the peak of its cultural development. (Design by ACA.)

suggests that these two are also spontaneously mummified bodies. All four of these are curated in the Museum of Pre-Columbian Technology in the National Agrarian University of Peru.

Initial Period and Early Horizon (1800–200 BC)

Early Horizon mummies

Introduction of ceramic arts ushered in the Initial Period, which also includes early development of irrigation. Within this milieu the beginnings of a religious tradition, Chavín, can be identified that gradually expanded its geographical zone (but with considerable regional differentiation) on Peru's northern coast during the subsequent Early Horizon. This cult was centered at the enormous highland temple complex Chavín de Huántar that was excavated in 1919 by the pioneer Peruvian archaeologist Julio Tello (1943) and later by Lumbreras (1971), Burger (1995a) and others. Of the few Early Horizon mummies that have been professionally studied and reported are two unopened funerary bundles housed at Berlin's Museum für Voelkerkunde. CT studies have provided insight into human corporeal preservation efforts at this early period (Herrmann & Meyer, 1993:126–129; Banzer *et al.*, 1995). One of these bundles contains a 16–18-year-old female (Fig. 4.23) and the other the body of an adult male, seated in a basket, whose antemortem tooth loss suggests a substantially older age-at-death. The interesting feature is that both of these bodies have been eviscerated and the abdominal and thoracic cavities have been filled with plant fibers and possibly some fur in the male. Unfortunately few provenience data are available for these bundles and the bodies have been dated only by the tubular form of the bundle that is surrounded by tightly spaced, horizontal, encircling, flat, braided bands – features characteristic of Chavín bundles. However, no artifacts within the bundle were identified by radiographical studies, so we have no features other than the bundle shape to link them firmly to the Chavín tradition. If subsequent radiocarbon dating establishes that they are, indeed, at least contemporary with the Chavín tradition, then to my knowledge this is the earliest evidence of anthropogenic mummification within today's political borders of Peru. One could argue that the Chinchorro culture in nearby northern Chile is several millennia older, and some bodies of that cultural group have been found as

Fig. 4.23. A Chavín funerary bundle.
The bundle's external pattern is characteristic of the Chavín cultural group. (Photo is courtesy of B. Herrmann & R. Meyer (1993) and Staatliche Museen zu Berlin.)

far north as the coastal city of Ilo in extreme southern Peru (Wise, 1991). However, that group played no significant role in the history of what is Peru today.

Paracas mummies

Although absolute chronology for this cultural tradition remains controversial (Paul, 1991), the influence of the Chavín religious tradition in Peru's northern area during the Early Horizon can be detected in a locally modified form in the iconography of the earliest phase of Paracas culture on the southern coast (Kauffmann-Doig, 1998b:69). The peninsula of the same name 200 kilometers south of Lima (about 14° S) provided an improbable and hostile home. Barren of vegetation and swept by recurrent dust storms, its environment even led some of the Paracas people to live in at least partially subterranean dwellings (Schreiber, 1996a:78–87). The site presents several clustered burial areas with differing features. The upper of several terraces contained vertical shafts up to 5 meters deep, with a centrally constricted diameter, beneath which a broader chamber (*caverna*) contained up to 25 mummified bodies (Fig. 4.24). Associated goods and relatively simple textiles suggest these represent principally the lower and middle socioeconomic classes of the Paracas population.

Fig. 4.24. Paracas "*caverna*" burial; ca. 400–100 BC. A lateral extension (*caverna*) at the bottom of a deep, funnel-shaped pit characterizes burials by members of the cultural group occupying the Paracas Peninsula in antiquity. (From Reiss & Stuebel, 1880–1887.)

At a lower terrace, however, now called the Great Necropolis of Wari Kayan, Peruvian archaeologists Tello & Mejia-Xesspe (1979) in 1927 extracted 429 individually buried, roughly triangular mummy bundles whose contents stunned the archaeological world. The bundles were huge, some of the largest weighing more than 220 kilograms. Unwrapping these revealed the nude, flexed, spontaneously mummified corpse in the center to be seated on a deerskin mat commonly set into a circular basket that had sides up to 20 cm high (Fig. 4.25). It was usually surrounded by a variety of new or previously worn clothing items including shirts, mantles and robes, while the head was often covered by one or more turbans held in place by a string. The body and basket were usually enveloped in one or more plain, cotton shrouds. These, in turn, were wrapped in up to as many as 15 further blankets, shrouds or other textiles many of which were woven of llama or alpaca wool (Fig. 4.26). It is the spectacular polychrome designs incorporated into these textiles and their incredibly fine preservation that has captivated their viewers the world over. Enclosed between the textile layers were other grave goods including weapons, often gold jewelry and ceramics (Schreiber, 1996a:78–81). A roughly circular, head-shaped bundle at the top of some bundles usually contained only additional textiles, the body itself being completely contained within the central, lower part of the bundle (Fig. 4.27). An example of this was on exhibit at Lima's

Fig. 4.25. Interior of a Paracas funerary bundle; SM, ca. 400–100 BC.
Lower and middle class burials are characterized by a simple, textile-wrapped burial surrounded by a layer of leaves within an outer wrapping. (From Reiss & Stuebel, 1880–1887.)

Fig. 4.26. An elite Paracas burial; SM, ca. 400–100 BC.
The huge bundle of textiles weighing hundreds of kilograms enveloping the spontaneously mummified body, topped by a textile pseudohead, is characteristic of elite Paracas and Nasca burials. (From Reiss & Stuebel, 1880–1887.)

National Archaeological Museum during my visit in 1982. A fantastic figure with two Picasso-like large eyes on the same side of a profile view was painted on such a "head" (Aufderheide, 1982a).

While the *cavernas* bodies include all ages of both sexes, those of the necropolis are composed mostly of older males (priests), and clearly represent the population's elite groups (Place, 1995a:30–31). Neither evisceration nor other evidence of deliberate attempts at soft tissue preservation has been identified in these bodies. Indeed, while the absorption of moisture from the skin surface by the porous, abundant textiles can be expected to have accelerated dehydration leading to desiccated, mummified bodies quite frequently, in fact this was achieved to a significant degree in only a moderate fraction of the bundles. Most reported bodies are skeletonized or consist of only partially skin-covered bones articulated by fragments of fibrous ligaments. The most ostentatious of these necropolis burials seem to date to the end of the Early Horizon. Interestingly, few Paracas bundles of this type have been found subsequently at this site or elsewhere. Curatorial efforts have had to deal with administrative apathy, political turmoil, profound economic depression, merging museums and unreturned exposition loans, but managed to achieve conservation of enough textiles to provide ongoing study opportunities (Daggett, 1994).

Iconography from the later Paracas culture also sug-

Fig. 4.27. An elite Paracas or Nasca burial; SM, ca. 400–100 BC.
Bundle similar to that shown in Fig. 4.26, but demonstrating a textile facial design on the pseudohead. (From Reiss & Stuebel, 1880–1887.)

gests the practice of "trophy heads" as represented by supernatural beings holding a knife in one hand and a severed human head in the other (Bruhns, 1994:136–137). The reality of this custom was verified in 1956 by the occasional headless burial and the finding of two caches totaling 15 human "trophy heads" in the Ica Valley by Pezzia (1968) of the Museo Regional de Ica. This feature is discussed in more detail in the section on *Nasca mummies*, below.

Early Intermediate and Middle Horizon mummies (200 BC to AD 1000)

The Nasca people

Culturally related features are spelled Nasca while geographically related ones are denoted Nazca (Richardson, 1994). The Nasca people are direct descendants of the Paracas population. Indeed, only the replacement of resin by clay slips on their ceramics is the archaeologically defined beginning of the Nasca culture. Most of their Paracas ancestors' burials are located on the small Paracas Peninsula, while the geographical core of the Nasca population lies about 60 kilometers inland from the coast in the Nazca and Ica valleys. These narrow valleys placed limitations on irrigation agriculture, and the Rio Grande River that they enclose was one of the most unpredictable and driest on Peru's south coast. These characteristics make the Nasca people's preoccupation with water availability and its related rituals more understandable. Three features distinguish this culture: ceramics, "trophy" heads and the Nazca lines. The beauty of their ceramics (Putnam, 1914; Uhle, 1914) can probably be matched only by those of the Moche people. Nasca pottery is illustrated principally with natural figures in the earlier phases and more abstract, supernatural ones in the later periods (Roark, 1965; Allen, 1981). Unlike Moche, however, Nasca iconography represents individual figures, not rituals, and thus provides substantially less information about their cultural practices between about AD 200 and 800. A Nasca nine-phase pottery sequence has been established and efforts have been made to use these time intervals as a model for all of Peru during the Early Intermediate Period (EIP) (Richardson, 1994:113–119). While much of their society suggests a chiefdom confederacy type of political structure, the sudden expansion to include all five valleys from the Chincha to the Acari during the Nasca 5 phase suggests a rather transient state organization, supported by contemporary militaristic iconography (Silverman, 1987).

Nazca lines

The well known Nazca lines are found near the modern town of Nazca. They consist of an often intersecting cluster of lines that extend in a straight line over hills and valleys. In addition, a small number of animal and even geometric figures are scattered about the plain, some of which measure as much as 0.5 kilometer in length, and are formed by a continuous line (Figs. 4.28 and 4.29). These geoglyphs were created by simply

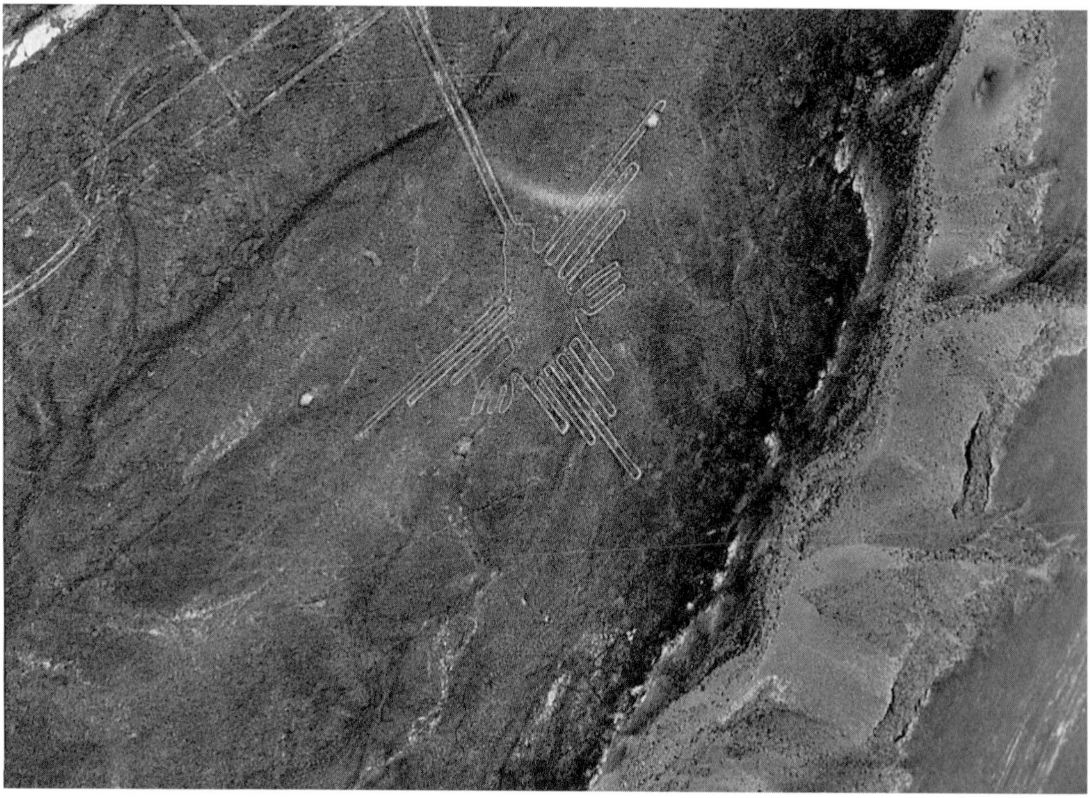

Fig. 4.28. Nazca lines, ca. AD 200–350.
The demonstrated hummingbird design is one of many designs comprising the elaborate, gigantic set of stone patterns called "Nazca lines" near the modern city of Nazca, Peru. (Aerial photo by ACA.)

inverting the darker, oxidized-surface stones that cover the valley floor and exposing the underlying lighter-colored sediments (Kauffmann-Doig, 1998b:69–70). Their dimensions, together with the fact that they are best (though not necessarily only) viewed from the air has led to the fantasy proposal that they were formed by alien astronauts (von Däniken, 1968). More realistically the recently deceased German archaeologist Maria Reiche, who made the valley her home for many years, has noted that some are aligned with astronomical bodies, suggesting that their function was calendrical. Currently the most popular view is that at least the straight lines represent sacred pathways to agricultural fertility ritual areas that relate principally to water (rain) needs. This interpretation appears to have support from ethnographical and archaeological sources as well as precedence from modern practices. The figures and occasional straight lines conceivably could bear astronomical relationships, though even these may represent water/fertility symbols (Reinhard, 1996a:55–56). Alternatively Aveni (1990) suggests that they are pathways for ritual processions that follow the lines of the animals.

Nasca mummies

These take two forms: entire bodies and isolated heads. The bodies were not dealt with in a manner differing greatly from that of their Paracas ancestors. The corpses were spontaneously mummified, flexed and wrapped in abundant textiles that obscured the body form. Some of these included a false, textile head at the

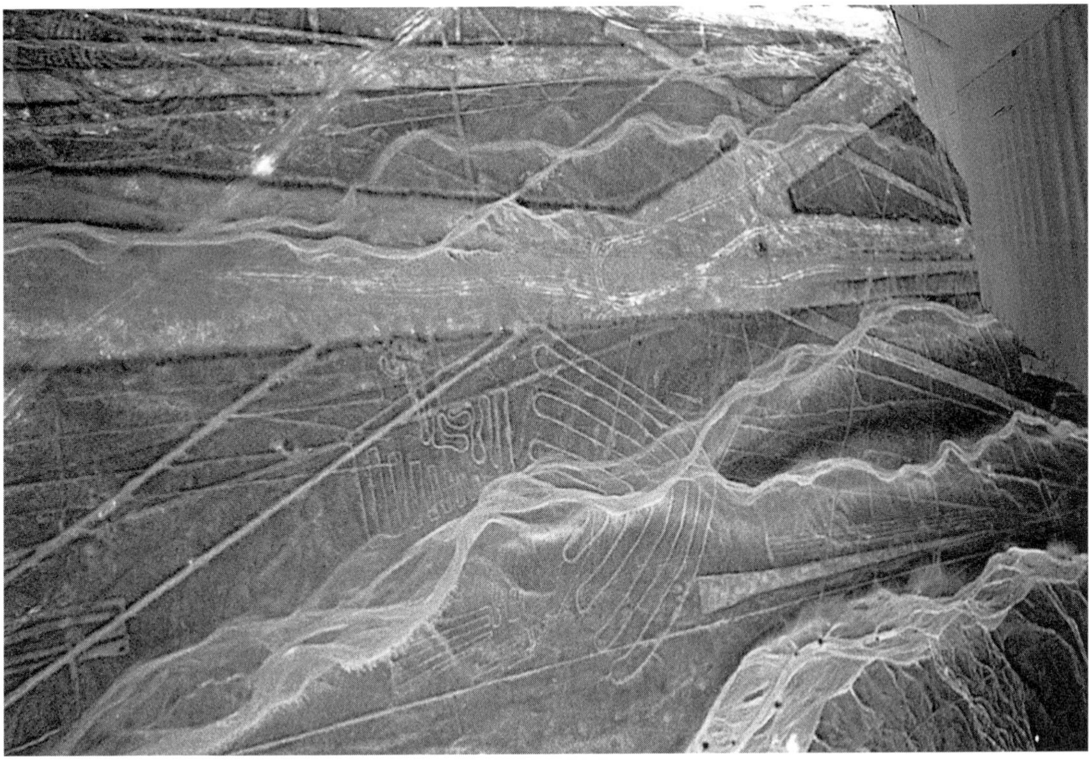

Fig. 4.29. Nazca lines, ca. AD 200–350.
Another geometric pattern similar to that shown in Fig. 4.28, but with superimposed straight lines. (Aerial photo by ACA.)

top of the bundle. In some areas, such as the principal Nasca ceremonial center Cahuachi, they were interred in simple, circular, shaft pit graves (Strong, 1957, cited by Lumbreras, 1974:132). In others (Chaviña) they were placed in underground, adobe brick-lined and often plastered cists. Tello (1929), cited by Vreeland (1998), describes a few mummies from the larger bundles, some of which had missing heads, that were eviscerated, some with defleshed extremities; these were then desiccated by fire. Subsequent to that mention, I could find no reports of other anthropogenically mummified Nasca mummies.

The significance of isolated head burials is less easily understood. While Paracas iconography is teeming with imagery of severed human heads, relatively few archaeological examples have been reported. However, Carmichael (1994) estimates that 5–10% of Nasca burials are isolated heads or headless bodies. Early Nasca ceramic images reveal a realistic human figure in warrior costume holding such heads. Late Nasca iconographic art is replete with supernatural figures holding severed heads bearing two or more features of "trophy" heads: suspension cord, bodiless presentation, eyes rolled upward (suggesting death), lips sealed with cactus thorns and locks of hair on the forehead (Richardson, 1994:113–114). Archaeologically the finding of such heads becomes more frequent, during and after the Nasca territorial expansion that may have been accomplished by militaristic means. Verano (1995) defines a Nasca "trophy" head as one that retains a significant amount of soft tissue, contains a midline defect in the frontal bone for purposes of cord suspension and shows irregular enlargement of the foramen magnum that he assumes was created for the

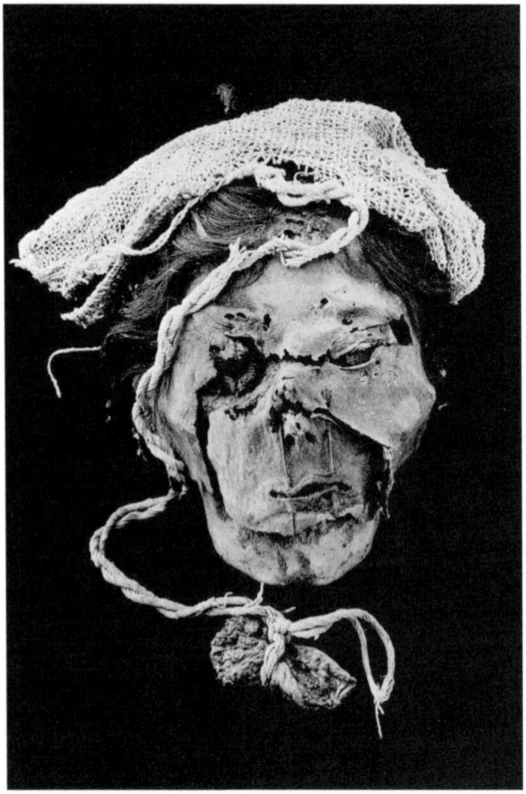

Fig. 4.30. A Nasca "trophy head"; SM, ca. AD 200–600. This spontaneously mummified Nasca head has the features described by Verano (1995) as characteristic of "trophy heads". (Photo by and courtesy of John Verano.)

purpose of excerebration (Fig. 4.30). Verano also notes that women and children account for fewer than 10% of such heads, the remainder being those of young adult males, suggesting that they are military conquest products. Readers, however, should note that the terms "trophy heads" (Kleiss & Simonsberger, 1984, in pre-Hispanic Brazil: Mundurucú) and "trophy skulls" (Sledzik & Ousley, 1991, in southeast Asia: Vietnam, 1971) have been employed for isolated heads acquired in different geographical areas or eras. These usually lack some of Verano's described features for Nasca heads, although circumstances strongly imply their acquisition for display as war trophies. Both single Nasca burials as well as caches of more than 40 heads have been reported. While occasionally some have been found in tombs with bodies, most are separately interred, often in relation to some architectural structure such as beneath a house floor. Cut-marks are present on some Nasca skulls that are devoid of soft tissues and it has even been suggested that some of them were made premortem (Baraybar, 1987; Drusini & Baraybar, 1991) on the basis of criteria with which many paleopathologists would not be comfortable. Few skulls show evidence of war violence.

Verano (1995) reconstructs the preparation of "trophy" heads as follows:

1. Sever head at neck.
2. Remove cervical vertebrae and soft tissues of neck and oral cavity.
3. Enlarge foramen magnum and excerebrate.
4. Tie jaw into position.
5. Stuff textiles or cotton into cheeks and oral cavity.
6. Remove scalp with attached hair, cover cranial vault with textile and replace scalp.
7. Pin lips and eyelids shut with cactus spine (*Prosopis* sp. or *Acacia macracantha*).
8. Create defect in midline of frontal bone and attach cord of textile or hair.

These observations have spawned the following differing interpretations (Table 4.2).

1. "Trophy" heads are militaristically acquired victory trophies. It was the German archaeologist Max Uhle (1914) who found the first of these heads and dubbed them thus. Verano (1995) notes that the age and sex distribution of the heads are consistent with military acquisition, and Proulx (1971) embraces this view, suggesting control of the slain individual's spirit as a motive. Drawing a parallel with Ecuadorian Jivaro beliefs, Allen (1981) feels the motive for such head-taking may reflect a perceived need to prevent loss of one's soul.
2. Ritual sacrifice of an enemy captive. Support for this concept is gathered from Late Nasca iconography demonstrating warrior-clad individuals holding severed heads, and the findings of cut-marks suggestive of deliberate induction of antemortem bleeding (Drusini & Baraybar, 1991).

Table 4.2. *A partial list of Paracas–Nasca "trophy" heads*

Reference	No.	Site	Phase	Comments
Paracas "Trophy" Heads				
Pezzia, 1968	2	Inca Valley	Late Paracas	Cerro Max Uhle; Cerro de la Cruz 1966
	13	Hacienda Oculate		
Riddel & Belan, 2000	5	Acari Valley Tambo Viejo	Late Paracas or Early Nasca	In pits under room floor; heads in ceramic (?)
Nasca "Trophy" Heads				
Browne *et al.*, 1993	48	Cerro Carapo	5	Burial
Carmichael, 1988:482	2	Cantalloc, Nazca Valley	2	Burial, juvenile
Carmichael, 1988:482	1	Majoro Chico, Nazca Valley	3	Burial
Kroeber & Collier, 1960	3	Cahuachi, Mound A	3	Best mummification burial Decapitated female body
Strong, 1957	1	Cahuachi, Burial Area 7	5–6	Urn burial #12,SA
Strong, 1957	1	Cahuachi	5	Burial 14
Neira Avedaño & Coelho, 1972:142	11	Chaviña, Acari Valley	4–5	
Doering, 1958, 1966:142	9	Cahuachi	8	Architectural content; burial tomb
Pezzia, 1969:146	6	Lower Nazca Valley	8	
Drusini & Baraybar, 1991	7	Cahuachi	3–4	
Doering, 1959	16	Cahuachi, Estaqueria, Las Trancas	?	

3. Ritual sacrifice of a Nasca individual. Iconographic portrayal of warriors could reflect a common feature of many Andean religions: ritual warfare performed by two sides selected from within the same Nasca group and presumably carried out in part to create bleeding, an act deemed necessary for certain rituals. The portrayed warriors commonly demonstrate cranial deformation as do intact Nasca bodies and Nasca "trophy" heads.
4. The head was severed from a Nasca body postmortem after the body had become mummified, presumably for use in a ritual. Carmichael (1994) describes a body with these features.
5. "Trophy" heads represent acts of ancestor worship. Guillén (1992) and Browne *et al.* (1993) suggest that the major effort made to preserve the heads' soft tissue, the retention of facial features, differing ages and sex and the near absence of war wounds all are most consistent with a supposition that these heads were memorials (even idols?) of highly respected ancestors.

Head-taking in South America has considerable antiquity, although the full constellation of "trophy" head features as itemized above is not commonly present outside of the Paracas–Nasca heads. Engel (1963) describes disembodied heads from 2000 BC, a period contemporary with separated heads among the Chinchorro people of northern Chile (Skottsberg, 1924; Guillén, 1992). Meighan (1980) describes headless burials from the Initial Period, and the Middle Horizon Period (following Nasca's Early Intermediate Period) has well-known separated heads among the Huari people (Brewster-Wray, 1982; and see below). McEwan (1987) reports a cache of ten skulls from a Middle Horizon Huari site in a valley south of Cuzco. The appropriate interpretation of "trophy" heads remains controversial.

For a world-wide perspective on head-taking, readers can be directed to the discussion by Kleiss (1998:210–215) who uses both archaeological and literary references to trace the practice from at least the Mesolithic Period to the present.

The Moche

On Peru's northern coast, this period is characterized best by the Moche culture. While it can be argued that the Moche structure lacked some features we normally associate with a kingdom, these people spread their influence (if not necessarily absolute control) over an area occupying at their maximum expansion nearly 640 kilometers of coastline (Richardson, 1994:112). The country is blessed with rich mineral deposits and immediate access to the sea, so exploitation of these resources resulted in complex, highly productive irrigation agriculture and marine harvests that, at their peak, probably supported as many as 50000 individuals. It also led to mastery of intricate metallurgical skills. Their proficiency with ceramics and the double-mold technique, as well as the "stirrup" forms, have resulted in a wealth of archaeologically recovered specimens. The imagery on these ceramic products has enabled reconstruction of many Moche ritual and other cultural practices. Furthermore, construction of some of their huge pyramids testify to their organizational skills. The Huaca del Sol pyramid exceeds 40 meters in height and was built with an estimated 100 million mud bricks (Kauffmann-Doig, 1998b:71).

While thousands of Moche burials have been looted, only about 350 have been excavated by archaeologically trained professionals (Richardson, 1994:108). Torrential, El Niño-induced rains contributed generously to the environmental disaster that is believed to be responsible for this civilization's collapse. These events and others that occurred several centuries later (particularly around AD 1100) undoubtedly resulted in the destruction of soft tissues of many Moche mummies (Nials *et al.*, 1979; Thompson *et al.*, 1985). As a consequence, most of the studied burials have been skeletonized remains (Verano, 2001).

Undoubtedly the royal burials in the pyramid near the community of Sipan have provided the most exciting finds that permitted verification of iconographic interpretations of Moche art. A common theme of that art is the "sacrifice ceremony", carried out by four individuals who are consistently represented by specific, identifying clothing and ornamentation as well as actions (Donnan, 1988). Excavations of the only partially looted tombs in Sipan's pyramid by Walter Alva (1988, 1990) revealed several obviously elite burials complete with human retainer burials (Verano, 1997, 1998) and a plethora of jeweled ornaments. Alva and his associates were astonished to find that the clothing and ornamentation of the richest burials corresponded precisely to the specific individuals (called Figure A and Figure B) portrayed in the mural and ceramic iconography of the Moche sacrifice ceremony! A third body (female), excavated by Donnan & Castillo (1992) wore the clothing characteristic of Figure C of that ritual. While two of the bodies had become skeletonized, several of the sacrificed retainer corpses demonstrated at least some persisting soft tissue. More informative of mummification practices for Moche royalty was a similar elite body, excavated by Strong & Evans (1952:150) from Huaca de la Cruz in the Virú Valley, whose remains revealed sufficient soft tissue retention to establish that the body had been eviscerated (Lumbreras, 1974:111). These bodies, as well as the many excavated skeletons, indicated that the common Moche burial position was an extended body, in contrast to the flexed position so common in later cultures. While most of the relatively few mummified bodies removed from the huge site at Pacatnamu (located at the mouth of the Jequetepeque valley in northern Peru) are from the Late Period, one spontaneously mummified was coincident with the Moche Period (Verano, 1995b).

The Maranga

This little-studied site, contemporary with the Moche culture in the north, is located in an area that was on the outskirts of Lima in the Early Colonial Period on the highway that connected Lima with the port city of Callo. Today it has become included within the urban limits of Peru's capital city. Its pyramid must have been one of the most important of the many in Peru's central and southern coastal areas in this period (Bruhns, 1994:199). This view is validated by a detailed study of the complex's structure and the road pattern involving

Fig. 4.31. A Maranga mummy; SM, AD 835.
The Maranga site has been overbuilt in the outskirts of modern Lima. This is one of about 40 burials from that site now curated in Quito, Ecuador, at the Instituto Jacinto Jijón y Camaño. (Photo by and courtesy of Felipe Cárdenas-Arroyo.)

it, as well as documents that refer to the site as the head of the Chayavilka lordship (Canziani, 1987). The pyramid measured 270 meters × 100 meters, was plastered and painted yellow, and was surrounded by smaller platforms. Except for infants (Fig. 4.31), burials were in the extended position with multiple ceramic offerings (Bruhns, 1994:199). The site has been studied by Middendorf (1894), Uhle (1910), Kroeber (1926) and Jacinto Jijón y Camaño (1949). The last of these was an Ecuadorian historian who transported more than 250 human crania and 40 funerary bundles from the Lima site to Quito, Ecuador. There they have become part of the permanent collections at the museum in Quito that bears his name (Cárdenas-Arroyo, 1993b). Currently these items are undergoing bioanthropological study (The Bioanthropology of Maranga, Peru Project) as well as an investigation of the elaborately decorated textiles that compose the *fardos* by the professional staff of the Museo Jacinto Jijón y Camaño in Quito and the Centro de Estudios en Bioantropología in Bogotá, Colombia.

The Highland Tiwanaku and Huari ("Wari") empires

While most of the groups noted above were principally involved with coastal or low valley Peruvian sites, social development was also evolving at Peruvian highland sites during the Early Intermediate Period. A collection of hamlet and village sites about 15 kilometers south of Lake Titicaca had developed an agropastoral existence detectable archaeologically as early as the later phases of the Early Horizon (Tiwanaku I and II) (Richardson, 1994:98,119). Their revolutionary discovery and development of raised-field agriculture technology (Tiwanaku III: AD 100–400) expanded their subsistence capacity from just a few thousand to about one-third of a million individuals (Kolata & Ortloff, 1996). Known for the size and quality of their monumental stone architecture and the fierce-appearing, feline-faced "staff" or "gateway" god, they expanded west to dominate the Moquegua Valley in southern Peru and south as far as what is now northern Chile. These other areas were particularly exploited when a severe drought about AD 600 threatened their highland-based maize-agricultural subsistence. A contemporary highland group (Huari) evolved from groups at the northern end of Lake

Titicaca, and matured in Peru's Ayacucho basin. Spurred by the highland drought, the Huari expanded to the north about AD 600, stopping short of Moche territory. They temporarily dislodged the Tiwanaku people from the Omo site in the Moquegua Valley and built a dramatic defense center nearby at Cerro Baúl (Goldstein, 1989:251–252). Stimulating the development of terrace agriculture in the mountainous areas, they also reached the coast in the Nazca Valley, but only 200 years later the Huari empire collapsed in about AD 800. The Tiwanaku empire survived at least two more centuries until a second major drought destroyed the base of that empire as well.

The Bolivian highland climate was not conducive to spontaneous mummification and most Tiwanaku burials at high altitudes are skeletonized (Ponce-Sanguines, 1966). Nevertheless, Arené *et al.* (1995) describe the eviscerated, desiccated body of a 2½-year-old Tiwanaku child whose tissues they softened sufficiently by immersion in a solution to allow radiological angiography of the leg vasculature by injection of a radiopaque dye into the aorta. Neither provenience data nor the composition of the injected fluid was offered. A flexed, desiccated mummified body known to be from Bolivian highlands but without further provenience data was found to have been stored at the Gakushuin University in Tokyo, Japan, for 100 years when it was examined in 1993 by Morimoto & Hirata. These investigators found it had been eviscerated via abdominal and anterior cervical incisions and then desiccated by an unknown method. The radiocarbon date was reported to be 100 BC. Three other Bolivian mummies were found to have been eviscerated abdominally; an additional seven were spontaneously mummified (Ponce-Sanguines & Iturralde, 1966, cited by Vreeland, 1998). Tiwanaku burial groups at low altitudes such as those from the Azapa Valley in northern Chile are small. However, from such we learn that burial of the bundle-wrapped, flexed body in rectangular, usually stone-lined cists was common (Fig. 4.32). Body preparation did not include any detectable efforts at soft tissue preservation, though spontaneous mummification by desiccation is a common result at these lower locations. Grave goods include characteristic ceramics and "snuff kits", small wooden trays used to pulverize hallucinogenic compounds that were sniffed nasally via a hollow bird bone "straw". Common also are coca leaf-containing bags or quids and polychrome textiles, especially the often-tasseled, geometrically patterned, four-cornered hat of the elite.

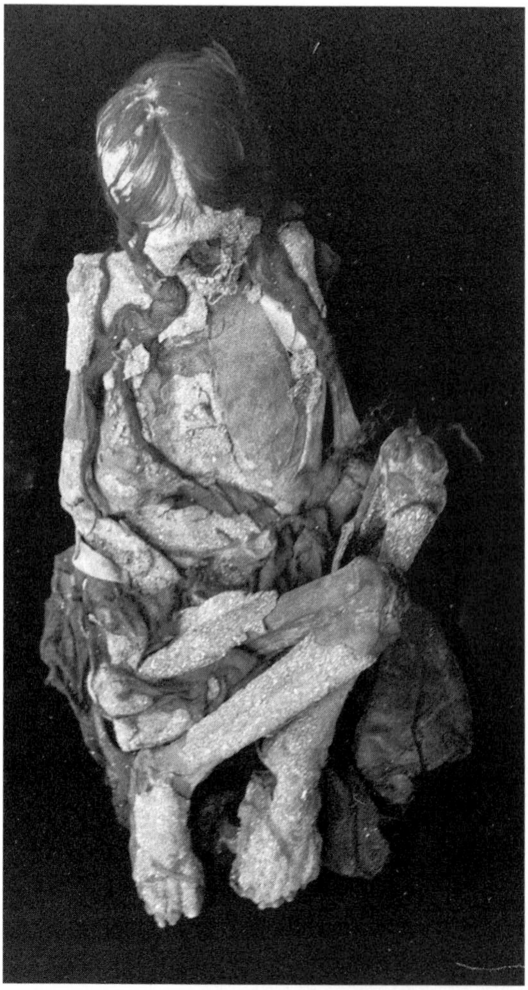

Fig. 4.32. A Tiwanaku Period burial from the Pica site near Iquique, Chile; SM, AD 835.
The seated position with maximal flexion of the legs is a feature of this cultural group's burial practices. Mummy no. IQU 95, T-3. (Photo by ACA, courtesy of Cora Moragas.)

The nature of the Tiwanaku and Huari presence beyond their circum-Titicaca homeland has not been easy to decipher. Based principally on the defensive structure on the Moquegua Valley's Cerro Baúl, it has become popular to view the Huari presence as controlling all aspects of an area, while the Tiwanaku areas suggest gradual growth in a new region, implying influence through trade. Later administrative representation only is evident and, in some areas, physical colonization. Thus, during Tiwanaku V in Peru's Moquegua Valley, a strong physical presence is suggested (Goldstein, 1989:251). However, in northern Chile's Azapa Valley the physical presence of small numbers of colonists can be demonstrated but only Tiwanaku "influence" in the tombs of local ("Cabuza") people is suggested by the inclusion of Tiwanaku items among the grave goods (Goldstein, 1995/1996:68) (see Mummies from Chile, below). At the 2436 meter altitude of San Pedro de Atacama in northern Chile's Atacama Desert, grave goods of tombs contemporary with the Tiwanaku empire show extensive influence but little evidence of Tiwanaku colonization.

The funereal bundles of four Huari children excavated from the near coastal site of Pachacamac, Peru, in 1896 by Max Uhle are now curated at the Pennsylvania University Museum. Two of these are bulky, textile-stuffed bundles with painted false heads, while two others are heavily padded tubular bundles held in position by cords. X-rays reveal the bodies of children in all four, aged 15, 12, 7 and 1 year at death; three are dated to tenth to eleventh centuries AD, and one to fourth to seventh centuries AD. In two the bodies are jumbled skeletons, a third appears spontaneously mummified as does the fourth, though decapitation and extremity disarticulation of the latter suggests a sacrifice victim (Fleming *et al.*, 1983).

Five Peruvian mummies, radiocarbon dated AD 800–1000, curated at London's British Museum and excavated from Ancón, Peru, are from the Tiwanaku V Period. They are now incompletely wrapped in their textiles, and were examined abdominally to recover coprolites for study (Holden, 1989).

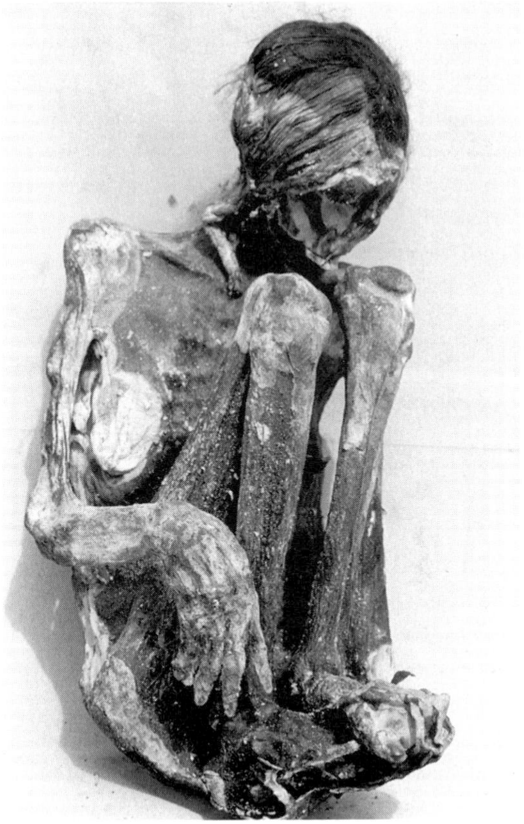

Fig. 4.33. A late Tiwanaku Period (Cabuza) burial, Azapa Valley, Arica, Chile; SM, ca. AD 800–900. This mummy is similar to that in Fig. 4.32 from a site about 200 kilometers north of that burial. Mummy no. AZ141, T-23. (Photo by ACA.)

Human heads were found buried under the floors of a Huari compound by Brewster-Wray (1982). At the time of my visit on July 25, 1982, to the Rafael Larco Herrara Museum in Lima, Peru, a Huari mummy bundle was exhibited, consisting of a large funereal textile bundle capped by a false head with a wig of artificial hair. An excavation of a Cabuza cemetery (AZ141) in which I participated in northern Chile's Azapa Valley near the seaport of Arica in 1982 included both local and Tiwanakoid artifacts; the burials were flexed bodies, spontaneously mummified, in vertical shaft graves without cists (Fig. 4.33). Similar findings are common in the Quitor and Coyo sites at San Pedro de Atacama in northern Chile's Atacama Desert, from which many similar mummies have been retrieved (Costa-Junqueira, 1988). In addition, three disembodied heads from cemeteries 2, 5 and 6 from the Quitor site at San Pedro de Atacama are also curated at the Museo Etnografico J.B. Ambrosetti, Buenos Aires, Argentina (Facultad de Filosofía y Letras, 1992).

Late Intermediate Period (AD 1000–1470)

The Chiribaya

Following the collapse of the Tiwanaku empire at the end of the Middle Horizon about AD 1000, the cultural groups occupying southern Peru and northern Chile coastal and valley sites realigned themselves with other regional groups during a period known locally as Regional Development. Some of these differentiated into archaeologically definable entities. Among these were the Chiribaya people. The largest of these clusters occupied the Osmore drainage basin, at whose valley mouth the modern Peruvian seaport of Ilo is located at about 17° 40′ S. Ceramic features unique to this group span the period of about AD 1050–1300, representing the earlier half of the Late Intermediate Period (Stanish & Rice, 1989).

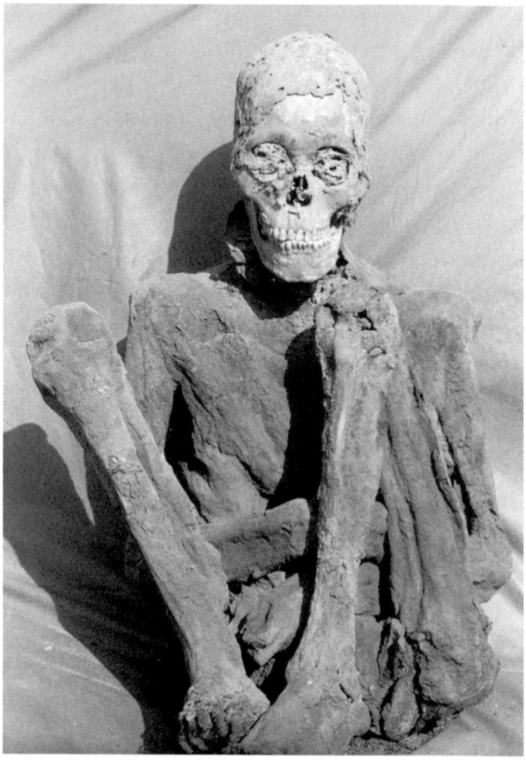

Fig. 4.34. A Chiribaya mummy; SM, ca. AD 1050–1250. This agricultural population in southern Peru's Osmore Valley buried their dead in stone-lined cists. Mummy no. 10. (Photo by ACA.)

In the late 1980s and early 1990s excavations were carried out on the Osmore Valley cemeteries as part of the Contisuyo Project that proved to be one of the most extensive, interdisciplinary projects in recent years. Mortuary analysis was under the direction of Dr Jane E. Buikstra. At least 134 mummified human remains were excavated, as well as skeletonized bodies more than twice that in number. The largest cemetery, Chiribaya Alta, was located atop a bluff of one of the valley's flanks near the sea; most of the bodies were found in the nine subdivisions at that site. These burials were placed, commonly clothed and in a sitting position, in underground cists containing capstones and stone-lined walls with a dirt floor (Fig. 4.34). Similar burials were also found at a site called Estuquiña, located about 75 kilometers up the valley near the modern city of Moquegua. That site also included a few, aboveground, "mini-chullpa" burials called collared tombs by the excavators. Many well-preserved, finely woven textiles with unique patterns characterized the grave goods, as well as distinctive ceramics (Williams *et al.*, 1989).

I had the privilege of examining these 134 mummies and noted that 40% were subadults less than 15 years old. Among adults, males and females were equally represented. Average stature was 158.5 cm (females) and 165.7 cm (males). Roughly one-half of the adults but nearly 90% of the subadults displayed intentional cranial deformation. Traumatic lesions included only one skull fracture and one abdominal stab wound. Osteoarthritis of extremity joints and the vertebrae was common. Two young adults demonstrated dilated colon, suggestive but not diagnostic of Chagas' disease (American trypanosomiasis). A 30–35-year-old woman revealed evidence of primary osteosarcoma of the distal right humerus, evidenced by a striking, radially oriented, spicular ("sunburst") periosteal reaction (Aufderheide *et al.*, 1999a). Infectious lesions included one example of osteomyelitis involving the knee of a young adult female and the maxilla and mandible of a 55–60-year-old male. Two fatal cases of pneumonia were identified (one adult and one child) as well as 27 mummies showing pleural adhesions that are evidence of a past episode of pneumonia now healed. Three adults revealed changes suggestive of tuberculosis. The lungs of one revealed the pattern of a Ghon complex. A sample of a partially calcified, hilar lymph node yielded a DNA sequence unique to the tuberculosis complex (Salo *et al.*, 1994). Another demonstrated the collapsed vertebra and "humpback" deformity of Pott's disease, while a third had characteristic tuberculous lytic lesions of the spine and claviculomanubrial joint. About one-third had pediculosis caputi. Both porotic hyperostosis and cribra orbitalia were common though, interestingly, nearly all were found in individuals with intentional cranial deformation (Aufderheide, 1990a; Walker & Guillén, 1995). Coca-leaf chewing practices were common as manifested in chemically demonstrable cocaine and its metabolites in mummy hair (Cartmell *et al.*, 1991) and the resulting periodontitis produced a high rate of antemortem tooth loss (Langsjoen, 1996). Subsequently a similar effect on dentition was demonstrated among modern coca-leaf chewing Andean highlanders (Indriati, 1998).

El Brujo

This site in the Chicama River Valley on Peru's north coast has yielded about 150 rather poorly preserved mummies buried in textile bundles in an abandoned Moche pyramid. They date to about AD 900–1100. They are still under study and unreported at the time of writing (J. Verano, 2000, personal communication).

Túcume

Excavations have unearthed some mummies at a large group of pyramids near Peru's northern coast (ca. 6° 30′ S, 79° 40′ W) inland and about 25 kilometers north of Lambayeque. This site, called Túcume, is being investigated by a group under the direction of Thor Heyerdahl and colleagues. The burials at this location span the periods of many Peruvian Pre-Hispanic cultures. Most are skeletons, but an unreported fraction appear to be spontaneously mummified remains wrapped within abundant textiles. Reports on examination of these have not yet appeared in scientific journals, though considerable information about the site, its excavation, ceramics and other artifacts has been made available (Heyerdahl *et al.*, 1995). The quality of the preservation as judged by a photograph of one of the two mummy bundles opened (Heyerdahl *et al.*, 1995, Fig. 56, p. 95) demonstrates what appears to be principally a cloth-covered cranium.

Chancay

While the Chirabaya differentiated during the Regional Development interval on Peru's southern coast following collapse of the highland empires, those on the central coast coalesced into an entity centered in the Chancay Valley and drew its archaeological definition from that region. Its origin can be traced ceramically to the fall of the Huari empire, though it flourished between AD 1100 and 1400. It was an actively trading group with coastal as well as trans-Andean routes. Their burials reflect a prominent level of socioeconomic stratification, both in grave structure and funerary goods. The most elaborate are large, rock-lined underground mausolea, filled with well-preserved handsomely embroidered textiles enveloping a seated, flexed corpse (Lumbreras, 1974:191–193). Such burials have been retrieved not only from the Chancay but also other central coast valleys including many from Ancón (Lothrop & Mahler, 1957). Some funerary bundles have a false head similar to Nasca burials; several such were on display at the Rafael

Larco Herrara Museum in Lima at the time I visited it in 1982 (Aufderheide, 1982b).

Mummies from Peru's central coast

The region around Huaral on Peru's central coast (ca. 11° 30′ S, 77° 12′ W) about 125 kilometers north of Lima was occupied by Chancay populations until about the middle of the fifteenth century AD. The museum at the seaport Huaral in 1998 acquired the spontaneously mummified body of a neonate of undetermined sex when a tomb looter had excavated and discarded it. Local newspaper photos reveal a disproportionately large head, suggesting the possibility of hydrocephalus. Similarly a flexed 15–18-year-old female mummy from the same region found in early 1999 and dating to AD 750–1100 is suspected of having been a sacrifice victim. Neither of these bodies has been studied by dissection and are known publically only through newspaper reports.

The Chachapoya mummies

In January 1997, two residents stood in the streets of the remote village of Leymebamba in northern Peru, arguing so heatedly that they attracted the attention of a police officer. When he learned that their disagreement dealt with the question of whether or not an object they possessed was gold plated, the policeman recognized the item as an ancient brooch. Investigating further, the police eventually recovered more than 200 such items from their homes and learned that the residents had looted a pre-Columbian site in the mountains. Thus the drama of the Laguna de las Momias began.

The Chachapoyas (a name derived from the Quechuan words for "Cloud People") at one time controlled about 40000 square kilometers of northern Peru's Amazonian Andes (*montaña*) between the Huallaga and Marañon rivers extending southward about 250 kilometers from Bagua in the upper Utcubamba Valley. While they are archaeologically identifiable as early as AD 800, the interval of their cultural efflorescence was the eleventh century, the period of Andean Regional Development following the collapse of Tiwanaku's highland empire. The foundations of their round, stone block houses are scattered throughout this area in clusters ranging from perhaps those of an extended family of 20 or so structures to cities of up to 400 such units. The largest of their structures, at Kuelap (3000 meters' altitude) was a platform 600 meters long with surrounding walls up to 19 meters high, made with about 100000 stone blocks, each weighing 100–200 kg. This has been variously interpreted as a fortress surrounded by habitations, or as a prestigious administrative center related to food production, with 400 food storage units for reserve and redistribution during El Niño crises (Kauffmann-Doig, 1996:128). Probably 200 years were required to complete its construction (Gessler, 1998). Some believe the quality of the Chachapoyas' streets, terraces and cities exceed even those of Machu Picchu.

Only a hardy people, however, could survive in the land they had chosen as their own – the *montaña*. Most of it lay at an altitude between 2000 and 3000 or more meters above sea level. Rain and fog dominated the weather patterns, as did the effects of these conditions: dense forest, lush vegetation and almost ubiquitous swamps, all superimposed on a mountainous, contorted terrain with often incredibly steeply sloped valley walls. Significant among Chachapoya cultural features was their social stratification, in which the majority of this probably multi-ethnic population consisted of farmers and herders whose surplus productivity supported an oligarchal aristocracy. Geographically this region offered passage between the eastern Amazonian lowlands and the western cordilleran slopes and coast at a much lower altitude than surrounding alternatives. This provided some favorable trade opportunities, but was also attractive to those wishing to use it for military purposes. Some feel this explains the need for walls 19 meters high at Kuelap. The nature of this terrain also helps us to understand why the Chachapoyas managed to resist invaders and maintain their independence until about

Fig. 4.35. The Laguna de los Condores (Laguna de las Momias).
This site, in the midst of Peru's rain forest, is a most unlikely site for preservation of human soft tissue. (Photo by and courtesy of Federico Kauffmann-Doig.)

AD 1470, when Inca forces paid a bloody price to gain control of this region (Kauffmann-Doig, 1997).

Peruvian archaeologist Federico Kauffmann-Doig, director of the Instituto de Arqueología Amazonica in Lima, Peru, however, was intimately familiar with these problems. He had already headed 15 expeditions into this region to study the Chachapoyas and excavated 20 of their burial sites, all of which had proved to have been looted to some extent, as had all other reported Chachapoya sites. Nevertheless when he heard about the Leymebamba finds he immediately organized still another expedition in association with the Ligabue Research and Study Center, authorized and registered at the Instituto Geografico Nacional del Peru. He arrived in Leymebamba at the end of May, 1997, with two other scientists and 13 men; they spent two days traveling the remaining distance. The second half, through rain, clouds and swamps could only be traversed on foot, and at times their path needed to be created with machetes.

Arriving at their destination, the Laguna de los Condores (renamed Laguna de las Momias because of the current absence of condors in this region) they were struck by its beauty and isolation. The lagoon's dark, smooth surface reflected the high, nearly vertical face of the cliff on the opposite shore (Fig. 4.35). In several areas bare of vegetation, a total of six stone-and-mortar structures clung as if glued to this precipice, though actually cleverly supported by log props and fitted with wood balconies. Dr Kauffmann-Doig recognized the pattern instantly, since he had encountered similar examples at other Chachapoya sites such as Revash, Ochin and Pajatén (Kauffmann-Doig, 1998c). These two-story mausolea were 5 meters tall,

Fig. 4.36. Chachapoya mummy mausolea.
The two-story, mummy-sheltering mausolea cling to a precipice overlooking the lagoon pictured in Fig. 4.35. (Photo by and courtesy of Federico Kauffmann-Doig.)

rectangular and fitted with lagoon-facing windows that were principally of trapezoid shape. Their plastered surfaces were painted with ocher of multiple colors inside and out, and the façade was trimmed with stone friezes (Fig. 4.36). In one area the cliff displayed geometrical patterns of rupestrine art. Later determined to be at 2800 meters altitude at about 8° 00′ S, 78° 15′ W, the site seemed to incorporate the elements of a fabled Shangri-La. Several hundred round stone foundations on the side of the lagoon opposite the mausolea suggested a probable sizable residence area (von Hagen & Guillén, 1998).

Making their way quickly to the nearest of the mausolea, Dr Kauffmann-Doig's team-mates boosted him to one of the windows and he scrambled through it. What he saw made him literally cry for joy. Not only was the chamber filled with mummy bundles but, while the Leymebamba looters had torn openings in at least 35 of them, more than 60 were intact (Fig. 4.37). Delighted that he and his colleagues were the first formally trained archaeological scientists to encounter such a hoard of ancient Chachapoya mummies, they promptly set about surveying and assessing these remains. They estimated that collectively the mausolea contained at least 100 mummified bodies; a later count exceeded 200 (Gessler, 1998).

Evaluation of the nature of these mummies became possible simply by viewing them in those bundles that had already been ripped open by the looters (Fig. 4.38). The bodies appeared to have become desiccated spontaneously and had been compacted by flexing the knees and hips maximally so that the bowed head rested on the knees, and the flexed elbows at the side of the trunk caused the hands to reach the face. Where the peri-

Fig. 4.37. Looted Chachapoya mummies; SM, ca. AD 1350.
Looters fragmented many Chachapoya mummies prior to the first archaeologist's visit. (Photo by and courtesy of Federico Kauffmann-Doig.)

neum, abdomen and chest could be seen, they presented no evidence of anthropogenic mummification efforts. Both women and men as well as children were represented. Many had quite light-brown skin and some even rather light-brown hair, leading some to speculate about the possibility of Vikings having reached this region (Gessler, 1998). Their dried eyes were intact in the orbits, the nose and lips well formed, the hair abundant, the genitalia recognizable and dentition preserved. Cotton had been stuffed into the mouth and nostrils of some. Earrings, necklaces and other ornaments were common. The outer layer of beautiful wool textiles enveloping the mummified bodies were commonly embroidered, often in the pattern of hand-sewn human faces. The archaeologists were astounded that the soft tissue had resisted decay for five centuries in such a humid atmosphere.

Examination of the artifacts provided further data. Many bundles included small bags filled with coca leaves, consistent with previous evidence of coca-leaf chewing practices by Inca and pre-Inca Andean populations (Cartmell *et al.*, 1991), though their placement in the tombs was probably intended as offerings. Abundant ceramics indicated that, while most of them were of Chachapoya origin, some had Inca patterns and a few even had Christian motifs, suggesting not only that some of the mummies had been buried after the AD 1470 Inca conquest, but also that others had been included during the early Spanish Colonial Period. The Inca presence also appeared to be

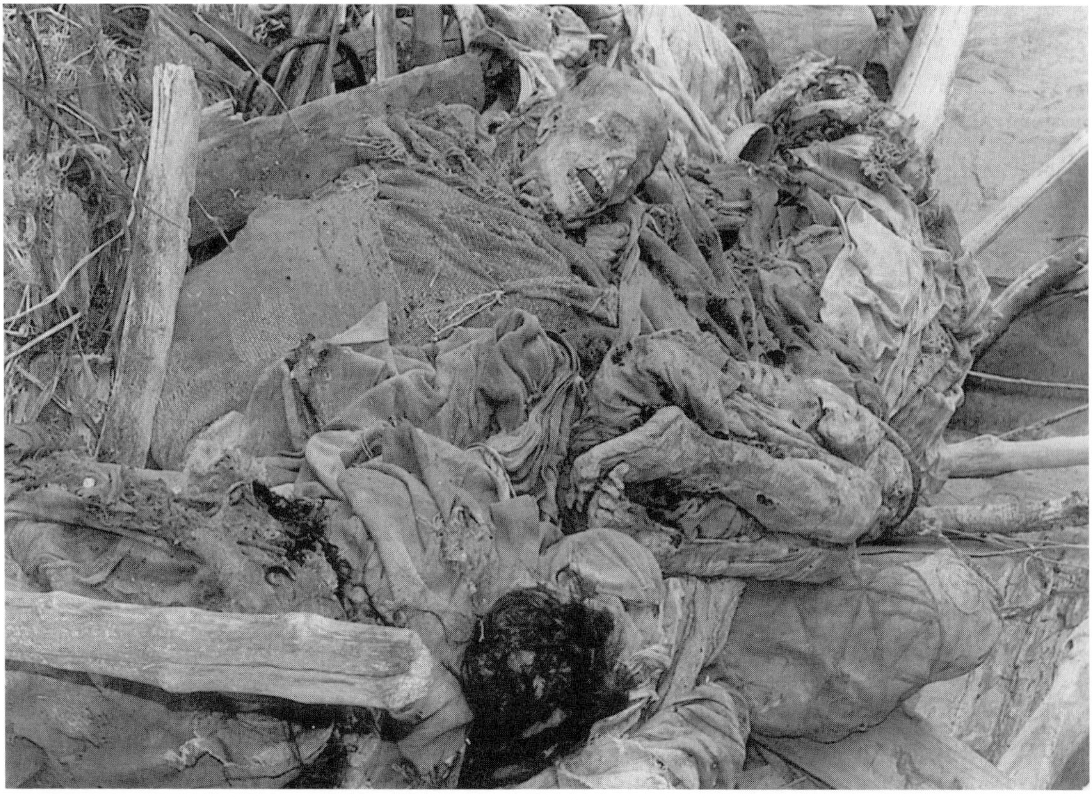

Fig. 4.38. Several Chachapoya mummies; SM, ca. AD 1350.
Looters had stripped wrappings from several mummies when Dr Kauffmann-Doig first visited the site. (Photo by and courtesy of Federico Kauffmann-Doig.)

confirmed by the finding of Inca *quipus*, corded and knotted devices used by Inca administrators to itemize tax and other business records. The finding of some Chimu ceramics was also consistent with the existence of trade with coastal groups.

Recognizing that unique local conditions were responsible for the mummies' soft tissue preservation and the danger of their rapid deterioration unless moved to where they could be conserved under appropriate conditions, Dr Kauffmann-Doig returned to Lima with only one intact mummy bundle for tissue sampling together with some artifact samples (Fig. 4.39). He intended to prepare an elaborate study protocol for sampling this group of mummified human remains and carry out a detailed, scientific investigation of these very special, fragile mummies after arranging for their proper storage. However, a subsequent dispute over privilege of access to these rare specimens has paralyzed this plan. Recent reports of external examinations, radiology and endoscopy of these mummies have been presented orally at scientific meetings. A recent visit to the site documented what appears to be evidence of further looting (Lerche, 2000).

Late Horizon (AD 1470–1534)

The Incas

Prior to about AD 1434 the tribe occupying the region centered around the modern city of Cuzco (altitude ca. 3300 meters) can not be distinguished easily from the surrounding tribes that appear to have been feuding

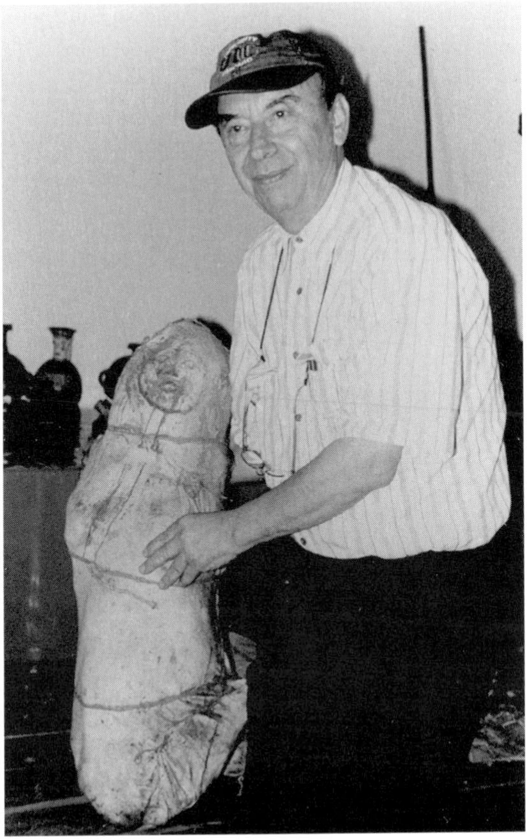

Fig. 4.39. Dr Federico Kauffmann-Doig with Chachapoya mummy (SM, ca. AD 1350).
Only this one Chachapoya mummy was retrieved by Dr Kauffmann-Doig during his first visit to the Laguna de las Momias. (Photo courtesy of Federico Kauffmann-Doig.)

perpetually with each other. However, in that year their eastern neighbors, the Chancas, sensing a lack of resolve in the Incas' reigning leader (Viracocha), attacked Cuzco. Viracocha, his heir Urco, and his royal retinue abandoned the city, leaving its defense to his generals and one of his sons, Cusi. The army routed the Chancas, after which Cusi usurped the throne, calling himself Pachacuti. The remainder of his career was characterized by successful military aggression until he had conquered all groups to the north as far as Cajamarca, as well as the western and southern parts of the province of highland Cuzco. Turning his attention then to the administrative aspects of creating Cuzco as his empire's capital, he appointed his son, Yupanqui, as military commander in 1470. Yupanqui was even more aggressive, pushed the conquests nearly to Quito, Ecuador, most of Bolivia and south beyond Santiago, Chile. By 1493 Huayna Capac had succeeded Yupanqui but in 1527 this leader died, probably of smallpox, whose contagious features caused it to precede the Spanish soldiers that brought this scourge to the New World. Before he died, Huayna Capac divided the empire between his two sons, Huascar and Atahualpa. The resulting civil war created political chaos, enabling Pizarro and his small cadre of soldiers to capture Atahualpa at about the time the latter had arranged successfully for the assassination of his brother in 1532. Since the Inca administration was an absolute monarchy, Pizarro's execution of Atahualpa paralyzed the Inca decision-making machinery. This allowed the Spanish sufficient time for re-enforcements so that they quite easily suppressed the inevitable rebellion several years later. If we accept AD 1470 as a time at which sufficient land had been conquered to justify the term "empire", then the Inca empire's duration is limited to a mere 62 years. If we measure the empire's duration as beginning with the Chancas' attack, it is still only barely a century. Yet the almost incredible civilization they created within that period owes much of its development to their genius for organization. If a conquered group had a well-organized structure, the Incas often retained it with minimal modification. On the other hand, entire uncooperative communities sometimes were transplanted deep within Inca territory, and commonly compelled to marry local people. Recognizing the disharmony inevitably resulting from multiple languages, they imposed their Quechua on vanquished populations. The maximum productivity of the land was calculated and obligations of annual crop-sharing and/or labor were imposed. Probably not since the pyramid-building pharaohs has any culture demonstrated so dramatically how much human labor can accomplish if it is effectively organized and integrated.

Yet they never developed a form of writing. In its absence we are left with many unanswered questions, including those of mummification. Torrential El-Niño-related rains reduced many Peruvian mummies to skeletons. One might anticipate that Spanish conquistadors would have recorded the details of a practice as unique as mummification, but they seem to have regarded at least the royal mummies as administrative impediments and disposed of them. Among the hundreds of Inca mummies excavated in Peru, only a few scientific reports about mummification methods can be found. Considering the brevity of the Inca empire, it is not surprising that mortuary practice varied throughout the conquered lands, particularly at the empire's periphery where the available time was insufficient for complete acculturation. We shall, therefore, concern ourselves to some extent with two subgroups – the royal mummies and the mountain sacrifices, followed by scattered other examples of mummies from the general population.

The royal Inca mummies

The deification of royal Inca mummies and the political employment of those mummies to enhance the centralization of power are discussed in some detail in Chapter 2. These mummified bodies were considered sacred, the very incarnation of the Incas' deity. Since the living, reigning Inca was a close relative (usually a son) of his now mummified predecessor, he could and did publicly also claim divinity. As a constant reminder to that effect, the royal mummies participated in public rituals, displayed in Cuzco's Temple of the Sun, paraded publicly at festivals and joined the living on the reviewing stand at graduation ceremonies of the military academy.

We look in vain, however, for details of their mummification methods. Illustrations by the chronicler Poma de Ayala (1613/1956) indicate clearly that the principle of royal mummification was desiccation, confirmed also by descriptions of their appearance (de la Vega, 1609/1987:307). But whether this was spontaneous, "spontaneous-assisted" or anthropogenic, remains an enigma. The altitude at Cuzco (about 3300 meters) is not high enough to result in mummification consistently and predictably by freeze-dry methods, and neither the soil nor the warm temperatures exist there that can regularly produce spontaneous mummification (though focal microclimates exist). Furthermore, chroniclers confuse the issue further with conflicting statements: Minelli (1993) cites Franciso López de Gómara (1522), saying that they embalmed the body, pouring purifying liquids down the throat of the corpse and then applying resin, while Fr. Jose de Acosta (1590/1940) wrote that the bodies were saturated in bitumen and orifices closed with plates of gold and silver. These statements are reminiscent of features of Egyptian types of mummification and probably simply reflect cultural preconceptions. Garcilaso de la Vega, himself not devoid of bias and national pride, describes evisceration, as does Prescott (1936:748), followed by embalming for royal mummies and by drying in Cuzco's cold, dry air for the general population. He does not speak of resin use for either royalty or bourgeois mummies.

We also have no opportunity to examine the bodies themselves. Since these royal mummies were revered as divine, they were clearly in conflict with efforts by the military conquerors to establish themselves as supreme and they also represented obstacles to the evangelistic efforts of the Christian clergy. Garcilaso de la Vega describes the facial appearance of several royal mummies about 25 years after Pizarro's conquest of Cuzco, but no record of their continued existence thereafter can be found. While rumors persist that suggest they remain hidden at a secret site in Lima, it is more probable that they were destroyed (burned?).

The mountain mummies

Human sacrifice, while not unknown, was not a common Inca event. True, they did not hesitate to execute recalcitrant conquered tribal leaders but they did not incorporate the practice into a major, celebratory event and ritual as did the Aztecs with war captives. The death of an Inca was viewed as a cataclysm and chroniclers indicate that nobles from each Inca province were expected to offer an adolescent as a

Fig. 4.40. Discovery of Inca mummy on the peak of Cerro el Toro; SM, ca. AD 1400.
Juan Schobinger (far right) and colleagues stand behind the mummy they found on top of Cerro el Toro. (Photo courtesy of Juan Schobinger.)

retainer for sacrifice at this occasion (Vreeland, 1998). While textiles, llamas and other animals sufficed for their routine sacrificial needs (Schreiber, 1996b), events such as famine, earthquakes and especially volcanic eruptions (Reinhard, 1998) that they perceived as mortal threats were viewed as calling for human sacrifice. Under these circumstances the sacrifice took place at sacred locales. The most sanctified of all such were the many Andean mountain peaks. A popular misconception states that the Incas worshipped the actual mountains. More specifically, they probably viewed deities as resident in the sky. Mountain peaks, approximating the sky more closely than any other site on their land, were considered the place where they could confront their deities to make their plea – an incarnation locale. Here then, is where they met their gods when they were desperate, mollified them with a human sacrifice and pleaded their cause. While they sanctified the mountains housing these shrines, their worship was directed at the gods that met the human representatives at these altitudes.

The isolation of these sites has resulted in sufficient survival of their architecture, artifacts and even some mummified remains of their human sacrifices to permit reasonable reconstruction of the event. Local rocks were used to create a small platform that appeared to serve as an altar. This was placed as near the peak as access would permit (Fig. 4.40). Such areas

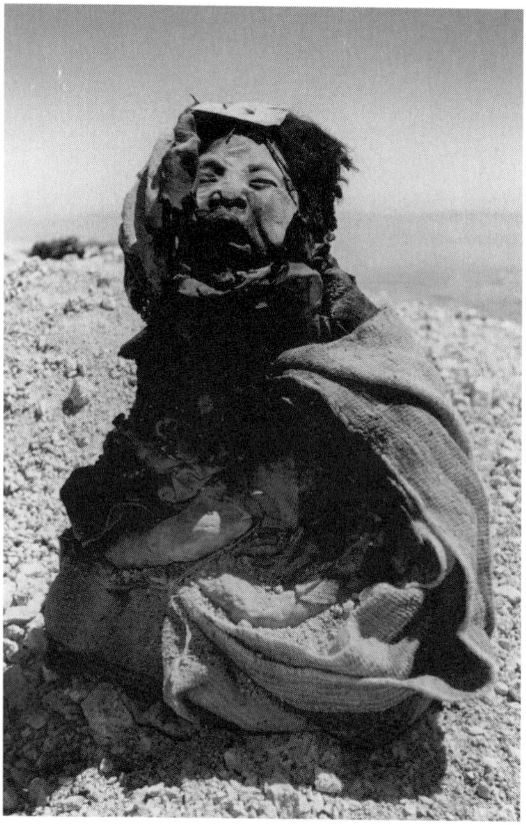

Fig. 4.41. A child's mummified body on Mt Llullaillaco, Argentina; SM, ca. AD 1400.
This 8-year-old female mummy was one of three mummified bodies excavated from the peak of Mt Llullaillaco by Johan Reinhard, Constanza Ceruti and expedition members. (Photo courtesy of Constanza Ceruti.)

Fig. 4.42. El Toro mummy; SM, ca. AD 1400.
The body of this young adult Inca male is also pictured in the foreground of Fig. 4.40. (Photo courtesy of Juan Schobinger.)

commonly do not accommodate a large number of individuals and in some examples a broader area at a slightly lower level appears to have served as a convenient encampment. We do not know how the individual was chosen but the bodies of sacrificed individuals have ranged from a 6-year-old child to young adults (Figs. 4.41 and 4.42). Metallurgical, lithic and textile artifacts suggest that these may have played a role in the ceremony. While earthquakes have dislodged the sacrificed bodies in some cases (Reinhard, 1996b), the most common disposal of the sacrificed individual seems to have been a somewhat superficial burial. In one a major skull fracture was found, but since that body had been dislodged by an earthquake and fallen to a lower ledge, the significance of that finding remains controversial (Fig. 4.43). Skull fracture has also been identified in several others (Table 4.3) but it is not certain that they occurred before death. In others the method is not obvious and has been speculated to be simple hypothermia (freezing to death) (Fig. 4.44), strangulation or even live burial (Fig. 4.45). Vomitus on the clothed body of a child (El Plomo mummy) together with evi-

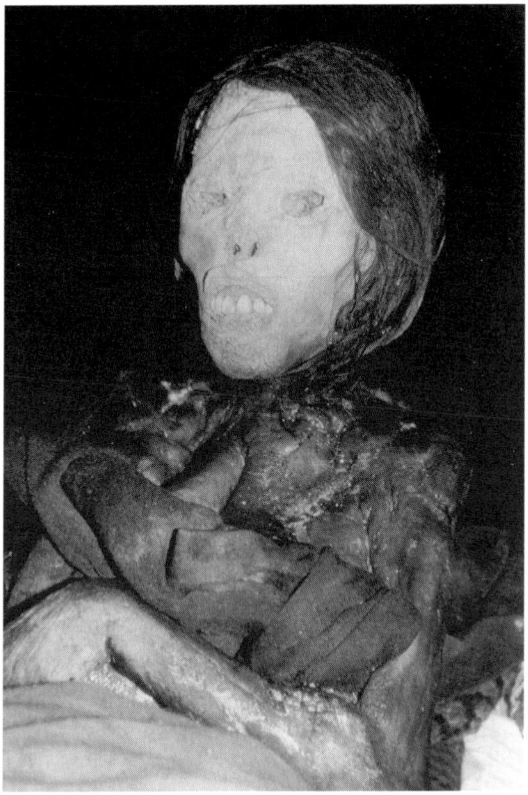

Fig. 4.43. "Juanita", the frozen body from the top of Mt Ampato, Peru; SM, ca. AD 1400.
This young adult female excavated by Johan Reinhard, Juan Chavez and expedition members on Mt Ampato had been dislodged from her burial site by an earthquake. (Photo courtesy of Johan Reinhard.)

dence of chicha (corn beer) suggests the possibility of alcoholic sedation (Fig. 4.46).

Some of the bodies (those from El Misti near Arequipa, for example) became largely skeletonized after death. Preservation of soft tissue, however, is quite common under these circumstances and is usually brought about spontaneously by simple freezing or freeze-drying (Figs. 4.47 and 4.48). A list of bodies recovered from such sites is presented in Table 4.3. To date, few invasive scientific studies have been carried out on these bodies (Fig. 4.49), though radiographical and other noninvasive examinations have been performed on several.

Mummies of the general Inca population

Lumbreras (1974:235) correctly notes that many Inca bodies were buried above ground in caves or under rock shelters. As noted above, however, local practices in various parts of the empire influenced these practices. The Cuzco Museum at the time of my visits in 1982 and 1985 was curating a very modest mummy collection of 10–12 bodies, though in 1935 when McCreery (1935) visited it he found 20 mummies there. These had been accumulated by a private collector so no detailed provenience data were available but the remains were believed to have been from the Cuzco region. They were fully flexed, wrapped in reed mats and appeared to have been naturally mummified through spontaneous desiccation. McCreery states that, in the Cuzco area, burials at Pisac in the Urubamba Valley were in tombs cut into the side of a gorge, some of which had beehive-shaped walls of adobe and rock.

An Inca body from the Cuzco area, located in the Museum of Anthropology and Ethnology in Florence, Italy, and acquired in 1875–1884 was studied by Fornaciari *et al.* (1992a) and Minelli (1993). The body had been spontaneously mummified, was flexed, wrapped in textiles and a vegetal fiber basket had been built around it. Subsequent dissection demonstrated the dilated cardiomyopathy, megaesophagus and megacolon characteristic of Chagas' disease (American trypanosomiasis) with immunohistological demonstration of the infective agent *Trypanosoma cruzi*. This diagnosis was further confirmed by extraction of a segment of ancient DNA unique to *T. cruzi* from a sample of this mummy's esophagus (Guhl *et al.*, 1999).

Five Inca mummies were also on display at the Rafael Larco Herrara Museum in Lima during my visits in 1982 and 1985. Textiles covered the lower half of the bodies but the exposed upper half showed no evidence of anthropogenic mummification. No provenience data were available.

Table 4.3. *South American mountain sacrifice mummies*

Province and country	Mountain	Altitude (meters)[a]	Approx. age (yr)	Sex[b]	Year found	Discoverers	Popular name	Reference	Latitude	Longitude	M, S, I[c]	Comments[d]
Ayacucho, Peru	El Toro	6380 6200	21	M	1964	Juan Schobinger; Beorchia Groch	El Toro	Schobinger, 1966	29° 10′ S	69° 45′ W	M	
Mendoza, Argentina	Aconcagua	6960 5321	7	M	1985	Juan Schobinger	—	Schobinger, 1991	32° 30′ S	70° 00′ W	M	
Santiago, Chile	El Plomo	5448 5421	8	M	1954	Grete Mostny	El Plomo	Mostny, 1957	33° 13′ S	70° 13′ W	M	
Arequipa, Peru	Pichu Pichu	5669 5630	15	F	1963	Johan Reinhard	—	Reinhard, 1992	16° 27′ S	75° 12′ W	I	Skull fx
Arequipa, Peru	Pichu Pichu	5669 I[e]	10	F	1996	Johan Reinhard; Jose Chavez	—	Reinhard, 1998	16° 27′ S	75° 12′ W	S	
Arequipa, Peru	Pichu Pichu	5669 I	8	M	1996	Johan Reinard; Jose Chavez	—	Reinhard, 1998	16° 27′ S	75° 12′ W	S	
Arequipa, Peru	Ampato	6309 6309	11	F	1995	Johan Reinard; Miguel Zarate	Juanita	Reinhard, 1997, 1998	15° 50′ S	71° 50′ W	M	Skull fx
Arequipa, Peru	Ampato	6309 5871	11	F	1995	Johan Reinhard; Miguel Zarate	—	Reinhard, 1997	15° 50′ S	71° 50′ W	S	Lightning
Arequipa, Peru	Ampato	6309 5871	13	M	1995	Johan Reinhard; Miguel Zarate	—	Reinhard, 1998	15° 50′ S	71° 50′ W	S	Lightning
Salta, Argentina	Cajon	5500 I	25	F	1924	I[a]	—	Vreeland, 1998	25° 03′ S	65° 53′ W	I	
Arequipa, Peru	Coropuna	6425 5150	Adult[f]	M?	1965	I	—	Vreeland, 1998	15° 30′ S	72° 39′ W	I	
Salta, Argentina	Chañi	5896 5896	6	M?	1905	Military	—	Ceruti, 1998	24° 00′ S	67° 07′ W	M	
Iquique, Chile	Esmerelda	905 905	9	F	1976	Construction crew	—	Schobinger, 1991	20° 10′ S	69° 10′ W	I	
Iquique, Chile	Esmerelda	905 905	18	F	1976	Construction crew	—	Schobinger, 1991	20° 10′ S	69° 10′ W	I	
Salta, Argentina	Quehuar	6130 6130	14	M?	1974	Antonio Beorchia	—	Reinhard, 1992	24° 20′ S	65° 46′ W	M	
Ayacucho, Peru	Sara Sara	5500 5500	15	M	1996	Johan Reinhard; Jose Chavez	Sarita	Reinhard, 1998	14° 48′ S	73° 36′ W	S	Skull fx
Arequipa, Peru	Ampato	6309 5872	Subadult	?	1997	Johan Reinhard; Jose Chavez	—	Reinhard, 1998	15° 50′ S	71° 50′ W	S	Lightning
Salta, Argentina	Llullaillaco	6739 6739	8	M	1999	Johan Reinhard; Constanza Ceruti	—	Reinhard, 1999	24° 42′ S	68° 32′ W	M	
Salta, Argentina	Llullaillaco	6739 6739	14	F	1999	Johan Reinhard; Constanza Ceruti	—	Reinhard, 1999	24° 42′ S	68° 32′ W	M	
Salta, Argentina	Llullaillaco	6739 6739	8	F	1999	Johan Reinhard; Constanza Ceruti	—	Reinhard, 1999	24° 42′ S	68° 32′ W	M	Lightning

Notes:

[a] The number on the left indicates the height of the peak; the number on the right identifies the altitude at which the mummy was found.

[b] M, male; F, female.

[c] M, mummy; S, skeleton; I, indeterminate.

[d] fx, fracture.

[e] I, indeterminate.

[f] Adult is defined as over 16 years of age.

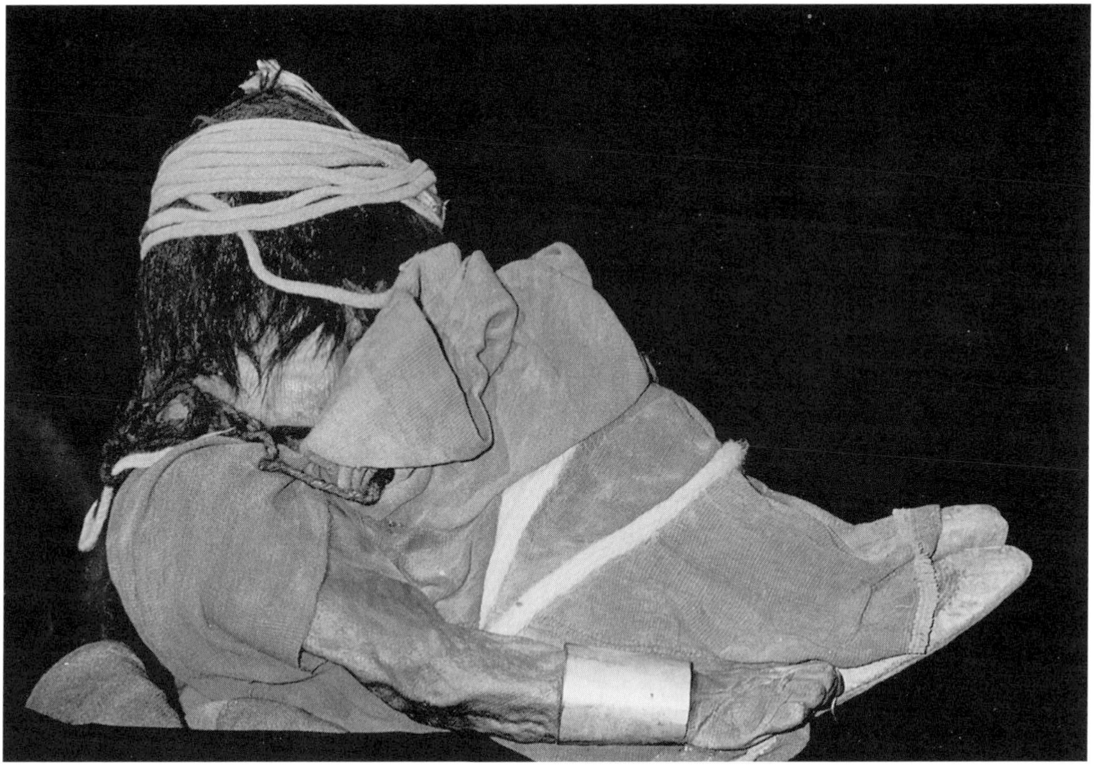

Fig. 4.44. Boy mummy from Mt Llullaillaco; SM, ca. AD 1400.
The frozen, mummified body of this 8-year-old boy revealed superb preservation of soft tissues. Excavated by Johan Reinhard, Constanza Ceruti and expedition members. (Photo courtesy of Constanza Ceruti.)

As many as 1800 mummies have been excavated from the Ancón site between 1945 and 1962 but detailed reports are lacking. At least some of them were from the Inca period, and Stothert (1979) described the opening of one of these. It contained the body of an adult female wrapped in four layers of textiles, each separated by stuffing the space between them with leaves. A wicker basket covered the body, which was enveloped in a shroud (Fig. 4.50). The body, however, was skeletonized, without any soft tissue preservation.

The Colonial Period (AD 1534–1821)

Relatively few mummies have been reported from this period, though of these few several are distinguished persons.

Viceroy of "New Spain"

In 1988 the crypt of Lima Cathedral was found to house the remains of about 300 individuals. Most of these were elderly Caucasian males, though about 30% were infants and children. Conservation methods varied: in some no effort to preserve soft tissues was made, while others were encased in lime with little resulting soft tissue preservation. Still others were eviscerated and their body cavities filled with earth. One unlabeled, elaborately decorated coffin contained the extended burial of an elderly male with clothing that clearly marked him as an aristocrat. He had been eviscerated and the thorax and abdomen had been filled with soil. The top of the skull had been removed to gain access to the cranial cavity, but at least part of

Fig. 4.45. Child body found at top of Nevado de Chuscha; SM, ca. AD 1400
This child, also known as Momia de los Quilmes, was found on a mountain about 20 kilometers south of Buenos Aires, Argentina. (Photo courtesy of Juan Schobinger.)

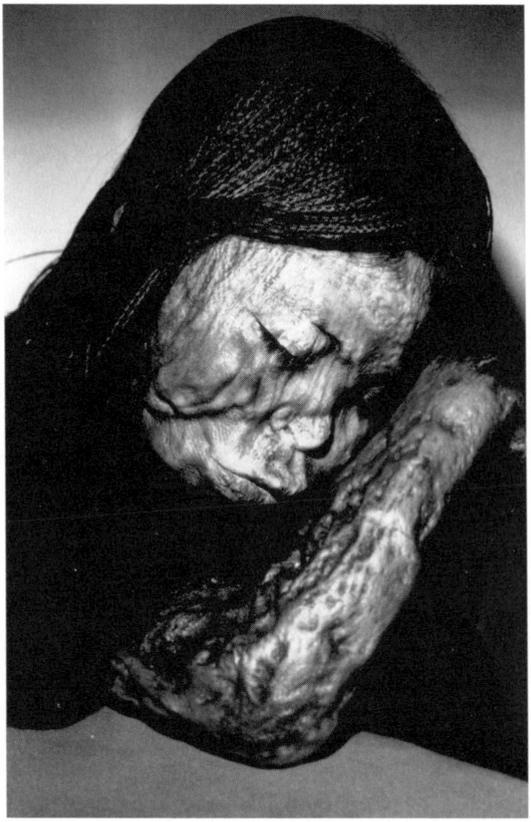

Fig. 4.46. The "Prince of el Plomo"; SM, ca. AD 1400. This child's frozen-dried body was found on Cerro el Plomo and is curated at the National Museum in Santiago, Chile. (Photo courtesy of Mario Castro.)

the brain was still inside. The bones were those of a robust, elderly male. The right forearm was missing, but was replaced by a prosthesis, attached to the humeral stump by a nail. The prosthesis was made of low-grade silver with an attached wooden hand; the proximal humerus demonstrated a healed, 14 cm wound. On the chest was a silver medal on the back of which the name Melchior was inscribed. The distinguished funerary goods, the unique prosthesis and the medal all supported the identification of Melchior Portocarrero Lasso de la Vega, Conde de la Monclova. This man had had a distinguished military career and lost his right forearm in the battle of Dumas in 1658. Later he was appointed Viceroy of New Spain, governed there for 16 years and died in Lima in 1705 at age 69 years (Guillén-Oneeglio *et al.*, 1995).

Francisco Pizarro

In 1532, accompanied by only several hundred soldiers, the Spanish conquistador Francisco Pizarro invaded Peru, captured and executed the Inca leader Atahualpa and took over the country. Almost inevitably, disputes arose about the fairness in distribution of the subsequently looted gold, silver and property. Nine years later several of Pizarro's dissatisfied officers,

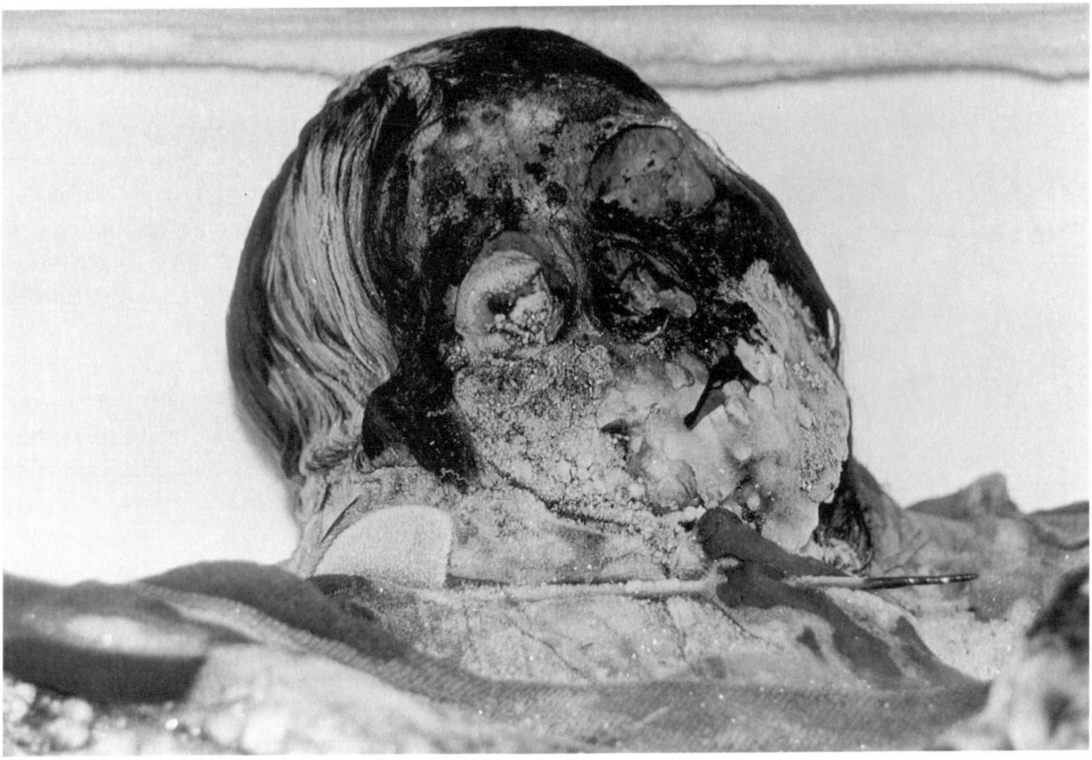

Fig. 4.47. The mountain sacrifice mummy Sara Sara; SM, ca. AD 1400.
Another well-preserved Inca mountain-top sacrifice. (Photo courtesy of Johan Reinhard.)

despairing of their chance for recovering what they regarded as their just share, attacked Peru's conqueror while he was entertaining at a dinner in his quarters at Lima. Most of the guests jumped from the windows while several tried to hold the assassins at bay at the door. Lacking time to assemble his armor, Pizarro wrapped a cloak around his left arm, grabbed his sword and, as his door guard fell, taunted the assassins, daring them to enter. As he slew the first, they hesitated. Sensing possible failure the attacking group's leader seized one of the other assassins, thrust him violently at Pizarro, who ran his sword through him. However, while Pizarro was struggling to extract the sword from the man's body, the leader struck Pizarro a mortal blow on his neck, felling him. One of the assassins then struck him on the head with a large water jug while the others repeatedly stabbed him. Subsequently they mutilated his body. Several hours later his body was wrapped in a shroud and placed in a crypt under the altar of Lima Cathedral.

Four years later his bones and sword were exhumed, placed in a wooden box decorated with the cross of Santiago and stored in a crypt of the main chapel at Lima Cathedral, as he had requested in his will. In 1606 it was moved again to a newly constructed area of the church, but three years later an earthquake required reconstruction of the chapel. Between 1623 and 1629 he was moved again. In 1661 a proclamation reported formally that these were the remains of Pizarro and that his head was contained in an accompanying lead box inscribed with a statement to that effect. In 1746 an earthquake again devastated the cathedral and by 1778 a virtually new church had been constructed (Maples *et al.*, 1989). During 1891, the

Fig. 4.48. The Aconcagua mummy; SM, ca. AD 1400. This 7–10-year-old boy was excavated in 1985 by members of an expedition led by Juan Schobinger. It was at the top of Mt Aconcagua in Argentina at an altitude of ca. 6283 meters above sea level. (Photo courtesy of Juan Schobinger.)

350th anniversary of his death was marked by a municipal commission that examined a body in a wooden box in the cathedral crypt beneath the main altar. This report noted that the hands and external genitalia were missing and that the head was attached to the body, though much of the neck's soft tissue had deteriorated over the anterior and lateral aspects. The dissectors then separated the head from the chest. They also noted that multiple lime pellets were present on the torso's surface, speculating that the earth in which he had been interred must have contained these. They also noted that the viscera were represented only by brown powder, that the perineum had been disturbed and that some brain tissue (mostly meninges) remained in the cranial cavity when it was exposed by removal of the calvarium. The body was then reconstructed by sewing the head back on the neck by its skin, the body varnished, replaced in the wood coffin and the casket was sealed. The following year the commission placed the body in a white marble sarcophagus with glass on three sides for display, including within it a glass bottle containing his brown, pulverized viscera together with a copy of their report (McGee, 1894).

In 1977 the founders of the Paleopathology Association, Aidan and Eve Cockburn together with James Vreeland visited Lima and viewed the body in the Cathedral, as had many thousands of tourists before them since 1892. Staring at the body they were puzzled by the head, since it appeared inappropriate for the body. A few days later they were surprised by a Lima newspaper's proclamation that a cathedral workman had found a lead box with the inscription that it contained the head of Francisco Pizarro! Thereafter the media were silent on this topic (Cockburn & Cockburn, 1977; Cockburn, 1978). In 1984 several physical and forensic anthropologists examined the contents of two wooden boxes. One of these contained the remains of two children, an elderly female, and the skull and postcranial remains of an elderly male. The postcranial skeleton of the latter articulated with the skull in the lead box that was inscribed with the statement that it contained Pizarro's head. Maples *et al.* (1989) meticulously examined the bones that clearly evidenced all the changes and traumatic lesions that the assassination's historical account would lead them to expect. They concluded that the postcranial bones in the larger of the two wooden boxes (which also matched the description of the 1891–1892 municipal commission report) were the remains of Francisco Pizarro, that the skull in the lead box was Pizarro's skull, and speculated further that the other remains may represent Pizarro's wife and children, and that the other complete skeleton of an elderly male was probably a cleric. Subsequently Stout (1986) examined a cross-sectional sample of the remains' rib, and used histomorphometry methods to estimate an age-at-

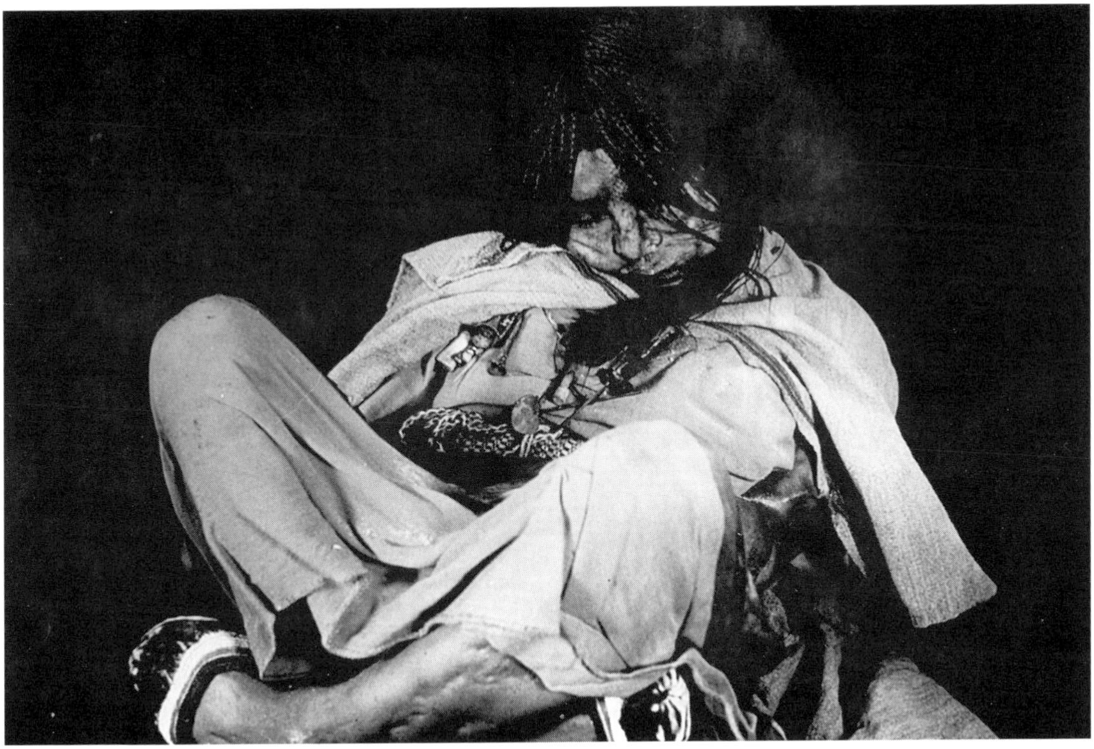

Fig. 4.49. Female child mountain sacrifice on Mt. Llullaillaco; SM, ca. AD 1400.
A 14-year-old female frozen body was one of three excavated from the top of Mt Llullaillaco (Argentina) by Johan Reinhard, Constanza Ceruti and expedition members. (Photo courtesy of Constanza Ceruti.)

death of 62 years. While Pizarro's exact birth date is not known, historical accounts suggest he was in his sixties when he died. Thus, after nearly half a millennium, it appears that the various surviving parts, now mostly skeletonized, of Pizarro's remains have finally been reassembled and housed again in Lima Cathedral (Fig. 4.51).

Harvard University mummy (HUM-1)

In 1980 a joint study by a team made up of representatives of the Paleopathology Association and Harvard University's Peabody Museum carried out a dissection of mummy no. N 18446, curated at the Peabody Museum. This had been collected and donated to the museum so provenience data were limited to the fact that it had come from Llactashica in the Rimac Valley on Peru's central coast. Later a radiocarbon date on the mummy's muscle tissue was reported as 140 ± 60 years BP, placing it firmly within the Colonial Period. The body was flexed and preservation was due to spontaneous desiccation (Fig. 4.52). External genitalia indicated that it was the body of an adult male and dental attrition suggested an age of 35 ± 5 years. Other findings included a circular 11 mm diameter, beveled complete cranial defect with evidence of early healing as well as two small (20 and 40 mm) fracture lines radiating from it posteriorly. Initially declared as a surgical trepanation wound of the right side of the frontal bone, later observers suggested this may represent a congenital defect. In addition, adjacent to the defect was a 20 mm long incised, healed wound of the outer table, and just lateral to it was an 18 mm × 12 mm × 2 mm sling ball wound impressed into the outer table. Soft tissue

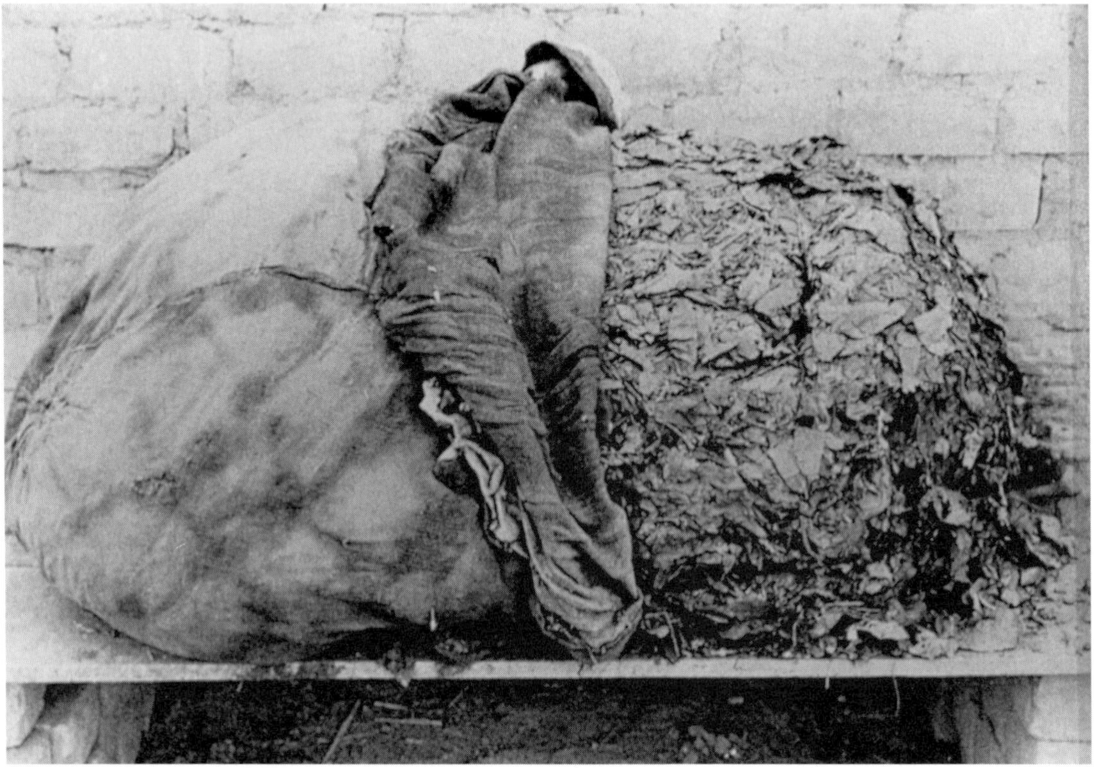

Fig. 4.50. Opened Inca mummy bundle; SM, ca. AD 1476–1532.
Four textile layers enveloped the skeletonized body. Each layer was separated from the next by a layer of dry leaves. Late Horizon Period. From Ancón, Peru. (Photo by and courtesy of Karen Stothert (1979).)

studies revealed severe anthracosis (Burger, 1978; Zimmerman *et al.*, 1981b).

Summary of mummies from Peru

Although several thousand mummified human bodies have been exhumed from a variety of Peruvian sites, most of these have been excavated by amateurs or adventurers. A few have been studied by professionals and most are now lost. Bodies of the Inca rulers have disappeared, presumably disposed of by the conquistadors for political reasons. The quantity of potential information permanently lost by these deplorable actions is staggering. In addition, nature (recurrent El Niño episodes with their related torrential rains) has conspired to destroy many of the remaining mummies. The result of these events is a surprisingly small group of bodies – too few to permit generalizations.

We know that the royal Inca bodies were mummified by desiccation, but we do not know how this was achieved or even whether they were eviscerated. Possibilities include desiccation by heat (fire), by smoking (improbable; no description of soot deposit on skin) or by a freeze-dry method such as controlled exposure at a high mountain site. Most of the nonaristocratic bodies that have become available are spontaneously mummified, often by burial in arid, sandy soil and frequently wrapped in highly absorbent textiles. These are features commonly known to contribute to spontaneous mummification. Finally we know that a few bodies have demonstrated evidence of evisceration with replacement of visceral structures by botanical elements.

Fig. 4.51. Pizarro's coffin in Lima Cathedral; SM, AD 1541.
Recently the separated head of Pizarro has rejoined the remainder of the conquistador Francisco Pizarro's body in his coffin kept in the cathedral at Lima. (Photo courtesy of *National Geographic*.)

This appallingly sparse database needs to be expanded by professional examination of hundreds of mummified Peruvian bodies from a large number of sites before we can understand both the details of mortuary rites as well as the impact of natural postmortem effects on Peruvian human remains.

Mummies from Chile

Geography and climate

A combination of physical factors in northern Chile interact to generate a unique climate favoring the production of spontaneous mummification of interred bodies. Principal among these is the effect of the Andes mountains. Tectonic action has caused this range to parallel and follow closely the Pacific coast of South America. At the latitudes of northern Chile these peaks commonly lie within a hundred kilometers of the shore. Their western slopes plunge steeply to a low level, then abruptly decline more gently over an alluvial plain that occupies most of the surface between the coast and the mountain base (Fig. 4.53). The Andes participated in the last Ice Age and at its end the glacial melt water descended the steep western slopes with such violence and turbulence that it lacerated that terrain into a succession of deep, V-shaped valleys (Fig. 4.54) while those of the coastal plain became half-filled

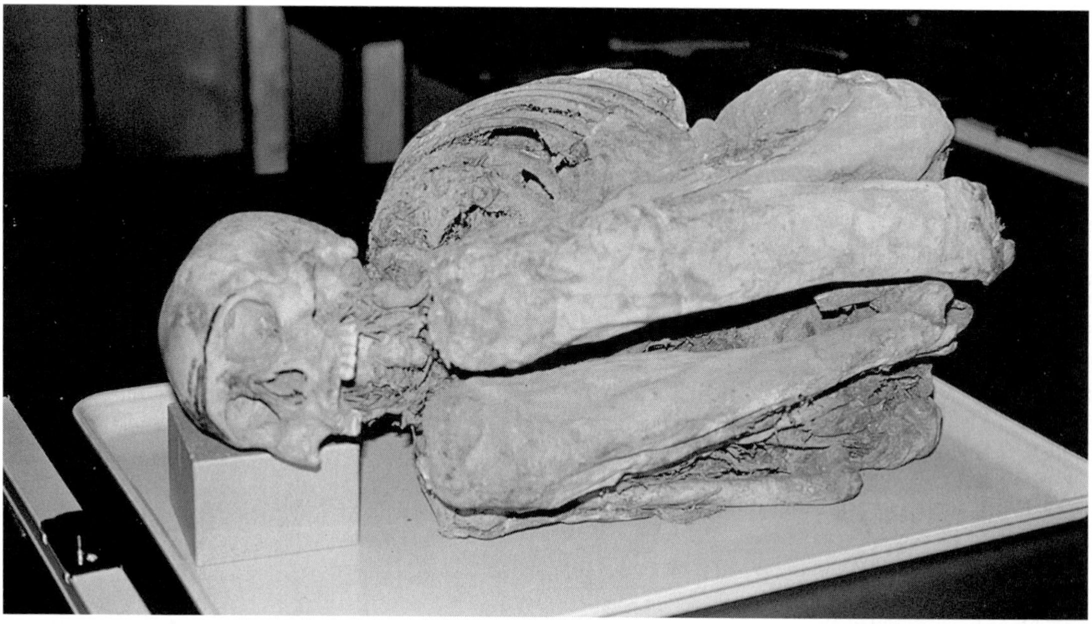

Fig. 4.52. Peruvian mummy no. HUM-1; SM, AD 1810.
The spontaneously mummified body of this Peruvian male from the Rimac Valley was being curated at Harvard's Peabody Museum when it was dissected by a team led by Michael Zimmerman and Erik Trinkaus. A slingball skull fracture was identified. (Photo by ACA.)

Fig. 4.53. Sketch of geography of northern Chile's Atacama Desert.
Note the Andes' high peaks at top of the sketch, separated from the lower, parallel range by the highlands (altiplano). The valleys traverse the steep western slopes of the lower cordillera, and continue across the alluvial plane to enter the Pacific Ocean at lower left. (Artist: Raul Rocha, Archaeological Museum, University of Tarapaca.)

Fig. 4.54. Atacama Desert at 3000 meters' altitude.
The arid, barren terrain reveals the steep-walled valleys carved by glacial meltwater at the end of the last Ice Age. (Photo by ACA.)

with alluvium. The latter has supported irrigation agriculture for the most recent populations. A somewhat lower cordillera west of the Andean Peaks results in a broad, highland plain (altiplano) at an altitude of about 4000 to 5000 meters above sea level (Fig. 4.55). Offshore a broad flow of cold Antarctic water, the Peru (formerly Humboldt) current, is pushed northward by predominantly southwest winds. The resulting hydrodynamic effect results in an upwelling that brings rich animate and inanimate nutrients to the surface, creating one of the world's most productive fishing areas. Periodic El Niño events do affect this current, with transient, profound suppression of marine fauna as well as avian populations, but northern Chile is spared El Niño's torrential rains that commonly affect the Peruvian coast.

When Atlantic air masses on the eastern side of South America are blown inland and rise to cross the Andes, they cool and their moisture condenses. The resulting rain produces all degrees of precipitation from well-watered plains (pampas) to the Amazonian rain forest on the Andes' eastern slopes. Only a small amount of moisture remains to be deposited as snow on the peaks more than 6000 meters high (Fig. 4.56). On the western side the Pacific winds need to cross the cold Peru current before reaching land, reducing their moisture content. When they arrive at the sun-warmed land the remaining moisture is vaporized and is not condensed again sufficiently to fall as rain until it rises to the level of the highlands. The consequence of these unique conditions is northern Chile's Atacama Desert, one of the world's most arid areas, where rain has not been reported below 3500 meters in some areas since records were kept (Fig. 4.57).

This narrow plain, compressed between the coast and the Andean mountains is generally less than 100 kilometers wide and extends from about 16° to 22° S. Most of its many east–west valleys, cut by glacial melt water, are dry (*quebradas* (gorges)) but some contain small streams (often only seasonally) originating in the highlands that irrigate coastal flat-bottomed, lower valley fields. Thus human habitation in this barren

Fig. 4.55. The Andes at latitude 22° S.
Andean highland plateau (altiplano) is flanked by the snow-capped peaks in the background and a lesser range in the foreground. (Aerial photo by ACA.)

Fig. 4.56. Andean highlands (altiplano).
Ground view of the Andean highlands shown in Fig. 4.55. The snow on the 6000 meter peaks, together with rainfall on the altiplano serve as the source of the streams that traverse the Atacama Desert. (Photo by ACA.)

Fig. 4.57. The Atacama Desert (about 3000 meters).
Below 3000 meters' altitude, rainfall in the Atacama Desert is essentially absent. (Photo by ACA.)

desert became possible by exploiting the rich marine resources (fish, shellfish and sea lions (*Otaria flavescens*)) or, as later populations did, practicing small-scale, irrigation agriculture in the low valleys near the coast (Fig. 4.58). Agriculture developed at these latitudes in the highlands, and was supplemented there by hunting vicuñas, guanacos, llamas and alpacas (collectively called camelids) and by pastoralism following domestication of the last two (Fig. 4.59).

Synopsis of prehistory and history of Chilean populations

The Early Archaic Period in the highlands (8000–6000 BC) was characterized by year-round hunters, but between 6000 and 3000 BC these were driven to lower altitudes by a warmer, drier climate and extensive highland volcanic activity that permitted only summer hunting parties in the altiplano. Full-time highland residence was restored in the Late Archaic, during which agriculture and camelid pastoralism evolved.

The relationship between the highlanders and the coastal people separated by the barren Atacama Desert interspersed with steep-walled *quebradas* has remained controversial. Northern Chile's coastline was populated by people (Chinchorros) of unknown origin about 7500 BC who maintained a preceramic, predominantly maritime subsistence until about 1500 BC. By at least 1000 BC highland influence (Alto Ramirez) from the Lake Titicaca area in the altiplano arrived and these migrants settled on the coast, introducing pastoralism and agriculture. About AD 400 the expanding Tiwanaku highlanders of Bolivia (locally called Cabuza) reached the north Chilean coast. When the Tiwanaku empire collapsed about AD 1000, regional groups tended to coalesce. On northern Chile's coast

Fig. 4.58. Irrigated low valleys in the Atacama Desert.
Small streams originating in the highlands flow through an alluvial plain of the Atacama Desert along their course to the sea. Small-scale irrigation of these permits limited agriculture. (Photo by ACA.)

Fig. 4.59. Camelids in the Andean highlands.
Feral (vicuñas, guanacos) and domesticated (llamas, alpacas) camelids permit hunting and pastoralism in the highlands. (Photo by ACA.)

Table 4.4. *Principal coastal cultures of northern Chile*

Culture	Time period
Chinchorro	7500–1500 BC
Alto Ramirez	1000 BC to AD 350
Cabuza	AD 400–1000
Miatas Chiribaya	AD 1100–1300
San Miguel and Gentilar	AD 1300–1450
Inca	AD 1450–1550
Colonial	AD 1550–1824

an archaeologically identifiable, primarily agricultural group termed Maitas Chiribaya evolved between AD 1100 and 1300, succeeded by the technologically more sophisticated San Miguel and the Gentilar groups. Inca influence (ceramic and textiles) is detectable on the coast by about AD 1450 but their physical presence probably was minimal (Rivera, 1977). These groups and their dates are listed in Table 4.4.

Mummification features of northern Chile

A large fraction of the bodies interred in the soil of the Atacama Desert underwent spontaneous mummification. Factors contributing to such mummification include the extreme aridity of the region's climate, the almost total absence of groundwater and the relatively sandy soil. Heat per se is probably not a major factor. Midsummer coastal daytime high temperature averages about 29 °C and midwinter about 22 °C. During one early fall day on excavation, I measured the air temperature in the shade and found it to be 25 °C, that of the sun-exposed sand at 2 cm depth was 47 °C and at a depth of 14 cm was 25 °C. Clearly the heating of the sand by the sun did not extend to the depth of the interred bodies. It has been alleged that the high nitrate content of the soil may have exerted an osmotic effect, contributing to removal of water. This is conceivable and certainly there are areas of the Atacama in which the nitrate content is so high that the 1879–1883 Pacific War was fought between Bolivia, Peru and Chile for its ownership (it was then the principal source of nitrogen for gunpowder manufacture until Germany learned how to extract nitrogen from the air during World War I). However, whether nitrate concentration is high enough to have such an effect at the specific burial sites in northern Chile has not yet been examined. Furthermore, the degree of spontaneous mummification varies enormously. A fraction (14–25%) of bodies from such a site have skeletonized, while an equal fraction demonstrates good preservation of the major viscera, with the remainder showing intermediate grades. These varying responses suggest that normally the margin between complete loss of all soft tissue and its excellent preservation is a narrow one in which small variations have large effects. Since burial methods usually are similar within a population it is also probable that the variables producing these differences are extra-sepulchral; for example, difference in time intervals between death and interment, season of death, the presence of bacteremia (bacteria in circulating blood) at time of death and similar other circumstances.

Only one group in this region practiced anthropogenic mummification – the Chinchorros. The details of their practices are discussed below. Unfortunately, climatic conditions in the highlands are not conducive to soft tissue preservation, so our observations regarding the mummification process need to be restricted to the coast and low valleys.

While the Chinchorros interred both their anthropogenically and spontaneously mummified bodies in an extended position, their successors (the Quiani people) were placed on their right side with semiflexed knees. Most of the subsequent groups, however, buried their bodies in an extreme "fetal" position, with hips, knees and elbows flexed, knees under the chin, and hands between the knees and chin. Chinchorros were placed directly into the sand without coffins, but usually the bodies were wrapped in a reed mat. Their successors commonly had textile wrappings and were buried vertically in shaft tombs, although some lined the tomb with stones.

The principal areas from which significant numbers of mummies have been retrieved in Chile are discussed individually below.

Fig. 4.60. Acha Man, one of the world's oldest mummies; SM, ca. 7000 BC.
This Chinchorro male was excavated from the Acha *quebrada* (gorge), one of the Atacama Desert's dry valleys. Portion of reed mat is adherent to facial skin. Radiocarbon dating of mummy muscle was about 7000 BC. (Photo by ACA.)

Coastal and low valley sites near Arica, Chile

Chinchorro (9000–1500 BC)

While the earliest evidence of this group is a shellfish midden near the port of Antofagasta, the oldest mummified body from this group (one of the oldest in the world) is Acha Man, whose spontaneously mummified body was radiocarbon dated as 7020±255 BC (Aufderheide *et al.*, 1993) (Fig. 4.60). Most of the bodies have been recovered from beach sites or locations within a few kilometers of the beach. The largest group (nearly 100) was a collection of anthropogenically mummified bodies clustered in the sand adjacent to a municipal water reservoir within the port city limits of Arica, about 1 kilometer from the sea, at a site labeled Mo 1. Later a smaller number of 69, mostly spontaneously mummified bodies were found nearby (Mo 1–6). Still smaller groups have been collected from a site at Chinchorro beach at the city's margin and other sites are scattered from Antofagasta to Ilo, Peru. The spontaneously mummified examples from Mo 1–6 had radiocarbon dates that clustered between 2000 and 1500 BC, while those from Mo 1 ranged from about 5000 to 1840 BC. Both groups were wrapped in totora reed mats. The spontaneously mummified bodies were in the extended position (Fig. 4.61), wore "loin cloths" composed of reed or camelid fur cords suspended from a waist cord (*faldellin*). No coffins were used; the bodies were simply interred in the sand, wrapped in reed mats

Fig. 4.61. Chinchorro mummy; SM, ca. 2000 BC. The Chinchorro mummies are extended bodies. Almost all subsequent ancient populations in northern Chile show some degree of flexion. Mummy no. Mo 1–6, T-22. (Photo by ACA.)

(Fig. 4.62). Found on the body or between reed mat layers were marine hunting artifacts such as harpoon foreshafts (Fig. 4.63) and bone or cactus needles, camelid or sea lion rib with a handle used to pry shellfish off rocks (chope), small segments of camelid and sea lion hide with attached fur, as well as feathered pelican skins and a small mesh net to transport collected shellfish (Fig. 4.64). Soft tissue preservation, including viscera was quite good in these bodies and about two-thirds of them had abundant, unstyled hair (Fig. 4.65). A small fraction revealed intentional cranial deformation but no evidence of cranial trepanation. Their health status appeared to have been at least satisfactory, with several showing changes suggesting estimated age-at-death in excess of 60 years. A moderate amount of evidence for interpersonal violence was apparent. About 25% had healed skull fractures. One adult male had been killed by two penetrating stab wounds of the chest, probably inflicted with harpoons (Fig. 4.66). That body also displayed unhealed skull fractures and facial lacerations. However, no patterns suggestive of battle death were identified.

The anthropogenically mummified Chinchorro bodies provide evidence of an astonishing degree of complexity that rivals that of Egyptian methods. In their most complex form they flayed the body, defleshed the skeleton completely and disarticulated it (absence of cut-marks in many suggests possible spontaneous decay of soft tissue), then reassembled the skeleton, reinforced the joints and stabilized major body parts with sticks (Fig. 3.14). The muscle and other soft tissue forms were simulated with mud and vegetals, after which they replaced the skin (or substituted animal hide for it) and sewed it up. A wig of adult hair and a clay mask over the face completed the restoration (Figs. 4.67 and 4.68), after which the body surface was painted with black manganese paint (5000–3000 BC) or with red iron oxide paint (3000–2000 BC). Less complex forms involved desiccation of the trunk soft tissues by placing hot embers into the eviscerated body cavities. At least half of the anthropogenically prepared bodies were those of children (Fig. 4.69). The end result had the appearance of a splinted doll (Fig. 4.70). Indeed, evidence that an occasional mask had been repainted led to a suggestion that these bodies may have been propped up in the sand for use in some type of ritual (M. Allison, 1987, personal communication). The effort involved in their preparation suggests it must have had profound meaning for the group. Considering that the oldest of these (about 5000 BC) was prepared at least 2000 years before the first pharaoh's body was effectively desiccated anthropogenically, their achievement becomes even more impressive. One wonders how Grafton Elliot Smith would have viewed this evidence of antiquity (he speculated that mummification originated in Egypt and the practice subsequently diffused

Fig. 4.62. Chinchorro mummies; SM, ca. 2000 BC.
Whether spontaneous or anthropogenic, these are almost all wrapped in mats made of totora reeds and buried without containers. Mummy no. Mo 1–6, T-22. (Photo by ACA.)

throughout the world, including Peru – see Chapter 1). The oldest body (about 7000 BC), that of Acha Man, is spontaneously mummified. The next oldest, about 5000 BC, was anthropogenically mummified in the complex manner. Progressive simplification occurred over time until about 2000 BC. Thereafter, little effort was made to preserve the body beyond encasing it in some kind of sand–cement mixture (Fig. 4.71), after which period only spontaneously mummified bodies have been found.

Many details of great interest can not be included in this abbreviated discussion, but the interested reader is referred to the Spanish article by Allison *et al.* (1984) that represents the primary scientific description of these methods, the Spanish book by Muñoz-Ovalle *et al.* (1993), documenting the body of Acha Man, the English description of Acha Man (and six other mummies) by Aufderheide *et al.* (1993) and especially the English book by Arriaza (1995c).

The origin of the Chinchorro people remains controversial. Until the nearly 9000-year-old, spontaneously mummified body of Acha Man was found, it appeared that the Chinchorros had burst upon the archaeological stage with a fully developed, complex form of mummification already intact. This, naturally, led to the suggestion that the evolution of this practice must have occurred elsewhere. Now that a spontaneously mummified body 2000 years older than the previously studied bodies has been identified (but no bodies between 7000 BC and about 5000 BC) it must be acknowledged that the 2000 year interval is time enough to evolve a local mummification method. An

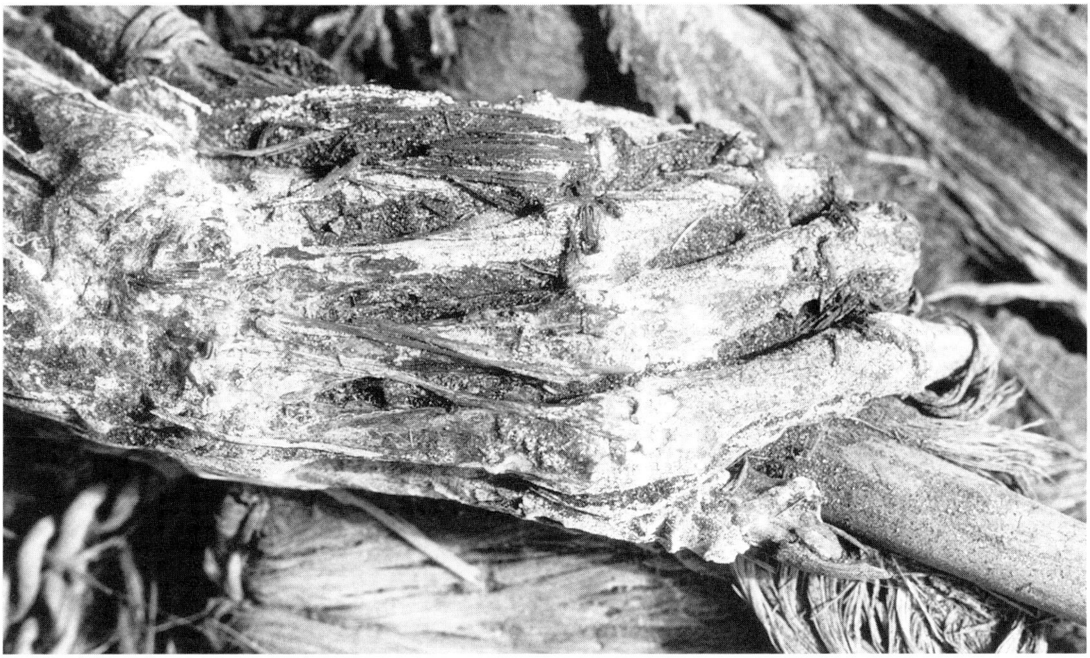

Fig. 4.63. Mummified Chinchorro hand grasping harpoon foreshaft; SM, ca. 2000 BC.
Artifacts are frequently bound to the hands of Chinchorro mummies. Mummy no. Mo 1–6, T-2. (Photo by ACA.)

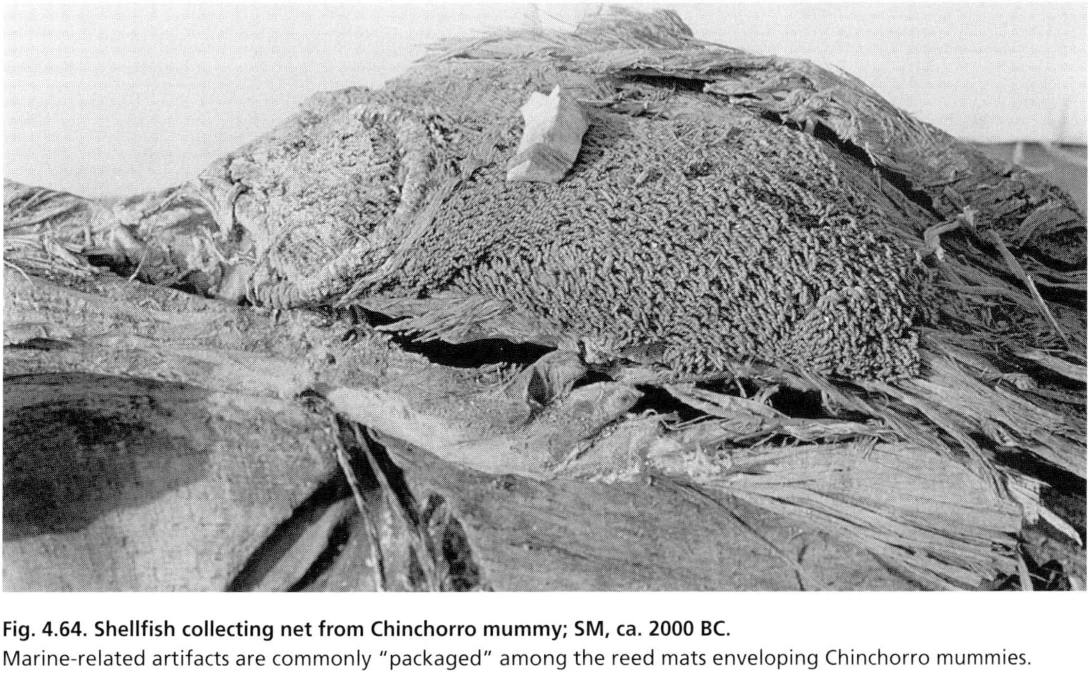

Fig. 4.64. Shellfish collecting net from Chinchorro mummy; SM, ca. 2000 BC.
Marine-related artifacts are commonly "packaged" among the reed mats enveloping Chinchorro mummies. Mummy no. Mo 1–6, T-53. (Photo by ACA.)

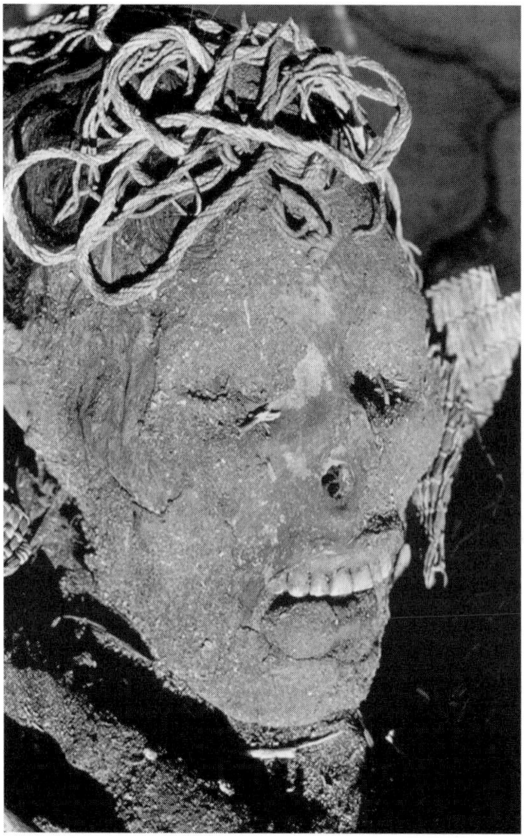

Fig. 4.65. Excellent soft tissue and hair preservation; SM, ca. 2000 BC.
Soft tissue preservation, especially hair, is often excellent in spontaneously mummified Chinchorro bodies. Mummy no. Mo 1–6, T-22. (Photo by ACA.)

Fig. 4.66. Penetrating chest wounds; SM, ca. 2000 BC.
Two chest wounds penetrated the thorax of this adult male Chinchorro mummy, one inflicted from the dorsal aspect and the other from the ventral side. Two harpoon foreshafts (artifacts from other mummies' wrappings) just fit comfortably into these wounds. Mummy no. Mo 1–6, T-18. (Photo by ACA.)

alternative hypothesis postulates that the Chinchorros migrated across the top of the Andes from the jungle on the eastern slopes. This is derived from suggestions of contact with the highlands in the form of artifacts accompanying the body such as camelid hide fragments, feathers from jungle birds and an occasional "snuff kit" (Rivera, 1977). However, these are scarce and some could have been obtained without contact (the highland guanacos commonly followed rivers to the coast during the winter season). Chinchorro coprolites contain ova of fish tapeworm (*Diphyllobothrium pacificum*), indicating that they consumed uncooked fish (Reinhard & Aufderheide, 1990) (Fig.4.72). Whatever the degree of contact between the Chinchorros and their highland neighbors, it was not sufficiently close to share highland foods and so become detectable by methods of chemical dietary reconstruction (Aufderheide, 1996). The problem remains an enigma. About 160 Chinchorro mummies have been retrieved from the coastal sites Mo 1 and Mo 1–6 within Arica and an additional 50–60 from Chinchorro Beach and other sites, giving a total in excess of 200 mummies.

Alto Ramirez (1000 BC to AD 350)

This interesting group represents the first of many subsequent highland populations that moved to the

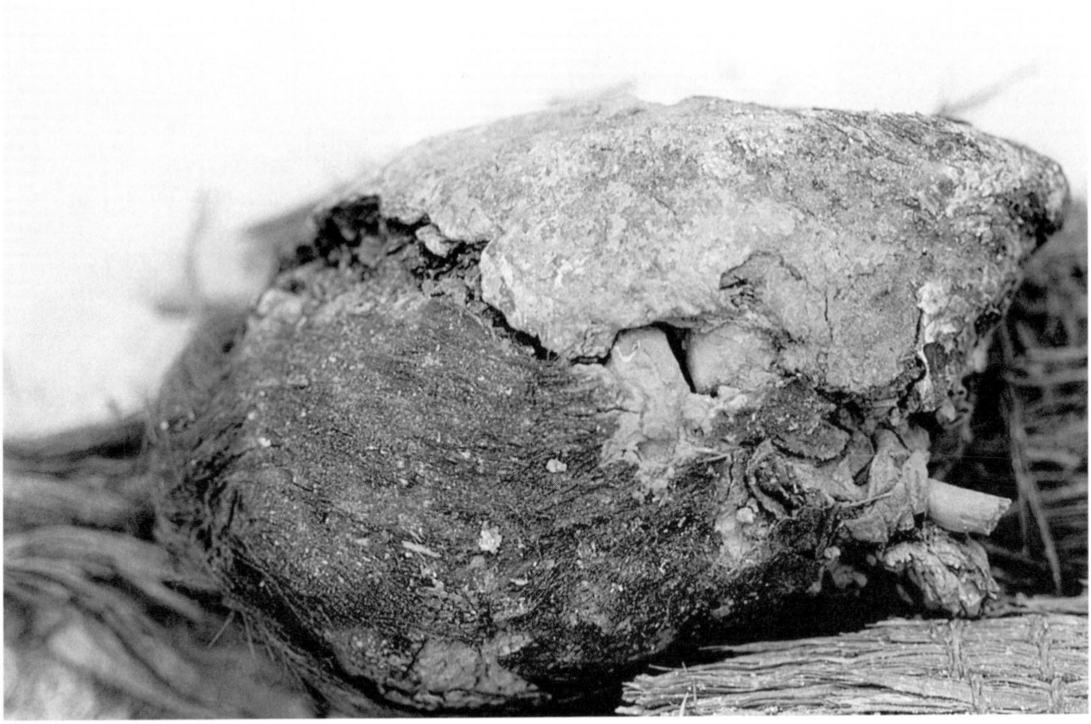

Fig. 4.67. Clay mask over a Chinchorro mummy face; AM, 4120 BC.
A clay mask overlies the facial area of an anthropogenically mummified Chinchorro body. Mummy no. Cam 17, T-1, C-3. (Photo by ACA.)

coast and became integrated with the coastal population. They are recognizable by their ceramics and subsistence articles among the grave goods. The first of these appeared in the Azapa Valley about 1000 BC and the last of them remain recognizable archaeologically about AD 350–400. Their presence is documented on the coast at river mouths and in low valleys from Arica to Iquique. They brought with them the agricultural technology and that of llama and alpaca pastoralism that they had evolved in the highlands. They successfully introduced the coastal Chinchorros to these techniques. Their integration into the populations pre-existing on the coast is epitomized in a small group of five bodies that I had the privilege of examining, excavated from the coastal site of Pisagua (PSG-7). The bodies at this beach site were buried in shallow sand pits on a cliff overlooking the ocean. They were semi-flexed, lying on their right side, similar to the Quiani populations which is a transition group between the coastal Chinchorro extended burials and the maximally flexed bodies of subsequent groups. Several infants were basket burials (Fig. 4.73) Grave goods include plain or decorated ceramics, geometric patterns on woollen textiles, wool turbans (Fig. 4.74), coil-type basketry and other items typical of Alto Ramirez grave goods of Azapa Valley burials. In addition, these same tombs also contained harpoons, fishhooks, chopes, shellfish collection bags of mesh nets and other marine items characteristic of Chinchorro artifacts. Furthermore, their chemically reconstructed diet was found to be predominately marine, as was that of the former Chinchorros. It is possible this was a small coastal colony that specialized in procuring marine foods to supplement the largely agricultural diet of a

Fig. 4.68. Chinchorro clay mask; AM, 4000–3000 BC. A black-painted clay mask is bound into position on an anthropogenically mummified Chinchorro child's body. (Photo by ACA.)

Fig. 4.69. Chinchorro mummy; AM, 1500 BC. The mask originally covering the clay head's facial area has been removed from this Chinchorro child's mummy. The skeleton has been excarnated, reconstructed with mud and vegetals and wrapped. Mummy no. Cam 15D, T-16, C-1. (From Aufderheide *et al.*, 1993. Photo by ACA.)

nearby low valley Alto Ramirez population inland from the coast (Aufderheide *et al.*, 1994). The label "Alto Ramirez" may encompass somewhat heterogeneous subgroups that differ in their burial structures (shaft and tumulus) and body positions (supine and lateral semi flexed) but have common highland origins, ceramics and subsistence modes. Rivera (1995–1996) has subdivided their 1400 year interval at coastal sites into Alto Ramirez I, II & III. At Azapa Valley sites AZ-14, 70, 71 and 115, about 130 Alto Ramirez bodies have been excavated and an additional 20–30 at other coastal sites including Pisagua (PSG-7). All Alto Ramirez bodies are spontaneously mummified.

Tiwanaku (Cabuza) (AD 400–1000)

The circum-Titicaca Tiwanaku people expanded their empire and reached coastal northern Chile by AD 400 and persisted there until the empire collapsed about AD 1000. Burials representing highland Tiwanaku

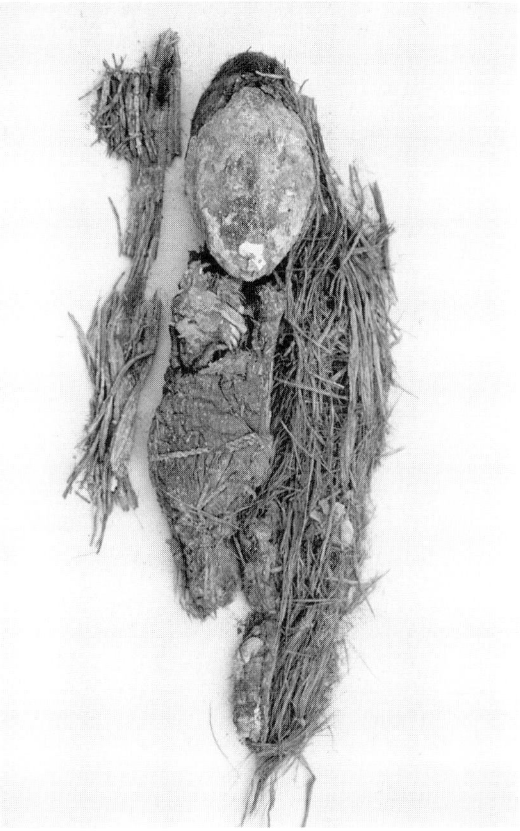

Fig. 4.70. Chinchorro mummy; SM, 4120 BC.
The completely reconstructed Chinchorro children's bodies simulate dolls constructed of natural materials. Mummy no. Cam-17, C-1, T-3. (Photo by ACA.)

administrators in coastal Arica are not easily separated from those of local but Tiwanaku-influenced Aricans. Nevertheless, consensus is that the empire stationed their representatives in Arica, but most Tiwanakoid graves there are those of local coastal people (Goldstein, 1995–1996). Their commonly stone-lined cist tombs contain spontaneously mummified, maximally flexed bodies (Fig. 4.75) that are accompanied by abundant Tiwanaku-style artifacts including distinctive ceramics, and especially spectacular polychrome textile patterns. At least 125 spontaneously mummified Cabuza bodies have been recovered from Azapa Valley sites near Arica labeled AZ-6, 17, 141 and 143.

Maitas Chiribaya (AD 1100–1300)

This group becomes recognizable archaeologically shortly after the time of the Tiwanaku empire collapse and probably represents a coalescence of regional groups. Their ceramics, for example, differ only minimally from those of the large excavation in the Osmore Valley at Ilo, Peru, described above. Most of the other mortuary features of this group in Arica are very similar or identical to those described for the Chiribaya of southern Peru. For that reason they will not be repeated here and the interested reader is referred to that section for details. Suffice it to say that more than 100 maximally flexed, spontaneously mummified bodies of this predominantly agriculture-based subsistence group were excavated, examined and reported from Azapa Valley sites AZ-71, 76, 140, and 141 (Fig. 4.76). All but 13 of these came from AZ-140. Although there are outstanding exceptions, the general level of soft tissue preservation is distinctly less for this group than in the previous ones, perhaps because at least some were buried in ambient space within a small, stone-lined cist.

San Miguel and Gentilar (AD 1300–1450)

These two are additional archaeologically recognizable groups that differentiated further in the period of regional development following the fall of the Tiwanaku empire. They differ from Maitas Chiribaya people largely in the high technological quality of their subsistence items, ceramics and textiles. They have carried agricultural subsistence to the maximum seen in this region but still always included consumption of marine foods to a significant extent. About 50 maximally flexed, spontaneously mummified bodies were recovered from Azapa Valley sites labeled AZ-6, 8, 71, 75, 76, 140, PLM-4 and San Lorenzo.

Inca (AD 1450–1550)

The cultural impact of the Inca empire is recognizable in the graves on the north Chilean coast (Fig. 4.77).

Fig. 4.71. Plaster-covered Chinchorro mummy; SM, ca. 2000 BC.
Just prior to discontinuing anthropogenic mummification practices, the Chinchorro people merely covered the body with a mixture of sand plus a bonding agent. Soft tissue preservation in these is poor. Mummy no. Mo 1–6, T-13. (Photo by ACA.)

These probably took the form of local chiefdoms. This area provides little evidence of a significant physical presence of Cuzco Incas, though a few local administrators could easily be overlooked archaeologically. On the northern Chile coast, excavation revealed the presence of a small, marine-oriented band at the mouth of the Camarones River about 80 kilometers south of Arica. Inca textiles and ceramics in these graves are mixed with characteristic local marine-related artifacts. The frequently patched textiles suggest a low socioeconomic level of this group. Burial patterns are characterized by simple shaft burials of spontaneously-mummified, maximally flexed bodies wrapped in a single layer of low quality cloth. About 30 such bodies have been recovered from this site labeled Cam-9 (Fig. 4.78). Chemical analysis suggests many of these had accumulated substantial amounts of arsenic in their tissues secondary to drinking from the Camarones River, whose high arsenic content threatens even modern populations in this area (Allison *et al.*, 1995–1996; Figueroa-Tagle, 2000).

Colonial mummies (AD 1550–1830)

Scattered bodies from the Colonial Period have been reported. In the Arica area of the north Chilean coast a group of 15 individuals were examined during a salvage excavation adjacent to a colonial church (site

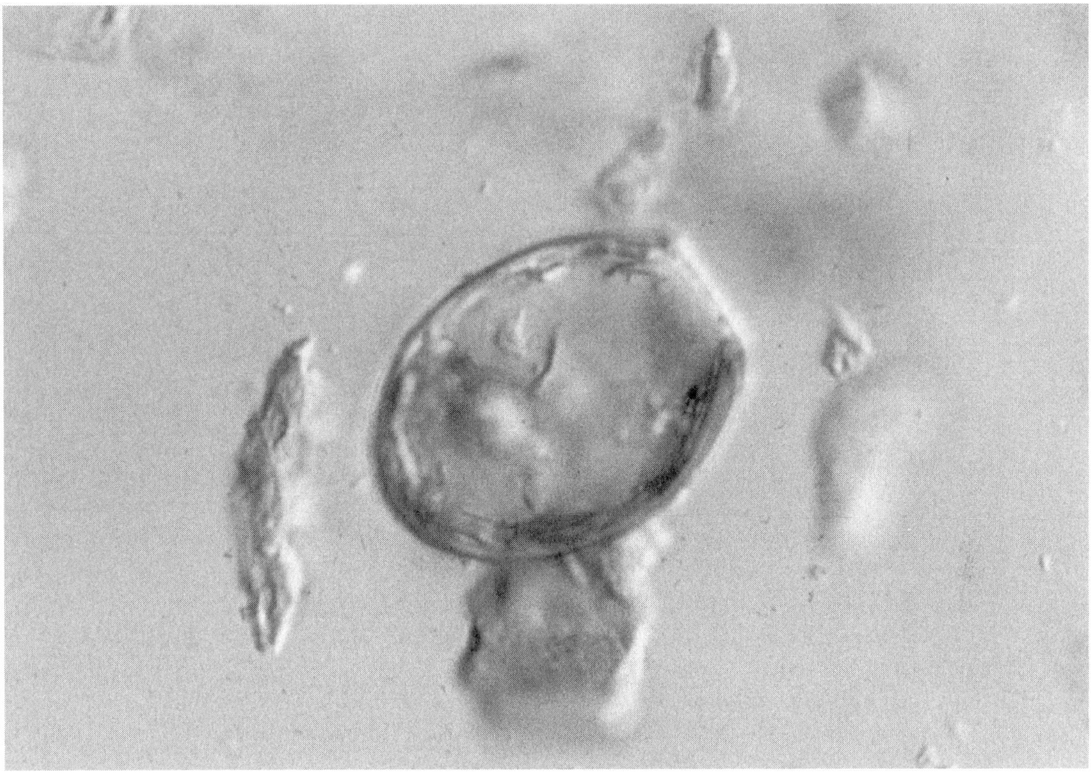

Fig. 4.72. Fish tapeworm ovum in Chinchorro coprolite; SM, ca. 2000 BC.
Finding of the ova of *Diphyllobothrium pacificum* in several Chinchorro coprolites from the Mo 1–6 group by Karl Reinhard (University of Nebraska, Lincoln) indicated the Chinchorro consumption of uncooked fish. (Photo by Karl Reinhard.)

AZ-142). They were in an extended position and skeletal analysis suggested both unmixed as well as mixed Caucasian, Negroid and Native South American features. Bodies had become mummified spontaneously and the generally poor soft tissue preservation was probably the consequence of irrigation agriculture as well as the practice of pouring water on the soil surface overlying the bodies adjacent to a church for the purpose of using it for recreational dancing.

Results of studies on these various groups from the Atacama Desert's coastal area are reported principally in Spanish language journals but English accounts include Allison *et al.*, 1979; Allison, 1985; Aufderheide, *et al.*, 1993; Arriaza, 1995c.

Summary of Arica area mummies

The coastal area of northern Chile was settled about 7020 BC (dated shell midden) by the Chinchorro people, who maintained a predominantly maritime subsistence for more than 5000 years. During the earlier phases they practiced a very complex form of anthropogenic mummification that was simplified over time until it ceased about 1500 BC. All mummification subsequent to that date in all cultural groups was the product of spontaneous mummification. About 1000 BC, highland migrants (Alto Ramirez) arrived on the coast and settled there, introducing agriculture and pastoralism to the coastal mariners. The Bolivian highland culture

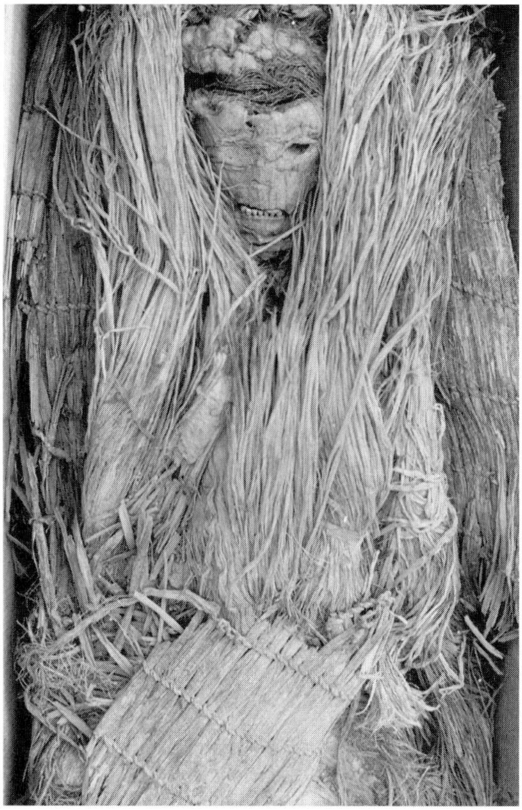

Fig. 4.73. Alto Ramirez basket burial; SM, 1000 BC. The spontaneously mummified body of this infant from an Alto Ramirez population from Pisagua, northern Chile, was buried at a sandy site, covered with dry grass. Mummy no. Psg-7, T-726 A. (Photo by ACA.)

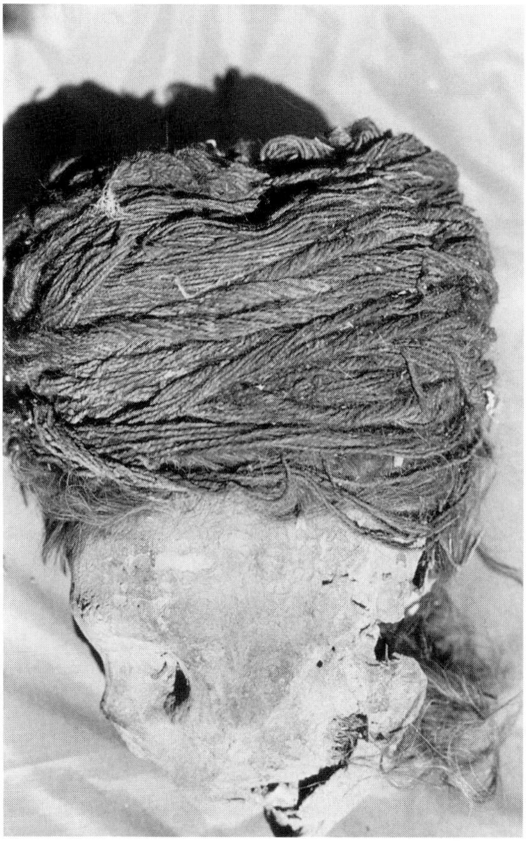

Fig. 4.74. Wool cord turban; SM, 1000 BC. The adult mummies of the Alto Ramirez population frequently reveal a turban of wool cords of llama or alpaca fur. Mummy no. Cam 15D, T-U2. (Photo by ACA.)

Tiwanaku (locally Cabuza) expanded to coastal northern Chile about AD 400 and dominated the region until its empire collapsed about AD 1000. Subsequently scattered groups coalesced into the definable archaeological entities of Maitas Chiribaya (AD 1100–1300), San Miguel (AD 1300–1400), Gentilar (AD 1400–1450) and finally the Incas. The latter arrived on Chile's northern coast about AD 1450 and influenced the area until the Spanish Conquest about AD 1534.

Excavation, examination and reporting of at least 1000 bodies from the region of the Azapa Valley at Arica to Antofagasta in extreme northern Chile can be documented easily from the cemeteries of these principal groups. An additional equal number can be estimated with confidence from multiple, smaller burial areas. Fewer than 10% of these are curated today in mummified form, though their disarticulated skeletal tissues are preserved. While this may seem shocking, the reality of the circumstances under which these studies were carried out is grim indeed. Most of these sites came to official attention as a result of illegal looting. In this region it became apparent quickly that,

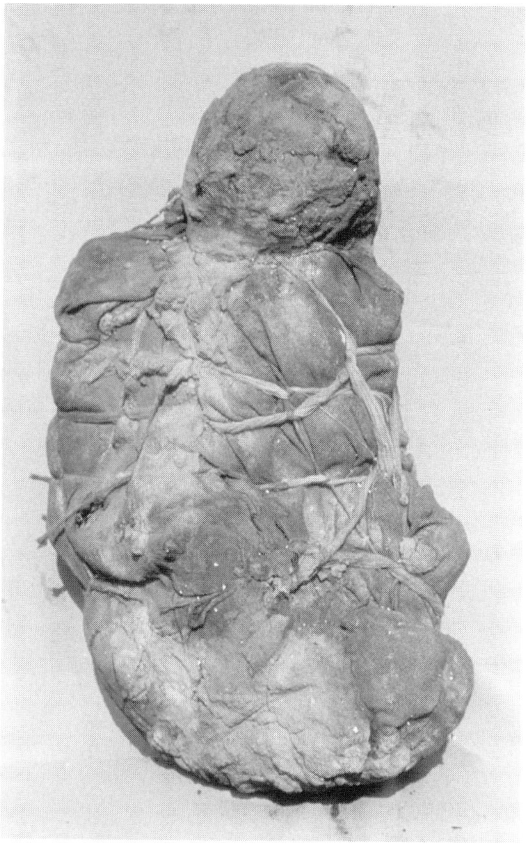

Fig. 4.75. Cabuza (Tiwanaku) funerary bundle; SM, AD 800–1000.
Northern Chile's coastal burials during the Tiwanaku Period consisted of maximally flexed body buried in a sitting position, often in stone-lined cists. Mummy no. AZ-141, T-43. (Photo by ACA.)

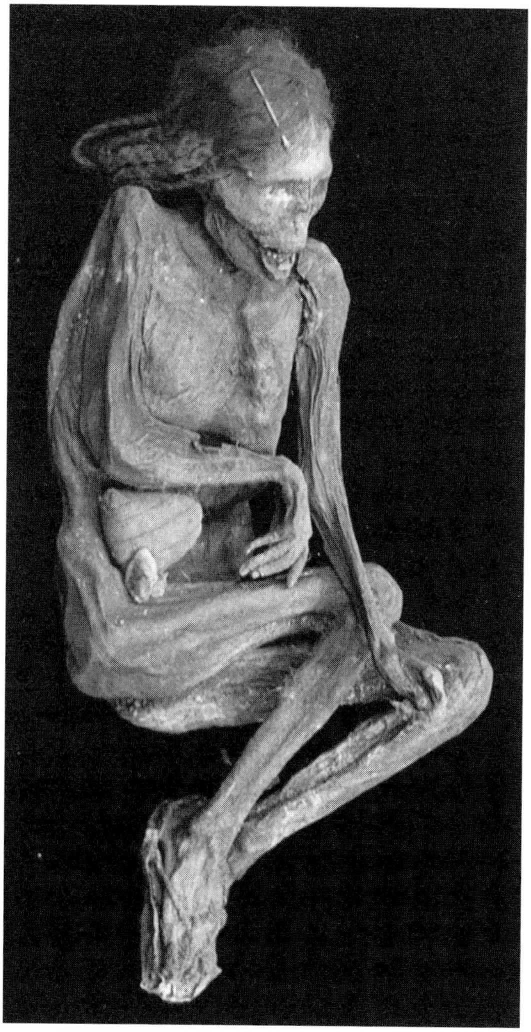

Fig. 4.76. Northern Chile mummy of Regional Development Period; SM, ca. AD 1250.
Coastal burials in northern Chile about AD 1250 are most commonly members of the Maitas Chiribaya population. Mummy no. IQU-95, T-6. (Photo by ACA courtesy of Cora Moragas.)

once looting was initiated, the entire site would be stripped and the mummies destroyed in the field within several months, because funding for continuous official site protection simply was not available. I know of no museum today that could, nor would even wish to attempt to, build the facilities to house and conserve at least 1500 mummified bodies excavated over about two decades, and train and permanently fund the personnel to provide for their perpetual care. During this period the affected institutions were operating under profoundly restricted fiscal conditions. The staff members that excavated them carefully, examined them in detail and continue to report results of their scientific studies are to be congratulated and admired for their contributions.

Fig. 4.77. Inca Period mummy; SM, ca. AD 1450.
Inca Period burials in northern Chile (AD 1450) are accompanied by artifacts relating to hunting (arrows), weaving (spindles) and agriculture (ceramics, maize). (Photo by ACA.)

Fig. 4.78. Inca Period northern Chile mummy; SM, ca. AD 1450.
A group of coastal burials of the Camarones-9 population at the mouth of the Camarones Valley had many Inca artifacts. Mummy no. Cam-9, T-14. (Photo by ACA.)

Mummies from San Pedro de Atacama, Chile

The community of San Pedro de Atacama is located at about 23° 00′ S and 68° 10′ W at an altitude of 2436 meters above sea level, adjacent to several large salt deposits (*salars*) that mark the remains of ancient lakes. At an altitude roughly midway between sea level and the highlands, it is an oasis in the midst of the hyper-arid Atacama Desert. Bodies interred in the soil there quite regularly become spontaneously mummified.

Arriving at this diminutive, remote community as a missionary in 1953, the Jesuit priest R.P. Gustavo Le Paige soon became entranced with the opportunity to study the regions' past by excavating its burials. When he died there in 1980 he had excavated hundreds of tombs, many of which contained human mummies, collected over 100000 artifacts, established a relationship with the nearest university and built a museum and archaeological laboratory. Unfortunately his archives of these items are not ideally documented. The archaeologists Agustin Llagostera, Lautaro Nuñez, physical anthropologist Maria Antoinetta Junqueira Costa and others have continued and expanded these studies and splendidly archived their excavated material. At this time probably more than 3000 tombs have been studied.

San Pedro de Atacama is a strategically positioned oasis. It found itself on an east–west route between the coast and the highlands and also one of the major

Fig. 4.79. "Snuff kit".
Many of the burials at San Pedro de Atacama dating to the Tiwanaku Period contain wood trays designed to pulverize hallucinogenic plants. Photo by ACA is that of a modern replica of such a tray. Note that the carved individual is holding a "trophy head".

north-south routes paralleling the coast. This crossroads location inevitably led to the development of trade activities there in antiquity. Among the many burial grounds of this area, some of which reach back to the Archaic Period, are those of Quitor (about AD 400–700) and Coyo (AD 700–1000). These are contemporary with the expansion, peak and finally collapse of the highland circum-Titicaca Tiwanaku empire. The tombs of these two major burial areas, each of which is subdivided into numerous smaller cemeteries, contain artifacts that quite clearly closely link this community with the Tiwanaku culture. In addition to characteristic Tiwanaku ceramics and textiles, several hundred "snuff boards" have been found in these tombs. These are small wooden trays about 16–18 cm long by 5–7 cm wide. On about two-thirds of their length the surface has been recessed to provide a convenient surface to grind dried plants containing hallucinogenic alkaloids into powder (Fig. 4.79). This can be "snuffed" (aspirated) into the nose with a "straw" made from a hollow bird leg bone. The remaining surface is usually ornamented with carvings of distinctive Tiwanaku themes (Llagostera *et al.*, 1988; Torres *et al.*, 1991). Since San Pedro was a major trade center, it is of great interest whether the abundance of Tiwanaku grave goods (greatest in Coyo) represents a dominating physical presence of Tiwanaku colonists or whether it reflects a prominent but more indirect influence. The question remains unresolved, though the latter currently is more commonly favored.

During my last visit there in 1985, about 50 intact mummies were located in the museum's storage building where the arid climate is quite favorable for soft tissue preservation, though too dry for ideal textile fiber conservation. All preserved bodies, of course, were spontaneously mummified as a consequence of the climatic and soil features discussed above. One of these is the mummified body of a young adult female whose well-preserved features are so strikingly attractive that local residents refer to her proudly as "Miss Chile" (Fig. 4.80).

Mummies from Iquique

This small coastal community was a thriving port at the turn of the twentieth century. Built on a narrow ledge at the foot of a steeply sloping cliff, it served as a major shipping port for the copper and nitrate mined from sites several kilometers inland. Dramatic reductions in mining activity and therefore in shipping commerce have impacted profoundly on the community's economy. During its earlier years, however, an archaeological sequence was established for the region that paralleled that described for the Arica area quite closely. During my visits in 1990 and 1995, a local museum was displaying five bodies from the

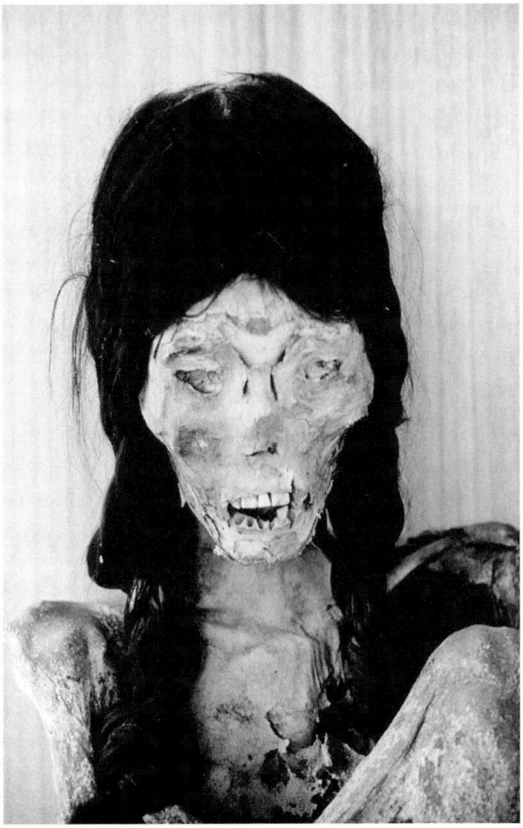

Fig. 4.80. "Miss Chile"; SM, ca. AD 1000.
This well-preserved Tiwanaku Period, spontaneously mummified adult female body has been embraced proudly by the local population, who have dubbed her "Miss Chile". Body is curated in the Museo "R.P. Gustavo Le Paige S.J." in San Pedro de Atacama, Chile. (Photo by ACA.)

Fig. 4.81. Late Tiwanaku Period burial near Iquique, Chile; SM, ca. AD 1000.
Multiple lytic areas in long bones, ribs and vertebrae of this burial at the Las Verdes near-coastal site indicates death by bacterial septicemia. Mummy no. Iqu-95, T-2. (Photo by ACA courtesy of Cora Moragas.)

Chinchorro culture (excavated from the nearby Patillo site), including a small carved wooden model of a mummy found on one of the bodies. In 1990, 15 Alto Ramirez mummies were excavated by a team under the direction of Dr Mario Rivera from a cliff-side site at Pisagua, and transported to Iquique, where I had the privilege of examining them as described above. Seven additional bodies, similarly spontaneously mummified, were made available by Dr Cora Moragas, Director at the Museo Regional, Iquique, from a salvage archaeology project near Iquique in 1995. Most of these proved to be from the period of Regional Development at the Bajo Molle site, dating from AD 1115 to 1310. One of these, from the Los Verdes site, died of septicemia, manifested by multiple, scattered, foci of osteomyelitis, probably secondary to a dental abscess (Fig. 4.81).

Mummies in Santiago museums

Most of the mummies housed in the National Museum were excavated from sites at Chile's northern latitudes. They include some mummies of Chinchorro and subsequent cultural groups excavated by Dr Max Uhle and some by Junius Bird. Others have been excavated more recently. The mummy now housed at the Museo Nacional de Historia Natural that has probably aroused the most public interest is that of the child mountain sacrifice "El Plomo" described above in the section devoted to such mummies in Peru.

Ancient Andean coca-leaf chewing practices

The custom of chewing cocaine leaves was well established at the time of the sixteenth century Spanish conquest of the New World. While its motivation was occasionally medicinal (for altitude sickness, febrile illness or alleviation of hunger), most commonly it functioned as a usually highly stylized ritual of social bonding. It still serves that purpose today. It was also commonly employed as a religious offering (Fig. 4.82). The cocaine in the few leaves chewed on such occasions was probably too little to produce the substantial mood alteration with which we are familiar today when a concentrated cocaine dose extracted from many leaves is injected intravenously for hedonistic purposes. However, as that small amount was absorbed and distributed by the blood in ancient Andeans, it passed through the hair follicles and entered the emerging hair shafts. In this dry environment the molecules of cocaine and its metabolite benzoylecgonine became stabilized. Larry Cartmell, M.D., of Ada, Oklahoma, who is a research affiliate of the Paleobiology Laboratory in Duluth, Minnesota, adapted modern, forensic, sensitive radioimmunoassay (RIA) and gas chromatography mass spectrometry (GC/MS) methodology to the substrate of mummy hair, which is commonly preserved in spontaneously mummified bodies (Fig. 4.83). He found that his methods effectively detected the presence of cocaine in the hair of ancient mummies. He then tested many samples of hair from the Chinchorro people as well as members of the cultural groups that succeeded them. Results indicated that cocaine was absent from all Chinchorro samples. However, when the first highland migrants (Alto Ramirez culture) arrived at coastal sites about 1000 BC, cocaine was detected in several hair samples of this group. Upon the subsequent arrival of the highland Tiwanaku-related people, locally known as the Cabuza culture, hair samples from a large fraction of that population were found to have cocaine as did samples from subsequent coastal cultures. Since infants less than weaning age revealed the same frequency of positive

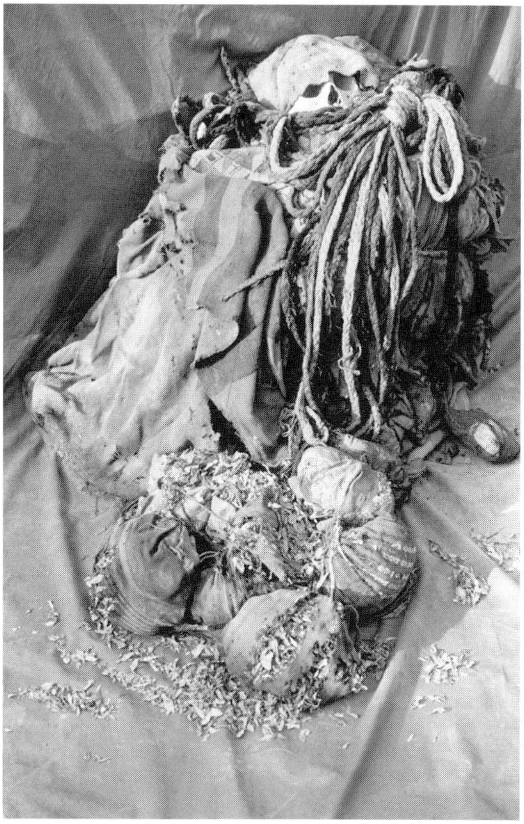

Fig. 4.82. Coca-leaf bag offerings from a Chiribaya mummy; SM, ca. AD 1050–1250.
Coca leaves were common offerings in coastal southern Peru and northern Chile burials after about AD 400. These took the form of oral coca leaf quids or bags stuffed with coca leaves. Shown here is a mummy of the Chiribaya population, Osmore Valley, Peru. (Mummy prepared by Sonia Guillen, photographed by ACA.)

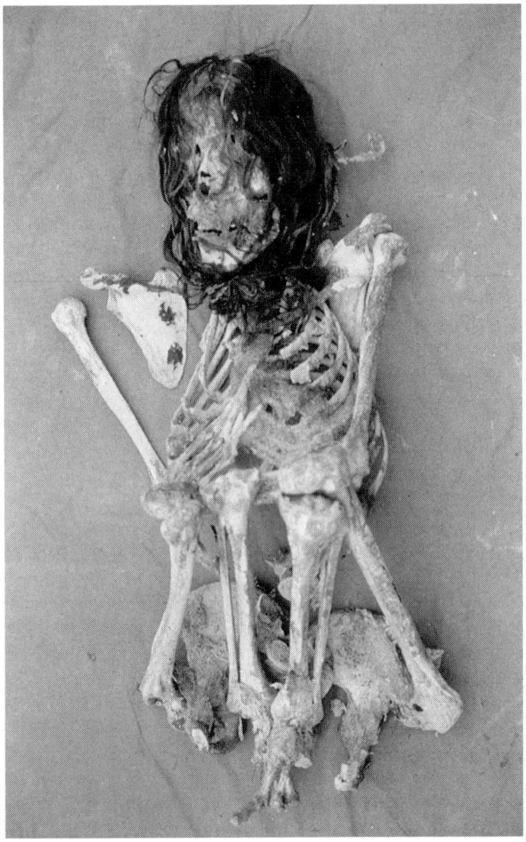

Fig. 4.83. Skeleton with hair; SM, ca. AD 1050–1250. Hair is commonly the last soft tissue to disappear after death. Hence, barely articulated skeletons, but with abundant hair are not uncommon in Atacama Desert burials. Mummy no. 70, Chiribaya Alta. (Photo by ACA.)

tests as their mothers, it appears the chemical was transmitted transplacentally and by nursing, though the small amounts detected are not likely to have had significant physiological effects (Cartmell *et al.*, 1991).

These findings can be translated into some behavioral predictions. This ritual of social bonding was so highly valued that, after conquest, neither the Church nor the state could extinguish it. Indeed, it persists today in the highlands, still used for such purposes. The total absence of evidence of coca-leaf chewing among the Chinchorros strongly suggests that they had not established such close relationships with highland groups. The findings do not exclude all contact, but certainly limit the degree of association. Interestingly, the study also found no statistical correlation between the presence of coca leaves in the tombs (in bags or in the oral cavity) and cocaine in the hair sample (the latter is evidence of antemortem consumption). This suggests that those burying these bodies placed the coca in the tombs as a spiritual offering, and that its presence in the tomb can not be used to predict its antemortem use by that specific tomb occupant; only the presence of cocaine in the hair or other body tissues can predict that.

Another, less direct line of evidence evolves from the findings by Odin Langsjoen, D.D.S., who is also a research affiliate of my Paleobiology Laboratory. He compared the degree of antemortem tooth loss (caused by advanced periodontitis) between the Chinchorros and the later Maitas Chiribaya culture. He found that the Maitas Chiribaya (who chewed coca leaves regularly) were more than 24 times as apt to lose posterior teeth from this cause before death than were the Chinchorros who did not chew coca leaves (Langsjoen, 1996). This devastating effect on the gingiva resulting in posterior tooth loss among the ancient people was then proved to be associated with chewing coca leaves by a study on living modern Peruvian and Bolivian highlanders (Indriati, 1998; Indriati & Buikstra, 2001). Again, the absence of such an effect on Chinchorro dentition supports the conclusion that they did not chew coca leaves.

An additional implication can be derived from this study. The coca plant is native to the eastern side of the Andes. To reach the western coast the plant must be transported first from the eastern side up to the altiplano, then down the western slopes to the coast to reach residents there. The fact that the practice was unknown to (or at least not indulged by) the Chinchorros would seem to diminish the probability that the eastern jungle represents the Chinchorros' locale of origin.

Mummies from Argentina

Mummies from the Museo de Arqueología y Antropología, La Plata

The high rainfall east of the Andean sierra is not conducive to spontaneous mummification. Consequently most mummies that are displayed in museums of such locations became mummified in some other areas. The museum at La Plata near Buenos Aires in Argentina is no exception. During my visit there in 1985, 11 mummies were displayed. Of these, four were from Egypt or Nubia and two from the Atacama Desert (San Pedro de Atacama) on the western slopes, one was from Ancón, Peru, and another from Tiwanaku in Bolivia. An extended, spontaneously mummified body wrapped in animal (goat) hides lying in a bed of reeds had been obtained from Tenerife, Canary Islands and represented a member of that island's aboriginal Guanche population (see Mummies from Spain, below). Only two bodies (both spontaneously mummified) had been found in Argentina. One was a nearly skeletonized body with ligamentous fragments attached to some of the bones. It was found in the extreme south of Argentina near Calafate, Lago Argentino (50° 12′ S, 72° 00′ W). The other was found near San Blas just south of Buenos Aires. This body was a member of a group that usually allowed the soft tissues to decay, then painted the bones in bright colors. Except for the Egyptian, Guanche and Nubian bodies, the others were all spontaneously mummified bodies.

Eva Perón mummy

The events relating to the corpse of Maria Eva Duarte de Perón is a superb example of the power that can be transferred by the cadaver of a revered leader to the individual or group that possesses it. While her husband, President Juan Perón, controlled the Argentinian army and the politics of that country between 1943 and 1955, it was his wife Evita that

Fig. 4.84. Pedro Ara.
The Spanish pathologist Pedro Ara was commissioned by the Argentinian president Juan Peron to embalm Perón's wife Eva. (Photo from Ara, 1996, with permission by Cristina Ara de Mackey.)

became endeared to Argentina's labor movement. She showered the workers with unprecedented benefits, to which they responded by their blind loyalty to both her and her husband, closing their eyes to reports of the Peróns' plundering of the state's treasury. When she died of cancer in 1952, the funeral cortège was surrounded by hundreds of thousands of her supporters.

In anticipation of her death, a distinguished Spanish pathologist, Dr Pedro Ara (Fig. 4.84), was retained to embalm her body. Dr Ara had established a prestigious reputation for effective, innovative techniques for postmortem soft tissue preservation. Eva had barely

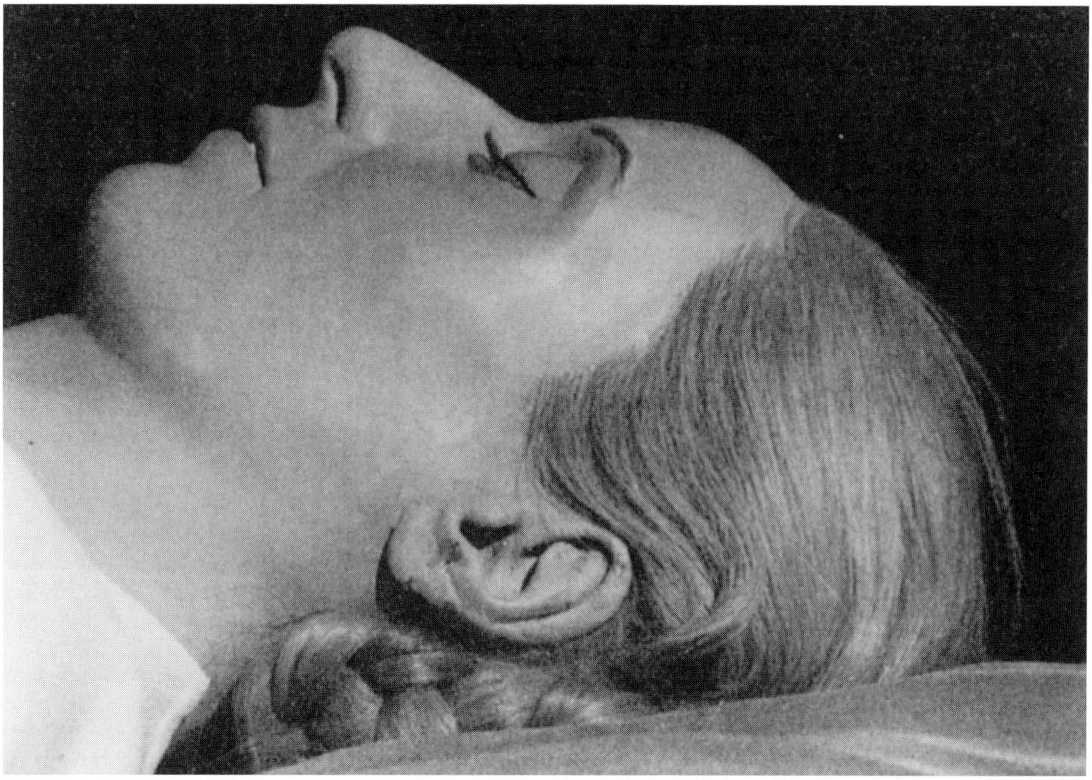

Fig. 4.85. Eva Perón; SM, AD 1952.
The popular wife of Argentinian president, Juan Perón, was affectionately called "Evita" by an adoring public. Her body was embalmed by Pedro Ara (Fig. 4.84), and survived considerable travel, multiple burials and exhumations as well as lying in state. (Photo courtesy as in Fig. 4.84.)

breathed her last, when Dr Ara was summoned and allowed to carry out his initial procedure alone in the room. As she had requested, her body was then transferred to the headquarters of the General Confederation of Labor where it lay in state briefly. Dr Ara never revealed the nature of his embalming process, though he continued to carry out further procedures during the next several months. The outpouring of national reverence for her body (Fig. 4.85) by the general public, especially Argentinian labor organizations, was unnerving to Juan Perón's enemies, and her body was moved several times for security reasons.

Without her guidance and participation, however, Perón found it increasingly difficult to maintain his power base. In 1955, three years after her death, an army general (Aramburu) led a successful coup against the government and Juan Perón fled to Spain in exile. Aramburu distrusted the labor organizations' still active identification with their memory of Eva Perón. Initially he transferred her body out of their headquarters and hid it. Labor protests became increasingly militant and they organized what could be termed a "cult of Evita", demanding to view her enshrined body. Fearful that their recovery of her body would lead to an insurrection, General Aramburu arranged to smuggle her body out of the country, shipping it to Italy under the assumed name of an expatriate being returned for burial there in 1957.

Rumors that "Evita lives" finally led to Aramburu's assassination by terrorists in 1969 and continuing political turmoil resulted in the restoration of Juan Perón to the presidency in 1973. Two years earlier, however, the Peronistas had learned the location of Eva's body, exhumed it and smuggled it to Madrid where her corpse joined Juan Perón and his new wife Isabelita. Upon opening the casket, her body was found to be unchanged during the 19 year interval since her death, except for superficial soiling. Dr Ara carried out corrective measures, restoring the body to it previous condition.

When Juan Perón and Isabelita returned to Argentina in 1973, they left Eva in Madrid. Without Eva, however, Juan Perón first faltered, then died less than a year later. Isabelita managed to succeed him as president, but soon realized her influence was eroding. In a desperate effort to restore her dwindling power, she brought the body of Eva back from Spain in a chartered plane. She displayed the bodies of both Juan and Eva (whose superbly preserved condition still pleased Dr Ara) lying in state, side by side. However, the Argentinian financial and political chaos had sapped the magic from the Peróns' corpses. Another military coup deposed Isabelita, and Evita's body found permanent rest in a glass-topped coffin within a steel vault located in the Duarte family's stone mausoleum, which is equipped with a burglar alarm (Barnes, 1978; Ara, 1996).

Mummies at the Museo Etnográfico "J.B. Ambrosetti" in Buenos Aires

The mummy collection at this museum is substantial. During a renovation project about a decade ago 53 mummified bodies were curated there. About half of these had been collected prior to 1920 and the remainder in the subsequent two decades. Inadequate conservation methods before 1990 had resulted in the deterioration and loss of a significant number of mummies. In 1990 the Faculdad de Filosofía y Letras at the University of Buenos Aires undertook a major rescue and conservation program that resulted in restoration and good preservation of the remainder. This includes a few anthropogenically mummified Egyptian bodies and five crania without bodies. The remainder are spontaneously mummified bodies from sites similar to those in the La Plata museum, such as northern Chile, Peru, Bolivia and a few from Argentina (San Juan, Jujuy and Pampa Grande provinces) (Faculdad de Filosofía y Letras, 1992).

In addition to these, several of the Inca mountain human sacrifice bodies were found on Argentinian peaks but discussion of these is included for convenience in the section Mummies from Peru, above.

Mummies from Brazil

Fundação Oswaldo Cruz, Rio de Janeiro, Brazil

This laboratory is very well known because of the staff's pioneering efforts in the field of paleoparasitology. They have established the presence of a large number of parasites in ancient populations by studying coprolites, many of which were taken directly from the intestines of mummified human remains. Of considerable interest has been their identification of hookworm in coprolites of a Brazilian mummy (Ferreira *et al.*, 1983) and fish tapeworm (*Diphyllobothrium pacificum*) in coprolites of the Chinchorro people, an ancient group of mariners from the coast of northern Chile (Araújo *et al.*, 1983). They have carried out an experimental program that established comparisons of ova measurements in ancient and modern samples.

MUMMIES OF EUROPE

Mummies from Spain

Canary Island mummies (*Guanches*)

The Canary Islands lie just off the northwestern corner of Africa opposite Morocco at about 24° 32′ N and 16° 07′ W. The island Fuerteventura is only about 120 kilometers from the African coast. Seven major islands are inhabited, all but two of which have volcanic, mountainous terrain. Tenerife's volcano ("Teide")

Fig. 4.86. Guanche cave on Tenerife.
It was in hidden caves like this one that the Guanche stored their mummified dead. (Photo by Steven Matzner.)

rises to 3718 meters and was active as recently as the nineteenth century, when a lava flow buried part of a seaside community. One of Teide's shoulders forms an island-long spine, whose slopes are riddled with lava-formed caves (Fig. 4.86), many of which served as burial sites. Moisture from northeast trade winds establishes a lush, semitropical climate on Tenerife's northern face but the southern slopes are so water-deprived they resemble parts of the Sahara Desert. The reliability of these trade winds was sufficiently well known in the fifteenth century that Columbus chose the Canary Islands as his departure site in his search for a western marine route to India.

Both linguistic and archaeological studies link Tenerife's aboriginal population (Guanches) with the Berbers – a pastoral group located in the Atlas mountain range that parallels the southern Mediterranean coast. However, we lack hard data about their arrival on the islands, though local legend suggests they represent a transplanted colony of rebels placed there during the second century BC by the Romans. Earliest radiocarbon dates of surviving mummies, though, reach back only to about AD 400. Spain spent much of the fifteenth century fighting the natives, conquering the islands one by one, completing its supremacy during the last decade of that century in bloody battles commemorated by such place names as "massacre valley" and "ambush site".

Following their victory the conquistadors and the various subsequent island visitors explored the burial caves and, in addition to many skeletonized bodies, they found a large number of mummified bodies. Many of the latter had not been buried but laid on wooden planks resting on rocks and some were even propped up in a nearly vertical position. In Tenerife's sixteenth century chronicles we find much conflicting information about the Guanches' mummification practices. Unfortunately, hundreds of these mummified bodies were taken away, most of them for curiosity's sake. Except for a few in scattered museums, the location of these today remains unknown, if they exist at all. The fate of those aboriginals who survived the military assault was not a happy one either. Many were eventually absorbed into the island's ever-increasing Spanish population, but a surprising number successfully fled the island or were enslaved, establishing a new life in the New World (Rodríguez-Martín, 1996).

By 1990 even most of the best-preserved mummies that had been collected at Tenerife's archaeological museum had disappeared. In that year I was fortunate enough to be invited to join Dr Rafael González Anton and Dr Conrado Rodríguez-Martín, the museum's director and paleopathologist, respectively, in a study of those mummies remaining in the collection there. We found that many of the 32 mummies were actually totally or mostly skeletons. They were still wrapped in goat skins, several tied to funerary boards (Fig. 4.87).

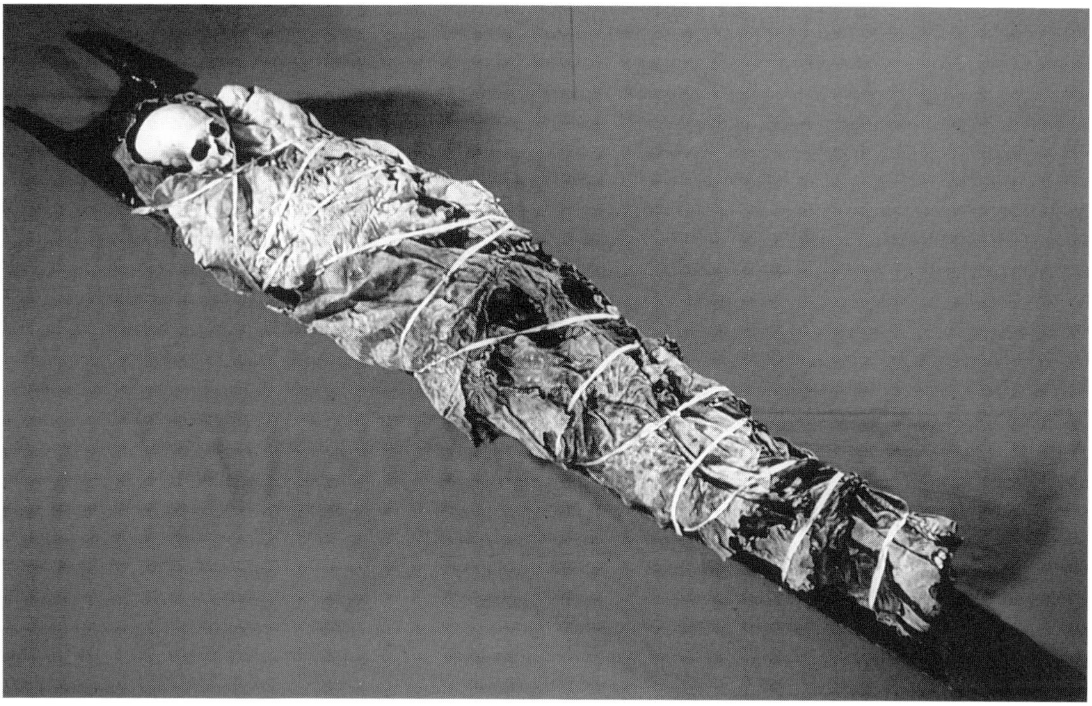

Fig. 4.87. Wrapped Guanche mummy; AM, AD 1375.
The Guanches wrapped their mummified bodies in goat hides and tied them to a funerary board. (Photo courtesy of Rafael González, Director of the Archaeological and Ethnographical Museum of Tenerife.)

Their radiocarbon dates ranged from AD 400 to 1400. Most were adults (Fig. 4.88), though a few subadults and even a fetus were represented. Two bodies were sufficiently intact to establish that they had been eviscerated and the body cavities stuffed with soil, pine needles, probably goat butter (*gofio*), tree bark, rodent droppings and other botanical, organic and mineral items (Fig. 4.89). The moss *Neckera intermedia* Brid was also found in the abdomen of a Guanche mummy curated at the Redpath Museum at McGill University in Montreal, Canada (Horne & Ireland, 1991). In several others, in which only body parts remained (e.g., pelvis and legs) we found no evidence of intentional, anthropogenic mummification efforts. Fifteen of the 26 in whom sex could be established were male. In three, hair was recovered (Aufderheide *et al.*, 1995a). Chemical dietary reconstruction indicated a predominantly terrestrial meat diet, probably mostly goats and sheep. Differing diet distributions in populations from adjacent valleys suggested that social interactions were carried out principally among residents within the individual valleys (Aufderheide *et al.*, 1995b). A Guanche mummy curated in the museum at Cambridge, England, was made available for our inspection on September 21, 1990. This demonstrated preservation of skin and underlying soft tissues over most of the body. Hair was present on the scalp. A horizontal sutured incision was obvious in the abdomen's right lower quadrant, while a left-sided anterior thoracic incision may represent the modern incision made to gain access to visceral organs. The right hand appears to have been removed and reattached with glue. Numerous incisions had been made in the skin over the right shoulder, neck, right chest and left abdomen (the mummy's back was not exposed for inspection). These incisions represent portals

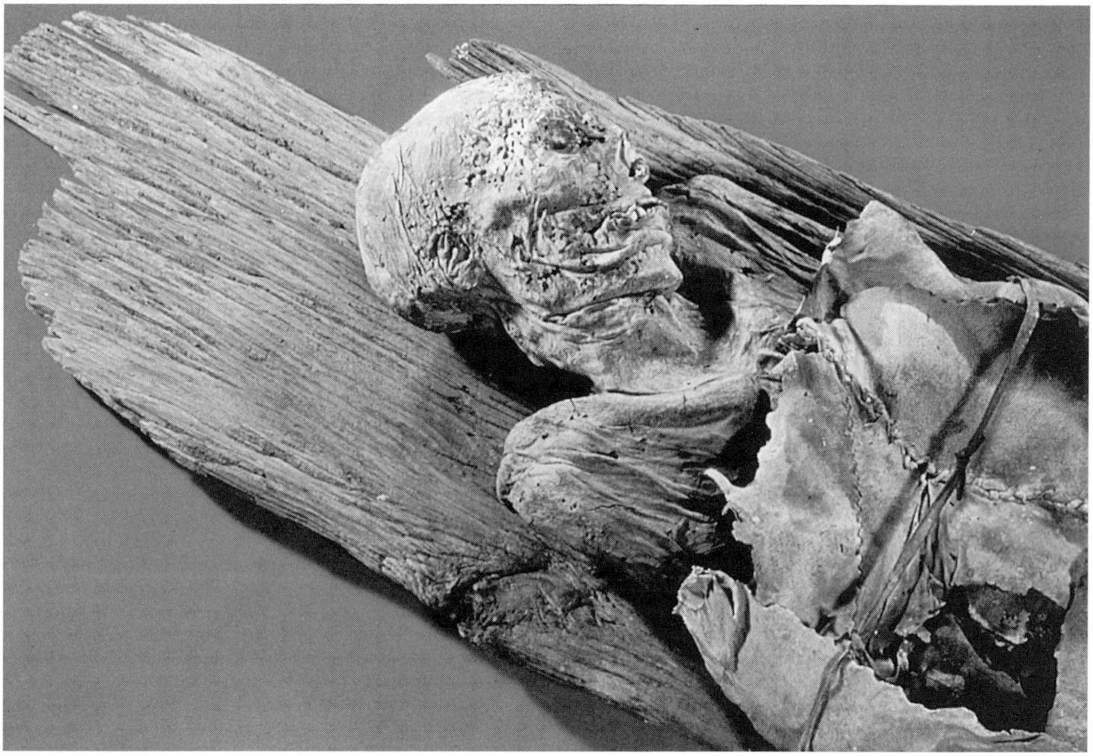

Fig. 4.88. Guanche mummy; AM, AD 870.
It was probably principally the Guanche society's elite members who were mummified. (Photo courtesy of Rafael González, Director of the Archaeological and Ethnographical Museum of Tenerife.)

employed to stuff soil and gravel into the subcutaneous areas. The mandible had sagged sufficiently to reveal an upper right maxillary fracture that has dislocated some teeth. An earlier study by Brothwell *et al.* (1969) had revealed extensive anthracosis. This mummy may be more characteristic of those removed from the Tenerife museum in the Canary Islands (Aufderheide, 1990b). The findings of these and other studies were presented at the First World Congress on Mummy Studies held in 1992 in Puerto de la Cruz on Tenerife, Canary Islands, Spain, and those proceedings were published by the Archaeological and Ethnographical Museum in Santa Cruz on Tenerife.

In addition to the Guanches, Spain has produced only a few other mummies. One of these was the body of a noblewoman, Inez Ruiz de Otálora. She had married the Secretary of the Royal House of Philip II in Valladolid. When she died in 1607 her body was shrouded and buried in the Church of the Franciscans of Valladolid. She was later moved to the family chapel in Mondragón, when the body was observed to have well-preserved soft tissue. The mechanism that resulted in desiccated mummification is not obvious, though clearly it was spontaneous. Skin is intact and some hair remains on the scalp. Over the years a belief arose that the body had therapeutic powers and thus it became an object of worship. Radiographic examination demonstrated renal and biliary lithiasis. The body has not been dissected for study except to retrieve the urinary calculi visualized by X-ray (Etxeberria & Herrasti, 1995) (Fig. 4.90A).

The Basque region of Spain also curates the bodies

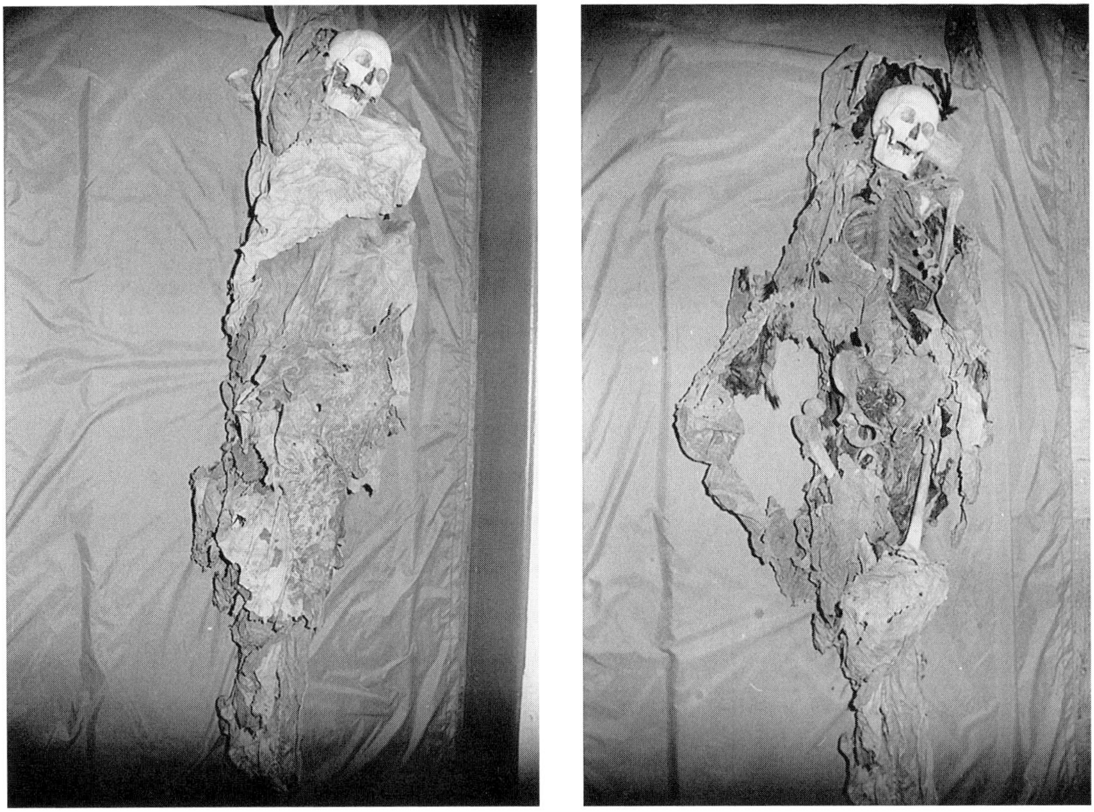

Fig. 4.89. Anthropogenic mummification of Guanche body; AM, AD 1015.
The left panel demonstrates the body wrapped in goat hides. The unwrapped body is skeletonized with abdomen filled with soil and botanicals (right panel). (Photo by ACA.)

of six other saints (Igorputz, Fausto, Felicia, Teresa, Leonor) as well as a male and female body from Durango. The origin of their veneration is indebted in part to the concept in the past that failure of postmortem soft tissue decay reflects divine recognition of the individual. Subsequently cults developed around these bodies, attributing curative powers for various ailments to them, including migraine, fecundity problems and others. All have become mummified spontaneously and vary in soft tissue preservation from poor to excellent. In several, missing body parts can be attributed to their transfer in other churches as relic items. Some have a colorful history, such as St Fausto, who, captured by Saracens and taken to Africa, returned to Spain after converting his master. In addition, the skeletons of four saints (Deodata, Fidel, Columba, Inocente) have been reconstructed using wire mesh and wax to refashion a body form, and stored in the chapels of the cathedral of Pamplona, Navaresse (Etxeberria *et al.*, 1994). Several other venerated but not yet sainted bodies are also preserved in various Basque religious structures (Fig. 4.90B,C,D).

The well-preserved, spontaneously mummified body of a young girl was found in a nineteenth century ossuary in Toledo, Spain. Histological study demonstrated the presence of large numbers of *Trichinella spiralis* in skeletal muscle sections (Gómez-Bellard & Abel-Cortés, 1990).

The spontaneously mummified body of a 5-year-old girl from the late nineteenth or very early twentieth

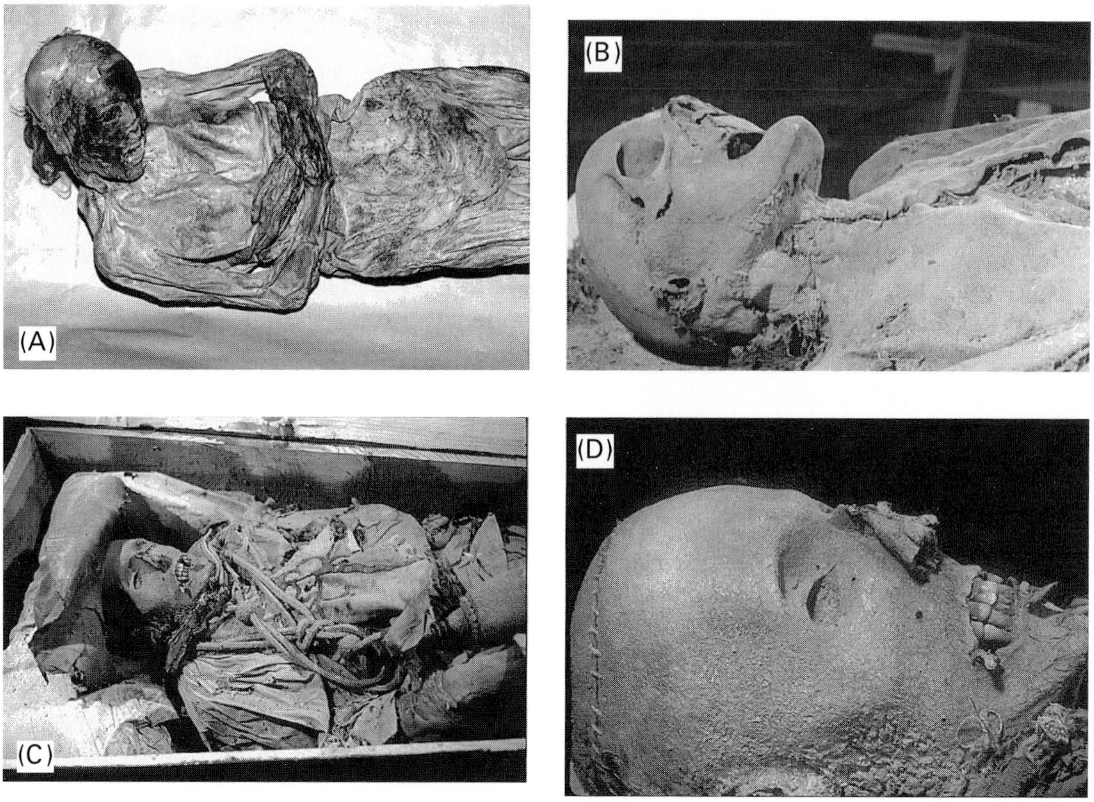

Fig. 4.90. Mummified Basque saints: (A) SM, AD 1607; (B) SM, ca. AD 1650; (C) SM, ca. AD 1720. (A) Inez Ruiz de Otálora; (B) Iñigo Fernandez Velasco; (C, D) Bernardino Fernandez de Velasco. (Photos by and courtesy of Francisco Etxeberria, Asociación Española de Paleopatología.)

century was found secreted in the cupola during restoration of a church in San Lorenzo, Spain, in 1987. Dissection and radiological examination demonstrated Sprengel's deformity of the right shoulder and hemithorax. Right shoulder dislocation was attributed to epilepsy at the time of death. Although the principal abdominal viscera were identifiable, advanced destruction of those organs had occurred prior to complete desiccation (Macias López, 2000).

Hydrocephalic mummy

The body of an intra-arterially embalmed infant, probably some time in the first half of the twentieth century, is believed to have become separated from other teaching display specimens of Barcelona's Pathology Museum during that country's civil war, ultimately coming to attention in the Archaeological Museum of Barcelona. There it was studied by Carvajal *et al.* (1995) by CT and found to be hydrocephalic (Fig. 4.91).

Miscellaneous mummies from Spain

A paleobiological study currently in progress at the Laboratory of Forensic Anthropology and Paleopathology in the medical school at Valencia, Spain, is studying the embalmed bodies of two bishops from the eighteenth century (the cranial cavity contained twigs and branches of juniper trees) and also some eleventh

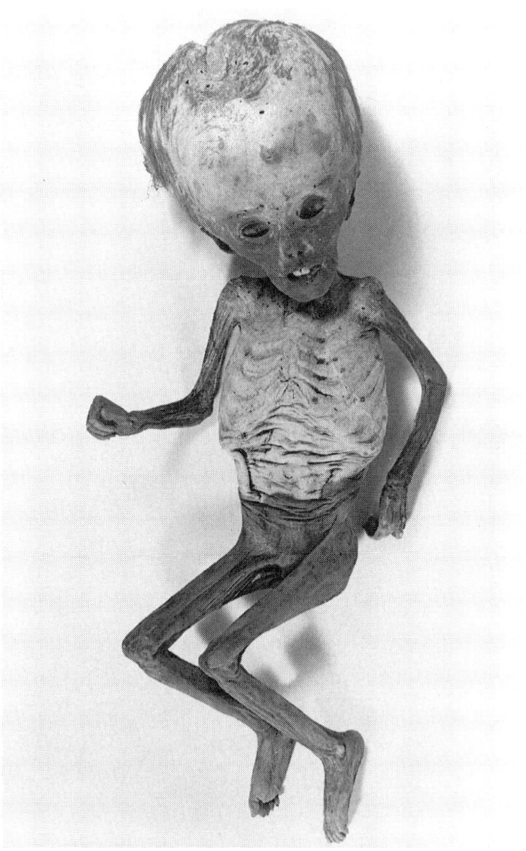

Fig. 4.91. Hydrocephalus; AM, ca. AD 1920.
This spontaneously mummified, hydrocephalic child is from Barcelona, Spain. (Photo courtesy of Domingo Campillo.)

century mummies, but their results have not yet been published (M.M. Feucht, 2000, personal communication).

Mummies from Madrid

In 1866 a Spanish commission returned from South America's west coast with 34 spontaneously mummified bodies excavated from Bolivia and the Atacama Desert. These were poorly conserved, and passed from one administrative unit to another over more than a century until they reached the Laboratory of Forensic Anthropology at the Universidad Complutense of Madrid. In 1990 they were rehabilitated there, photographed and described. At the time of writing (2002) they remain without reported dissection or other forms of laboratory analysis other than radiology and blood type determination (Reverte, 1986).

Mummies from Portugal

Portugal's climate is not particularly conducive to spontaneous mummification, yet this country venerates the mummified remains of at least three of its clergy. Here, as throughout medieval Europe, the failure of the usual postmortem decay process was viewed as a spiritual signal of divine recognition. In the now almost deserted village of Vila Chã da Beira, a small road leads to a chapel housing the spontaneously desiccated body of the priest Reverend (now Saint) Julião. The village of Penafiel, near the shipping center of Porto on Portugal's northern coast proudly preserves the dehydrated body of Saint Tarouca.

Although Isabel, daughter of Peter III of Aragón, was not a native Portuguese, she married its ruler D. Dinis in 1282 and ardently embraced her adopted country. She became so popular that the public attributed many miracles to her following her death. In 1612 the naturally preserved state of her body was confirmed by opening her tomb, and 13 years later she was canonized.

All three of these bodies appear to have been sheltered within the stone religious buildings under conditions that appear to have induced spontaneous mummification.

Mummies from France

John Paul Jones

Few careers demonstrate better the fickle nature with which fate sometimes deals with the lives of the great than that of John Paul Jones. Born in Galloway, Scotland, in 1747, this gardener's son at age 12 years signed as cabin boy, seeking better opportunities on an American-bound freighter. The next 7 years found him advancing his position in the slave trade until, at age 19, he brought his ship into a British port after the captain and mate had died of malaria. By 1775 the American Colonial Congress asked him to create a navy and the following year made him a captain. He moved into high military and political circles after he carried the good news of the colonies' military successes in the revolution to Benjamin Franklin in Paris. His several successful naval battles included the defeat of the British warship *Serapis*, during which he replied to their demand for his surrender with the defiant retort "I have not yet begun to fight" (Horne, 1978).

The end of America's revolutionary war, however, also ended the colonies' perceived need for his services. In the political tumult characterizing efforts to organize themselves as an independent, political state, the government offered Jones little more than a citation, gold medal and cash grant. The latter was not paid until 50 years after his death (Cohen, 1996).

This frustrated commander then sought further opportunities in France where Louis XVI had earlier acknowledged his contributions with a medal and cash, but found no further encouragement there either. Ultimately he responded to Catherine the Great's invitation to build up the Russian navy. There, too, he had stunning victories over the Turks in the Black Sea, but soon again became the innocent pawn in a political tangle. He escaped with his life only through intervention by the French ambassador. In 1792 he returned to Paris, a seriously ill, embittered man of only 45 years of age. Suffering from a chronic respiratory problem (asthma?), he became progressively anorexic, finally jaundiced and edematous and died July 18, 1792, impoverished and spiritually broken. He was buried in Paris with charity funds, though he was placed in a lead coffin to facilitate his return in the event Americans requested it: they did not (Dale,1952; Lasky,1982).

A century passed, as did the tumultuous course of events in both Washington, DC, and France. When the then President Theodore Roosevelt sent General Horace Porter to Paris as America's ambassador, the newly appointed representative of the USA was thoroughly knowledgeable about Jones' contributions and embarrassed by his country's shabby treatment of the father of its navy. He launched a search for his body that was to span a 5 year interval. He scoured French records endlessly, pored over old municipal maps and interviewed civic employees and historians. Gradually the net he cast closed in on his quarry when he discovered Jones had been buried in a public cemetery that had been abandoned in 1840, the surface paved and industrial buildings erected over the unmarked grave. Cemetery maps narrowed the possibilities to a focal area. Porter bought the property with his own funds and on February 3, 1905, hired workmen to dig five shafts beneath the buildings in different directions. Five lead coffins were retrieved; the third was the one he had sought for the previous 5 years. On April 7 it was opened (Porter, 1905).

The coffin was indeed composed of lead; the overhanging lid had been soldered and the seal was broken with difficulty. A strong odor of alcohol was detected immediately. A hole 2 cm in diameter was noted in the lid, also sealed with solder, that probably served as an adit for the introduction of the alcohol and exhaust for air before being sealed. Removal of the lid indicated that all internal space not occupied by the body was stuffed with straw and hay. Tinfoil enveloped part of the body, removal of which disclosed that the body was clothed but not with his uniform. His nose was flattened by the lid, but the body's soft tissues had been "wonderfully preserved" (Fig. 4.92). After being photographed, the facial features were found to resemble closely those of a bust that had been prepared by the sculptor Houdon during Jones' life. This, together with his initials on the clothing, satisfied Porter, the anthropologists and the pathologists he had summoned that identification had been established. Dr J. Capitan then proceeded to perform an autopsy.

He opened the thorax posteriorly and found additional alcohol still within it. Left pleural adhesions overlay an area of focal fibrosis in the left lower lobe but later microscopic examination failed to identify acid-fast bacilli, and the lesion was finally interpreted as the residuum of an episode of nontuberculous bronchopneumonia. The heart, stomach, intestine, liver and

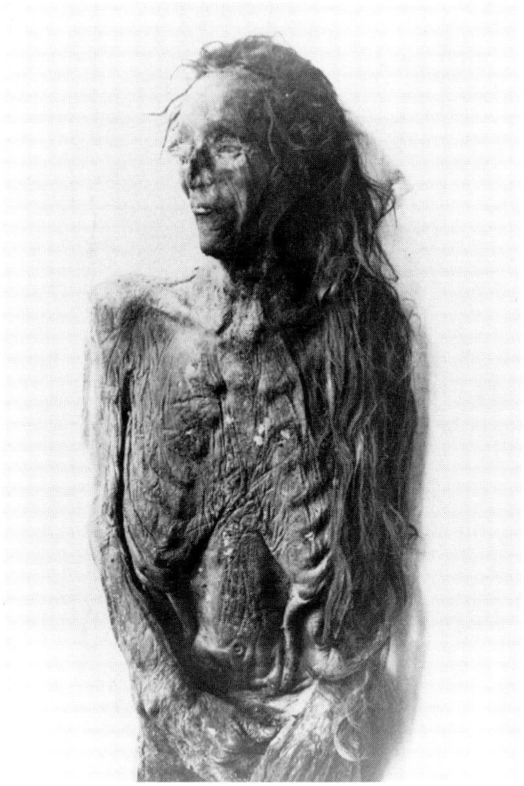

Fig. 4.92. Mummy of Admiral John Paul Jones; AM, AD 1792.
The lead casket in which the body was buried was filled with sand and alcohol. Body exhumed about 113 years after initial burial. (Photo courtesy of Horace P. Mende, Kusnacht, Switzerland and the United States Naval Historical Center Negative no. NH78750.)

gallbladder were without evident pathology, though contracted (by the alcohol?). The spleen was thought disproportionately enlarged, though its maximum length was estimated as only 6–7 cm. Kidneys, however, were small and hard. Their histology revealed thick-walled arteries, glomerular sclerosis and interstitial fibrosis. Dr Capitan's interpretation was that John Paul Jones died of what today we would call end-stage renal disease, but then termed chronic interstitial nephritis (Macfarlan, 1915; Eckert, 1982).

Porter arranged for a formal public carriage transfer of Jones' body to a waiting American warship convoy that returned Jones to the USA in July, 1905. There more celebratory services were held and his public contributions lauded. A special shrine bearing his body was created in the chapel of the US Naval Academy, where it rests today.

Although America's medical profession has heralded this as the oldest autopsy (113 years after death) ever performed (Macfarlan, 1915), Dr Pettigrew (1834) might challenge that claim with his Egyptian mummy dissections, as could several that have examined Andean mummies.

Miscellaneous mummies from France

Within the church of St Thomas in Strasbourg the 600-year-old, coffined, spontaneously mummified bodies of Count de Nassau and his daughter were found. That of the count appeared to have retained all of its principal soft tissues, while that of his daughter demonstrated decay of principally the facial features (Gannal, 1838/1840:25).

Although France enjoys a rich tradition of Celtic culture, most of the biological evidence of human remains is skeletal. The statuary art and literature places such a prominent emphasis on the head that this cultural feature is often termed a "cult of the head" (Ross, 1967). Though both Diodorus Siculus (ca. 50 BC/1947, V: 29) and Strabo (ca. 10 AD/1977, IV: 4, 5) describe a Celtic practice of embalming the heads of their enemies in cedar oil, preserving and displaying them proudly in wooden containers, I have been unable to find a modern archaeological report to document that custom.

Mummies from Germany

In addition to the many bog bodies found in Germany (discussed below) the spontaneously mummified remains of about 25 bodies were found in a church vault at an old monastery ("Kreuzberg") situated on a high hill in the village of Popplesdorf near Bonn. No doubt both the sheltering effect of the church, together with its well-drained hilltop location, must have been a major factor contributing to the mummification process. The bodies are believed to be more than 400 years old (Gannal, 1838/1840:25).

One of Germany's more interesting mummies is that of the nobleman Christian Friedrich von Kahlbutz. He sired 11 legitimate children and 30 illegitimate. The latter were the result of exercising a feudal lord's right of sleeping with a subject's bride the night of their marriage. When the fiancée of one of his shepherds denied Kahlbutz this privilege, the nobleman had the shepherd killed. Accused of assassination, he swore innocence and predicted that his own body would not decay postmortem if he was not innocent. About a century after his death in 1702 his burial urn was opened during reconstruction of the Kamdehl church in Prussia. His dark-red, desiccated, shrivelled, spontaneously mummified body was found intact within the urn. Speculation that the fiancée may have poisoned him with arsenic was disproved by chemical analysis. His body was exhibited to the public, eventually even on television. Factors believed to have contributed to mummification are anaerobiasis (the urn had a double wall), the coffin's tannin-containing oak wood, the church's arid atmosphere, absorbent cloth in the coffin and perhaps the effects of some fatal disease. Tissue examination, however, established no definite cause of death. No doubt some still see his mummification as the hand of divine justice (Kleiss, 1995).

Fig. 4.93. Catacomb mummy; SM, ca. AD 1700. Ekkehard Kleiss is holding a spontaneously mummified body from the catacombs beneath St. Michael's Church in Vienna. (Photo courtesy of Ekkehard Kleiss, Merida, Venezuela.)

Mummies from Austria

Austrian catacomb mummies

Between the late sixteenth and eighteenth centuries members of the clergy and the laic elite were buried in catacombs and crypts of various churches in Vienna. Arid conditions there, together with apparently adequate ventilation resulted in spontaneous mummification of many of these. When Kleiss (1975, 1977, 1995) viewed some of these beneath the churches of St Francis and of St Michael he found the full range of soft tissue preservation represented, from mere skeletons to what he termed "nearly perfect mummification" (Fig. 4.93). Apparently the church of St Stephen had similar mummies, but they were no longer being displayed in the 1970s. Kleiss noted neither odor from the bodies nor evidence of rodents in the catacombs.

Fig. 4.94. The "air-dried chaplain"; SM, AD 1746
The spontaneously mummified body of the vicar of St Thomas in Blasenstein is well preserved after two centuries of burial in sandy soil of a Vienna chrchyard. (Photo courtesy of Ekkehard Kleiss, Merida, Venezuela.)

He was particularly impressed with the degree of preservation of a cleric called Reverend Franz Xavier Sidler von Rosenegg, vicar of St Thomas am Blasenstein (later dubbed "the air-dried chaplain"). This man died in 1746 at the age of 37 years and was buried in the sandy soil of the churchyard. During relocation two centuries later his coffin was opened. When it became evident that spontaneous mummification had occurred, the wood coffin was fitted with a glass panel and he was placed on display (Fig. 4.94).

The practice of "catacomb burial" was discontinued in Austria in 1784 when Emperor Joseph II, in his efforts to improve the population's health status, issued several regulations, one of which forbade interment within the city limits of Vienna.

All these "catacomb mummies" are in well-ventilated cirucmstances whose mummification mechanisms are discussed in Chapter 3.

Hallstatt's "salt mummies"

Large areas of salt (sodium chloride) are deposited in certain of Austria's mountainous regions. In the area of Hallstatt the concentration of raw salt ore averages 45%, with some strata having as much as 85%. These deposits were mined even in antiquity, and were also employed as burial sites. At about the beginning of the seventeenth century (AD 1593 and again in 1616) in the area of the Hazel Mountains several tombs became exposed by weather conditions. When the coffins were opened the local people of Hallein and Hallstatt were astonished to discover that the bodies inside had startlingly realistic preservation of the soft tissues. These probably were the products of the hyperarid atmosphere within the salt mines, but the frightened populace insisted on prompt reburial with additional efforts to create effective seals on the new tombs. A similar experience occurred in AD 1734, when the salt-preserved body of a man wearing mountain clothing was exposed. Fear also resulted in his immediate reburial without further study or observations (Kleiss, 1967; Wells, 1980; Bogucki, 1995).

Tyrolean glacier bodies

In the alpine regions of central Europe it is not unusual to find intact or fragmented bodies emerging from glacial ice (Kaufman, 1996b). They are usually the remains of mountaineers who fell into a crevasse many years earlier, became incorporated into the glacier's ice mass before they decayed and were then transported by the movement of the ice down the mountain. Eventually the ice descends to levels low enough to melt, at which time the bodies become exposed and found. Bereuter *et al.* (1996b) describe recovery of the remains of a 62-year-old male and a 28-year-old female whose bodies had been enveloped in glacial ice for 57 and 29 years, respectively. Both cadavers demonstrated evidence of adipocere formation whose composition was investigated by those workers.

Undoubtedly the most well-known glacier mummy is that of "Oetzi" (the Iceman). While Austrian scientists recovered and studied it, a site survey revealed the site to be within Italian territory, so it is discussed in the section Mummies from Italy, below.

Mummies from Switzerland

Several bodies of individuals dating to the medieval, Renaissance and even the Industrial Revolution periods have survived the ravages of time in Switzerland. Each of these appears to have been placed in a burial container that was sheltered within a religious building or elite private mausoleum.

Franciscan friars at Basel

These were all anthropogenically mummified, though the details of the method are not always obvious. One sixteenth century body, now in that order's museum, was placed nude on a bed of wood chips and cloth fragments. Facial tissues had decayed but most of the other soft tissues were found intact. Some red spots on the skin surface proved to be cinnabar (mercury). Dissection including viscera that were exposed via a posterior incision as well as some of the extremities demonstrated skeletal changes suggestive of treponematosis (Kaufmann, 1996a).

Other mummified clerics

One preserved body from the charter house in Basel had died in 1556. Five children's remains from the Church of St Leonard and mummified bodies of three adults from the Church of the Dominican Friars have also been recovered.

Mummies of historical figures

The Baron of Sennwald, Johann Philipp of Hohensax, figured prominently in local history. He died in 1596 and was buried as a protestant. When the church burned, his body was returned to Basel in 1979. Allegedly his nephew had stabbed him to death but when his coffin was opened at time of transfer it was found to be quite well preserved. However, no knife wounds were evident at an autopsy that was then carried out. Desiccated viscera were found *in situ*, and a massive skull fracture was obvious. Though the desiccation mechanism of mummification in this body is clear, how it was brought about is not. It should be noted that a common practice at the time of this person's death was by hanging a body in a windy, well-ventilated area such as a church tower until it dried. A circular impression around this body's neck was consistent with such a cause. The body has been extensively restored and preserved.

An adult female, the Viennese Hapsburg Queen Anna, died there in 1281. She was eviscerated and "embalmed" with some indeterminate material (lye?), wrapped in cerecloth covered by her daily wear and placed in a beechwood coffin. Her grave was opened more than two centuries later (1510) and the body's soft tissues appeared intact, though when it was again opened during a transfer to a monastery at the end of the eighteenth century only skeletal tissues remained (Kaufmann, 1996a).

In addition, Swiss glaciers occasionally release bodies of mountaineers that have been trapped in the ice. Between 1988 and 1992 various body parts of an adult female came to light in this manner near the village of Bergün; she appeared to have died between 200 and 400 years earlier, based on a changing sequence of teeth eruption that had been established for that locality, and on clothing items (Kaufmann, 1996b).

Bog bodies of northern Europe, England and Ireland

One moonlit night about 2000 years ago deep in the heart of what is now called Lindow Moss (Bog) near Manchester, England, members from the highest level of the Celtic priesthood were performing their most serious ritual. Threatened by the advancing military horde from the continent, they felt that only direct intervention by their god would save them and their people from conquest and extinction. Invoking a form of appeal reserved for only the most desperate circumstance, they were substituting a living human for the small animals normally employed as an offering that accompanied their usual, less vital requests. Standing submissively at the consecrated position was a 40-year-old man, stripped of his clerical garments except for a fox-fur arm band, silently displaying his blemish-free body. At the appropriate point in the ceremony a priest standing behind him abruptly struck the sacrificial victim by a blow to the top of the head with a club wielded with sufficient force to fracture the skull and stun him. A second similar blow to the back of his head as he lay on the ground stilled the victim's movements. A second priest quickly placed a cord around the victim's neck and twisted it with a small stick until the tourniquet compressed the neck structures sufficiently to prevent respiration. A few minutes later a third priest slashed the soft tissues of the neck deeply enough to bisect the major blood vessels and collected the flowing blood in a container for an offering to their deity. Completing their exhortations for salvation, they placed the sacrificed individual's body in a watery location where it quickly sank out of sight.

The above is an outline of only one (Ross & Robins, 1989) of several perhaps equally plausible (Briggs, 1995) but differing reconstructions that have been proposed as explanations for the physical findings presented by ancient bodies excavated from the peat bogs of northwestern Europe. They offer fascinating challenges to explain how these water-logged bodies, placed there as long ago as the Neolithic, can retain their soft tissue structures for many centuries and how they acquired the often gruesome wounds they display so frequently. My review of the "bog bodies" phenomenon will start with a brief note about the structure of peat bogs and an overview of the number, distribution and chronology of bodies recovered from such bogs. The chemical details and mechanisms of bog features that result in soft tissue preservation are presented, and the effects of these will be noted here. Physical characteristics of bog bodies are then examined, followed by a brief discussion of the purposes for which bodies were placed in bogs.

Peat bogs

Peat is a term used to describe the accumulation of partly decomposed plants in water. Common adjectives modifying peat describe their principal vegetative components: bog peat (mosses), heath or meadow peat (grasses and sedges), forest or wood peat (trees), and sea peat (seaweeds) (Neilson *et al.*, 1939). Peat bogs are also commonly separated on the basis of their location: highmoor and lowmoor. Lowmoor peat is flushed continuously by groundwater draining higher land. Thus its water contains buffering dissolved salts that result in pH values of about 5.5 to 6.5 (slightly acid), not too different from that of other soils that have no tissue-preserving qualities. Highmoor peat bogs are watered by rainfall and commonly have a lower pH, ranging from 3.2 to 4.5 (Painter, 1994). It is important to note that moss of the genus *Sphagnum* can initiate growth and often subsequently dominate a bog at any time, probably because of its ability to grow in water and soil with low concentrations of nitrogen and phosphorus (Richardson, 1981). This moss grows on the surface at a rate of about 15 mm per year, though compression from its weight reduces the rise of its surface to about 6 mm (Fischer, 1998). Sphagnum bogs are thus called "raised bogs". Virtually all bodies with preserved soft tissues have been found only in Sphagnum-raised lowmoor bogs.

About 1% of the world's surface is occupied by bogs (Richardson, 1981). Although their distribution is global, it is not uniform; for example, about 15% (1 million hectares) of Ireland's surface is peat bogs (Gmelch & Gmelch, 1980). In developed countries extensive drainage has reduced the area occupied by bogs. As a fuel, peat's high caloric value places it intermediate between coal and wood. It has been dried and burned for fuel since antiquity. In modern times it is exploited for that purpose particularly during periods of economic stress such as World War II, and consequently its preserved bodies have been found at greatly increased frequency during such intervals.

Structurally the growth of *Sphagnum* peat bogs is at the surface, but its weight there compresses it so densely that molecular oxygen is limited to a depth of no more than 30–50 cm from the surface; below that level virtually anoxic conditions prevail. Bacterial density of the surface areas is about 10^8 per gram dry weight but in the anoxic area it is profoundly reduced (10^4–10^6).

The preservative feature of peat bogs has traditionally been attributed to these characteristics: the acid pH and the anoxic nature of peat bog's deeper water. However, the pH of most bogs yielding mummified remains averages 5.5 to 6.5, only mildly acidic. Some are as low as 3.2, a value that Painter (1991) points out is the same as is found in wine or yogurt production. Alternatively, he also notes that anaerobic bacteria can destroy human proteins and carbohydrates just as effectively as do aerobic microbes and, while the anoxic layer of peat has a reduced microbial density, it is far from sterile. Neither the anaerobiasis nor the acidity of peat, therefore, can account for its preservation qualities. Because peat had been found to be capable of absorbing fourfold greater quantities of water than does cotton, it has been used by local residents as bandages for centuries. It was natural, therefore, that it was widely believed to contain an antimicrobial substance. As early as 1913 Czapek cited an 1895 publication that Winterstein had isolated such a substance and called it sphagnol. Many competent investigators, however, were unable to confirm his work. Since tannic acid, used by taxidermists to "tan" hides is found naturally in oak-containing bogs, some had attributed the mummies' preservation to this compound until it was noted that tannic acid usually was absent in *Sphagnum* bogs.

We are indebted to a research chemist on the British Museum's staff, Terence Painter (1991, 1994, 1995) and colleagues for combining previous observations (Moore & Bellamy, 1974:122–128) with those of their own investigations and focusing their results on the effects of the compound they call "sphagnan" on human bodies immersed in such bogs that result in soft tissue preservation. The following is my attempt to summarize those contributions. Sphagnan is a polysaccharide component of the cell wall in *Sphagnum* moss and is released when the plant dies. It has a chelating property (removes metal ions from a solution) that is responsible for the decalcification of bones and teeth, an effect very commonly observed in bog mummies. Sphagnan also removes other metal ions (iron, copper, zinc, etc.) that are vital for bacterial growth, as do humic acids (a sphagnan product) and even intact holocellulose (of which sphagnan is a component). This, not the acid pH or the anoxia, is the principal mechanism that suppresses microbial growth in the anoxic layer of peat bogs. Sphagnan also inactivates secreted bacterial enzymes, further neutralizing the effects of the already suppressed bacteria. Sphagnan gradually forms humic acids that are resistant to bacterial attack and that incorporate the peat water's nitrogen content, making it unavailable to bacteria, further inhibiting their growth. Finally, sphagnan attaches to collagen fibers that are principally responsible for maintaining the structure of soft tissues, altering them in a manner which, while chemically different from that produced by tannic acid, has a "tanning" effect that makes the collagen fibers less reactive to putrefactive digestion.

While certain conditions can overwhelm one or more of these mechanisms, the usual ecology of *Sphagnum* peat bogs permits sufficient operation of sphagnan's effects to result in the preservation of animal and human soft tissues that are immersed in them so frequently. Unusual heavy rains, for example, could flood a lowmoor bog area with enough metal ions to exceed its chelation capacity; if the drainage area is agricultural, the nitrogen-binding potential could be swamped. The result of such episodes could be skeletonization of a body or even complete dissolution. In England only 15% of bog bodies are mummified (Brothwell, 1996). The finding of skeletons in bogs that also produce mummies suggests that such episodes may well occur periodically but that their effects are temporary. If so, then overall statistics could suggest that the average age of bog skeletons would be expected to be older than those of the mummies. While our radiocarbon-dated database for bog bodies is still too small to validate such a speculation, several skele-

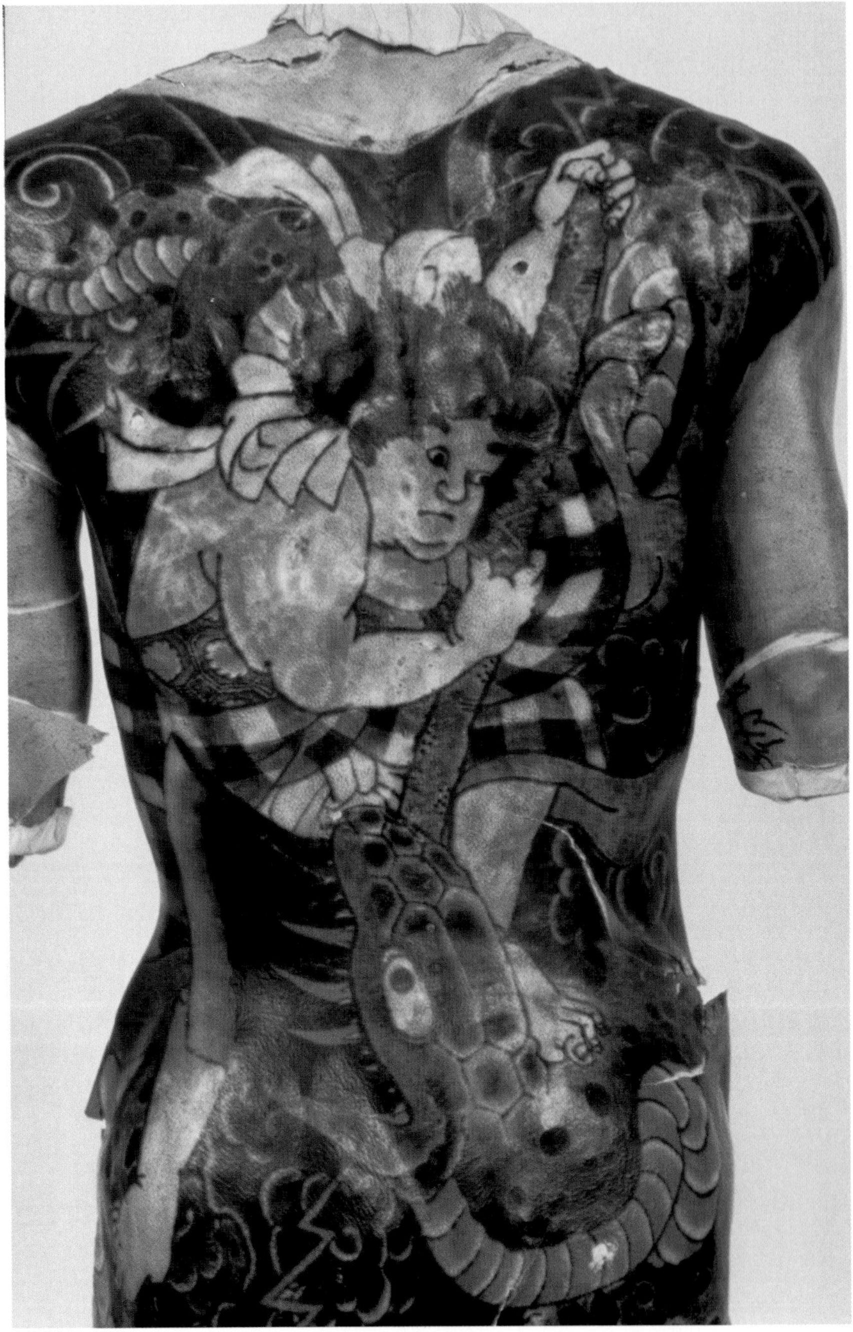

Plate 1. Japanese tattoo; AM, probably nineteenth century.
Although provenience data for this specimen are not available, its kaleidoscopic nature and theme images suggest a Japanese origin. The skin (probably preserved by a form of tanning) is mounted on a wooden form. (Specimen from National Museum of Health and Medicine, courtesy of Paul Sledzik.)

Plate 2. Tollund Man bog body; SM, 220 BC.
Note the flaccid appearance of the extremities due to loss of bone mineral secondary to the leaching of calcium by the chelating effect of sphagnan. (Photo courtesy of Christian Fischer, Silkeborg Museum.)

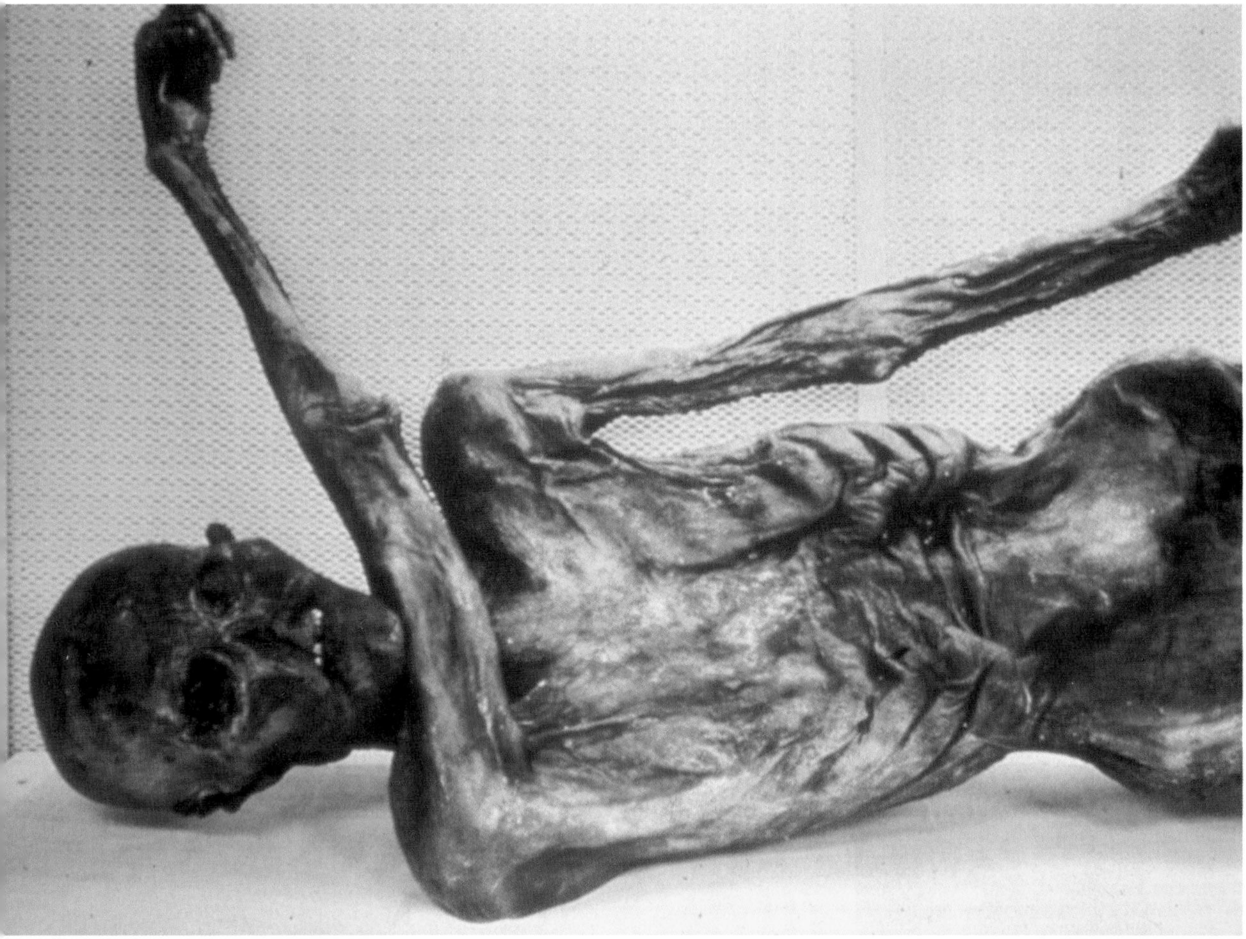

te 4. Mummy of the Tyrolean Iceman.
s spontaneously freeze-dried body, preserved by the glacial ice of the Tyrolean Alps, is more than 5000 years old. (Photo courtesy Luigi Capasso.)

ate 3. The mummified remains of Charles Francis Hall, AD 1871.
ne wooden coffin housing this body was buried at a shallow level on Greenland's west coast. Recurrent summer aws permitted some decay but the underlying permafrost is probably responsible for the partial soft tissue eservation. (Photo courtesy of Chauncey Loomis (2000).)

Plate 5. Chinese jade burial suit; SM, 220 BC.
Although enormous expense and effort were expended to create this suit that enveloped the deceased Chinese ruler Liu Sheng, in the hope of long-term soft tissue preservation, it was all in vain. Only a skeleton was found inside the suit. (Photo courtesy of Robert Harding Picture Library.)

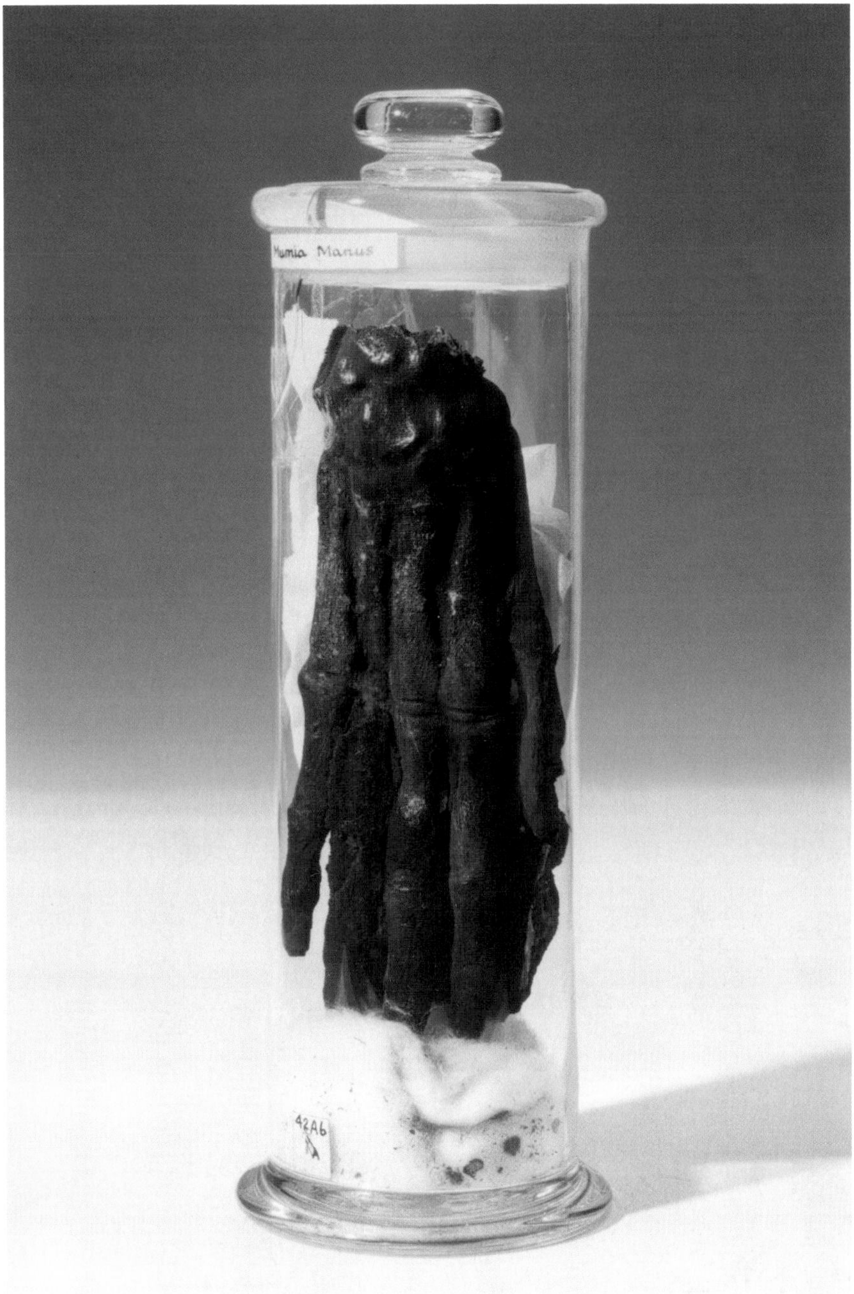

Plate 6. A sixteenth century British pharmaceutical jar.
Body parts from Egyptian mummies were kept in stock by European pharmacists during the early Renaissance Period to respond to physicians who prescribed powdered "mummy" as a medical therapeutic for patients. (Courtesy of Museum of the Royal Pharmaceutical Society of London.)

tons from the Danish bog Bolkilde were so dated to be more than 5000 years old (Bennike *et al.*, 1986).

Preservation features of peat bog mummies

While almost every bog mummy differs from others in some way, many share certain features. Skeletal and dental decalcification is probably the most obvious. This may be so marked that bones and even entire limbs can be grossly deformed by soil pressure on such hypocalcified extremities. Often all enamel has been removed, limiting study of the dentition. The decalcified skull can be deformed by soil pressure in a manner that can be mistaken for antemortem trauma. Sufficient calcium may not remain to produce high quality radiographs; MRI (magnetic resonance imaging) may provide better images for study. However, we have only one report of such an application – Lindow Man – where it produced images that provided no additional information beyond that generated by CT scans. In contrast, the high content of sphagnan-tanned collagen fibers in the dermis and keratin in the hair, finger and toenails results in excellent preservation of these structures. Such effects have enabled fingerprints to be prepared from a bog body of a type and quality indistinguishable from those made from living individuals (Tollund Man: Fischer, 1951). The anatomy of antemortem wounds and lacerations are also well-preserved because of sphagnan's tanning effect. Facial skin integrity preserves the countenance, causing the finders to comment on its appearance by describing the face "as if asleep" or other similar comments. Eyelashes, pubic and axillary hair may all be preserved, though gentle excavation techniques are commonly required to prevent their detachment by simple abrasion. However, the frequently noted presence of "beard stubble" suggesting that the individual had not shaved for several days before death is most commonly the consequence of skin retraction as normally occurs in modern forensic bodies that are unattended for several days after death. In bog mummies the retraction is probably the result of sphagnan's tanning effect on the collagen fibers, a process that shrinks the fibers. Pollen and helminth ova in the stomach and intestine are not altered by bog water, thus validating the results of coprolite analysis.

Fig. 4.95. The Weerdinge couple; SM, 30 BC
This double bog body burial proved to be two men, found in Bourtangermoor Bog in the Netherlands. The body on the right reveals a stab wound with prolapsed intestines. (Photo courtesy of Drents Museum, Assen, Netherlands.)

The viscera do not fare as well as the skin. Structures in which the majority of the volume is composed of epithelial cells, such as the liver and kidney, are commonly reduced in size, deformed by pressure or simply unrecognizable. Supporting structures, with a predominantly collagenous component, such as the pericardial sac and ligaments, are more apt to retain their form. Lung and intestinal wall (but not its lining epithelium) are probably the most commonly preserved, recognizable viscera (Fig. 4.95). Since this pattern is also seen in spontaneously mummified bodies in deserts, it may be occurring in bog bodies because the high content of lysozomal enzymes in epithelial cells enhances rapid postmortem destruction of such cells in the liver and kidney. Not only do fibrocytes have fewer lysosomes, but collagen is also less susceptible to destruction by those enzymes. It is also possible, though never demonstrated, that sphagnan's tanning of the skin impairs the diffusion of sphagnan-rich bog water through the skin, limiting the access of sphagnan to the viscera. The presence of particles of *Sphagnum* in a sample of stomach contents of a bog body has been cited as evidence that the victim drank bog water with her/his last meal (Fischer, 1998). That is certainly possible, but if the individual was alive when entering the bog, it is common for water to be gulped into the stomach in the process of drowning. If a corpse is placed into a bog, the weight of bog water on a body immersed to a depth of 2.5 meters can be expected at times to be sufficient to force at least a small amount of bog water via the mouth and esophagus into the stomach. Adipocere, though not common, has been found in several, including the medieval mummy from Meenybraddan, Ireland (Delaney & O'Floinn, 1995), and the Danish Huldremose Woman (Brothwell, 1996).

The number and chronology of bog mummies

The first record of a bog body find is one from Shalkholz Fen in AD 1640 (Nield, 1986; Brothwell, 1996), followed by two more in 1747 and in 1781 (Gmelch & Gmelch, 1980). The most recent represent World War II soldiers in Russia (Fischer, 1998). Bog bodies are most commonly found while peat is being dug for fuel. Hence, bog body finds tend to coincide with periods of economic stress. The total number of bodies that have been removed from peat bogs will never be known with accuracy. Probably the most major effort expended to identify them was made by Alfred Dieck (rivaled only by that of van der Sanden, 1995), who began by gathering information about German bog bodies but eventually expanded his studies to create a world-wide registry. In 1958 he listed 465 examples, by 1972 he had reached 1354 and when he died in 1986 his total exceeded 1850 (Dieck, 1986; Turner & Briggs, 1986). He has, however, been criticized for excessively liberal criteria for inclusion in these lists, and efforts to find firm documentation for many of them have failed. Fischer (1998) notes that many of the listings are only body parts, and van der Sanden (1995) points out that Dieck never examined a bog body himself, basing many of his entries only on local lore. Nevertheless, while stricter criteria for eligibility for inclusion would certainly reduce the numbers, his reports remain useful to provide perspective. Turner & Briggs (1986:158, Fig. 53) present the distribution in graphic form of the 1354 bodies listed by Dieck in 1972. They include Switzerland 7, Austria 5, Czechoslovakia 9, Poland 6, Greece 2, Denmark 418, Norway 10, Sweden 55, Finland 45, England 57, Scotland 10, Ireland 60, France 5, Belgium 4, Netherlands 49, Germany 602, and Russia 10. These recovered bodies were not uniformly distributed according to their dates of death. All of Scotland's bodies are from the medieval period or later. Of the nine Danish bodies whose dates have been verified by radiocarbon methodology, one is from the Bronze Age and eight from Denmark's Iron Age (about 400 BC to AD 400); if we include dated Danish skeletons we find 18 bodies from the Neolithic Period (Bennike *et al.*, 1986). In Ireland and England, however, the range is from 900 BC to AD 1800 (Brothwell, 1996). Worldwide, the range extends from 8000 BC to about AD 1945 (Fischer, 1998).

Cultural nature of bog mummies

The purpose of placing bodies into bogs has been and remains highly controversial. It would, of course, be naive to expect a common motive to apply to all of the approximately 2000 bog bodies discovered to date. Furthermore, if textual material was interred with these bodies, they did not survive the bog milieu. Hence, we must look to the bodies themselves, their environment, grave goods that did survive and literature relating to mortuary practices to make our judgments. Based on modern experience with documented cases, we can expect some of the recovered bodies to represent accidental drownings. Many of the paths through such bogs are of ancient origin and some have hazardous segments even today. Such bodies, if preserved, would be clothed and without grave goods.

Modern experience also leads us to anticipate that bogs may have been employed to dispose of bodies by murderers. An excellent example is the finding of an isolated head (dubbed Lindow I) in 1984 near the location of the complete body we now know as Lindow Man or Lindow II. When that head was recovered and the police interrogated a local resident whose wife had disappeared mysteriously 20 years earlier, the suspect admitted murdering her, disarticulating the body and hiding the parts in the bog. Even though the subsequent radiocarbon date indicated the head was about 1740 years old, the man was tried and convicted of murder (Turner, 1986, 1995c). It is unlikely that he was the first to have employed a peat bog for that purpose nor was this the only isolated head to have been found in a bog (Garland, 1995).

Local topography may well dictate that bogs be used as burial sites for routine deaths by residents of villages juxtaposed to peat bogs. Such burials would conform to general mortuary practices of a region and thus be quite apt to contain clothing as well as other grave goods. Turner (1995a) describes such burials in Scotland, some of which have historical support.

Superstition may be the basis for a subgroup of atypical bog burials. These are bodies that appear to have been tied to wooden stakes, some of which also have a stake driven through the chest. This has been interpreted as an effort to immobilize the deceased's perceived malevolent spirit, preventing it from wreaking vengeance on the living by symbolically "killing" it. Similar practices were in vogue in New England and other parts of North America as late as the nineteenth century (Sledzik & Bellantoni, 1994). Some of the isolated heads may be products of this activity.

In addition to routine inhumation, accidents and murder victims, several other possible explanations for

Fig. 4.96. The bog body Borremose Man; SM, 700 BC. Note the stout rope around the neck. (Photo courtesy of National Museum of Denmark.)

bog bodies have been offered. Of these, the most difficult to exclude would be that they represent acts of punishment. A small fraction of the studied bodies have showed physical signs of violence, including changes suggestive of death by hanging (Fig. 4.96). The term "hanging" is used here to include both asphyxia by body suspension at the neck from a rope, and asphyxia by strangulation with a cord around the neck used as a tourniquet to compress the trachea (garroting). In some cases lacerations or stab wounds are also present. Those that adhere to the punishment explanation point to the historical use of hanging as a judicial form of execution, to the commonly prone position of such bog bodies (suggestive of social disapproval), and to the fact that children are rarely represented. They also point to the description of the Gauls (Celts in today's France) by the second-century Roman historian Tacitus (98 AD/1981: Chapter 12) that their retribution for military or civilian misbehavior (battle cowardice, certain disapproved sexual acts, treason *et al.*) was to "drown [the guilty] in swampy bogs . . ." (Tacitus 98 AD/1981:106–111; Fischer, 1998). Those critical of this explanation note that it is virtually impossible to determine whether the wounds were inflicted just before or just after death (Briggs, 1995) and also point out that the function of describing Rome's enemies was to glorify Rome. Denigration of Rome's enemies was a part of that process.

The very same physical findings on the bodies described above are also employed by those who postulate that they reflect religious rituals that include human sacrifice (Glob, 1969). These features include (1) fatal or potentially fatal lacerations (commonly of the neck) or stabbing (often the chest), (2) severe blunt blows that inflict fractures of the skull, ribs and/or extremities, and (3) asphyxia by hanging or neck tourniquet. Additional findings may include nudity, the high male/female ratio, hair shaved close to the scalp and often a prone position. Of the 50 English and Irish bog bodies, 19 revealed at least one and commonly several of these features (Turner & Briggs, 1986). They also mobilize support for the concept from ancient customs. They note the common Celtic practice of using watery sites as places for leaving votive offerings to their deities. These include weapons (often military), ornaments, art objects (an inscribed silver cauldron in one case), a chariot and many others. Thus bogs themselves could be viewed as sacred sites for some ancient populations, or at least that revered sites were often located in bogs (Bradley, 1990:161–177).

It seems improbable that the relatively slow sphagnan preservation processes described above would be able to preserve a body placed into warm bog water in summer. Proponents of the sacrifice hypothesis note that pollen and seeds in the stomach and intestinal contents of bog mummies have all indicated winter or early spring as the season of death. This time of year coincides with fertility festivals that were held by many Iron Age Danish groups just before the agricultural season began. They call attention also to another comment by Tacitus that "a human sacrifice is made by those assembled to celebrate the horrible beginnings of their barbaric cult . . .". The smooth fingernails unmarred by scratches and the absence of calluses suggest that these individuals were not common criminals but selected from among the elite. Indeed, Ross & Robins (1989), who participated in the extensive study of Lindow Man, have reconstructed an elaborate Druidic ritual to explain his death as a human sacrifice at a time of dire population stress. While that proposed scenario could apply to Lindow Man and others from the Danish Iron Age, a recent review of Danish bog skeletons found a rope around the neck area present in a sequence of skeletonized bodies traceable back to the Early Neolithic (Bennike *et al.*, 1986). This places the practice in a much broader context. While the sacrifice theory is certainly currently the most popular, the precautions noted by Briggs (1995) and the suggestions of Bennike *et al.* (1986) require an explanation that can accommodate a practice that appears to have begun 3000 years before the Iron Age and its related fertility cult evolved.

Overview of some well-known bog mummies

Although bodies had been recovered from north European bogs for many years, it was Glob's (1969) publication and his article in *National Geographic* (1954) that generated world-wide interest in this phenomenon. Both of these publications focused on Tollund Man and Grauballe Man. The exquisitely preserved facial detail of these bodies together with the dramatic evidence regarding the nature of their death captivated the imagination of professionals and the general public. Demographic data of some of the more well-known bog bodies are detailed in Table 4.5.

Tollund Man (Plate 2)

Tollund Man was found in Bjaeldskovdal bog near Silkeborg, Denmark, by local peat cutters in 1950. Except for a peaked cap and a belt, the body was nude. A leather thong encircling his neck was sufficiently tight to compress a groove into the soft tissues of the

Table 4.5. *Demographic data on selected well-known bog bodies*

Reference	Country	Body name	Date found	Age/yr	Sex	Death period[a]	Injury	Comments
Thorvildsen, 1947	Denmark	Borremose Man, 1946	1946	Adult	M	840 BC	A, T	
Thorvildsen, 1952	Denmark	Borremose, 1947	1947	Adult	I	475 BC	T	Infant bones too
Thorvildsen, 1952	Denmark	Borremose Woman, 1948	1948	Adult	F	770 BC	T	
Thorvildsen, 1951	Denmark	Tollund Man	1950	Adult	M	220 BC	A	
Fischer, 1951	Denmark	Elling Woman	1938	25–30	F	AD 35	A, S	
Munck, 1956	Denmark	Grauballe Man	1952	Adult	M	310 BC	T, S	
van der Sanden, 1996	Netherlands	Yde Girl	1897	16	F	AD 200	A	
van der Sanden, 1996	Netherlands	Weerdinge Couple	1904	Adult	M, M	100 BC	S	2 bodies
Turner & Scaife,1995	England	Worsley Man	1958	20–30	M	AD 240	A, T	Isolated head
Turner, 1995a	England	Lindow Man II	1984	30–50	M	AD 10	A, T, S	
Delaney & O'Floinn, 1995	England	Meenybraddan Woman	1978	25–35	F	AD 1220	N	Adipocere
Glob, 1969: 85	Germany	Windeby Man	1952	Adult	M	AD 10	A	
Gebuehr, 1979	Germany	Windeby Woman	1952	14	F?	AD 10	N	
O'Floinn, 1988	Ireland	Gallagh, Co. Galway	1821	Adult	M	843 BC	A	Stake by body
van der Sanden, 1995	Denmark	Huldremose Woman	1879	Adult	F	AD 100	S	Adipocere

Notes:

F, female; M, male; I, indeterminate; A, asphyxiated (choking, hanging); T, skeletal trauma; S, stab wounds; N, none.

[a] Dates listed represent the mean if more than one value was obtained.

front of his neck. Though arms and hands were almost skeletonized the rest of his externally visible soft tissues appeared well preserved (Thorvildsen, 1951; Glob, 1969). The bones and teeth were largely decalcified. In the body cavities the heart, liver, and some lung and bowel structures were recognizable. The short-cut scalp hair was intact. The stomach and intestine contained enough pollen and seeds to suggest death occurred during the winter season and revealed the presence of *Trichuris trichiura* ova. A radiocarbon study dated his death to about 220 BC, the pre-Roman Iron Age. Ultimately the body was preserved by placing it in a succession of fluids (formalin, acetic acid and alcohol), after which the body was left to dry while the head was submerged in toluol, saturated first with paraffin and then beeswax (Fischer, 1998).

Grauballe Man

The body was found in a small bog also near Silkeborg, and preservation of soft tissues was roughly similar to that of Tollund Man. It was even possible to make fingerprints of his thumb that resembled modern patterns. Though this body lacked evidence of asphyxia, the neck tissues had been slashed deep enough to sever the esophagus and trachea, and he had suffered a skull fracture (Fig. 4.97). The reader is again reminded that it is usually not possible to differentiate between wounds of this type that were acquired immediately before from those that were inflicted shortly after death, so that they are commonly described with the nonspecific adjective "perimortem". A calibrated radiocarbon date of a hair sample was reported as 265 ± 40 BC. Like Tollund Man, coprolite analysis also revealed *Trichuris trichiura* ova. In contrast to Tollund Man, the entire body of Grauballe Man was immersed in a solution of tannin (containing oak bark) at the museum for 18 months. Subsequently a problem of continuing oozing of oils and fats with subsequent evaporation and skin cracking was minimized by placing the body in a nitrogen-filled, sealed, glass container (Glob, 1954; Fischer, 1998).

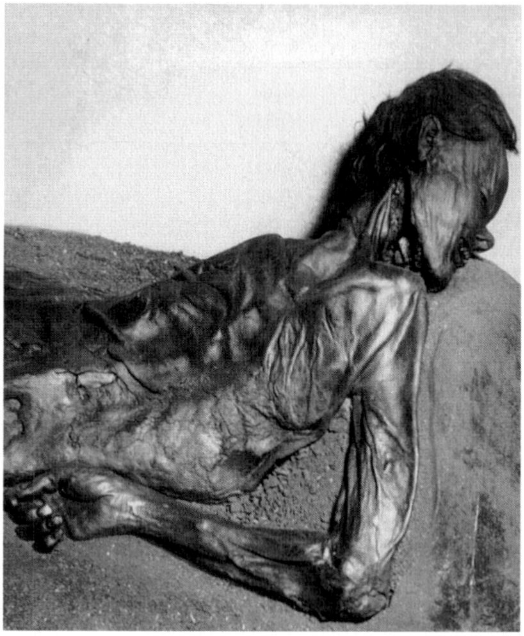

Fig. 4.97. Grauballe Man; SM, AD 410–560.
The deeply slashed throat is apparent. He was found in a bog near Silkeborg, Denmark. (Photo courtesy of the Moesgård Museum from Glob (1969) Aarhus, Denmark.)

Lindow Man

In 1983 the isolated female head (called Lindow I), largely a skull, was found in Lindow Moss, a peat bog a short distance south of Manchester, England. The following year most of a body of a 30–50-year-old man (Lindow Man II) was identified at nearly the same spot. The peat-cutting machine had amputated several body parts, bisecting the body at the level of the fifth lumbar vertebra. The missing left thigh was recovered in 1988 (Turner, 1995c). In 1987 the peat-cutting machine produced many fragments of an adult male (Lindow Man III). My discussion will limit itself to Lindow Man (Lindow II). The body was transported to the British Museum still encased in a block of peat (Fig. 4.98), where the excavation was completed and subsequently subjected to a still continuing series of laboratory studies including all forms of radiology and imaging, medical dissection, histological examination, coprolite analysis and many others (Stead *et al.*, 1986). The soft tissue was well preserved as were gastric and intestinal content, but no other viscera were recognizable. Bones and teeth were extensively decalcified. Principal findings included skull fracture, cervical vertebral fracture-dislocation, laceration of the right neck soft tissues deep enough to transect the jugular vein and a neck ligature in a groove impressed onto the neck tissues (Fig. 4.98). The ligature was only 1.5mm thick and the groove in which it lay was visible only in front and on both sides, but not on the back of the neck, suggesting that it was not used to asphyxiate the victim but to enhance the blood flowing from the severed jugular vein. The body was eventually conserved by freeze-drying it.

Meenybraddan Woman

This 25–35-year-old female body was uncovered by an Irish peat cutter in 1978 near Inver in Donegal, Ireland. The body was supine in the bog, and nude, but partly enveloped in a cloak. No other artifacts were present. Most of this somewhat obese body was intact, including tissues and scalp hair. After recovery the body was frozen. In 1985 the body was thawed and studied. Teeth and bones were profoundly decalcified. The lower extremity soft tissues were absent. No external pathology was evident. Endoscopy revealed liver and lung tissue that were confirmed histologically. No skull or other fractures were noted nor was either a neck ligature or neck lacerations or other wounds apparent. Subcutaneous tissue of the trunk and breasts demonstrated extensive adipocere formation (Fig. 4.99); chemical analysis revealed increased palmitic and decreased oleic acid composition. The body was eventually transferred to the British Museum, where it was conserved by freeze-dry methods.

For readers interested in more details about bog bodies, the gazetteer of bog burials from Britain and Ireland will be useful (Briggs & Turner, 1986). It lists about 150 bodies from 120 sites in England, Wales, Scotland and Ireland. This information is presented again in 1995 with some omissions of the less well-documented items but arranged in a more convenient, tabulated form (O'Floinn, 1995; Turner, 1995b).

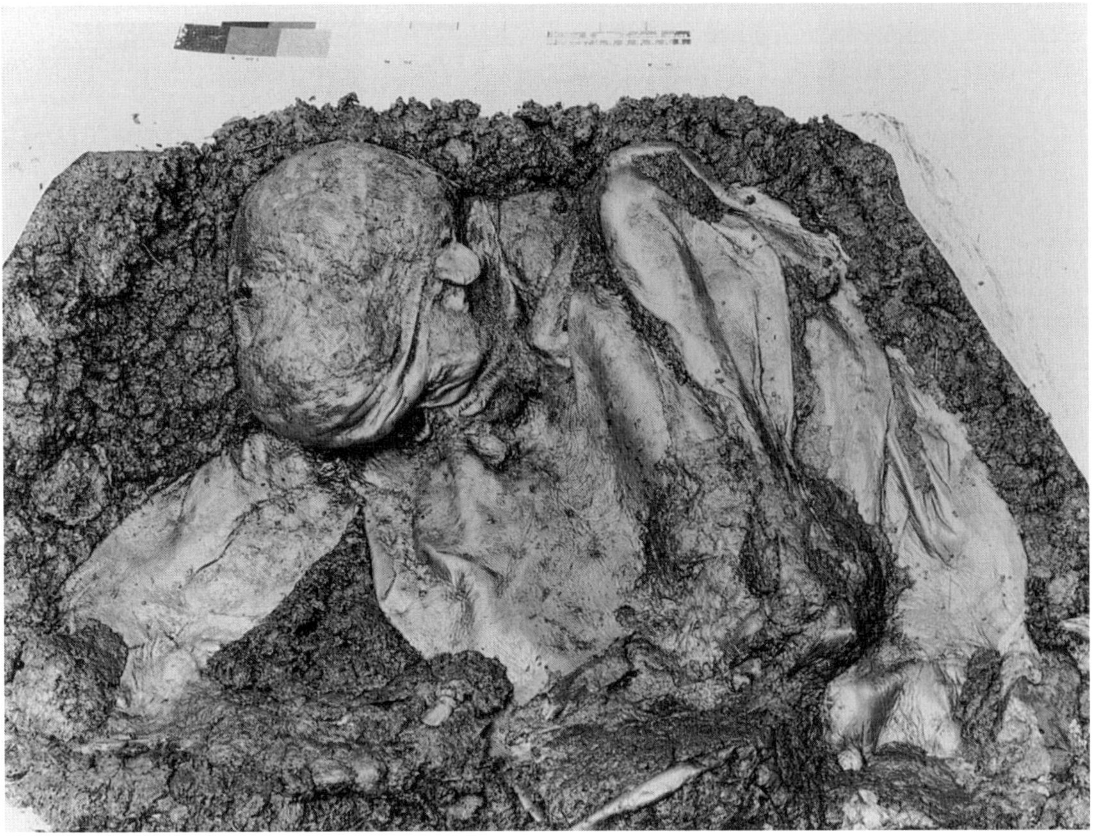

Fig. 4.98. Lindow Man *in situ*; SM, AD 410–560.
Still encased in peat, Lindow Man was transported to a museum for meticulous excavation under controlled conditions. (Photo courtesy of the British Museum of Natural History, Negative no. PS151984.)

Mummies from Ireland

In addition to bog bodies, Ireland also houses some catacomb mummies. The tombs below St Michael's Church in Dublin also contain the spontaneously mummified bodies of late medieval crusaders (Fig. 4.100).

Miscellaneous mummies from Scotland

The Department of Anatomy in the medical school of the University of Maryland at Baltimore houses a unique collection of human bodies. In the early nineteenth century both European and American medical schools shared a common problem: availability of human cadavers that were necessary for the teaching of anatomy. Only the bodies of executed criminals could be acquired legally for such purposes, so the need was met by suppliers carrying out clandestine retrieval of recently buried bodies from graveyards. Furthermore, embalming methods employing intra-arterial injection of tissue fixatives that would result in long-term preservation of tissues had not yet reached a stage satisfactory for teaching anatomical dissection (the mid eighteenth century Hunter brothers' technique required evisceration).

It was within this milieu that Allen Burns, an anatomist at Glasgow's Andersonian Institute, solved at least part of the problem. He developed a method of long-term human whole-body tissue preservation,

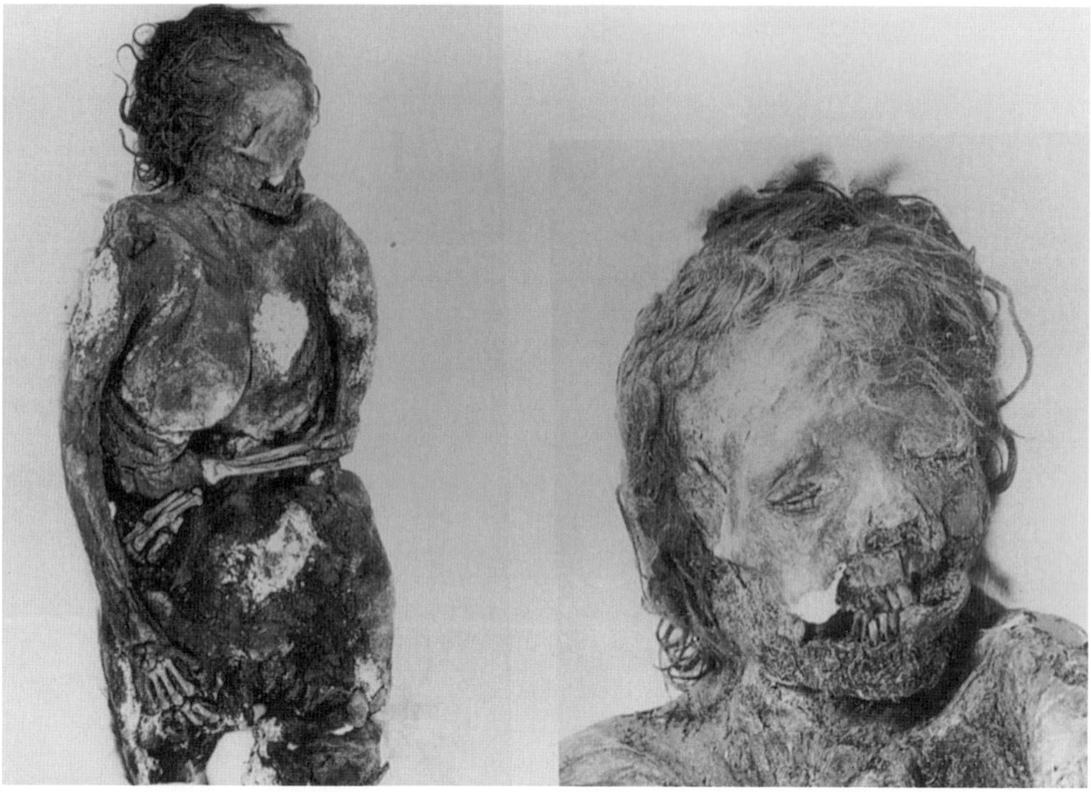

Fig. 4.99. Meenybradden Woman; SM, AD 1220.
Widespread adipocere formation is evident in this Irish bog body. Thawed after 8 years of frozen storage following discovery in 1978, the right panel reflects her reconstructed appearance. (Photo courtesy of the National Museum of Ireland.)

dissected the prepared bodies and enhanced the preparation by using colors to define structures for clarity. These proved to be completely satisfactory for use as permanent demonstrations in his anatomical teaching lectures. Given the fact that he probably acquired at least some of those bodies from sources other than executed criminals, it is not surprising that he shared little of the details of his technique with any one. We do know that the tissues were at least partly desiccated, and that he employed the principle of osmosis, using salt and sugar to achieve the dehydration process (Fig. 4.101). The ancient Egyptians had employed the same principle effectively for several thousand years using carbonate salts in the form of a locally available ore (natron). The composition of that ore, however, and the technique of its use were not widely known in the early 1800s. It is, therefore, likely that Burns accomplished his results by independent invention. He probably achieved it by exposing the area of interest, dissecting it and then packing the tissues in his powdered mixture of dry salt and sugar, replacing it as needed until the exposed tissues had become desiccated. The dried, colored arteries and muscles resulted in a permanent anatomical exhibit (Fig. 4.102).

When his successor, Granville Sharpe Pattison, left Scotland to assume a similar post at the University of Maryland in 1820, he carried the Burns Collection of anatomical specimens with him, persuading that institution to build a museum to house them. Many of these specimens are still stored in usable condition at the medical school there (Wade, 1998) (Fig. 4.103).

Fig. 4.100. Irish crusaders; AM, AD 1190.
Spontaneously mummified bodies of Irish crusaders were found in the catacomb beneath Dublin's St Michael's Church. (From L. Mercer McKinley, *J. Pathology* **122**: 27–29 (1977), copyright 1977, courtesy of John Wiley & Sons, Ltd, reproduced with permission.)

Mummies from Denmark

The Mound People Mummies from the Bronze Age

During the Bronze Age (about thirteenth century BC) the elite among the people occupying what is now Denmark buried their dead in wood coffins on stone platforms, then covered them with a large deposit of earth. These "Mound People" made the coffins from a segment of a huge oak tree's trunk that had a diameter of up to or more than 1 meter. These were bisected longitudinally, hollowed out and the clothed bodies laid therein. The two halves were then rejoined and fixed into position. Excavation of some of these near Egtved in Jutland, Denmark, in the early 1920s revealed that the soft tissue of many of these bodies had been preserved, presumably consequent to the tanning effect of the tannin in the oak (Kaner, 1996). Some of these are now on display at the National Museum in Copenhagen (Fig. 4.104).

The preserved brains from Denmark's medieval monastery

Between AD 1236 and 1540 the clergy and laity of Svendborg on the Danish island of Fyn buried their dead in the cemetery adjacent to the local Franciscan monastery. Excavation of about 10% of those burials in the 1970s revealed the presence of residual brain tissue (but no other soft tissue) within the cranial cavity of 56 of 74 skulls. Their appearance varied from

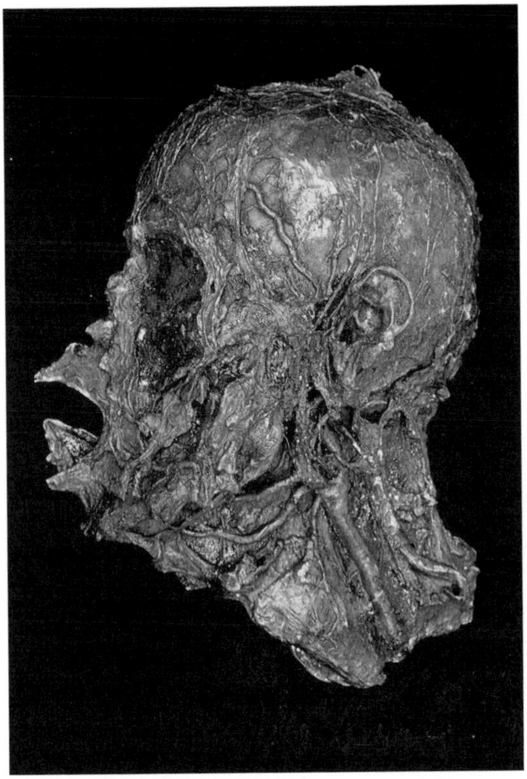

Fig. 4.101. A Scottish "sugar mummy"; AM, AD 1800.
An anatomical specimen, prepared 200 years ago by the Glasgow anatomist Allen Burns using salt and sugar, remains an effective teaching aid today. (Photo by Mark Teske, R.B.P., and courtesy of Ronald Wade.)

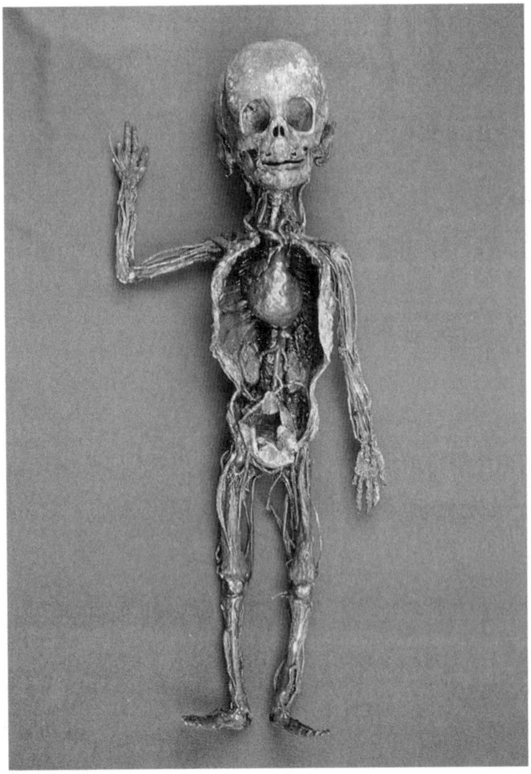

Fig. 4.102. A 200-year-old "sugar mummy"; AM, AD 1800.
This body, prepared as in Fig. 4.101, can be seen today in the Burns Museum, Anatomical Services Division of Baltimore's University of Maryland School of Medicine. (Photo by Mark Teske, R.B.P., and courtesy of Ronald Wade.)

small, irregular, shapeless masses to about half-size but otherwise normal brains demonstrating surface convolutions. Gross, histological and electron microscopic anatomical studies showed stainable axons and clear separation of gray and white matter, but displayed "ghost" cells without nuclear or cytoplasmic structure. Preservation by adipocere formation was established chemically (Tkocz *et al.*, 1979).

Greenland mummies

Charles Francis Hall

During the international fervor concerning exploration of North American arctic regions in the late nineteenth century, the American explorer Charles Francis Hall found himself, his crew and their ship overwintering off the northwestern coast of Greenland. Returning from a two week reconnaissance of the area, he became ill promptly after he was served and drank a cup of coffee, suffering abdominal pain. He died several days later and was buried ashore. Earlier in the expedition he had complained of being poisoned, particularly suspecting the ship's surgeon, with whom he frequently quarreled. In 1871 a ship's surgeon's medical kit contained arsenic for use as a medical therapeutic.

On August 8, 1968, Hall's remains were exhumed by Dr Chauncey Loomis, a Dartmouth College faculty

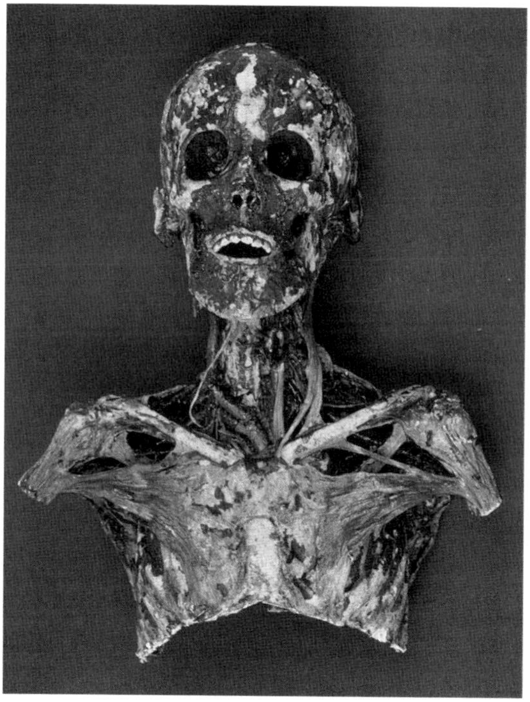

Fig. 4.103. "Sugar mummy" of the Burns Collection; AM, ca. AD 1800.
Note the high preservation quality of the dissected arteries, veins and nerves of the neck areas. (Photo courtesy as in Figs. 4.101 and 4.102.)

member, and colleagues (Loomis, 2000). Having been buried in winter, the top of the coffin was found only 18 inches (46 cm) below the surface. At 81° 46′ N the soil is permafrost in which only the surface thaws during the summer. A body so superficially buried would probably remain frozen throughout most years, but could partially thaw during the occasional warmer, longer summers. When the coffin was opened, the body, clothed in a military uniform, retained sufficient soft tissue to resemble an adult body, though a significant amount of postmortem decay had occurred. The lower half of the body was encased in ice, suggesting that sufficient thawing had occurred to at least partially fill the coffin with meltwater at some occasion(s?) after death (Plate 3). The viscera revealed advanced postmortem changes. Samples of hair, tissue, fingernail, bone and soil were acquired for analysis. Analytical results indicated that the arsenic content of the soil was 22.0, bone 11.0, and hair 15.3 micrograms (μg) per gram. When hair was sectioned into six individual segments linearly and each segment analyzed separately, the pattern of arsenic concentration demonstrated the highest value in the hair root sample (29.4), progressively decreasing to the hair tip (10.6). The authors noted that these findings are consistent with the administration of a higher dose of arsenic shortly before death (Paddock *et al.*, 1970).

However, more recent research results have made us aware of variables that mean such findings may be less conclusive. The gross findings at time of exhumation indicated clearly that the body had become thawed at least several times since inhumation and that during these intervals groundwater (containing 22.0 μg of arsenic per gram) pooled in the coffin, soaking the clothing and the body. Under those conditions the probability that arsenic-containing water reached the sampled tissues is high. Indeed, it is virtually certain that it soaked the hair. The sulfur in disulfide bonds in hair constitutes about 4% of the hair's weight. Such sulfur would combine irreversibly with arsenic to produce an insoluble precipitate that would resist washing of the sample. The progressive change of hair segment concentrations might be explained by other variables such as the tendency for hair structure to disintegrate nearer the hair ends. Given what we know today about heavy metal binding to sulfur in hair, I am not certain that modern forensic scientists would indict the ship's surgeon, on the basis of the tissue arsenic content.

Pisissarfik mummies

In addition to the mummified body of Hall, several spontaneously mummified bodies of Greenland natives have been found. The least studied of these includes three caribou hide bundles located beneath several large rocks covering a small burial chamber near Nüük (formerly Godthaab) in southwestern Greenland at the foot of a nearly vertical cliff called Pisissarfik. Within these were the poorly preserved, spontaneously mummified bodies of three infants wrapped in sealskins

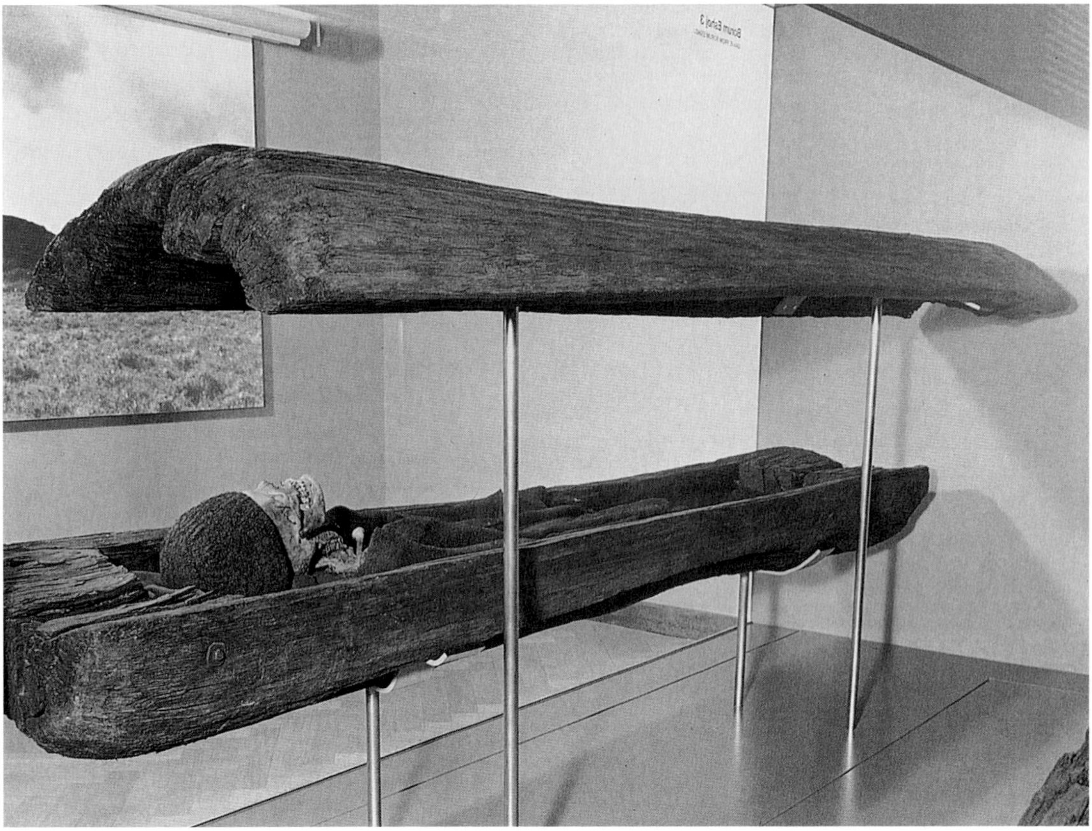

Fig. 4.104. Danish log coffin burial; SM, ca. 1000 BC.
Soft tissue preservation in these burials is attributed to the tannins in the oak coffin wood. (Photo courtesy of Nationalmuseet Danske Afdeling, Copenhagen.)

(Fig. 4.105). Not yet scientifically dated or studied further, they are definitely from a pre-Christian period.

Qilakitsoq mummies

Better-preserved, more spectacularly clothed bodies were found by Inuit hunters in 1972 at a northwestern coastal Greenland site (about 71°N, 52°E) called Qilakitsoq on Nuussuaq Peninsula. Covered with large rocks and sheltered by an overhang, the frozen bodies of six women and two children were identified. The caribou and sealskin clothing was exceptionally well preserved, particularly that of an infant (Fig. 4.106). The freeze-dried bodies were spontaneously mummified. One of the tattooed, adult bodies (about 50 years old at death) had suffered extensive cranial destruction by a nasopharyngeal carcinoma (Fig. 4.107). The remainder of her viscera were also preserved well and revealed pulmonary anthracosis with fibrosis, undoubtedly secondary to the use of seal oil as lamp fuel. These bodies and their clothing are now curated at the Greenland Museum in Nüük. Radiocarbon dates of the bodies cluster about AD 1475.

Unartoq mummies

Three infants buried in a kayak in a coastal cave on the island Uunartoq were found in 1934 by the Danish

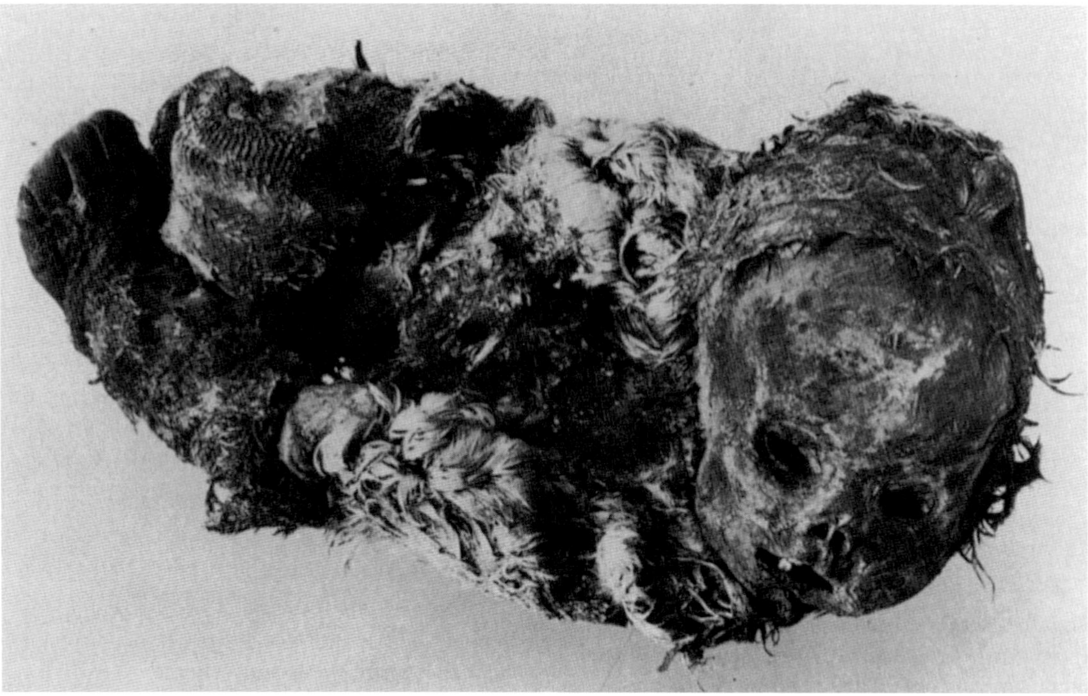

Fig. 4.105. Greenland mummy; SM, ca. AD 1650.
The freeze-dried, spontaneously mummified body of this Greenland native was found on Greenland's western coast, and is now curated at the Greenland Museum in Nüük. (Photo courtesy of the National Museum of Denmark and the Greenland Museum.)

archaeologist Therkel Mathiassen in South Greenland. These bodies were removed to Copenhagen's National Museum (Hart Hansen *et al.*, 1991).

Mummies from Sweden

Freeze-dried mummies of the Arctic balloonists

To understand the tortured logic of attempting to reach the North Pole by balloon in 1897 it is necessary to note the state of international politics at the turn of the twentieth century. Germany, Italy and several other European countries had gathered their various provinces and welded them together into nation states during the previous half-century. Even the USA was emerging from an internal focus on consolidation of its own territory and was beginning to concern itself with its status from a global perspective. National pride was effervescent and demanding international recognition, but more than two decades would pass before the mechanism of military conquest by such newly formed states would be employed in an attempt to achieve it. Interest in the unexplored area of the high arctic regions had acquired world-wide interest resulting from efforts to forge a commercially feasible Northwest Passage through arctic North America. As a succession of (principally) British expeditions began to make it clear that an arctic route would never reach practicality, interest shifted to reaching the North Pole. This was believed to be possible only by overcoming the formidable natural hazards through the applications of the most modern available technology. All realized that the successful country would gain profound esteem

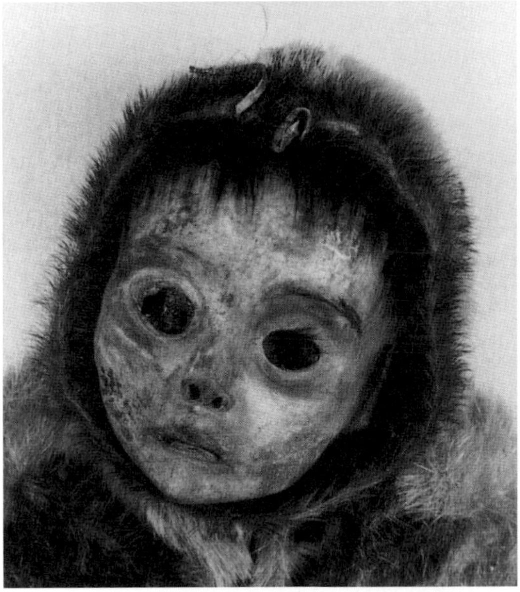

Fig. 4.106. Inuit child mummy; SM, ca. AD 1475.
This spontaneously freeze-dried mummy of a Greenland native was found on Greenland western coast at the Qilakitsoq site on Nuussuaq Peninsula. (Photo courtesy of Greenland Museum, Nüük.)

and admiration from the remainder of the industrialized world. The North Pole became an incredibly coveted prize. No lure of commercial gain could have driven the newly emerging nations to expend greater energy and resources than they employed in pursuit of the otherwise useless feat of being the first to reach the North Pole.

Into this bizarre milieu stepped a Swedish engineer from Stockholm's patent office. Entranced by the dazzling potential touted by developers of the hydrogen-filled balloon, Salomon Andrée convinced the Swedish government to back his proposal for reaching the North Pole by air, using this most recent product of the Industrial Revolution. Arriving in Spitzbergen, a remote island more than 800 kilometers north of Norway's most northern tip, he and his two crew members launched their balloon into a strong wind from the south, never to be seen alive again. Several optimistic messages were received at the launching site via homing pigeons, and then – silence.

Thirty-three years later a Norwegian fishing boat landed at White Island to retrieve several walrus they

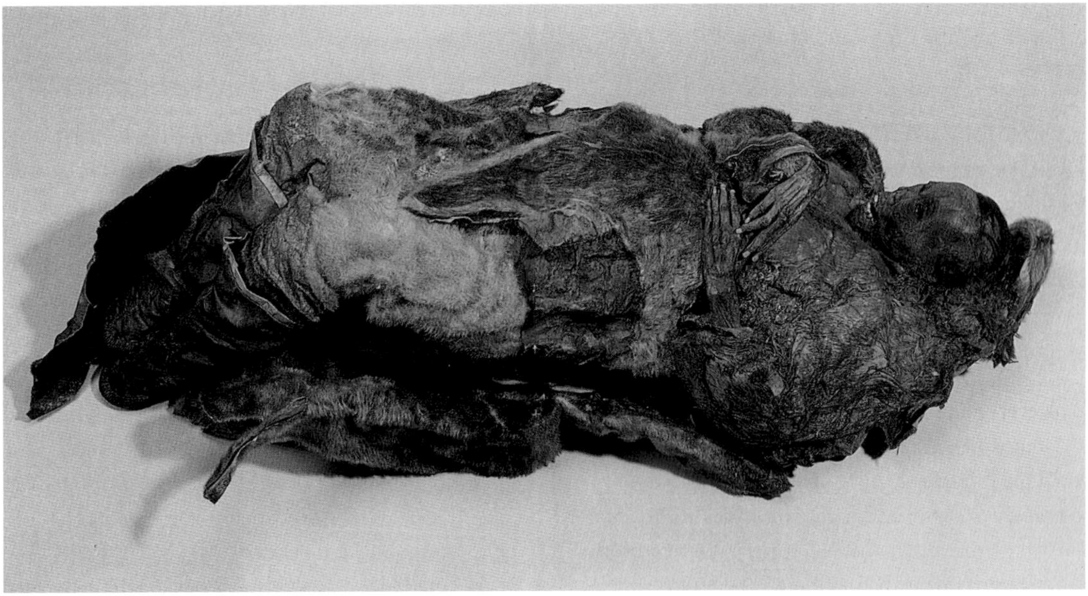

Fig. 4.107. Inuit adult mummy; SM, ca. AD 1475.
This adult is from the same site as Fig. 4.106. She died of nasopharyngeal carcinoma. (Photo courtesy of Greenland Museum, Nüük.)

Fig. 4.108. End of the *Eagle* flight; SM, ca. AD 1897.
Icing forced the landing of the balloon that Andrée and two companions had hoped would take them from Spitzbergen to the North Pole. (Photo courtesy of Lennart Bäck, Chairman of the Swedish Society for Anthropology and Geography.)

had shot. This island is about the same latitude as the launch site, but on the opposite (eastern) side of Spitzbergen. Most years this island is completely locked in ice that fails to melt at all. But the summer of 1930 was unusually warm, permitting the Norwegian ship access. To their astonishment they found a boat, several sledges, a crude camp containing fuel, stove and abundant food. Nearby were the frozen bodies of the three balloonists, and their diaries. The Norwegians took everything aboard and returned them to Sweden.

Reconstructing the sequence of events as recorded in the diaries, they learned that the balloon had accumulated heavy ice after about 48 hours of flight, forcing them down and terminating the flight (Fig. 4.108). During an exhausting trudge across the continually shifting and moving ice floes, they had arrived at White Island nearly two months after abandoning the balloon, dragging a boat and several heavily laden sledges behind them. Food had not been a problem because they shot several polar bears during their trek, the meat of which was savory enough to be eaten raw (Fig. 4.109). In spite of what seemed to be adequate nourishment, they suffered progressive weakness and agonizing muscle pains during the latter part of the ordeal. Within a few weeks all three men were dead, even though they had shot enough seals, birds and polar bears to support them through the entire winter.

The mystery of their ailments and death was solved by autopsy and histological study of tissues. The muscle of the polar bear meat was laden with the larvae of the parasite *Trichinella spiralis* (Fig. 4.110). Even today most polar bears harbor these organisms. If their meat is eaten without cooking these organisms invade the consumer's tissues, especially the muscle. Each new meal of the contaminated meat sent another wave of worms throughout the body muscles. The heart

Fig. 4.109. Polar bear (SM, AD 1897) shot by Andrée and other members of the expedition.
This is one of several polar bears shot by Andrée and his companions during their traverse over the ice to White Island. This and Fig. 4.108. are from film found in their cameras and developed after their discovery. (Photo courtesy of Lennart Bäck, Chairman of the Swedish Society for Anthropology and Geography.)

muscle is no exception, and in such extensive infections the heart may weaken, causing heart failure or cardiac arrest. Thus, after 33 years, the freeze-dried mummified tissues yielded the secret of their deaths: advanced trichinosis, acquired by eating uncooked, infected polar bear meat.

Mummies from northern Scandinavia

Mummies from Lapland

The Lapps of northern Scandinavia were reindeer herders, and thus moved with them in a seasonally related cycle. Despite such movements, they established a base area where their movements originated and at which their cycle was completed. Their principal burial grounds were located at such locations, usually adjacent to a stream or lake. The nature of routine burials was influenced by the location of such sites. In forested areas they often took the form of tree burials, while in tundra areas they were interred in the soil or, in eastern areas, simply on the surface. If death occurred during a season when they were distant from such permanent burial sites, it was common practice to achieve temporary mummification by suspending the body in the wind without evisceration. While some accounts suggest that the resulting desiccation was often hastened by smoking the body, this remains unsubstantiated. Their purpose in producing temporary mummification apparently was simply to prevent decay until the seasonal cycle movements would bring them back to where disposal in their permanent burial grounds could be carried out (Stora, 1971:109; Barber, 1995).

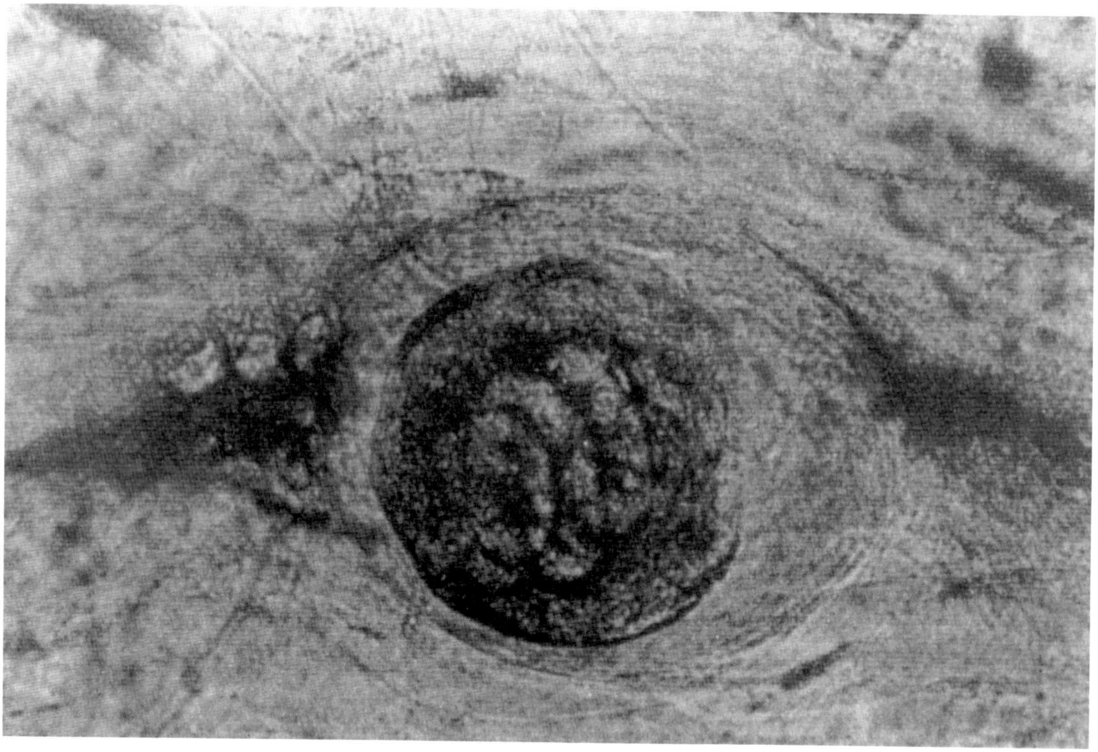

Fig. 4.110. The polar bear's revenge.
Tissue histology revealed that the muscle tissue of polar bears eaten by Andrée and his team members (SM, AD 1897) was riddled with the larvae of *Trichinella spiralis*. (Photo courtesy of Lennart Bäck, Chairman of the Swedish Society for Anthropology and Geography.)

Mummies from Hungary

During a 1989 visit to Hungary I found only Egyptian mummies being curated. The majority of these were housed at Budapest Fine Arts Museum. On display were two Egyptian heads as well as a mummified left hand from the New Kingdom Period. There was also an unwrapped New Kingdom adult with a gaping, 10 cm long incision in the right lower abdominal quadrant through which linen could be visualized inside the abdomen. Arms were extended along each side of the body with palms facing the lateral thighs. Hands and feet had been broken off and reattached with glue. Another intact, Ptolemaic Period adult was completely wrapped with a metal plate resting on the abdominal wrappings. A Roman Period completely wrapped adult female was present within an open wooden coffin. A wrapped adult male with exposed face, nasal tampons in place, arms crossed on the chest with hands on shoulders was labeled with a first century AD date. Another wrapped adult revealed the surface contours of a face but labels did not indicate sex or age period.

In addition to these a bundle wrapped to resemble a first century AD child mummy was accompanied by its X-ray film that showed the wrappings enveloped only an adult lower leg. In addition an unwrapped Egyptian head with abundant hair bore no labels for sex or age period.

Animal mummies included one adult and five immature crocodiles, one wrapped cat with exposed head, one headless ibis (X-ray), a falcon and a snake (both wrapped), an ichneumon and another ibis within a ceramic pot.

The museum's curator also informed me that only one other Hungarian city had mummies: three Egyptian mummies in the Déry Museum in Debrecen. One of these was a wrapped adult mummy from the Late Period but this one had been subjected to radiological study. Another 30–35-year-old Egyptian female mummy had also been similarly studied and found to be bearing a fetus *in utero*. The cartonnage mask of the third adult mummy suggested a date of Dynasty 30 (4 BC) but its coffin was dated to an earlier period.

The directors of these museums knew of no mummified bodies of native Hungarians.

Mummies from the Czech Republic

What no doubt is one of the largest accumulations of Egyptian mummies outside of Egypt is found in the Czech Republic, most of them located in the Náprstek Museum, affiliated with the National Museum of Prague. Conserved mummies native to the Czech Republic, however, are not numerous and come principally from the Renaissance era. The manuscript by Subert *et al.* (1993) presented at the International Mummy Symposium held at Innsbruck University in 1993 provides some information about such bodies.

Those authors note that the bodies of socially and politically important individuals during the sixteenth and seventeenth centuries in the Czech Republic were anthropogenically mummified. The bodies were eviscerated via an abdominal incision, though the brain remained in the cranial cavity. The body cavities were filled with herbs. Some bodies were then placed in a large metal tank filled with resin solution (resin type not specified) that also contained a mixture of potassium chloride and potassium sulfate. Several kings were so preserved, including Premysl Otakar II, Venceslav I, Charles IV and his father John of Luxemburg. Such bodies were commonly then placed in churches, cloisters or crypts. Anthropogenic mummification in the Czech Republic was forbidden at the end of the eighteenth century, after which the bodies were simply placed in the crypts without special preparation, though immersion in a disinfectant solution, usually a salt solution in vinegar or extracts of fruits and juniper needles was practiced during epidemics (Subert *et al.*, 1993).

Subert *et al.* also examined a group of 42 corpses from crypts at Klatovy in 1977. These were principally elite bodies from the seventeenth century as were four bodies of the aristocratic Bokuvka family from a crypt in the cemetery chapel in Postrelmoy–Sumperk region. The latter were Moravian Knights. The investigators felt that ventilation contributed generously to soft tissue preservation of these bodies and provide a valuable and useful explanation of architectural structure designed to maximize air exchange. Their mathematically derived estimates of ventilation volumes and rates for the Gothic and Baroque crypt styles are of interest. The origin, chemical nature and use of various waxes employed for conservation of these mummies is detailed. One body was treated with gamma ray radiation to destroy the insects. They estimate that more than 200 mummified crypt bodies are present in the Czech Republic.

Mummies from Italy

The mountainous nature of much of Italy's terrain is not widely appreciated because many of Italy's popular tourist sites such as Rome, Naples, Venice or Pompeii are located at or near the sea. The Appennine range, for example, extends from the toe of Italy's boot, the full length of its leg, swings westward and in its extreme northwestern course joins a southern branch of the Alps at an altitude of up to 4000 meters. During most of its length the peaks range from 2000 to 2500 meters, while the mean height is distinctly less than 2000 meters. Its northwest–southeast orientation results in a latitudinal range from nearly 37° to 47°N, roughly equivalent to that between Tunis and Boston. The consequence of these impressive differences in temperature, daylight hours and humidity can be expected to enhance spontaneous desiccation in some regions but not in others. The distribution of Italian mummies, therefore, comes as a bit of a surprise. Spontaneous mummification occurs in the southern Italian seaport of Naples (annual mean temperature about 20 °C) as well as in the alpine community of Venzone in the far northeast, which has a climate similar to that of Zurich (annual mean temperature about 10 °C).

In addition to some 142 Egyptian mummies and mummy parts curated at Turin and excluding the bodies at the Capuchin monastery in Palermo, Sicily, about 500 mummified bodies have been identified in Italy. Most of these date to the late medieval era or more recent times. How did this country of relatively modest size accumulate so many mummies, particularly since the routine mortuary practice of Romans included cremation as well as inhumation? The answer is found in influences from both the socioeconomic and the theological realms. The majority of Italian mummies are associated with this country's large churches, cathedrals and monasteries. Had the Vatican become based in some other country, most of these religious structures would probably have been built elsewhere. Furthermore it was the reactivation of trade leading to the Renaissance and ultimately to the Industrial Revolution that fueled the concentration of wealth. This supported the construction of many of these edifices together with their associated mausolea and sarcophagi that preserved the bodies of the elite.

My review of Italian mummies will begin with the "catacomb mummies", a term I use for the spontaneously mummified bodies preserved in subterranean chambers, often related to a religious structure. These will be followed by descriptions of those bodies recovered from mausolea and sarcophagi, and will conclude with the relics of saints.

"Catacomb mummies" from Venzone

Located in alpine foothills in a pass connecting the Friulian plain in Italy's northeastern region near Trieste with transalpine areas, the small community of Venzone proudly displays its 15 spontaneously mummified bodies to the general public. In 1338 the residents consecrated the small Church of St Andrew, which they have expanded subsequently into a grand cathedral. The original level of the older church served as a stone platform for the new cathedral's subfloor tombs designed for the bodies of the cathedral's clerical and related staff. Later renovations expanded these tombs beyond the area occupied by the original Church of St Andrew, and these more recent subterranean tombs had only a floor cut into the native limestone. In 1679, construction works opened one of the latter tombs and the body was found to be spontaneously mummified. Later burial of additional bodies identified similarly preserved corpses until, in 1850, a total of 27 such spontaneously mummified bodies had been found. All of these were in tombs without a prepared floor but protected from rain and groundwater by the overlying cathedral. Flooding following a heavy rainfall led to detection of standing water in several tombs, followed by the removal of the catacomb mummies from those tombs. A widely held ancient European belief viewed the failure of normal postmortem decay as a divine suspension of natural laws and celestial affirmation or even sainthood. It is not obvious today whether this motivated the local residents, but they clearly value their mummified ancestors. They retrieved them from the tombs and dried them. They have conserved them meticulously since that time, exhibiting them in the baptistry adjacent to the cathedral. Over the years several bodies have become separated from the group and a catastrophic earthquake that collapsed the baptistry in 1976 reduced the surviving mummies to a total of only 15.

During my visit to the incredibly hospitable community of Venzone in 1983 I viewed (but was not allowed to sample) these 15 bodies. They were all males and most had been identified by name. The bodies were in an advanced state of desiccation (Fig. 4.111). One elderly individual demonstrated a spine deformity consistent with Pott's disease (collapse of a vertebral body secondary to its destruction by a tuberculous abscess; Fig. 3.22). However, recent radiological study of this body suggests that this is a postmortem artifact secondary to forcing the body into a container too small

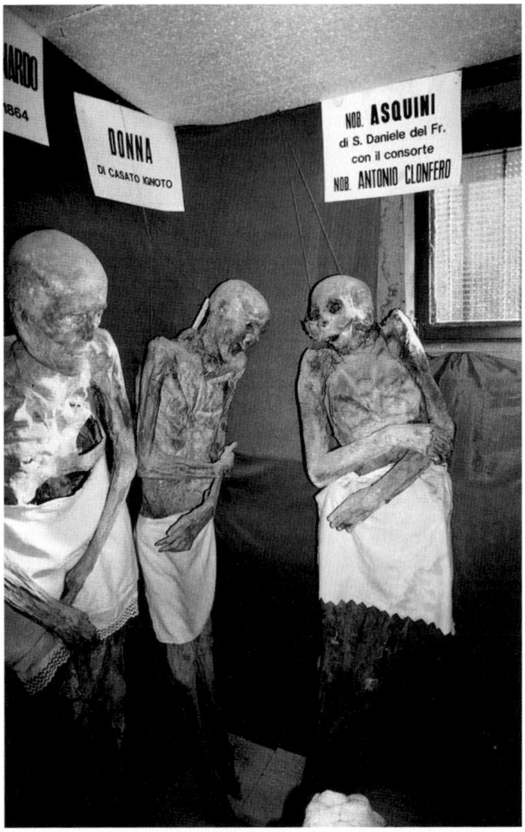

Fig. 4.111. Mummies of Venzone, Italy; SM, ca. AD 1400–1800.
These bodies became spontaneously mummified due to the aridity in the limestone catacombs beneath the cathedral. (Photo by ACA.)

to accommodate it properly (Baggieri *et al.*, 2001). Facial features and external genitalia were preserved in most of them.

In 1829, B. Biasoletto, director of the Trieste Botanical Garden, examined the mummies and suggested that a white fungus that he called *Hypha bombycina* was responsible for the mummification. With the help of John Rippon, Ph.D., from the Section of Dermatology at the University of Chicago a reference to this classification was found on p. 65 in *Mycologia Europaea*, Vol. 1, published in 1822 by Persoon. It was also listed on p. 1192 in *Sylloge Fungorum*, Vol. 14, published by its author P.A. Saccardo in Pavia, 1899. In both it was described as a white fungus with sterile mycelia commonly encountered in caves. The description is insufficiently detailed to separate it from many other common white fungi in the modern fungus taxonomic system. Our cultures from the mummies, soil, coffins and tombs at Venzone did not reveal the presence of any fungus fitting that description. The underground tombs from which the mummies were removed had a natural limestone floor, the tomb itself being cut out of limestone that underlies most of the city (Mainardis, 1976). Such regions are well drained and the tombs themselves were protected from the usual rain and groundwater by the church below which they were located. While obviously not totally immune to flooding occasionally, these conditions can be expected to have been sufficient to bring about corporeal desiccation. It is difficult to assign a major role to *Hypha bombycina* or any other fungus for the mummification process (Aufderheide & Aufderheide, 1991)

Mummies from Urbania, central Italy

A group of "catacomb mummies" similar to those described above from Venzone are curated in the community of Urbania in central Italy. The Church of the Dead there was built in AD 1380 and the monks' duties included assisting the dying and burying the dead. Little information is available about the details of their mortuary practices, but they appear to have been buried in underground tombs related to the church. In 1836 a semicircular wall was built in a chapel called "Cemetery of the Mummies" to house mummified bodies transferred there from some of the tombs. Twenty such bodies are placed in wall niches; they date from the fourteenth to the eighteenth century.

During my visit to this community in 1983, local officials were kind enough to open a tomb beneath the stone floor of the "Sepulcher of the Friars" building. This tomb had been last opened 400 years earlier. Four largely decayed, clothed bodies had been placed on board shelves. The boards had decayed and collapsed, the skeletonized bodies now lying in a jumbled heap on the floor. No mummies were present. The small

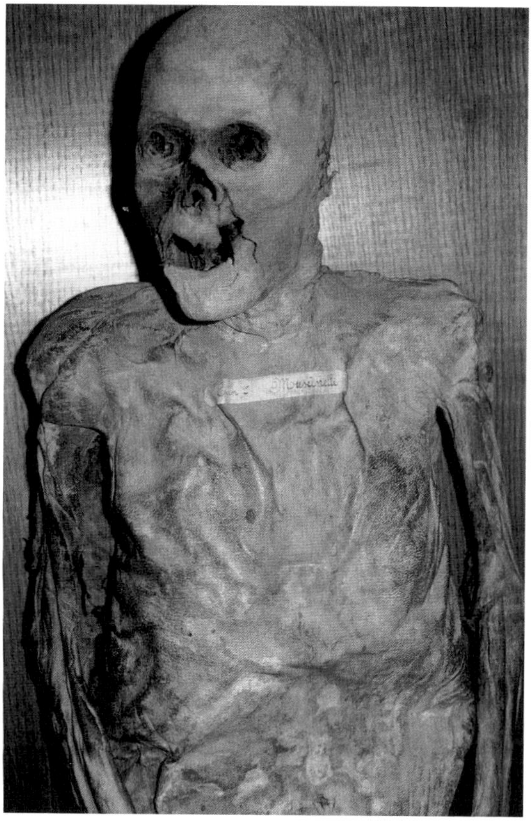

Fig. 4.112. Mummies of Urbania, Italy; SM, ca. AD 1390–1834.
Mummification method is identical to those of Fig. 4.111. Mummy no. 9. (Photo by ACA.)

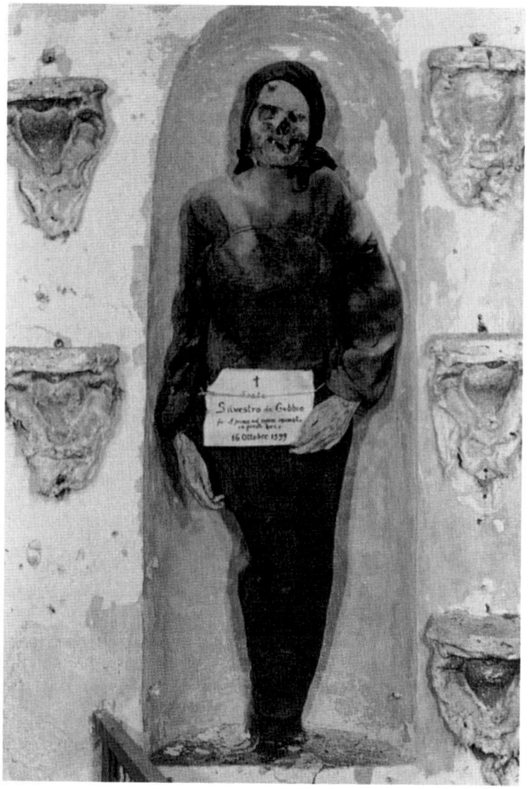

Fig. 4.113. Mummy in Palermo's Capuchin monastery; SM, AD 1599.
This is the first of thousands of bodies mummified (some spontaneously, most anthropogenically) by the monks of Palermo, Sicily. (Photo courtesy of Padre Superiore della Confraternita de Cappacchini, Palermo.)

chamber (about 3 meters × 2 meters × 3 meters) had limestone walls. No white fungi were grown from multiple samples.

Fornaciari (1982) describes observations made on the mummified bodies in the chapel during his 1982 study. Two poorly preserved corpses had been treated with lime, but the remainder were desiccated bodies characteristic of spontaneously mummified "catacomb mummies" (Fig. 4.112). Although he did not carry out major dissections, interesting pathological observations included a knife wound in a young adult male, another with mongoloid facies, a young adult female with evidence of a cesarean section incision and a fetus in the uterus, and another adult with evidence of rickets.

"Catacomb mummies" from Palermo, Sicily

Beneath the sixteenth century Capuchin monastery built in Palermo, Sicily, huge subterranean chambers connected by spacious halls were constructed out of the massive limestone deposit supporting the building. These were created as burial sites for deceased monks. The first was placed there in 1599 (Fig. 4.113) and the last in the late nineteenth century. The bodies of deceased monks from other areas were also transferred there. During this interval the families of the monastery's laic supporters were granted permission

Fig. 4.114. Capuchin monk of Palermo; AM, ca. AD 1600–1739.
One of many of the monks described in legend of Fig. 4.113. (Photo by ACA.)

to place their deceased in these catacombs. The practice became ever more popular. Today this extensive network of chambers beneath the monastery has housed as many as 8000 bodies, of which it is estimated that nearly 2000 retain at least part of their soft tissues (Matranga, 1983).

During my 1983 visit I found the bodies placed in a variety of positions. Many are in wall niches, but after these became completely occupied, the available space was maximized by placing bodies in a vertical position, many suspended from a wall hook (Fig. 4.114). Coffins and a few sarcophagi are scattered among these bodies, which have been grouped by occupation (professors, priests), age (children) and sex. Almost all are clothed, including hats and shoes and even gloves on some. Many of these expose only the face (Fig. 4.115); among such, bare skulls are common. Where soft tissue is apparent, the mummification principle is desiccation.

Discussion with curators and monastery administrators revealed that most have been anthropogenically mummified. Some bodies are said to have been dipped in lime or arsenic if death occurred during an epidemic. The clothed body of a very young child, one of the last to be buried here, is displayed in a glass-covered container and appears to be remarkably well preserved in the face and hands (the only exposed parts of the body) (Fig. 3.10). She is known to have been intra-arterially injected with some chemical (almost certainly an arsenic solution that was popular in the mid nineteenth century). Most of the rest are said to have been laid on a ceramic grid in a cell that was sealed for about nine months. They were then washed in vinegar, dressed and exposed to air.

Officials were kind enough to allow us entrance to

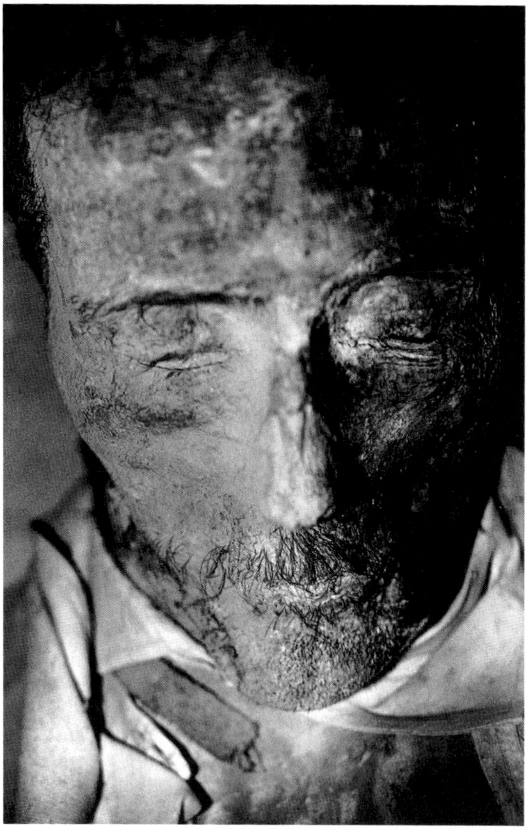

Fig. 4.115. Another of Palermo's monks; AM, ca. AD 1600–1739.
The range of soft tissue preservation of these bodies is evident in this and the preceding two illustrations. (Photo by ACA.)

one of these cells – the last one that had been used. It was a rectangular room about 4 meters × 5 meters with a vaulted, stone-lined ceiling. A ventilating tube about 8 cm in diameter placed high on one wall traversed the rear wall of the room to reach the outside air. The floor was dirt (limestone). A low wall about 1 meter high extended around three sides of the room, placed about 0.5 meter from, and paralleling, each room wall. The space thus created between the room's wall and this low wall was filled with limestone, ground into gravel-like consistency. The top of the low wall was joined to the room wall by horizontal, ceramic tubes about 10 cm in diameter, spaced along the length of the walls about every 15 cm. Lying supine on the grid so created were two unclothed human female bodies (Fig. 4.116). Most of the soft tissues of the back, the perineum and the lower abdomen were missing. In one the anterior abdominal wall was markedly distended. All remaining soft tissue was dry and desiccated. Straw had been stuffed into the areas of soft tissue loss. Clearly the arrangement was designed to allow fluids oozing from the body to drip into the limestone gravel.

An interesting reference to this process can be found in Pettigrew (1834:241–244). While titling his book *History of Egyptian Mummies*, the author includes a final chapter addressing several non-Egyptian mummy collections, including the Palermo Capuchin mummies. Here he quotes a certain Captain Smith, who observed them in 1824 and was told that the mummies were prepared by drying the bodies over a slow fire in an oven. Additionally he cites a Captain Sutherland, who described the mummification as being brought about by drying the bodies "in a stove heated by a composition of lime which makes the skin adhere to the bones . . .".

While reports by "chroniclers" are notoriously unreliable, it is an interesting exercise to attempt reconstruction of the mummification process that would be consistent with the descriptions of these two travelers, that of the monastery officials and my own observations. One possibility occurs to me: if the limestone gravel beneath these bodies had been heated sufficiently before use to have converted the carbonate of limestone to carbon dioxide (which would evolve into the air) it would leave behind the oxide of calcium (CaO), commonly known as lime and used commercially as lime for a mortar or cement ingredient. Lime combines with water in an exothermic reaction emanating heat. Thus fluids dripping from the body would drip onto the lime and combine with it. The resulting heat would thus accelerate the desiccation of the body on the grid above the lime. The extensive loss of tissue from the back of the bodies would be consistent with such a poorly efficient desiccating mechanism. However, while this would be consistent with the descriptions offered, it remains a purely personal speculation.

Fig. 4.116. Desiccating chamber at Palermo.
Walls are limestone, as is the floor. The body on the ceramic grate is one of the last bodies mummified (AM, AD 1739) in these chambers more than 100 years ago. (Photo by ACA.)

Mummies from Savoca, Sicily, and Ferentillo, Tuscany

Twelve similar spontaneously mummified, desiccated "catacomb mummies" have been found in Savoca, Sicily. To date I have not been able to identify reports of their scientific study (Fornaciari & Capasso, 1996). An additional 25 mummies were found in Ferentillo in Italy's Umbrian region (Fulcheri *et al.*, 1991). About AD 1500 the Church of St Stefano was constructed on top of the ruins of a church that had been built more than three centuries earlier. The original church then became a cemetery used until 1871. Excavation of the chambers beneath the floor of the current church in the late nineteenth century identified a crypt containing the human remains of many bodies. The room measured 24 meters × 9 meters × 25 meters carved out of the solid rock of the mountain. One windowless wall was formed by the warm rock of the mountain, but another contained four unshuttered windows that were perpetually open, providing excellent ventilation. One corner contained a heap of completely skeletonized bodies, but 25 spontaneously mummified bodies were found near the windows. The bodies had been buried in wooden boxes in soil at a depth of about 1 meter. Favetti & Pennacchi (1992) abstracted a series of reports by C. Maggiorani and A. Moriggia (Fulcheri *et al.*, 1991) who described the mummies' skin (desiccated), and opened the abdomen of one. The principal viscera were identifiable in a desiccated state. Maggiorani and Moriggia also performed chemical analysis of the soil, reporting it as composed principally of the calcium salts of limestone. They felt that the limestone provided excellent drainage, the overlying church protected the area from rainfall and the windows contributed ventilation. Probably influenced by the local views of the Venzone residents, they also felt that, in addition to the physical factors noted, fungal growth (Torulaceae?) may have contributed to the cadavers' desiccation. The illustrations of the mummies in the current publication by Favetti & Pennacchi reveal what appear to be very well-preserved, spontaneously desiccated mummies whose scientific study can be expected to yield substantial bioanthropological data.

"Catacomb mummies" from Navelli, central Italy

Another group of mummies from the Tuscany highlands in central Italy was identified in the small community of Navelli in the province of L'Aquila, Abruzzo. There, in five rooms beneath the Church of San Sebastiano, are at least 200 (probably more) mummified bodies. They were discovered when a part of the church floor collapsed, revealing the body-filled rooms beneath the floor. They probably represent a municipal burial site. If so, they may represent a priceless collection of enormous paleopathological and historical value, beginning with the church construction in the thirteenth century. Their desiccated state suggests that they are spontaneously mummified but to date there has been no report of scientific studies of the Navelli mummies (Capasso & di Tota, 1991).

Mummies from Arezzo, Tuscany

During recent restorations of the floor in the Basilica of San Francesco in the Tuscany community of Arezzo, nine wooden coffins were discovered beneath the floor. The basilica had been constructed in 1377 above an earlier church built at the end of the thirteenth century. The rooms of the latter had been filled with soil and used as tomb sites. Thus protected from rain and groundwater, the dry soil that filled five of the coffins appears to have enhanced spontaneous desiccation of the bodies. These represent wealthy laic supporters of the church. All were dressed in later sixteenth century clothing. Currently they remain under study (Fornaciari, 1997). However, radiological examination (including total body helical CT with three-dimensional reconstruction) was carried out on a female body, aged 20–30 years, clothed in a lace-adorned Renaissance dress. The study demonstrated not only renal lithiasis and cholelithiasis but also swollen breasts, pubic diastasis and a large abdominal mass. Laparoscopic biopsy with histological study confirmed the latter as uterus. These findings strongly suggest recent pregnancy, with death occurring as the consequence of a puerperal complication. The Basilica's archives include the burial of a woman who died shortly after delivery, a report consistent with this scenario (Ciranni *et al.*, 1998, 1999).

Mummies from the Abbey of St. Domenico Maggiore, Naples

In addition to the mummification of bodies in clerical catacombs, some Italian bodies became spontaneously mummified and others remained anthropogenically mummified when placed in coffins or sarcophagi within churches. Probably the best example of this phenomenon is in the St Domenico Maggiore Abbey in Naples, southern Italy. This monumental structure includes a passageway suspended about 5 meters above the church floor that was constructed specifically to house 38 coffins containing bodies of the Neapolitan kings, queens and nobility from the fifteenth and sixteenth centuries. Their scientific value is enhanced substantially by the fact that many are individually identifiable and were so well known that we have much historical data about them with which paleopathological findings can be correlated (Fornaciari, 1985).

Fourteen of the 27 bodies that became available for study were anthropogenically mummified, 7 mummified spontaneously and 6 were completely skeletonized. Soft tissue preservation after death in Naples is probably dependent on the season of death and the microclimate of the coffin's location in the church. Naples' summers (May-September) have a mean daily high of about 26 °C, but only 16 °C the rest of the year

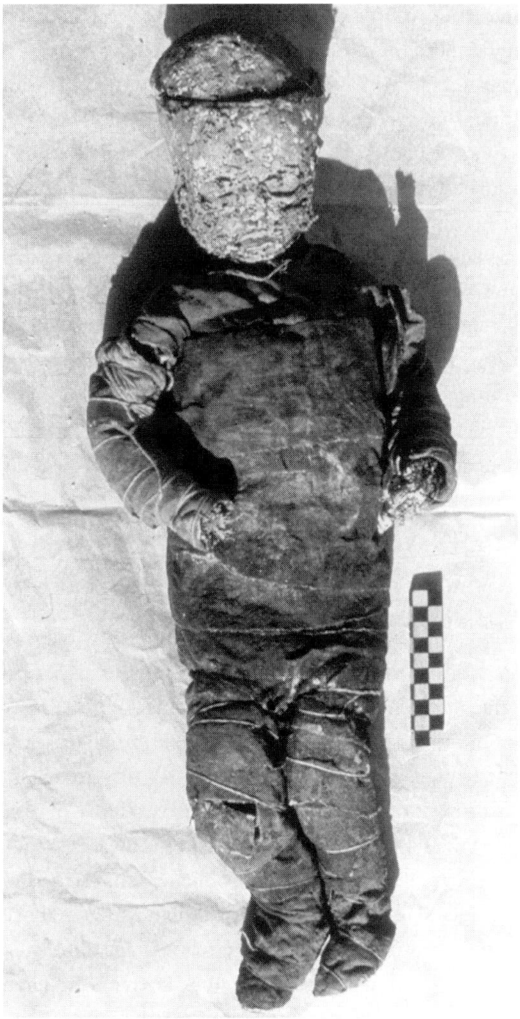

Fig. 4.117. Child mummy of Naples; ca. AD 1530. Evisceration and mercury salt insertion in body cavities were employed to preserve this child member of the Naples' elite family in the seventeenth century. Mummy no. 2 of San Domenico Maggiore Abbey. (Photo courtesy of Gino Fornaciari.)

(Hall, 1981:236). Placement of a coffin suspended in the upper part of the church will subject the contained body to a temperature substantially higher than those recorded figures, and can be expected to enhance soft tissue desiccation. The skeletonized bodies of these individuals may have died during a cooler season or the coffin may have been placed initially in a different location within the church, whose temperature was less conducive to mummification.

While the general principles of anthropogenic mummification were similar for the bodies in which significant human effort at preservation was exerted (evisceration, decerebration, etc.), the details often differed widely. The evisceration incision ranged from small to torso-length; its direction was oriented craniocaudally in most, but in two it was transverse. Following evisceration the body cavities in some (five) were filled with resin-soaked cotton-wool wadding, with earth (one), with lime (one), with mercury-containing substances (two) (Figs. 4.117 and 3.11) or with lime and aromatics (three). Similar materials were placed into the cranial cavity following craniotomy and brain removal. Considerable excarnation was obvious in four bodies.

Fornaciari & Capasso (1996) also remind us that, in Europe, bodies of deceased, distinguished individuals were commonly placed in a sitting position over a jar that collected the fluid dripping from them for several months. Thereafter they were washed with vinegar and placed in a wood box that was then filled with dry soil for several more months. With desiccation completed it was then transferred to its permanent coffin. The appearance of the spontaneously mummified Neopolitan bodies was consistent with such treatment.

Subsequent study of these bodies has generated some striking findings. The ulcerated skin lesions in a 2-year-old boy revealed a virus structure consistent with smallpox when studied by electron microscopy, confirmed with immunochemical staining (Fornaciari & Marchetti, 1986). A pelvic tumor in the abdomen of Ferrante I of Aragon, King of Naples (1431–1494), proved to have the histological structure of a colon adenocarcinoma; molecular biological studies demonstrated a mutation in the K-*ras* gene as expected in modern colon adenocarcinoma (Marchetti *et al.*, 1996). A bandaged arm ulcer of Mary of Aragón (1503–1568) revealed spirochetes in the gumma-like lesion studied histologically and by electron microscope. The diagnosis was confirmed to be treponematosis by indirect immunofluorescence using a human anti-*Treponema pallidum* antibody (Fornaciari *et al.*, 1988a). An invasive naso-orbital tumor in the Duke of Gravin (died 1549) revealed a histological structure of a malignant epithelioma (Fornaciari, 1985). Other studies on these bodies are ongoing, but the conditions already identified are dramatic examples of the research potential in the scientific study of mummies.

Mummies of Italian saints

In addition to the mummies noted above, Italy also curates another group of cleric-related mummified whole bodies or body parts. These represent the mummies of individuals sanctified by the church. Fulcheri (1996) has provided us with a superb review of this phenomenon and the interested reader is urged to read this easily accessible article in detail. In this volume I can only abstract some of the principal features. Of these, Fulcheri notes that postmortem preservation of soft tissues is viewed as evidence of divine recognition and even today is an item of consideration when one evaluates the qualifications for granting sainthood status. Initially, spontaneous mummification was the desirable form of evidence but later some efforts to enhance preservation of tissue was made. This included evisceration and use of balsams and aromatics in the case of martyrs or for other reasons that led to consideration of possible sainthood status at the time of death. Anatomical dissection of exhumed bodies was sometimes carried out to confirm the integrity of the body. When death occurred a long distance from home during the Crusades, a practice evolved that consisted of excarnation and boiling of the body for purposes of returning the skeletal remains. This was subsequently practiced for possible saints as well. The resulting mummification permitted multiple parts of the body to be processed and buried or installed in various churches as relics by several different devoted groups. Indeed, it was not unusual to carry out anatomical dissection and mummification for the specific purpose of obtaining relics that would be used for veneration.

Fulcheri documents several specific examples of such mummies, including one in whose examination he had participated (Santa Margherita of Cortona); the body had undergone many incisions of the limbs as well as evisceration and the application of balsams. In other cases anatomical evidence was sought in the form of some feature that could be considered as a symbol of the divine (Fig. 4.118). Santa Chiara da Montefalco's body was dissected by one of her convent's nuns who is alleged to have recognized clearly the presence of a cross (perhaps postmortem clot or embolus?) within the heart. He also notes paleopathological findings in modern studies of some of these saints' bodies. Multiple fractures of tibia and fibula in one individual may reflect crurifragium in a martyr. The finding of congealed lead along the esophagus and trachea in the martyred bishop San Ciziaco probably reflected the occasional contemporary practice of pouring hot, molten lead into the mouth of heretics. Fulcheri estimates that Italian churches probably curate more than 300 relics and at least 25 mummified bodies of saints (Fulcheri, 1987a,b). Tables within this valuable article document the names, dates and references for both spontaneously and anthropogenically mummified saints. His interest led to a conference held in L'Aquila, Italy at which these bodies of the saints were declared to be a national heritage, deserving of support for both scientific study and conservation. Fulcheri (2001, personal communication) has also initiated a registry for mummies of Italian saints.

The mummy of Saint Zita

Saint Zita died in 1278 at the age of 60 in central Italy and her venerated, spontaneously mummified body was kept in the Basilica of Saint Frediano in Lucca, Italy. In 1988 Fornaciari *et al.* carried out an interdisciplinary study on this body that included endoscopy, radiology, histology and atomic absorption spectroscopy (Fornaciari *et al.*, 1989b). When they encountered

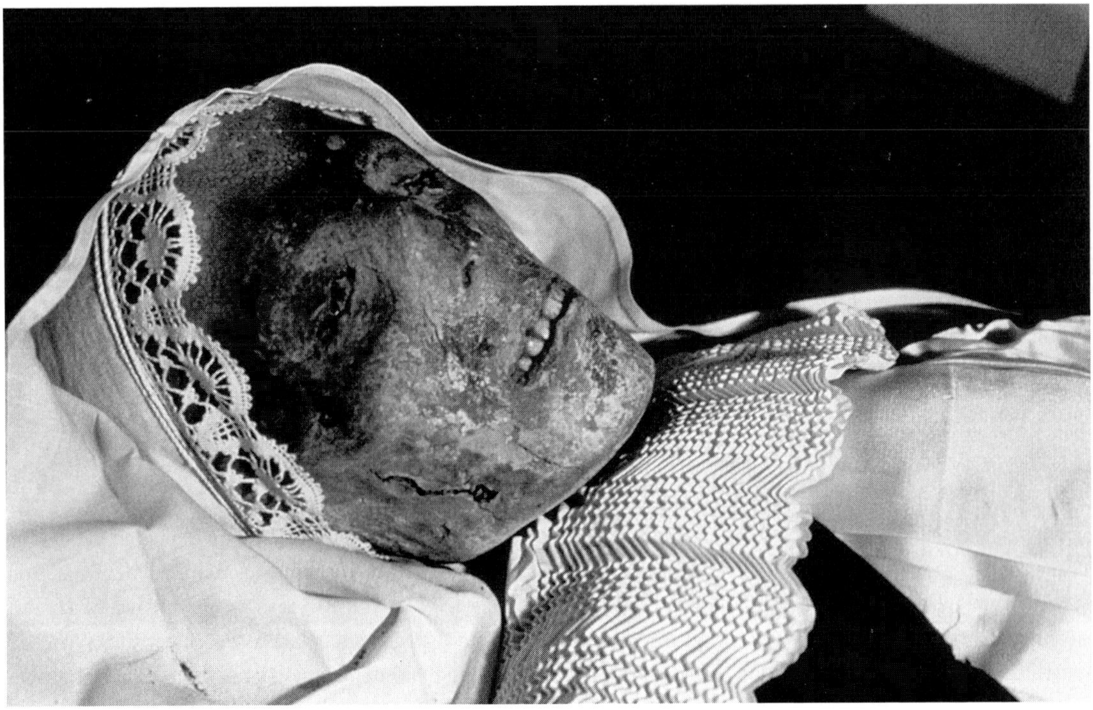

Fig. 4.118. Renaissance clerical mummy; ca. AD 1320.
The body of Beata Margherita Vergine di Città in Italy was anthropogenically preserved by methods including evisceration. (Photo courtesy of Ezio Fulcheri.)

numerous areas of black material embedded in focally fibrotic areas of pulmonary tissue, they employed Fourier transform infrared microspectroscopic methodology to compare that tissue with normal lung and also with coal, confirming the histological diagnosis of anthracosis with fibrosis (Fig. 4.119).

Saint relics are not always treated with such reverence. Mould (1996a:97) published a sketch of skeletal relics of St Blaise, a martyred bishop. Drawn in their individual, appropriate anatomical positions, the result was the portrayal of a five-headed skeleton with six arms and legs (Fig. 4.120).

Miscellaneous Italian mummies

Several individual mummies in Italy have features of interest. They include the following.

The Grottarossa mummy

The Romans practiced cremation as their principal mortuary ritual, so the finding in AD 1964 of an 8-year-old female's body that had been prepared by anthropogenic mummification in a grave near the Via Appia 8 kilometers from Rome startled the archaeological community. No evisceration wound was present and the brain was intact. No evidence of natron or bitumen use could be detected, but chemical tests identified the presence of coniferous resin and myrrh pollen on the body. The sarcophagus was clearly Roman, second century AD. The face had Caucasian anthropometric dimensions, but the wrappings were linen. A lump of embalming resin was found on the skin. Many of these features are characteristic of Egyptian type mortuary practice during the Roman

Fig. 4.119. Saint Zita; SM, AD 1278.
The body of this saint reached a desiccated state spontaneously in a church in Lucca, Italy in 1274. (Photo courtesy of Gino Fornaciari.)

Period (50 BC to AD 400), though the body is clearly Caucasian (Fig. 4.121). Why Roman parents would employ Egyptian mortuary techniques remains speculative (Ascenzi *et al.*, 1996)

"Mummies" of Pompeii and Herculaneum

In AD 79 Mount Vesuvius erupted, burying the nearby cities of Pompeii and Herculaneum. The falling hot ash first smothered the fleeing residents, then hardened around them. This process suspended both the communities and the inhabitants archaeologically at a single moment in time, analogous to a single frame of a long motion picture film. Recent excavations reveal that at least some of the bodies' skeletons but no soft tissue had survived. Nevertheless, the soft tissue had remained intact while the ash that covered them hardened. Subsequent soft tissue decay created a cavity within the hard ash that retained the form of the body at the moment of death. Recently excavators breaking into such a cavity carefully filled it with plaster, then removed the surrounding hardened ash, recreating the human body form as a plaster sculpture. These were so detailed that the method of the fleeing individual's desperate effort to escape the volcano's effect is patently apparent (Figs. 4.122 and 3.15). Several of these "plaster" mummies are currently on display in the modern museum of excavated Pompeii (Place, 1995b).

The Varese mummy

In the northern Italy community of Varese, the local museum curated the spontaneously mummified body of a boy of about 11 years of age, and of unknown

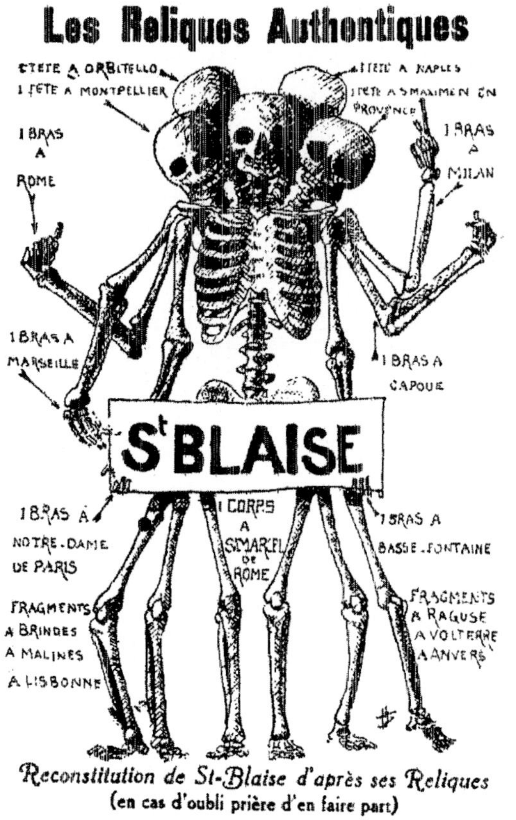

Fig. 4.120. Relics.
A diagram of relic bones curated at various clerical institutions (as indicated), all claiming to be from the body of St Blaise. (From Mould, 1996a. Courtesy of the Wellcome Library, London.)

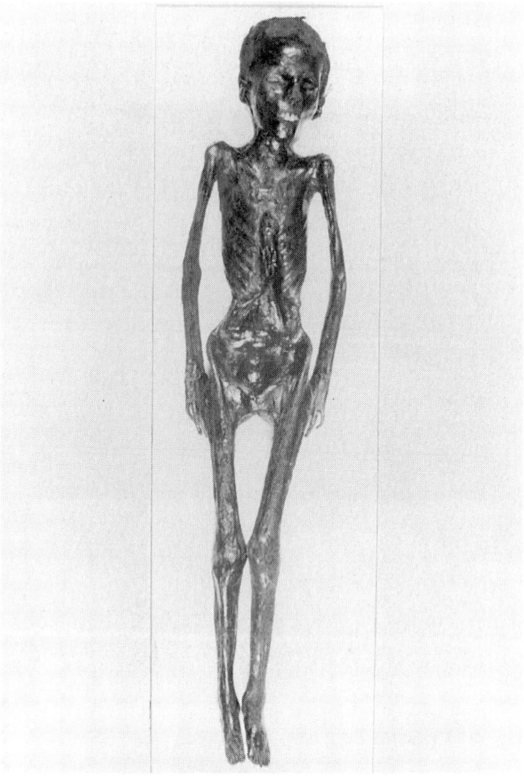

Fig. 4.121. Grottarossa mummy; AM, ca. AD 150–200. This child's body, found in Rome, had mummified features commonly used in Egyptian mummification. (Photo courtesy of Antonio Ascenzi.)

provenience. Radiology revealed multiple Harris lines of bone growth arrest, and a canine tooth showed enamel hypoplasia. Using a posterior approach, Fornaciari *et al.* (1993) dissected the thoracic and abdominal cavities and concluded that, following a succession of severe illnesses, the boy succumbed to an acute respiratory infection.

The mummy of Pandolfo III Malatesta

This Italian nobleman who died in AD 1427 was a well-known historical figure. Known as a valiant soldier and avid horseman, his military feats led him to become prince of Brescia and Bergamo. Historically he died of a fever at Loreto in central Italy. His sarcophagus was recently identified in Fano (Marche, central Italy). When endoscopy of the sarcophagus revealed a well-preserved body, Fornaciari & Torino (1995) carried out an autopsy. Findings included a renal staghorn calculus, benign prostatic hypertrophy, robust bones with prominent muscular attachments, right hand exostoses likely to be secondary to sword use and bowed, calloused knees – the probable result of a lifetime indulgence in horseback riding.

Fig. 4.122. Pompeii "ash mummy"; SM, AD 79.
These represent plaster casts created by archaeologists who poured plaster into defects found during excavation of the ash that covered Pompeii during Mt Vesuvius' eruption in AD 79. (Photo courtesy of Sylvia Horwitz. See also Fig. 3.15.)

Egyptian mummies in Italian collections

Fulcheri (1996) notes that Italy has at least 28 different collections of Egyptian mummies, totalling 142 mummies and 214 body fragments. The collection at Turin represents 96 mummies and 85 fragments. These are included in the section devoted to Egyptian mummies discussed separately in this volume.

The Tyrolean "Iceman" or "Oetzi"

In the early fall of 1991 two vacationers were hiking in the Oetztal region of the Tyrolean Alps when they came upon a human body partly frozen in a meltpool at a glacier's edge (Plate 4). After the find was reported, the body was viewed by several visitors, but it was four days before it was transferred by helicopter to the University of Innsbruck. There, anthropologist Konrad Spindler recognized that this was not just another of the "glacier bodies". These are bodies of mountaineers who fell into a crevasse many years earlier and finally were extruded from the ice as a result of glacial ice movement. One or more emerge from Tyrolean glaciers at irregular intervals nearly every summer. Among the body's associated artifacts were numerous tools and weapons including a foxhide quiver with arrows, fragments of a grass cloak, a lithic blade and the wooden frame of a backpack. It was, however, the body's axe that caught Spindler's eye. It was covered by a dull patina but its shape was obvious. If the material beneath the patina was bronze, it was at least 3000 years old; if simple copper, at least 5000 years. Later radiocarbon dating and chemical studies established the latter.

Responsibility for study and preservation of this unique find instantly became controversial as soon as its age became known. The site was close to the Austro-Italian border. A surveying team soon determined that it was about 100 meters within Italian territory. Eventually a compromise was reached: the Austrians would retain possession of the body, preserve it and study principally the artifacts scientifically for about 5 years, after which it would be transferred to Italy for permanent custody.

Spindler and colleagues plunged into a rapidly established research protocol. The body had apparently lain in a small, sharply defined rocky depression that permitted the glacial ice to move over the body without disturbing it. Thus it was not subjected to the sometimes shearing action of moving glacial ice. Its exposure in 1991 was attributed to recent climatic conditions that brought reduced snowfall and warmer temperatures. The body itself was that of an adult male of 40–55 years of age at death. Initial research focused on the associated artifacts. His equipment was that of an experienced mountain hunter, though some cereal grain kernels found on his clothing suggested his membership in an agricultural-based community. The body itself had become freeze-dried, and was essentially intact except for some tissue lost during the process of extracting the body from the ice. At the University of Innsbruck it was preserved in a specially constructed cabinet that kept the tissues frozen at controlled relative humidity.

Research on the artifacts permitted successful reconstruction of the clothing. This included a fur cap and shirt, covered by a grass cloak, fur leggings suspended from his belt by a leather strap, and leather shoes. In addition to 14 arrow shapes, the quiver also included a small arrow repair kit. Ultimately some of the body's viscera were sampled by needle biopsy. DNA studies suggested that this material had undergone extensive degeneration, although some short fragments could be amplified. After about 6 years and a hundred different studies on samples from the body and its artifacts, the frozen mummy was transferred to Italy, where a special refrigerated room had been constructed with a glass window for the public to view it (Spindler, 1992, 1994; Nedden *et al.*, 1994; Gorman, 1999). The first step in the study of the body was to thaw it and carry out radiography. This established the probable cause of death when it revealed a lithic point embedded in the shoulder.

Overview of Italian mummies

In view of the fact that Romans cremated their dead, it is not surprising that most of the Italian mummies are from the medieval era and the Renaissance Period. Their principally elite status offers very exciting opportunities, since extensive historical information can be correlated with paleopathology findings. Both the investigators and the mummies' curators deserve kudos for their willingness to exploit these opportunities by conducting bioanthropological and paleopathological research on such bodies. This field of study probably has no better examples of the profound increase in diagnostic specificity and resulting new information contributed by the application of modern laboratory methodology to ancient human remains.

Mummies from Greece

Greeks during the Classical Period and the Hellenistic Age did not mummify their dead. An exception was Alexander the Great, whose mummification process is discussed in Chapter 3. However, some soft tissue preservation occurred beneath the copper mask and pectoral artifact of a body of a male in one of the shaft graves (V) at Mycenae, ca. 1500 BC (Stone, 1975; Dickinson, 1977:49, 57; Steel, 1996). This is most probably the enzyme-inhibiting effect of copper leached from the mask by groundwater. Hans Schliemann, who was in passionate search of

Agamemnon's body, summoned a local pharmacist who painted the entire body with a mixture of resin dissolved in alcohol, designed to solidify the soft tissue sufficiently to enable removal of the body from the tomb intact and preserve the surviving skin and subcutaneous tissues (Stone, 1975:384–394).

Mummies from Russia and Ukraine

Mummies from the Siberian steppes

A belt of more or less unbroken high plains grassland (steppe) extends from about Kiev, Ukraine, in the west to Siberia's Lake Baikal in the east at roughly 50° N. In the first millennium BC this was occupied by a succession of tribes with overlapping territorial occupation (Artamonov, 1965). While they either raided or traded with each other a good deal, they shared a language family and common cultural features. Theirs was a pastoral, yurt-housed, semi-nomadic subsistence centered around the horse. In the area north of Iran they were called Scyths or Scythians by Herodotus (ca. 450 BC/1981, Book IV:I, 27) in antiquity. However, that appellation in modern times has been applied often to the steppe groups as a whole, even though it is improbable that those tribes near Iran ever had contact with the eastern tribes north of eastern Mongolia (Barber, 1999). West of the Urals modern historians may use the phrase "horsemen of the Eurasian steppes" while east of that range the term "Scythio-Siberian" may be employed (Polosmak, 1994). For convenience, this discussion will use the term Scythians to apply to all the steppe tribes unless otherwise indicated, particularly since Chinese documents from the third century BC report that an earlier Indo-European people transmitted Scythian art styles to Mongolian Huns and Chinese (Artamonov, 1965).

Scythian mortuary practices for their chieftains included underground burial in log chambers covered by a gravel mound (kurgan). The Russian word *pazyryk* means mound but subsequently Pazyryk has been used not only as a name for the ancient people, but also for the area. These kurgans were often looted both in antiquity and later. During the seventeenth and eighteenth centuries sufficient Scythian art objects became available to permit Czar Peter the Great to accumulate an enviable collection. Most Scythian burial excavations have occurred in the Altay province, a region many Russians proudly claim as Asia's geographical center (about 49° 20′ N, 87° 30′ E). In much of this area at an altitude of about 2500 meters the ground more than a meter below the surface grassland is permanently frozen (permafrost). V.V. Radloff excavated two burials in this region as early as 1865 and found distinctive artistic artifacts closely resembling those of Scythian burials north of Iran. Rudenko (1970) found several such mounds near Schibe in central Altay in 1924, trenched them in 1927, and excavated one in 1929. Interrupted by World War II he returned in 1947 in a collaborative excavation by the Institute of History of Material Cultures of the USSR Academy of Sciences and the Hermitage Museum. Four more burials were excavated in 1947–1949. All had been looted in antiquity (Rudenko, 1970: xxxiii).

Rudenko's reconstruction of these tombs indicates that they were built in spring and summer. Aided by melting of the exposed surface, a vertical, rectangular depression was excavated to a depth of 4–5 meters and lined with logs. After preparation, the body was placed in a hollowed-out, longitudinally bisected log, often accompanied by that of a female (queen? retainer?) in a similar coffin. In at least one tomb a four-wheeled wagon was found as well as horses' bridles and saddles. Two or more horse bodies were placed immediately outside the log chamber's walls. Numerous other items including felt wall hangings, rugs, clothing, wood carvings, and items of metallic art were often found. A log roof was then constructed, the remainder of the shaft filled with soil and a large mound up to 45 meters in circumference and up to 3 meters high, consisting of gravel and rocks, was built over the shaft. The first rain after completion trickled through the rocks, down the shaft and eventually filled the log chamber. Buried in permafrost and insulated by the rock mound this now ice-filled chamber did not thaw again until the tomb was opened by looters or archaeologists (Bogucki, 1996).

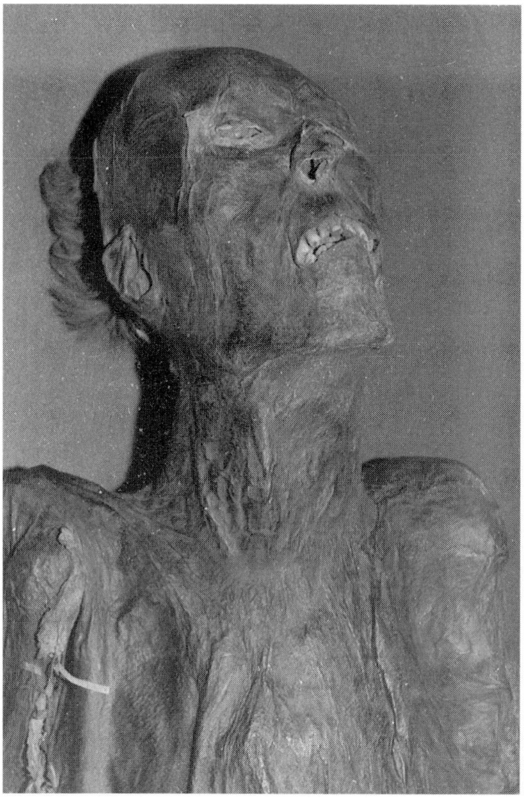

Fig. 4.123. Pazyryk mummy; AM, ca. 425 BC. Mummies from this central Asian plateau were eviscerated initially for temporary preservation, but interred in permafrost about 1 year after death. (Photo courtesy of Leonid Sergeevich Marsadolov from The Hermitage, St Petersburg, Russia.)

Scythian mortuary ritual also included preparation of the body in a manner that would minimize soft tissue decay so the body could be transported to many areas to be viewed by the populace and until the above-described permanently frozen state was achieved. Evisceration was carried out via an abdominal incision (Fig. 4.123), and excerebration through a defect created in the occipital bone (rarely via the temporal bone). Herodotus (ca. 450 BC/1981, IV:71) states that the body cavities were filled with chopped cypress, frankincense, parsley and anise seeds. Observations also indicate that posterior incisions were made along the length of the spine from occiput to sacrum and from shoulder to shoulder as well as medially along the length of the extremities. Rudenko (1970:279–287) speculates that these, together with many stab wounds over the back and the buttocks served as entry ports for the introduction of tissue preservatives, though he admits he has no supportive evidence for this assumption. To me they resemble skin incisions to drain tissue fluids. In some cases excarnation (Barrow 5) of at least the extremities was carried out, with replacement by sedge grass, probably via such incisions. These incisions were then sewed with continuous sutures. Bodies were buried only in spring and summer, probably to enhance excavation in permafrost. Presumably the relatively elaborate anthropogenic mummification procedure was for individuals who died during the other seasons and was designed to prevent putrefaction while the body awaited construction of the kurgan and lay in state in various communities. Tufts of hair and fingernails in small leather bags were found in several kurgans. As in some other, unrelated burials throughout the world, these could reflect beliefs that the deceased's spirit may be resident in such tissues.

Their weapons, devotion to horses and habit of raiding their neighbors has caused history to paint a Scythian image of a fierce, aggressive, unfeeling people. This inappropriate view, however, ignores the obvious evidence that among them were at least some more sedentary individuals who were highly sensitive to the beauty of their environment, and expressed this in their artifacts, particularly those relating to horses. In some of the royal kurgans almost every square centimeter of the bridles' and saddles' leather surfaces are emblazoned with geometric and animal figures and patterns (realistic, stylized or fantastic), joined together by metal rivets with impressed faces. These demonstrate an extraordinary harmony of symbolism, elegance and function (Gryaznov, 1950). That this appreciation was not limited to exceptional, creative artists may be evidenced by the observation that the identical saddle patterns are sometimes duplicated in the form of tattoos on the mummified skin of the rider in the tomb. In addition to their aesthetic features, these patterns (together with linguistic clues) have been viewed also as a form of information-transmission writing (Sadovsky, 1985). They have been employed to link the Pazyryk people to the Canadian Inuit and the modern Native Americans of Ventura County north of Los Angeles, California.

The role of the horse in these people of the steppes can not be exaggerated. This is most obviously exemplified by a kurgan in the Turan-Uyuk Valley near Kysil (Kyzyl) at about 52° N and 95° E. The man and woman interred in this Pazyryk-type kurgan, named after the nearby village of Arzhan, were surrounded not only by the bodies of 15 dignitaries, but also by those of many horses. The ritual significance of these equine bodies became clear when excavation demonstrated individual underground wood cells, 180 of which were arranged in linear alignments radiating 360° from the central tomb. At least some of these are believed to have been constructed with astronomical relationships. Many, though not all, of these cells contained the body of a horse. These are viewed by excavators as reflections of world views conforming to nature's annual cycle of life and death, with each horse representing one day (the total = one-half year). This arrangement is duplicated by another steppe kurgan at Ulug-Chorum in southern Tuva province, and the arrangement is similar to some of the Native North Americans' stone circles (Marsadolov, 1993).

More recent excavations of Altay kurgans from the Ukok plateau have recovered several female eviscerated, mummified bodies. One of these is an 18–20-year-old female about 163 cm tall who has been dubbed with several informal names: Ukok Maiden (Campbell, 1994), Ukok Princess (Bogucki, 1996). Her excavator, however, referred to her simply by the dignified term "The Lady" (Polosmak, 1994). She was enveloped in silk, crowned with a half-meter-high headdress, with a Scythian-pattern tattoo on her left arm and accompanied by six horses. Inclement weather, a transporting helicopter crash, and administrative problems delayed her evacuation from this mountainous area (Campbell, 1994). She was subsequently conserved by Sergei S. Debon, the Russian biochemist who also embalmed Ho Chi Minh and helps to maintain Lenin's body. She has even undergone facial reconstruction by Dr Mikhail Gerasimov, who also carried out a similar sculpted portrait of Ivan the Terrible (Stanley, 1994). A total of about two dozen such bodies have been recovered from kurgans in the Altai region. Bioanthropological studies on these bodies have not yet been reported.

Russia's royal Romanov mummies

After more than three centuries, Russia's Romanov family, one of the last representatives of Europe's absolute monarchies yielded its hegemony by submitting to arrest by a delegation from the Bolshevik revolutionaries in March, 1917. Imprisoned in a house in Yekaterinberg, Siberia, the members of this family were held in captivity there until July 16, 1918, when Lenin ordered an execution squad to their house. Czar Nicholas II, his wife Alexandra, hemophilic son Alexei, four daughters, a valet, two other servants and the family physician were led into a basement room and summarily shot and/or bayoneted. Their bodies were trucked to a nearby forest with intent to place them in a deep mine shaft. Instead, the truck bogged down in a muddy area, so an attempt was made to burn the bodies of Alexei and his sister Anastasia. Dissatisfied with the result, the squad dug a shallow pit, disfigured faces with sulfuric acid and buried the remaining nine bodies.

Although the bodies were discovered in 1979, their exhumation had to await an appropriate political climate that evolved in 1991. Considerable skeletonization had occurred but the czar and his wife could be identified by dental comparisons. Bullet and bayonet body wounds were consistent with witnesses' descriptions. Facial reconstruction assisted identifications also, but the most definitive evidence was achieved by DNA analysis with comparisons to specimens acquired from the body of the czar's deceased brother and other, living relatives including the Duke of Edinburgh (Bahn, 1996). The bodies of Alexei and Anastasia were not recovered. Seven years later the remains were interred, amidst some controversy, at Peter and Paul

Fortress in St Petersburg, Russia, taking their place there with the bodies of their Romanov predecessors (Quinn-Judge, 1998).

The mummy of Nicholas Rubinstein

Europe's mid nineteenth century music world clasped the Russian Rubinstein brothers, Anton and Nicholas, to their heart, disputing only which of them was the world's most accomplished pianist. There was, however, no doubt which was the most enjoyable social companion – not the serious, reserved Anton but rather his boisterous, gregarious brother Nicholas. When 45-year-old Nicholas died of tuberculous enteritis in 1881 the continent mourned as it buried his unembalmed body in Moscow. The many accomplishments and escapades (in time embellished with additional imaginary exploits) of this brilliant pianist and composer gradually metamorphosed into legend, remaining vivid in the Russian public's memory.

In October of 1934 that Moscow cemetery was relocated and Nicholas' coffin was opened. To everyone's astonishment, they found Nicholas himself staring out at them, his body so well preserved that he was easily recognizable. Considering the peoples' legendary view of this man's life, it was probably inevitable that they interpreted this "uncorrupted" body preservation as a form of divine recognition via suspension of natural laws by the deity, believed to be celestial acknowledgment of a saintly life. Whatever validity such a claim has, it is not unreasonable to suspect that the basic pH of the well-drained chalk soil in which he was buried may have made a significant contribution to his body's preservation, especially since several other bodies buried nearby in this same soil had been preserved spontaneously in a similar manner (Bowen, 1939:278).

The mummy of Vladimir Ilych Lenin

Lenin provided intellectual guidance for Russia's Bolshevik Revolution of 1917. He died at his country estate in Gorki, following a succession of strokes, on January 21, 1924. An autopsy was performed the next day, after which he was embalmed "in the normal manner". He was then taken by train and lay in state at Moscow's House of Trade Unions. The massive public response to view the body overwhelmed officials, who then decided, against his widow's wishes, to prepare the body for permanent display. He was moved to a wood mausoleum in Moscow's Red Square and shown in a glass-topped, refrigerated unit, kept at 0 °C and low humidity. This resulted in gradual skin discoloration, wrinkling and drying. Noting the unrelenting long lines of viewers, officials then assigned responsibility for long-term preparation and maintenance of Lenin's remains to Vladimir Vorobyov (professor of anatomy) and Boris Zbarsky (biochemist). Details of their methods were kept as state secrets until recently but the general principles are now known to have included the use of a solution containing glycerine and potassium acetate (perhaps with formalin). A mixture of paraffin and vaseline is believed to have been injected into the skin to smooth facial wrinkles. The body was then kept at 16 °C at 70% humidity (Quigley, 1998:28–37) (Fig. 4.124).

During the next 70 years, throughout the reign of the communists, Lenin's body remained a public icon. The wood mausoleum was replaced by stone, the body was provided regular, periodic maintenance and the viewing lines never abated. The body survived two bombing attempts that shattered the bodies of the bombers and some visitors but not that of Lenin. In 1934 Boris Zbarsky yielded his maintenance duties to his son Ilye; Vorobyov died 5 years later. Between 1941–1945 Lenin was removed to western Siberia to prevent German confiscation. Today the maintenance team includes 27 scientists and 33 technicians (Quigley, 1998:28–37).

The secrecy relating to their methods has led to many rumors, some suggesting that the "body" in the casket is a wax mannequin, others that only the head and hands are real. Periodic efforts to refute such allegations have still left at least some critics unconvinced. Egyptologist Bob Brier examined Lenin in 1999 and

Fig. 4.124. The mummy of Lenin; AM, AD 1924.
The nature of embalming solutions used for preservation of the body of Vladimir Ilyich Lenin has never been revealed. (Photo courtesy of the Russian Embassy, Washington, DC.)

confirmed the fact that it was, indeed, a human body (R. Brier, personal communication).

When Josef Stalin died on March 6, 1953, his body was also assigned to Lenin's curators, who presumably used those same methods to prepare and maintain him. For 9 years the bodies of both Lenin and Stalin lay side by side on exhibit in Red Square. They were inspected twice weekly, immersed for a month in a glycerine acetate solution at 18 month intervals and enjoyed regular cosmetic applications. But political climates change, and in 1962 Stalin's body was removed from the mausoleum and presumably buried.

Miscellaneous mummies from Europe

Cromwell's head

European mummies also include several oddities among which the perambulations of Cromwell's head is perhaps the most curious. The effects of the tumultuous events that characterized the career of Britain's Oliver Cromwell followed him beyond his grave. When he died in 1658 an autopsy was performed (his cause of death was declared to have been "the ague", a term at that time applied to malaria). After the autopsy the body was embalmed (by methods unknown) and promptly buried. Nor did his regime survive. Supporters of Charles I, whose execution Cromwell had guided, recalled Charles II from exile and avenged themselves on Cromwell by exhuming his body, decapitating it and impaling his embalmed head on a metal spike. They then fixed this to a pole that they mounted atop Westminister Hall. There it remained for more than two decades, its sightless eyes overlooking the comings, goings and doings of the parliament he had so cavalierly dissolved.

When a gale-driven storm finally tore it from its fastener one dark night, a soldier retrieved it as a souvenir. Twenty-five years later it reappeared in a museum and disappeared again when that firm closed in 1738. In 1775 a jeweler, who had purchased it from an actor, sold it, after which it passed through several private collections. The last known of these collectors can be dated to about 1950, when he offered the head to Cromwell's alma mater, Sidney Sussex College in Cambridge, England whose administrators then buried it at a nearby undisclosed location (Mould, 1996b).

It is unfortunate that the reinterment occurred just before our modern analytical skills became applied to ancient specimens, for surely few of our modern embalmers could achieve as durable a specimen as Cromwell's head.

MUMMIES FROM EGYPT AND THE REST OF AFRICA

Anthropogenic mummification evolved in Egypt over more than 3000 years. The stepwise nature of this process is presented below, correlated with the contemporary sociopolitical changes. Readers interested only in the most fully developed stage of Egyptian mummification can proceed directly to the detailed description in the section The New Kingdom.

Geography and climate of Egypt

Egypt forms the northeastern portion of the Sahara – that vast, arid, rocky desert whose difficult traverse undoubtedly played a major role in the development of its ancient civilization by providing Egypt with protection from neighboring invaders. Traditional regionalization of this country employs the northward-flowing River Nile to separate its eastern and western portions. Additionally, the immense marsh the Nile has created at its entry into the Mediterranean Sea has been named the Delta Region, while the almost detached, triangular peninsula wedged between Egypt and Arabia is named the Sinai. Finally, the fertile Nile itself, with its consequent densely adherent population clustered along its banks, merits separate regional recognition.

Originating in the mountains surrounding Lake Tanganyika and Lake Victoria as the Kagera River in the heart of Central Africa's most remote forest area, the River Nile (in successive segments called the Victoria, Albert and White Nile) wanders northward through jungle country, emerging from swamp and forest and flowing through the southern part of the Sahara Desert as far as Khartoum, a city in what is now Sudan (Fig. 4.125). Here the White Nile is joined by the Blue Nile, a river of formidable size. Thereafter these merged waters are termed simply "the Nile" throughout the remainder of its northward course. The Blue Nile is the Nile's most important tributary and contributes four-fifths of the Egyptian Nile's water. Annual spring monsoons inundate Ethiopian mountains, high plateau and Lake Tana, generating torrents whose violent turbulence wrenches the region's rich, black soil from those slopes drained by the Blue Nile. Disgorging this black, silt-laden deluge into the Nile causes the river's downstream course to flood through drainage canals cut into its high banks, annually depositing from July until October a fresh, highly fertile layer of silt on the lower fields constituting the flood plain adjacent to the river. Crops were planted following the waters' recession in October and harvest began in January, continuing until April. It was the grain grown on this narrow strip of productive soil adjacent to the river that provided the energy necessary for the construction of Egypt's pyramids and the economy that made possible the rise of Egyptian civilization. The recent construction of the high dam at Aswan and

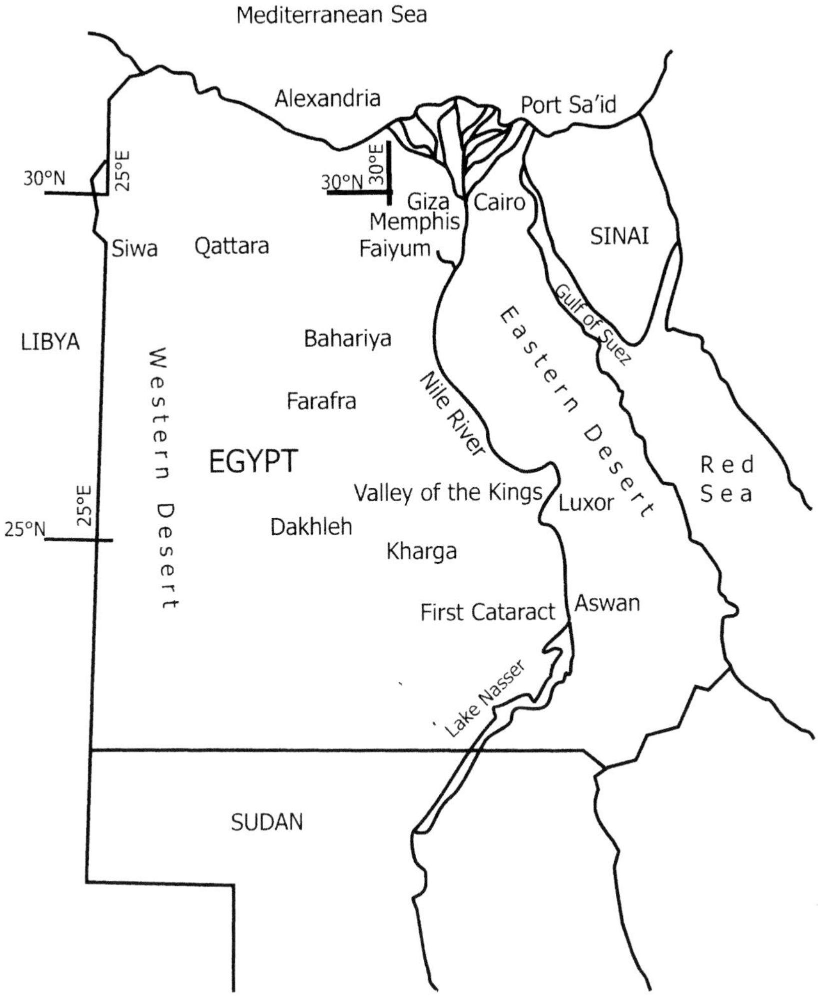

Fig. 4.125. Map of Egypt.
The Egyptian segment of the Nile threads its way through otherwise desolate terrain. It provides a narrow ribbon of cultivable land that supported most of the ancient Egyptian population clustered along its banks. The formidable Sahara Desert enveloped and protected the Nile Valley's people from hostile neighbors.

other river level controls as far as the delta now choke much of this flow of silt, forcing Egyptian farmers into the same dependence on synthetic fertilizers as are agriculturists in developed countries today. It will be interesting to see whether the hydroelectric energy generated by the dam will effectively offset the negative ecological, cultural and economic impact caused by cessation of the annual delivery of fertile soil.

While the Nile has a complex geological history, it is fair to say that the perpetual erosion by the Nile's water augmented by fluctuations in the Mediterranean Sea level over many millennia has resulted in a progressive lowering of the river bed to the point that it is now flanked by high banks (in some sections as much as 400 meters) throughout much of its course. Its average width where it enters Egypt is about 4 kilometers,

Fig. 4.126. Egyptian irrigation agriculture.
The ox-driven water wheel (*saqia*) raises water from a Nile canal into irrigation ditches. This device is still in use today. (Photo by ACA.)

though an 800 kilometer long section beyond Aswan in some areas reaches a width of 15 or more kilometers. The nutrients derived from its sources support an abundant population of fish, including the Nile perch, which was an aquatic staple for ancient Egyptians and made a significant contribution to their economy.

It was grain, however, that provided the food surplus which was eventually coveted by the larger of the then world's nations such as Rome. Egypt raised principally emmer (*Triticum dicoccum*) and barley (*Hordeum vulgare*). Earlier the flood plain (developed and controlled by cutting channels through the river's high banks) defined the perimeter of land that could be used for agricultural purposes, but, as generations passed, Egyptians developed the *shaduf* (bucket on a pole) and later even animal-powered devices (saqia) to lift water from the river into irrigation canals (Fig. 4.126). This permitted them to broaden the strip of arable land, but even today in upper Egypt it exceeds a width of a few kilometers only in certain areas. Unlikely as it might appear, that method of farming was so productive that Syrians, Hittites, Assyrians, Persians, Greeks, Romans and others at various times strove for control of the food surplus it generated.

The above generalizations, of course, include those contributions achieved by farming the delta's swamps. More than 150 kilometers long and twice that width, its wetlands and areas of brackish waters created a host of microecological areas that could be exploited for both nutritional and industrial horticulture and, of course, agriculture. Modern canals and controls of water levels have doubled and in some areas even tripled the delta's yield, lending truth to Herodotus' comment that Egypt (i.e., its civilization) is the gift of the Nile. Equally important, its transportation contribution was of enormous value. The Nile is navigable even by the smaller modern freighters as far as Aswan. Between Aswan and Khartoum are six sets of shallow rapids or small falls called cataracts, but these are short enough that goods could be transported around them by land (about 10 kilometers at Aswan) (Hamamsy & Jones, 1984). In the Middle Kingdom, channels were cut through the rapids.

The area between the Nile and Libya is called the Western Desert. Rupestrian (rock) art such as that from the Tassili and Messak Sattafet in southern Libya suggests that the Sahara at one time was a well-watered grassland supporting such animals as giraffes, elephants, rhinoceros, horses and, probably later, also camels and cattle (Lutz & Lutz, 1995). Precise chronology is difficult but, with the help of hydrologists and climatologists, it has been suggested that a humid period near the end of the last Ice Age may have filled the water basin underlying part of the Sahara, with further contributions during the early Neolithic. Drilled wells can reach these waters, creating artesian water sources even today. Exactly when desertification began again thereafter is also uncertain but it seems quite probable that its arid status deteriorated further through the late fourth and much of the third millennium BC (Baines & Malek, 1983:13). Certainly at the time of the Old Kingdom it was still a royal hunting site for grazing animals but since the New Kingdom it has maintained its present hyperarid state (Brewer & Teeter, 1999:23).

The Western Desert is the largest and probably most desolate of Egypt's modern regions. Relatively flat with few wadis, it receives so little rainfall now that most of its seminomadic people are settling into villages. Largely a rock-exposed desert with level terrain (except in its extreme southwest corner), it does embrace several areas of sand dunes, some of which reach a height of 150 meters in its northern and central portion, especially south of the Qattara Depression.

Depressions are a singular feature of the Western Desert. They represent focal areas, averaging perhaps 600 to 800 square kilometers, whose surface is depressed as much as 350 meters below that of the surrounding desert. In some cases such as at Dakhleh, the northern periphery is characterized by a nearly vertical escarpment (Fig. 4.127). The consequence of such a dramatic difference is that the depression's surface may be at or below the waterline, and the spontaneous exudation of water on the surface may create a localized grassland (oasis). A succession of these begins at the Dungus Oasis a short distance west of the Nile Valley near Aswan and, with the exception of the Fayum, extends in an approximately northwest direction to Kharga, Dakhleh, Farafra, Bahariya, Qattara and, finally Siwa near the northeast corner of Libya. Greco-Roman cemeteries containing large numbers of mummies have been found in Kharga (Dunand & Lichtenberg, 1991), Dakhleh (Aufderheide *et al.*, 1999b) and Bahariya (Hawass, 2000a). These oases are separated from each other by roughly several days of desert animal caravan travel. Eventually trading activities became centered in these, though they were actively exploited by the Egyptian state principally in its later periods. Earlier dynasties seem to have viewed the oases as being of value for deterring invasion by Libyans. Though probably no part of the Sahara is immune to the dry, hot, southern winds (*khamsin*) that commonly arrive at the end of February and remain until June, the Western Desert appears to be particularly vulnerable to them (Hamamsy & Jones, 1984:448). These can create choking, sun-obliterating, blinding dust storms that can be fully as threatening as a snow blizzard on the North American plains. This region's low rainfall permits the use of unfired mud brick as the principal architectural construction method.

The Eastern Desert lies between the Nile Valley and the Red Sea. It differs profoundly from its western equivalent in that terrain is higher and often sculpted. A ridge of discontinuous mountains parallels the Red Sea coast. Some of these "Red Sea hills" have peaks approaching more than 2000 meters, though their mean height is probably only half that. Their eastern slopes descend quite steeply down to a narrow coastal plain and are extensively dissected by wadis (normally

Fig. 4.127. The escarpment at the southern edge of the Dakhleh Oasis.
The depth of the Dakhleh Oasis floor in the foreground is made evident by the 330 meter high escarpment in the background. (Photo by ACA.)

dry ravines that drain the occasional torrential downpours to which the Eastern Desert is susceptible). Such wadis also lacerate the generally less precipitous inclines on these hills' western side, where they drain into the Nile. These wadis make north–south travel in the Eastern Desert difficult, but many wadis served as east–west routes between the Nile Valley and the Red Sea in antiquity. In addition, the Eastern Desert was extensively mined for precious stones such as lapis lazuli and gold.

The Sinai Peninsula's southern area resembles the Red Sea hills. Its wadis lead into the Red Sea as well as to the north into the Mediterranean. However, its northern portion is composed of several broad, irregular, arid sandstone plains. The Suez Canal separates the Eastern Desert from the Sinai Peninsula.

The Fayum was a lowland area, just west of the Nile, fed by a small Nile branch (Bahr Yusef) originating near Asyut; today it is a canal. This low area was a dead end for the Nile overflow, creating a large lake (Moeris) in antiquity that occupied most of the Fayum early in the Pharaonic Period. Much of that land was reclaimed and planted, beginning with Dynasty 12, but most of it by the later Ptolemies.

Nubia

The land between Aswan and Khartoum in antiquity was a picturesque corridor but of low agricultural productivity. Today much of it lies below Lake Nasser, created by the high dam at Aswan. Pharaohs valued the region's accessible sandstone for construction, as well as its role as a corridor for central African products, such as ivory and ebony. The degree of control over

Nubia by ancient Egypt's rulers fluctuated throughout its history. As a generalization, Egyptian domination of Nubia can be viewed as maximal during Egypt's stable periods (Old, Middle and New Kingdoms) but far more tenuous during the Intermediate Periods. Indeed, Dynasty 25 included Nubians as pharaohs (Baines & Malek, 1983:20). However, during much of Egypt's ancient period Aswan marked Egypt's southern boundary.

Sources of our knowledge

The elaborate mummification of the dead by this geographically isolated population is inseparably linked to their religious credo, political governance, subsistence strategy and environmental features. An understanding of that unique custom can be attained only by appreciating its integration into these other aspects of their life. I hope the interested reader can recognize such integration in the following considerable condensation of Egypt's three-millennial history that represents my response to that daunting challenge. While I have drawn heavily on the outstanding presentations by Hoffman (1979: Predynastic), James (1974: History), Brewer & Teeter (1999: Chronology and History), Gardiner (1961: Chronology and History), David (1982: Religion), Ikram & Dodson (1998: Mummification) and others, the reader must understand that all Egyptologists rely largely on documents and inscriptions recorded on papyri, ceramic fragments (*ostraca*) and animal hide (vellum, parchment) or on inscriptions that appear on monuments, tomb walls, coffins or even mummy wrappings. We are all indebted to Young (Peacock, 1855) and especially Champollion (Hartleben, 1906; Davies, 1987:47–68) for their contributions to the decoding of hieroglyphic script, thereby unlocking 3000 years of ancient Egyptian history. For predynastic Egypt we must depend on archaeological excavations and interpretations, buttressed by the results of investigations by biophysical scientists. Central to such efforts is the list of Egyptian kings prepared in the third century BC by the Egyptian priest Manetho, who himself was dependent on many of these same sources. King lists also have been recovered from Abydos, Saqqara, and several other sites. Most of such sources were not originally created for literal historical use and they can generate confusion and contradictions when employed for that purpose. Others have been altered over time. Manetho's list, for example, is available to us only in the form of fragmented copies laden with transcriptural errors. The following, therefore, represents what appears to be the most plausible of interpretations, subject to modification by the continuing flow of new scholarly data. Following Manetho's custom, subdivisions of the 3000 year span are presented as dynasties, each of which is composed of a succession of kings presumed to have been derived for the same family or at least region (not always true). These dynasties are further grouped according to modern Egyptological custom into three major intervals (Old, Middle and New Kingdoms) followed by a Late Period, each of which is separated from its predecessor and successor by a numbered Intermediate Period (Table 4.6). Because of the historical vagaries cited above, given dates are best viewed as approximations, particularly during the earlier half of the 3000 year interval. In the main, though not exclusively, I have followed the chronology presented by Ikram & Dodson (1998:4–12).

The Predynastic Period

Sociohistorical features of the Predynastic Period

Considering how much detail we have about the Egyptians during the interval of their 3000 year history, the realization that we still do not know today who their predecessors were comes as a bit of a surprise. At the time of Egypt's unification about 3050 BC we are already dealing with a settled, agricultural population with sufficient cultural sophistication to have just developed a form of written communication. We do know that, following a drought of many millennia, a period of increased rainfall (the Neolithic

Table 4.6. *The political and regnal chronology of Ancient Egypt*

Many of these dates, especially the older ones, are approximations. The dates I selected are the principal ones used by Ikram & Dodson (1998: 8–12), though I chose to end the Middle Kingdom after the death of Sobknefru in 1781 BC and also to begin the Late Period with the Saite Dynasty 26 even though at least the earlier kings from that dynasty were Assyrian vassals. Dynasty 26 rulers centered their government at the delta city of Sais and that dynasty's period (664–525 BC) is known as the Saite Period (or Dynasty). Some published chronologies initiate the Late Period with the Saite Dynasty (as I have done here), while others begin it with Dynasty 25. Persian occupation occurred twice (525–404 BC and 341–332 BC). Arabs conquered Egypt about AD 640 and ruled until the early fifteenth century. The Ottoman Empire then held sway until the early nineteenth century.

Period	Dynasties	Years
Predynastic	—	5500–3050 BC
Archaic	1–2	3050–2663 BC
Old Kingdom	3–6	2663–2195 BC
First Intermediate	7–11a	2195–2066 BC
Middle Kingdom	11b–12	2066–1781 BC
Second Intermediate	13–17	1781–1550 BC
New Kingdom	18 – 20	1550–1064 BC
Third Intermediate	21–25	1064–656 BC
Late	26–31	656–332 BC
Ptolemaic (Greek)	—	332–30 BC
Roman	—	30 BC to AD 395
Byzantine	—	AD 395–640

Subpluvial) about 7500–7000 BC turned the parched surface of the Sahara Desert into a rich grassland populated by grazing animals. Furthermore, the people who hunted this game soon began to herd domesticated cattle on the Saharan savanna. Whether Saharan cattle pastoralism reflects a Mesopotamian influence or a local, independent invention remains speculative (Schwabe, 1978:30) (Fig. 4.128). Though archaeological evidence is tantalizing but uncertain, these cattle herders may have brought their culture and practices to the people populating the Nile Valley about 5500 BC (Hoffman, 1979:218). Sufficient archaeological data have now been accumulated, so that we may begin our cultural history of Egypt at this point.

Fig. 4.128. Stone statue from Persepolis (529–330 BC). This massive statue of a bull (his removable horns have been lost) from the Mesopotamian site at Persepolis reflects the veneration accorded this animal there. Cattle pastoralism, originating in Mesopotamia about 6000 BC, may have reached Egypt's Western Desert from the Fertile Crescent. (Photo courtesy of John Larson and the University of Chicago's Oriental Institute. Negative no. (N-1)A 24065.)

Excavations near Hierakonpolis (Nekhen) south of modern Luxor have provided some understanding of events leading to development of the Egyptian state. During the Badarian Period (5500–4000 BC, Table 4.7), Nilotes (people of the Nile Valley) began to cultivate barley and raised sheep, pigs and cattle. The subsequent Naqada I (Amratian) Period (4000–3500 BC) includes evidence of a ceramic industry. The Naqada II (Gerzean) Period (3500–3150 BC) followed by Naqada III (3150–3050 BC) demonstrate a response to decreasing rainfall, with increasing aridity. Additionally, the supply of trees as fuel for the ceramic kilns had become exhausted, forcing relocation closer to the Nile's shores. With the wealth created by the for-

Table 4.7. *The chronology of the Predynastic Period*

Period	Years BC
Paleolithic	Ends about 5500
Badarian	5500–4000
Naqada I (Amratian)	4000–3500
Naqada II (Gerzean)	3500–3150
Naqada III	3150–3050

Notes:
Gerzean is often subdivided into Early (3500–3300) and Late (3300–3050).
Source: Modified from Hoffman, 1979:16.

merly thriving ceramic industry, local chieftains shifted to agriculture as an industrial alternative, developing an irrigation system to increase the size of arable lands. Their increasing social status is reflected in larger, grander tombs and funerary furniture, suggesting increased regional conflict with expanding areas of control (kings?). By the time of Naqada III regional warfare had become common, setting the stage for the consolidation of political control over all of Egypt as a single region. This view of a gradual process of cultural homogenization is supported by a progressive decline in regional differences relating to physical, archaeologically identifiable cultural features such as ceramics, agricultural technology, architecture etc. (Hoffman, 1979:343). Of all these factors, Hassan (1988) believes that the most influential was the struggle for power.

The above sequence of events was outlined by Hoffman (1979:35–78) and, following an additional decade of study, was amplified by Hassan (1988), who notes evidence of food production in the Nile Valley by at least 5000 BC. He feels that the valley was settled by native Nilotes, by a mix of desert nomadic herders fleeing the desertification of both the Eastern and Western deserts, as well as by migrants from the Levant. He traces evidence of division of labor and suggests that the need to control disastrous Nile floods led to regional cooperation and appointment of representatives which, in turn, led to regional conflicts among them. The trade that began with ceramics and that subsequently included agricultural surplus and exotic goods further enhanced wealth concentration and the militarism that culminated in the formation of the Egyptian state. Currently scholars have come to view the process as principally a sociological one in which the military elements become dominant primarily as the final step of unification (Kemp, 1989).

Thus, archaeological evidence suggests that this was not a static segment of Egyptian prehistory. Rather, almost every investigated feature demonstrates continuous change, beginning slowly in the first, accelerating in the second and reaching efflorescence in the third segment as the Archaic Period is approached. Castillos (1982) notes that predynastic mortuary practices began with shallow, often unlined, oval pits little more than 1 meter deep often capped by a limestone rock, progressing later to rectangular tombs. Large tombs with abundant goods reflecting wealth were rare in the first interval but reached 13% by the end of the Predynastic. Coffins did not appear until the middle of the Predynastic Period. Excavations at Hierakonpolis in Upper Egypt indicated a very rapid evolution of elaborate, adobe-lined tombs just prior to Dynasty 1.

Mummification features of the Predynastic Period

The almost universal body position during the Predynastic was one of partial lower extremity flexion and lying on its (usually left) side (Fig. 4.129). The common fate of most such buried bodies was complete skeletonization. Junker's excavation in the late 1920s of 125 burials at the delta site of Merimde (founded ca. 5000 BC) produced only skeletons. Similar findings were recorded from excavations at Ma'adi, a predynastic community near modern Cairo. However at Naqada, a Predynastic site just north of Thebes (modern Luxor), the famed British archaeologist W.M. Flinders Petrie opened more than 2000 tombs. While burial methods were similar to those at Merimde and Ma'adi, at Naqada many of the bodies had become

Fig. 4.129. Predynastic burial; SM, ca. 4000 BC.
Bodies buried prior to 3050 BC were generally lying on their side with flexed knees. Many of these became mummified spontaneously. (Photo courtesy of William Pestle and Nina Cummings at The Field Museum, Chicago, Negative no. CSA71913.)

mummified spontaneously. It was the appearance of such mummies that caused Grafton Elliot Smith to speculate that the Old Kingdom rulers' awareness of the bodies' soft tissue preservation under these conditions led them to develop the anthropogenic mummification methods after they buried their bodies in stone sarcophagi. Unfortunately, reports of scientific studies on such bodies are rare, and only a few of the bodies survive today in museums, including at least one ("Ginger") in the British Museum. Petrie's Tomb 5 at Naqada and others from Gerza also revealed skeletal tissue with teeth marks and evidence that the marrow had been scooped out of smashed long bones. Petrie interprets these as evidence of cannibalism. Hoffman (1979:116) suggests the possibility of human sacrifice. He notes that the latter was an occasional royal experiment that ceased abruptly after the end of Dynasty 2 about 2700 BC. Junker (1944:113–117) offers the same explanation for similar findings at Giza. However, other more mundane explanations, including secondary burial or soft tissue decay have been offered for such observations, and the suggestion of cannibalism remains controversial.

The Archaic (Early Dynastic) Period

Sociohistorical features of the Archaic Period

Egyptian history begins about 3050 BC, approximately coincident with the appearance of that unique combination of ideographic and phonogramic symbols we call hieroglyphic writing (though some scholars propose at least a prelude to writing as early as about Naqada II). This not only provides us with the names and chronology of Egyptian rulers but is also contemporary with a conformity of governance (dare we call it control?) common to the full length of the Egyptian Nile inclusive of its delta areas. This political fusion (usually termed "unification") is commonly presented as a single, triumphant military event achieved by a single leader called Narmer (Menes). The evidence for this scenario is heavily dependent upon interpretation of a stone stele ("palette") upon which is inscribed the image of a king wearing a crown of Upper Egypt who is obviously subjugating his antagonists, while the other side presents him in an equally victorious status, but wearing the crown of Lower Egypt (Brewer & Teeter, 1999:32). As described above, the predynastic archaeological evidence would suggest that much of the unifying process probably came about at a much more gradual pace, though the final step could have been one similar to that pictured on the Narmer palette.

The features characterizing this interval focus principally on the availability of a literary record, and the organization of the populace's activities directed by a central authority that was centered at Abydos and the Memphis area. Administration of that authority was achieved by dividing the entire Nile Valley and delta into districts (nomes) supervised by governors (nomarchs) appointed by the king, each of which was directly responsible to the royal representative, the vizier. Initially most of these appointees were probably blood relatives of the royal family. Nomarchs oversaw a host of scribes and others with appropriate skills to carry out the usual, necessary civil services. This structure enabled royal decisions and policies to flow smoothly through the bureaucratic structure, providing a uniformity of practices at all levels throughout what had now become a kingdom. Much more archaeological information dealing with the centuries immediately preceding the Archaic Period needs to be generated before we can determine just exactly how such remarkable centralization of authority was brought about. Once achieved, however, it remained the dominant feature of Egyptian history. As readers pursue the subsequent unfolding of Egypt's historical events, they will understand them best if they measure them against the perpetual efforts of royalty to maintain its status even in the face of sometimes transiently successful challenges to that authority during the Intermediate Periods.

Table 4.8. *The kings of the Archaic Period*

Dynasty 1 (3050–2813 BC)	Dynasty 2 (2813–2663 BC)
Narmer (Menes)	Hotepsekhemwy
Aha	Raneb
Djer	Ninetjer
Djet	Weneg
Den	Sened
Anedjib	Peribsen
Semerkhet	Neferkasokar
Kaa	?
	Khasekhemwy

Kings of the Archaic Period

We know little about any of the eight pharaohs of Dynasty 1 and the nine of Dynasty 2 (Table 4.8), and enough to interest the reader about only two.

Narmer is probably identical to the individual whom Manetho called Menes; we know principally only what the Narmer palette informs us. While we are certain that Narmer was a specific, historical individual, we are not sure whether the name Menes is a synonym for Narmer or a symbol for a series of chieftains that brought about the creation of a unified state. In addition to the Narmer palette, Narmer's name also appears on several, scattered inscriptions but these provide no further historical information about him.

Of the 17 kings composing the first two dynasties, only the body of Djer (Itit) has been found (by Petrie at Abydos). Of Djer's body (or that of his queen, see below) only a wrapped arm remained in his plundered tomb and even that is no longer extant. Thus we can study today not even the skeletal remains of any ruler during the Archaic Period.

While the other 15 kings were responsible for establishing the governing principles upon which the subsequent three millennia of Egyptian life were based, we can recognize the tangible products of their ideas and efforts, but cannot attribute features to specific rulers.

Mummification features of the Archaic Period

Anthropogenic Egyptian mummification as we know it did not develop simultaneously with pharaonic claim to divinity. Monumental tomb architecture expanded rapidly during this interval, but the bodies found in tombs of the Archaic Period had decayed to the status of skeletonization. Typical of these were the mid Dynasty 1 graves at Abydos reported by Habachi (1939). These tombs consisted of clay coffins placed about 0.5 meter below the surface, surrounded by ceramics. Bead necklaces and bracelets adorned the markedly flexed skeletons.

We have only one finding that hints at the possibility of interest in retention of at least body form during the Archaic Period. In the looted tomb of Djer, the third king of Dynasty 1, a disarticulated arm surmised by Petrie to have belonged to Djer's queen was found jammed into a wall niche, presumably by ancient tomb robbers who had been disturbed during the looting episode. This arm had been wrapped very tightly, as if an effort had been exercised to prevent distortion by postmortem bloating (David & David, 1992:41). Spencer (1982:34), however, notes that the arm is probably that of Djer's body because both kings and queens wore bracelets (Fig. 4.130). Petrie's photo reveals only bones without soft tissue. Subsequent curatorial management of the arm resulted in retention of the jewelry but the arm was discarded.

This tomb is also often cited as evidence of ritual human sacrifice during this early period. Djer's Abydos tomb was surrounded by hundreds of smaller tombs that many feel were attendant retainer burials. However, each of these contains only one burial whose features are simply those of other burials not directly associated with royal burials. Only those buried underneath the royal tomb could be interpreted as having been entombed at the same time as the king. Thus the others could as well be interpreted as members bearing some administrative or personal relationship to the king and were privileged, at the time of their later death of natural causes, to be buried near him. Evidence of such probable human sacrifice is not manifest after Dynasty 2. Djer's queen (if that arm really was hers), however, seems to have been buried with the king when he died (Hoffman, 1979:275–279).

Bodies during this period were usually placed in coffins of wood or clay. Their position, as in the Predynastic Period, continued to be flexed, lying on their left side. The tomb chamber, however, now was larger and often had mud brick walls. Several other Dynasty 2 burials at Saqqara demonstrate whole bodies, still contracted, tightly wrapped in linen that was impregnated with resin. The resin became rigid after cooling, retaining the body contours even though all soft tissue inside had decayed (Spencer, 1982:35). A large fraction of both Predynastic and Archaic tombs were looted in antiquity.

Smith (1912a) examined a flexed body from the end of Dynasty 2 or early in Dynasty 3 excavated by Quibell from a mastaba (rectangular, bench-like superstructure) tomb at Saqqara. It was wrapped in more than 16 layers of bandages, and a space occupied by only corroded linen was found between the bandages and the skeletonized body. Smith felt some kind of material was applied to the body surface in what proved to be a vain effort to prevent decay. He felt that this was the earliest evidence of attempted mummification.

Thus only little more than a century had passed following Egyptian unification when we encounter the first evidence of pharaonic perception that postmortem body preservation could enhance the royal public image by joining it to the expanding mortuary architecture. Had we only Djer's wrapped arm to support it, this statement would be vulnerable to the criticism of extrapolation beyond the evidence. However, the employment of resin in the wrappings by the end of the second dynasty seems to allow no interpretation other than an attempt to retain at least body form.

Fig. 4.130. Bejeweled arm from Djer's tomb; SM, ca. 3000 BC.
Though skeletonized, this arm found in the Dynasty 1 tomb of Djer was tightly wrapped. Many Egyptologists regard this as the earliest evidence of concern for retention of soft tissue body form in Egypt, though others find the Dynasty 2 body excavated at Saqqara by Quibbel more convincing. (Photo courtesy of Petrie Museum of Egyptian Archaeology and Egypt Exploration Fund.)

The Old Kingdom

Sociohistorical features of the Old Kingdom

The 26 kings of this period's four dynasties ruled for 500 years. Those of the first of these (Dynasty 3) were occupied particularly with the consolidation of central power. Especially notable was the step pyramid (often considered to be the first major stone building in the world's history) built by King Djoser's architect Imhotep (later vizier and physician, ultimately deified) at Saqqara. The vital role of mortuary ritual and religion in the reinforcement of power centralization is detailed in Chapter 2. Here it will be noted simply that the epitome of mortuary emphasis was reached during Dynasty 4 – an interval often called the "Age of Pyramids". The most massive of these structures, built by the pharaoh called Cheops was composed of more than two million blocks of stone, each weighing more than two tonnes. Nearly 5000 years later its awesome

Table 4.9. *The kings of the Old Kingdom*

Dynasty 3 (2663–2600 BC)	Dynasty 4 (2600–2470 BC)	Dynasty 5 (2470–2350 BC)	Dynasty 6 (2350–2195 BC)
Nebka	Senefru	Userkaf	Teti
Djoser	Cheops	Sahure	Pepy I
Sekhemkhet	Djedefre	Kakai	Merenre (Nemtyemsaf I)
Setka	Khephren	Shepseskare (Isi)	Pepy II
Khaba	Mykerinos	Neferefre	Merenre (Nemtyemsaf II)
Nebkhare	Shepseskaf	Niuserre (Ini)	
Huni		Menkauhor (Ikauhor)	
		Djedkare (Isesi)	
		Unas	

proportions still serve as one of the world's most visited tourist attractions in the Cairo suburb of Giza. The many thousands of peasants (or, in Arabic, *fellahin*) who supplied the labor for its construction were probably recruited during the period of the Nile's inundation, when farm duties were least demanding. While pyramid-building continued for 1000 years after Cheops died, their dwindling size paralleled progressive erosion of royal political and economic control (Table 4.9).

The numerous daily religious rituals (as well as civil services) carried out in the adjacent elaborate temples and other mortuary structures relating to the numerous pyramids and surrounding burials of the king's family members served as constant reminders to the populace of royal power over commoners. Additional features serving a similar function added during this period include further emphasis on the pharaoh's divine status and his incarnation as the god Horus. Particularly important also was the accent on the role of Osiris, presented during Dynasty 5 as the embodiment of the deceased previous ruler who, after death, served as gatekeeper to the land of eternal life. Dynasty 5 also witnessed the elevation of Re, the sun god, to the apex of the Egyptian pantheon. These latter changes have often been viewed as evidence of an erosion of central power, since they offered hope of participation by the general public in the enjoyment of postmortem pleasure that had previously been restricted to the divine rulers (James, 1974:467). However, the fact that each individual's access to eternal bliss was controlled by Osiris, who would judge them and who was the deceased father of the divine, currently living ruler could be expected to deter the populace's challenge to the central authority's power. Institution of the Sed festival, an odd ritual designed to rejuvenate the physical and spiritual power of an aging king, can be viewed in similar perspective. Though the arts and trade flourished, stimulated by military campaigns into Nubia, there can be no denying the archaeological evidence reflecting progressively increasing power of the nomarchial and probably clerical nobility. These factors, together with possible expanding expectations of the populace and the negative impact on the economy imposed by pyramid construction and by increasing climatic aridity led eventually to collapse of monarchial absolutism at the end of Dynasty 6.

Mummification features of the Old Kingdom

Experimentation with body preservation during the Archaic Period reflects those rulers' increasing efforts to employ expanded royal mortuary practices in order to enhance centralization of political power. During the subsequent period of the Old Kingdom these efforts in the area of tomb architecture were maximized as the progressively enlarging mastaba (bench-like) rectangular tombs yielded to the awe-inspiring pyramids of

Dynasty 4. Simultaneously, the frequent failure of soft tissue preservation using the methods of the Archaic Period led to much more successful modifications. Indeed the combination of evisceration, desiccation with powdered natron and use of hot, liquid resin to discourage microbial growth and prevent rehydration as initiated during the Old Kingdom ultimately became so effective for body preservation that their employment endured throughout the subsequent three millennia of ancient Egyptian history. Experimentation did not, however, achieve instant success. Detailed below are some of the excavated mummies whose features have enabled reconstruction of mummification's evolutionary chronology during the Old Kingdom.

The tomb of King Djoser, founder of Dynasty 3 which initiated the Old Kingdom, had been plundered in antiquity when it was opened in 1900. No complete body was present but scattered bones and body parts betrayed the presence of ancient looters. One of these was an isolated foot wrapped in linen bandages. Removal of the outer layer demonstrated that the foot was enveloped by a resin-impregnated linen cast. The superficial layer of the cast was covered with a resin layer thick enough to permit its creative artist to sculpt the tendons and other normal, superficial anatomical structures of the foot. However, recent radiocarbon dating suggests that the alleged remains of Djoser are actually at least a millennium more recent than Dynasty 4 (Strouhal *et al.*, 1995). The significance of this finding is so important that verification of this date seems desirable, since the mixture of bitumen with resin (which results in an incorrect, older radiocarbon dating value) is known to have occurred during the later periods (Nissenbaum, 1992).

In a rock tomb dated to late Dynasty 3 or early Dynasty 4, Garstang (1907) found a flexed body in which the arms were wrapped separately in linen. Some Egyptologists regard this as one of the earliest pieces of evidence of attempts to preserve at least body form of soft tissues.

Later, other complete bodies were found in which the entire body was encased in resin-soaked wrappings and with particular sculpting carried out in detail over the face, genitalia and breasts. One such is a Dynasty 4 body of a man called Renefer excavated at Meidum. He was wrapped in many layers of linen, the resin over the outer layer being sculpted with facial features that were also painted with different colors (Fig. 4.131). The penis had been so faithfully molded that circumcision was recognizable. Petrie sent this body to the Museum of the Royal College of Surgeons in London where unfortunately it was destroyed by a bomb in an air raid during World War II. In a Dynasty 5 body (Waty), prepared in a like manner and found at Saqqara, the outer layer of resin was replaced by plaster that had been similarly sculpted. Iskander (1980) feels that any mummy displaying the features of mortuary preparation found in Waty or Renefer can be assigned to the Old Kingdom with certainty.

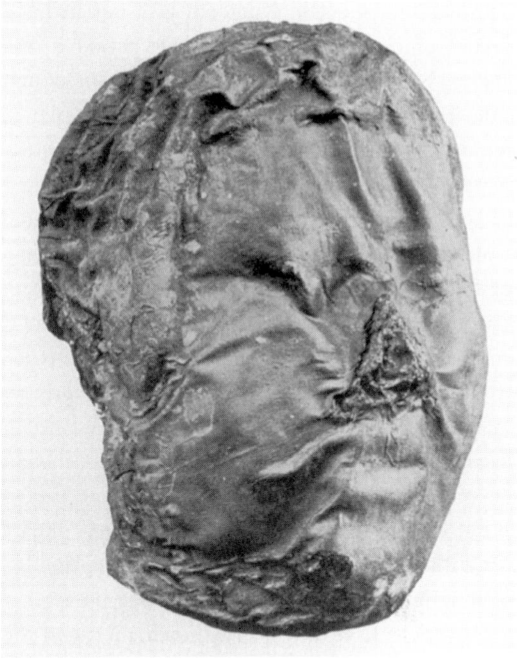

Fig. 4.131. Renefer's facial modeling; AM, ca. 2660–2200 BC.
Saturation of outer linen wrappings with resin and subsequent modelling of facial features (preserved after resin hardens) is a feature of Old Kingdom mummies. (Photo courtesy of the Royal Philosophical Society of Glasgow at www.royalphil.org.)

In addition, the Old Kingdom also initiated the practice of evisceration. During this period it was achieved by a vertical transabdominal incision located in the left flank. A special "Ethiopian stone" knife was employed for that purpose, carried out by an individual who was then chastised (ritually) for having violated the body. The embalmer then thrust his arm into the abdomen via the created defect and removed the viscera with a knife. The heart, together with the proximal aorta was allowed to remain *in situ* or replaced if removed inadvertently. Commonly, though not regularly, the retroperitoneal structures of the urinary bladder, prostate and both kidneys were also retained in the body. The incision was usually not closed, though sometimes a wad of resin-soaked linen was stuffed between the gaping wound lips; in only a small minority was the wound sutured or covered with a plate.

A tomb attributed to Queen Hetepheres, the mother of Dynasty 4 King Cheops was associated with the huge pyramid her son built at Giza. When Reisner (1928) opened the sarcophagus in her tomb he was astonished to find it empty. However, in a tomb recess he found a four-chambered alabaster container. Removing the lid he saw it contained several flat, linen packets. Subsequent experience with later tombs has taught archaeologists to recognize that these linen packets house at least part of the removed viscera. Importantly, in three of the four chambers the packets were covered with liquid. Later analysis by Lucas (1932) revealed it to be a 3% solution of natron. Thus, even in the absence of a whole body, these findings identify the practice of evisceration as early as Dynasty 4. It is also the earliest use of natron for soft tissue preservation that has been documented chemically, though in this case we can only be sure it was used for viscera preservation, not necessarily for the whole body. The body of a Dynasty 3 king (Setka) is described by Derry (1942) as having the abdominal cavity stuffed with linen but subsequently all soft tissues decayed, resulting in an eviscerated skeleton enveloped in linen.

Strouhal & Gaballa (1993) provided some evidence indicating probable use of natron in the Old Kingdom mummy of Dynasty 5 King Djedkare (Isesi). The looted tomb in his Saqqara pyramid was littered with his mummified body parts. Many skeletal fragments were covered with soft tissue (Batrawi, 1947); their surface revealed several layers of linen covered by plaster. Adherent to the inner aspect of a rib was a scrap of resin-soaked linen, implying evisceration followed by stuffing of the abdomen with textiles. Histological preparations demonstrated excellent preservation of tissue microstructures. These authors believe that such an unexpectedly high quality of preservation implies the application of dry natron for desiccation of the body. Radiocarbon dates were consistent with Dynasty 5. The use of natron is discussed in detail below in the section devoted to use of materials in mummification. Suffice it to say here that the only chemically documented use of natron during the Old Kingdom is that found in Queen Hetepheres' tomb; in the other bodies the evidence is indirect, based on such features as histological demonstration of unusually good preservation of tissue architecture. With the exception of the undissected body of Merenre (Nemtyemsaf II) in the Cairo Museum (last king of Dynasty 6), we have no intact, complete royal bodies with soft tissue preservation from the Old Kingdom.

The First Intermediate Period

Sociohistorical features of the First Intermediate Period

The century and a half following the end of the sixth dynasty were years of sufficient political turmoil that records of regal succession are difficult to validate. The most plausible surmise suggests that regional nobles engaged in power struggles that evolved into a competition between those in the north centered in Herakleopolis and those clustered in the Thebes area. After a century of battling nomarchs, princes and usurpers, Egypt was again reunited into a single kingdom from the delta to the second cataract in Nubia in a manner identical to its original unification: by mil-

Table 4.10. *The kings of the First Intermediate Period*

Dynasty 11a (2160–2066 BC)
Mentuhotep I
Inyotef I
Inyotef II
Inyotef III

Notes:
Dynasties 7–11a (2195–2066 BC): Inadequate documentation, except for the four listed.

Table 4.11. *The kings of the Middle Kingdom*

Dynasty 11b (2066–1990 BC)	Dynasty 12 (1990–1781 BC)
Mentuhotep II	Amenemhet I
Mentuhotep III	Sesostris I
Mentuhotep IV	Amenemhet II
	Sesostris II
	Sesostris III
	Amenemhet III
	Amenemhet IV
	Sobeknefru

itary means. This incarnation of Menes/Narmer was Mentuhotep II (Tepya), who soon displayed the administrative skills necessary to maintain his achievement. He and his successors moved the capital from Memphis to Thebes, and replaced the Old Kingdom principal deity (Re) with Osiris, thus weakening the Re priesthood (Table 4.10).

Mummification features of the First Intermediate Period

We have little evidence related to mummification that can be dated firmly to the First Intermediate Period. Whether this can be attributed directly to the political chaos of this interval is attractive but speculative. Ikram & Dodson (1998:112) describe the fluid-filled sarcophagus of Ptahshepses, analysis of which detected the presence of natron. They note, however, that the most plausible source of the fluid was groundwater that may have leached residual dry natron from the body. They also describe a body from Dynasty 9 that was desiccated and eviscerated but lacked the molded surface features of Old Kingdom mummies. The face and chest, however were covered by a cartonnage mask with sculpted facial structures, and linen ears were sewed to the outer wrappings. With these exceptions, one reasonably consistent practice characterizing the First Intermediate Period is the termination of molding body surface features onto an outer layer of resin covering the mummy's wrappings.

The Middle Kingdom

Sociohistorical features of the Middle Kingdom

Following the death of the weak Mentuhotep IV, last of the Dynasty 11 rulers, the throne was usurped by Amenemhet in 1994 BC. Restless nomarchs were dealt with by both military and diplomatic methods, restricting their growing power and redefining their regions. Lower Nubia was occupied and fortified, and Libyan expansion suppressed. Amenemhet was eventually assassinated, probably by palace guards (Brewer & Teeter, 1999:37). The remainder of the Middle Kingdom was a time of peace in which both trade and the arts again flourished and the Fayum was developed. Royal mortuary structures of this period were of more modest size but built to resist thievery, partly for reasons of economy and partly because the wealth in the royal tombs had been plundered during the First Intermediate Period. During the latter part of the Middle Kingdom, however, the increasing trend of Asians to move down the Levant and settle in the eastern delta, led the kings to initiate construction of fortifications in that region to retard this movement. This action proved to be too little too late. Cooperation among the "Hyksos" (Egyptian term for "rulers of foreign lands") led initially to economic and ultimately to political dominance to usher in the Second Intermediate Period (Table 4.11).

Mummification features of the Middle Kingdom

If, however, mummification practices reflect social and political order, the stability and uniformity of the early part of the Old Kingdom was incompletely restored. Mummies from this period demonstrate considerable variation in practices. For example, the tombs of Mentuhotep's six queens and princesses were found adjacent to and below his temple at Deir el-Bahari. These had all been similarly prepared and buried without evisceration. Viscera remained in the abdomen. Soft tissue preservation had been achieved by desiccation, probably by application of natron to the body surface, since natron-filled bags were found with some of the tombs, as well as on one other body and natron was identified chemically, adherent to the surface of an embalming table associated with the tombs. Jewelry imprints on the skin suggested incomplete dehydration when they were wrapped. Resin stains were found on only two of the wrappings and a body of another woman was covered with a layer of beeswax. Visceral prolapse via the rectum and vagina was attributed to postmortem abdominal gas formation (Derry, 1942). The bodies of several nobles had similar features. Evidence of pine wood resin acids and high sodium concentration on skeletal tissue of a Middle Kingdom high official also implies use of resin and natron, though no soft tissue survived in this body (Koller *et al.*, 1998). While Derry interpreted these findings as a decline in the quality of mummification techniques established in the Old Kingdom, it should be noted that their soft tissue preservation far exceeded that of Old Kingdom bodies.

Mummification, originally limited to pharaohs and their immediate family, was extended also to many members of the middle class during this period (David, 2000), heralding the complete democratization of this practice evidenced in the New Kingdom.

Most mummies from other areas during Dynasty 11b reveal that they were eviscerated through a left abdominal incision and evidence of natron desiccation is often overt. Two Dynasty 12 mummies Quibell found at Saqqara were so prepared (the male had a beard and mustache) as were two more from the same dynasty at Rifeh by Petrie and one from Lisht (Smith & Dawson, 1924:78–82). The flexed forearms so common in the Predynastic Period yielded to arms extended along the sides of the body during the Old Kingdom when evisceration was initiated and this was generally maintained, with exceptions, during the Middle Kingdom. Typical of this period's variations were 12 bodies from Dynasty 12 at Dahshur, most of whom had extended arms but in a few they were crossed over the chest (Ikram & Dodson, 1998:116).

Near Mentuhotep II's temple a group of battle casualties were buried, though the event probably occurred during Dynasty 12. They were apparently exposed on the battlefield for some time before burial, as evidenced by the stripping of muscle from tendons – a characteristic of vulture feeding. Though manifesting considerable soft tissue decay, the remainder was preserved by desiccation probably by placement in sand, as reflected by sand adherence to the skin. Arrow wounds and other forms of trauma testify to the violent nature of their death (Winlock, 1945).

Resin applied to the body (not to the wrappings as in the Old Kingdom) is also a feature characteristic of the Middle Kingdom mummies, though occasional exceptions are reported. An extreme example was reported by Mace & Winlock (1916) in which a thick layer of resin was poured over the mummy of Senebtisi after the body had been placed into the coffin (Fig. 4.132).

Another mummification feature is one whose introduction, formerly attributed to the New Kingdom but recently applied to the Middle Kingdom, is brain removal via the nasal passages – a method I shall call transnasal craniotomy. It was achieved by pushing a metal rod up through a nostril, usually the left, forcing it upward to create a passage through the thin bony layer of the ethmoid plate into the lumen of the cranial cavity. A rod with one fishhook-shaped end found with embalmers' tools has been assumed to have been inserted through such a passage to remove the brain.

Fig. 4.132. The tomb of Senebtisi; AM, ca. 2000–1800 BC.
After the body was laid in the coffin, huge quantities of hot, liquid resin were poured on this Middle Kingdom body, fixing the body to the coffin. (Photo courtesy of the Metropolitan Museum of Art Photographic Library, Negative no. L(6–7)806.)

However, it would appear that the brain is too soft and mushy to have permitted such an instrument to be effective. Conceivably, however, at least part of the meninges, falx cerebri and tentorium could yield to such applications. We do not know exactly how the brain was extracted, though several experimental efforts have demonstrated that repeated insertions of a straight rod manipulated with a twisting motion can macerate the brain. If wiped after each withdrawal, enough tissue was found to adhere to the rod to result in ultimate removal of the bulk of the brain (Sudhoff, 1911; Leek, 1969). If this is how it was actually carried out, it could account for the apparent absence of the brain among the viscera, because the macerated brain wiped off with a rag would no longer be recognizable. More likely, at several days after creation of such a passage, the remaining brain must have liquefied spontaneously and could be poured out of the cranium via that passage. Indeed, no effort at brain removal other than creation of the passage need have been performed at all. It has also been postulated that removal of mashed brain tissue may have been enhanced by insertion of a hollow tube into the cranium and flushing the cranial cavity with water. Because this would leave no physical evidence of its performance, this suggestion remains speculative.

Transnasal craniotomy is now firmly established in at least seven, well-dated bodies from the Middle Kingdom (Strouhal, 1995). In addition, Strouhal suggests re-examination of several reports of this procedure in Old Kingdom bodies as early as Dynasty 4 whose antiquity has been considered too insecurely dated for validation.

Most of the tombs (principally small pyramids) of the Middle Kingdom rulers were plundered so thoroughly in antiquity that archaeologists found no trace of pharaonic remains in the burial chambers. The British Museum houses a partial mummy of what appears to have been Mentuhotep II. That of Sesostris II was recovered, but its present location is uncertain (Ikram & Dodson, 1998:317–318).

The Second Intermediate Period

Hyksos kings in the delta and Theban princes ruled simultaneously. Names and regnal periods are not well known. Dynasty 17 ends with expulsion of the Hyksos initiated by Seqenenre (Taa II) (1558–1553 BC) and completed by his sons Kamose (1553–1550 BC) and especially Ahmose I.

Sociohistorical features of the Second Intermediate Period

Borders between heavily populated areas of two countries commonly carry the potential for conflict. Though protected from its neighbors by the Eastern and Western Deserts, Upper Egypt's Nile Valley junction with Nubia was the site of periodic hostilities. In Lower Egypt the delta's flanks were particularly vulnerable, and were exploited from the west by Libyans during Egypt's later periods. Its eastern flank, however, faced the Levant. Unrest in Mesopotamia, Syria or elsewhere in the eastern Mediterranean could be expected to be reflected by a southward migration of refugees leading into the eastern delta. Indeed, by the time of the Middle Kingdom, inscriptions and papyri document that many individuals with Asiatic (especially Palestinian) family names had already been incorporated into Egyptian society (Trigger *et al.*, 1983:155). Acceleration of this process, accompanying irresolution and disorder of Egypt's centralized government in Thebes during the latter part of Dynasty 12 led first to concentration of such foreigners at multiple foci within the eastern delta and subsequently into a loose military federation among them that gradually gained control of the delta. Though the transition was probably incremental, most historians identify the end of Dynasty 12 as the end of the Middle Kingdom.

During Dynasty 13, Egyptians called these Asiatics "Hyksos" ("rulers of foreign lands"). They may well have constituted an ethnically heterogeneous group, though many modern scholars believe they were principally a Semitic people, probably predominantly Palestinians. They eventually extended their domination upstream as far as Nubia. Their capital was at Avaris in the eastern delta, probably at the site of subsequent Tanis (or ruins at Tell ed-Dab'a) near the modern community of San el Hagar at about 30° 58′ N and 31° 53′ E. During their reign, much of Egypt's infrastructure appears to have remained intact and functioning, though the paucity of surviving records has forced the exercise of at times a somewhat arbitrary reconstruction of king lists. Trigger *et al.* (1983:152) cite the recording of 175 reigns on the Turin papyrus between the end of Dynasty 12 and the beginning of Dynasty 13, many of which may have occurred simultaneously in different parts of the country, with the Hyksos dominating the delta area and natives controlling Upper Egypt. By the end of Dynasty 17 (the dynasty number used to refer to Theban rulers during this period) the ruling monarchs of Thebes began to coordinate their activities sufficiently to mount serious military resistance to the Hyksos, probably beginning with the rule of Seqenenre. His son and successor Kamose is credited with major successes in this revolt and his brother Ahmose I completed the rout of the Hyksos, initiating Egypt's golden age of Dynasty 18 about 1550 BC (David & David, 1992). To guarantee royal bloodline purity, Ahmose I married his own sister, a practice initiated during the late Dynasty 17.

Mummification features of the Second Intermediate Period

Few of this period's bodies, either royal or nonroyal have survived to our day. Whether this is the product of politico-economic instability or merely the humidity of the delta is not clear. Of the six royal tombs found so far, only one contained a reasonably intact body. This was the mummy of King Hor of Dynasty 13, found in his pyramid at Dahshur near the pyramid of Amenemhet III, who may have ruled less than 1 year. While the body was broken, canopic jars implied that he had been eviscerated. Another royal tomb at Dahshur was that of Ameny-Qemau also of Dynasty 13. This largely destroyed pyramid contained a sarcophagus but no mummy, although this tomb also included canopic jars (Brier, 1994:252). A Dynasty 17 body of nonroyal status excavated by Petrie from the Qurna necropolis near Thebes demonstrated much soft tissue decay, but the abdomen was stuffed with linen, indicating that evisceration had been performed (Spencer, 1982:117). The tomb of King Khendjer (Dynasty 13) was identified in his pyramid at south Saqqara but was empty. Several royal bodies were recovered but were lost through neglect. The body of Kamose with a dagger strapped to his arm was found intact by Mariette and Brugsch while they were trying to "seed" the tombs along the expected route of a cousin of Napoleon III during his proposed visit in 1857. Unfortunately their interest focused on artifacts and they abandoned the body in the excavation debris. The well-preserved body of Seqenenre's wife suffered a similar fate, while that of Nubkhepere (Inyotef VI), a Dynasty 17 king, was reported in 1827 to have been fragmented and discarded by local looters in search of jewelry and artifacts (Brier, 1994:254).

Without doubt, however, the most well-known mummy of this period is that of King Seqenenre. This late Dynasty 17 Theban ruler is believed to have initiated active, effective revolt against the Hyksos. His quite poorly preserved mummy reveals devastating cranial trauma that includes two horizontal frontal lacerations with cranial fractures, nasal and facial bone fractures from a blunt blow, a laceration of the left cheek made with a pointed instrument and another like it over the left side of the head below the ear (Fig. 4.133). Smith & Dawson (1924:83–86) describe the rest of the body as a disarticulated skeleton enveloped by dark-brown skin. A left abdominal, gaping eviscer-

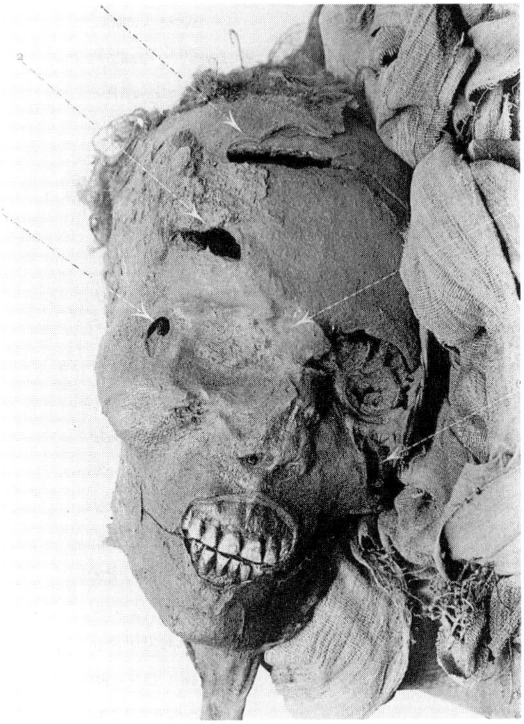

Fig. 4.133. Seqenenre's wounds; AM, ca. 1553 BC. Several axe wounds are evident in the right side of the frontal bone. A laceration below the left eye can also be seen, as well as a puncture wound to the right of the right eye. (From Smith, 1912b.)

ation wound was obvious and the abdominal cavity (but not the thoracic) was stuffed with linen. No transnasal craniotomy had been performed. Aromatic sawdust had been sprinkled over the body. They felt that the frontal lesions had been inflicted with an axe and the wounds at the side of the head by a spear. The poor preservation of the soft tissue was attributed to hasty embalming, probably carried out on the battlefield where he had suffered the injuries. This explanation has been perpetuated since they expressed this view. A recent review seemed to lend support by noting that the lesions were consistent with but not diagnostic of wounds produced by Palestinian weapons. However, the lesion near the ear has been shown to reveal radiological changes associated with a healing process, and the horizontal orientation of the frontal lacerations, together with the absence of trauma elsewhere on the body have led to the suggestion that they were inflicted by assassins while Seqenenre was sleeping (Ikram & Dodson, 1998:118).

In summary, embalming methods of this period essentially represent a perpetuation of those carried out in the previous period of the Middle Kingdom. Evisceration, extended body with extended arms and wrappings that enveloped the limbs separately all had precedence in the period just prior to the First Intermediate Period. Transnasal craniotomy was practiced, but irregularly. Effectiveness of mummification covered the entire range from nearly complete skeletonization to well-preserved soft tissue. No surviving mummies have come to us from the Hyksos' central capital in the eastern delta. Scientific studies of those found in or near Thebes are rare.

Table 4.12. *The kings of the New Kingdom*

Dynasty 18 (1550–1300 BC)	Dynasty 19 (1300–1187 BC)	Dynasty 20 (1187–1069 BC)
Ahmose I	Ramses I	Sethnakhte
Amenhotep I	Seti I	Ramses III
Tuthmose I	Ramses II	Ramses IV
Tuthmose II	Merneptah	Ramses V
Tuthmose III	Seti II	Ramses VI
Hatshepsut	Amenmesses	Ramses VII
Amenhotep II	Siptah	Ramses VIII
Tuthmose IV	Tawosret	Ramses IX
Amenhotep III		Ramses X
Akhenaten		Ramses XI
Smenkhare		Hrihor
Tutankhamun		
Ay		
Horemheb		

The New Kingdom

Sociohistorical features of the New Kingdom

With the Hyksos successfully expelled and the country reunited one might expect that the subsequent period would be one of peace and prosperity. For the general population, this was probably so and, in retrospect, this interval is usually viewed as Egypt's golden age. However, at least for the leaders (Table 4.12) it was not simply a return to Middle Kingdom ways. Egypt was now very conscious of its relationship to other Mediterranean populations. It carried on a thriving trade with at least the eastern Mediterranean countries that was highly profitable and fueled its expenses. The Hyksos experience had heightened Egyptians' distrust of foreigners. Internally the struggles for power among Theban princes had made them conscious of the need to consolidate their newly won positions (Wente, 1974:471–481).

Expression of these concerns took several forms that can be understood best if we remember the kings' claims to divinity and identity with their principal god. An early major "religious" change was the fusion of the Theban god Amun with the Old Kingdom sun god Re; this Amen-Re was then identified as the principal god to which all other gods were subordinate. This was accompanied by a simultaneous restatement of the pharaoh as the product of a royal but human mother and Amen-Re as the father (David, 1982:121–124). Success in the Hyksos expulsion effort was attributed directly to Amen-Re, enhancing the king's image as a powerful military leader. Exploiting this further, many pharaohs of the New Kingdom carried out foreign military campaigns. Thutmose III established an "Asiatic Empire" that reached as far as the Euphrates River, gathering tribute from Hittites, Assyrians and even Babylon. Later he turned south and gained control of Lower Nubia.

During the latter half of the New Kingdom, however, central power declined. Bureaucratic positions, once entirely appointed by the rulers, gradually became hereditary, including many of those of the priesthood. Lavish gifts of land to the temples by early kings of Dynasty 19 (and the wealth they generated) empowered the clergy. The death of every pharaoh initiated still another cult whose temple and priests needed to be maintained in perpetuity. The construc-

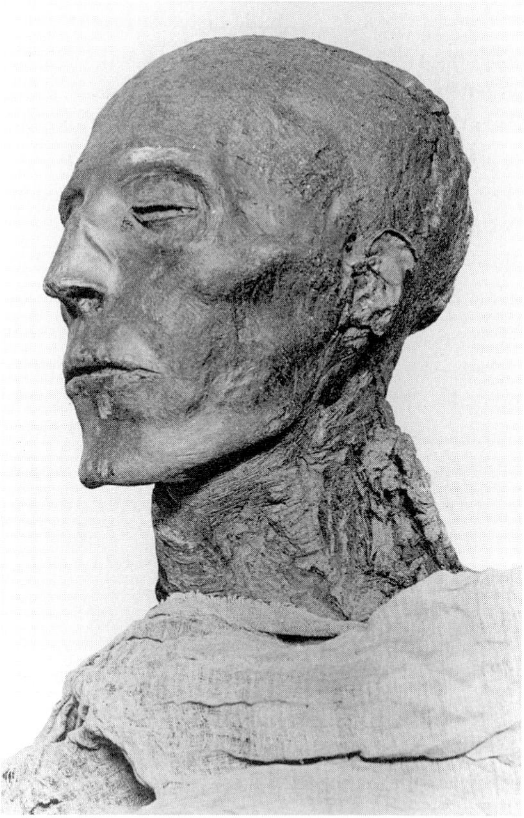

Fig. 4.134. Sety I; AM, ca. 1279 BC.
The serenity of this mummy's expression reflects the high level of embalming expertise in Dynasty 19. (Photo from Smith, 1912b.)

tion and maintenance costs of these architecturally lavish edifices, together with the loss of revenue under the king's control gradually sapped the power of the throne. A desperate effort to free the government from the clerical burden was carried out by Akhenaten (Amenhotep IV) when he elevated a sun-god (symbolized as a solar disk) to the level of supreme god and eventually declared all other gods at first subordinate and finally invalid. All other temples and cults were eventually suppressed and their priests discharged. He claimed absolute identity of the pharaoh with this new god, elevating himself from the former status as merely the son of a god to this "new" status (actually equal to that claimed by early Old Kingdom rulers) as the very incarnation of the supreme god – now called Aten. This claim required no priesthood to mediate between the king and Aten. The experiment, however, was short lived when it was overturned by his successor, the youth Tutankhamun.

In addition to Akhenaten and Tutankhamun, other well-known kings of the New Kingdom include Sety I (Fig. 4.134) and especially Ramses II (who ruled for 67 years and was probably responsible for a larger number of statues and monumental architecture than any other pharaoh) and his successor Merneptah, whom some regard as the pharaoh of Israel's Exodus (Wente, 1974:475). The latter half of the New Kingdom following Ramses II was characterized by contraction. Most of the buffer states in the Levant were lost to the "Sea People". It became increasingly difficult to prevent Libyan intrusion into the delta from the west and military invasion by Syrians and Assyrians from the east (Trigger *et al.*, 1983:229). By the end of Dynasty 20 the traditional government had become so weakened that most Egyptologists view Dynasty 21 as the onset of the Third Intermediate Period.

The kings of the New Kingdom

Of the many New Kingdom rulers whose exploits are documented and shamelessly exaggerated in monumental inscriptions, the reader needs to know at least the five discussed below.

Tuthmose I

He expanded the military conquests of his father to the Euphrates River, Syria, and deep into lower Nubia. He initiated the rock tomb burials of what we now call the Valley of the Kings. His tomb, discovered in 1899, was empty. His coffin was found in the Deir el-Bahari royal mummy cache, but was occupied by the body of Pinudjem, the Dynasty 21 high priest at Thebes. Another, unlabelled body in the cache was identified as Tuthmose I on the basis of facial family resemblance.

Hatshepsut

As the daughter of Tuthmose I she married Tuthmose II but bore only a female child. Her husband sired an heir, Tuthmose III, by a concubine. Hatshepsut's husband, Tuthmose II, died when the heir was only a child, so Hatshepsut served as regent to her stepson Tuthmose III. Two years later she declared herself pharaoh. During her reign she constructed a spectacular, terraced and colonnaded façade to her tomb in the Valley of the Kings. When Tuthmose III became an adult he eventually succeeded his stepmother. His emotional relationship to Hatshepsut remains controversial, but we do know that, after her death, he erased her name from most monuments. Her body was not in her tomb when Howard Carter discovered it in 1903 and opened it. Several other unidentified bodies found elsewhere have been suggested to be those of Hatshepsut but to date none have been confirmed.

Akhenaten (Amenhotep IV)

This "heretic king," born Amenhotep IV, elevated the god Aten to supreme divine status, and first demoted and then purged all other gods. This essentially divorced the politically powerful priesthood of the god Re from all religious and associated political status. He built a new capital at Amarna with revenues from the closed temples. His statues suggest a unique physical morphology suggestive of Marfan's (Burridge, 2000) or Froelich's syndrome, but those of his wife, Nefertiti are accepted as the very model of Egyptian beauty. Some historians have suggested his monotheism became a model for the Jewish Jehovah, others that he was the pharaoh of the Exodus (David & David, 1992). Firm identification of Akhenaten's mummified body has not been established.

Tutankhamun

This "boy king" died at about age 20 years. He is usually regarded as the child of Akhenaten but not with Nefertiti, though some believe him to be Akhenaten's brother. Smenkhare, probably his older brother, may well have ruled briefly before he died and was succeeded by Tutankhamun at the age of 8 or 9 years. Tutankhamun retained Akhenaten's sun god Aten, but he also restored the gods (and their priesthoods with those temple-related revenue sources) that Akhenaten had dethroned. While his abbreviated reign did not allow time to distinguish himself in any unique manner, his death certainly did. His was the only Egyptian pharaoh's tomb in the Valley of the Kings that archaeologists have opened and found intact with all its incredible treasures. Thereby this otherwise obscure pharaoh has become the most well known of all his fellow rulers of all ages. His body continues to remain in the sarcophagus in his tomb today, though two mummified fetuses (his offspring?) have been removed.

Ramses II

This is the national hero of Egyptian history. His monumental inscriptions present him as a fierce warrior, suppressor of Hittite power and Libyan invasion, though archaeological evidence suggests lesser levels of military achievement. His domestic policies generated a long interval of order and economic stability. He fathered more than 100 children with a half-dozen queens and numerous concubines. Today he is best known for his almost incredible creation of monuments, buildings, colossi and temples. Among these are major structures at Karnak, his funerary temple and the Ramesseum at Thebes. In addition he created the gigantic figures of the Abu Simnel temple near Aswan that were disassembled and reconstructed in a salvage effort that rescued them from submersion by the Aswan dam in modern times. He died as a near-centenarian after a 67 year reign. While his tomb in the Valley of the Kings was empty when opened, his body was included in the cache of royal mummies at Deir el-Bahari. It is a well-preserved body (Fig. 4.135). When this body, housed in the Cairo Museum, was found to be suffering the effects of microbial action in 1977 it was sent to France for sterilization by total-body gamma radiation.

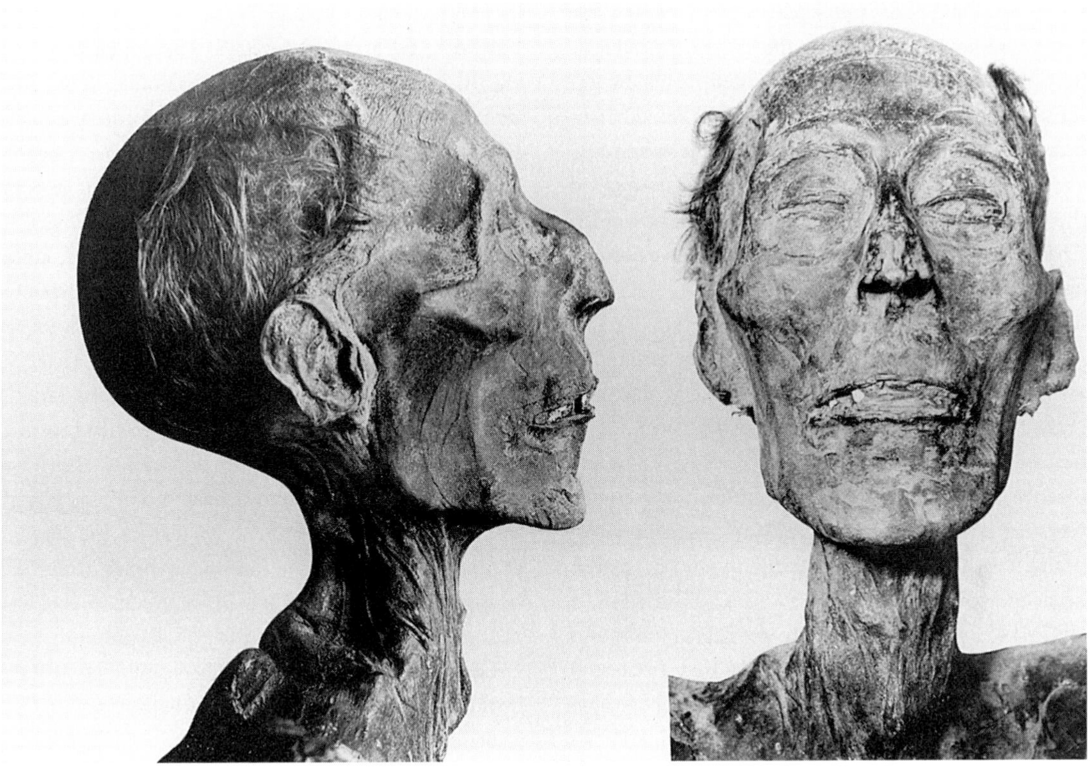

Fig. 4.135. Ramses II; AM, ca. 1212 BC.
This king ruled 67 years, dying at an age probably older than that of any other pharaoh. (Photo from Smith, 1912b.)

Mummification features of the New Kingdom

Our study of Egyptian mummification practices during this period has, ironically, been enhanced by the actions of tomb robbers. Relaxation of governmental supervision during the Second Intermediate Period rendered royal and noble tombs vulnerable to looting. The practice continued in the New Kingdom. At some later date apparently, the violation of many royal tombs was discovered by Theban priests. These then secreted as many of the New Kingdom royal mummies as they could find, repairing their bodies that had been broken during the looters' search for jewelry and then hiding them in two nondescript tombs that could be more easily guarded. In 1871 one of these was discovered near Queen Hatshepsut's tomb at Deir el-Bahari and found to contain 40 relocated royal and aristocratic bodies of which 10 were those of kings. The other was found in 1898 in the tomb of Amenhotep II at the Valley of the Kings. This included 14 bodies of which 9 had been kings. Together, the mummies of most New Kingdom rulers had been recovered in the caches of these two tombs. An additional, still undiscovered cache containing the remaining missing bodies may exist. Recently a body included among a small collection of Egyptian mummies originally acquired by a Montreal adventurer has been suggested (but not confirmed) to be that of Ramses I (Gibson, 2000; Holden, 2000; Waldie, 2000). Identification problems resulting from placement of some of these bodies in coffins inscribed with names of other individuals have persisted to date. Yet the mummification features of many of the more intact bodies could be reconstructed.

Grafton Elliot Smith was asked to examine all these bodies and he describes his findings in a 1912 publication (Smith, 1912b). In his book he comments on features of the individual mummies but not in sufficient detail to carry out desired statistical studies regarding variations. Derry (1942) has also examined some of these as well as the mummies of many nonroyal nobles. These, together with reports by individual excavators, constitute the basis for reconstruction of New Kingdom mummification methods.

These mummies demonstrate that the principles initiated during the Old Kingdom had finally been refined to the point that effective soft tissue preservation had become the rule. While it is often suggested that the Egyptians' "secret" mummification methods remain a mystery, in fact the techniques can now be detailed with a great deal of certainty and have even been verified experimentally (Notman & Aufderheide, 1995; Brier, 1997). The process was quite simple and pragmatic from a mechanical point of view. Because it was a religious procedure, it was made complex only because the ritual demanded intensive participation by a large number of individual subspecialists who were required to carry out prescribed prayers, incantations and procedures for each minor step of the mummification process. The process is described here in detail as well as tabulated in Table 4.13 and only subsequent changes are discussed for later periods. Though Iskander (1973) has described it as a 13 step procedure, it can be thought of as a four stage process.

Operative stage

Shortly after death, relatives contracted with professional embalmers to perform a selected form of mummification, after which the body was transferred to the "House of Beauty" where initial body cleansing and purification was carried out after which it was moved to "The Pure House", the embalmer's workshop. Archaeological identification of such establishments is so rare that it has been suggested it may have been of portable and perishable construction, such as a tent. This seems reasonable, since the volatile gases emitted by a putrefying body could be expected to seep into the walls of a permanent structure whose stench would eventually make the room uninhabitable. The fifth century BC Greek historian Herodotus (ca. 450 BC/1981, Book II:89) notes that, in some cases, a female corpse was kept at home for up to four days to discourage violation of the body by male embalmers. If true it was most likely an infrequent practice, since Egyptian summer temperatures would be expected to produce devastating degrees of putrefaction in an untreated body during such an interval.

In this workshop the initial step was excerebration. The usual method of brain removal was to create a passage from the outside into the cranial cavity by inserting a metal rod into a nostril and pushing it through the skeletal ethmoid plate into the cranial cavity, as described in the section Mummification features of the Middle Kingdom, above. Such a procedure became routine (though not universal) in New Kingdom mummies, yet it had been initiated at least during Middle Kingdom and perhaps even in Old Kingdom times. Several days later, after the then liquefied remains of brain tissue were drained, some hot, liquid resin was poured into the cranial cavity. In others dry linen was stuffed into the cranial cavity, while in still others neither was used. In some the nasal passage created was left open, while linen tampons closed it in others. Resin sometimes reached the cranial cavity in a retrograde route via the spinal canal when liquid resin was poured into the desiccated abdomen, seeping into the spinal canal via the intervertebral foramina.

The next step was removal of the viscera from the thoracic and abdominal cavities. Most commonly this was carried out through an incision in the left side of the abdomen. This was a vertical incision approximately in the left midclavicular line beginning several centimeters below the left rib margin and extending caudally about 10–15 cm (Fig. 4.136). After the reign of Tuthmose III this incision began at about the same level but more laterally and carried medially and obliquely, parallel to Poupart's ligament, extending to a point near the pubic symphysis (Smith, 1912b:32–36). This was performed with a flint

("Ethiopian") stone by a designated subspecialist who was then ritually and publicly chased down the street for desecrating the body. The embalmer then inserted his hand into the abdominal cavity through the wound, removing most of the viscera. Lungs were reached by incising or excising the diaphragm. Allegedly embalmers cut their attachments with a wooden instrument (hard to believe). Special care was exerted to permit the heart to remain in situ, for it was believed to be the seat of thought, emotion and memory and thus needed for Osiris in the underworld to judge the quality of the individual's behavior during life. In mummies of royal or important nobles a scarab inscribed with a prayer was sometimes placed adjacent to the heart. Kidneys were also often retained as were pelvic structures; their retroperitoneal location may have made their removal more laborious.

During this period the viscera were preserved and placed into special containers called canopic jars by Egyptologists, a term referring to Kanopus, the Homeric name of Menelaeus' helmsman who was worshipped in the form of a jar with a bulbous abdomen. By the time of the New Kingdom the heads of three of the four human-headed jars had been replaced by those of three specific animals representing the sons of Horus to denote the visceral content of each jar. Imsety, the human-headed, stored the liver; Duamutef, the jackal-headed, held the stomach; Hapi, the baboon-headed, contained the lungs, while Qebehsenuef, the falcon-headed, preserved the intestines. Actual inspection of their content, however, indicates a good deal of variation in the distribution of the viscera. Preservation of the viscera was usually achieved by the use of natron for desiccation after which they were wrapped in sometimes resin-soaked linen and placed in the jars. The jars were placed in the same tomb as the mummy. For royal bodies the viscera were sometimes placed in elaborately designed containers, such as the gold miniature coffins in the tomb of Tutankhamun.

Following excerebration and evisceration the body cavity was cleaned and wiped with palm wine (13% ethyl alcohol) and spices. The presumed intent was, no doubt, to sterilize the body cavity. It is improbable that such treatment reached that goal, though it would be expected to reduce its bacillary content substantially. Iskander (1973) describes the abdominal insertion of linen packets containing only wads of linen in some to absorb moisture, natron in others for more effective moisture removal and aromatic resins to counter the odor of putrefaction in the body cavities. The head was then commonly but far from invariably shaved close to the scalp. After this the body was transferred to the desiccation table.

Arm positions changed during the New Kingdom. They were extended with hands on the sides of the thighs at the beginning of Dynasty 18 but beginning with Tuthmose II were flexed at the elbows resulting in arms crossed over the chest. Thereafter, until Dynasty 21 all males maintained this position while the arms remained extended in females. With the onset of Dynasty 21 all arms were again extended with hands in a prepubic position or on the lateral thighs. During the Ptolemaic Period they were again flexed but returned to an extended pattern during the Roman Period. Numerous exceptions have been noted (Waldron, 1995).

Desiccation stage

The temporary stuffing was removed from the body cavity several days later, and replaced with natron-containing packets stuffed snugly into the body cavity to occupy all available space. A linen band was placed around the head and jaw to hold the mandible in position. The arms, hands and legs were tied into the desired position. The table was covered with a layer of pulverized natron 5–10 cm deep. The body was laid on this, after which a huge mass of this same, dry natron was poured over the body until it was completely covered everywhere to a depth of perhaps at least 5 cm. Probably natron-filled linen packets were placed in the axillae, between upper thighs, external genitalia and other areas to prevent skin-to-skin contact. Modern experimental experience suggests that frequent changes of the natron that was in direct contact with

Table 4.13. *Egyptian mummification chronology*

	Predynastic	Archaic	Old Kingdom	IP 1	Middle Kingdom	IP 2	New Kingdom	IP 3	Late Period	Ptolemaic	Roman	Christian
Arms	F	F	F,P,T	P	P,F,T	P,T	P,T,C	P,T	P,T,C	T,C	T	T
Canopics – chest	0	0	W,S,P	W,S,P,C	W	W	W,S	0	0	0	0	0
Canopics – jars	0	0	S,P	W,S,P,C	W,S,P,C	W,S,P,C	W,S,G,F	W,S	0,W,S	0,W,S	0	0
Canopics – stoppers	0	0	U	H	H	H	A	A	A	A	0	0
Cavity packing	N	N	N	N	L,S,R	L	L,S,R,F	L,S,R	L,S,R,M	L,R	L,R	N
Cranial cavity	E	E	B	B	B	E	R,E,S,I	R,L,S,I	R,L,I	R,A,P	R,V,L	B
Evisceration	N	N	N,Y	Y	Y	Y	Y	Y	N,Y	Y	N,Y	N
Evisceration incision	0	0	0,F	F	F,P	?	F,I,U,S	F,A,S	F,P,U	F,P,U	0,F,P,U	0
Excarnation	?	0?	0?	0	0	0	0	0	0	0	0	0
Excerebration	0	0	0,E	0	0,T	0,T	T	T	T	T	0,T	0
Eyes	G	G	G	G	R,G	G	O,L	C,Q,W,P	C,Q,W,P	G,L,R,P,S	G,L,R,P	G
Facial painting	0	0	0	0	0	0	P	P	P,0	P,0	0	0
Legs	F	F	F,E	E	E	E	E	E	E	E	E	E
Molding	0	B	B,F	0	0	0	0	0	0	0	0,B	0
Mummy portrait	0	0	0	0	0	0	0	0	0	0	P	0
Natron	0	?	B?	B	B	B	B	B	B	0,B	0,B	0,B
Resin and beeswax	T	?	0,L	L	T,L,P,W	P	P,L,W	P,L	P,L,B	P,L,B	P,L,B	0
Restoration	0	0	0	0	0	0	S	0	S	S	S	0
Salt	S?	0	0	0	0	0	0	0	0	0	0	S,P
Subcutaneous packing	0	0	0	0	0	0	0,P	P	0	0	0	0
Viscera disposal	0	0	S	S	S	S	S,C	C	C,B	B	S,C	0
Viscera modelling	0	0	G	?	0	?	G	P	G	0,G	0,G	0
Wrappings	G,R	L,M	R,P	R,P	R	L	R	R	R,H	R,H	P	S

Notes:

IP, Intermediate Period.

Abbreviations:

1. Arms: F, flexed; T, hands on thigh; P, hands prepubic; C, chest crossed.
2. Canopics – chest: 0, none; W, wood; S, stone; C, cartonnage; P, pottery.
3. Canopics – jars: 0, none; W, wood; S, stone; C, cartonnage; P, pottery; F, faience; G, gold.
4. Canopics – stoppers: 0, absent; U, unmodelled; H, human-headed; A, animal-headed.
5. Cavity packing: N, no; L, linen; S, sawdust; M, mud; R, resin; F, fungus.
6. Cranial cavity: E, empty; V, resin vapor; R, resin; L, dry linen; S, resin-soaked linen; I, insects; B, brain present; A, asphalt; P, plants.
7. Evisceration: N, no; Y, yes; I, indeterminate.
8. Evisceration incision: 0, none; F, left flank; I, inguinal; A, midabdominal; P, perineal; S, sutured; U, unsutured.

9. Excarnation: 0, none.
10. Excerebration: 0, none; E, extranasal; T, transnasal craniotomy.
11. Eyes: O, onion; G, globe present; L, linen pad; R, resin in orbit; S, shell; P, pupil; Q, stone; C, colored; W, wax.
12. Facial painting: 0, none; P, facial painting.
13. Legs: E, extended; F, flexed.
14. Molding: 0, none; F, facial; B, whole body.
15. Mummy portrait: 0, none; P, portrait on wood.
16. Natron: 0, none; B, present with body.
17. Resin and beeswax: 0, none; T, in tomb but not on body; P, present on body; B, bitumen also present; L, present on linen; W, beeswax.
18. Restoration: 0, none; S, sticks.
19. Salt: 0, absent; S, symbolic; P, preservation effort.
20. Subcutaneous packing: 0, absent; P, present.
21. Visceral disposal: 0, non; S, separate from body; B, body surface; C, body cavity.
22. Visceral modelling: 0, none; G, modelled genitalia; P, present.
23. Wrappings: G, goat skins; L, linen; M, modelled; S, shroud; R, resin-soaked wraps; H, herring-bone wrap; P, painted shroud.

Source: Modified from Bertoldi & Fornaciari, 1997.

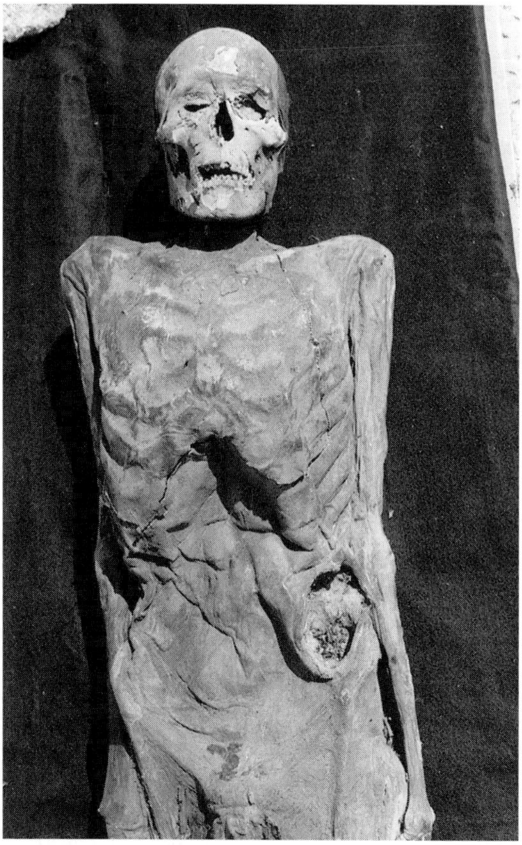

Fig. 4.136. Evisceration incision; AM, ca. 30 BC to AD 395.
Prior to the reign of Tuthmose III the incision was placed as shown here, but thereafter it was made just above and parallel to Poupart's ligament. (Photo by ACA and Michael Zlonis of mummy no. 3 of Dakhleh Oasis (Kellis site) mummies of the Roman Period.)

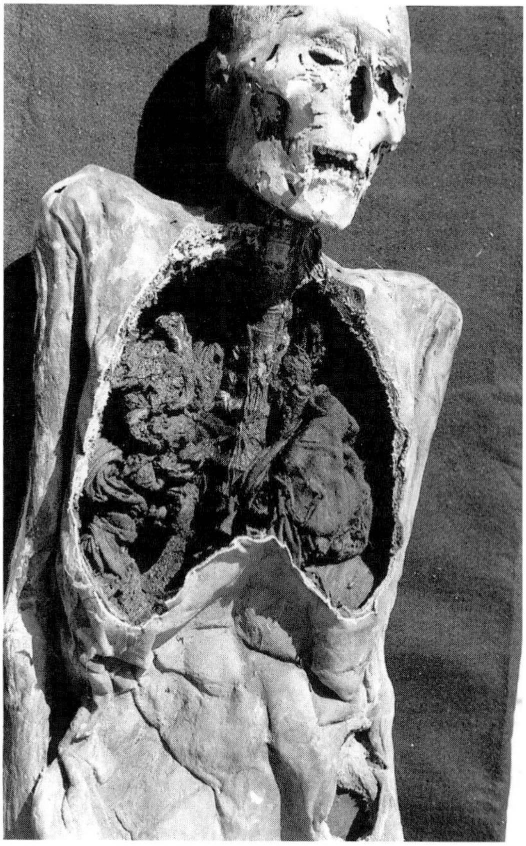

Fig. 4.137. Stuffing of the body cavities; AM, ca. 30 BC to AD 395.
Resin-saturated linens were rolled up and placed into the eviscerated and desiccated body cavities during the latter half of the history of the ancient Egyptian kingdom. (Photo is by ACA and Michael Zlonis of the same mummy pictured in Fig. 4.136.)

skin was necessary to maintain the osmotic movement of moisture through the skin (Notman & Aufderheide, 1995).

Waterproofing and cosmetic stage

Following about 40 days of natron desiccation the natron was removed from both the skin surface and the body cavity. That cavity was again rinsed with palm wine, wiped dry, and the entire cavity surface was painted with hot, liquid resin, commonly mixed with a substantial amount of beeswax. After this had cooled and hardened, hot liquid resin in generous quantities was poured into the body cavity not only to sterilize it, but to seal all openings. The final treatment of the body cavity was then applied most commonly in the form of stuffing it with hot, resin-soaked linen (Fig. 4.137). The evisceration wound was usually not sewed shut in this period; in some it was left gaping while in others the opening between the wound's lips was filled by stuffing a resin-impregnated linen rag into it. In only a few cases, was a metal plate placed over the wound.

This was also the stage when the finishing procedures over the entire body's skin surface were carried out. In royal or noble bodies the hair was sometimes arranged in the desired coiffure – braided, curled, or plaited. Body openings (nares, anus, external auditory meatus, vagina) were often closed with a linen tampon sometimes soaked in resin. The eyes were frequently allowed to remain in the orbits but depressed, covered with resin or resin-soaked linen patch and eyelids pulled over them. Metal plates under the eyelids in some helped to support the lids. The transnasal craniotomy defect was sometimes sealed with a resin-soaked linen patch. Fingernails and toenails were tied to the fingers and toes to prevent their loss; in royal bodies gold thimbles were occasionally employed for this purpose. The skin surface was anointed with cedar oil or other ungent, and often aromatics such as myrrh were applied to the skin. Finally the entire skin surface of the desiccated body was made water-repellant by painting it with a layer of hot, liquid resin. While most resins are water soluble, the thick layer of hardened resin on the skin and in the abdomen effectively protected the dried soft tissues, since the desert climate provided only occasional exposure to air of high humidity and virtually never liquid groundwater, unless the tombs were below the Nile's level of seepage water. Further water-repellant effects were contributed by the beeswax often mixed with the resin. The body was then adorned with necklaces, bracelets, hair ornaments, arm bands, rings, scarabs, and other jewelry and amulets.

Wrapping stage

Linen was employed almost exclusively. While some of this in elite burials was specially prepared for this purpose and was inscribed with religious messages, more commonly the inner wrappings appear to represent used, even cast-off clothing, table covers or other utilitarian items. One Roman Period burial at Dakhleh Oasis included a linen tunic, complete with head and arm holes, made to fit a child of about 4 years of age, which had been used to wrap an adult mummy's arm. During the New Kingdom the wrapping procedure was elaborately prescribed with individual prayers accompanying every step. Limb wrapping began with the fingers and toes, which were individually wrapped. The wrapping then progressed to the hands and feet. Broader sheets were then first applied to the arm in a longitudinal orientation. After several layers of this, circular bands, held in place by daubs of hot resin, enveloped the arm. Lower limbs were similarly wrapped. The penis was individually wrapped and, if excessive shrinkage or damage had occurred during desiccation, it was modeled with resin-impregnated linen. Large linen sheets were commonly then used to envelop the already wrapped arms with the trunk and similarly another to enclose both lower extremities. Thereafter further packing of linen in localized areas was employed to simulate a desirable body form. Such packs were commonly employed in women to shape the appearance of breasts. The outer shroud, held in position by circular bands, was painted red in some mummies. Jewelry, amulets, ushabti, scarabs, and other items were included between linen layers during the wrapping process. Finally, in some, facial features (eyes, eyebrows, mouth) were painted on the outer layer of linen. The rituals associated with wrapping the body were so elaborate they required 15 days to complete (el-Mahdy, 1989:68).

Nonroyal mummies from the New Kingdom

Probably the best measure we have of nonroyal bodies from the New Kingdom are those from Deir el-Medina, the village of workmen in the Valley of the Kings. While not royal nor even aristocratic, neither were they *fellahin* (peasants). Though empty canopic jars were found in their tombs, these bodies were not eviscerated neither were they excerebrated. Funerary tomb inclusions were abundant and often luxurious. The bodies had been dried with natron and revealed external application of resin in generous quantities (Ikram & Dodson, 1998:124). In the same tomb as that

Table 4.14. *The kings of the Third Intermediate Period*

Dynasty 21 (1064–940 BC)	Dynasty 22 (940–743 BC)	Dynasty 23 (Thebes) (867–724 BC)	Dynasty 23 (743–715 BC)	Dynasty 24 (731–717 BC)	Dynasty 25 (752–656 BC)
Smendes	Shoshenq I	Harsiese	Pedubast II	Tefnakhte	Piye
Amenemnesu	Osorkon I	Takelot II	Osorkon IV	Bokkhoris	Shabaka
Pinudjem I	Shoshenq II	Pedubast I			Shabataka
Psusennes I	Takelot I	Iuput I			Taharqa
Amenemopet	Osorkon II	Osorkon III			Tanutamun
Osokhor	Shoshenq III	Takelot III			
Siamun	Shoshenq IV	Rudamun			
Psusennes II	Pimay	Peftjauawybast			
	Shoshenq V				

of Hatnufer, the wife of Hatshepsut's principal architect, were seven other skeletonized, unembalmed bodies, one of which was wrapped as a mummy. The latter contained the remains of a female and two children with features suggestive of secondary burial. Derry (1942) suggests that this may represent the tomb of Hatnufer's family, and that the bodies of previously deceased members were transferred into her tomb at the time of her death.

The parents (Yuaa and Thuiu) of Thiy, one of Akhenaten's wives, are excellent examples of New Kingdom mummies. The evisceration wound in Yuaa's mummy is covered with a gold plate and that of his wife Thuiu is partly closed by a suture. Soft tissue preservation is superb even though no resin was applied to Yuaa's skin (Quibell, 1908:68–73).

The Third Intermediate Period

Sociohistorical features of the Third Intermediate Period

When Ramses XI died at the end of the New Kingdom, the centralized control of Egypt under a single king once again collapsed. Though the Ramesside rulers had kept their religious center in Thebes, their secular duties were carried out from the delta city Tanis (Table 4.14). After the death of Ramses XI the governor of Tanis was declared king, where the kings of Dynasty 21 ruled in the eastern delta in an uneasy relationship with Theban rulers. Libyan military leaders in the delta (Dynasty 22) re-established control of Palestine temporarily, but, while the delta rulers were occupied with warding off trouble from the Levant, Nubian leaders exploited the political turmoil by consolidating their forces and extending their influence northward until, by 716 BC they controlled not only Thebes but the entire country. During Dynasty 25, Egypt was ruled by Nubian pharaohs. The Assyrians, however had expanded and recovered control first of Palestine, then the delta and finally drove the Nubians out of Thebes back into Ethiopia in 667 BC. Continuing domestic problems weakened the Assyrian grip on Egypt, but they perpetuated their influence more indirectly through one of their vassals from Sais who regained control of all Egypt in 655 BC, initiating the Late Period under Dynasty 26, the Saite Dynasty.

Mummification features of the Third Intermediate Period

During Dynasty 21 (which introduced this period) mummification reflects a very distinct break with the past, reflecting major efforts by embalmers to make the body after desiccation appear more life-like. Smith & Dawson (1924:111) attribute this to the fact that ancient embalmers, as a consequence of an ancient investigation into tomb-robbing, had an opportunity to visualize a mummy's appearance after a long interval of burial. Several papyri document how the looting of royal tombs during Dynasty 20 had come to royal attention and provided detailed testimony of the arrested culprits. During the investigation, Smith & Dawson believe that contemporary embalmers were motivated (or royally commanded?) to improve their skills beyond the production of the shrunken, shriveled scarecrows that were revealed when the looters stripped the wrappings from the mummies.

The information about mummification in this period is derived principally from nine royal mummies and those of more than 40 priests and priestesses from Thebes. Smith & Dawson (1924) cite the mummy of Queen Nozme, wife of Hrihor, the last ruler of Dynasty 20, as an example of early embalming experimentation in which the wrappings were elaborately padded to simulate the body form of the living individual, but whose actual mummified body's shape was not different from those of earlier Dynasty 20 mummies. Thereafter the principal method used to give the desiccated bodies the rounded contours they possessed during life was the process of forcing foreign material into a space created at the time of evisceration by dissecting the subcutaneous plane between the skin and underlying muscles. Smith & Dawson (1924) detail this both in their narrative and in their illustration (their Fig. 31 between pages 88 and 89). They began their dissection by incising the fat layer at the edge of the evisceration wound (Fig. 4.138), inserting their knife-containing hand into the cut and enlarging it by forcing the hand in the desired direction. They then stuffed a variety of materials into the space to force the skin outward. Stuffing materials included linen, mud, resin and sawdust. Sticks were employed to force these materials beyond the reach of the arm. Additional skin incisions were employed to stuff parts of the body that could not be accessed from the evisceration wound (Fig. 4.138). Excessive zeal in the face area sometimes resulted in the bursting of the skin there during desiccation, as occurred with the body of Queen Henttawy (Fig. 4.139). Ikram & Dodson (1998:126–128) cite the Rhind papyrus as prescribing a total of 17 such incisions, though the mummified bodies usually demonstrate less than half of that. Interestingly, no effort was made to stuff the breasts, in spite of the fact that they commonly shrank dramatically during desiccation. The New Kingdom Dynasty 18 mummy of Amenhotep III was packed in subcutaneous areas identical to the special methods of Dynasty 21 mummies three centuries later. Rulers during that 300 year interval were mummified in the usual New Kingdom manner. This oddity has been variously explained as an early experiment that was abandoned in Dynasty 18 and revived in Dynasty 21 (Smith & Dawson,

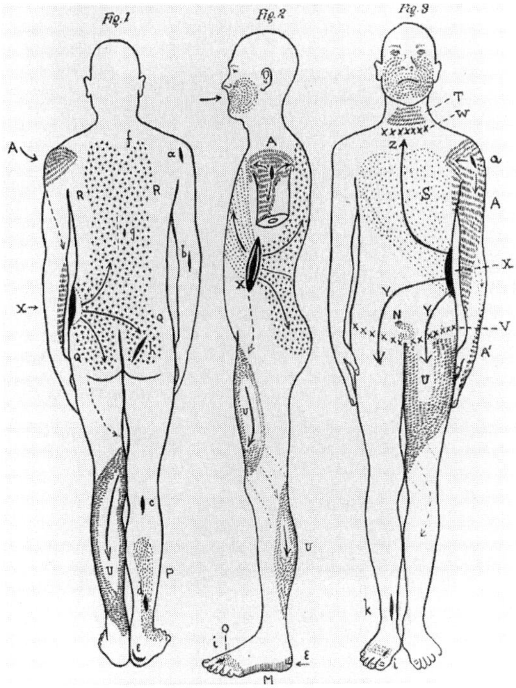

Fig. 4.138. Subcutaneous packing; 1064–940 BC. To produce a desiccated body whose form more closely approximated that of the living, embalmers of Dynasty 21 dissected the subcutaneous tissue in a pattern shown here, then filled the space with sawdust or soil. The lettered areas represent skin incisions through which packing material was inserted into a subcutaneous space created by manual dissection. The dotted areas indicated the extent of such subcutaneous packing. (Photo from Smith & Dawson, 1924: Fig. 31 and p. 115.)

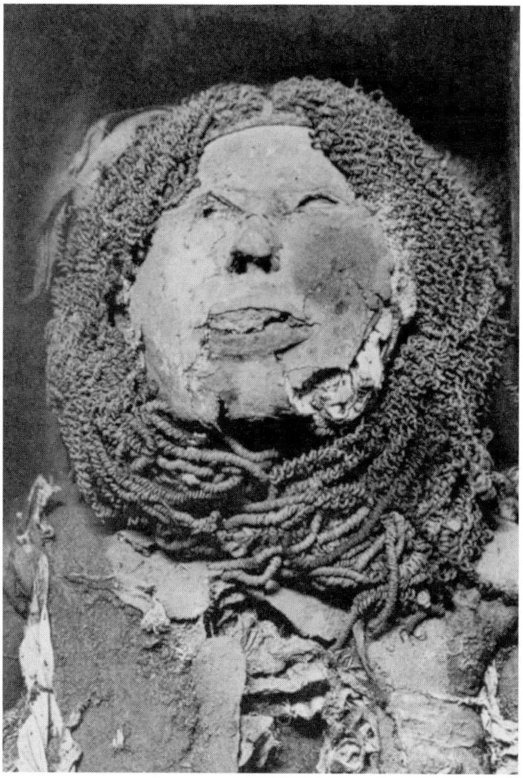

Fig. 4.139. Facial subcutaneous packing; AM, ca. 1026 BC.
The face, packed as indicated in Fig. 4.136, had been stuffed so excessively in the Dynasty 21 body of Queen Henttawy that the skin burst during desiccation. (Photo from Smith, 1912b.)

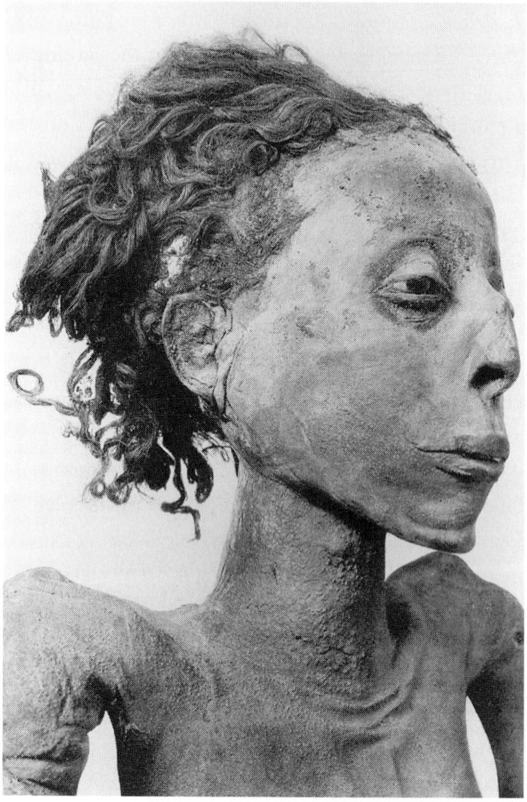

Fig. 4.140. Artificial eyes; AM, ca. 1064–940 BC.
Some royal or elite Egyptian bodies were made to appear more lifelike by implanting mollusk shells in the orbit, simulating the pupil with a round piece of black obsidian, as in this Dynasty 21 mummy of Nesitanebasher. (Photo from Smith, 1912b.)

1924:111) or as evidence that the body in the coffin of Amenhotep III is really that of a later person (Brier, 1994:267–269).

Further modifications included stuffing the neck via a thoracic incision with packets of a mixture of fat and soda. Sawdust was the preferred filler for the eviscerated body cavities. The employment of white stones with insertion of round, small, dark centrally placed stones to simulate a pupil inserted into resin-filled orbits produced a startlingly life-like effect (Fig. 4.140). Further efforts to simulate a life-like statue were employed by painting not only the face, but the entire body (red for male, yellow for females). Wigs were commonly prepared, and the cheeks were rounded by filling the oral cavity with sawdust. In Queen Nozme's mummy, real hair was pasted above the orbits to simulate eyebrows.

However, such efforts to seek perfection of the mummification process did not survive beyond Dynasty 21. Already by Dynasty 22 the quality of mummification had begun to regress. Water gained access to the silver coffin in which Shoshenq I was buried at Tanis in the delta and caused much loss of nonskeletal tissue. However, sufficient soft tissue remained to establish his mummification method as resembling that of Dynasty 18 more than that of

Table 4.15. *The kings of the Late Period*

Dynasty 26 (Saite) (664–525 BC)	Dynasty 27 (Persian I) (525–404 BC)	Dynasty 28 (404–399 BC)	Dynasty 29 (399–380 BC)	Dynasty 30 (380–342 BC)	Dynasty 31 (Persian II) (342–332 BC)
Psammetikhos I	Cambyses	Amyrtaios	Nepherites I	Nektanebo I	Artaxerxes III
Nekho II	Darius I		Akhoris	Teos	Okhos
Psammetikhos II	Xerxes I		Nepherites II	Nektanebo II	Arses
Apries	Artaxerxes I				Darius III
Amasis					
Psammetikhos III					

Dynasty 22/23 (Derry, 1939). An Egyptian mummy in Edinburgh's Royal Scottish Museum reveals that the subcutaneous packing efforts had been abandoned by Dynasty 24 (Dawson, 1927a). The emphasis shifted from the condition of the body to that of the mummy's wrapped, exterior appearance that was characteristic of the Late Period.

The Late Period

Sociohistorical features of the Late Period

The peaceful era that initiated this period was soon followed by recurrent tumult. Babylonian occupation of Palestine drove not only Egyptians but Greeks and other nationalities back into the delta where Egyptian commerce had already accumulated a mixture of nationalities. Civil war followed and during the frequent shifts of power the Persians twice occupied Egypt (524–404 BC and again in 343–332 BC). The spectacular Macedonian sweep through the Levant and the first defeat of Darius III's Persian army by Alexander the Great caused the Egyptian populace to embrace this conqueror as a liberator, leading Egypt into 300 years of Greek rule (Table 4.15).

Mummification features of the Late Period

Surprisingly few mummies from this period have become available from the Nile Valley. Indeed, Grafton Elliot Smith (Smith & Dawson, 1924:121) laments that "our information from Egypt is so meager, that for later periods we have to rely very largely upon the data afforded by the examination of a very large series of mummies from Nubia . . .". The series to which Smith refers are mummies excavated from Nubia during a major salvage archaeological survey carried out by George Reisner in 1907–1908 (Reisner, 1910) just prior to construction of a dam that subsequently permanently flooded a substantial area of Nubia. Only a small fraction of mummies buried in those now flooded areas could be recovered, but those which were rescued represent invaluable scientific treasures. They were examined by both Smith and F. Wood Jones and later by D. E. Derry, though it was principally Wood Jones who is responsible for the detailed field notes and their subsequent expansion and publication (Smith, 1910:181–220). I will summarize their findings as representative of mummies from the Late and Ptolemaic Periods, though the reader should remember that Nubian practices undoubtedly were not identical to those of contemporary Nile valley embalming methods. Even Smith (1910:195) warns that "the bulk of our specimens, therefore, cover a period which probably extends from the time of the Late New Empire to Persian and Ptolemaic times, and since archaeological separation of these periods is not precise (in our series) I shall deal with the whole of this series as one group. . .". Even today we can cite only a few additional, independently reported Egyptian Nile Valley examples. However, a huge number of mummies from this period were found in the Bahariya Oasis in 1999 (Hawass, 2000b) and may prove to be an auxiliary source of information after they have been

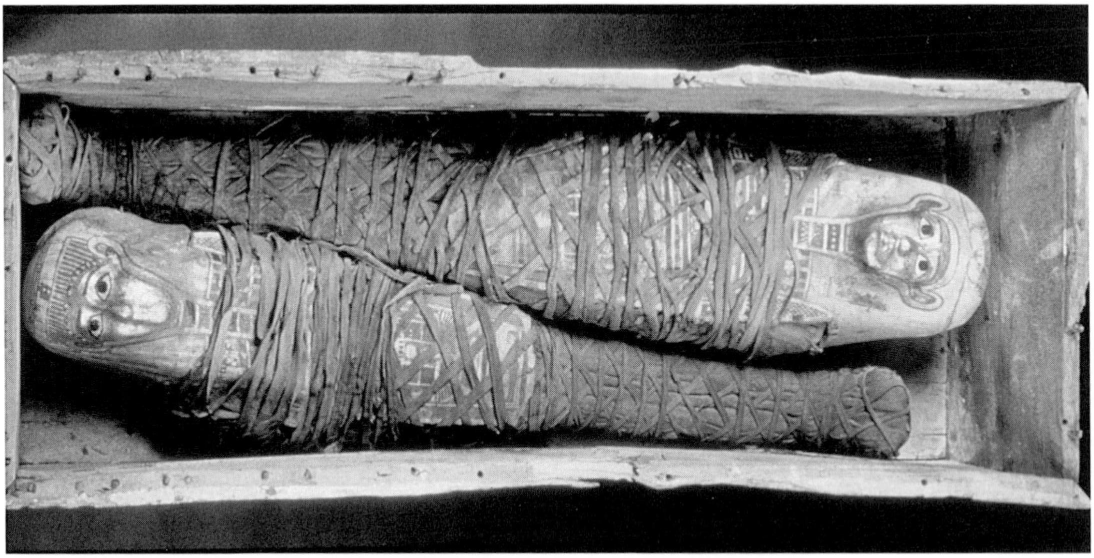

Fig. 4.141. Late Period mummies; AM, ca. 525–332 BC.
In the Nile Valley, superficial wrappings of elite bodies frequently assumed a complex, geometric pattern. (Photo courtesy of William Pestle and Nina Cummings at The Field Museum, Chicago, Negative no. A110660c.)

examined and reported on. In the following summary, principal emphasis is directed at features that differ from the previous period.

The wrappings of the Nubian mummies are described by Wood Jones as reflecting only a casual interest in either appearance or effectiveness, and Smith interjects his observation that this is a point of difference in the Nile Valley, where the outer layers were arranged in more complex patterns (Fig. 4.141). The initial layers next to the body were applied with the help of generous quantities of resin, and often considerable resin was poured over the outer layers. The hair of Nubian mummies was not cut before burial and was often quite long on female mummies, though frequently it was inadvertently removed during unwrapping because it was so adherent to the bandages. The mouth was commonly agape and filled with linen and/or resin. Resin-soaked linen tampons often seal the nares, anus, vagina, and external auditory meatus. Collapsed globes of the eyes were present in 80% of the orbits, covered by resin, but no artificial eyes were identified. Transnasal craniotomy was performed in fewer than half, and were especially infrequent in children. Resin was present within the cranium in only one-third of these.

Uneviscerated body cavities were the rule in children but found in only a few adults. The abdomen of most eviscerated adults was emptied via the same left flank incision as in the Nile Valley mummies, though visceral disposal was quite variable; in one example they were found bound to the abdominal skin. In many they were returned to the abdomen wrapped in resin-soaked linen packets. The evisceration wound was never covered, and in many a resin-soaked linen pad was stuffed between its gaping edges.

On pages 213–215 of Smith & Wood-Jones' (1910) report is described an unusual practice in which a stick is placed in the eviscerated body cavity, emerges at the neck and to which the head is fixed with bandages. The head in these had been excerebrated via the foramen magnum. In others some body parts or bones were either missing or present in excess (composite mummies, made up of parts from several different bodies), all of which had been wrapped so as to suggest the normal exterior appearance of a single enclosed body. Similar reconstructions have been described in Greco-Roman mummies from the Dakhleh Oasis (Aufderheide *et al.*, 1999b).

Descriptions in some individual reports of Late Period Nile Valley mummies tend to be largely consistent with the principal features of the Nubian mummies listed above. Dawson (1927b) describes a Dynasty 26 mummy donated to the British Museum by the Duke of Sutherland; this conformed to the Nubian pattern. In another example (Dawson, 1927d) he describes the well-known mummy of the Persian period dissected and reported by Granville (1825), in which there was no abdominal incision but some of the viscera had been extirpated via the perineal route and replaced with a packing of mud and resin. Gray (1966) X-rayed several Late Period mummies in museums and confirmed the observations of Smith & Wood Jones (1910) by identifying surplus bones. Pahl & Parsche (1991) note changes suggestive of defleshing in several skeletal remains from the delta not unlike those Wainwright (1910) had reported from Meidum. A mummy from Dynasty 26 was reported by a museum in Middletown, Connecticut, that demonstrated extensive padding of the wrappings to provide a living body contour but otherwise resembled the treatment reported by Smith & Wood Jones (Dyson, 1979). The mummy of Amentefnakht was unique in that the dorsal part of his body within the sarcophagus was immersed in a brown liquid. Chemical analysis was used to reconstruct his mummification. This appeared to consist of depositing his body into the sarcophagus, the bottom of which contained a layer of natron. His body fluids then dissolved these salts without evaporation in the plaster-sealed sarcophagus (Zaki & Iskander, 1942). With a few minor exceptions these mummies are consistent with the basic features found in contemporary Nubian mummies. It must be admitted, however, that most of the above-described mummies have never been radiocarbon dated. Hence, they may have been labeled as "Late Period" mummies simply because they did resemble the Nubian mummies from that time.

Egypt's Greek Period

Kings of the Ptolemaic Dynasty

Alexander III ("The Great") of Macedon defeated the Persians and took control of Egypt in 332 BC. He assumed the position of pharaoh, though he left Egypt in the charge of an administrator, Cleomenes, and led his army in a series of astonishing conquests that overran Babylon and even extended into India. When he died of an illness in Babylon in 323 BC, one of his generals (Ptolemy) led part of Alexander's army back to Egypt. He pronounced himself pharaoh, successfully fought off his competing generals' armies and established the Ptolemaic Dynasty that persisted for 300 years in Egypt: Ptolemy I to XV (Caesarion, son of Julius Caesar and Cleopatra). Cleopatra VII who inherited Egypt from her father, Ptolemy XII, married her older brother Ptolemy XIII, who drowned in a delta battle against Julius Caesar. Cleopatra then married her younger brother Ptolemy XIV, but bore a son with Caesar who became Ptolemy XV. However, after Caesar's assassination she joined Anthony against Octavian (Caesar's nephew) in a struggle for supremacy in Rome. When the Anthony–Cleopatra fleet lost the battle of Actium and were pursued to Alexandria by Octavian, the pair committed suicide. Octavian (thereafter known as Caesar Augustus) dispatched the child Caesarion (Ptolemy XV), terminating the Greek Ptolemaic Dynasty. He then claimed Egypt as a Roman province in 30 BC, initiating the Roman Period in Egypt that endured for the subsequent four centuries until Rome fell in AD 395.

Sociohistorical features of Egypt's Greek (Ptolemaic) Period

Alexander visited the desert oasis at Siwa where the oracle declared his divine origin. Leaving Cleomenes to organize the Greek administrative structure of Egypt and begin construction of a new delta capital (Alexandria), Alexander returned to complete his campaign. He died after a brief illness in 323 BC and his generals divided the empire among themselves. Ptolemy returned to Egypt, where he ultimately declared himself pharaoh (Ptolemy I) after successfully defending himself against other ambitious generals.

The first several Ptolemaic rulers were good administrators. They established order, keeping the culture intact but overlaying it with Greek principles. International trade and commerce flourished. Additionally, Ptolemy I and Ptolemy II also initiated and expanded an intellectual center (museum and library) that included the largest collection of books in the then Western world and attracted the best poets, historians and scientists as "artists-in-residence" (Canfora, 1987:37–44). Nevertheless, Roman need of grain found Egypt irresistible. Desperate diplomatic and personal manipulations by the last of the Ptolemaic pharaohs, Cleopatra VII, to maintain Egyptian independence collapsed when the combined fleets of Anthony and Cleopatra fell to Octavian (thereafter called Caesar Augustus) in the battle at Actium, followed by their suicides in 30 BC (Grant, 1972:226).

Mummification features of Egypt's Greek (Ptolemaic) Period

Because mummification practices during this interval are transitional between those of the Late and the Roman periods, they will be included in the discussion of the following section.

The Roman and Greco-Roman Periods

Sociohistorical features of the Roman Period

The treatment of Egypt as a Roman protectorate ceased with Octavian's victory. Thereafter Egypt was formally a Roman province. Roman dependence on Egyptian grain led Augustus to administer it himself so jealousy that no high-ranking Roman could visit it without his permission. The political tumult of subsequent imperial Rome, however, was also reflected in Egypt and, after the formal division of the Roman empire in AD 395, Egypt became a Byzantine province. By this time Christianity had become the predominant religion in Egypt even though it was often in administrative conflict with Constantinople.

The remainder of Egyptian history, including the Byzantine Period, its inclusion within the sweep of Islam across northern Africa, the Mameluke rulers and their defeat by the Ottoman empire, Napoleon's aborted conquest, the oscillating French and British influences, the modern Egyptian Kingdom, and finally Egypt as an Arab Republic, comprises fascinating historical Egyptian chapters, but mummification declined rapidly following the fall of Rome. Hence this will end our historical review with the termination of the Roman Period.

Mummification features of the Greco-Roman Period

Mummies from the Greco-Roman Period reflect a dramatic shift in emphasis within Egyptian mortuary practices. During this interval embalmers expended much less emphasis on preservation of soft tissues and more on the external appearance of the wrapped, mummy bundle. Narrow linen bands were often arranged in a diagonal fashion in such a manner as to create diamond-like patterns. In some of these a flat gold leaf was visible within each diamond-shaped enclosure, creating the illusion that the entire mummy was enveloped in gold leaf (Brier, 1994:101). It was also within this period that linen was woven for the express purpose of mummification; prior to this the bandaging was composed principally of cast-off clothing and rags. These new linens were commonly inscribed with prayers or inked patterns that further enhanced the external appearance of the bundle. Excessively generous use of resin, sometimes filling the abdomen, with a surplus poured on the bandaged surface also characterizes Roman Period mummies. Internal features became quite variable. At the Roman Period necropolis of Douch, in the Kharga Oasis, French excavators found that only two-thirds of the mummies revealed evidence of transnasal craniotomy, while a mere 15% had been eviscerated (Dunand & Lichtenberg, 1992:199–223).

At the Roman Period site of Kellis in the Dakhleh Oasis, only the minority of the bodies were prepared by evisceration with generous applications of resin. However, many of the bodies that had been buried without, or with only minimal effort to effect, soft tissue preservation became spontaneously mummified. These tombs were looted regularly, often resulting in body fragmentation. Subsequent efforts to repair such damage often involved employment of wooden rods and resin to fix disarticulated body parts into position before rewrapping them. Resin was also inserted via atypical portals such as the back muscles, orbits and others (Aufderheide *et al.*, 1999b).

A unique feature introduced during the Roman Period was the use of portraits of the deceased. Ancient Egyptian graphic art invariably presented the face in profile, but in the art of Roman Period mummies, the subject is facing the viewer directly (Fig. 4.142). The realistic detail of these has led to the suggestion that they constitute the world's earliest surviving portraits (Berger, 1999). Although some had been recovered earlier (in the seventeenth century), they came to public attention in 1887 when the Austrian merchant Theodor Graf asked his agent to purchase as many as possible, principally from Bedouin salt miners near er-Rubayat in the Fayum (Thompson, 1976:7). He offered these for sale in a traveling exhibition, but they were so radically different from conventional Egyptian art that they were suspected of being a hoax. However, Petrie excavated 150 of these during the next 2 years from the Hawara site in the Fayum, not only verifying their authenticity but generating the term "Fayum mummy portraits" by which they are still commonly known today. This is, however, a misnomer, since a moderate number of additional ones were excavated from multiple sites between the Fayum and Thebes (modern Luxor) subsequently, especially at Antinoopolis. A total of nearly 1000 are known to have been recovered to date from a variety of sites in the communities scattered along much of the Nile Valley's length (Jucker, 1984).

Most of these portraits were painted between AD 50 and 250, a few later but none after 350. They are all bust views; no full length postures have been found. Excavators such as the French archaeologist Gayet believe they see a great deal of similarity between the portrait and the mummy within the wrappings, though at least one discrepancy has been reported in which the nonconcordance of sex was noted. One wrapped mummy was subjected to CT (computed tomography) scanning, the results of which were employed to create a plaster model of the skull. A manual technique was then used to create facial reconstruction on the plaster model. Comparison of this reconstructed face with the appearance of the mummy portrait for that body

Fig. 4.142. Mummy portrait; ca. 525–332 BC. During the Roman Period in some areas (especially the Fayum), realistic portraits were painted on boards and then incorporated into the wrappings to make them visible externally. (Photo by ACA of an item at the Milwaukee Public Museum, courtesy of Lupton Carter.)

revealed a striking likeness (Macleod *et al.*, 2000). Thompson (1976:12), however, feels that the very latest group consists of a few general stereotypes with variations. While some, especially those of children may have been painted post mortem, most were probably painted while the subject was still living. Support for this impression is gained not only by the intensity of the subject's appearance but also by the fact that some are still in a wood frame that includes the string on the reverse side, used to hang the portrait on the wall. Sometimes a decoratively painted stucco frame housed such portraits (Thompson, 1976: Fig. 14). When the subject died, the portraits were removed from their frame, the four corners rounded and the portrait included within the bandaging in a manner that permitted the painting to remain visible. These portraits replaced the cartonnage masks of the former periods.

Thompson (1976), Riding (1997) and Berger (1999) describe how these were prepared. The painting was made on a flat slab of wood (cypress, oak, cedar, linden, lime, or fig) that was first sealed with gesso (gypsum and glue). Pigments (gold, black, red and several others) were mixed with molten beeswax and applied while still warm – a technique termed encaustic. Later the colors were mixed with eggwhite (tempera). Both brush (camel, cat, squirrel or even human hair) and palate knife or spatula were used for their application. A minority, especially those from Antinoopolis, were painted on a linen shroud.

Both Berger (1999) and Riding (1997) call attention to the fact that, in spite of their intensity, the usual purpose of portraits (memorialization) was not the intended use of these paintings. These had a religious purpose – to facilitate, by means of facial recognition, the transition from this earth through the various passages of the underworld after death. Berger & Riding also note that artistically these were the last realistic portraits until the Renaissance.

Mummies after the Roman Period

Since Christianity included an expectation of corporeal resurrection, the concept of mummification was not unacceptable to members of this sect. However an edict in AD 392 by the Christian Roman Emperor Theodosius I forbade embalming. While we have few Egyptian bodies thereafter that demonstrate substantial anthropogenic soft tissue preservation, there is evidence of some degree of postmortem body manipulation. Salt is present on the body in quite a few during this interval. Most commonly, however, it occurs as only a small quantity sprinkled on the body surface, more reminiscent of ritual than a utilitarian embalming effort. Only a few demonstrate quantities great enough to suspect efforts at whole-body preservation (Prominska, 1986). However, myrrh, frankincense, unctions, and shrouds were commonly employed. Even these vestiges of former Egyptian mortuary practices ceased after the Arab conquest of AD 642.

Use of Egyptian mummification methods as a dating system

The evolution of anthropogenic mummification of Egyptian bodies has been detailed in the foregoing section from its initiation in the Archaic Period through a millennium of experimentation to reach its acme in Dynasty 21, after which it deteriorated quite rapidly, disappearing altogether in the Byzantine Period. Many, particularly early workers, have felt that the evolving changes in mummification techniques were sufficiently consistent and unique to lend themselves to a system allowing prediction of the time period in which an individual body was mummified. Indeed, Bertoldi & Fornaciari (1997) have prepared a useful chart of features designed for such use, which I have expanded (Table 4.13).

Readers employing such an approach, however, need to be aware also of its limitations. Of primary importance here is the quality of the database. While most of the references used for the creation of such tables today are reasonably current, they are often dependent on observations reported as much as a century ago. For example, Grafton Elliot Smith's (1912b) seminal comments on the bodies of royal mummies are not detailed, standardized records lending themselves to modern statistical evaluations, because he does not provide us with the presence or absence of specific features for every body. He does comment on every individual royal body available to him at the time, but in narrative form. We can not assume that his failure to comment about the presence of a feature implies the absence of that feature. In fact, some of the bodies had not even been completely unwrapped when he examined them, and apparently he was allowed only minimal dissection privileges. Furthermore, the placement in antiquity of some bodies of the royal caches into coffins designed for others and so labeled has resulted in uncertainty regarding the identity of some of the royal mummies. The mummy identified as that of Amenhotep III, for example, is prepared with the elaborate unique method of subcutaneous insertion of "filling" materials, a feature not appearing again until two centuries later. It is not at all certain whether this is a "failed experiment" or whether it represents misidentification of a Dynasty 21 body. While the features Smith originally associated with certain dynasties have been supplemented by observations from more recently excavated bodies, we do not today have a set of features whose individual statistical "weight" (for both presence and absence) of specific features has been established on a large group of both royal and nonroyal bodies whose ages have been firmly identified by reliable criteria other than those of Smith-defined mummification criteria for the various periods.

These limitations do not declare that all features have equal value or vulnerability. The molding of facial detail using the outer layer of plaster or resin-impregnated linen wrapping certainly appears to be an Old Kingdom practice that is not easily confused with the subcutaneous stuffing and other forms of attention paid to the body's face in the Late Period's Dynasty 21.

Assessment of the presence or absence of a feature such as perineal evisceration in an individual body can also be difficult, because the perineum's dependent position in a supine body often results in the accumulation of such a large quantity of resin there that its removal by dissection often destroys the diagnostic tissue changes. For decades transnasal craniotomy was considered to be a feature limited to the New Kingdom and subsequent periods, but dissection has now demonstrated its presence in the Middle and perhaps even the Old Kingdom (Strouhal, 1986).

For some features such as arm and hand positions our information is derived principally from royal or aristocratic bodies; these data should not be extrapolated to the mummies of commoners. Gradual accumulation of information about this unmistakably assessable (even by X-ray in wrapped bodies) feature on well-dated bodies can help to define this to a satisfactory degree (Gray, 1972). Instructive are data from a Roman Period necropolis of Douch in the Kharga Oasis (Lichtenberg, 1994) in which arms were extended in 91% of 47 mummies but crossed in 8%, legs were parallel in 87% but crossed in 13%. Transnasal craniotomy had been performed in only two-thirds, circumcision in about one third and evisceration in 13%. Derry (1942:261–262) notes that some features such as materials employed for stuffing an eviscerated abdomen were widely variable among royal bodies (lichen in Siptah and Ramses IV, sawdust in Ramses V and natron with fat in Merneptah). This feature, therefore, probably has lesser predictive value. Zimmerman (1986b) also calls attention to such variations. Furthermore, changes in embalming practices can not be expected to have occurred simultaneously in all regions throughout Egypt.

Summarizing this section, I do not wish to present an iconoclastic view that overemphasizes the constraints that variations of the features in Egyptian mummification place on the employment of mummification methods to predict the dynastic origin of a mummy. However, it is important to be aware of the vulnerability of such a practice. Perhaps identification of some of the database's weaknesses may be of value to those who attempt such predictions and hopefully might motivate someone to improve its quality.

Egyptian mummification materials

Listed below and grouped according to usage are some of the materials employed by Egyptian embalmers to achieve their mummification goals. Much of this information was extracted from a study by Baumann (1960),

Lucas & Harris (1962), Majno (1975), Ikram & Dodson (1998), Nicholson & Shaw (2000). While the chemist A. Lucas pioneered this field, recent applications of subsequently available instrumentation are enabling not only confirmation of many earlier studies that employed less specific methodology but also generating new findings not previously measurable. Dramatic examples of the latter include Nissenbaum's (1992) identification of Dead Sea bitumen in Roman Period mummies and the demonstration by Mejanelle *et al.* (1997) that Egyptian embalmers employed a vegetal tanning agent.

Body cleansing agents

Palm wine

The volatility of this agent frustrates a simple chemical test to demonstrate its presence in an ancient body. However, the date palm (*Phoenix dactylifera*) is native to Egypt. Wine's liquid nature could provide a mechanical cleansing effect after removal of the abdominal and thoracic viscera, and its 13% ethyl alcohol could contribute an antibacterial influence.

Myrrh

This was used to wipe the inside of the body cavity walls following the cavity's cleansing by palm wine. Myrrh is a gum resin derived from *Commiphora opobalsamum* acquired from Somali and Southern Arabia. Its resinous exudation emerging from an incision into its bark dries in the form of small accretions ("tears") usually less than 3 cm in diameter. Known also as the "Balm of Gilead", it has many medical internal and external applications. It was also a common perfume ingredient; the *Oxford English Dictionary* cites Psalm 45:8 as "All thy garments smell of myrrh and aloes and cassia . . ."

It is an aromatic substance with a turpentine-like odor whose use in wounds is more than 4000 years old. It was not, however, until the mid 1970s that Majno (1975:217–219) demonstrated formally what the ancients had suspected, that myrrh exerted a powerful antibacterial effect on common causes of wound infection such as *Staphylococcus aureus* and the putrefactive family of *Clostridium* spp. Thus its use to wipe the body cavity following evisceration and again after natron desiccation was bacteriologically sound.

Body stuffing materials

Myrrh

This supplied further antibacterial action, as described above, when employed for temporary stuffing, together with natron.

Frankincense

This fragrant aromatic is obtained in a manner similar to that of myrrh from the tree *Boswellia papyrifera*. Egyptian sources indicate that Queen Hatshepsut sent an expedition to the land of Punt (Somalia) to acquire this material. Baumann (1960) considers it the most commonly used incense throughout the world. Lucas & Harris (1962:91) identified it in material from the tomb of King Tutankhamun. While Majno (1975) suspects it has antiseptic qualities, these have not been demonstrated scientifically.

Lichen

Only a few Egyptian mummies have revealed the presence of lichen in the body cavity. This was *Evernia (Parmelia) furfuracea*. The Egyptian embalmers were probably trying to exploit its absorptive qualities (Baumann, 1960). Siptah's abdomen was packed with this material (Smith, 1907).

Sawdust

It was used as abdominal packing both alone (as was done with Ramses VI: Smith, 1907) and also mixed with sand or resin. Examination of occasional mummies reveals that it was sprinkled over the skin when mixed with an aromatic substance. Other applications include its mixture with natron in canopic jars. In all of these its large surface area no doubt enhanced its absorptive effect.

Resin

This is a term applied to the substances that exude from a tree spontaneously or following an injury such as bruising or incising the tree bark. This viscous liquid usually hardens after it is exposed to air, forming globs (sometimes called tears) of hard, commonly translucent, often brownish spherules. The well-known substance we call amber is an excellent example. Resins usually have a low melting point, but harden at room temperature. Trees from which resins are often extracted include pines (most common), spruce and, much less frequently, cedars. They have a broad range of applications including their use in paints and varnishes. Some resins (gum resins) contain a gelatinous material with mucilaginous qualities that hardens upon evaporation of its water content. While references dealing with ancient Egypt frequently refer to the cedars of Lebanon (*Cedrus libani*), this tree, whose wood has many highly valued qualities as timber, produces only small amounts of resin and was probably not a common source for embalming needs. Egyptian embalmers probably obtained most of their resin from the Cilician fir tree (*Abies cilicica*), the Aleppo pine (*Pinus halepensis*), the umbrella pine (*Pinus pinea*) and the oriental spruce (*Picea orientalis*), all obtained from Syria and/or Asia Minor. Less frequently employed were the turpentine-related resins from the nonconiferous tree *Pistacia terebinthus* found on the Greek island Khios just off the eastern Mediterranean coast of Turkey (Lucas & Harris, 1962:316–320).

These identifications were principally dependent upon solubility and fluorescence tests. Far more informative are the results of gas chromatography/mass spectrometry (CG/MS) methodology. Less than a dozen studies to date have been carried out, principally on Late or Greco-Roman Period mummies (Rullkoetter & Nissenbaum, 1988; Connan, 1991; Connan & Dessort, 1991; Proefke at al., 1992). These have identified conifer resins with frequent minor components of bitumen from the Dead Sea, Iraq or unknown sources, as well as beeswax. One sample from a Dynasty 19 mummy suggests an earlier use of bitumen than previous, cruder tests had led us to believe (Columbini *et al.*, 2000). What we have learned from the application of the GC/MS method is that, to understand best the pattern of resin use, the data as defined by methodology available to the era of Lucas & Harris must be refined by further examination using more sophisticated methodology. Perhaps a reasonable summarizing statement of our present knowledge about "resin" use in ancient Egypt is that the principal component was resin from several genera of conifers acquired from various areas at different times and that bitumen from the Dead Sea, Iraq and other sites as well as beeswax were added as minor components at least during the later periods (Nicholson & Shaw, 2000:466).

Resin lumps not associated directly with the body have been found commonly in Egyptian predynastic tombs containing spontaneously mummified bodies, where they can most logically be interpreted as playing a religious role, probably representing an offering. The various qualities of these materials were actively exploited during the Dynastic Period by Egyptian embalmers. They used resin as a heated liquid material to paint the body cavity wall following natron desiccation, also frequently pouring it in generous quantities into the eviscerated, dried body cavities. In later periods the body's skin surface was painted with resin. In all but the earliest periods it was employed to attach the initial layer of wrappings to the skin surface and daubed onto the various layers of wrappings to hold them in position. Resin-soaked linen rags were used as tampons for various body orifices, or stuffed between the gaping lips of the abdominal evisceration wound. Sometimes it was poured into the excerebrated cranial cavity or was mixed with the most superficial wrappings of the mummy, permitting the molding of facial features on Old Kingdom mummies. Resin-impregnated linen was also used to wrap desiccated viscera before they were placed in canopic jars or returned to the abdomen.

In some Roman Period mummies I examined at the Dakhleh Oasis it was a simple matter to trace the route of the hot, liquid resin from the abdominal cavity into which it was poured, entering the vertebral canal via the intervertebral foramina, then passing via this canal up into the cranial cavity. In several cases the resin did not reach the cranial cavity in the liquid state but was hot enough to vaporize, becoming deposited on the cranial walls as a feathery mesh several millimeters in thickness. In several bodies the spinal canal contained generous quantities of resin, a small quantity of which had "spilled" over the foramen magnum brim, draining into the otherwise empty cranial cavity as a hardened trickle several centimeters long. Resin reaching the cranial cavity via this route may account for the frequent finding of resin within the cranial cavity of an Egyptian mummy in which the well-defined transnasal craniotomy channel shows no evidence of resin stains; to date such a finding has been explained by the postulated use of a funnel for the introduction of the resin into the cranial cavity, though no such resin-stained instrument has ever been reported.

The term "kyphi" has been applied to the use of a mixture of resins poured on top of an already desiccated and wrapped mummy after it was placed into a coffin. It was used in amounts that would spill off the mummy, gluing the body to the coffin's interior. Frequently an additional amount was poured onto the coffin lid as well. Gum mastic from Khios was a common ingredient of kyphi (Baumann, 1960).

Aside from its employment as mucilage for wrapping linens, the nature of resins' use by embalmers suggests it had preservative action. Though Majno (1975:217) only formally tested myrrh for antiseptic activity, he generalizes the resins as probably exerting antibacterial effects. Nevertheless, in spite of resins' surprising ability to penetrate tissues (even reaching the joint spaces in some mummies), external application alone without natron desiccation in Egypt's later periods resulted in poor soft tissue preservation. It is certainly conceivable that exposure of the tissue surfaces (such as the eviscerated body cavity lining) to the intense heat of liquid resin would have a purely thermal bactericidal effect and help to reduce the bacterial load of that tissue. However, I suspect that Egyptian embalmers were using resin application principally as a moisture repellant for their desiccated, mummified tissues in order to prevent their subsequent rehydration through moisture absorption. The beeswax and bitumen that were often mixed with the resins surely contributed further water-repellant effects.

Bitumen

The naturally occurring hydrocarbon, sometimes called asphalt and sometimes bitumen, has a long history. A major source in Persia was exploited for medicinal purposes in medieval times, as detailed in Chapter 10 (Mummy as a drug). Because in its hardened state bitumen is black and shiny, it can closely resemble embalming resins that, over millennia, become altered to acquire a black, glistening appearance. Hence the modern appearance of ancient resin in Egyptian mummies has been mistakenly identified as bitumen in the past. Such an assumption gained credence from Diodorus of Sicily's (ca. 50 BC/1947) comment that Dead Sea bitumen was shipped from Palestine to Egypt for embalming use. The Dead Sea in modern Israel must have an asphalt-oozing cleft in its floor, because periodically since antiquity and even until today large globs of bitumen emerge at its surface and drift ashore. However, while some trading with Palestine occurred in antiquity, Egypt's relation to that country during most of its history was not one of close cooperation. Nevertheless, trade involving Dead Sea asphalt between these two countries may have occurred as early as 900 BC and was actively promoted early in the Ptolemaic Period. Hence it has been suggested that the less expensive Dead Sea asphalt may have been substituted for resin (or used as a diluent) by Egyptian embalmers primarily after about 300 BC. Attempts to establish the presence of bitumen in mummy samples, based on bitumen's solubility and of its content of vanadium, molybdenum and nickel led to conflicting results, some of which suggested that bitumen might be present in mummies of Ptolemaic or younger periods (Lucas & Harris, 1962:303–308). Spielman (1932) used ultraviolet fluorescence and spectrographic analysis to identify bitumen in "resin" of Ptolemaic mummies. Application of modern methods of GC/MS recently have demonstrated elegantly that material in a mummy from the second century AD clearly contained Dead Sea asphalt while that from about 900 BC revealed no trace of it (Nissenbaum, 1992). However, a survey of 13 "resin" samples from Egyptian mummies ranging from Dynasty 12 to the Roman Period studied by pyrolysis/GC/MS revealed no evidence of petroleum bitumen (Buckley & Evershed, 2001). Now that a specific chemical identification method for bitumen is available, a larger-scale study from different periods is indicated to define the demographic pattern of Egyptian use of bitumen for embalming in greater detail.

Anointing materials

Storax is a balsam oleoresin acquired from the tree *Liquidambar orientalis* that is found in Asia Minor, used not only for anointing the body after its desiccation by natron, but also medicinally and as a perfume ingredient. Pettigrew has suggested that colocynth oil, squeezed from the seeds of *Citrullus colocynthis* was also employed (Baumann, 1960). Lucas & Harris (1962:329–332) note that Ben oil from *Moringa pterygosperma*, *M. oleifera* or *M. aptera* is reported to have been used as an anointing oil and also sesame oil (which Majno (1975:53) believes is an antiseptic). Castor oil and balanos oil were also employed. Their chemical detection, however, on a mummy's body would be a formidable challenge even with modern technology. Myrrh, however, was commonly used also as an ungent ingredient for anointing and spices are mentioned as well. The nature of these spices is difficult to identify. Cinnamon (from *Cinnamomum zeylanicum*) and cassia (from *C. cassia*) have been recorded by Diodorus and Herodotus, respectively, though Baumann (1960) notes that their source in India would probably not have made them available before 300 BC.

Desiccating agents

Natron

The Wadi Natrun near Cairo is the source of most of the natron used in Egypt. Rains wash these soil ingredients into stagnant, shallow lakes or pools where subsequent evaporation concentrates them. Natron is a naturally occurring mixture of salts in which sodium carbonate predominates, accompanied by lesser amounts of sodium bicarbonate (baking soda), sodium sulfate and sodium chloride (common table salt). Variations in the concentrations of these components are evident in analyses of both modern and ancient samples (Garner, 1979). Its desiccating action is osmotic. However, sodium carbonate is employed in Ruffer's solution to break down protein bonds (especially those of collagen) to permit better penetrance of liquid reagents into the tissue. Thus it is conceivable that this action would enhance further the transfer of water through the skin with which it is in contact. Chemical, physical and historical evidence abounds to confirm its central role in the desiccation of Egyptian mummies. For years the question of how it was employed was debated. Early workers like Grafton Elliot Smith insisted that the body to be mummified had been immersed in a liquid natron bath. These workers were initially misled by a faulty translation in Herodotus, and further by an assumption that the frequent absence of the epidermis in mummies was the product of immersion in liquid (as indeed frequently does occur in immersed bodies, but not exclusively so). That misimpression was enhanced further by the finding of viscera in a canopic jar in the Old Kingdom tomb of Queen Hetepheres, Cheops' mother. These viscera were lying in a liquid solution whose analysis by Lucas was found to be a 3% solution of natron.

However, experimental mummification of small animals by Garner (1979) and Lucas & Harris (1962:292–293) could not confirm the effectiveness of liquid natron for soft tissue preservation. Subsequently alternative explanations for the epidermal loss became apparent. Application of dry natron to body skin often results in firm skin adherence of the epidermis to the natron (or the cloth if the natron is applied in bags) when the natron or cloth is peeled off the body. In addition I would like to suggest that all internal water must exit from the body during the desiccation process via the skin and this water is commonly enzyme-laden by the time it reaches the skin surface, rendering the skin susceptible to digestion. This suggestion is supported by the observation that the epidermis is found to be absent in anatomical sites overlying bulky soft tissue such as the thigh or anterior abdominal wall, but the epidermis on the external ear or other sites with minimal underlying soft tissue is most commonly still intact. In addition, the above-mentioned action of sodium carbonate in natron may also make a contribution.

Furthermore Queen Hetepheres' canopic jars may also be otherwise explained. Viscera normally were dried with dry natron before being placed in canopic jars with additional dry natron. It is conceivable that during the early phase of mummification's evolution in the Old Kingdom the viscera may have been placed in the jar without initial desiccation. If then they were covered by an amount of dry natron insufficient to desiccate them, the water content of the viscera may have dissolved the natron, creating a 3% solution without the addition of more added water, thus simulating a deliberate liquid bath. Another canopic jar filled with a pint of liquid was found at the Lahun I site by Brunton (1920:20). No organs were present but a mixture of mud and resin was found in all the canopic jars, including the one in which the fluid was present. Lucas (1908) found the principal ingredients of natron in the liquid of that jar, as well as humic acid. The most appealing explanation of these findings is the suggestion that they represent the residua of a flooded tomb.

Today few adhere to the "natron bath" concept. No container large enough to immerse a body for that purpose has ever been excavated. Considering the number of mummies in Egypt, such bath containers ought to be numerous for they would not be expected to have been as disposable as the drying board needed for dry natron application, several of which have been found (Winlock, 1922; el Amir, 1948). Also excavated have been a large number of natron-filled bags in refuse piles near tombs, including that of Tutankhamun. Most convincing of all are the experimental mummification procedures on the bodies of a large dog (Notman & Aufderheide, 1995) and a human (Brier, 1997) that were based on the application of dry natron alone.

Cosmetic materials

Henna

The leaves of the plant *Lawsonia alba* and *L. inermis* were extracted and the resulting reddish-orange pigment used to paint fingernails, toenails and even hair of both the living and the mummies. In Dynasty 21 it was employed as a dye for the outer shroud on male mummies. Cultivated in Egypt, it was also an ingredient of ointments and perfumes.

Onions

The eyes were left in the orbits after body desiccation, but in the latter half of Egyptian history they were commonly compressed and a pad of linen, often resin-impregnated was placed on them to occupy the space. The eyelids were then drawn down over the linen. Sometimes an onion was substituted for the linen pad. The onion does have some antibacterial influence, but the fact that an onion is sometimes placed also in such improbable locations as the sole of the foot, the external auditory meatus or even on the outside of the body suggests it also may have played a religious role.

Beeswax

This had multiple uses including employment as a paint vehicle, as a protective for painted surfaces and others. In mummification it was used to fix curls in hair and to fashion Horus figures molded to resemble the lids for canopic jars during Dynasty 24 when the viscera were returned to the abdomen (Lucas & Harris, 1962:337). Its presence has been documented in modern analyses of ancient "resins" using GC/MS methodology (Nissenbaum, 1992). A study employing similar analytical techniques on "resins" of 13 mummies between the New Kingdom and Roman Periods revealed the presence of beeswax as a resin component commonly used in later periods and was the dominant element (87%) in a Roman Period sample (Buckley & Evershed, 2001).

Paint

In Dynasty 21 every effort was exerted to achieve a lifelike body appearance. This included painting not only the face but often the entire body. Henna, discussed above, was also used to provide color to the face, palms and soles. Other pigments were also from natural sources such as iron oxide (used as facial rouge). Red ocher was also used to paint lips with a brush. Soot was employed to blacken areas, and both metal ores as well as vegetal sources provided other needed colors.

Artificial eyes

Prosthetic eyes were used in later New Kingdom mummies and especially in the Late Period. These were usually made of white stone, but it was during the special efforts made by Dynasty 21 embalmers that the black "pupils" were added to produce an effect more closely resembling the eyes of a living person.

Textiles

Mummy wrappings are virtually exclusively linen. The production of linen from flax was a major industry in Egypt for both domestic and foreign markets. Its use can be documented in the Nile Valley several millennia before the unification of Egypt. In predynastic times *Linum angustifolium* was the flax that was cultivated. Eventually this was replaced by *L. usitatissimum*. Since flax was also the cloth from which clothing for the living was fashioned, it is not surprising that the same cloth was commonly used for funeral purposes. The reason for that is evident when the mummy textiles are examined: they are usually composed of used, often patched sheets commonly torn into strips or clothing that is no longer fit for wear (Baumann, 1960). Only the most elite of burials employed previously unused linen, specifically manufactured for use on mummies and often inscribed with prayers from the Book of the Dead (Caminos, 1982).

Disposal of the viscera (canopic apparatus)

Methods of disposing of the viscera changed dramatically and almost continuously throughout the history of ancient Egypt. Furthermore, nonroyal burials attempted to emulate the fashions of royalty, but expense and tradition caused them to trail the changes in royal practices and only rarely even approximated their extravagance. Those readers desiring the details of such changes will be well rewarded by consulting the superb and well-illustrated presentation by Ikram & Dodson (1998:276–292) under the heading "Canopic Equipment". I will present here an abstract of this broad subject, drawn in part from their description and in part from a broad base of articles beginning with those of Smith & Dawson (1924) and David (2000), as well as several other texts.

The term "canopic jars" is often used as a synonym for visceral containers, but the latter were not always jars. Hence in this presentation I will use a term implying a wider range of structures: canopic apparatus. Even this, though, is an artifactual label. While some have related the origin of the name for the vases simply to the town of Canopus situated on the western end of Alexandria's major harbor near modern Abu Qir, Strouhal (1992a:261) describes how Kanopos (hereafter spelled in this volume as Canopus) was the name of the helmsman for Menelaus, husband of Helen and brother of Agamemnon. Canopus was revered as a form of Osiris at Abu Qir, and symbolized by a globular vase. When the viscera-containing jars were found in tombs by Egyptologists, scholars associated their shape with the Canopus vases; hence the term "canopic jars".

Predynastic and Archaic

Because anthropogenic mummification is not detectable during these intervals, it is not surprising that none of the canopic apparatus elements can be identified in tombs of these periods.

Old Kingdom and First Intermediate Period

The earliest evidence of visceral disposal becomes apparent in the tomb of Queen Hetepheres, wife of Snefru, already met above in the discussion of the use of natron for desiccation, as well as in the pursuit of the earliest evidence of Egyptian mummification. This Dynasty 4 tomb's sarcophagus was empty but a wall niche contained a rectangular stone chest with four compartments. No jars were present, but within each compartment was a linen-wrapped mass resembling viscera (Iskander, 1980:5) though to my knowledge these were never examined histologically. The chest was not decorated and had a flat lid.

By the end of Dynasty 4 it has become evident that such chests were usually placed near the corpse's feet, that some were inscribed with the occupant's name and that jars began to appear quite regularly thereafter within the chests' compartments in royal tombs. Both coffins and chests remained wooden in nonroyal tombs and in these, viscera were wrapped and placed in compartments without jars.

Middle Kingdom

During this interval, the stoppers of canopic jars became shaped to resemble human heads. In royal tombs the chests containing the jars were adorned with two goddesses on opposing sides, alternating with the symbols of gods on the other two. In occasional tombs the viscera were wrapped in a bundle enveloped by a mask with a human face. By the Second Intermediate Period some chests bore a figure of a jackal, representing Anubis, the Egyptian god of mummification (Ikram & Dodson, 1998:278–280).

New Kingdom

In Dynasty 18 the chests expanded into shrines, still of rectangular shape but having a curved roof. Goddesses placed at each corner linked their outstretched arms, encircling the structure. Canopic jar stoppers now carried the symbols of the four sons of Horus, each of which usually housed specific visceral organs. Imsety had a human head and stored the liver; Hapi (ape) the lungs; Duamutef (jackal) the stomach and Qebehsenuef (falcon) the intestines (Fig. 4.143). In 1837 Pettigrew examined a Dynasty 21 Egyptian mummy whose visceral distribution did not follow the above pattern. What Pettigrew described was accepted as the norm, causing great confusion until Smith (1911b) published the above-noted pattern. Nevertheless, it has become apparent subsequently that, while Smith's pattern is the rule, Egyptian embalmers frequently took liberties in the distribution of the viscera. The visceral bundles were first dried, then wrapped and placed in the jars. In royal tombs they were placed into mummiform miniature coffins, often of solid gold.

Change in mortuary practices, however, were rapid, as was change in religious and governmental structure during this and immediately subsequent periods. By Dynasty 20, canopic jars, which had become fixed into position by being incorporated into a stone chest carved from a single block, now returned as mobile, individual entities, though their location within the tomb became variable. Indeed during Dynasty 21 the viscera were simply wrapped and returned to the body – either within the abdomen or between or on the legs and feet. Canopic jars, however, continued as empty symbols, occasionally even containing carved substitutes for the viscera (Brier, 1994:85; 1997).

Late and Ptolemaic periods

During the early part of the Late Period some burials reveal a short-lived return to the use of canopic jars for

Fig. 4.143. Canopic jars; ca. 1187–940 BC.
With caps shaped to simulate the four sons of Horus, these jars were employed to store the bodies' principal viscera. (Photo courtesy of William Pestle and Nina Cummings at The Field Museum, Chicago, Negative no. A111049c.)

visceral storage. Thereafter canopic jars disappear, never to return. The chest that had formerly housed them grew into a heavily adorned, enlarged structure bearing carvings or paintings of the figures formerly dedicated to canopic jar stoppers. Few of these contained viscera and by the end of the Ptolemaic Period even the chests had disappeared (Ikram & Dodson, 1998:291–292).

Mummies from northern Africa

Mummies from Libya

While much of Libya's terrain is indistinguishable from adjacent Egypt's Western Desert, the extreme southwestern region incorporates part of the Tadrart Acacus range of mountains. Within these slopes a rock shelter is located called Uan Muhuggiag whose walls display a large array of rupestrian art. Excavation of its floor in 1959 uncovered the mummified body of a 2½-year-old child. Wrapped in an animal hide (whose radiocarbon date was 3455 ± 180 years BC), the body demonstrated evidence of evisceration after which it had undergone desiccation (spontaneous? natural-assisted?) (Mori & Ascenzi, 1959; Ascenzi, 1998).

Plaster burials from northern Africa

The bodies of Christian burials in Algeria and central Tunisia during the first four centuries AD were frequently encased in gypsum or lime. These "plaster burials" were described by Green (1977) and discussed in Chapter 3 (Anaerobiasis and plaster mummies), so are only noted here.

MUMMIES FROM THE MIDDLE EAST AND ASIA

Mummies from Jordan

Given its climate, the dearth of reports dealing with mummified human remains in the Middle East and northern Africa (excluding Egypt) is surprising. A rescue operation was carried out in the cemetery of Qirbet Qazone, at the southern tip of the Dead Sea. Tombs of the Early Bronze Age I–III were identified there as well as those from AD 0–200. The cemetery was being looted extensively. The first find in 1994 was the body of a spontaneously mummified adult male, reported in 1997 (Gruspier, 1997). Subsequent excavation encountered the full range of soft tissue preservation from none (skeletonization) to spontaneously fully mummified remains. Later examinations of that first mummy included radiology, radiocarbon dating (AD 1450), chemical dietary reconstruction (principally vegetal diet with minor terrestrial meat and of C3 and C4 plant components), osteological study and skeletal histology (Lucas *et al.*, 2000). Bioanthropological studies of the mummies excavated later have not yet been published (Fig. 4.144).

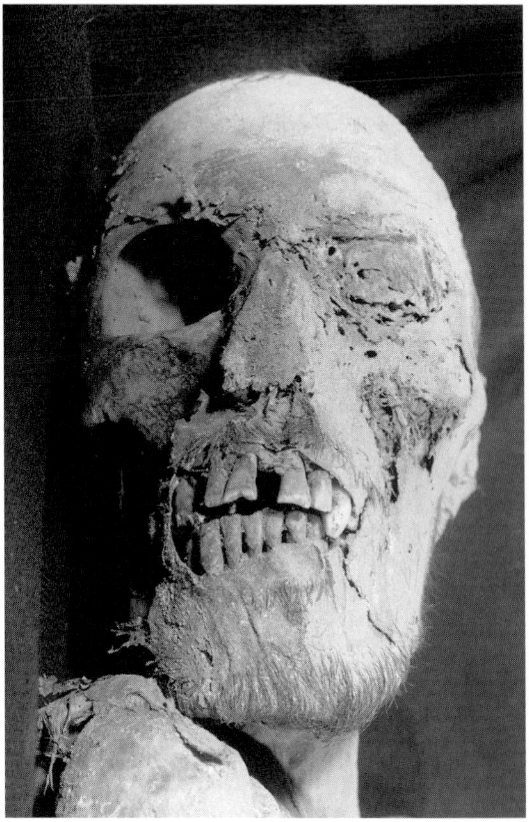

Fig. 4.144. A mummy from Jordan; SM, ca. AD 100. This spontaneously mummified body is that of an adult male from the Qazone site in Jordan. (Photo by Belal Degedeh and Hussein Debajeh, courtesy of Jerome Rose and Mahmoud el-Najjar.)

Mummies from Lebanon

In the thirteenth century AD the Mamelukes had gained control over Egypt, Syria and Palestine, and now focused their efforts on Lebanon. The region about Mount Lebanon was of special interest to them because the Maronites, a Christian, mountain, pastoral people had sided with the Crusaders who had bases there. The rugged terrain presented a formidable challenge, but by AD 1283 the Mamelukes laid siege to the village of Hadath el Gibbet that lay in the Qadisha Valley, a refuge for persecuted Christians. Some of the villagers had fled high into the mountains. Among them were a handful that had found a cave today called Asi al Hadath Grotto. This was situated on a precipice 700 meters high and of such difficult access that they felt as secure there as circumstances permitted. For the next 7 years they endured the siege undetected. The siege ended in AD 1290 (Hourani, 2000). Exactly seven centuries later, July 1990, a group of young speliologists (calling themselves Group d'Études et de Recherches Souterraines du Liban) were exploring the cave when they began to find thirteenth century artifacts (Freese, 1996). Excavation soon identified the human remains of ten individuals: three adult females, one adult male, five infants and a fetus. Eight of these bodies retained soft tissue, presenting as mummies. The arid cave atmosphere had combined with the low temperature to retard postmortem decay. The cave altitude is about 1300 meters above sea level and is located at about 34° 15′ N, 36° 04′ E. The bodies were wrapped in multiple layers of cotton shrouds. They were spontaneously mummified (Abi-Aoun *et al.*, 1994:35–57).

Dealing with the fragile bodies and textiles with meticulous care, they exhumed them, crated them and successfully transported them through the countryside in the midst of a civil war, delivering them intact to the National Museum of Lebanon.

Five bodies showed varying degrees of soft tissue preservation. A four-month-old infant wore three dresses, a headdress, necklace and earrings; her body appears intact. A few were little more than skin and bones, while the remainder reflected intermediate stages of soft tissue retention. An 18-month-old infant was placed at the left shoulder of an adult female, a pattern to which the local tradition still adheres for a mother–daughter relationship. Long strands of black hair between the four-month-old infant's toes most likely are the mother's, since local custom even now dictates that a grieving mother pull out her hair while kissing her dead child's feet (Hourani, 2000).

It would be of great interest to use the methodology of molecular biology (DNA profiling) on tissue samples to identify family relationships between these bodies, as well as the many other studies that could enlighten our knowledge from this period in Lebanon. Unfortunately no scientific studies on these bodies have been reported. Ms Guita Hourani, Chairwoman of the Maronite Research Group (Mari), has organized a group of scientific investigators interested in carrying out such a project, but local administrative instability and priorities have prevented implementation of the study. Laboratory studies on ancient tissues always tax the scientists' skills, and mummified tissues are known to deteriorate even under the most ideal conditions. The bodies now have been removed from their mummifying environment for a decade. Hopefully the remaining obstacles to initiation of the project can be overcome soon and scientific investigation initiated before further decay frustrates any studies.

The mummified patriarch

Finding himself in a village near Beirut, Lebanon, in the fall of 1815 the British physician Charles Lewis Meryon was attracted by a group surrounding the local monastery. Entering it he found the lifeless body of the local patriarch slumped in his chair, his hands still grasping a bible and other accouterments of his profession. Early putrefaction odors together with the directionless milling of the populace made it clear that this was a situation beyond their experience. When Dr Meryon learned that it was the tradition to embalm church patriarchs, and that such skills were not available locally, he volunteered his services. This was eagerly accepted, and the body carried to the church cellar and laid on the floor. A local pharmacist supplied compounds prescribed for such use, found recorded in church records. When many of the simple requests Dr Meryon made were not fulfilled, he began to suspect that the urgency to identify and install the patriarch's living successor carried a higher priority than disposing of the deceased one.

Nevertheless, he kneeled on the floor, incised the abdomen and removed the visceral organs with all the dignity the situation permitted. He resisted the temptation to incise and dissect the individual organs to seek the cause of death, fearing this might give offense to the supervising elders. Though unaware of its nature he smeared the prepared compound on the lining of the emptied, cleaned body cavities, and filled them with bran. He then removed the brain, filling the cranial cavity with powdered drugs. At his request a blue cotton thread was supplied, which he used to close all incisions with meticulously placed sutures. Five hours after he began the process, the now clothed, embalmed patriarch was back in his armchair as his funeral service began. Relieved that he apparently had not inadvertently violated any local religious taboos and that the bystanders admired his sutures, he decided not to incur further risks and rode back to the city without attending the final service (Meryon, 1815/1987).

The mummy of Alexander the Great

Although Alexander was a Macedonian native, he died in Babylon. How his body was preserved for about 2 years before it was transferred from Babylon to Egypt remains controversial. One version suggests it was submerged in a vat of honey (Budge, 1889) while others suspect it was prepared by Egyptian embalmers. The available evidence is described in Chapter 3, to which the interested reader is referred. This mummy is listed here solely because the procedure most probably was undertaken at Babylon, the Asian location of Alexander's death.

Mummies from Myanmar (Burma)

The paucity of reported mummification from this area is frustrating, and the little available is primarily anecdotal. Dawson (1928a) comments, as an aside, that the bodies of Burmese rulers and priests were "elaborately embalmed and then cremated . . ." but says nothing about the technique itself. A small degree of confirmation can be derived by a photo, unaccompanied by text, whose legend alleges that it represents a Buddhist abbot from "Queen Suphavarlat's golden monastery lying in state, embalmed and covered in gold leaf . . .". The photo's date is not supplied, though the book's title suggests the illustration must have been prepared prior to 1925 (Singer, 1993).

Mummies from Vietnam

The mummified bodies of two Buddhist monks in Vietnam were examined by Cuong (1984). The monks had served as the principals (superiors) for a local religious unit (pagoda) in a village 23 kilometers south of Hanoi during the Le dynasty in the seventeenth century. Both were in a sitting position with crossed legs, and presented a statuesque appearance. This was apparently achieved by covering the entire body surface with a thin layer (2–4 mm thick) of earth mixed with sawdust and paint; this, in turn, was covered with a 5 mm layer of now-hardened resin. Thereafter a final coat of shiny, enamel paint was applied. Apparently much of the soft tissue decayed, liquefied and evaporated over time (the body now weighs only 7 kg). The applied coat is supplying rigidity to maintain the shape of a human body. Radiology of one of the bodies revealed that the bones of the hands, feet and skull were in their usual location but the ribs and vertebra have become detached and collapsed into a heap inside. The outer layer of the other body contained so much powdered material that it was radiopaque. Thus it seems that the bodies have largely decayed but the material applied to the external body surface is strong enough to maintain a shell that retains the form of the original, desiccated, mummified bodies (Fig. 4.145).

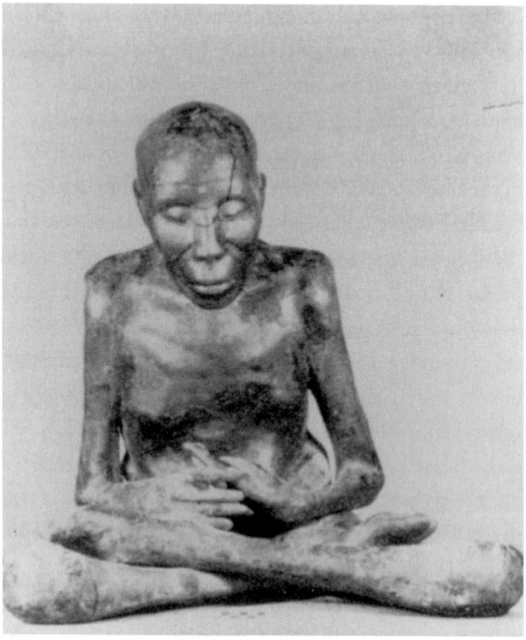

Fig. 4.145. Buddhist monk mummy from Vietnam; AM, ca. 1650.
This "mummy" survives as a lacquer shell around a decayed body. (Photo courtesy of Nuguyen Lan Cuong.)

Mummies from China

Mummies from the Western Han Dynasty

Abolishment of the feudal system together with other changes led to formation of the first, national Chinese state by the short-lived Chin dynasty in 221 BC, followed by the more stable Han Dynasty that endured for nearly half a millennium (206 BC to AD 220)

(Garraty & Gay, 1972:12). Several excavations from this period have uncovered some noteworthy mummies. Of these, the Mawangtui tombs are probably the most dramatic. Certainly they are eligible for inclusion in a short list of the world's largest tombs. Of the three that have been reported in European languages, we will detail the first, and comment on the others to the extent to which they differ.

The structure of the Mawangtui tomb no. 1 was massive. Located near modern Changsha in Hunan province, it was excavated in April, 1972. An earth mound 4 meters high with a circumference of 100 meters covered it. Beneath this a square, funnel-shaped pit was dug, the upper end of which was nearly as broad as the overlying mound and whose stepped walls narrowed downward to reach the roof of the burial chamber 20 meters from the top of the mound. This chamber had walls of white clay, the inner lining of which was composed of a layer of charcoal 40cm thick. Interior to this were wooden walls that formed the burial chamber itself. Within the chamber's center were four nested wooden coffins, the innermost containing the tomb's body (Peng, 1995). Regal grave goods filled the space between the coffins and the chamber's wooden walls. These goods included not only clothes, furniture, ceramics, silk designs, wood carvings and many musical instruments, but also food – fish, eggs, fruit and meat. Documents identified the body as being from the family of Litsang, Chancellor of Changsha and Marquis of Tai of the Western Han Dynasty, placing the date of burial at about 120 BC. The excavators believe the body to be that of the Marquis' wife.

The 154cm long body weighed 34.3kg and was enveloped by 20 shrouds and dresses, most made of silk with a few of linen, and it was lying half immersed in a solution. Later tests indicated that this was acidic and included low concentrations of mercury, giving it weak bactericidal properties. Most remarkable was the appearance of the body. It was that of a female of about 50 years of age. The body was corpulent and the skin surface uneven, the product of focal areas of subcutaneous adipocere formation (Fig. 4.146). Most unusual was the fact that much of the skin was relatively soft

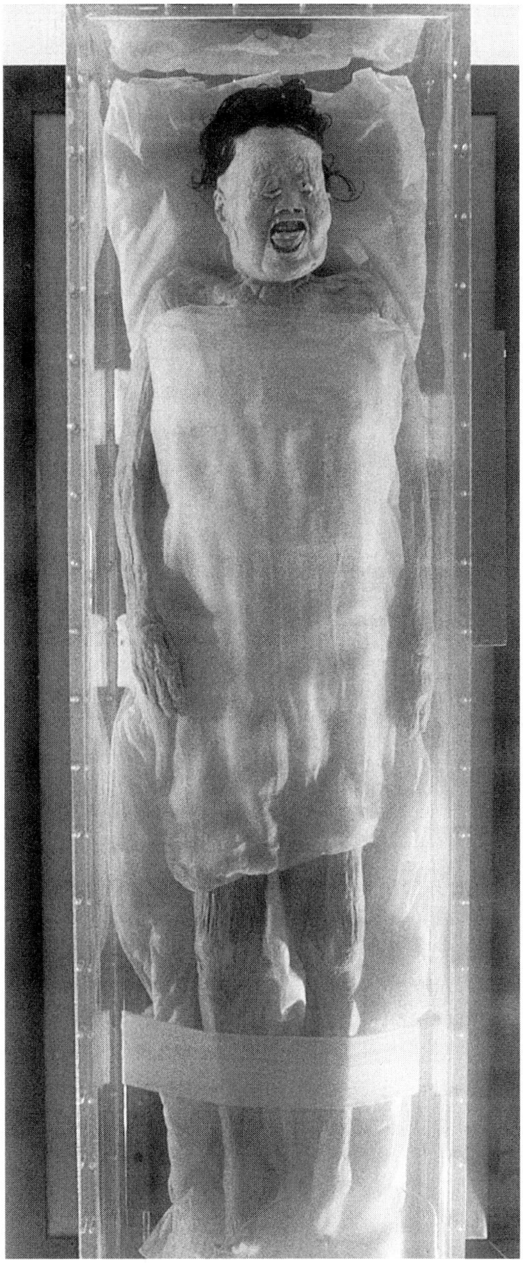

Fig. 4.146. Chinese mummy from Mawangtui tomb no. 1; SM, ca. 150 BC.
Factors contributing to this body's remarkable preservation include anaerobiasis and mercury, but some remain enigmatic. (Photo courtesy of Zhang Yang at the Cultural Relics Printing House, Beijing, Peoples Republic of China.)

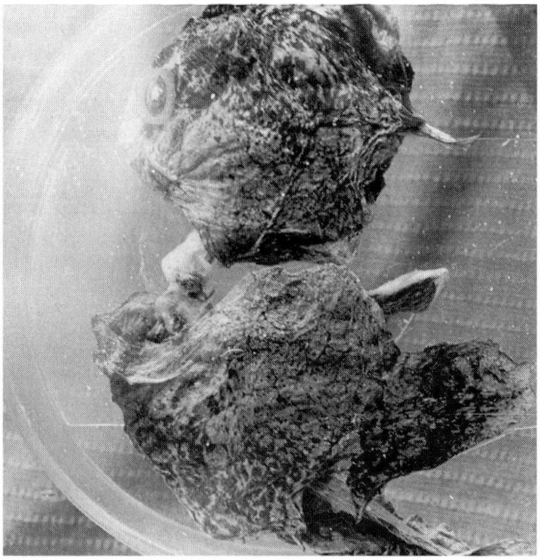

Fig. 4.147. Lungs from mummy in Fig. 4.146; SM, ca. 150 BC.
These lungs appear well preserved and did not collapse after death. The dark streaks represent anthracotic pigment (inhaled soot). (Photo courtesy as in Fig. 4.146.)

and elastic, a feature later found to be shared by the internal viscera. Furthermore, joints were easily movable. There were signs that some postmortem putrefaction had begun, but became arrested: sunken eyes, mildly protruding tongue, rectal prolapse and others. Yet, the body cavities revealed retention of the principal organs, which, though shrunken in size, were easily recognizable. Autopsy dissection revealed a brain one-half normal size and of "bean-curd" consistency. Hair was intact. The heart demonstrated an atherosclerotic plaque in the left main coronary artery that reduced the artery's lumen to 25–50% of its original size. Gallstones were found in the common hepatic duct and in the ampulla of Vater. Nodular pulmonary calcifications were present in the upper lobe of the left lung (Fig. 4.147). The esophagus, stomach and intestine contained 138 musk melon seeds. A degenerated, herniated, lumbar intervertebral disk was evident, and a perineal scar was noted (Wei, 1973).

X-rays revealed malunion of an old fracture of the right distal radius and ulna. Histology demonstrated intrahepatic bile duct dilatation, small myocardial scars and pulmonary anthracosis. Most epithelial cells had autolyzed but collagen and elastic tissue were intact. Parasitological studies found that she had been infected by *Schistosoma japonica* in the liver and rectal wall, and colon content included ova of *Enterobius vermicularis* and *Trichuris trichiura*. Chemical analysis showed evidence of chronic lead and mercury accumulation. Examiners felt that biliary colic had stimulated spasm in her atherosclerotic coronary artery, leading to myocardial ischemia with ventricular fibrillation causing sudden cardiac arrest. Following completion of the examination, the body was conserved by injection with and immersion in 4.5% formaldehyde (Hunan Medical University, 1980). It is now being curated at Junan Medical College (Hall, 1974).

It is difficult to be certain regarding the mechanism of preservation for this body prior to excavation. The examiners make a major issue of anaerobiasis, brought about by decay of the food left in the burial chamber, and maintenance of the anaerobic state by the depth of the burial (more than 20 meters) and the method of sealing the wood chamber (clay was surrounding 60 tonnes of charcoal that covered the burial chamber on all sides). They admit that these conditions would not impede anaerobic putrefactive action and, indeed, note that some findings suggest that this had begun early in the postmortem interval. They attribute the arrest of this process, however, to the accumulation of putrefactive products, as evidenced by the acidic nature of the coffin fluid.

Little experimental evidence is available to evaluate this proposed scenario. In an as yet unpublished study carried out by Anya Lukasewycz and Saras Aturaliya in my laboratory, rat fetuses were harvested aseptically and stored in sealed glass vials at room temperature. Within 48 hours the pH of the liquefied tissue dropped to about 5.5 where it remained thereafter during the more than two months of the study. Tissue liquefaction, however, continued throughout the observation period. Thus lysosomal action itself as well as nonenzymatic hydrolysis even without additional bacterial enzymes would appear to require inhibition in order to effect tissue preservation.

The Mawangtui body, however, was half immersed in fluid within the coffin. This acidic fluid (actual pH is not included in the report) did contain traces of mercury, though its ability to inhibit bacterial growth was stated to be "weak". The report's authors suggest that the coffin fluid's source was condensation of vapor from the soil and clay surrounding the wood chamber. Absence of such fluid in the chamber and in the other five coffins makes that suggestion appear unlikely. More attractive is the contemporary tradition that they cite of cooling the body by placing ice under it within the coffin at time of burial. It is conceivable, though not probable, that such ice was prepared from a solution that included some tissue-preserving material not yet identified in the fluid.

Two other tombs of similar, massive construction were excavated. One of these contained a skeleton of a 30-year-old male with artifacts and documents identifying his date of death as 168 BC. The remaining tomb housed a male body with seals, jade articles, documents and 30 books, identifying him as Li Tsang, the Marquis of Tai, who died in 186 BC. The books contained much cultural information that was helpful in reconstruction of the sociology of the Han Dynasty (Hudson, 1996a).

In addition to the Mawangtui graves, another Han Dynasty body was excavated from tomb no. 168 on Phoenix Hill in Jingling county of Hubei province in 1975 by Dr Wu Zhongbi of Wuhan Medical College. The tomb was found to contain the spontaneously mummified body of a 60-year-old male with preservation features similar to those of Lady Tai in the Mawangtui tomb no. 1 described above. Documents indicate his death in 167 BC and examination revealed that he had died of acute peritonitis following a ruptured gastric ulcer (Cheng, 1984) (Fig. 4.148). Parasitological study identified *Schistosoma japonicum*, *Clonorchis senensis* and *Taenia* sp. (Dexiang *et al.*, 1981).

Mancheng burials and jade suits

During China's formative period Chinese philosophical world view was flexible enough to embrace the concept that humans could increase their lifespan by natural means. By the time of the Han Dynasty (206 BC to AD 220) the idea of longevity had been expanded to include material immortality. Supported by the principles of evolving Taoism, Chinese alchemists were busily seeking the "immortality elixer" by preparing various mixtures of gold, silver, arsenic and other natural ores. Jade, also a natural ore, eventually became included as a candidate for the elusive mixture that would provide resistance to body tissue aging and decay. In his book Pao Phu Tzu, written about AD 320, Ko Hung describes the concepts underlying these efforts. Extrapolating from perceived effects of chemicals on cadavers, he wrote "When gold and jade are inserted into the nine orifices, corpses do not decay . . . So when men ingest substances which are able to benefit their bodies and lengthen their days, why should it be strange that [some of these] should confer life perpetual?" (cited by Needham, 1974:284). Thus, the concept that jade possessed anti-decay properties was an outgrowth of the search for an "elixer of longevity" in China and had become well established by the time of the Han Dynasty, when imperial efforts to achieve immortality through preservation of the body almost rivaled those of Egyptian pharaohs.

At Mancheng, a tunnel 50 meters long was chiseled into the solid rock of a mountain leading into a tomb chamber measuring 15 meters × 13 meters × 7 meters. The tunnel was blocked by a stone wall which was reinforced by an iron plate (Thorp, 1991), while in another similar one 16 boulders each weighing 7 tonnes obstructed it (Harrington Spencer, 1996). Exhaustive effort was also expended directly on the bodies. These were centered around the conviction that jade prevented tissue decay. While literary sources first document this concept about AD 320, the archaeological record provides evidence of such convictions at least half a millennium earlier (Thorp, 1991). Jade is also called nephrite, possibly because it was thought to have therapeutic value for persons with kidney disease in late medieval times. It is a rock composed of calcium magnesium silicate, and today has no known, demonstrated, tissue-preservative or healing value. Royal tombs of the Western Han Dynasty of the late second century BC are located at Mancheng near Xuzhou in

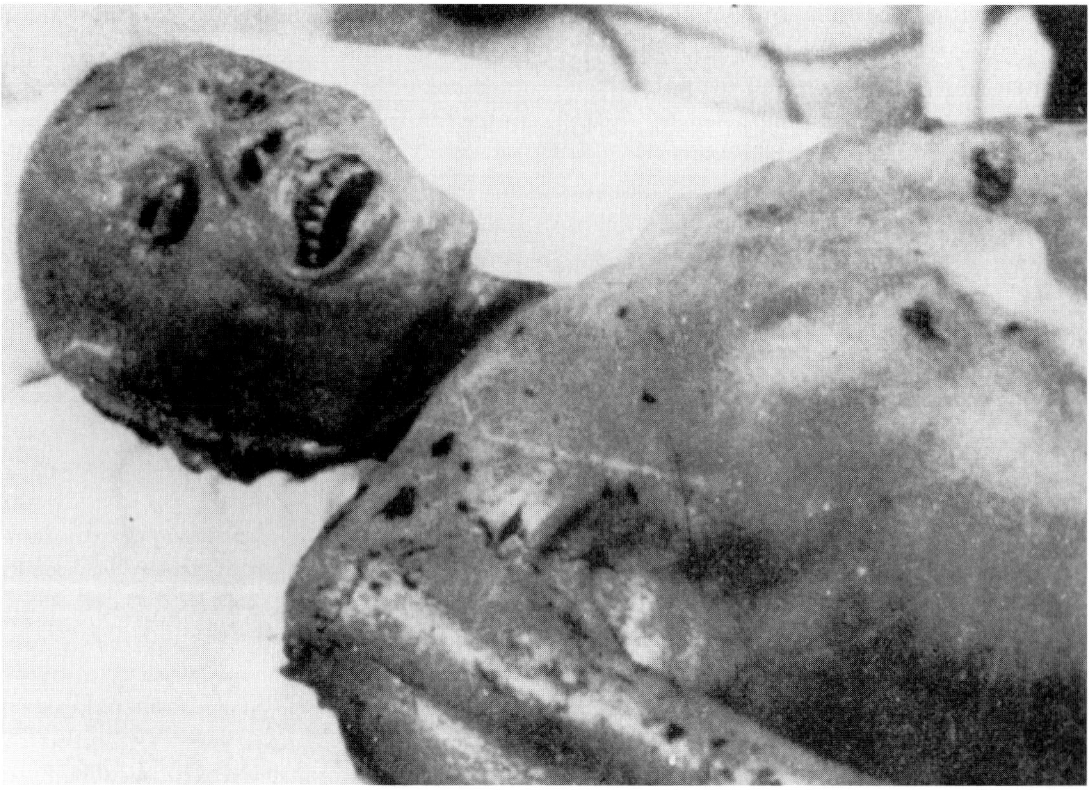

Fig. 4.148. Mummy of China's Western Han Dynasty; SM, 167 BC.
The spontaneously mummified body of this 60-year-old man was buried 167 BC. He died of a perforated gastric ulcer. (Photo courtesy of Tsung O. Cheng and *Annals of Internal Medicine* **101**(2): 714, 1984.)

Hebei province, about 34° 15′N, 117° 15′E. That dynasty's third king, Liu Sheng, and his consort Dou Wan were buried there and excavation revealed that both bodies were encased in a jade "suit" (Fig. 4.149). These were created by sewing small, thin, rectangular, peripherally perforated, jade plaques with gold thread onto a cloth backing. The facial area included eye slits and a rudimentary nose and eyes. The torso had been prepared in an upper and lower segment, to which sleeves for two arms, hands, legs and boots were attached (Plate 5). The amount of work to create such a unit must have been staggering, and have been possible only in a central imperial workshop. Their use seems to have been limited to the imperial family, and Brier (1998:83) describes imprisonment of a family who violated that restriction. The jade burial suits were also considered a valuable imperial gift. Fewer than two dozen such completed suits have been recovered (Thorp, 1991), probably because of their expense. Emperor Wen of Wei in AD 223 forbade their production because they encouraged tomb looters who burned the suits to recover the gold threads (Brier, 1998:84). This decree probably also contributed to the dearth of jade suits. In reality, however, the occasional intact, unlooted Mancheng tomb reveals a wealth of coins, ceramic figures, royal chariots with their horses and other grave goods. Though not as rich as earlier tombs, these treasures were nevertheless sufficiently attractive even without jade suits to provide incentives for plundering by looters.

Fig. 4.149. Jade "mummy"; SM, AD 220.
In the belief that jade imparted soft tissue preservation, this ancient Chinese body (Liu Sheng) of the Han Dynasty was covered with jade plates. However, the body inside the jade "suit" was found to be a skeleton. (Photo by William MacQuitty made available by Edmund Capon.)

However great the efforts to prevent tissue decay by the use of jade, they were not successful. In all cases to date, the excavators have found only skeletons without preserved soft tissue within such jade suits.

Xinjiang mummies

The Orient's Himalayan mountain chain creates the world's most dramatic terrain. Its overall shape somewhat resembles that of a giant fishhook whose shank is formed by the east–west segment of the chain north of India. The convex portion of the fishhook's curve faces west, while that part approaching the barb (Tien Shan) extends west–east, rapidly declining in height to meet the desert floor below sea level at Turpan. The height of these mountains blocks all air masses except those from the east from reaching the land area they envelop, forcing the air to an altitude that wrings out all of the moisture. The result is the Taklimakan Desert, a part of which is known as the Tarim Basin. This is one of the world's most arid regions, a major central portion of which is occupied by giant, shifting sand dunes. Small wonder, then, that it is here, in Xinjiang province, that China's oldest, spontaneously mummified bodies have been found. It is an area of temperature extremes ranging from about −7 °C in January to almost unbearable summer heat. Habitation and travel in this desert is possible only along the foot of its surrounding mountains, or along one of the streams, such as the Tarim, that spring from the mountains to enter the desert from its periphery but soon vanish into the sand. As ancient caravans from southern and eastern China approached Dunhuang at the edge of this segment of what eventually came to be part of the Silk Road to the Mediterranean, the near absence of water in the desert's center forced them to make a choice. The southern option skirts the desert along its southern edge at the foot of the steep Kunlun slopes descending from Tibet's high plateau. Alternatively, the northern route passes through Hami and those communities lying along the Kongi and Tarim rivers that lead to Loulan and Lop Nor. It is along these routes that mummies from the Tarim Basin have been found.

Several bodies have been excavated from graves near Loulan, a site that once bordered a still-shrinking lake fed by the Kongi River. Among these is the body of a young female with remarkably well-preserved facial features, whose radiocarbon date indicates that she died about 1200 BC. The Xinjiang residents of Turkish descent have embraced her, calling her the "Beauty of Loulan". She is one of the few, very well-studied (1979–1980) mummies in China. A multidisciplinary team, mostly from Shanghai, has published its results in a Chinese monograph, but thoughtfully included English summaries of almost every chapter (Zhiyi & Yongqing, 1982). Bioanthropological findings included anthracosis and pulmonary silicate deposits, blood type O, head lice, trace elements in the hair and many others. The body was heavily clothed with furs and many artifacts of everyday use were found buried with her.

Subsequent excavations of multiple sites in the Tarim Basin during the past four decades have opened more than 500 tombs. Of these, several hundred included mummies. These have been found near Hami in the desert's northeastern quadrant (Fig. 4.150), the Lop Nor area (Tarim Basin) and near Cherchen along the desert's southern edge. Most of these mummies have been distributed among many other museums, some as distant from Xinjiang as Shanghai, but only two (Loulan, Hami) are reported to have been studied in detail. Dr Wang Bing Hua, director of Ürümxi's Archaeological Research Institute, attributes the spontaneous mummification to three factors: arid climate, salty soil and shallow, winter burials (Wang, 1996). Average salt content of the desert soil near Turpan is about 10 g/liter but in the very surface layer it can be five times greater. At Hami the soil contains layers of gypsum and at Cherchen actual salt blocks are obvious within the soil, especially near the surface. Most burials are only about a meter below the surface. An infant 51 cm long dating to 800 BC was excavated near Tiebanhe. It was accompanied by a feeding "bottle", made of ewe skin and also a cup carved from an ox horn (Palmer, 1998). Several other mummies were excavated from the Hami area in the late 1970s, dating from 1060 to 771 BC (Vescia, 1980). A male mummy from Subaski had a neck incision sewed up with horsehair (Hadingham, 1994).

Fig. 4.150. Xinjiang Province mummy; SM, 1200 BC. This mummy examined by Victor Mair (shown) was excavated at the Qizilchoqa site near modern Hami in the Taklimakan Desert of China's Xinjiang province. (Photo courtesy of Victor Mair, University of Pennsylvania.)

Fig. 4.151. Ruins of Gaochang City.
Gaochang, also known as Khocho, was the capital of an Uyghur area settled AD 850–1250 at the northern edge of the Taklimakan Desert in China's Xinjiang Province near modern Turpan. (Photo by ACA.)

During my visit to Ürümxi, Xinjiang province, in 1989 (Aufderheide, 1989a) we found six mummies, most from the Tarim River area, in that city's Archaeological Research Institute. Two were males, two females and two infants. Several adults had distinctly Caucasoid features. All were extended, supine and spontaneously mummified. I then travelled into the desert to Turpan, a city located 154 meters below sea level. Five mummies were curated at the Turpan Archaeological Museum. Three were from the sixth and two from the eighth centuries AD. All were from the local area and all were spontaneously mummified. None had been studied bioanthropologically. Modern Turpan is an outgrowth of the ancient city called Gaochang (Fig. 4.151). This was a trading center on the Silk Route, and was particularly active during the sixth to tenth century AD. Visiting the Gaochang ruins adjacent to the modern city, I was surprised at the high degree of preservation of these buildings. Some excavations had been carried out at the nearby cemetery (Astana) used by Gaochang residents. These tomb chambers were about 2.5 meters square and about 4 meters below the surface; some had plaster walls. One was decorated with wall paintings of animals on plaster. In one of these, a male and a female body lay on a wood plank (Fig. 4.152). Their desiccated, extended, supine bodies were spontaneously mummified. The degree of dental attrition suggested that the female was a very young adult, while the male had died about in the fourth decade of life. Although soft tissues externally appeared well preserved, the bodies had not yet been studied formally.

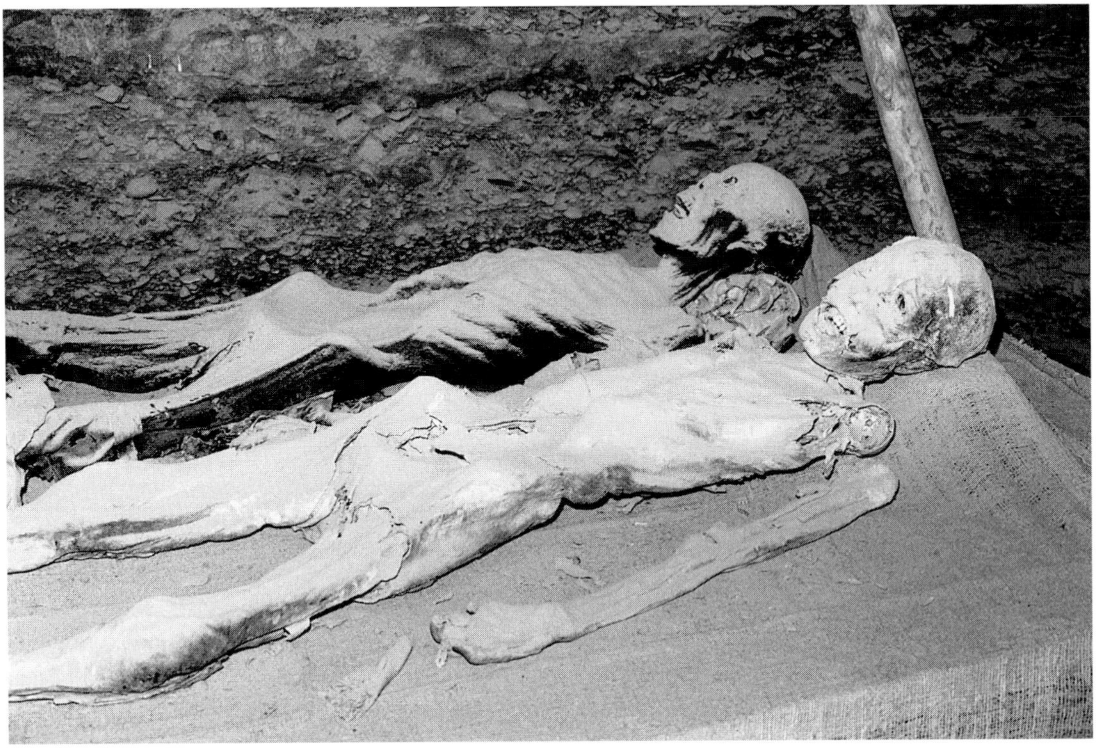

Fig. 4.152. Mummies from the Taklimakan Desert; SM, ca. AD 850–1250.
The spontaneously mummified bodies of a male and female were excavated from the Astana burial ground adjacent to Gaochang City (see Fig. 4.151). (Photo by ACA.)

In the Shanghai Museum of Natural History in 1989, four mummified bodies were on display, one of which was a young adult female from a site near Hami in the Tarim Basin. Radiocarbon dating indicated death at about 1200 BC during the Tang Dynasty. She had light brown hair, was in the supine, extended position and appeared spontaneously mummified and well preserved (Aufderheide, 1989a).

Two years prior to my visit, Dr Victor Mair of the Department of Indo-European Studies at the University of Pennsylvania had seen these mummies and was struck by their Caucasoid appearance. About 100 years earlier, documents had been found in this area written in Brahmi script but in a language (termed Tocharian) that suggested it had Indo-European roots. To date, however, archaeologists have unearthed no Caucasoid bodies to confirm the presence of Caucasians in that region. Although the Tocharian documents were dated largely from the sixth to the eighth century AD, the most recently excavated mummies were almost all much older (some as old as 3000 BC). The idea that they might represent Tocharian ancestors was a concept so exciting that Mair has pursued opportunities to study them since his first visit. He found ancient paintings of the Yuezhi (about 500–50 BC) people with blue or green eyes, long noses and full beards, and felt that they too might be part of the descendant chain leading to the Tocharian-speaking people (Mair, 1995).

The same conditions that prevented the bodies' soft tissues from decaying acted equally well to preserve not only their textile fibers but also the pigments. This brilliantly colored clothing has already supplied another clue consistent with Mair's hypothesis. Four bodies were removed from a 3 meter deep tomb near Cherchen, a site along the Cherchen River at the foot of the steep Kunlun slopes descending from the Tibetan plateau. Their radiocarbon dates clustered around 1000 BC. A male body ("Cherchen Man") was 2 meters tall, had light-brown hair and was spectacularly garbed in white deerskin boots, brightly colored trousers and shirt as well as felt leggings (Kamberi, 1994). The weave pattern of some of Cherchen Man's clothing has revealed a unique, "long-hop" twill type (weft of over three, under two, offset by one warp/row) that has not been identified in other areas for this period except at a nearly contemporary Celtic site at Hallstatt (Barber, 1999). Such findings suggest possible Celtic contact and exchange during the first millennium BC. In addition, an initial report of mitochondrial DNA analysis of a sample extracted from a rib segment of a mummy from Hami demonstrated near identity with that of a Cambridge European reference sample, a finding also consistent with biological links between the Xinjiang mummies and Europeans (Francalacci, 1995). Studies of 200 skulls from seven Silk Road sites dating to 1800 BC by the physical anthropologist Han Kangxin from the Ürümxi Archaeological Research Institute indicated measurements conforming to the Caucasoid pattern (Cro-Magnon Man) (Wenkan, 1995). Another tantalizing finding was the painting of a sun-ray design on the head of a spontaneously mummified male from Zaghunluq near Qiemo on the Qarqan river along the Taklamakan's southern route. This design is similar to that of an ancient Iranian deity (Hudson, 1996a). While these results represent principally preliminary reports, collectively they are all supportive of possible early contacts of this region with western Caucasians and, at minimum, provide a powerful incentive to expand the scientific study of these Tarim Basin bodies in a multidisciplinary manner.

Mummies from Tibet

Findings of mummies from Tibet are not reported. Goodman (1986:29) notes that a body is only of importance to Tibetans while the soul is within it. After death, therefore, the bodies of common Tibetans are defleshed and the soft tissues together with the pulverized bones are laid out to feed the vultures or placed in streams for consumption by fish. The current, exiled Dalai Lama indicates that this practice is motivated by compassion for such animals (Goodman, 1986). Cremation is limited to high status Tibetans such as teachers because of the paucity of firewood at those altitudes. The living Dalai Lama confirms that only the bodies of the elite, the Dalai and Panchen Lamas were embalmed. We lack details of the embalming method, though Bell (1946:446) comments that repeated application of salt was part of the process. These mummies were then placed in specially constructed, small Buddhist shrines (*chörtens*; *chaityas*; *dagobas*; *chokdens*) that were placed in accessible locations for veneration by the living.

However, in his published autobiography, T. Lobsang Rampa (1957) describes his participation in the mortuary rites for a high status Tibetan lama. Following evisceration, the body cavities are painted with a special lacquer, then filled with lacquer-saturated textiles. The body surface is then painted with a different lacquer. Thereafter the body is wrapped in silk and placed in the center of a special room that is then filled to the ceiling with salt. This specially built room is then subjected to a high level of heat from an intense fire for one week. After cooling the body is removed, its silk covering peeled off and his entire body covered with gold leaf. It is then placed on a golden throne in a sacred room reserved for similar predecessors where they can be venerated. I have not been able to find a report of an intact mummy from Tibet in the scientific literature.

Mummies from other Chinese regions

During our visit to the Shanghai Museum in 1989 (Aufderheide, 1989a) we found three mummies on exhibit from various other Chinese sites. One was an adult male from the Ming Dynasty, excavated from Gaochun in Jiangsu province near Nanking. The body

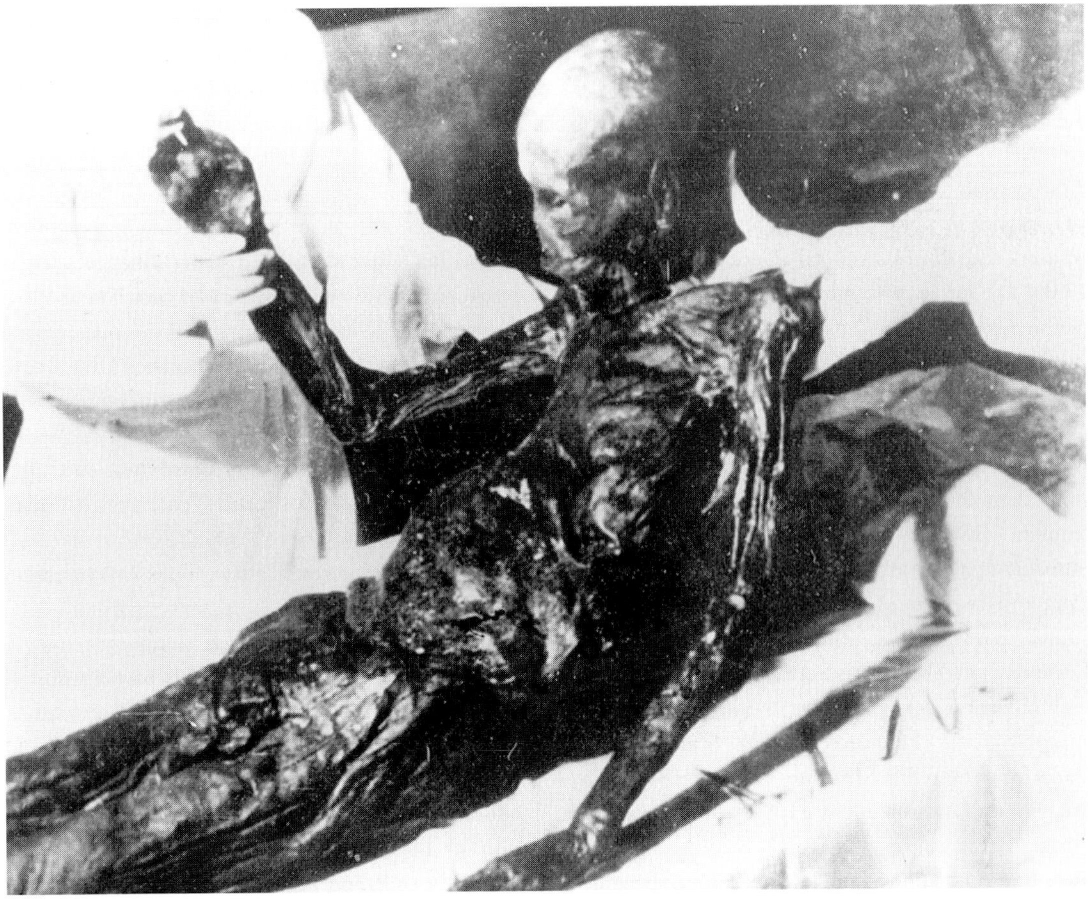

Fig. 4.153. Chinese Southern Sung Dynasty mummy of Chou Yü; SM, ca. AD 1127–1279.
Excavated in the Jindan county of Jiangsu province in China, the tissues of this spontaneously mummified body remained remarkably flexible. (Photo courtesy of Li Guo Ping, Director of Jintan Museum.)

was exhibited in a metal case immersed in a colorless liquid and demonstrated a sutured incision from the suprasternal notch to the pubis, testifying to an autopsy. Bilateral hammer toes were evident, as was a normal penis but huge scrotum (evidence of probable postmortem fermentation or antemortem hernia). The label on the display stated that many Ming Dynasty bodies have been found, especially in Jianqou and Zhejiang province, with uniform burial services and prevalent antiseptic methods at the time. Good preservation of soft tissues, very valuable for study, was noted in this body.

The second was an aged male from Jianxi. The coffin label stated his name (Zhouyu) and that he was a student of the Imperial College in feudal China during the Southern Sung Dynasty. His body had been found soaked in a yellow liquid in the coffin with movable joints and elastic skin. This body, too, had been subjected to autopsy, as evidenced by a sutured skin incision similar to that described above.

A third body was that of an adult male from the Southern Sung Dynasty excavated in Jindan county of the Zhenjung region in Janqiu province. Liquid was found in his coffin (Figs. 4.153 and 4.154). Only one

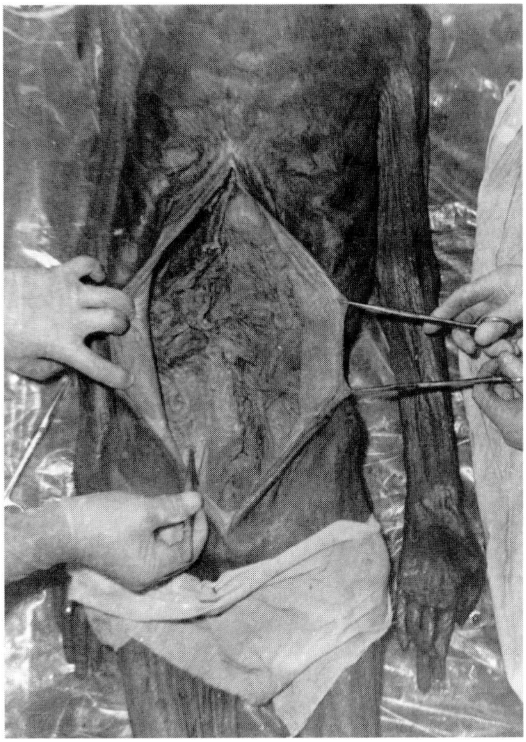

Fig. 4.154. Autopsy of Chou Yü; SM, ca. AD 1127–1279.
The photo illustrates the pliability and flexibility of the mummified tissue of Yü (see Fig. 4.153). Autopsy reveals well-preserved viscera. (Photo credit same as Fig. 4.153.)

other Sung (Northern) Dynasty mummy has been found. He had unstyled gray hair, with dark, shiny skin and advanced dental attrition. The chest had a "barrel" shape, while the abdomen was profoundly scaphoid. Position was extended, supine.

Reports of mummy studies in the Chinese language not available to me probably exist, but the above were the only English language publications about Chinese mummies I could find.

Mummies from Japan

The humid climate and acid soil combine to create unfavorable conditions for soft tissue preservation in Japan. Indeed, exhumation of even skeletal tissue of great age is unusual there. I could find reports of fewer than three dozen mummified bodies from this area (Table 4.16).

Mummies of political leaders

In the twelfth century AD, the northern portion of Japan's largest island, Honshu, was ruled by chieftains of the Fujiwara people. Bodies of the first three rulers are preserved in Chusoni, Hirazumi, and Iwate prefectures, as is only the head of the fourth (Tadahira). Although the ages at death of the first three range from 54 to 73 years, Tadahira was beheaded as a young adult of 23 years when the Kamakura shogunate conquered the Fujiwara capital in AD 1189 (Hudson, 1996b). These bodies were studied by a multidisciplinary team in 1950 who found them to be quite well-preserved, desiccated bodies. They found no evidence of evisceration, though rats had consumed the viscera during the nearly eight centuries of postmortem interval. Tadahira's head showed sword cut-marks that had been responsible for the decapitation as well as amputation of the nose and both external ears. A long iron nail had been driven into his head. Whether the desiccation was brought about completely spontaneously, or whether applied heat had contributed to the dehydration process could not be ascertained with certainty. A skeletal study suggested that this population more closely resembled modern Japanese than the indigenous Ainu.

Mummies of Buddhist priests

Almost all of the remaining Japanese mummies are those of Buddhist priests. Prior to the arrival of Buddhism in Japan in the sixth century AD, corpses were considered to be polluted. By the time of the Heian Period (about AD 800–1200) a new concept had evolved that death was the first step toward entering paradise. Among a subgroup of Japanese Buddhists this view evolved further into an idea that an ascetic life can lead directly to becoming a Buddha, and preservation

Table 4.16. *Mummies from Japan*

Mummy	Location prefecture	Died	Mummification method[a]	Evisceration	Age-at-death
Kochi	Niigata	1363	I	0	82
Tetsumonkai	Yamagata	1829	A,S	0	62
Tetsuryukai	Yamagata	1868	heat, lime	+	62
Bukkai	Niigata	1903	I	0	76
Shungi Shōnin	Ibaragi	1686	I	0	78
Zenkai Shōnin	Niigata	1687	I	0	85
Chūkai Shōnin	Yamagata	I	A,S	0	58
Shinnyōkai Shōnin	Yamagata	I	A,S	0	96
Enmyokai Shônin	Yamagata	1822	A,S	0	55
Myokai	Yamagata	1863	A	0	44
Shukai	Niigata	1780	N	0	62
Homyokai	Yamagata	1683	I	I	61
Sinnyokai	Yamagata	1783	I	I	96
Tansei	Kyoto	1613	I	I	63
Myoshin	Gifu	1817	I	I	I
Kankai	Niigata	1878	I	I	I
Komyokai	Yamagata	1854	I	I	I
Junkai	Niigata	1636	I	I	78
Yutei	Fukushika	1686	I	I	92
Gyojun	Nagano	1687	I	I	45
Chiei	Nagano	1736	I	I	35
Fujiwara Kiyohira	Iwate	1128	I	I	73
Fujiwara Motohira	Iwate	1157	I	I	?
Fujiwara Hidehira	Iwate	1187	I	I	66
Fujiwara Yasuhira	Iwate	1189	I	I	23
Miscellaneous (5)	U. Toronto	19C	S,a	0	I
Tadamune	Miyagi	1658	A,M	0	59

Notes:
19C, nineteenth century; I, indeterminate; A, anthropogenic; N, spontaneous; S, smoke-dried; M, mercury; a, adipocere; +, present; 0, absent.
All except the last six listed (Fujiwara family, Tedamune and those at the University of Tokyo) were Buddhist priests.
The table was constructed from information in: Morimoto, 1993; Yamada *et al.*, 1990, 1996; Hudson, 1996a; and Sakurai *et al.*, 1998.

of the body became linked to this belief (Group for Research of Japanese Mummies, 1993; Sakurai *et al.*, 1998). Such convictions encourage mummification to which the individual can contribute while still living, a process called self-immolation (*nyujos*, entering into nirvana; *sokushin-butsu*, becoming a Buddha in this body). The concept of conscious efforts aimed at self-mummification is very ancient in Asia, certainly much older than Buddhism. In the section describing mummies from China, I quote Needham's (1974:284) citation of an AD 320 reference embracing the concept of ingesting appropriate chemicals to achieve immortality of the body. It would seem only a small step to extend that idea to deliberate self-mummification. This can be achieved by deliberately refraining from the inclusion of a variety of proscribed grains in the diet. Progressive weight loss commonly occurs over a period of 2–4 years, with disappearance of much of the body's fat and much of the muscle. When a state of profound debilitation is achieved, water intake is then also gradually diminished until the individual dies in a dehydrated, cachectic state (Morimoto, 1993).

The postmortem handling of the body varies. It may be buried and exhumed after several years or it can be conserved immediately in a place of honor in a Buddhist temple where it may be worshipped. As Yamada *et al.* (1990, 1996) suggest, it is highly probable that at least some additional desiccation was required (what they call "finishing touches") in many cases. In others we have specific records indicating that such assisted dehydration was achieved by drying with candles. There is no reason to question the validity of these processes. Immediately after death, decay processes compete with desiccation effects. It is usually a close race, and if most body fat, much of the muscle mass and at least some of the water has been removed before death, sufficient desiccation to arrest decay may be achieved with modest or, in some cases, very little anthropogenic postmortem effort. Because such efforts commonly are not detectable upon later examination they cannot be excluded from consideration.

Morimoto (1993) details the *sokushin-butsu* process by listing the five cereals from which such individuals abstain initially (rice, barley, soybeans, red beans, sesame seeds), and at a later time they added five more to the list (millet, broomcorn, panic-grass, buckwheat and corn). After death, initial burial of the body was in a double coffin. Within the inner one the body was surrounded by charcoal. The coffin was placed in an underground stone chamber fitted with a drain. Several years later the body was exhumed, forced into a cross-legged, sitting posture and completion of dehydration accomplished. This was achieved by exposure to wind, the sun's heat or smoking it with burning candles, incense and mugwort (rarely soaked in sea water). Thereafter the body was clothed in priestly garb and displayed. Morimoto (1993) describes several mummies believed to have been mummified in this manner: Zenkai Shōnin from the Niigata region (Fig. 4.155), Chūkai Shōnin from the same region (Fig. 5.156) and Enmyōkai Shōnin from the Yamagata region (Fig. 4.157). The bodies of the latter two may have been smoked after death.

Only a single subgroup of Buddhist priest bodies are known to have been anthropogenically mummified.

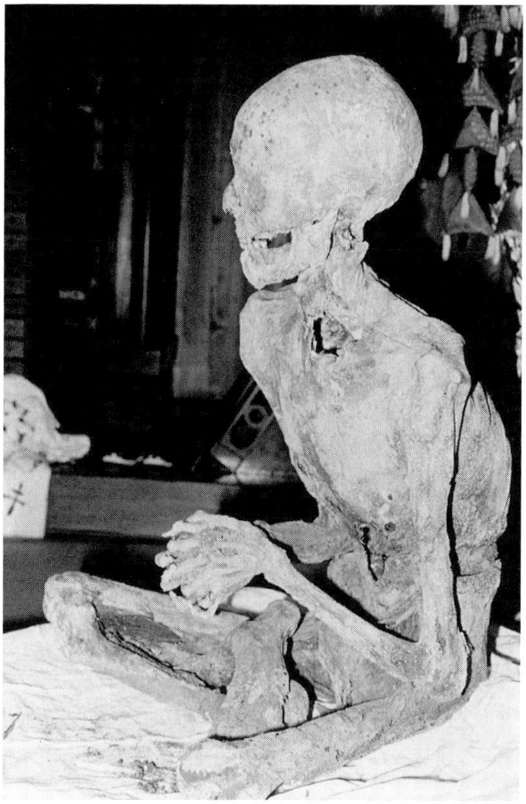

Fig. 4.155. Buddhist priest mummy from Japan (Zenkai Shōnin); AM, AD 1687.
Immediately after death the body's desiccation was aided by ventilation, sun exposure and heat by fire. (Photo courtesy of Iwataro Morimoto.)

They are so indicated in our Table 4.16. The first of these, dating to about AD 1300, was that of a priest named Rinken. This is believed to be the earliest example of anthropogenic mummification in Japan (Sakurai *et al.*, 1998). Sakurai *et al.* (1998) provide interesting details about these bodies. At least one of these was eviscerated and the abdomen filled with lime. All have been exposed to some heat source to bring about desiccation, and many have been deliberately exposed to smoke. The extremities were often painted a black-brown color with paint probably composed of persimmon tannin. Neither viscera nor brains were extracted (Sakurai *et al.*, 1998).

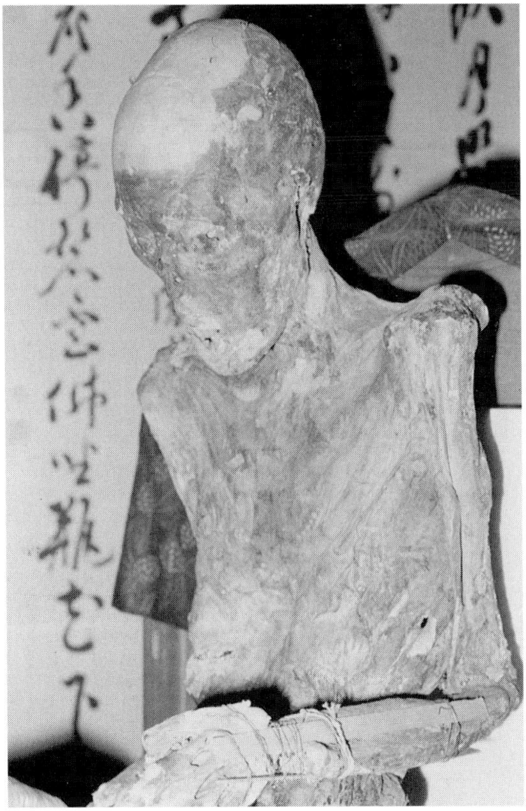

Fig. 4.156. Japanese Buddhist priest mummy (Chūkai Shōnin); AM, AD 1755.
Slow starvation and terminal water restriction were major soft tissue preservation factors for these mummies. (Photo courtesy of Iwataro Morimoto.)

Although Hudson (1996b) cites a nineteenth century Japanese explorer's comment that the Ainu mummified their chiefs on the now Russian-owned island of Sakhalin, I have been unable to find descriptive reports of such recovered bodies in the English literature, nor could I find data about the mummies conserved at the University of Tokyo except that four of the five showed adipocere formation.

A list of all individuals discussed above are itemized and tabulated by Yamada *et al.* (1996) together with their dates at death, age at death, current curatorial location and other data.

Mummies from the Philippines

The heat and humidity of most of the Philippines normally would lead to low expectations of finding mummies there. In 1906 Dean Worcester of the United States Philippine Commission described cultural features including mortuary rites of the many tribes located on Luzon. He recorded a broad range of burial practices but mummification was not among them at that time. Yet in the higher and cooler altitudes of some Luzon regions, mummified members of the formerly headhunting Ibaloi people have been found in caves. Morimoto (1988) states that these bodies had been mummified by evisceration via the anus. Several of the bodies bore tattoos and two revealed intentional cranial deformation. They were buried in wood coffins, though what procedures other than evisceration were carried out has not been reported.

More recently, however, more data have become available about these mummies. Between about AD 1100 to 1500 this form of mummification was practiced on bodies of their elite in the province of Benguet, about 300 kilometers north of Manila. The bodies demonstrate maximally flexed knees and hips (Fig. 4.158). Coffins are of pine wood, often consisting of hollowed tree segments. The burial caves are located at about 2000 meters' altitude, sheltered from rain and not sealed. The coffins are clustered in family groups. Orlando Abinion, the curator of the National Museum of Manila, described the mummification legends among present tribal members (Donnet & Duclos, 1999). According to such lore the process actually began before death with a drink of concentrated salt solution by the dying person. As bizarre as this may sound, it is consistent with normal human physiology. Ingestion of such a concentrated saline solution would result in a major shift of extracellular water into the intestinal lumen. The sheer bulk of this much water would trigger peristalsis. Such a laxative effect would expel not only the water (contributing to total body dehydration), but also the colon's content of feces. This would deplete substantially the intestine's bacterial

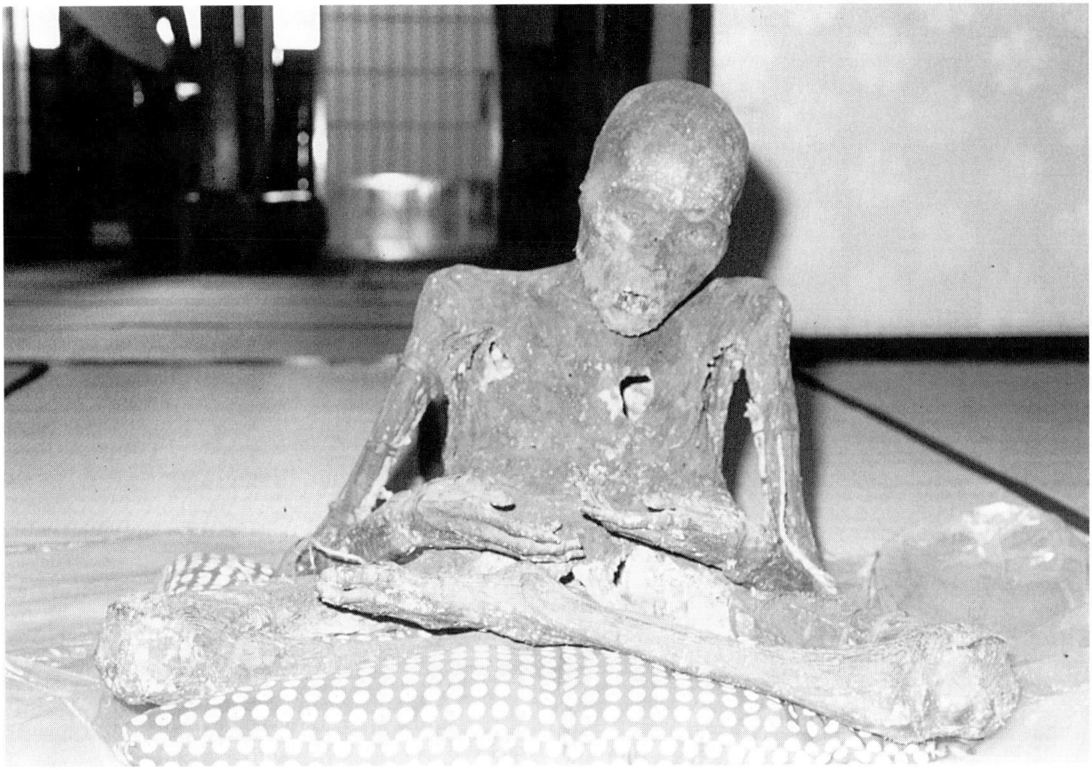

Fig. 4.157. "Self-mummified" Japanese mummy; AM, AD 1822.
The body is that of the Buddhist priest Enmyōkai Shōnin. (Photo courtesy of Iwataro Morimoto.)

load upon which subsequent decay is partly dependent, delaying onset of the tissue-destructive enzymatic process.

Tobacco smoke was blown into the mouth of the body shortly after death. The body was then seated over a low fire followed by completion of body desiccation through sun-drying. The influence of sixteenth century Spanish priests was probably responsible for cessation of these practices (Fonda, 2000).

During the first quarter of the twentieth century some of these mummies were stolen. One of these eventually arrived at the National Museum of Manila, where it was exhibited for some years before it was returned to the Ibaloi in 1999. That body had been displayed lying on its back. During return of the body to the tribe from which it had been stolen, a well-preserved tattoo was found on the back. The natives did not agree to sample tissues for study nor dissection of any kind before restoration to the burial cave. In contrast to the report by Morimoto (1988) cited above, Orlando Abinion states that evisceration is not a feature of these mummies (Donnet & Duclos, 1999).

MUMMIES FROM OCEANIA

Introduction to mummies from Oceania

Oceania includes the vast area occupied by the southern portion of the Pacific Ocean that is littered with islands ranging from flat, diminutive, uninhabitable atolls to Australia, an island of continental magnitude.

Fig. 4.158. Philippine mummies; AM, AD 1500–1800.
The Ibaloy people of Benguet province at more than 2000 meters altitude placed the uneviscerated, fire-dried bodies of their elite in log coffins in a cave. (Photo courtesy of Gamma Agency, France.)

Oceania's roughly rectangular dimensions include north and south latitudes of about 23° (tropics of Cancer and Capricorn), while its western border (western Australia) is about 120° E and its eastern extension is marked by Easter Island (Isla de Pascua) at about 110° W. This immense area, encircling one-half of the world, contains 8000 to 10 000 islands. The relative isolation of these insular populations has provided ideal conditions for maximal cultural and linguistic evolution, manifested by almost 2000 languages spoken by only about five million inhabitants. Steep-walled valleys on the many volcanic islands further enhance separation of subpopulations, leading to cultural practices so individualistic they almost defy classification of subareas within this immense expanse. Nevertheless, in addition to the Australian continent, its area has been divided into three, quite unequal-sized units, based on linguistic and cultural features as well as human physical characteristics: Melanesia, Micronesia and Polynesia (Force, 1984).

Reconstruction of Oceania's prehistory necessarily relies principally on a modest body of linguistic studies, most of which is supported by a disappointingly small archaeological database that depends a good deal on "Lapita" pottery. While enough inconsistency remains to make any particular chronology vulnerable to controversy, the most common working model indicates that New Guinea was already occupied prior to 35 000 BP (Gorecki, 1996). Furthered by their evolved sea travel skills and by the lower sea level during the glacial Pleistocene Period, people from New Guinea also reached Australia (the Torres Strait area's current mean depth is only about 30 meters). Recent evidence (Zimmer, 1999) suggests an even older date:

60 000 BP in Australia. Oceania's languages at contact can be traced to one of two families: (1) Papuan, involving New Guinea and the clusters of Melanesian island masses both directly east and west of it; and (2) Austronesian, including almost all the remainder of Oceania (though Australia's 200+ languages can not be linked yet to either of these) (Force, 1984). Tryon (1996) provides an excellent synopsis of how the Austronesian-speaking people are believed to have had their home in the South China area, moving to Taiwan, whence about 6000 years ago a Malayo-Polynesian speaking subgroup migrated to the Philippines. Southern extension brought a subdivision of these to northern New Guinea and New Britain about 200 BC while further southeast movement reached Vanuatu (New Hebrides before 1980) about 1200 BC and Fiji 200 years later. After 1000 BC colonization radiated therefrom, arriving in the Marquesas about AD 400, Hawaii AD 650 and Easter Island a century later. The Society Islands were reached about AD 750 from which the trail led to the Cook Islands AD 900, reaching New Zealand in AD 1000–1400. More recently, however, genetic markers (mtDNA) studies generally support the proposed pattern, though findings in several Melanesian populations were in conflict (Hagelberg *et al.*, 1999). Mitochondrial DNA studies using a 9 base pair deletion indicated that Polynesians did not arise from the Melanesian population (Merriwether *et al.*, 1994; 1995; Merriwether, 1999). Still more recent studies, however, involve also mutations in several loci of the Y chromosome. These suggest the possibility that Austronesian speakers may have arisen not in Taiwan but in southeastern Asia and expanded slowly enough to incorporate Melanesian features (Gibbons, 2001). Only broader studies will resolve these conflicts.

The resulting cultural diversity that evolved among these many insular populations is also reflected in mortuary practices, including mummification. I will review the sparse information available on this topic for Australia first, then direct attention to the area of Torres Strait and Queensland, ending with the practice of overmodeling skulls in the Vanuatu region.

Mummies from Buang Tribe in Papua New Guinea

Tribal groups' mortuary practices in New Guinea vary enormously. We have little reason to believe that a mummification rite practiced by one group is representative for even a region, much less the whole island. Nevertheless, descriptions by professional anthropologists or archaeologists of mummies from New Guinea are so sparse that it seems appropriate to include at least one. An eyewitness account of mortuary practices in the Snake River Valley near Salamaua in Papua New Guinea includes the production of some mummies. Local practices vary among tribes and even within tribes, based on age and social status among the Buang natives. Flexed extremities are tied into place and multiple skin incisions created to permit fluid drainage. After being wrapped in tapa cloth, the body is placed in a pit that is filled with grass until the body is covered. Earth is then used to fill the remaining space. After several months the then-desiccated body is exhumed, wrapped in a grass-filled bundle and moved to a well-aerated ledge on a limestone cliff. Vail (1936) found many such bodies to have lost much muscle, the skin appearing to be applied almost directly to extremities' long bones. The desiccation mummification appears to have been the product of fairly effective conductance of water away from the skin surface by the grass envelope and windy ledge location. Because the many dry limestone caves themselves were not used as burial sites, Vail suspects that mummification was not a goal but rather an incidental occurrence. However, the skin incisions would seem to have no purpose other than to encourage water conductance out of the body.

Mummies from Australia

Since more than 200 languages are spoken in Australia, it is not surprising that mortuary practices vary substantially in the different portions of this continent. These languages, however, do not appear to be part of either the Papuan or Austronesian families. Hence Australian aboriginal death rituals should not be expected to resemble those of the remainder of the Oceanic groups, and indeed, most differ substantially. The one feature that most Australian aboriginal groups share with each other as well as with the people of Papua New Guinea is the concept that adult male death is not a normal, biological process. Given this conviction, it is natural that the structure of their mortuary rituals would focus on identification of the human individual who perpetrated the sorcery, assuming that the deceased's spirit will not rest until the death is avenged (Elkin, 1948:253).

Methods of sorcerer identification do vary greatly. Most, but not all, incorporated the aid of the group's shaman. Signs that were interpreted as indicating the guilty party often involved manipulation of the corpse. In the Bloomfield River area, for example, the corpse was exhumed after several days' burial and searched for bruises, while natives of the lower Tully River opened the abdomen and retrieved the stomach, looking for divination signs that would lead the shaman to identify the one responsible for the death. For all these many variations in detection of the "murderer", the body needed to be available for a period of time long enough to permit completion of this vital portion of the funeral process. Such preservation could be achieved by temporary burial, by storage on a tree platform or many other techniques. Among some groups, the body was carried from camp to camp for weeks or even months, seeking certain signs that would betray the sorcerer. For such practices the body must have undergone more formal preservation methods that would guarantee soft tissue conservation adequate to complete the rite. These involved mummification of sufficient quality that some of these cadavers have survived to the present. Dawson (1928a) and others have deplored that textual reference to Australian aboriginal mummification is common, but few mummies have been identified and described. However, it must be remembered that the goal of such body preparation in most cases was temporary preservation for ritual completion. These usually were too ineffective for long-term survival and, after completion of the divination function, most were buried, cremated, placed in a hollow tree trunk or disposed of in other ways that resulted in soft tissue decay (Roth, 1907). Furthermore, mortuary services for women were much simpler, usually consisting of little more than wrapping the corpse in a shroud composed of tea-tree bark and burying it promptly after death, as happened to the bodies of children.

Mummies from the Queensland Area

Of the few described mummies that survived time, Dawson (1928a:plate VIII) illustrates two from the Queensland area, both similarly prepared. They are from North Australia, Trinity Bay on the eastern coast of Queensland. The mummies had been eviscerated and thoroughly desiccated by the heat of fire and/or sun. Hips, knees, elbows and neck were maximally flexed and bound tightly into position with cords and netting after being painted with red ocher. This produced a bundle of minimal size and weight that was sufficiently portable to be carried from camp to camp over long intervals.

Hamlyn-Harris (1912a), director of the Queensland Museum, also published photos of three Queensland mummies, one of which (*ibid.*, Plate 4) is the same photo of the Trinity Bay mummy included in Dawson's (1928a) article. The other two are those of an adult from the Johnstone River area, and that of a child from Cairns. All of these had been prepared in a manner identical to that described by Dawson. This article also offers four plates (8–11) demonstrating rare examples of partial or complete skeletons wrapped in tea-tree bark bundles.

In addition to these a Queensland mummy was acquired in 1905 by a German traveler (Klaatsch, 1905). Near Cairns (about 16° 50′ S, 148° 40′ E) he came upon a village still mourning the death of an older indigenous chief (called Narcha) whom they had mummified after his death eight months earlier. He comments that the early stages of such mummification must involve some loathsome practices, since the local authorities are attempting to suppress the ritual on hygienic grounds. The body was buried for several days, following which it was exhumed. After removal of the hair and putrefying epidermis, the body was eviscerated. It was then placed on a wood rack beneath which a fire was kept burning until complete desiccation was achieved. Drippings of fat were collected and smeared into the survivors' hair as part of the mourning process. The corpse was then tied tightly into a humped, sitting attitude that he described as an attempt to reproduce the fetal position. He also remarks that such mummification practices seem to be limited to the Cape York zone between Cooktown (about 15° 10′ S) and Townsville (19° 10′ S).

Using methods that would embarrass most of us today, Klaatsch acquired this mummy and, together with another that he does not describe, forwarded these (as well as 45 skulls and other excavated bones) to the Berlin Society of Anthropology, Ethnology and Ancient History. To date I have not been able to determine whether these mummies have survived and, if so, where they are being curated. His report included a photo of Narcha's mummy.

A brief but interesting description of funeral rites carried out by the Munkan, a Cape York, Queensland, tribe occupying a locale facing the Gulf of Carpentaria in northeastern Australia was described by a social anthropologist. Ursula McConnel (1957:17) lived among these people between 1927 and 1934, collecting folk myths and other data. She was an eyewitness to at least part of such a funeral ritual (the cremation ceremony). Characterizing their mortuary rites as a blend of those practiced in Papua New Guinea and Torres Strait, she notes that the body of the deceased may be buried for a while or eviscerated promptly, after which it is placed on a platform or on forked sticks. Dehydration of the corpse is accelerated with a fire. The desiccated mummy is then tied to a pole, wrapped in bark of a tea-tree, tied with strands of vegetal cord to forked sticks, after which it may be kept in this condition for several years, depending on wishes of the family. Eventually, following a final feast, the mummy is cremated as part of an elaborate ceremony.

Mummies from Torres Strait

Between the southern coast of New Guinea and the tip of the Cape of York projecting northward from northeastern Australia's Queensland district lies Torres Strait. Only 200 kilometers wide, this constricted passage overlies the shallow continental shelf bearing the northernmost extension of Australia's Great Barrier Reef. Cluttered with islands and reefs, it represents a navigational nightmare for mariners, as attested by the hulls of many a sunken derelict.

Cultural practices on these islands are closely related to those of nearby Papua New Guinea and to the rites of Australia's Cape York. Even within this island cluster, however, substantial variation was common, principally in the most eastern islands of Darnley (Erub), Stephens (Ugar) and Murray (Mer) (native names in parentheses). Overmodeling of skulls was practiced more widely as described below (Macintosh, 1949; Moore, 1984). Whole-body mummification was designed to serve a temporary purpose–appeasement of the deceased spirit (Moore, 1984:32), a process involving a series of ceremonies carried out over a period of more than six months. Thereafter it was assumed that the spirit had been effectively transferred permanently to its spiritual residence. Following these efforts at appeasement of the spirit the mummified body had no further religious significance, and was disposed of in a variety of ways dictated by convenience and secular interest to the survivors.

We have bodies of only nine mummies from which to reconstruct the anthropogenic mummification process employed to achieve soft tissue preservation for the 6–12 months required to complete the mortuary rituals. Fortunately Hamlyn-Harris (1912b) has recorded the observations of a long-term resident of one of the eastern islands, Murray (Mer), whose statements are consistent with inspections of the mummified bodies. In addition we have a large body of data recorded by the Cambridge University Anthropological Expedition led by A. C. Haddon (Haddon,

1935). This investigator visited New Guinea and the Torres Strait Islands initially in 1888–1889 to study the fauna there but became very interested in native life, spending half of his time with the local people. Ten years later he returned with a team of scientists that generated an enormous ethnographic and biophysiological database, including mummification practices. The following reconstruction is based primarily on these two sources, supplemented by others as indicated.

1. Within 24–36 hours after death the body was placed on a platform about 3 meters above the ground.
2. A small fire was built below it for ritual reasons, not to induce desiccation.
3. Several days later the blistered skin was removed (on some islands, carried out in a canoe on the sea), and the body cavities were eviscerated via a lateral abdominal incision; in several bodies the perineal route was used. The tongue was excised, eyes pierced to remove the fluid, and palms and fingernails (with attached skin) were stripped to be worn as a necklace by the widow. Excerebration was achieved with an occipital incision and insertion of an arrow that was twisted to macerate the brain.
4. The body was laced to a wood frame and erected vertically (beyond the reach of dogs) and supported by sticks in the abdomen, legs and head.
5. The evisceration wound was closed by a continuous or interrupted suture.
6. The body was painted with red ocher.
7. The body skin and joints were pierced and dripping fluid collected in coconut bowls.
8. The body was massaged several times daily.
9. When nearly dry it was decorated: nautilus shells placed into orbits to simulate eyes and the body painted again with ocher and coconut oil.
10. When completely air-dried, the body was moved to a beehive-shaped hut and tied to the center post.
11. After serving its ritual functions, the head eventually decayed, was removed and given to the widow. The remainder of the body was left to decay, be buried or cremated.

Plate 2 of Hamlyn-Harris' (1912b) article has photos of two such adult mummies curated at the Queensland Museum. They are lashed to a wood trellis, and the bodies have features described above. The trellis was formed by two long, upright poles with eight cross-pieces tied into place (the uppermost was double). To this the body was attached with vegetal cords, suspended by the shoulders, hands, knees and feet, while the separated head was tied into position around the forehead, with the mandible fastened to the head by a cord. The right abdominal evisceration wound was also employed to insert several pieces of wood to replace the viscera and then the wound was closed with a suture. A large shell covered the genitalia (Sengstake, 1892). These two correspond to my Table 4.17, nos. 1 and 2. In addition, Figure C of plate 8 in an article by Pretty (1969) also demonstrates a child similarly mummified and is described (and illustrated, his Fig. 11) by Dawson (1928a). This child mummy probably originated at Darnley (Erbu) Island and came to England's Royal College of Surgeons Museum via the Leicester Museum (Pretty, 1969:35) by unknown transporters about 1900.

A mummy of this same type from Darnley Island is of historic interest. In Chapter 1 I noted that the famed Egyptologist Grafton Elliot Smith (1915) propounded a theory stating that a set of cultural features characterized by sun-centered religion, megalithic monuments and anthropogenic mummification originated in Egypt. He believed it then diffused culturally from that center to other, globally distributed sites, including Torres Strait and Peru. In pursuit of supporting evidence he traveled to the latter two regions to examine their mummies. In 1914 in Australia he studied a mummy curated at Sydney's Macleay Museum and felt that this mummy's features were unmistakable evidence of direct Egyptian influence. Decades later, however, Pretty noted the following features: it had no transnasal craniotomy, no sign of natron desiccation and the abdominal incision was sutured. The mandible was fixed in position by a cord through the internal nasal structures, the skin supported by a wood prop

Table 4.17. *Mummies from Australia and Torres Strait*

Mummy no.	Age	Sex	Collector	Year	Source	Curation[a]	References
Mummies from Torres Strait							
1,2	Adult	M,F	W.F. Pettard, A.H. Palmer	1898	Stephen's Is. (Ugar)	Queensland Museum	Hamlyn-Harris, 1912b
3	Child	M	I	1900	Darnley Is. (Erub)	R.C.S. Museum	Dawson, 1927b; Pretty, 1969
4	Adult	M	Wm. Macleay	1875	Darnley Is. (Erub)	Macleay Museum	Pretty, 1969
5	Adult	M	C. Lemaistre	1872	Darnley Is. (Erub)	R.C.S. Museum	Flower, 1879: 392
6	Child	I	Capt. O. Stanley, T. Huxley	1849	Darnley Is. (Erub)	Merseyside Co. Museum	West, 1984
7,8	Child	M,F	James Chalmers	1880	Stephen's Is. (Ugar)	Voelkerkunde Museum	Haddon, 1935: 137
9	Adult	M	John Douglas	1884	I	British Museum	Dawson, 1924
Mummies from Australia							
10	Adult	M	H. Klaatsch	1905	Cape York	I (Berlin Soc. ?)	Klaatsch, 1905
11	Adult	M	George Grey	1845	Adelaide	R.C.S. Museum	Flower, 1879: 393
12	Adult	F	N.B. Tindale, C.P. Mountford	1936	Cape Jervis	National Museum Australia	Tindale & Mountford, 1936
13	Infant	I	H.L. Sheard	1927	Murray River	National Museum Australia	Sheard *et al.*, 1927
14	Adult	I	U. McConnel	1927	Cape York	Cremated	McConnel, 1957: 17

Notes:

M, male; F, female; I, indeterminate.

[a] R.C.S. Museum, Museum of Royal College of Surgeons, London; Queensland Museum, Brisbane; Macleay Museum – Sydney; Merseyside County Museum, Liverpool; Voelkerkunde Museum, Berlin; British Museum, London; Berlin Soc., Berlin Society for Anthropology, Ethnology and Prehistory; National Museum of Australia, Adelaide.

and skin drainage punctures were present. Fingernails and toenails had been deliberately removed, no cloth or bark wrappings were present nor were resin or aromatics used, and the entire body was lashed to a wood frame (Pretty, 1969). During my visit to that museum in August, 2000, I did not have an opportunity to evaluate even those features that are observable without radiography or dissection. These are all features that differ from those often present in Egyptian mummified bodies. Smith's views fueled controversy among his colleagues and, of course, subsequently have been refuted: anthropogenic mummies from the western coast of South America have been radiocarbon dated to about 2000 years before the time of the first pharaonic dynasty (Allison *et al.*, 1984).

An almost identical mummy from Darnley Island (Table 4.17, no. 5) was acquired by the Royal College of Surgeons as early as 1872 and is described by Flower (1879), who unfortunately chose to strip the soft tissue from the bones after his inspection in order to add the skeleton to the museum's collection (as he did with the Adelaide mummy described above). His illustration has an appearance very similar to those described above. A headless child's body from Darnley Island otherwise mummified in like fashion was acquired by a Liverpool museum in 1849 (Table 4.17, no. 6) and two other children's mummified bodies from Stephen's Island (Ugar) are in Berlin's Museum fuer Voelkerkunde (Haddon, 1935, and my Table 4.17, nos. 7 and 8). An adult (Table 4.17, no. 9) suspended from a wood frame is curated at the British Museum (Dawson, 1924) but its island origin is indeterminate.

To my knowledge, none of these mummies has been studied by bioanthropological methodology.

Mummies from South Australia

The desiccated, flexed body of an elderly female was found wrapped in kangaroo skin and buried in the salty, sandy soil of a cave (Kongarati) in a 128 meter high sea cliff. This was located near Second Valley, South Australia, 8 meters above sea level, in a rock tomb lined by slabs of slate. Ethnological data about the Ramindjeri tribe from that area confirmed that desiccation was normally carried out by exposing the body to smoke over a slow fire. Desiccated viscera were visualized through a skin defect over the right back area (Tindale & Mountford, 1936).

A 2-year-old infant's flexed body was found on a ledge of a cliff on the lower Murray River. Much of the body was skeletonized but skin covered the extremities, right flank and foot. An irregular mass anterior to the spine was probably viscera. Parts of the body were charred, suggesting exposure to fire. Ethnological information suggests stillborn infants were burned in this region. Preservation of soft tissue in this body was probably desiccation by fire, with contributions by sun exposure and eventual coverage by blown sand and animal excreta (Sheard *et al.*, 1927).

A mummy from the Adelaide area in southern Australia was also described by Dawson (1928a). After evisceration the adult body had been tied in a hyperflexed position and desiccated. This body was curated at the museum of London's Royal College of Surgeons, where it was described by Flower (1879). Unfortunately, having described and photographed it, Flower then skeletonized it.

Maori mummies from New Zealand

That New Zealand aboriginals (Maoris) frequently decapitated corpses and mummified the disarticulated head is well known but commonly misunderstood. At the time of European contact by Tasman in 1642, intertribal warfare among indigenous tribes was so common that nearly every tribal residence area included a fortified shelter. Revenge for earlier raids was the most frequent motivating condition for such aggression. The immediate goal of a battle was the death of the enemy's chief or chiefs. Because it was usually impractical to escape with the entire corpse, the cadaver was commonly decapitated and only the enemy chief's head carried back to one's own territory as a symbol of success: a trophy head. Such heads were impaled on pointed sticks, publicly ridiculed and even physically abused at prescribed rituals. Which heads were thus acquired and brought home depended on the victim's status. Chiefs or high-ranking enemies were preferred, as were unusually fierce combatants.

Individual motivation for the taking of such heads included such factors as personal status (Sinclair, 2000) or occasional employment of heads by tribal shamans in rituals designed to enhance agricultural success. Community benefits were the avenging of previous tribal victims, status (and therefore security) of the tribe among regional communities, their use as memorials and particularly their value in peacemaking efforts, where the mutual exchange of such heads was an integral feature (Orchiston, 1967).

As might be expected, warriors of the slain chief exerted every effort to prevent the slaying tribe from acquiring the head of their dead leader. If they were successful, they also found it difficult to transport the entire body home, and so decapitated it. This head was displayed at their home base with the highest respect, mourned over and venerated in every possible manner (Birket-Smith, 1957:218).

Finally, heads of individuals dying of natural causes sometimes were kept in a mummified state either as a community memorial if the person was among the elite or simply as a cherished personal memory of a beloved family member.

The mummification method for all of these is similar in principle, though Orchiston (1967) cites six descriptions that vary in detail. In many cases the following mummification steps were carried out before return to the home village after the battle. The soft tissues of the neck were kept attached to the disarticulated head.

Excerebration was carried out, after which the eyes were usually removed. In some cases the eyes, ears, nose and lips were closed with sutures. The nose was commonly splinted to retain its shape. Sometimes the cranial cavity was stuffed with flax, as were the orbits. The head was then heated by several different methods. In the simplest form it was simply placed in boiling water. In others it was heated by a crude oven (heated stones in an earthen pit). In still others the top of this oven was fashioned so that the skin of the neck fitted snugly over an opening at the oven's top, forcing the steam to traverse the head tissues. The skin surface was wiped frequently to prevent wrinkling during shrinkage. If soft tissues separated from the skull, wood splints were inserted into the space to retain contours. Heat was controlled at moderate temperatures to permit gradual desiccation. Eventually the steaming or boiling was discontinued and heat without water was applied (baking) and commonly smoke by placing appropriate leaves in the "oven". This latter process hardened the skin. The entire process required about 24 hours.

Because such heads commonly were adorned during life with deeply pigmented tattoos in striking patterns, they attracted the interest of visitors and tourists. Gradually a significant market for these heads evolved. They can be found today in globally distributed museums (Ackerknecht, 1944) where, however, they are now only rarely displayed.

Hongi (1916) comments that occasionally the entire body was mummified and provides a witnessed account. In that example the body was placed over a prepared hole in the ground and eviscerated via the perineal route. The viscera were allowed to fall into the hole and the hole refilled with earth. The body was then desiccated, presumably by heat from a fire. He also comments that the space between the lips and the upper and lower jaws was filled with wax. Hongi's own sister was so mummified.

In addition, Hongi provides an ethnographic account from an old Maori aboriginal woman who told him that such mummification efforts involved puncturing the body in many places, including the skull, to drain body fluids and drying the body in an oven-containing hut constructed for that purpose. At times the body cavities were filled with flax. Although Hongi has met with dispute in the literature relating to the question of whether ancient Maoris carried out whole-body mummification, Orchiston (1968) has also documented several reasonably convincing examples. During an 1820 visit to New Zealand, Cruise (1823) viewed the body of a local chief's brother who had died "a considerable time earlier . . ." and whom the chief had just transported back to his home. The body was maximally flexed and wrapped in multiple vegetal mats. The legs' soft tissues were shriveled and the abdomen collapsed, although no abdominal incision was apparent. A local chief also showed Cook's surgeon the desiccated body of a neonate (Beaglehole, 1955: Appendix IV, 584–585). A missionary's report refers to the New Zealand Maoris practice of "embalming the dead and putting the corpse into a box . . ." (Elder, 1934:90).

Perhaps most tellingly, Polack (1838:375, 1840:127) describes how a local chief preserved the body of his 4-year-old child by applying the same "boiling" methods New Zealand Maoris used to preserve the heads of their enemies. In addition Reischek (1930:215) was shown two adult mummies in a New Zealand cave near Kawhui and transported two of them to a Vienna museum. Firth (1931) describes these in detail and illustrates a photo of one of them. These reports suggest that, while Maoris probably did not routinely mummify the entire body, they did so under certain special circumstances such as the need to transport the body a long distance (Orchiston, 1968). Some of such bodies, especially those of children, acquired the mummified state spontaneously, while in others evisceration followed by "boiling" was employed.

Mummies from the Cook Islands, Tahiti and Samoa

Hongi (1916) cites a book by Dr Wyatt Gill, *Life in the Southern Islands* (London, 1876), in which is described how ancient residents of the island of Mangaia smeared oil over the deceased's body, wrapped it in a cloth and buried it or, more commonly, placed it in a cave. At varying intervals the body was unwrapped, exposed to the sun and rewrapped. The author goes on to describe a cave where he saw hundreds of such mummies. He emphasizes that they were not eviscerated but desiccated. These mummies have not been described, studied or reported subsequently in the scientific literature.

For Tahiti, Service (1958) describes a similar practice for deceased chiefs, though there evisceration was carried out and the sun is noted to have been the desiccating agent. This appears also to have been temporary mummification, because he goes on to explain that, after the body disintegrates, the skull is kept in a special shelter. McHargue (1972:105) quotes Sir James Cook as describing that Tahitian mummification was carried out partly by sun-drying and partly by squeezing the body to express fluid that escapes via skin incisions; later the body was anointed with oil. In spite of subsequent, extensive social activity in Tahiti by individuals from other continents, I could find no scientific reports of such mummies.

Samoa had similar funerary practices restricted to the populations' leaders. Women there prepared the body by evisceration, removal of all body hair and anointing the body with oil repeatedly while incising and compressing the skin to squeeze out fluid. After several months the abdomen was stuffed with tapa cloth, clothed and kept in a dedicated building. Repeated oiling extended the mummified state for a long time (Murdock, 1934; James, 1957:125). Nevertheless, the absence of reports in the scientific literature suggests that this method also proved to be temporary mummification.

Mummies from Indonesia

Mummification in these islands was widespread and produced by desiccating the body in the sun and desiccating them further by smoking. This also appears to have been temporary mummification because, following appropriate rituals, the body ultimately was placed in a hollow tree, burned and buried (McHargue, 1972).

Chapter 5. Soft tissue taphonomy

Introduction

In 1940 Efremov defined a new paleontology study area, calling it taphonomy. It dealt with all the various processes that operated to form a fossil as we find it today ("the science of laws of embedding . . ."). Later the term was also employed to describe changes in skeletal tissues examined by physical anthropologists (Bonnichsen, 1982). Fossil proteins have also become a special area of interest for some (Behrensmeyer, 1984). The term taphonomy, therefore, also seems applicable to the field of mummy studies, since interpretation of our analyses are in part dependent on the effects of postmortem processes that altered the decay mechanism sufficiently to result in mummification. Hence I will label the discussion of these effects "soft tissue taphonomy". Though observations on postmortem changes in mammalian bodies have been recorded as early as 1819 (Davy), the field has certainly not progressed in proportion to its importance.

The taphonomy of proteins

Considering how much we know about the behavior of proteins within the living body, it is surprising how little the postmortem fate of proteins has been studied. Because proteins form such a large fraction of the body's structural as well as functional components, they house information of great interest to bioanthropologists. The sequence of amino acids in a protein reflects the structure of the gene that directed its synthesis; blood globulins include antibodies that can provide information about infectious diseases suffered during life; polymorphic antigen systems can offer a measure of biological distance and evolutionary descent; isotopes of carbon and nitrogen in protein can reflect dietary histories. In view, then, of the importance of preserved protein to those attempting to reconstruct the life of an ancient body, the meager status of information available on this topic at the time of writing comes as a disappointment.

Normal protein structure

Amino acids are the building blocks of proteins. These are relatively short, linear molecules of carboxylic acids (those containing a -COOH group) that have an amino group (-NH_3) attached to the carbon atom (alpha carbon) adjacent to that of the carboxyl group. Twenty such amino acids found in proteins are coded in genes; nine are "essential", i.e., the human body can not synthesize them, so they must be included in the diet. The liver joins amino acids by linking the alpha amino group of one amino acid to the carboxyl group of another (via a peptide bond) to form short chains of amino acid "residues" (peptides). Some peptides themselves have biological function. Commonly, however, longer chains are formed (polypeptides).

When the chains exceed a molecular weight of 10 000 they are called proteins. These proteins commonly include as many as a thousand amino acid residues. Such protein chains are principally linear covalent links but noncovalent interactions can result from contact or at least proximity of one point in a chain to that of another segment of the same chain. One common form is that of a spiral or helix (fibrous proteins). At such contact sites interatomic forces may result in mutual attraction or repulsion. In other molecules the folding process may be far more extensive, resulting in a very compact molecule of spheroidal shape (globular proteins). The combination of the molecule's sequence of amino acid residues and the manner in which it is folded determine chemical behavior. Fibrous proteins are most stable (e.g., tendons) and are used for structural purposes while globular proteins are commonly functional (e.g., hormones). Even though fewer than two dozen amino acids are available for construction of a molecule of 1000 amino acid residues' length, the possible sequence permutations number in the millions (Morrison & Boyd, 1973:1132–1163).

Certain features of protein structure are of special interest to bioanthropologists. Considerable energy is required to break the chemical bonds, those that cross-link between chains as well as the peptide bonds linking amino acids. This lends stability to the protein molecule. Degradation is more apt to occur via microbial enzymatic action than from nonenzymatic influences of the postmortem environment such as thermal or pH effects, though both degradation methods occur. Removal of water from its immediate environment (desiccation) may also stabilize a molecule. In some proteins peptide chains are linked to nonproteins, such as sugars, to produce glycoproteins (as occurs in the ABO red blood cell antigens); these bonds may or may not be as strong as linkages between amino acid residues. The actual amino acids present also play some role in resistance to postmortem protein degradation because some amino acids are more vulnerable than others to postmortem environmental action. Protein sequestered in some protective locations (bone mineral, teeth) may be physically sheltered from contact by destructive agents (Ambler & Daniel, 1991; Eglinton & Logan, 1991). Finally, humic acids or other environmental exogenous nonprotein substances may covalently link with proteins, introducing "foreign" carbon and nitrogen atoms that could confuse protein isotope determinations (Ambler & Daniel, 1991).

Methods of protein degradation

Hydrolysis

We know that postmortem protein breakdown is commonly initiated through hydrolysis carried out by release of intracellular enzymes, subsequently accelerated by microbial enzymes and later by those of insects. But nonenzymatic degradation also plays a role. We need to know exact chemical mechanisms and rates under a broad range of postmortem environments in order to predict and assess a sample's quality and therefore interpretation of tests performed on it. By far the most common mechanism of loss of a protein molecule's integrity after death is via hydrolysis. As the term implies, this involves insertion of a molecule of water (H_2O) into the protein chain at a point that results in the molecule's fragmentation – the hydroxyl group (OH) remaining on one fragment and the hydrogen atom (H) on the other. Additional hydrolytic action may eventually create free amino acids that may be lost by diffusion. As indicated above, this can be brought about quite rapidly by microbially produced enzymes, or much more slowly by nonenzymatic action of environmental agents. Folding of protein molecules makes the bonds of some parts of a protein molecule less accessible to bond-disrupting agents, stabilizing the molecular structure of the protein. This is an operating influence in the various fibrous structures in the body such as collagen (Eglinton & Logan, 1991). In the presence of water, light (especially ultraviolet) can result in photooxidation and peptide bond disruption. Terminal amino groups of proteins or even individual amino acids can combine with carbohydrates to produce brown, insoluble compounds by a process termed the Maillard reaction (Karpowycz, 1981). These products are relatively resistant to further chemical reactivity. Evershed *et al.* (1997) recently demonstrated that the Maillard reaction also contributed to plant preservation in the Nubian Desert.

Denaturation

The three-dimensional, highly specific structure of proteins is responsible for their chemical behavior. One mechanism by which proteins may escape postmortem degradation is denaturation. This is a term used for a change in protein conformation so major that the molecule no longer behaves as expected. In most cases denaturating agents make proteins more susceptible to hydrolysis. Denaturation of globular proteins by heat usually acts by disruption of secondary and tertiary bonds, resulting in an uncoiling of the protein molecule that may terminate in irreversible protein precipitation. When denaturation leads to insolubility, hydrolysis slows, a process that contributes to preservation; the insoluble proteins in hair and nails are such examples. Other common agents bringing about denaturation include strong bases or acids and even certain compounds used for tissue fixation prior to sectioning, such as formaldehyde. The latter acts by forming methylene bridges between denatured proteins (Morrison & Boyd, 1973:1150).

Miscellaneous chemical taphonomic mechanisms and effects

Oxidation, nonenzymatic cross-linking and modification of amino acid side chains are additional methods of protein degradation. The unique mechanisms operating in peat bogs to preserve soft tissues are discussed in detail in Chapter 4 (Bog bodies of northern Europe, England and Ireland), and are noted here briefly for readers' convenience. Holocellulose (cellulose plus hemicellulose), a component of *Sphagnum* moss in peat bogs, includes a product called sphagnan that comprises much of the hemicellulose. It has a chelating quality, exchanging various bivalent ions for calcium ions in peat water and thus sequestering them; humic acids have a similar property. This is not only responsible for decalcification of bog bodies' bones, but also effects a bacteriostatic action on calcium ion-dependent bacteria, suppressing their growth. The carbonyl (C=O) groups in sphagnan also build bridges between collagen molecules, preserving them from enzymatic degradation ("natural tanning"). The cross-linking of collagen fibers causes collagen to expel water from its interstices, preventing rehydration and resisting microbial attack (Painter, 1991). The holocellulose of *Sphagnum* moss also binds and inactivates proteases and other destructive bacterial enzymes (Painter, 1995). All these actions modify protein structure in such a manner as to contribute to their preservation probability.

Applications of protein degradation principles

While the forensic literature contains many descriptions of postmortem protein changes (Coe, 1977), most of these data deal with the immediate postmortem interval of only a few days and are designed to help to estimate time of death. Allison & Briggs (1991) have collected a considerable body of information about re-entry of proteins into the geosphere, only a modest amount of which is directly applicable to the needs of bioanthropologists. Because collagen forms 90% of bone protein, we have the largest body of experience with this substance. Collagen composition is ideally adapted for its purpose in mammalian bodies. It is composed of a recurring triad of amino acid residues in which every third residue is glycine. Proline and hydroxyproline are frequent in the other two positions. Three alpha chains are arranged in an intertwining helical arrangement, resulting in regularly spaced interchain hydrogen bonds that produce an insoluble material of great tensile strength and stability, and a unique "banding" pattern visible by electron microscopy (Collins *et al.*, 1995b) (Fig. 5.1).

As early as 1957 Abelson, using principally a paper chromatography method then available, extracted

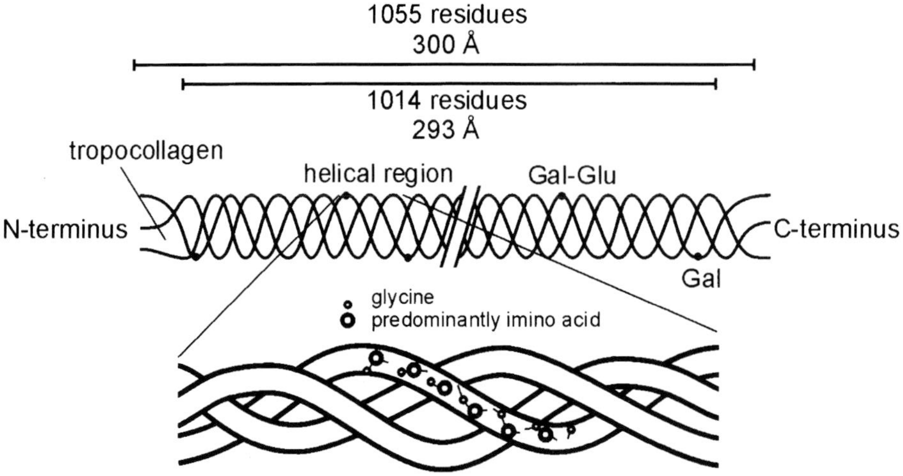

Fig. 5.1. Structure of collagen molecule.
The triple-helical structure of collagen is evident, composed of three polypeptide chains. Repetitive amino acid sequence results in cross-linking, a feature that contributes to its stability and tensile strength (1 Å = 0.1 nm). (Figure drawn by Christine Jeans (University of Newcastle), reproduced here from M.J. Collins *et al.*, *ABI Newsletter*, November 1995, with permission from Natural Environment Research Council, United Kingdom.)

ancient bone samples and found a range of amino acids similar to that in modern bone. He acknowledged that contemporary methodology did not permit him to separate original from diagenetic (contaminant) sources of these amino acids. That same year Ezra & Cook (1957) studied fossils up to 6000 years old but found only a few amino acids remaining in some and none at all in others. About a decade later Ho (1966, 1967) qualitatively identified a wide range of amino acids in Pleistocene mammal fossils, though they retained little hydroxyproline (an amino acid unique to collagen). Miller & Wyckoff (1968), however, did find hydroxyproline in a million-year-old fossil, and Gurley *et al.* (1991) found evidence of protein that probably was not collagen in a 150 million-year-old dinosaur. Abundant hydroxyproline was also found in a Dynasty 12 Egyptian mummy bone (Lausarot *et al.*, 1972). Just as Abelson had earlier found more amino acids in the more highly mineralized fossil shells than in bone, so Ho noted better preservation in dentine than in bone. Later workers confirmed that a mineralized matrix offers significant protection from protein-degrading agents (Ambler & Daniel, 1991; Collins *et al.*, 1995; Bada *et al.*, 1999). Collins *et al.* (1995a) even feel that the enzyme molecules are so large that they can not penetrate collagen, acting primarily on its surface and that therefore most collagen degradation in nature is probably brought about by nonenzymatic, chemical mechanisms. Barraco (1980) was able to demonstrate that much more protein had been hydrolyzed into polypeptides and free amino acids in a spontaneously mummified Egyptian body than in one preserved by the traditional natron technique, and that preservation of protein and lipid molecular structure varied in direct proportion to the tissue's sodium salt (presumably natron) content (Barraco *et al.*, 1977).

Modern methodology including gas chromatography with or without mass spectrometry has enabled more sensitive and specific identification as well as quantification. In addition, polarimetry methodology has permitted racemization measurements. The latter is based on the fact that, with the exception of glycine, all amino acids include a carbon atom that has four different substituents. This renders them "optically active", capable of rotating a beam of plane-polarized light (right (D) or left (L)). After death, the amino acids of body proteins alter their optical activity ("racemize") at a slow but temperature-dependent rate (see Chapter 6: Amino acid racemization). These and other methods have been employed to identify features of the protein in a specimen that could serve as predictors of the degree of original protein preservation in a sample. Deliberate exploitation of protein degradation leading to racemization for the purpose of dating an ancient specimen is of major interest to bioanthropologists and is discussed more extensively in Chapter 6.

Masters (1987) notes that the following features of collagen diagenesis may be so employed to assess a sample's degree of diagenesis: decreased content of amino acids and nitrogen; composition suggestive of high noncollagenous protein; increased D/L racemization ratio; a carbon/nitrogen concentration ratio outside normal range (2.9–3.6); loss of organic content (as reflected in lower levels of nitrogen content), and a relative increase in the acidic amino acids, aspartic and glutamic, because of their absorption to mineralized matrix. Pfeiffer & Varney (2000), however, demonstrated an absence of correlation between the quality of preservation as judged by bone histology and the retention of collagen's molecular structure, indicating that histology is of questionable value as a preservation marker for collagen in bone. Subsequently Bada *et al.* (1999) found that in a satisfactory sample the D/L racemization ratio of aspartic acid should not exceed 0.10–0.15 and that alanine should be less racemized than aspartic acid, noting further that preservation of extractable DNA was predicted by these same criteria. Examination of dinosaur bones revealed mixed results with respect to these racemization criteria. They also noted that D/L ratios of amber-embedded insects were the same as those of modern insects, and attribute the lack of racemization to the anhydrous state of those trapped in amber (Fig. 5.2).

Only a few reports of analysis of proteins other than collagen in ancient bodies are available. Aseptically collected human blood stored in sealed ampules at 1 °C and examined 24–42 years after collection was found to contain intact hemoglobin that demonstrated a normal hemoglobin absorption spectrum, oxygen dissociation curve, oxygen capacity and effectively functioning enzymes catalase, glyoxalase, choline esterase and carbonic anhydrase (Keilin & Wang, 1947). Similar results were noted in sterile dried blood samples 8 years old, which also revealed chemically intact albumin at about one-third of normal amounts (Sensabaugh *et al.*, 1971). The enzyme Cu_2Zn_2 superoxide dismutase has been extracted from a 3000-year-old spontaneously mummified body and found to be functional (Weser *et al.*, 1989).

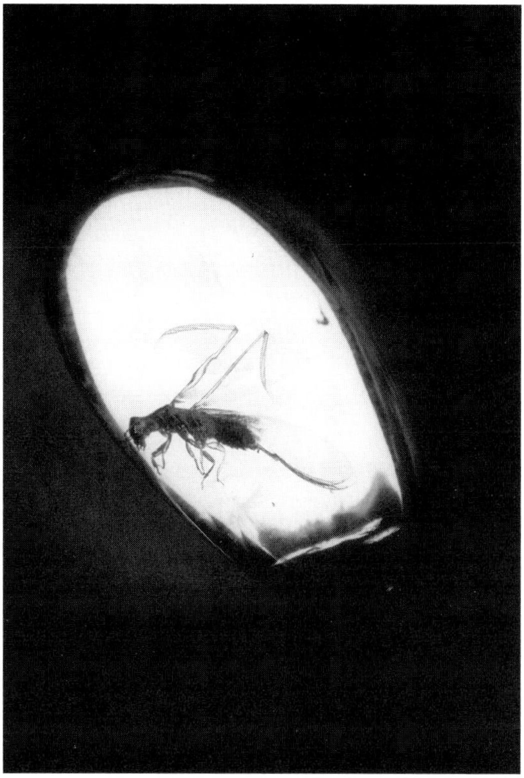

Fig. 5.2. Amber-embedded insect; SM, 25–40 million years old.
Soft tissue taphonomic processes are almost suspended when insects are embedded in amber. Termite in amber from the Dominican Republic. (Photo by and courtesy of Jon Kramer of the Potomac Museum Group.)

Immunological studies that employed monoclonal antibodies successfully identified proteins secreted into the blood by *Plasmodium falciparum* (malarial parasite) and *Schistosoma* sp. in ancient Egyptian mummies (Howard *et al.*, 1986; Deelder *et al.*, 1989, 1990; Miller *et al.*, 1994; Rabino-Massa *et al.*, 2001). Rutherford (1999) used an antibody directed against an antigen on schistosome ova to identify them in tissue sections of Egyptian mummies. These successful efforts demonstrate sufficient retention of at least some antigenic determinants of protein molecules (epitopes) to provide useful research data.

Recently Kolman *et al.* (1999) reported their spectacular recovery of an IgG antibody and demonstrated its intact functional status against the antigen of *Treponema pallidum pallidum* from a 200-year-old Easter Island human bone that revealed a characteristic "saber-shin" change. Recovery of the immunoglobulin required intensive efforts to increase its concentration in the extract. The finding was confirmed by extraction and identification of a segment of DNA unique to the syphilitic spirochete from the same sample. Both these findings are unprecedented, and probably emphasize that we still need far more research before we understand protein degradation to a satisfactorily predictable degree in the many different postmortem environments normally encountered.

If the specimen to be tested is from ancient tissue, Tuross & Stathopolos (1993) note the need to modify the techniques that were designed for use on fresh samples collected from living or recently deceased subjects. Using fresh human bone samples as controls, Kaup *et al.* (1994) isolated a functional 200 kDa protein characteristic of alkaline phosphatase from an Egyptian mummy radiocarbon dated to the Ptolemaic Period (ca. 323–30 BC), and the following year Weser *et al.* (1995) again extracted structurally and functionally intact alkaline phosphatase from the clavicle of a 4000-year-old Old Kingdom Egyptian mummy. Immunoreactivity for type 1 nitric oxide synthase has been demonstrated in sural nerves of Old Kingdom Egyptian mummies (Appenzeller & Aufderheide, 1999) (Fig. 5.3).

Of special interest to bioanthropologists and paleopathologists is the question of activation of ancient proteases that destroy proteins during rehydration of desiccated, mummified tissue samples. The gross and microscopic preservation of tissue structure in bodies whose postmortem tissues are exposed to high concentrations of heavy metal ions (arsenic, lead, copper, mercury) suggest that these ions inactivate tissue proteases (Snow & Reyman, 1977); Green *et al.*, 1981; Yamada *et al.*, 1990; Schulting, 1995). However, direct demonstration of such a phenomenon and its quantification in ancient tissues under controlled circumstances is still wanting. Tuross (1991) provides a useful review of this topic and also presents her own contribution to the demonstration of several preserved bone and serum proteins (osteonectin, IgG and albumin) in ancient Native American bones. Ascenzi *et al.* (1991) employed an immunological method to demonstrate the presence of hemoglobin subunits in ancient bones in an effort to diagnose thalassemia.

So, what can we conclude from this dreary litany of isolated observations and profoundly limited experimental studies dealing with the chemical mechanisms of postmortem protein degradation?

1. Protein degradation can occur enzymatically (rapidly) or nonenzymatically (slowly).
2. Its rate is temperature dependent.
3. Molecular structure is a major variable in a protein's vulnerability to degradation.
4. Fibrous proteins resist degradation better than globular proteins.
5. A mineral matrix affords some protection from degradation to its included protein.
6. Desiccation retards or stops protein degradation.
7. Hydrolysis is the most common form of postmortem protein degradation.
8. Microbial enzymes can greatly accelerate protein degradation.
9. Some proteins survive intact after a very long postmortem interval.
10. The degree of some proteins' integrity may be predicted by determination of several, easily measured chemical parameters.
11. The D/L racemization ratio value for aspartic acid is predictive of the degree of collagen integrity.

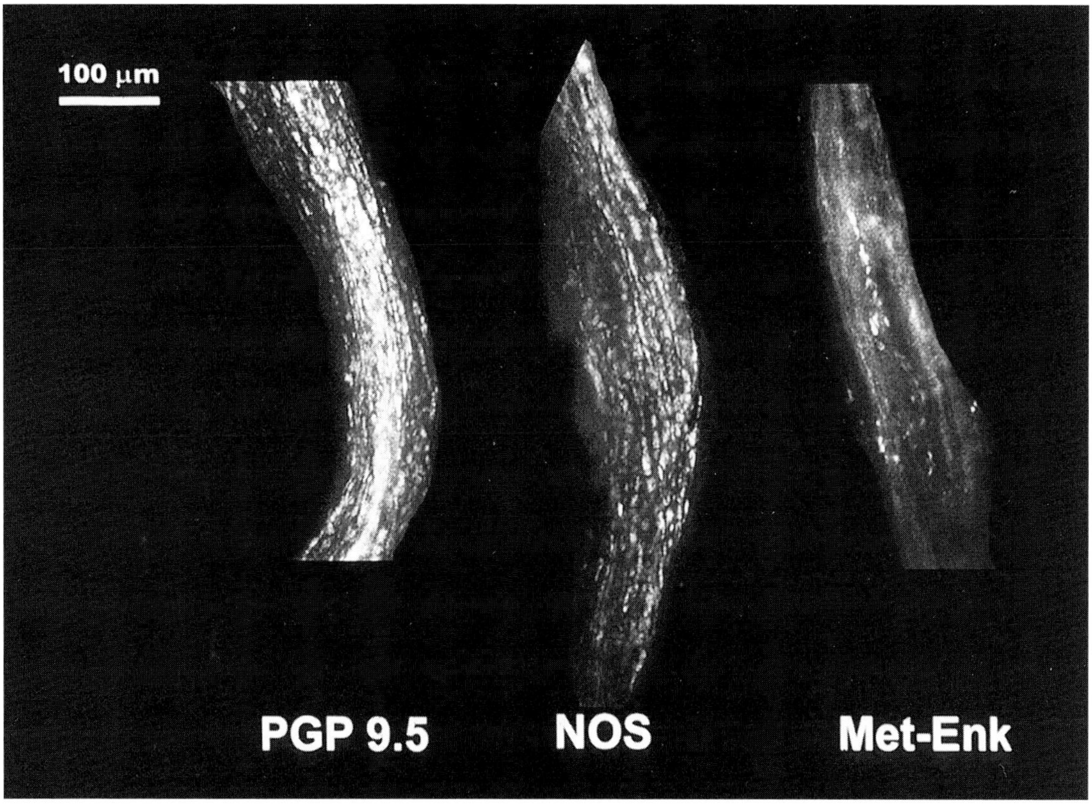

Fig. 5.3. Preservation of functional enzyme in mummy tissue; SM, 1000 BC.
The immunoreactive enzyme nitric oxide synthase (NOS) was demonstrated in a sural nerve dissected from a 3000-year-old Andean mummy. PGP, protein gene product 9.5; Met-Enk, met-enkephalin. (Photo courtesy of C.H.V. Hoyle and O. Appenzeller.)

The above, broad generalities will pacify few bioanthropologists' apprehensions. They are based on individual observations, often in the uncontrolled circumstances of field findings, or in experimental studies commonly highly focused on specific proteins (especially collagen). Clearly we need far more studies under the widely different postmortem circumstances of climate, soil, burial depth, coffins and other mortuary taphonomic variables in order to interpret our laboratory measurements with satisfactory certainty.

Fortunately, as has become evident in other areas of mummy studies, we can turn for help to our colleagues in other disciplines. The geochemists have already created a database whose integration into the field of mummy studies can be expected to make major contributions to our understanding of postmortem chemistry. In their investigations of sediments, both marine and terrestrial, they have recognized the orderliness of stratigraphy. The deepest layers are oldest, with the result that the early degenerative changes in organic molecular structure are found in the superficial layers while the lowest levels reflect the end stages of chemical taphonomy. As related so elegantly by Eglinton (1998), on this basis it has been possible to recognize certain predictable patterns of molecular breakdown that appear to be unique to various classes of organic molecular structure. Admittedly, the milieu of the grave may not mimic those of sediments, but certainly the patterns established in sediments could serve as general guidelines for future research into human soft tissue taphonomy.

The taphonomy of nucleic acids

Early studies suggested that nucleic acid would be at least as susceptible to denaturation as most proteins and certainly more than collagen. As expected, ancient DNA (aDNA) from archaeological bones was found to have suffered extensive molecular damage from hydrolytic as well as oxidative degradation, and was reduced quantitatively to a profound degree (Pääbo, 1985). While the polymerase chain reaction has been effective in amplifying short segments of DNA extracted from archaeological specimens, it still remains difficult even today to identify nuclear DNA consistently from such specimens. Mitochondrial DNA (mtDNA) is present in several hundredfold greater number of copies than is nuclear DNA; hence most reported aDNA studies deal with mtDNA. Even so, segments exceeding 200 base-pair (bp) or nucleotide lengths are difficult to recover.

Applications of these principles have resulted in amplified sequences from mtDNA of an extinct (since AD 1883) zebra-like animal – the quagga (Higuchi *et al.*, 1984). More recently Krings *et al.* (1997) have succeeded in isolating a mtDNA segment from a Neanderthal fossil, approaching the theoretical limit of 100 000 years for the recovery of aDNA. Others, however, have suggested that DNA sheltered in a more protective matrix such as amber or lithics might survive much longer. In 1990 Golenberg *et al.* reported recovering aDNA from a 15 million-year-old magnolia plant fossil. Pääbo & Wilson (1991) greeted this report with cautious optimism, and the following year DeSalle *et al.* (1992) recovered aDNA from a fossil termite encased in Oligo-Miocene (25–30 million-year-old) amber. More recently Cano & Borucki (1995) reported recovery of aDNA from a spore of *Bacillus sphaericus* embedded in a 25–40 million-year-old sample of Dominican amber. In 1998 Lambert *et al.* recovered a novel organism resembling a *Staphylococcus* from amber of similar age and locale. Such reports, however, have been received with considerable skepticism (Lindahl, 1993). Lambert et al. (1998) reported culturing from Dominican amber a *Staphylococcus* with unusual DNA structure. However, several failures to extract aDNA from amber have been documented (Austin *et al.*, 1997). Lindahl (1997) regards those very ancient recovery reports as a "fiasco" of experimental artifact but accepts validity of the Neanderthal report by Krings *et al.* (1997). Nevertheless, Cano (1998) cites the unique mechanisms by which bacterial endospore structures protect the integrity of bacterial DNA, rendering them of great value in the study of bacterial evolution.

Because racemization of amino acids is affected by the same factors as spontaneous degradation of DNA by depurination, Poinar *et al.* (1996) have established a correlation between the easily measured D/L racemization values for aspartic acid and recoverable aDNA from a specimen; this simple procedure can predict the probability of success in aDNA recovery from a given specimen (for details, see Chapter 6: Biochemical methods).

The taphonomy of lipids

Normal lipid composition of the human body

Lipids are found in a variety of chemical forms in the human body. Those detectable by modern laboratory instrumentation and of archaeological and anthropological interest include the following.

Neutral fat

This consists of fat representing stored energy. It is located within the cytoplasm of fat cells and distributed widely throughout the body but most abundantly in the subcutaneous tissues. Molecules of this lipid form are called triacylglycerols (triglycerides), composed of the three linked carbon atoms of glycerol each of which is attached to an esterified fatty acid chain. Differences in the nature of the fatty acid chains are common in various body locations. Incorporation of water molecules into neutral fat splits the triacylglycerol into glycerol, freeing the fatty acids – a process termed hydrolysis. Hydrolysis can occur slowly and spontaneously in an alkaline environment, but enzymes secreted by certain bacteria can accelerate the process enormously.

Phospholipids

These are 1,2-diacylglycerol molecules with a base attached to the glycerol core's third carbon atom by a phosphodiester bridge. They have an ionized ("polar") end that is hydrophilic and an opposite, nonpolar end that is hydrophobic. They are incorporated into the body's membranes where they are structured in a back-to-back arrangement, their resulting hydrophilic ends residing on the outer and inner cell wall surfaces and their hydrophobic ends between these two surfaces within the membrane. They represent a vital component of the wall of virtually every cell membrane and thus constitute a significant fraction of total body fat.

Fatty carboxylic acids

These are usually of linear chain structure, with a carboxyl group (monocarboxylic) at one end and a hydrocarbon chain commonly composed of even-numbered carbon atoms. The most abundant of those forming human triglycerides have 16 or 18 carbon atoms. In normal human fat, free fatty acids are unusual; about 95% are linked to a glycerol core. The most common fatty acids in humans are oleic, palmitic and linoleic acids. When all the hydrocarbon's carbon sites are occupied ("saturated") with hydrogen atoms, the fatty acid is termed a saturated fatty acid; if even a single double bond is present, it is said to be unsaturated. In the living state most fatty acids containing triacylglyceride moieties have one or two double bonds. Oleic acid contains one double bond and comprises roughly half of the lipids' fatty acids. Saturated fatty acid molecules pack together more tightly, increasing their intermolecular forces, raising their melting point and increasing the total molecule's chemical stability (Morrison & Boyd, 1973:1055–1067).

Sterols

These compounds have three six-sided carbon rings and an additional five-sided cyclopentane ring. Most of the body's hormones are of this structure (steroids). Relevant to this discussion is the "mother compound" of steroids – cholesterol, which has a similar ring structure with a long side chain on the cyclopentane ring and an alcoholic hydroxyl group. Among its multiple roles (as steroid derivatives) is the production of bile salts, whose excretion into the intestine facilitates fat digestion. A similar compound produced by plants can be separated on the basis of its methyl and ethyl groups in the C-24 side chain. Plant sterols are very unlikely to migrate into human tissue, even when the human tissue is immersed in a watery bog, because of their water insolubility. Their identification in a human tissue sample of a body in such an environment is evidence that the analyzed human tissue was physically contaminated by an intact plant fragment (Evershed, 1993). Eglinton (1998) notes the stability of sterols in sediment and calls attention to the potential of extracting intact sterols from ancient sediments to identify the source of their carbon atoms via stable isotope ratio analysis in such single preserved compounds.

Postmortem alterations of human body lipids

Initial studies

Except for those relating to the production of adipocere (see below) our understanding of the taphonomy of human lipids is characterized by an appalling paucity of such directed studies. Mair (1914) examined some brain tissue from Egyptian mummies given him by Grafton Elliot Smith in 1914, and found that the chloroform extract consisted chiefly of fatty acids without phosphorus or cerebroside. Lack of better laboratory instrumentation, however, restricted the extraction of further useful information. Sixty-three years later, Barraco *et al.* (1977) studied both skin and visceral samples from several Egyptian mummies and noted the routine absence of demonstrable phospholipids and gangliosides (brain and nerve lipids). Thin-layer chromatography (TLC), however, permitted separation of six components among the neutral lipids and 11 in the total lipid extract. These were, however, not individually characterized in their initial report. In Egyptian mummy samples, Barraco *et al.* (1977) found lipid was extractable, but that gangliosides and phospholipids could not be demonstrated, though cholesterol was probably intact. Proof that lipids did survive in archaeological bone and that they were not soil contaminants was demonstrated by Evershed *et al.* (1995).

The dearth of such reported studies makes the findings of a recent investigation (Evershed, 1990) of great interest even though the studied ancient bodies had all been extracted from peat bogs. Skin samples from seven such bodies from Dutch sites as well as Lindow Man from a British site were examined by the modern methods of TLC, gas chromatography (GC) and gas chromatography with mass spectrometry (GC/MS). In Lindow Man unaltered cholesterol was abundant, as were fatty acids with the same length distribution as is found in normal, fresh human tissue, but neither triacylglycerols nor phospholipids were detectable. Samples from the seven Dutch bodies also revealed no identifiable phospholipids, the phosphate and alcohol groups having apparently been split off and diffused into the peat water. However, TLC did reveal evidence of some degree of triacylglycerol and/or steryl fatty acyl ester molecule preservation, but specific identification of these by GC/MS was deferred for future studies. Saturated free carboxylic acids were common. Particularly important was the total absence of oleic acid (the principal such acid in the neutral fat of living humans); however, 10-hydroxyoctadecanoic acid (derived from oleic acid by hydroxylation) was observed. Intact cholesterol was also found, as were some of its degradation products including coprostanol, produced by microbial action of intestinal bacteria. Evidence that this was achieved by stereospecific conversion of oleic acid almost assures us that it was carried out by microbial action. Thus, Evershed's (1990) observations indicate that, in the peat bog environment, hydrolysis of human neutral fat resulted in freeing the fatty acids and subsequent hydroxylation of at least the most common of these fatty acids (oleic) by microbial action.

These findings need to be measured against those of spontaneously desiccated bodies in a dry environment. In later work from the same laboratory, Buckley *et al.* (1999) studied two Egyptian mummy skin samples, one of which was spontaneously mummified, and found the fat extensively hydrolyzed, with oxidation and hydroxylation of the fatty acids. Bereuter *et al.* (1996a) studied three adult males, two of which were found dead in their apartment 3½–4 years after death and the third who was found hanging from a tree three months after committing suicide. In each case the body had become mummified spontaneously via desiccation. Studies were carried out by GC but not MS. The dominant fatty acid in the hanged man was palmitic acid while in the other two oleic acid still predominated. Hydroxy fatty acid was not identified in any of the three. In addition, Gülaçar *et al.* (1990) studied a 4000-year-old Nubian mummy whose fat samples were characterized by absence of unsaturated fatty acids, abundant hydroxystearic and hydroxypalmitic acids, and some oxo fatty acids. They felt that bacteria had contributed to the monocarboxylic acid and sterol distribution, but that most of the substituted carboxylic acids were produced by a physicochemically induced oxidation or oxidative degradation process.

In a later study Evershed *et al.* (1995) extracted lipids from bone and found a low level of triglycerides, but identified detectable intact cholesterol as well as its diagenetic degraded products, suggesting its stable carbon isotope ratio might be predictive of the nature of the paleodiet. In a different study, Evershed *et al.* (1999) were able to extract individual lipid compounds from archaeological ceramic containers and, by measuring their stable carbon isotope ratios ($\delta\,^{13}C$), predict that the lipids were derived from milk fats rather than from adipose tissue. This made it possible to verify the practice of a dairying subsistence in an ancient population by chemical analysis. These studies demonstrate that the isolation and subsequent study of a chemical substance that remains intact but is encountered in a sample demonstrating extensive postmortem diagenesis can still provide accurate information on premortem practices.

On the basis of such meager evidence it appears that in mummified tissue triacylglycerol hydrolysis leads to release of free fatty acids, but that their subsequent modification varies with the postmortem environmental circumstances. Under natural conditions, however, most lipids are degraded at a slower rate than are carbohydrates and proteins (Allison & Briggs, 1991:11).

Adipocere formation

While the postmortem development of lumps of material of "whitish consistency" in the orbits of a medieval human cadaver (St Cuthbert) had been described earlier (Battiscombe, 1956), it was Antoine François de Fourcroy, in 1791, who joined the two gross features of this material by naming it adipocere (adipo, fat; cere, wax) (Moulton, 1905:87; Skinner, 1961:9). He found this material in bodies exhumed about 15 years after burial in coffins at a Paris churchyard. His primitive laboratory examination led him to believe it was composed of ammonia soap with calcium phosphate and fat, a putrefaction product of soft tissues (Evans, 1963:40–60). The nineteenth century documented endless numbers of case reports and occasional subject reviews of adipocere formation. These made it clear that the development of adipocere in a cadaver was not a rare event, and that it could be restricted to a local area or involve the entire body. It was also noted that it occurred most commonly where fat is most abundant, such as the breasts of obese women or the abdominal subcutaneous tissue of obese individuals of either sex. Furthermore, it was established that early lesions were most apt to be superficial, that it almost never was seen in children (Wetherill, 1860), that it was common in mass burials and that it occurred in humid environments. The interval between death and identification of adipocere development in these case reports ranged from six weeks to many years.

Lack of the necessary laboratory instrumentation, however, retarded understanding of adipocere's chemical nature and development. Fourcroy had determined that the average melting point of adipocere samples that he studied was 52.5 °C (cited by Evans, 1963:40–60), but he did not recognize that it was entirely a product of altered fat. By 1823, however, Chevreul (cited by Mant & Furbank, 1957) had found that adipocere was composed almost entirely of fatty acids derived from neutral fat.

By the mid twentieth century two valuable reviews of the adipocere topic emerged: Mant & Furbank (1957) and Evans (1963). These summarized previous reports and provided a database that served as a starting point for the subsequent elegant studies made possible by availability of GC and GC/MS methodology. A virtual cascade of reports from the laboratories of Takatori *et al.* (1977, 1986, 1987, 1996; Takatori & Yamaoka, 1979; Gotouda *et al.*, 1988) detail their elucidation of the chemistry of adipocere formation. They first documented that fatty acids in fresh, normal human fat were all part of triglycerides, but in adipocere from ocean-recovered bodies they were all free fatty acids. The methyl-10-hydroxyoctadecanoate they found was optically inactive and had a melting point 6 degrees Celsius lower, suggesting its formation occurred by nonenzymatic organic synthesis in this case. Subsequent studies consistently identified long-chain 10-hydroxy- and also 10-oxo-fatty acids in adipocere, whose substantially increased melting point added chemical stability to the adipocere. They also revealed that the 10-oxo-fatty acids were produced from the 10-hydroxy-fatty acids. They eventually achieved such conversion *in vitro* with enzymes extracted from sonicated preparations of the bacterium *Flavobacterium meningosepticum*. By 1986 they had demonstrated that a variety of bacteria, most of them Gram-positive, contributed enzymes that effected these hydrolytic and oxidative conversions. A report by Tomita (1984) made it clear that the sequence of events in experimental adipocere formation in mice involved initial hydrolysis of triacylgycerols that freed the fatty acids and that subsequent hydration of oleic acid (the most common fatty acid freed from neutral fat) resulted in the production of 10-hydroxystearic acid, the most common fatty acid found in adipocere. Using the *Clostridium perfringens* bacterium, he was able to convert oleic acid to 10-hydroxystearic acid *in vitro* (but not from a substrate of palmitic and stearic acids). Tomita found that these compounds diminished in number after reaching their peak. The finding by Pfeiffer *et al.* (1998) that common Gram-positive bacteria are capable of degrading adipocere may be relevant to that observation. Simpson (1965) points out that *C. perfringens* proliferates rapidly at 37 °C, but virtually ceases at 21 °C. The importance of temperatures appropriate for proliferation of such organisms in order for adipocere to develop was demonstrated well by Cotton *et al.* (1987). These workers had found adipocere formation in two bodies that were recovered after 5 years of immersion in fresh water. By reconstructing temperature records kept by a nearby electric power plant during those 5 years they demonstrated that the bodies became submerged during the only month of the year (August) when the water was warm enough to support growth of *Clostridium* sp. As early as 1926, Den Dooren de Jong had demonstrated that conversion of an unsaturated fat to a saturated state made it more resistant to microbial attack, stabilizing it chemically (Den Dooren de Jong, 1961).While Wallen *et al.* (1962) and Davis *et al.* (1969) found several Gram-negative pseudomonads capable of producing 10-hydroxystearic acid from oleic acid, this is an aerobic oganism and therefore may not play a major role in the usually anaerobic milieu of a corpse's tissues.

A more recent study of fat (skin, intracranial, intrathoracic) in a Nubian mummy from Kerma in modern Sudan, however, suggested a more complex postmortem process. Bacterial contributions to the postmortem fat alterations were obvious in monocarboxylic fatty acid and sterol distributions. Most of the substituted carboxylic acids were produced by physicochemical-induced oxidation or an oxidative degradation process (Gülaçar *et al.*, 1990).

Another sophisticated study by Evershed (1992) on adipocere from an Irish bog body (Meenybradden) dating to about AD 1570 found it to be composed entirely of fatty acids without triacylgycerols, and in which 10-hydroxystearic acid is predominant. A very low content of unsaturated acids remained, suggesting that a microbial reduction step in conversion of oleic to palmitic acid occurred. Evershed also notes that demonstration of residual intact triacylgycerols in some preserved fat from peat bogs ("bog butter") implies that hydrolysis in peat bogs may sometimes be quite slow, implying a nonenzymatic, nonbacterial mechanism.

In a series of studies by Bereuter *et al.* (1996a) and Mayer *et al.* (1997) on bodies retrieved from glaciers these workers found greater variation in fat alterations, including adipocere composition. Thus it is evident that, while fatty acids predominated in some glacially produced adipocere, substantial oleic acid remained intact; in others stearic and palmitic acids as well as 10-hydroxystearic acid predominated; calcium soaps were also common. In fat from nonadipocere sites, some intact triacylglycerols were demonstrable. In fat from the Tyrolean Iceman (>5000 years old), 5% of the fatty acids were unsaturated, 16% of these consisting of odd carbon-number acids. They also suggested that, at mountain glacier altitudes, ultraviolet radiation may catalyze conversion from saturated fatty acids to hydroxyl fatty acids. In that event the hydroxyl groups could be almost anywhere on the carbon chain. While the mechanisms involving adipocere formation occurring under the usual circumstances have been clarified a good deal during the past several decades, it is obvious that we still have much to learn about its formation under unusual environments such as glaciers.

Summary of the taphonomy of human lipids

Human lipid breakdown initiates with death. The steps involved are occasionally carried out by nonenzymatic chemical mechanisms, in which event they progress more slowly. In most cases, however, conditions support intestinal bacterial proliferation. In these circumstances the breakdown of the larger molecules is enhanced by bacterial enzymes, enormously accelerating the process. The very first phase involves cellular death accompanied by rupture of the cytoplasmic lysosomes. The exposed phospholipids of the membranes suffer cleavage of the phosphate groups, disrupting these lipids until they are no longer detectable. Neutral fat may also suffer effects of endogenous lipase, but the alkaline milieu preferred by those lipases is short lived, if it exists at all. Neutral fat undergoes hydrolysis, fracturing the triacylglycerols into glycerol and free fatty acids. The majority of these are unsaturated, straight-chain C_{16} and C_{18} fatty acids, oleic acid being the most common.

If conditions for adipocere formation are appropriate, subsequent hydroxylation of oleic acid generates 10-hydroxystearic acid. This process may continue until oleic acid is no longer identifiable. This product can form the substrate for oxidation to oxo-fatty acids. Both 10-hydroxystearic and oxo-fatty acids have higher melting points than the original oleic acid from which they are derived. Hence they are solid at room temperature, producing the physically and chemically stable substance with fatty and waxy features we call adipocere. The bacteria contributing enzymes that drive this process to completion, sometimes within one month, are principally Gram-positive organisms that commonly form part of the normal intestinal flora. These processes are enhanced by an abundance of fat, by warm temperatures conducive to bacterial proliferation, by humid atmosphere (though normal tissue water content is sufficient for such reactions), by clothing and by limited aeration. While local adipocere development often occurs, it may involve much of the fat throughout the body. Organs enveloped and protected by adipocere are commonly preserved in a dehydrated state. The lowered pH generated by free fatty acids may ultimately act to constrain further change by inhibiting bacterial degradation of adipocere, though some further breakdown of longer chain fatty acids may occur before stability is achieved. The lesser studied sterols may survive better in intact form. Their extraction, isolation and study may provide valuable data for isotope analysis. These processes are all potentially operational after death. However, which are activated and at what rate depend heavily on the circumstances of the cadaver's environment. Hence, variations among these changes can be expected in bodies located in varying milieus, such as glaciers, bogs or cemetery burials.

Mummification by adipocere formation (Fig. 5.4)

Undoubtedly the two most well-known bodies preserved by adipocere formation are the ones commonly called the "Soap Man" and "Soap Lady". In 1875 they were exhumed from a Philadelphia cemetery during road construction. Eventually the Soap Lady became conserved in the Mütter Museum in that city while the Soap Man found a permanent home at the National Museum of Natural History, a unit of the Smithsonian Institution in Washington, DC. Although museum records indicate that these individuals died in 1792, recent radiographs demonstrate artifacts establishing that interment could not have been earlier than 1824, and the family name and alleged death by yellow fever also appears to be false information (Conlogue *et al.*, 1989 (female), 1997 (male)).

The taphonomy of carbohydrates

The basic unit of carbohydrates is a monosaccharide – a simple sugar that can not be hydrolyzed into smaller units, though it may be metabolized to carbon dioxide, water and, in amino sugars, to ammonia. Such monosaccharides can be connected to each other to form linear chain polysaccharides. One of the most abundant polysaccharides in the human body is glycogen, which is composed of a succession of linked glucose residues with similar side branches. It serves as a storage product that is mobilized for energy needs. Another polysaccharide of importance to humans is starch, whose long chains are arranged in a helical manner. These chains may be linear (amylose) or branched (amylopectin). Cellulose of plant origin is also composed of very long linear chains of glucose that provide stability by being bound to adjacent, parallel chains (Logan *et al.*, 1991). Glycogen, the starches and cellulose are homopolymers of glucose differing in linkage types and numbers.

Carbohydrates of other than dietary function in the human body are linked to a polypeptide or protein to

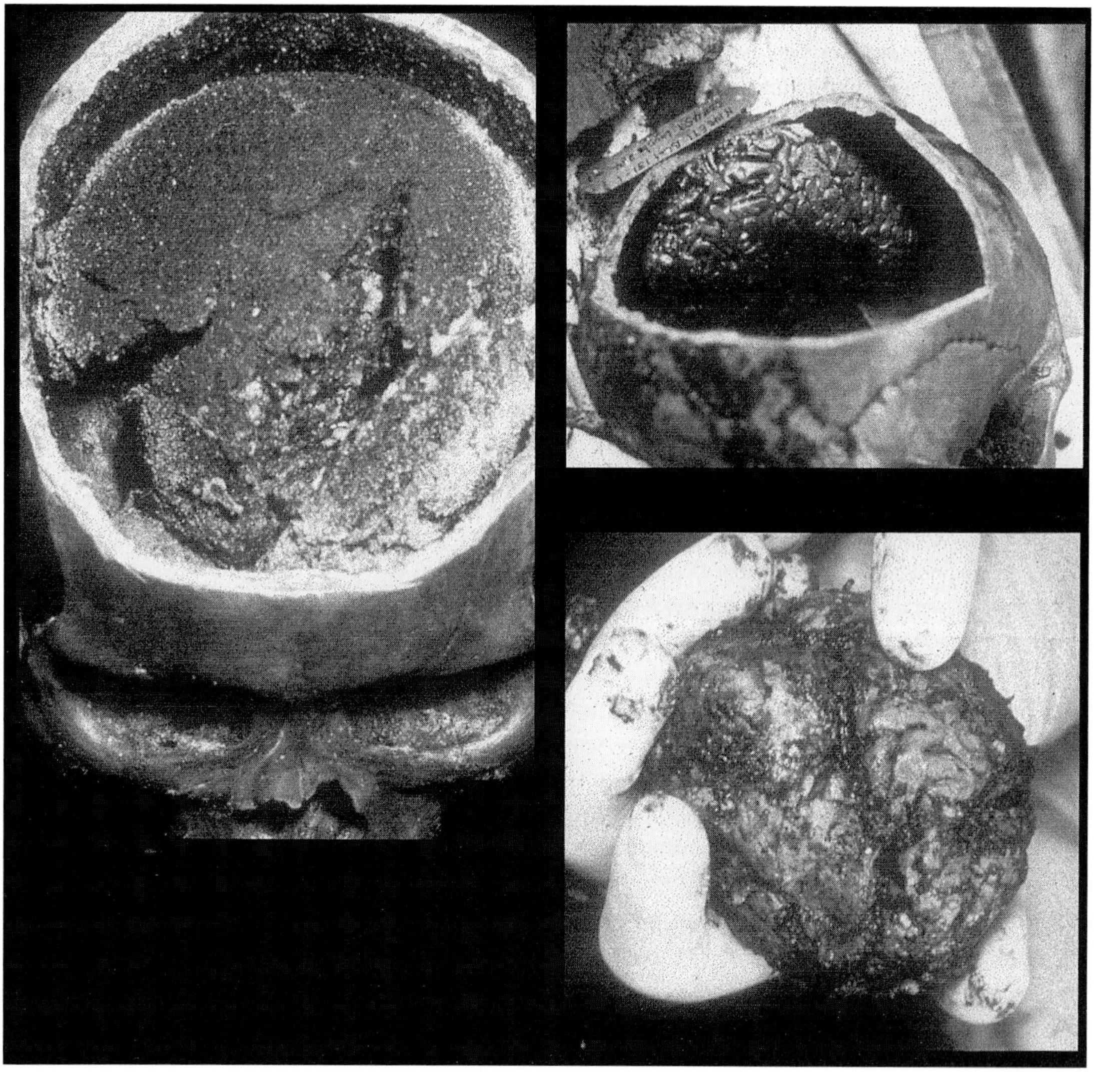

Fig. 5.4. Adipocere formation; SM, ca. 6000 BC.
Several Paleoindian skulls recovered from a southern Florida peat bog contained preserved brain tissue, probably at least in part due to adipocere formation. (Photo courtesy of Glen Doran, Dept. Anthropology, Florida State University (Doran *et al.*, 1986).)

form glycoprotein and to lipids to form glycolipids. These function in a broad array of sites including epithelial and red blood cell membranes (ABO blood group antigens on lipids and proteins), hormone structure and others.

Like proteins, polysaccharides are degraded by hydrolysis. In the living, this is carried out enzymatically. After death, some of these enzymes can function in the autodigestive process initiated by lysosomal enzymes, augmented subsequently by those from bacteria or insects. Nonenzymatic hydrolysis also occurs. Polysaccharide susceptibility to hydrolysis, together with their high water solubility, contribute generously to their common postmortem loss of chemical integrity. The sugars released by hydrolysis are readily metabolized for energy by microorganisms. Laves

(1950) as well as Logan *et al.* (1991) note that the final metabolic products are CO_2, H_2O and NH_3.

The literature records relatively few efforts to identify carbohydrate molecules in ancient human remains. The red blood cell antigens of the ABO blood group are an exception, and methods to identify these have been reported since the 1930s (Boyd & Boyd, 1933; Candela, 1939). These early tests were applied to bone, but by 1957 it was becoming apparent these were not always reliable (Thieme & Otten, 1957). Later studies attempted to improve results using agglutination inhibition or mixed cell agglutination techniques. Bioanthropological interest in such methods rested primarily on the expectation of provision of information regarding biological distance between studied populations. However, before completely satisfactory methodology to identify these antigens directly could be established, the more promising technique that utilized mitochondrial DNA evolved. Subsequently further interest in ancient ABO blood groups subsided.

While few other efforts to identify carbohydrates in mummified human remains have been documented, intact saccharides have been found in very ancient sediments (Moers, 1989), cited in Logan *et al.* (1991), and a review by Vallentyne (1963) notes that pectins have been identified in fossil algae and glucosamine in an Eocene insect fossil. These basic findings provide optimism for a search for other carbohydrates in mummified human remains.

The taphonomy of bone mineral

The focus of this book is primarily soft tissue mummification. Nevertheless, alterations in the skeleton's protein matrix will also affect bone mineral preservation. We have noted above that bone mineral provides protection from destructive postmortem agents, be they enzymatic or nonenzymatic. Protein matrix degradation, however, exposes the mineral to postmortem agents as well; these are principally environmental. Because bone mineral houses dietary information, a brief review of such changes seems appropriate.

In the earlier stages of bone mineral analysis for chemical dietary reconstruction Parker & Toots (1970) note that the strontium in dietary vegetal was reflected in the strontium of bone mineral and was useful to differentiate the vegetal fraction of the diet from the strontium-free meat portion. They felt that postmortem alterations of bone mineral (diagenesis) was not a problem. Numerous reports during the subsequent decade, however, made it clear that, at least under certain circumstances, diagenesis of a degree sufficient to affect diet interpretations did occur (Aufderheide, 1989b). Research focus then targeted two goals: (1) "markers" that could identify diagenetic change in a specimen, and (2) manipulation of the specimen to recover the biological signal from an altered specimen. The development of inductively coupled plasma methodology permitted measurement of multiple trace elements simultaneously. A host of such studies provided a wealth of data on bone mineral samples from which it became evident that certain elements including aluminum, iron, manganese and others were common in soil but were infrequently found in living bone. These could be employed as diagenetic markers, though we still lack sufficient quantitative data on each of them to predict the loss or distortion of the biological strontium content; thus dietary interpretation is thwarted.

Application of physical and physicochemical methods have led to an understanding of the action of environmental taphonomic agents on the hydroxyapatite crystal of bone mineral. In acid soils, such "apatite" crystals may dissolve and the ions be leached by groundwater, resulting in a loss of strontium from the bone mineral. Alternatively the space in archaeological bone (created by degeneration of the cells and vascular structures) can be filled by deposition of groundwater ions in the form of nonbiological, purely chemical apatite crystals; these may or may not contain strontium ions. Finally, both may occur – the initial dissolution of biologically formed apatite crystals (that contain the dietary signal), followed by their mixture with groundwater ions and resulting recrystallization of apatite crystals composed of a mixture of these two strontium ion sources. To date no method has been found capable of recovering the biogenic dietary signal from this latter, mixed mechanism (Tuross *et al.*, 1989). However, Sillen (1989) has noted that nonbiologically formed apatite crystals are "more crystalline" (i.e., have fewer substituted carbon ions and fewer structural defects); this results in solubilities that differ at certain pH values. Exploiting this difference he has provided data that suggest successful recovery of the original biogenic signal in fossilized samples that demonstrate considerable diagenetic change but in which the original apatite crystals were not lost, and in which the nonbiogenic crystals had merely been added. In the hands of others this method has been equally successful in some but not all specimens, perhaps because in some samples one of the alternative mechanisms of diagenesis mentioned above may be the predominant form of chemical alteration.

In sites from hyperarid deserts devoid of groundwater (Figs. 4.54 and 4.57), chemical dietary reconstruction by trace element analysis still appears to be effective if coupled with strontium isotope studies to identify the fraction of strontium derived from vegetal resources (Aufderheide & Allison, 1995a), but at most sites the problem of diagenesis remains the major constraint in chemical dietary reconstruction studies by trace element methodology.

The interested reader may supplement this relatively superficial review of the chemistry of bone mineral taphonomy with the following references: Price, 1989; Sanford, 1993; Lyman, 1994.

Soft tissue taphonomic agents: water

That water can act as an agent of taphonomic change is self-evident. In this discussion we will not discuss the role of water that is so important for geologists and paleontologists – i.e., water as an agent of transportation. Instead, we will focus particularly on the role of water in providing the medium or matrix within which other biological agents (enzymes, microbes, chemical changes) express their influences, with special emphasis on those features leading to spontaneous desiccated mummification.

After death the sudden drop in tissue oxygen tension initiates cellular autolysis. Organelle membranes rupture, including those of the lysosomes. The lysosomal hydrolytic enzyme content spills into the cytoplasm, breaking down protein and carbohydrate molecules until the cell structure, including its outer membrane, is destroyed. In the meantime that same oxygen tension decrease has acted to suppress the colon's aerobic bacterial population, providing the gut's minor population of anaerobes a milieu in which to proliferate and become the dominant bacterial form. Invading the visceral veins, these bacteria follow the blood flow route to the liver, vena cavas, right heart, lungs, left heart and then the remainder of the body via the route of the systemic arterial circulation (Aturaliya *et al.*, 1995).

Arrival of these bacteria at any given point in the body initiates the second phase of decay orchestrated by bacterial enzymes. In a body protected from insects and scavengers, these bacteria carry out the bulk of soft tissue destruction. Focal areas of bacteria initially tend to accumulate in the most dependent areas, carried there by gravity-directed blood seepage. There the anaerobic bacteria ferment the tissues in such localized areas, the gases formed accumulating as swollen, bloated regions. Early foci are almost exclusively within skeletal muscles, not within the peritoneal cavity. The latter is involved only after some of the initial foci dissect into the abdominal cavity. Pressure resulting from the bloated areas commonly compress the gut, forcing dark fluid, stained with hemoglobin and myoglobin breakdown products, out of the mouth, nose and anus, a process termed "purging" (Aturaliya & Lukasewycz, 1999).

If insects have access to the body the third stage of decay begins. Flies will lay eggs in the moist areas (eyes, nose, mouth – and genital area if exposed) whose several waves of larvae emerge over the period of several days. These larvae are voracious and, after several cycles, will be present in such large numbers that they will be the principal agents of soft tissue destruction and consumption. As these numbers of larvae increase they attract a variety of beetles most of whom, with a few exceptions such as the carrion beetle, feed on the larvae, not the decaying body (Haskell *et al.*, 1997). These processes can continue until skeletonization is achieved, although commonly large animal scavengers will disarticulate body parts and may carry them to a new environment where the sequence may be altered (Haglund, 1997).

The rate of these processes can be influenced profoundly by several variables including temperature, pH, protective clothing, soil cover and others. It will be apparent, though, that virtually all of these steps require the presence of an aqueous medium. While these activities are proceeding, water is being lost – some by evaporation, some by tissue liquefaction, some by seepage. In the usual environments, the decay processes are so rapid that the soft tissue disappears before tissue dehydration becomes influential. However, we know that some bodies reach desiccated mummification spontaneously. Death, therefore, initiates a race between those factors that accelerate and those that retard the decay processes. Since water is central to the action of nearly all decay mechanisms, much of what follows is directed at the postmortem factors influencing the decaying body's soft tissue water content.

Initially the above-summarized sequence of events was observed and reported by Aturaliya *et al.* (1995) in our Paleobiology Laboratory at the University of Minnesota, Duluth. The next step was to study factors affecting the rate of tissue water loss. The findings were reported by Aturaliya & Lukasewycz (1999) and will be summarized here. The study involved rodent carcasses under controlled conditions. Initially the fraction of visceral tissues' weight that was represented by unbound water was determined by simple desiccation methods. In living tissues about 78% of the weight was found to be removable water; in samples of these same tissue types removed from spontaneously mummified Andean mummies, residual removable water was less than 1% of its calculated original weight at the time of death. Exposing the "clothed" and nude bodies in a temperature- and humidity-controlled chamber revealed that the "clothed" body lost 25% more weight than did the nude body over a period of 69 days. Burying both "clothed" and nude bodies in sand again resulted in the "clothed" body losing more weight, though the difference was less (11%). Isolated fresh soft tissue (liver) blocks' weight losses when buried in sand were compared with those exposed to a free flow of air. The air-exposed tissue reached a dry state (about 72% weight loss) in 24 hours while the sand-exposed had lost only 60% at the end of 7 days. These results indicated that clothing enhanced water loss from the body ("wicking" action?) in both air-exposed and buried bodies, and that such water loss rate was far faster in air-exposed than in buried bodies.

These findings are consistent with one feature: accelerated removal of water emerging from the skin surface speeds up the rate of water removal from the whole body. All water beneath the skin (internal water) can leave the body only by penetrating the skin and presenting at the skin surface. The rate of water transfer from the saturated internal structures to the skin surface can be expected to be related to the difference in water concentration ("water gradient"). This gradient can be maximized by keeping the skin surface as dry as possible at all times. Thus the conductance of water away from the skin surface as effectively as possible will maintain a high gradient and result in an accelerated rate of body dehydration. By presenting a large surface area, porous clothing can accelerate such water conductance away from the skin surface. In air-exposed situations an increased flow of air results in even more rapid evaporation, contributing further to more rapid removal of water from the skin surface.

Our working hypothesis, therefore, can be that maintenance of a high water concentration gradient (by maintaining continuous, effective water removal from the skin surface) will maximize total body dehydration rate. Now let us review observations on spontaneously mummified bodies from the Andean area where I have dissected many such mummies. These mummies often present sharply localized areas of skin loss and maceration, commonly located on the back or palms of hands and on the chin or lower face. These are precisely the points of contact in flexed bodies where the knees are drawn up to the chin, the hands placed on the knees (palm down) with the forward-flexed head's chin resting on the back of the hands. This position is maintained with ropes, and the whole body is then wrapped. In such a situation where skin areas are in contact with each other, water emerging from the skin surface can not be removed rapidly. This keeps the skin surface wet, delaying the rate of internal water removal at such sites and providing additional time for the enzyme-laden water to digest the tissue. Because Andean burial body orientation is commonly in the sitting position, precisely the same explanation can be employed to interpret the reason for the commonly macerated pudendal body areas. These effects are exaggerated by the fact that the water at the skin surface is rich in enzymes from the decaying internal tissue. Another marked effect is the amount of tissue internal to any given skin area. For example it will surely require a much longer time to transfer the water in heart tissue to the skin surface than it will to remove the water from the muscle of a finger. The best-preserved soft tissues in spontaneously mummified bodies are commonly the fingers, toes, ear lobes and other skin-covered areas with little underlying soft tissue. Skin in such areas often reveals preservation of even all epidermal layers, while the full thickness of epidermis over the anterior abdominal wall or that of the thigh has usually been lost. Overall differences in soft tissue preservation among such bodies may be due to general differences affecting the whole body, such as temperature (season of death), delay in initiation of mortuary rites, etc.

Many groups practicing anthropogenic mummification employ methods to enhance water removal from the skin surface. Some of these simply suspend the body in such a manner that the water can drain away from the body by gravity as described by Fornaciari (1985) among bodies from the elite of Naples as recently as the nineteenth century. In Colombia the body was dried in a hammock above a small fire, the exuding body fluids being allowed to drip into clay vases (Cárdenas-Arroyo, 1995; see also Chapter 4: Mummies from Colombia). The bodies of the catacomb mummies of the Capuchin Friary at Palermo were suspended in a supine position to permit accumulated fluids to drip into a limestone gravel-filled trough below them (Matranga, 1983). Other groups make multiple skin incisions in the dependent areas and even compress the skin around the incisions to squeeze out the fluid. The many mortuary practices involving anthropogenic mummification include frequent wiping, washing or "anointing" of the body during the process that would involve removal of fluid from the skin surface. All these variations have in common the goal of leading water away from the skin surface by one method or another.

Soft tissue taphonomic agents: heavy metal ions

Most biochemical agents are highly adapted, i.e., their molecular structure is designed to function only within a specific chemical reaction occurring in a defined environment (optimum matrix, temperature, pH, etc.). After death many of these environmental factors change: the body temperature usually approaches ambient levels, tissue acidity increases and decay mechanisms break down macromolecules. Additionally, in buried bodies groundwater may gain access to the body, exposing it to exogenous agents, some of which have chemical effects.

Tissue enzymes are exquisitely sensitive to heavy metal ions often present in groundwater. These include ions of mercury (Hg), lead (Pb), arsenic (As) and copper (Cu). All of these can "poison" (disable) biochemical enzyme action. Indeed, the toxicity of these substances when introduced into a living body is well known, and some (arsenic, lead) have a record of homicidal applications (Jolliffe, 1993).

Arsenic (As)

Paralysis of the enzymatic decay processes by heavy metal ions can result in excellent postmortem tissue preservation. The use of embalming solutions incorporating high concentrations of arsenic was so effective that its employment in the nineteenth century had to be terminated by legislative action because of its perceived hazard to the living resulting from environmental contamination (though the loss of evidence such embalming caused in cases of suspected homicidal arsenic poisoning also played a role in stimulating such legislation – see Chapter 3: History of modern embalming practices). The excellent state of both microscopic and gross soft tissue preservation by arsenic in the case of the body of Elmer McCurdy described in Chapter 4 is such an example (Snow & Reyman, 1977) (Figs. 4.8–4.10). The high concentrations of arsenic in the soil of Greenland as well as the arsenic probably ingested premortem may have contributed to soft tissue preservation of the late ninteenth century arctic explorer Charles Francis Hall (described in Chapter 4: Mummies from Denmark) who was buried there in 1871 (Paddock *et al.*, 1970). However, whether an entire human body could be preserved by passive postmortem perfusion of arsenic-containing groundwater without intra-arterial injection of an arsenic solution has not been established.

Mercury (Hg)

Mercury is a very effective enzyme inhibitor. It is broadly disseminated by the atmosphere, generating a regional environmental hazard when used indiscriminately, such as in gold mining as was done during South America's Spanish colonial period (Martínez-Cortizas *et al.*, 1999) and still practiced in the Amazon today (Nriagu *et al.*, 1992). Mercury levels in ancient sea mammal tissue were found to be lower than in modern similar animals and humans, but were still at undesirably elevated concentrations even in antiquity (Smith *et al.*, 1978). In the sixteenth century, mercury was used as a therapeutic agent for syphilis. Unfortunately, Hg ions were retained on the tooth surfaces, subsequently combining with sulfur and resulting in black staining of tooth enamel (d'Errico *et al.*, 1988). Parts of the thoracic viscera, lying in a pool of about 1000 grams of mercury within the chest remained preserved in a 320-year-old Japanese burial even though the remainder of the body, not included in the mercury pool, was completely skeletonized (Yamada *et al.*, 1990). It is probable that the mildly acidic pH resulting from lysosomal enzyme action ionized part of the metallic mercury that, in turn, impaired enzymatic decay. The superbly preserved condition of the soft tissues found in the 2100-year-old body of a Chinese woman (Wei, 1973; Hunan Medical University, 1980) has been attributed in part to the mercury salts found in the water present within the sarcophagus (Peng, 1995). A mercury paste applied to the body of a late medieval child of three months of age in Naples, Italy, probably prevented the soft tissue loss that is so commonly seen in this age group (Fornaciari, 1985) (Fig. 3.11).

Copper (Cu)

While copper is an ion essential for the functioning of some mammalian enzymes, it is clear that a significant surplus of copper ions will impair the action of many enzymes. Estes (1993) demonstrated the antibacterial effect of copper in the laboratory (Fig. 5.5). It has been the experience of many veteran archaeologists that an occasional body is found completely skeletonized except for a local area of soft tissue immediately beneath a copper bracelet or other copper artifact. This can be highly convincing evidence of the soft tissue preserving effect of copper ions. A similar finding resulted from copper necklaces, pendant and bracelet on the 200-year-old body of an infant Native American from British Columbia, Canada (Schulting, 1995), in whom the green-stained copper compounds were evident thoughout the soft tissue and completely penetrated the bone, as confirmed by X-ray fluorescence analysis. However, probably no other body demonstrates this more dramatically than that of "the Copper Man" currently curated at New York's American Museum of Natural History (Fig. 3.8). Copper Man's green-stained body was originally found in a collapsed ancient mine shaft in northern Chile's Atacama Desert, together with his axe and other mining tools where he had apparently been gathering copper ore when his hammer blows led to the collapse of the cave's ceiling. The area is a huge, commercial, active copper mine (Chuquicamata) today. The soft tissues of this body are extremely well preserved. Chemical analysis, however, reveals that, while the surface of the rock ledge around the body has a copper content high enough to keep it sterile, the copper ions had not penetrated the body's soft tissues beyond the depth of the skin. While surely the excellent skin preservation was at least partly due to the copper ions, that of the deeper tissues and viscera could not have been a copper ion effect. Nevertheless, this area of the Atacama Desert is hyperarid, and bodies buried in its soil normally desiccate so rapidly that spontaneous preservation of soft tissues in such bodies quite commonly rivals that found in Copper Man. Hence, while copper probably contributed to the superb skin preservation, persistence of the remainder of his body's soft tissues is most logically attributable to the Atacama Desert's climate. Moreover, several bodies with retention of considerable soft tissue found in a Tennessee "copperas" cave are often cited as evidence of copper ions' ability to preserve soft tissue. However, copperas is an older term for ferrous sulfate ore, and the soft tissue retention of those bodies is more probably due to the arid cave air maintained by the cave's gypsum (calcium sulfate) and alum (potassium aluminum sulfate) content (Miller, 1812).

Lead (Pb)

Lead, too, is often cited as an enzyme inhibitor, and in the living individual absorption of toxic quantities of lead is known to impair the enzymatic process of hemoglobin synthesis. Little evidence exists for the deliberate use of lead as a soft tissue conservation agent except for its employment as a burial container, either lining a coffin or for use in the creation of the entire coffin. Here we must be careful in assigning motive to the choice of lead as material for coffin construction. Certainly of all metals, lead requires the least elaborate effort and skill for melting and casting the container's shape. If the original cast was not sufficiently precise to meet waterproofing desires, the lid could easily be lead-soldered to achieve that goal. Smaller containers, such as those designed to contain human tissue relics or smaller body parts (Pizarro's head: see Chapter 4: Mummies from Peru) could be kept dry even more easily by lining them with a thin layer of lead. Thus any lead container for human tissue may well have been chosen principally because of the above-mentioned features, rather than to exploit lead's enzyme-inhibiting action. For example, the exhumed lead coffin containing John Paul Jones' body (that was preserved sufficiently well to be recognizable 113 years after his death: Fig. 4.92) contained sufficient alcohol to be easily recognizable by its strong and distinctive odor. It probably escaped from the coffin very slowly through two, small, identified cracks. Clearly, those burying this body did not expect the lead to prevent the body's decay.

An adult body placed in a lead coffin and buried in a late Roman cemetery (Poundbury) in present-day England was exhumed in 1971. The body was found to be completely skeletonized except for a mass of hair. No lead was found in the hair and only a trace of lead could be found in scrapings from the skull. As might be expected, if a lead container is sealed sufficiently well to exclude groundwater totally, then lead ions from the container's lining would have little opportunity to saturate the body's tissues sufficiently to preserve them.

Heavy metals for protease suppression

Unpublished work in my laboratory in 1992 dealt with efforts to design a better method of rehydrating desiccated tissue. Fresh animal tissues were placed for 48 hours in solutions containing up to 10% concentration of As, Cu and Pb ions in an effort to inhibit any protease in the tissues. The histological appearance of the tissues so treated differed very little from those placed in a solution of 0.9% saline as a control. Higher concentrations might have been successful but were not attempted because, even if effective, a method requiring such high concentrations of toxic compounds would be undesirable for routine use (Lukasewycz, 1992). These observations suggest that little soft tissue preservation should be expected when only small amounts of these heavy metal ions are available to the tissues. However, copper artifacts in direct contact with soft tissues may have a local conservation effect.

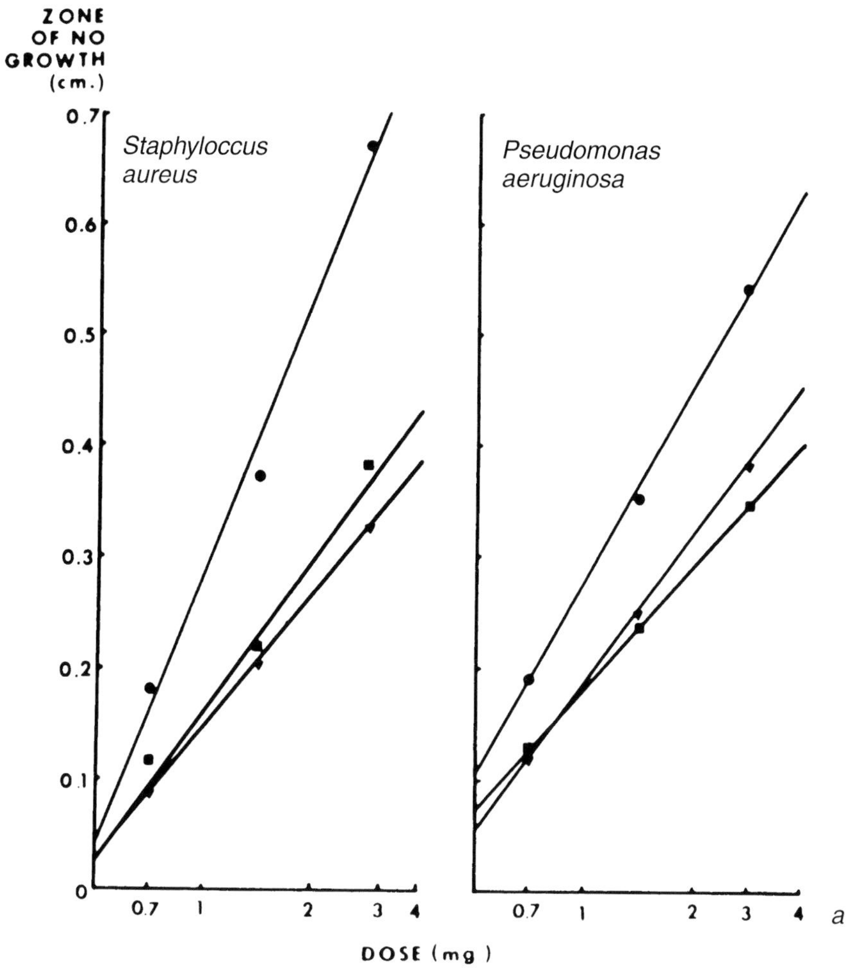

Fig. 5.5. Bacteriostatic effect of heavy metal ions.
The antibacterial action of copper ions is evident. Cupric carbonate (circles), sulfate (squares) and acetate were tested against the two bacteria indicated. Growth suppression was correlated with the concentration of copper ions. (From Estes, 1993: 67 (revised edition) by courtesy of publishers Science History Publications.)

Summary of antimicrobial effect of heavy metals

If the apparent tissue-preserving effects of these ions really is operating by impairing bacterial enzymes, then they might also impair enzymes vital to the bacteria's own vital metabolic sequences. Majno (1975:113) tested the antibacterial effects of several copper ores (malachite, chrysocolla and verdigris) and found them to have easily demonstrable bacterial growth-suppressing effects. Estes (1993:66–68) repeated this work using the soluble chemical salts that such ores represent in crystalline form (cupric carbonate, cupric sulfate and cupric acetate, respectively), confirming their antibacterial action and documenting that the degree of their effectiveness was proportional to the number of copper ions in the ores (Fig. 5.5).

Soft tissue taphonomy agents: bacteria

Bacteria are uniformly present in soil samples. Various species of the genus *Clostridia* have been found in soil samples whenever they have been sought (Hatheway, 1990). They are also present almost as commonly within the normal human colon, and especially within the appendix (Barza, 1993). Thus virtually every human body is exposed to bacteria after death, almost all of which include anaerobic bacteria of the genre *Clostridia* or *Bacillus*. In soil presenting suboptimal conditions these species commonly survive by spore formation. In the human colon their uninhibited growth is suppressed during life by microbial competition from aerobic, Gram-negative bacteria and by substantial levels of tissue oxygen concentrations maintained by the tissues' circulating blood. While Gram-negative bacteria can certainly proliferate during at least the initial part of the postmortem period, the rapid decrease in tissue oxygen levels that develops when blood circulation ceases favors proliferation of the anaerobic, Gram-positive members of the colon's bacterial fauna, especially *Clostridia* and *Bacillus*. The Gram-negative colon bacteria do survive a long time, but the oxygen tension and pH of their postmortem environment progressively depart from values optimal for these organisms and approach the optimum for the anaerobic, Gram-positive rods. Eventually the latter become the predominant organisms.

After death, bacteria can penetrate the full thickness of the intestinal wall, appearing on the serosa (the external intestinal wall surface) in an interval of no more than 2–15 hours, depending on the temperature and the nature of the bacterium (Kellerman *et al.*, 1976; Melvin *et al.*, 1984). Their entrance into the small veins of the bowel wall and mesentery occurs after a similar interval. Initially the aerobic bacteria that produce powerful proteases appear on the intestinal surface but after about 48 hours the *Clostridia* and *Bacillus* predominate. Thus the usual aerobic flora of the living colon is suppressed and largely replaced by the anaerobic postmortem flora two to three days after death.

The postmortem spread of bacteria through the body

An experimental study by Aturaliya *et al.* (1995) has defined the spread of bowel organisms throughout the body. At room temperature bowel bacteria invaded the mesenteric veins, reaching the liver and spleen (via the splenic vein that joins the mesenteric vein) and the portal vein by 48 hours after death, the heart and lungs by 72 hours and the kidney (and probably all other body organs) by 96 hours. This spread of organisms follows the venous channels from the colon to the liver (and retrograde via the splenic vein to the spleen). Twenty-four hours after the bacteria reach the liver, they arrive via the inferior vena cava in the heart and promptly progress to the lungs. An additional 24 hours are required to traverse the lungs, spreading through the left atrium, left ventricle and aorta to reach the kidneys and probably other peripheral tissues.

In the above-described sequence, bacteria were detected by the very sensitive technique of culture on agar. In some of these, only a single bacterium or small numbers were present. In a study of infected burns of living humans, it became evident that microscopic evidence of tissue destruction could be attributed to enzymes contributed by bacteria only if these organisms were present in numbers of 10^5 per gram of tissue or more (Volenec *et al.*, 1979). If we can translate these observations to postmortem tissue changes, the study by Aturaliya *et al.* (1995) indicates that about two weeks are required to effect such changes, since tissue sections of the viscera did not reveal bacteria during the first two weeks after death at room temperature. Observation of the decay of unburied human corpses at about 25 °C by Rodríguez & Bass (1983), however, indicated that bloating was evident after 4–10 days. Bloating also occurred in the rats studied by Aturaliya *et al.* at about similar intervals. Since such bloating is usually the consequence of anaerobic, Gram-positive bacterial putrefactive action, this is probably closer to the time interval between death and widespread bacterial dissemination, with sufficient subsequent proliferation to produce significant tissue destruction in cadavers exposed to these circumstances. In addition, these rates and time intervals proved to be profoundly temperature sensitive. Elevation of ambient temperature to 43 °C generated a similar but greatly accelerated sequence of events while exposure at 4 °C virtually arrested bacterial proliferation.

All of these described changes, however, were noted in unburied animals. Rodríguez & Bass (1985) point out that in buried human bodies these changes are retarded enormously, the extent of deceleration being directly proportional to the depth of burial. They also note that the buried body's temperature exceeds that of the surrounding soil by 3 to 10 degrees Celsius, beginning weeks after burial. Whether or not bacterial proliferation was the cause of this phenomenon was not clear to them. However, whatever the cause, this degree of temperature elevation is sufficient to enhance bacterial growth above the rate expected from the soil temperature.

Bacterial effect on postmortem tissue structure

Because of initial postmortem delay of bacterial dissemination, histological changes in many tissues during the first few days at 25 °C can be attributed to the effect of endogenous lysosomal enzymes. All tissues revealed detachment of endothelial and epithelial cells from their respective basement membranes. The degradation and fragmentation of such cells prior to the arrival of bacteria seem to parallel those cells' degree of metabolic activity during life. In the kidney, for example, proximal convoluted tubule epithelial cells, but not glomerular structure, demonstrated extensive fragmentation during the initial 72 hours after death before bacteria arrived. Peripheral skeletal muscle revealed only minimal histological change during that interval. Thereafter anatomical disruption correlated very positively with bacterial proliferation in the tissues.

Spleen and kidney proved to be especially susceptible to bacterial destruction. At 43 °C these organs were no longer identifiable four days after death. The initial gross evidence of bacterial tissue destruction was focal evidence of gas accumulation (bloating). Contrary to common assumption, these areas of bloating involved the peritoneal cavity only late after the anterior abdominal wall had become at least partially digested. It was clear that these areas originated in peripheral skeletal musculature. Their distribution was gravity dependent, perhaps because blood tended to drain preferentially into such locations. As they grew in size, some fused with adjacent involved areas and superficially some of these provided an external appearance of intra-abdominal distention, but dissection did not confirm intraperitoneal involvement until abdominal wall digestion resulted in complete defects.

Clostridia are capable of producing a variety of toxins. One of these is targeted at skeletal muscle, producing myonecrosis in living individuals. Others include a hyaluronidase that would be expected to enhance bacterial spread through tissue, a deoxyribonuclease (DNase), a hemolysin capable of lysing red blood cells and lecithinase, which can destroy any membrane structure (Barza, 1993). While Gram-positive bacterial rods predominate in such bloated foci, Gram-negative organisms, though usually in the minority, are also often present and may contribute other unique enzymes. Collectively, bacterial enzymes appear to be able to account for the rapid tissue destruction in such bloated lesions and in viscera.

Summary of bacteria as taphonomic agents

From these observations we can conclude that bacteria are responsible for the bulk of body tissue degradation in most postmortem circumstances involving bodies protected from insect access. In exposed bodies at room temperature this begins a few days after death. The rate of bacterial proliferation is profoundly temperature sensitive. Gram-positive, anaerobic bacteria (principally *Clostridia* sp. and *Bacillus* sp.) normally present in the human colon produce the bulk of the destruction with the production of their powerful proteases, though other organisms commonly make lesser contributions. The rate at which the bacterial destruction occurs varies significantly with environmental circumstances, temperature constituting the major variable.

Soft tissue taphonomic agents: fungi

Considerable information has accumulated that indicate some fungi can produce invasive disease in living humans. Examples include *Coccidioides* and *Histoplasma* (lung infections), *Sporotrichum* (subdermal lymphatic infections), *Trichophyton* (superficial skin infections) and many others. Given access to appropriate tissues by inhalation, skin pricks and other methods, these organisms are capable of producing progressive infections, some even fatal. With the recent increase in impairment of the human immune system of iatrogenic origin such as cancer chemotherapy programs or acquired viral infection (acquired immune deficiency syndrome, AIDS), fungi that normally are not pathogenic for immunocompetent adult humans (*Mucor, Candida,* etc.) can also produce human infections, and we find that our available database about these fungi is far less comprehensive. If we focus on the fungi that grow on decaying bodies, we are dealing with a group of microbes that have enjoyed lesser interest from mycologists, unless the fungi are involved with commercial or household activities such as preparing baking goods.

Fungi are ubiquitous in soil and their spores are often distributed very effectively by air currents. Cemetery locations devoid of fungi are rare indeed. This often frustrates our ability to determine whether a fungus identified in or on a mummy is of ancient or recent origin. The paleopathology literature is littered with observations in which such findings are variously interpreted, often unaccompanied by evidence supporting the rationale. Few will deny that fungi can be destructive taphonomic agents, but most will have difficulty describing criteria to define when an identified fungus is acting principally as an incidental passenger and when it is clearly a significant contributor to decay. Museum curators will need to continue to use arbitrary preservation approaches and the discussion in this volume will assume that the presence of any fungi anywhere in or on a mummy represents at least a potential for local destructive effects.

Location of fungi

Superficial fungal growth

We find fungi on mummies most commonly limited to exposed surfaces. The skin surface is the usual point of interface between the fungus in soil or air and the mummy's body. Furthermore study often demonstrates that this is its *only* location. Because most mummies retain their soft tissue as a consequence of its desiccation, fungi limited to the skin surface can be most parsimoniously explained by the fact that the skin surface, being in contact with environmental air or soil containing moisture, is the only part of the body where the mummy's tissue is apt to contain the minimum concentration of water (13%) necessary for fungal reproduction (Matossian, 1989:7).

During the nineteenth and twentieth centuries bodies placed in chambers carved out of limestone beneath the cathedral floor in Venzone, Italy, sometimes mummified spontaneously by desiccation. While generally protected from rain by the cathedral above it, the chambers were occasionally flooded during particularly severe rainstorms. A white fungus (then called *Hypha bombycina*) commonly grew on the mummies' skin surface. Although local lore attributes the mummification to the fungus, its presence more likely reflects such flooding episodes that punctuated the otherwise arid atmosphere (Aufderheide & Aufderheide, 1991).

Aspergillus fumigatus or *A. niger* has limited invasive potential in an immunocompetent adult, though it may colonize the bronchial surface and even the lining of a lung abscess in a living person. After death it may proliferate in mummified bodies, though the extent to which it is capable of invading beyond such locations after death has not yet been documented well (Horne *et al.*, 1996).

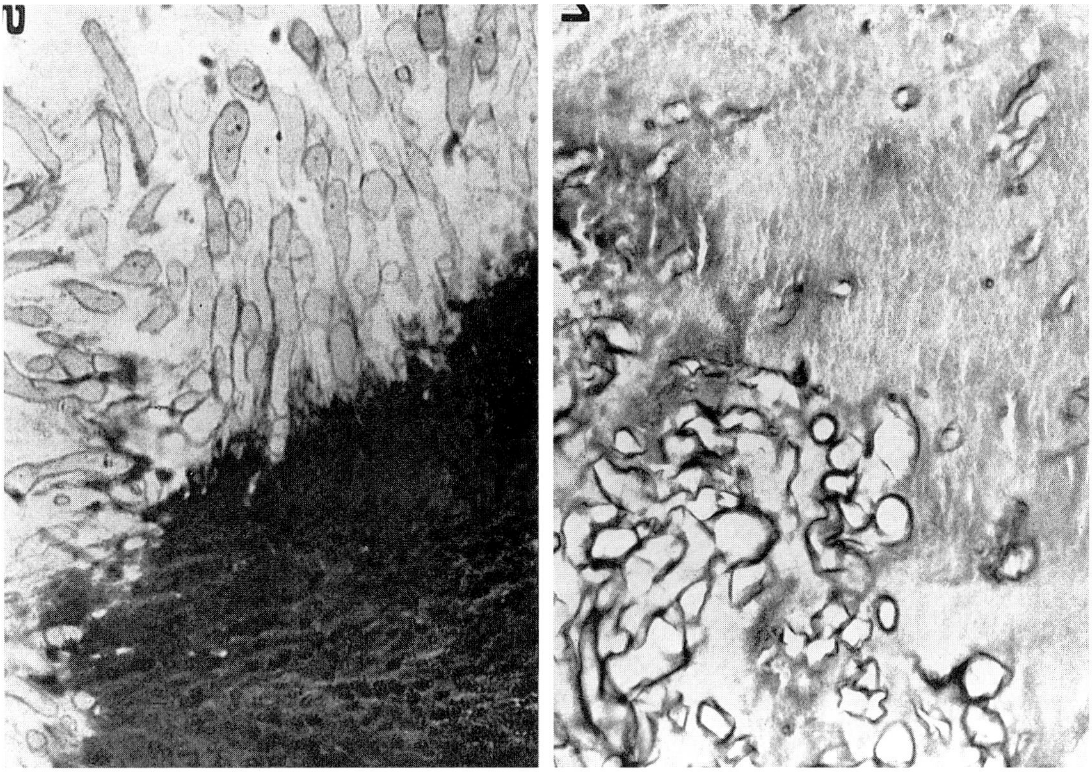

Fig. 5.6. Fungal osteoclasia.
The photo on the left reveals hyphae of *Mucor* sp. penetrating a trabecula, forming pits in the normal, calcified bone. In the right photo fungi (*Mucor* sp.) fill the medullary space, forming canalicular tunnels seen well in the right upper area of the photo. (Photo from Marchiafava *et al.*, 1974, by permission of author and © Springer-Verlag GmbH & Co. KG.)

Invasive fungal effects

If available moisture continues, such fungi are capable both of following existing tissue defects or of creating such channels themselves. The best-known example of such fungal behavior is termed fungal osteoclasia. In such cases fungi gain access to Haversian canals or other normal channels in bone and proliferate there. Proliferation is accompanied by resorption of bone mineral as well as its organic, protein matrix. The end result is the creation of a new maze of tunnels independent of the various normal skeletal tissue channels (Fig. 5.6). Viewed via a fine-ground transverse section of long bone, this tunnel pattern is reminiscent of the tunnels apparent in a glass-paneled ant farm. Since the edges are sharply defined, it is presumed that some substance, possibly secreted by the fungus, is responsible for the chemical resorption (Marchiafava *et al.*, 1974). Because fungi are often found within these channels microscopically, Perotti *et al.* (1998) warn that their presence could confound chemical, immunological or other tests, including radiocarbon dating. Caution, however, must be exercised in interpretation of some microscopic structures, because some earlier reports mistakenly identified some fungal elements as red blood cells (Horne & Jarzen, 1993).

Fulcheri *et al.* (1999) have urged assessment of mummy tissue preservation by a systematic monitoring of conserved mummies using histological tissue section techniques that can demonstrate such fungi in tissues.

If sufficient moisture is available, fungi may also invade soft tissues. A cluster of Native American Inuit spontaneously mummified bodies about 500 years old were found at a rock beach at high latitude on Greenland's northwestern coast (Hart Hansen *et al.*, 1991). Not only had their fur clothing and their skin surface been colonized by the rare fungus *Sporotrix fungorum* (Bodenhoff *et al.*, 1979), but this organism had also invaded some of the soft tissues such as the eye (Andersen, 1994:65–66). Another example is the more than 4000-year-old body of Egypt's ancient pharaoh Ramses II. This anthropogenically desiccated body had been plundered several times in antiquity and, after its nineteenth century recovery, had been housed until recently under circumstances that allowed little control of temperature and humidity. Fungus growth, both surface and invasive, became so extensive that, in a desperate effort to salvage the body, in 1976 it was exposed to 1.8 megarads of gamma radiation to sterilize the body without disfiguring it (Bucaille, 1990). A total of 89 species of fungi were cultured from various parts of his body before irradiation (Balout & Roubet, 1978).

Spontaneously mummified bodies in a church in Guanajuato, Mexico, remained well preserved as long as they were kept in a large, open room, but only months after several were placed in a glass display unit they became covered by a green fungus.

Fungal contamination of cereal grains

All flour prepared from cereal grains contains fungi. Closely monitored storage conditions in developed countries usually keep the moisture content below 13%, preventing fungal proliferation. In antiquity as in some modern countries this was not always carried out. If the cereal (usually rye, though maize ("corn smut") and others can be similarly involved) is stored under more moist conditions, fungal proliferation can occur. Some such fungi secrete toxins injurious to humans that ingest them. *Claviceps purpurea* produces a substance (ergot) toxic to nervous tissue, causing convulsions and delirium; in addition it also causes spasms of arteries leading to gangrene of the extremities, a condition called ergotism (Haller, 1993). The fungus *Aspergillus flavus* under such conditions secretes a chemical termed aflatoxin, which, when ingested, can cause lethal liver damage (Patten, 1981; Aufderheide & Rodríguez-Martín, 1998:323) while a toxin (T-2) produced by *Fusarium tricinctum* results in ulceration and bleeding of the oral cavity and intestine. Surely all of these effects must have been experienced by many ancient populations, particularly during the early period of agricultural experimentation.

Another striking, though not harmful, example of an effect of fungal secretion in stored grain is that of tetracycline production by *Streptomyces* sp., a soil fungus. This compound is a powerful antibiotic, but also has another, unusual quality: it becomes bound by bone's organic matrix that is forming at the time the tetracycline is ingested. Such deposition can be made visible as a fluorescent band if sections of bone are examined under the microscope. When Bassett *et al.* (1980) examined bone sections in that manner from a Nubian population about AD 350, they were startled to see such fluorescent bands. Certain other fungi including *Mucor, Candida* and others also cause bone matrix fluorescence in buried bones, but that pattern is diffuse, not the banded form that tetracycline causes when ingested antemortem. The authors concluded that tetracycline exposure had occurred when the ancient Nubians had eaten bread whose flour had been contaminated by *Streptomyces* sp., which proliferated when the grain had been stored in moist conditions and produced the tetracycline. Similar findings were observed in bones from the Roman Period site Kellis at Egypt's Dakhleh Oasis (Cook *et al.*, 1989).

Vreeland (1978) found spores of *Nigrospora* sp. and *Lasiodiplodia theobrumae* in the anal region of a Peruvian mummy within a bundle dated to about AD 800. Though harmless to humans, these are plant pathogens capable of infecting maize, providing us with archaeological information about that agricultural problem in antiquity.

Soft tissue taphonomic agents: insects

An undisturbed human corpse normally undergoes a series of progressive, grossly apparent changes that lead from death to a completely skeletonized state. The medicolegal applications of detailed knowledge of this process was appreciated as early as 1898. In addition to studying the insects involved in 150 human burials, Motter (1898) also buried animals to investigate the sequence in which the various animals arrived at the corpse. Forensic pathologists have subdivided this continuing process into four reasonably discretely definable stages and named them according to their most obvious morphological features (Rodríguez & Bass, 1983; Goff, 1991): (1) Fresh, (2) Bloated, (3) Decay, and (4) Dry. Concerned with the need to estimate time of death, they have also documented the arrival time and behavior of the many insects that exploit the corpse as a food source. While the details of that sequence are of great interest to forensic specialists, I will note here only those broad features that fall within the realm of anthropological applications.

The first (Fresh) stage begins at death and persists until the first sign of bloating becomes apparent. Blow flies (Calliphoridae) may arrive within minutes, certainly within the first few hours. The moist areas of the nares, eyes and genitalia (in a nude body) are highly attractive sites for them to deposit their eggs so that the emerging larvae can feed on moist tissue. Later very large larvae ("maggots") evolve that are voracious. In a well-established colony those in this stage can destroy a large mass of soft tissue in 24–48 hours. After gorging themselves they often migrate to the soil where they form a casing (puparium) from which an adult fly will emerge several days later and will deposit a mass of eggs in about a week to repeat the cycle. Many other species of fly participate in this process.

The second (Bloated) stage begins with the first observable evidence of putrefactive bacterial gas formation (bloating) and terminates when bloating ceases. This stage also marks the arrival of beetles. The earlier-appearing ones are usually carrion beetles that feed directly on the fly larvae (Haskell *et al.*, 1997).

The third (Decay) stage is aptly termed. Most of the soft tissue removal occurs during this stage. The flies have usually established large colonies by now (indeed, the earlier maggots may have left the body already to pupate in the soil), and their larvae are capable of consuming a very impressive fraction of the cadaver in a surprisingly short time. Most of the beetle species that arrive at this time, often dominated by the dermestids that prefer drier tissue, feed on the body itself. The dermestids were described by Hope as early as 1834 in Egyptian mummies' cranial cavities.

In the final stage (Dry) before skeletonization only now do desiccated fragments of soft tissue hang from skeletal parts. These are not attractive to fly larvae and the flesh-eating beetles finish the remainder of the soft tissue destruction. Mites are capable of imposing extensive superficial damage to a dry body, particularly in spontaneously mummified, curated bodies.

Throughout this process flies, especially the Calliphoridae, dominate the insect population of a decaying corpse and their larvae are responsible for most of the soft tissue destruction. Beetle and mite fragments have been found microscopically within natural spaces of dry bones (Schwartz & Schultz, 1994).

The described sequence of events is an idealized cycle occurring in an undisturbed body at summer temperatures exposed on the earth surface. The rate of destruction is obviously temperature-dependent and, under ideal conditions, near skeletonization has been achieved in as little as a week (Payne *et al.*, 1968). However, it can be retarded five- to tenfold in winter temperatures that remain above freezing, or cease altogether if the corpse freezes (though a well-established colony of larvae can generate a warmer microclimate). Fly larvae can penetrate soil, and even natron (Garner, 1986), to some depth to reach buried carrion, but burials more than half a meter deep are usually beyond their reach.

From the above we can extract some useful generalities. Fly larvae carry out most of the insect destruction of soft tissue in a decaying, exposed body. On a heavily clothed, partly destroyed body, the face may well be the site of maximal destruction by insects. Insects usually make only a minor or no contribution to the destruction of a deeply buried body. Rodríguez & Bass (1985) found no insects on bodies buried deeper than 0.3 meters at the end of 1 year. Those readers seeking more detailed data on this topic may consult Nuorteva (1977), Smith (1986) or Haskell *et al.* (1997).

Soft tissue taphonomic agents: thermal effects

Thermal changes in ambient temperatures can alter postmortem decay rates, whether the change involves elevations or depressions. However, the bulk of the available data was generated by forensic scientists whose principal interest lies in the application of such information to the estimation of time of death. For that reason most of the observations are short term and few were carried out under circumstances where variables could be controlled. What follows here, therefore, is largely a listing of such observations, few of which have been verified or quantified by experimental methods.

Cooling as a taphonomic agent

Assuming normal room temperature is about 20–25 °C, the term "cooling" will be considered to mean temperatures less than room temperature. The study by Aturaliya *et al.* (1995) dramatically demonstrated the deceleration of bacterial proliferation and its resulting decrease in decay rates. Bacteria can be very specific regarding their optimum temperature for proliferation, and hospital laboratories have learned to alter incubation temperatures, depending on the expected organism's growth requirements.

Most of the data relating to tissue cooling come from experience with frozen tissues. In most circumstances bodies of living humans and animals become frozen (less than 0 °C) at a relatively slow rate. Under such conditions ice crystals form throughout the body tissues, initially in the extracellular tissues and water within cells. This type of freezing is very tissue disruptive. As extracellular ice crystals form, water shifts across the cell membrane to maintain water concentration equilibrium. Progressive intracellular dehydration occurs and if the cell membrane does not rupture as a consequence of osmotic disequilibrium, the formation of subsequent formation of intracellular ice crystals commonly brings about structural cell disruption mechanically. Certainly red blood cells suffer hemolysis regularly by one of these mechanisms when an intact body freezes. Furthermore, if a cell's structure has survived the initial freezing sequence, the osmotic changes are often exaggerated during any subsequent thawing event, rendering the cell even more vulnerable to destruction during that phase.

It is also true that the tissue state may not be static during the frozen period. Every cook knows that meat maintained at the average home freezer temperature of about −29 °C will eventually become too dry to be used as food. The transfer of water in the frozen state to the air without first melting is called sublimation. Its rate is determined largely by the temperature, the relative humidity and the physical state of the interface between the surrounding air and the frozen tissue. It often is not realized that air bubbles trapped in ice surrounding a frozen body may act in the same manner, though the rate of exchange, of course is profoundly less. Nevertheless, the frozen-dried state of the glacier-entrapped Tyrolean Iceman found in 1991 (Plate 4; Spindler, 1994) and the Prince of el Plomo (Fig. 4.46; Horne, 1996) testify that 500–5000 years is sufficient to bring about desiccation of a frozen body. Furthermore, nonenzymatic breakdown of protein structure remains active at these temperatures; it is necessary to lower temperatures to the range of −80 °C to arrest that process. Such values are quite easily established in laboratory freezers, but not often found in nature where human bodies are apt to be found.

Some archaeological finds have been made where human bodies have been found in a frozen but not freeze-dried state. Probably the most well-known are the bodies of three crewmen of the mid nineteenth century British Arctic expedition to the Canadian Arctic led by Sir John Franklin seeking a Northwest Passage. These three died aboard ship in 1846 and were buried on the gravel beach of Beechey Island (74° 28′N, 92° 00′W). The permafrost (permanently frozen soil) at this latitude reaches from a few centimeters below the surface down to a depth of many meters. The bodies had been buried at a depth of about 2 meters by removing chunks of gravel laboriously with pickaxe, sledge and shovel. The clothed bodies were placed at that depth in crude wooden coffins and the gravel replaced. Arctic temperatures rise above freezing in the summer, so the next time it rained, water trickled down through the gravel, encasing the bodies in ice. Because of the permafrost conditions, they did not thaw again until they were exhumed in 1984, 138 years after their interment. As these bodies had never thawed since their burial, they demonstrated superb soft tissue preservation without evidence of dehydration (Beattie & Geiger, 1987) (Figs. 4.1 and 4.2). The Pazyryk bodies from the Russian steppes are similar, though those had been eviscerated (see Chapter 4: Mummies from Russia and Ukraine) (Rudenko, 1970; Campbell, 1994).

The process of freezing also inhibits subsequent bacterial growth after the frozen body is thawed. Micozzi (1986) demonstrated that frozen–thawed animal bodies in the field underwent the same sequence of decompositional changes as those placed in the field immediately after death without prior freezing, but at a much slower rate. Numerous mammoths have been found in the frozen state in Siberia. While they appear grossly to have excellent soft tissue preservation, the few studies of tissue preservation that have been carried out demonstrate evidence of protein structure deterioration. Unfortunately the discovery and retrieval circumstances of most of these did not preclude tissue thawing during that part of the process (Goodman *et al.*, 1979; Johnson *et al.*, 1981; Micozzi, 1991:12; see also Chapter 7: Mammoths).

Heat application as a taphonomic agent

Chapter 4 identified many ancient populations who employed heat as a method of soft tissue preservation after death. While sun exposure may have been employed as part of the mummification process in some desert regions, in most cases heat was applied in the form of an actual fire. This was certainly the case in the Darien area of Central America, the islands of the Torres Strait, the overmodeled skulls of Vanuatu in Melanesia, the Benguet in the Philippines, and others. In most of these the mummifiers attended the heated corpses closely, continually wiping moisture off the exposed skin surface. By so doing they maximized the gradient of tissue water content between that of the deeper, internal tissues and that of the skin, enhancing the conductance of water out of the body.

In some cases, however, the body was suspended on a platform or hammock directly above the fire. In some such cases the fire was too small to transfer a substantial quantity of heat to the body. In those circumstances the probable taphonomic agent was not so much the fire's heat as the smoke generated by the fire. Smoke contains many aldehydes whose carbonyl groups form polymeric bridges between the free amino groups on different collagen molecules, providing these with a high degree of stability that is one form of tanning (Painter, 1995).

In all forms of heat application for the purpose of soft tissue preservation, it must be kept in mind that increasing the temperature will also increase bacterial proliferation, an effect that would normally accelerate tissue destruction. In order to overcome this problem the temperature can be raised high enough (>50 °C) to denature protein or the body can be exposed to smoke that has both tanning and antibacterial effects.

Cremation is the extreme form of heat application. Its goal, of course, is not soft tissue preservation but rather its destruction. It comes somewhat as a surprise to learn how much heat is required to reduce a human body to ashes. Modern human crematoria usually achieve this by exposure of a human body to a temperature of about 900 °C for one hour; obese bodies require more time or greater heat. Viscera commonly have the highest water content and may be the last of the soft tissues to burn (Evans, 1963:83–87). Cremation heat also denatures and coagulates protein. If they escape burning, such denatured proteins are more resistant to subsequent environmental degradation. Complete cremation temperatures could have been achieved by the funeral pyres carefully constructed by some ancient populations. The amount of heat required to reduce a body to ash is often underestimated by modern murderers attempting to hide evidence of violence by arson. Quite frequently such burned bones still retain evidence of firearm or other forms of trauma and residual desiccated soft tissue may even permit identification by DNA technology (Eckert *et al.*, 1988).

This text focuses principally on soft tissue, so no discussion detailing changes in heated skeletal tissue will be offered here. Readers interested in that topic should know that major changes occur. For example the glycine/glutamic acid ratio is lowered by cremation temperatures and ammonia content is relatively increased; these changes can be employed to separate cremated from noncremated bones (Taylor *et al.*, 1995). While substantial alterations in crystallization are effected by cremation temperatures, Grupe & Hummel (1991) were able to recover the *in vivo* strontium/calcium ratio values by the use of regression equations, permitting their employment as paleodietary indicators. Discussions of other skeletal alterations by cremation can be found in the following references: Shipman *et al.*, 1984; von Endt and Ortner, 1984; Stiner *et al.*, 1995; Krogman & İşcan, 1986:7–39; Mayne-Correia, 1997.

Anatomical taphonomic effects

Probably the best information we have on the decomposition of a human cadaver is that produced at the research facility in the Department of Anthropology at the University of Tennessee, supplemented by the field observations of forensic scientists. The sequence of changes in an undisturbed body placed on the surface outdoors has been noted in the section above dealing with insects as taphonomic agents. Rodríguez & Bass (1983) gave these stages descriptive terms (Fresh, Bloating, Decay and Dry), and their duration at midsummer temperatures were approximately 5, 5, 13 and 20 days. Temperature was the major variable and insects (principally fly larvae) were responsible for most of the tissue destruction.

Rodríguez & Bass (1985) reported on decomposition of buried bodies. The most significant differences from bodies decaying on the surface was the markedly reduced insect involvement, with more superficial burials (0.3 meter deep) and absent insect activity below 0.5 meter. In addition, the body temperature decreased in direct proportion to the depth of burial. The consequence of these factors was a very substantial decrease in the rate of decomposition, which in the buried bodies was being achieved principally by bacterial proliferation. Other features they observed included the relatively rapid loss of breast fat, resulting in the collapse of these structures to the point at which they often became unrecognizable in four to six weeks. Hair also tended to slough off the head as an apparent intact mass after only a few days in hot weather in surface-exposed bodies. Factors accelerating decomposition included increases in temperature, humidity, insect access and clothing (maggots shun sun) while deeper burials and embalming prolonged decomposition time (Mann *et al.*, 1990). Postmortem collagen degeneration accompanied by abdominal wall tension secondary to abdominal bloating can cause dramatic postmortem wound dehiscence that can easily be mistaken for antemortem wound infliction (McGee & Coe, 1981). Bloating commonly applies pressure on the body cavities sufficient to cause rectal or uterine prolapse as well as protrusion of the tongue.

An alternative source of information about postmortem taphonomic changes is my own experience with mummified bodies. Because relatively little has been reported about soft tissue loss or preservation under conditions that result in at least partial preservation (spontaneous mummification), I will present such alterations in considerable detail here. The model for this presentation will be a group of 57 mummies dissected in 1987 at the University of Tarapaca's Museo San Miguel in Arica, Chile, under the direction of Marvin Allison, Ph.D., who was at that time serving as a full-time faculty member and Director of Biological Resources (Aufderheide & Allison, 1994). This cultural group was a Preceramic coastal population called Chinchorro that carried out a subsistence strategy involving extensive exploitation of marine resources, operating as small groups at the band level centered at the mouths of rivers on the coastal area of the hyperarid Atacama Desert in northern Chile. Radiocarbon dates for this particular subset (archaeologically labelled Mo 1–6) ranged from ca. 2350 to 1750 radiocarbon years BC. They had been buried without coffins or grave goods on the sandy slopes about 75 meters above the valley floor. While their ancestors had employed elaborate anthropogenic mummification methods, soft tissue preservation in this late subgroup was entirely spontaneous.

Table 5.1 indicates the B, ST, STI and Organ scores obtained for this Chinchorro population (see Chapter 6: Performance of mummy dissection). Standard deviations were approximately ±25 for each of the mean values listed in the table. Such high values are commonly obtained in large populations of spontaneously mummified bodies because they reflect the fact that, while many bodies retain most of their soft tissues, others are nearly skeletonized.

The soft tissue index ranges from as high as 74% in the adults to as low as 53% in infants. Subadults (<16 years) constituted 40% of the 69 bodies in this population and infants (<2 years) were 28% of the total. The higher water content per square meter of body surface for infants may be responsible for the lower STI. Such values are characteristic of spontaneously mummified bodies excavated from hyperarid deserts such as the Sahara or Atacama.

For those interested in research studies to be carried out on samples from various body organs and tissues, preservation rates of the various body structures are of great importance. Table 5.2 provides a listing of those most apt to be of interest (Zlonis, 1995). Annotations

Table 5.1. *Preservation of bone, soft tissue and viscera*

Age (yrs)	B	ST	STI	Organ
Adults[a]	73	54	74	31
11–15	80	51	64	35
3–10	49	32	65	10
<2	32	17	53	11

Notes:
B, estimate (%) of preserved fraction of skeleton; ST, estimate (%) of preserved soft tissue covering the bones; STI, Soft tissue index = ST/B = fraction (%) of preserved bone still covered by preserved soft tissue; Organ, estimate (%) of preservation of 10 viscera (listed in text).
[a] Adult is defined as over 16 years of age.

that reflect my experience with these and other populations follow.

Head

Only about half of the bodies had attached heads. Those removed from the other half were not encountered at this burial site nor have any caches of heads been found at any of the many sites excavated in this region. Further north in Peru and 1000 years later the "taking of heads" became a prominent feature among the Paracas and Nasca cultures and until the twentieth century also among the Jivaros in Ecuador. The usual explanation by archaeologists for such a pattern is that this reflects the practice of "ancestor worship", but the cultural significance of the missing heads of the Chinchorros remains controversial.

Hair

The low overall frequency of hair preservation (54%) is entirely the result of the missing Chinchorro heads. Hair was noted in 90% of the heads present. Hair must be the last soft tissue to yield to decay in most circumstances. In coastal areas of Peru and Chile it is quite common to excavate skeletons with hair. In all probability the low water content of hair, its high content of insoluble

Table 5.2. *Frequency of preservation of individual body structures*

Body part	% of total[a]	% when body part is present[b]
Head	52	100
Hair	54	90
Eyes	25	93
Nails	68	86
Breasts	7	56
Heart	19	32
Lungs	44	55
Esophagus	14	22
Stomach	7	13
Ileum	30	54
Colon	33	59
Coprolites	30	50
Liver	42	63
Gallbladder	1	1
Spleen	0	0
Kidney	13	23
Uterus	0	0
Brain	14	30
Pancreas	0	0
Prostate	0	0
External genitalia – male	77	88
External genitalia – female	28	38
Urinary bladder	7	15

Notes:

[a] % of total: represents the fraction of all bodies in which the indicated structure was identifiable by gross examination.

[b] % when total body part is present: refers to the fraction that the indicated structure was identifiable among only those mummies in which the major body part housing the structure was present. For example: brain tissue was present in 30% of the crania, but heads were associated with only 52% of the mummies, so brain tissue was noted in only 14% of all mummies examined.

Source: From Zlonis, 1995.

fibrous protein (keratin), together with its location on the external surface of the body and its high surface area, all contribute to the high frequency of its preservation. Indeed, braided and even often elaborately pleated hair styles can be found on bodies of some groups, though not this Chinchorro population. Long braids provide the possibility for studying the individual's dietary or drug history (Cartmell *et al.*, 1991; White, 1993) by cutting the braids into short lengths (1.5–2.0 cm, equivalent to about one month's growth of scalp hair) and carrying out individual chemical tests of each segment. Pubic hair can be present in amounts useful for diagnostic testing, but usually is not available in such quantities.

Eyes

In spontaneously mummified bodies the eyes are commonly collapsed within the orbit, anchored there by the still identifiable optic nerve. Surprisingly, in view of their availability (preserved in 93%) little serious study of ancient eyes has been carried out (Andersen, 1994).

Nails

These are also regularly preserved when hands or feet survive (86%). Their keratinous structure provides the same type of preservation as does hair, but its slower growth reflects a different (longer) time interval. Attribution of mystical qualities to fingernails and/or toenails was not uncommon among ancient cultures or even some modern ones. It may also have been a feature of Chinchorro cultures because one-third (33%) of them had fingers tied with vegetal cords and in 20% toes had been similarly treated.

Breasts

Postmortem hydrolysis releases free fatty acids from the breast's neutral fat that seep away, causing the breast to collapse. It comes as a surprise to many that such collapse may flatten a breast to the degree that it can no longer be recognized, especially if the chest was tightly wrapped. Hence, even when the anterior chest was intact, breasts could be recognized in only about half (56%) of the bodies among the Chinchorro adult females. Even in some of these it was only the larger areola and/or nipple that differentiated a female from a male chest. In breasts that were lactating at the time of death, the increased water content often resulted in loss of skin over the breast surface. Breasts were identifiable in bodies of postmenopausal age just as frequently as in those of premenopausal women.

Heart

The heart could be identified in only about one-third of the bodies (32%) even when the other thoracic cavity soft tissues were intact. At least part of the pericardial sac could be recognized more frequently. This appears to be due largely to the myocardium's high lysosomal content. Indeed, it is not unusual to open an intact pericardial sac only to find the heart has been reduced to a black stain (hemoglobin breakdown products) on the sac's inner surface. Since, however, some cardiac structures such as the valves, chordae tendineae, interatrial septum, visceral pericardium and major arterial trunks have a major fibrous or elastic component, and since such tissues survive the postmortem environment better than does the myocardial fiber, it is not uncommon to find the heart's basic structure defined by these residual tissues even though no myocardium remains (Fig. 5.7). Both heart and pericardial sac may be digested, liquefied and disappear if enzyme-laden abdominal fluid has digested the diaphragms and gained direct contact with the mediastinal structures. If the skin over the thorax contained a gross defect or even numerous insect perforations, heart tissue could be identified in only 15% of the cases, but when the chest wall was completely intact, heart tissue was identified in 89%.

Fig. 5.7. Desiccated fibrous "skeleton" of the heart; SM, ca. AD 200.
All myocardial tissue has decayed, leaving only the fibrous structures (large arterial trunks, valves with attached chordae tendineae) of the heart, enveloped by the partly opened pericardial sac. Mummy no. AZ-75, T-57 from the Azapa Valley near Arica, northern Chile. (Photo by ACA.)

Lungs

Except for major bronchial and vascular structures in the hilum region, much of the space in an expanded, living lung is taken up by capillaries suspended in air with only a thin alveolar epithelial covering. After death the blood dries and the air is resorbed. As the lungs collapse in a body in the supine position, they contract toward the hilum and come to rest on, and conform to, the curvature of the inside of the posterior chest wall. Because in most such lungs bronchial and alveolar epithelium is lost, the mass of collapsed tissue is made up principally of thin fibrous membranes reflecting the visceral pleura and those that subdivide the lobes into their smaller subunits. The trachea is normally easily identifiable in the neck (though a normal thyroid is often undetectable), and both it and the major bronchi usually can be dissected easily to at least the level of their tertiary divisions. Of course, if lung disease such as pneumonia has caused a portion of the pleura and lung to adhere permanently to the inside of the chest wall, such adhesions will prevent those portions of lung from collapsing after death. Intact lungs can be found in at least half (55% of adult Chinchorros) of the spontaneously mummified bodies when the thoracic wall soft tissues are intact.

Gastrointestinal tract

The esophagus can usually be identified in a transected neck but in only about one in five bodies (22% of the Chinchorros) can the mediastinal segment be identified in its normal, collapsed state. Occasionally postmortem gas formation in the stomach will distend it, as will certain premortem conditions including American trypanosomiasis (Chagas' disease) in South American bodies. If the esophagus is identifiable, the esophagogastric junction can be used to identify the stomach. If not, the stomach may not be easy to dissect (identified in only 13% of Chinchorro bodies) because it may retreat under the rib cage and other bowel loops (both ileum and colon) may overlie it and dry in a position that causes them to adhere to the stomach. Loops of ileum and colon are so numerous and large that at least some parts of their course remain identifiable (50–59%). Coprolites may be found in parts of the colon about half the time (50%). In these bodies a coprolite-laden cecum may come to rest near the anus, where it is easily mistaken for rectum. The much lower frequency of coprolites in the mummies of children suggests that diarrhea may have been a feature (perhaps even a cause?) of their fatal illness.

Liver

The liver, weighing 1500 g in life, usually is also the largest and most obvious of the surviving viscera. It can be identified in about two-thirds of spontaneously mummified bodies (63% among the Chinchorros). Anchored by its incorporation of part of the right diaphragm into its capsule and also by the biliary tract leading to the duodenum, it is usually present in the right upper abdominal quadrant. Nevertheless in extensively decayed bodies I have found it detached, lying loose elsewhere, including in the pelvis and left thoracic cavity. However, more than 90% of a normal liver's volume is composed of epithelial cells that are very susceptible to autolysis by lysosomal enzymes and those of bacteria. Hence its volume in these spontaneously mummified bodies is often only one-fourth to one-fifth of its living size. The thin gallbladder is usually collapsed and rarely identifiable unless it is thickened by chronic infection or its lumen filled with gallstones. It was recognizable in only one Chinchorro mummy; in that one the wall was fibrous, about 1 cm thick and the lumen filled with gallstones.

Spleen

This organ is composed chiefly of blood cells. These fragile cells autolyze readily after death, so much so that it is not an unusual experience to fail in its identification. I could not recognize the spleen among any of the 57 Chinchorros. Since I have been able to demonstrate it frequently subsequently, part of the reason for its apparent absence among the Chinchorros may have been my inexperience at the time of dissection.

Kidney

As indicated above in the section describing our experimental taphonomy studies, renal microscopic structures autolyze for several days before bacteria arrive that accelerate their breakdown. It is, then, not surprising that intact renal structures were apparent in only 23% of the Chinchorros. Sometimes a reniform cavity was all that remained of the kidney. Careful dissection at the pelvic brim can often reveal the ureter, which can then be traced retrogradely to the kidney. This is especially useful on the right side, where the liver may overlie the kidney.

Bladder and prostate

In one Chinchorro the urinary bladder dried in a greatly distended state as a thin, delicate, translucent saccular membrane. In four others it was only 2–4 cm in diameter. All were in males. In the others it had dried in a collapsed state and was so obscure that it was unrecognizable in any of them. The prostate was not identified in any of the Chinchorro bodies.

Uterus and adnexae

While inexperience at the time of this dissection undoubtedly played a role in my failure to identify a uterus in any of the female Chinchorro bodies, it is also true that this part of the pelvis is the most dependent in a female cadaver. Accordingly, as the viscera decay, the enzyme-laden fluids tend to accumulate in the pelvis and digest tissue to a greater extent there than elsewhere. Among the Chinchorros, this tissue destruction had penetrated to the level of the labia majora and in some males the perianal area, destroying evidence of both internal and external genitalia in many of these (see below).

Pancreas and adrenals

Autolysis is so rapid in these enzyme-rich organs after death that I have never been able to identify them in either a spontaneously or anthropogenically mummified body.

External genitalia

These were more regularly apparent among males (88%) than females (38%) among Chinchorro mummies. A significant reason for this is because the penis usually lay above that part of the pudendal area that had been digested as described above. In some bodies the penis is so well preserved that the easily identified prepuce indicates whether circumcision had been carried out. In some males the scrotal folds had flattened against the skin of the thigh sufficiently to be barely perceptible, and testicles were not identified in any of the Chinchorro mummies.

Brain

Tissue derived from the brain was found in 30% of adult Chinchorro crania. In none of them was a mass present larger than an estimated 20% of its original size. In only one was a pattern suggestive of cerebral convolutions noted on the surface. In most bodies the brain was represented only by a dark brown, formless mass of pasty consistency without grossly recognizable anatomical structures. In several the brain had apparently liquefied after death and then subsequently dried, its solutes precipitating into minute pebbles of 1–2 mm diameter, resulting in 10–15 cubic centimeters of coarse, gray, sand-like material. The anatomist and Egyptologist Grafton Elliot Smith recognized and described all these forms of spontaneously preserved brain tissue in Egyptian bodies as early as 1902. Tkocz *et al.* (1979) found preserved brain tissue in 76% of 74 medieval skulls in a Danish cemetery. Anatomical and chemical analyses established that they were preserved by adipocere formation.

Chapter 6. Mummy study methodology

Introduction

In spite of recent emphasis on biochemical or other nonanatomical methods of analysis, paleopathology remains rooted firmly in morphology. Most diseases generate tissue alterations at either the gross or the microscopic level. In living and even recently deceased individuals the pathologist can usually translate such changes into one of the major disease categories such as inflammation or neoplasia and commonly is even able to identify a specific etiological diagnosis. Taphonomic effects on archaeological remains, however, frequently obliterate many of the anatomical clues on which such specific disease taxonomy is dependent. Soft (nonskeletal) tissues usually disappear quite promptly after death, deleting the evidence of such common diseases as lobar pneumonia, cancer of the breast and others. In addition, bone has such a limited range of responses to disease agents that similar bone changes can be caused by a wide variety of conditions. Even when soft tissue is preserved, epithelial cells have usually degenerated beyond recognition. For the diagnostic paleopathologist this is a serious loss, for it is these very cells that commonly contain the diagnostic clues. Furthermore, in contrast to bone, soft tissues shrink, deform, and lose both color and texture before desiccation finally arrests the decay process. These changes introduce further variables into the process of disease recognition.

Hence paleopathologists must learn the unique appearances (and sometimes disappearance) of diseased soft tissue in mummified human remains (Fig. 6.1). Their diagnostic expectations must be more restricted when dealing with ancient tissue, and they must become particularly familiar with taphonomic changes so as to avoid misinterpretation of pseudopathological postmortem alterations as antemortem disease. Simultaneously they must maximize their recognition of what information the usually better preserved structures of mesodermal origin can offer.

Despite these limitations, morphology remains central to recognition of disease in ancient human remains. The biochemical, immunological, biophysical and other nonanatomical methods transposed from other disciplines and focused on anthropological questions during the past several decades derive their utility from their specificity. This feature makes them invaluable supplements to the fields of physical anthropology, paleopathology, bioanthropology and archaeology. Because, however, these methods are usually focused on narrowly stated questions, they remain largely ancillary aids. The primary models still need to be fashioned by morphological techniques. It is appropriate, then, to begin our review of mummy study methods with those related to the *in situ* normal and diseased morphological structure of mummified remains: gross and microscopic tissue examination, endoscopy and radiology.

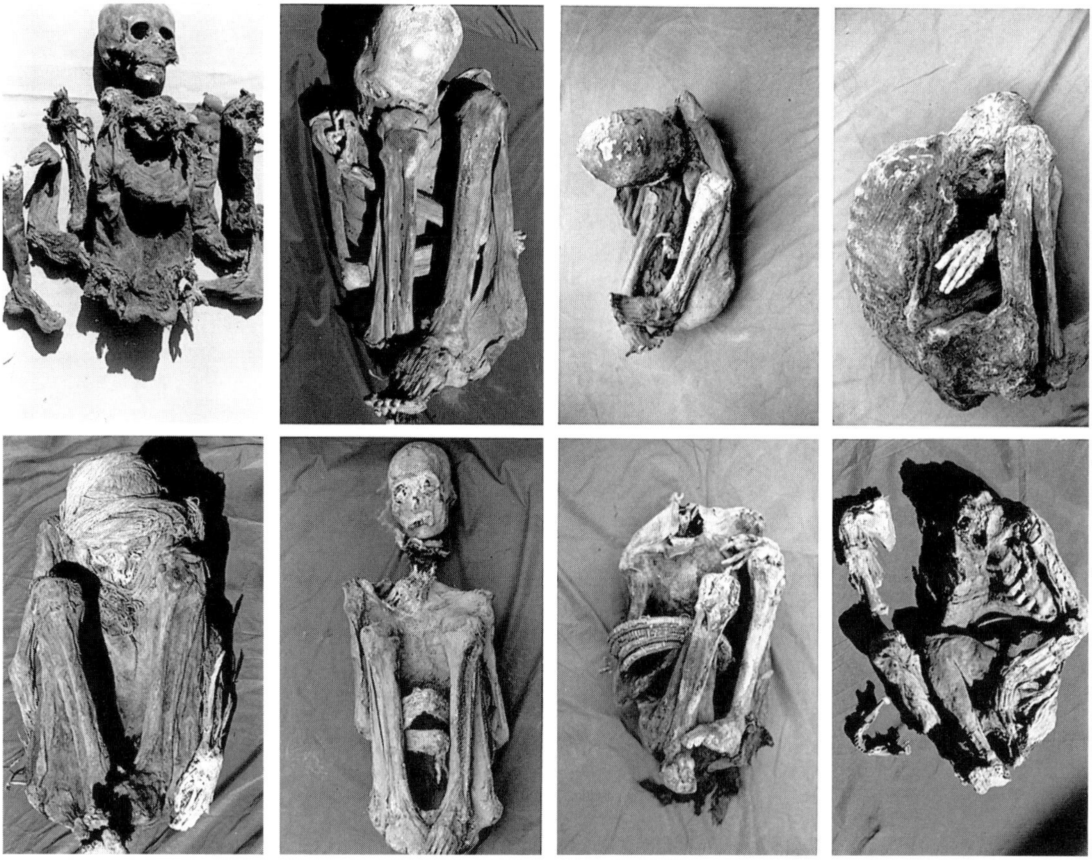

Fig. 6.1. Variations in soft tissue preservation; SM, ca. 50 BC.
These eight Andean mummies from northern Chile's Atacama Desert were all excavated from a 2000-year-old cemetery. In spite of their similar age, they demonstrate substantial variation of soft tissue preservation. Cemetery AZ-75, Azapa Valley. (Photo by ACA.)

Morphology-based methods: anatomical dissection methods

The study of morphological changes in the tissues of mummified human remains is still the cornerstone upon which the science of mummy studies is based. Such observations substantially increase the range of diagnoses that can be established beyond those possible by osteological study alone. This is true in spite of the dissolution of histological integrity in epithelial tissues. The synthesis of a mummy study database begins with analyses of anatomical changes.

Purposes of anatomical dissection

Anatomical dissection seeks to supply information in the following areas:

1. *Cultural data* involving subsistence activities as judged from: tool and product assemblages often associated within wrappings or clothing; ritual forms reflected by spiritual offerings, some of which may even be placed within the body's oral cavity or other orifices (drugs, sacrificial items); social status (body decorations, jewelry, hair styles; clothing and textiles enveloping the body); population identity as

revealed in body alterations (including tattoos, cranial deformation, alterations in the shape of ears, teeth, feet and others).

2. *Health and diseases.* Health is evaluated by demographic statistics and nonspecific physiological "stress" markers: skeletal transverse lines of bone growth arrest, dental enamel hypoplasia, Wilson's bands, stature. Specific diseases are commonly identified by patterns of gross morphological alterations in organs. These are usually supplemented by certain specialized methods including histological identification of intestinal parasite ova or recovery of certain DNA sequences unique to a specific infectious agent. Examples include the tuberculosis bacillus (Salo *et al.*, 1994), and viruses (1918 influenza virus: Taubenberger *et al.*, 1997).
3. *Demographic data.* Determination of sex, age-at-death, and similar features characterize the population. The recorded observations can then be used to derive other information of value such as infant mortality rates, survivorship curves, stature and similar characteristics useful in population comparisons. While much of this is obtained from osteological observations and measurements, the data are enhanced by soft tissue observations such as external genitalia in subadults, internal sex organ differences, age changes in organs, etc.
4. *Dating.* The most reliable specimens for radiocarbon dating of mummies are the viscera of the mummies. This source eliminates potential errors arising from skeletal samples or from noncontemporaneity of body and alternative samples such as charcoal or an artifact. El-Najjar *et al.* (1985) have reported that massive recent fungus growth within soft tissue (muscle) can introduce a major source of error, though this remains controversial, since such saprophytic fungi may be consuming ancient tissue.
5. *Sample procurement* for biophysical and biochemical studies. Some of these (for molecular biology, isotope ratios, trace element studies, etc.) require special handling and preservation.
6. *Basic research.* Knowledge about the postmortem changes that archaeological remains undergo (soft tissue taphonomy) is essential for proper interpretation of our findings. The effect of diagenetic alterations influencing chemically quantified stable isotope ratios in buried bodies would be an example. So little is known in this area that every studied mummy can make a contribution. Appropriately selected samples can also provide study material for other disciplines such as background radiation levels in antiquity, pollen or phytolith data from coprolites, comparative environmental toxicology and others.

Extent of dissection

This is governed primarily by three factors: ethical and cultural concerns, curatorial concerns, and scientific goals. Ethical and cultural concerns have already been addressed (see Chapter 1: Introduction). It is inappropriate for this discussion to be persuasive for any particular view, but the following observations can be helpful to those entering the field.

Curatorial concerns

Curators and governing board members of museums view their relationship to the institution's acquisitions as embracing three major activities: preservation, public education, and research. While investigative dissection of mummies usually generates new information useful to the second of these functions (public education), dissection is often seen to be in conflict with preservation interests. Loss of a museum item's integrity by a deliberate action is often more distressing to a curator than its gradual deterioration through conservational neglect. The investigative scientist, therefore, can anticipate considerable hesitation on the part of museum officials to approve anatomical dissection research studies on curated mummies. This instinctive compulsion for preservation may even be applied to inadequately conserved, continually decaying human remains. It is important for all parties involved to be aware of the following:

1. The type of information that dissection can be

expected to produce and how this could be used to amplify the museum's educational role. It is important to be realistic about the nature and degree of this expectation. If it is *excessively* optimistic, the final results may be disappointing to those granting permission. Examples from experience will help to translate this from an abstract concept to the actual mummies considered for dissection. Production of an educational video (for public use) focusing on the project itself has been a popular success in several museums.

2. Unless ideally maintained (an unusual status), the mummies are already undergoing gradual but continual loss of integrity (with associated loss of potential research and even public exhibit value).
3. Bulk samples of various dissected tissues are much more easily properly preserved than is an entire human body. Many such specimens can be stored for infinite periods of time, unaltered and therefore useful for future research.
4. Modifications of dissection methods may be designed to meet both the investigator's goals and the curatorial concerns. Some possible alternative dissection approaches are detailed at the end of this section's discussion (see Modifications of anatomical dissection techniques).
5. Access to mummies for dissection is rarely gained unless the dissector is genuinely interested in incorporating the museum's curators into the project as active and contributing collaborators, including participation in the preparation of subsequent publications.

The above items are of common concern when one is dealing with one or a few mummies that have already been excavated and curated. These problems become much more complex in the event of a major expedition proposing to excavate a large number of mummies. Few museums, for example, could, or would wish to, expend the major resources necessary to house the 135 mummies I examined during a single summer's excavation in South America in 1990. In some remote areas, where most such burial sites come to professional attention because of looting, the plundering of tombs compounds the problem. Unless such sites are excavated reasonably promptly after the looting has come to attention, the contents (including mummies) are soon irreparably destroyed (Fig. 6.2). Security is rarely sufficient to prevent looting of the remainder of a partially excavated cemetery, yet traditional curation and conservation of all of the excavated mummies is not possible (Fig. 6.3). In such situations only compromises are realistic possibilities (also discussed under Modification of anatomical dissection techniques, below). Although only a few days elapsed between discovery and recovery of an ancient body from a Tyrolean glacier in 1991, Spindler (1994:3) notes that handling of that body and the site items by nonprofessionals had already resulted in loss of certain scientific opportunities and information.

A plan to deal with large numbers of human mummies could include the following: (1) select (by whatever criteria are relevant for the museum) a subgroup of the mummies that the museum wishes for long-term retention and is willing to provide resources to carry out proper conservation; (2) select another subgroup for investigative study, including the possibility of extensive dissection; (3) after external study and sampling (preferably including viscera examination) rebury the remainder under circumstances that provide security as well as environmental conditions that will maintain their mummified state.

Scientific interests

The amount of information harvested from a mummy dissection is directly related to the extent of its dissection. Maximal information is derived by interdisciplinary study of a planned dissection and sampling of all soft tissues followed by a complete osteological examination. Studies of a more focused nature, however, frequently can be completed without violation of preservation concerns. Evisceration of a spontaneously mummified body with detailed study of each organ usually can be carried out in a manner that will still permit public exhibition of the body following the study. We have even used clinical endoscopy techniques

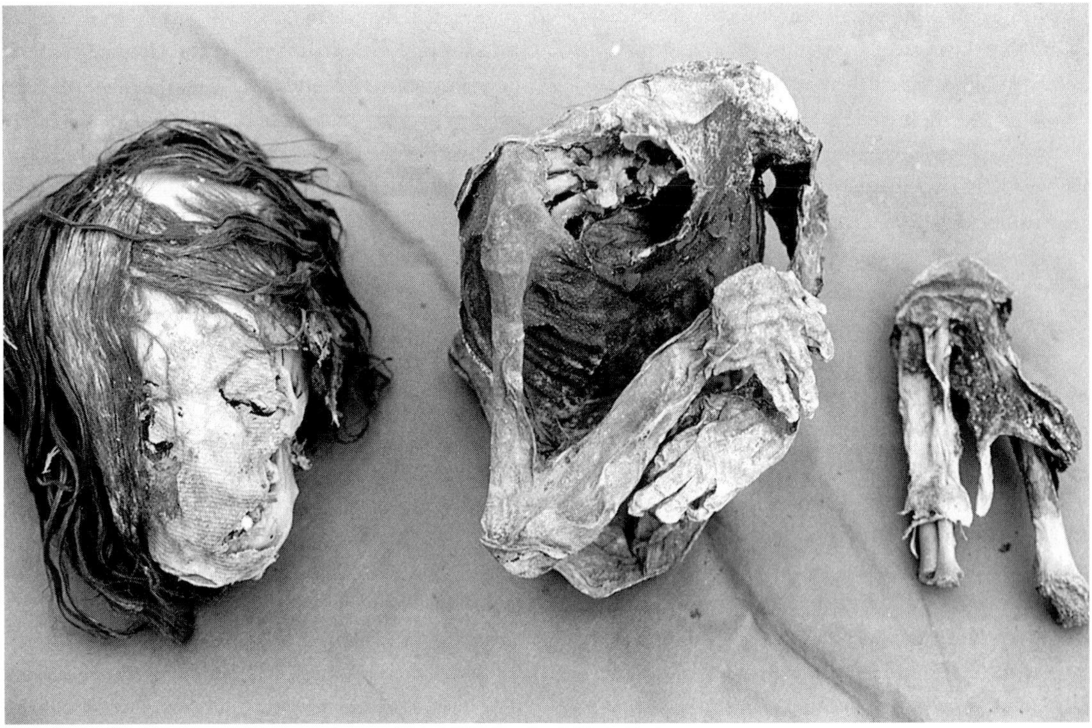

Fig. 6.2. Mummy fragmented by looters; SM, AD 1150–1300.
Human bones and mummified body parts littered the surface of the looted burial ground of the ancient Chiribaya population near the modern seaport of Ilo, Peru. Chiribaya Alta site. (Photo by ACA.)

(via a created aperture of less than 2.0 cm diameter in a rib) to obtain tissue that settled the question of the nature of an intrathoracic structure observed in a radiological computed tomography (CT) scan of an Egyptian mummy (Notman *et al.*, 1986).

Dissection procedure

Factors governing the approach to dissection include the following:

1. *Desiccated or hydrated mummy?* Dissection of a fluid-saturated "bog mummy" can permit the use of scalpels and other instruments commonly employed in routine hospital autopsies. The cardboard-like consistency of tissues in a desiccated body, on the other hand, requires saws and sturdier equipment.
2. *Mummification method.* In anthropogenically ("artificially") mummified bodies special efforts need to be made to reconstruct the sequence of events and materials used to achieve body preservation. I have found it useful to start with extremities of one side first, carefully dissecting successive layers, in one area *in situ*, then transecting an extremity and verifying observations by viewing it in cross-section. Limiting dissection to one side initially may even be useful in facial analysis, though it is rarely practical within the body cavities. In spontaneously desiccated bodies the thoracic and abdominal cavities commonly provide the principal useful observations and usually are approached directly.
3. *Extent of dissection.* This is established by agreement before initiation of dissection and obviously

Fig. 6.3. Salvage excavation of mummy burials.
More than 100 containers, each occupied by a spontaneously mummified body, fill the work space at the San Miguel Museum in the Azapa Valley near Arica, Chile. These mummies were excavated from a nearby ancient burial site (Cabuza culture, AD 400–1000) when it was found that the cemetery was being looted daily. (Photo by ACA.)

defines the limits and therefore the approach by dissection.

4. *Body position*. The extended body provides the largest number of dissection options (Fig. 6.4) while a tightly flexed body imposes limitations (Fig. 6.5). For example, if dissection is to be limited to evisceration, the extended body provides no obstruction to access, while in the maximally flexed body the legs, arms and sometimes even the head are rigidly interposed between the operator and the body cavities.

General features of anatomic dissection

The following discussion will relate to a spontaneously mummified, desiccated body. Departures from this approach necessitated by differing circumstances are discussed under Modifications of anatomical dissection techniques, below.

1. Because desiccated tissues are brown, rigid and brittle, the loss of guidance that is normally provided by color and texture in tissues of recently deceased individuals means that identification of tissue structure in desiccated bodies is often dependent on anatomical position. To the extent possible, therefore, dissection should proceed from a known anatomical structure. Normal anatomical structures may be difficult to recognize due to postmortem (often position-influenced) deformation. Complete disappearance by decay is common in certain high water content organs such as thyroid or spleen.

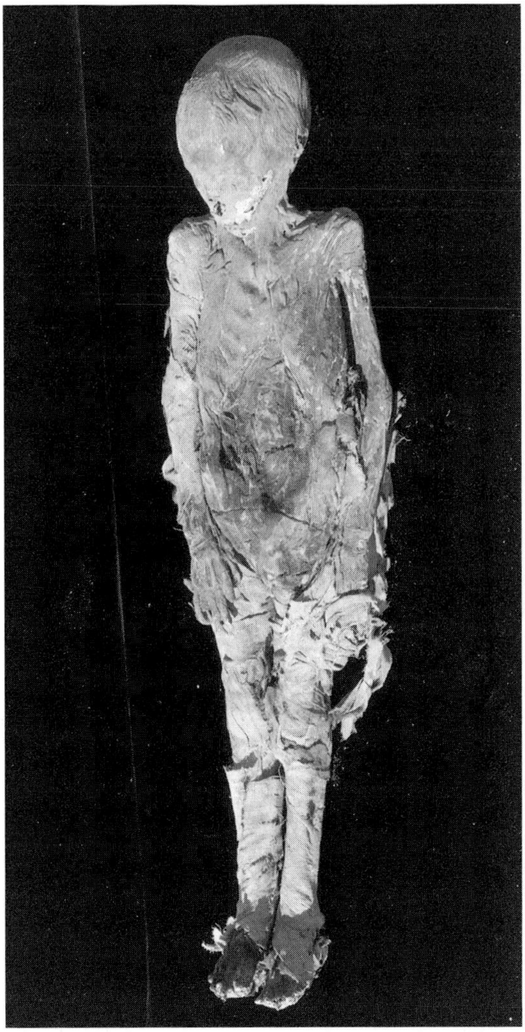

Fig. 6.4. **Extended body position; SM, 30 BC to AD 395.**
Access to body cavities is simplified in extended bodies. Spontaneous mummification from Kellis 1 cemetery at the Dakhleh Oasis in Egypt's Western Desert. Roman Period. Mummy no. 102. (Photo by ACA.)

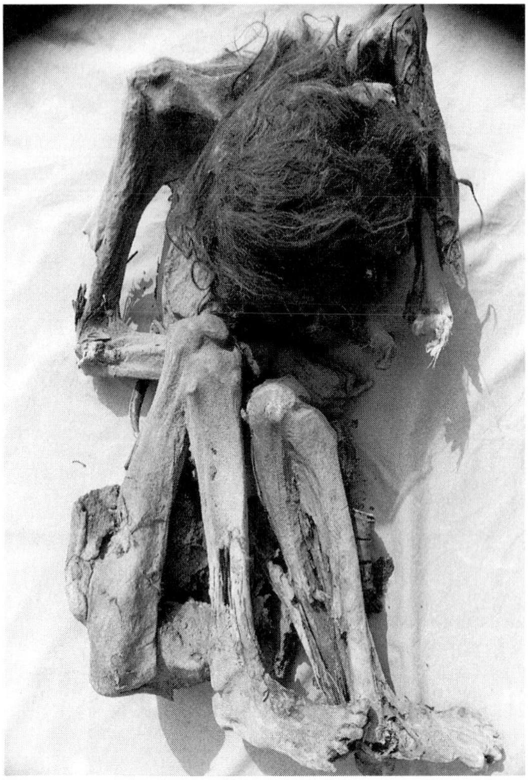

Fig. 6.5. Flexed body position; SM, AD 1150–1300.
Attempts to dissect visceral organs in the body cavity of flexed mummies are impaired by the arms and legs overlying the ventral trunk. Mummy no. 92 is an Andean adult Chirabaya from Ilo, Peru. (Photo by ACA.)

Similarly, structures with high enzyme content such as pancreas, kidney cortex or myocardium can disappear after death. Thus a continuing search for dissection guidance from fixed, known anatomical structures throughout the procedure is desirable.

2. Defects and perforations from insect or rodent action, decay areas and postmortem trauma can admit sand, soil, pebbles, vegetals and other nonhuman material that must be recognized to prevent misinterpretation.
3. Instruments: Table 6.1 provides a complete list of field instruments useful for dissection of a desiccated mummy, a few general comments are appropriate here. The hard, brittle nature of the desiccated tissues rarely yields to a scalpel. My preferred cutting instrument is a Swiss Army™ pocket knife (preferably a model that includes scissors) and a portable sharpening device to maintain its keen

Table 6.1. *Field dissection instruments list*

Aluminium foil (double strength, sterile)
Anthropometric instruments
Autopsy protocol
Backdrop cloth (velvet? plastic?)
Battery-operated dictation device
Chisel
Copy of Ubelaker (1989) manual
Dental pick
Duct tape
Fiber-point pen
Flashlight
Hacksaw
Hammer
Keyhole saw
Large garbage bags
Long, flat-bladed knife
Measured labels
Measuring tape
Oscillating saw (if electricity is available)
Paint brushes
Pencils
Pocket knife (Swiss Army type) and manual, diamond-dusted sharpener
Rubber gloves
Scissors
Shipping tags with wire attachments
Sieve
Surgical hemostatic clamps
Specimen bags
Teaspoon
Thumb forceps

edge. If electricity is available, a Stryker™ (or similar) autopsy or cast-cutter saw with an oscillating blade is ideal for cutting bone, both for dissection and sampling purposes. Frequently dried viscera are too fragile to be biopsied by scalpel or scissors. To avoid the crushing effect from routinely used biopsy instruments, the use of a small electric drill (the type commonly used by model-builders, such as the Dremel™) fitted with a thin, rotary saw blade bit has been found to produce satisfactorily thin slices of even the most fragile, dried tissue. A simple teaspoon facilitates recovery of crystalline or powdered precipitates as well as pulverized intestinal content. Because the dissection environment is rarely ideal, a flashlight with replacement batteries commonly becomes a treasured item. Rubber gloves are desirable, not to protect the operator (desiccated mummy tissues very rarely contain viable pathogens) but to prevent the operator's own DNA from contaminating the specimens selected for molecular biology studies. A kitchen sieve is useful for straining sand to recover small tissues such as loose fingernails.

4. Sample containers: I have found it convenient to create a permanent storage system in my laboratory that will accommodate specimen-filled field containers, obviating the need for repackaging the specimen after return to the laboratory. The specimen container around which the system is built is the Whirlpack™ 6 oz plastic bag that is sterile, waterproof and airtight (Fig. 6.6). For storage of mummy specimens the bag must *never* be opened by blowing one's breath into it! The lips of the bag's mouth must be opened by pulling gently on the paper tabs attached to it for that purpose. After the two lips of the mouth have been separated, one side is grasped with one hand while the other hand grasps the bottom of the bag. Now a single, sharp tug (avoid repeated tugs) will open an ambient air-filled space extending the full length of the bag that will permit even the lightest tissue fragment placed into it to drop to the bottom of the bag. The air is then kneaded out of the bag and the container is sealed by folding the lips. These are held in place by bending its enclosed wires. This will seal the specimen into the bag that requires a minimum of storage space. After labeling on a strip of freezer tape applied to the bag's other end, the specimen is ready for permanent filing without further handling. If the excavation takes place in a desert (the usual location of spontaneously mummified bodies), the residual air in the bag will be virtually free from moisture. Thus the sterile, air-tight and waterproof nature of the bag provides a satisfactory environment for long-term storage. For permanent

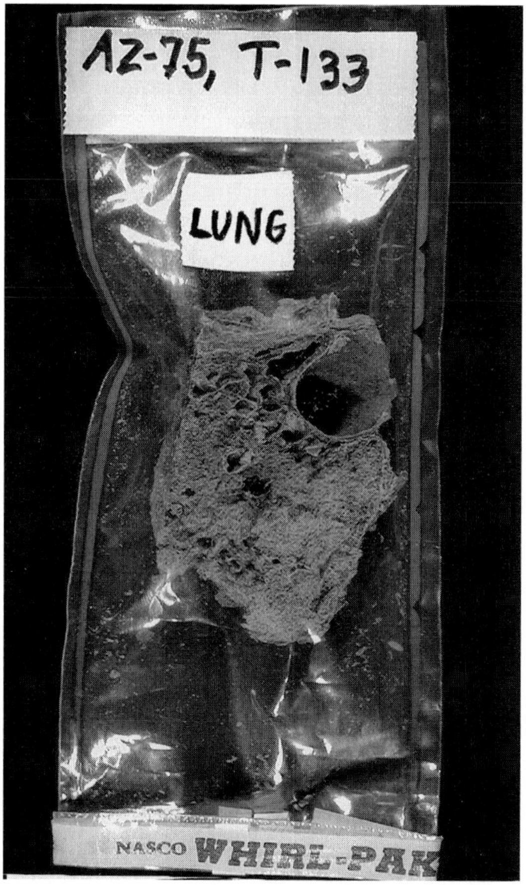

Fig. 6.6. Plastic specimen container.
A desiccated lung sample will remain well preserved in this waterproof, airtight, sterile plastic container, and permanently mounted in custom-fashioned, plastic boxes containing a desiccant. (Photo by ACA.)

filing these specimen-filled bags can be stored in larger plastic sealed bins that include a desiccant.

An average of 25 of the 6 oz bags per mummy has proved to be adequate. For field convenience the 6 oz specimen-filled bags from an individual mummy are placed into larger (24 oz) similar bags. Usually all of the collected containers from one autopsy can be put into one or two of the larger bags. Additionally, several larger bags are also desirable for bulky specimens. For estimation of needs, 25 6 oz and five 24 oz bags suffice for one mummy's specimens. Since Whirlpak™ bags have been sterilized prior to sale, these are satisfactory also for specimens intended for DNA analysis. Specimens acquired for radiocarbon analysis can be wrapped in several layers of heavy-duty, sterilized aluminum foil. *After* they have been well wrapped in such foil, the foil-enveloped mass can be placed into a plastic bag for security.

Performance of mummy dissection

A current copy of my mummy dissection protocol is reproduced in Figures 6.7–6.10. It is in a state of perpetual revision and is altered to meet the unique and variable needs for each field trip. Its entries are arranged approximately in the sequence in which the operations and observations are performed.

External examination

The complexity of unwrapping a mummy depends on the nature of its wrappings. Fur and even hide wrappings commonly retain enough flexibility to permit separation of these items from the skin in a reasonably intact manner if meticulous care is used. Problems in the perineal and other dependent areas are common. Conditions sufficiently arid to desiccate a body prior to its decay will also desiccate textiles. Vegetal fibers, however, become very brittle in such a desiccated state. Fiber breakage is common upon separation efforts, even in flat, even areas, and is predictable on curved contours. Practical solutions to this problem are uncommon. If textile preservation is highly desirable, some form of moisturizing before attempted removal is essential. This, however, will hydrate at least the mummy's more superficial tissues, inviting microbial growth. If a soil specimen was not obtained in the field, the soil commonly found between the folds or layers of the wrappings can be harvested at this time, though that adjacent to the skin is commonly permeated by the chemicals found in liquids of a decaying body. It is advisable to carry out the unwrapping stage with the body lying on a large cloth or plastic sheet, so that all soil and small, loose items (teeth, bones, nails, artifacts, etc.) can be collected and the soil sieved for such structures. I have found the number of mummies in whom

1

MUMMY AUTOPSY PROTOCOL
PALEOBIOLOGY LABORATORY
UNIVERSITY OF MINNESOTA DULUTH
SCHOOL OF MEDICINE
DULUTH, MINNESOTA 55812

Date________________ Site Name______________________________
Cemetery No.____________ Tomb No.________________ Body No.__________
Project Director____
Autopsy Dissector____
Age________ Based On____
Sex________ Based On____
Mummification Type Anthropogenic_____ Spontaneous_______ Indeterminate______

Sketch of Body

Description of Funerary Bundle

____
____
____
____
____
____
____
____
____
____

Fig. 6.7. Protocol for complete mummy dissection.
This began as a copy of that prepared by Marvin Allison, Ph.D., and subsequently was adapted for use for non-Andean mummies as well. In addition to the basic four pages printed here, I also use a skeletal sketch for manual indications of lesions and another page for items unique to an investigated population (hair styles, cranial deformation classification, etc.). Dental pages (Figs. 6.29 and 6.30) are also included routinely.

2

External Examination

Total Body Length (Crown-heel)____________ Head Circumference______________
Cranium: Length_______ Width_________ Index________ Deformation?___________
Hair Presence____________ Color__________ Style__________ Insects__________
Ears Presence_______ Lobe Perforation________________ Canal Exostosis________
Eyes Presence________ Orbit Content____
Nose Tampon_________ TNC*___________ Beard Presence_______ Color________
Breasts Presence___________________ Description______________________________
Penis Presence____________ Circumcision____________ Scrotum______________
Labia Presence____________ Prolapse________________Tampon_______________
Nails_______________ Cranial Deformation Presence______________ Type______
Comments____
____
____
____

*TNC, transnasal craniotomy.

Internal Examination
Organs & Pathology Identified

Heart____ Esophagus______ Prostate________
Lung_____ Stomach_______ Ovary, Rt.______
Liver_____ Ileum________ Ovary, Lt.______
Spleen______ Colon_________ Uterus____
Kidney, Rt.______ Rectum_______ Brain_________
Kidney, Lt.______ Bladder_______

	Head	Chest	Abdomen	Arms	Legs	
Bone:	____	____	________	_____	____	x 4 = B
Soft Tissue:	____	____	________	_____	____	x 4 = ST

STI = $\frac{ST}{B}$=_____=

Visceral Index

Value	Male	Female
10	Heart	Heart
10	Lungs	Lungs
10	Liver	Liver
10	Spleen	Spleen
10	Kidneys	Kidneys
10	Bladder	Bladder
10	Intestine	Intestine
10	Prostate	Uterus
10	Penis	Breasts
10	Hair	Hair

Total:

Specimens Collected

____
____
____
____
____
____
____
____

Fig. 6.8. Mummy autopsy protocol, page 2. TNC, transnasal craniotomy; STI, soft tissue index.

3

AUTOPSY FINDINGS

Fig. 6.9. Mummy autopsy protocol, page 3.

4

OSTEOLOGICAL MEASUREMENTS

Humerus Rt Lt
Length _____ ______
Head Diam. _____ ______
Midshaft Circ. _____ ______
Radius Length _______ ________
Ulna Length _______ ________

Stature__________________________
Based on Formula ____________________
Cribra Orbitalia Rt______ Lt______
Porotic Hyperostos _________________
Other Osteopathology

Femur Rt. Lt.
Length _____ _______
Head Diam. ______ _______
Midshaft Circ. _______ ________
Tibia
Length _______ _______
Head Diam. _______ _______
Midshaft Circ. _______ _______
Fibula:Length_______ _______

FINAL DIAGNOSES

Fig. 6.10. Mummy autopsy protocol. page 4. Rt, right; Lt, left; Circ, circumference.

fingernail or toenail specimens could be found was tripled by using this precaution.

Superficial features

An estimate of tissue preservation is desirable. The terms "well-preserved" or "poorly preserved" are too variably employed to be useful. I have created a simple system of estimates resulting in numerical values without the need for instrumental measurements. This is described in more detail elsewhere (Aufderheide *et al.*, 1992) but can be outlined here as follows. The percentage of skeletal tissue present can be estimated simply by assigning five numerical units ("points") to the major body parts (head, chest, abdomen, both arms and legs – each 5 points for a total of 25 points); multiply this total sum by 4 to obtain an estimate of the percentage of skeleton present (completely preserved body with a head, complete trunk and all extremities = 25 points x 4 = 100 points = 100% preservation of bones). Bodies with missing parts can also be estimated easily: subtract $2\frac{1}{2}$ points for a missing whole arm, $\frac{1}{2}$ point for a missing hand or foot, 1 point for a missing forearm or lower leg, etc. The goal is not absolute precision but a rough estimate without required measurements. After this is recorded, exactly the same procedure is repeated for soft tissue preservation estimates. For this purpose soft tissue is recorded as "present" if sufficient skin and/or muscle remains so that the bone is not exposed. The same numerical values are applied to the same body parts to generate a "percentage soft tissue preservation value" (ST) in the same manner as for bone (B). The value for a body part only partially covered by skin and/or muscle can be reduced roughly by the extent apparent. With these two estimates completed, the fraction of the preserved bone that is covered by soft tissue is expressed as the soft tissue index (STI)=(percentage soft tissue preserved/percentage bone preserved) $\times$ 100. Example: preservation of only an isolated head completely covered by skin and/or muscle would have a soft tissue index = 5/5 = 1 $\times$ 100 = 100. That value indicates that the only body part that remained contained all of its bone and soft tissue; the percentage bone and percentage soft tissue values reflect how much of the body's total bone and total soft tissue remain. When these indices are calculated for an entire population of mummies, the standard deviation values reflect the range of tissue preservation in the group. Field experience indicates that these estimates can be made reproducibly in less than one minute merely by looking at the body, yet convey far more information than the usual descriptive adjectives. The quantitative limitations of these estimates are self-evident and subsequent interpretations of these values should not exceed these constraints. In spite of this, the minimal time and effort involved in creating these values results in transmission of a far greater amount of, and more reproducible, information than the traditional "well preserved" or similarly nebulous adjectives.

A similar approach to visceral organ preservation is useful. Ten major organs that are recognizable grossly are listed for males and females in Table 6.2. Each is assigned 10 points if present. Total sum equals the percentage of visceral preservation ("visceral index"). Hair, although not a visceral structure, is included because of its importance as a specimen for analysis.

Table 6.2. *Visceral index calculation*

	Visceral index	
Value (units)	Male	Female
10	Heart	Heart
10	Lungs	Lungs
10	Liver	Liver
10	Spleen	Spleen
10	Kidneys	Kidneys
10	Bladder	Bladder
10	Intestine	Intestine
10	Prostate	Uterus, Ovaries
10	Penis	Breasts
10	Hair	Hair
Total: 100		

Fig. 6.11. Nodular surface of skin over dorsum of right flank; SM, AD 1450.
The nodular skin surface was shown by dissection to be the site of crystals ranging from 1 to 10 mm diameter. Energy-dispersive X-ray analysis demonstrated their nature: sodium chloride. Burial site: marine sand beach. (Photo by ACA.)

Pseudopathology

Several common hazards include the following:

1. Cords used to hold flexed extremities or wrappings in position may have been pulled so tightly that they leave depressions in the skin. These are usually easily detected but they may cause the skin between the depressions to protrude above the level of the surrounding skin to a degree that can be mistaken for a soft tissue tumor.
2. Dry coastal sites, even in areas where rain is rare, may still be subject to marine inundation secondary to tsunamis (tidal waves) or even the common, wind-transported surf spray. This may result in permeation of sea water into the dermis sufficient to create salt crystal formation upon subsequent desiccation. These crystals can range in size up to 10 mm, producing nodular elevations resembling skin eruptions (Fig. 6.11).
3. Facial expressions: variations in the degrees of contraction or relaxation of the many different facial muscles result in distortion of facial features. These alter the shape of the lips, wrinkling the skin of the forehead and about the eyes and raising or lowering of the mandible and eyelids. The resulting patterns produced by these combinations are called expressions. While expressions in the living are largely under voluntary control, commonly they are produced subconsciously. Thus they become the products of often unarticulated thoughts and reflect our emotional response to our introspections. That is to say, they express those unspoken emotions and

hence are termed expressions. During infancy we have learned to "read" expressions on others' faces and translate the messages that such expressions transmit. Indeed, the voluntary creation of an expression consciously assumed to radiate a specific message is not only employed frequently by most of us, but is a technique older than humans. Even some fish and amphibians deliberately alter their facial appearance to threaten approaching predators or to invite courtship.

The perpetual interpretation of others' expressions during the human interactions that are part of our daily social intercourse, therefore, lead us inevitably to read and respond emotionally to the expression we see when viewing the face of a mummified corpse. The facial features of a mummy constitute some kind of an expression similar to that of a living individual. Small wonder, then, that a mummy's "expression" evokes a powerful emotional response within us (see the poem by Phyllis Janik Chapter 10: Mummy as language, literature and film). These are honest and often moving confessions of the authors' or artists' personal reactions to mummies' expressions and they merit admiration and appreciation for their sharing of such feelings with readers and viewers.

This is, however, quite a different matter if we attempt to predict the mummy's emotional state at the moment of death, based on its expression. We must recall that a mummy's expression is involuntary and entirely the product of decay effects. Following an initial muscle contraction shortly after death (rigor mortis), skeletal muscles undergo decay. This degeneration of muscle tissue results in muscular relaxation. The degree of such muscle lengthening before the desiccation process fixes it into a permanent position varies enormously with environmental circumstances. Advanced decay of the face's masseter (chewing) muscle results in a sagging mandible associated with a gaping mouth. Rapid desiccation, on the other hand, can fix facial muscles into their tensed state, generating a strained expression. Varying combinations of these effects in different muscles shaped further by wrappings are capable of creating a range of apparent expressions wide enough to include almost all those seen in living persons.

In brief, expressions on mummies represent postmortem artifacts. Much nonsense has been written in efforts to relate a mummy's expression to the emotions experienced by that individual at the moment of death. Figure 6.12 represents a random selection of mummies' countenances, ranging from those that resemble a smile to those that simulate agony. Yet all are artifacts of decay processes and are useless as guides to the mummies' agonal emotions.

Fractures

All fractures must be examined carefully. Identification of an obvious postmortem fracture does not exclude the presence of an antemortem fracture elsewhere. Absence of the patina on the fractured surface that is evident elsewhere on the bone usually is sufficient to localize the time of the fracture to the excavation period or thereafter (Fig. 6.13). The presence of a local hematoma can be used as evidence that the fracture was incurred at or immediately preceding death. It must be kept in mind, however, that one or more blows (especially to the head) may cause death with cardiac arrest so quickly that subsequent further violence may not be accompanied by substantial local hemorrhage.

Wounds

Criteria similar to those described for fractures can often be applied also to skin defects to ascertain their antemortem or postmortem nature. Lacerations over bones are quite easily recognizable (Fig. 6.14) but the most difficult are those sharp-edged, apparent lacerations in areas of skin unsupported by any underlying soft tissue. Such defects over the cheek, thoracic cage or anterior abdominal wall can present major problems, since stab wounds in these locations may lack the hematoma expected from a similar wound in a larger muscle mass such as the thigh. Furthermore, relatively

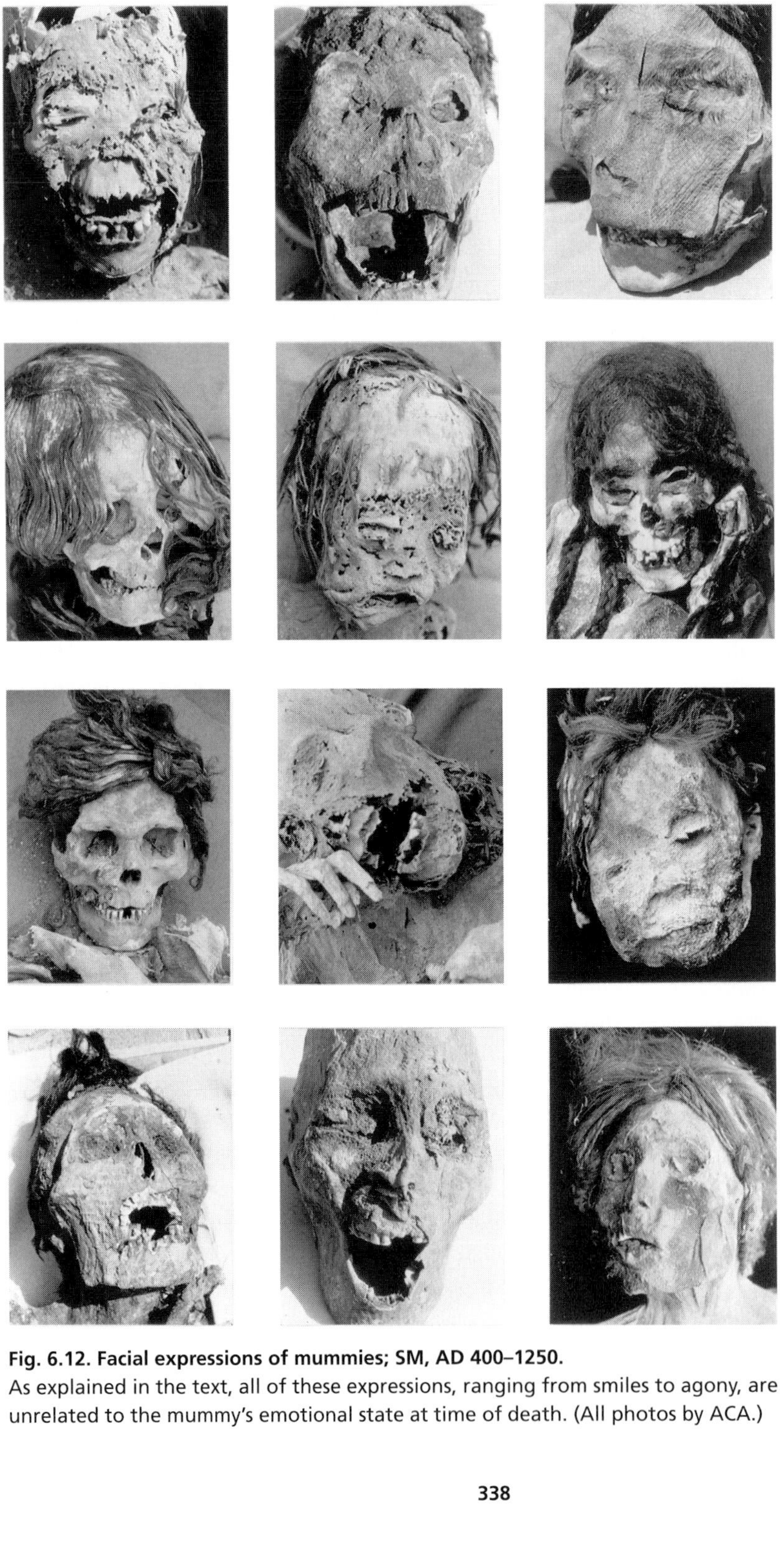

Fig. 6.12. Facial expressions of mummies; SM, AD 400–1250.
As explained in the text, all of these expressions, ranging from smiles to agony, are postmortem artifacts, totally unrelated to the mummy's emotional state at time of death. (All photos by ACA.)

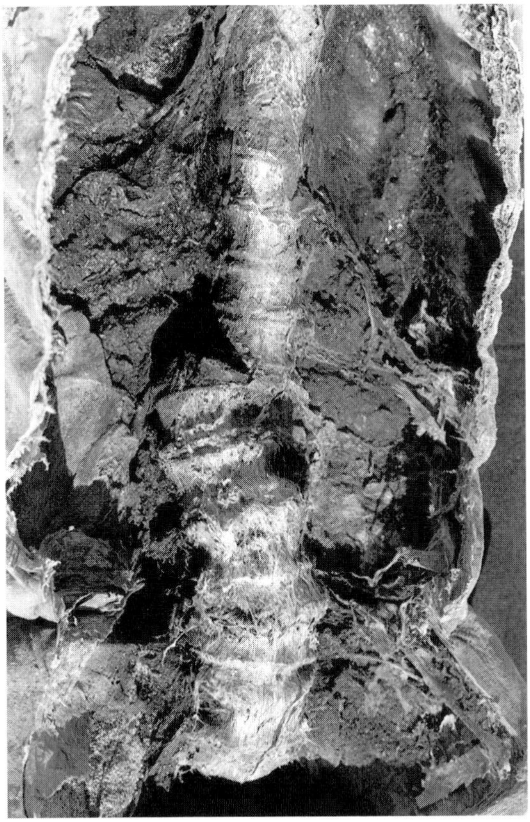

Fig. 6.13. Postmortem spine fracture; AM, 30 BC to AD 395.
The vertebral fractured surface was not resin-stained as was the remainder of the abdominal cavity in this anthropogenically mummified Egyptian body of the Roman Period at Egypt's Dakhleh Oasis. This is evidence that the fracture was not antemortem. Mummy no. 5. (Photo by ACA.)

minor pressure on dry, brittle skin, unsupported by tissue beneath it may easily disrupt the skin's integrity during postexcavation handling of the body.

External measurements

It is my practice to carry out several body part measurements before initiating dissection. If the body dried in an extended position these can be performed reproducibly, provided that discrete anatomical points are consistently utilized. Often, however, both upper and lower extremities are maximally flexed, and the spine is bent into a profoundly kyphotic position with the neck flexed to the point of chin contact with the sternal area (Fig. 6.5). Under these conditions measurement of total body length and certain other direct measurements at a useful level of accuracy may not be possible.

Head

Attachment of the head to the trunk is heavily dependent on paraspinal ligaments. Consequently spontaneous detachment of the head is common when either decay processes affect these ligaments or, more often, if advanced desiccation renders them so brittle that they fracture under conditions of even minor bending forces at the neck. Additionally, in some cultures, the head may serve a unique ritual function ("trophy heads") and for that purpose may already be disarticulated and processed or buried separately (Browne *et al.*, 1993). In most populations looting, either ancient or recent, is the most common cause for separation of the head from the body.

Intentional cranial deformation is common in certain cultural groups, particularly in the Andean area (Soto-Heim, 1987). It is most easily evaluated after soft tissues have been removed from the skull (Fig. 6.15), though the more marked degrees of deformation are identifiable without defleshing. For the southwestern coastal areas of South America, Allison *et al.* (1981) have proposed a useful taxonomy of such deformation types, based on lateral view silhouettes. These patterns result from textile bands that encircle the head with pads or boards placed in strategic locations to achieve the desired elongation or compression effects. Some elaborate ones also involve facial compression. Such an apparatus is termed a "cranial deformer" and is applied during the first months after birth. It is progressively tightened to force skull growth in chosen dimensions. It is usually removed before the third birthday. Some of the subgroups in the classification of Allison *et al.* (1981) may be the consequence of perhaps inadvertent variations in applied tension, with resulting effects on

Fig. 6.14. Antemortem laceration in a mummy; SM, 2000 BC.
The skin overlying the mandible has been lacerated in the perimortem period as evidenced by separation and retraction of the wound edges. The chest was also perforated by two harpoons (see Fig. 4.66). Chinchorro population, 2000 BC. Mummy no. Mo 1–6, T-18. (Photo by ACA.)

growth rates or overriding bones at cranial sutures. Adult skulls demonstrate that sometimes an astonishing degree of cranial shape distortion is compatible not only with life but with normal cerebral function. Nevertheless, undesirable and occasionally even fatal complications can occur secondary to this process, including premature synostosis of cranial sutures or soft tissue and even bone necrosis with resulting infection. The purpose(s) of such deformations remains speculative. It is most probably related to those of other body deformations (enlarged earlobes of Inca nobility, tattoos, etc.) to mark the individual as a member of certain groups (tribes, clans, other) or subgroups (socioeconomic or political status; profession such as shamanism, etc.). In groups among whom intentional cranial deformation is common, infants are often

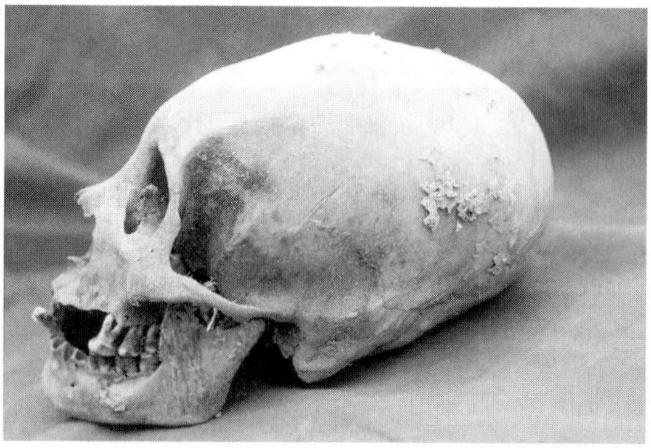

Fig. 6.15. Cranial deformation; SM, AD 1150–1300.
The deformation of this cranium was achieved by placing a textile band around the head that was continually tightened during the first 2 years of life. Chirabaya population, Ilo, Peru. Mummy no. 61. (Photo by ACA.)

Fig. 6.16. Head band ("turban"); SM, 1000 BC to AD 400.
In adults, head bands on mummies are common (to indicate status? Control hair? Other?). Those from the Alto Ramirez population of northern Chile commonly were built up in turban fashion. Such bands bear no relationship to cranial deformers. Mummy no. Cam 15D, T-U2. (Photo by ACA.)

buried with such deformers in position. The interpretation of the identification role of such deformers is supported by the fact that newborns and even probable stillborns also are buried wearing deformers; even a few adults have been excavated in the Andean areas with deformers intact. Turbans, often formed by many circular cords of wool threads, are found in certain cultural groups such as the Alto Ramirez people of the Azapa Valley in northern Chile (Fig. 6.16). These simple head bands should not be confused with cranial deformers.

Hair

Scalp hair is more resistant to decay than most soft tissues, probably because of its low water content, high surface area, protein structure and virtual absence of lysosomes. In conditions that are marginal for spontaneous mummification by desiccation to occur, it is not unusual to find tombs occupied by skeletons with abundant scalp hair but little other preserved soft tissues (Fig. 6.17). It is useful to record hair styles, if any, since they may supply tribal and/or sex information. Arriaza *et al.* (1986) have created an Andean hair style taxonomy based on the number and styles of braids and pleats. The classification has demonstrated that the most complex patterns were most common in males (Fig. 6.18). Practices in other areas may require construction of locally applicable hair style taxonomies.

When recording hair color it is important to know that sun exposure, modern fungus growth or other

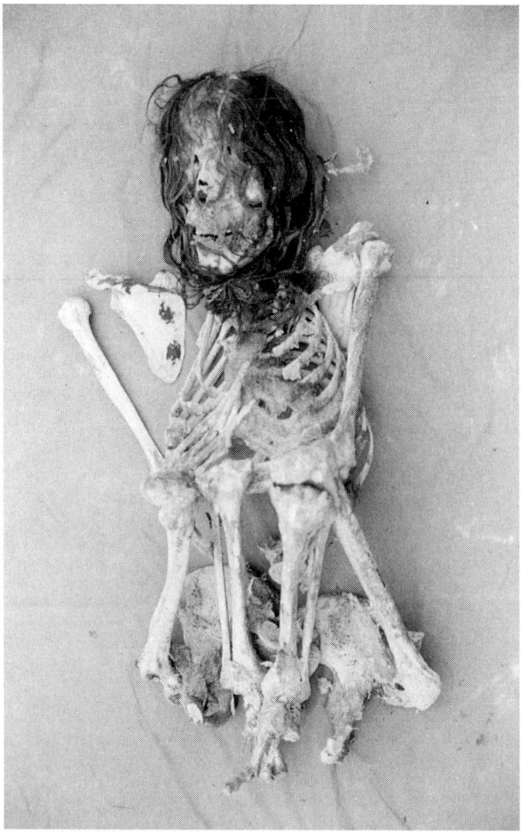

Fig. 6.17. Hair preservation; SM, AD 1150–1300. Hair is usually the last of soft tissues to undergo decay and disappearance. Among spontaneously mummified bodies, it is common to find some that are little more than skeletons but retain abundant hair. Chiribaya population, Ilo, Peru. (Photo by ACA.)

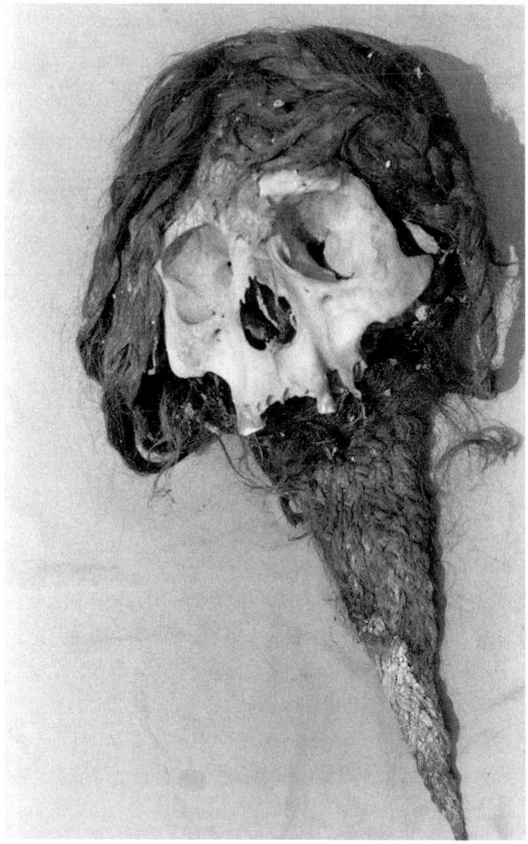

Fig. 6.18. Complex hair styles; SM, AD 1150–1300. In unlooted Andean tombs in the Atacama Desert of northern Chile, hair styles are preserved. Complex hair preparation including both braids and plaited styles are common among males. Chirabaya population, Ilo, Peru. Mummy no. 70. (Photo by ACA.)

environmental conditions may bleach a dark brown or black hair color to that of a blonde or, more commonly, reddish hue. Since this may be a focal effect, cognizance of such potential postmortem alterations of hair color will prevent misinterpretation. Hair color depends on the mixture of black or brown eumelanin with red or yellow pheomelanin granules. Eumelanin is more susceptible to degeneration under most burial conditions. As it fades, the reddish color or pheomelanin becomes more evident (Wilson *et al.*, 2001). Unawareness of such a taphonomic effect may be responsible for the disproportionate frequency of reported "red heads" among mummies.

Mummy hair should always be inspected for the presence of dried adult lice (*Pediculus capitis* or *P. pubis*) or, more commonly, the presence of their egg casings ("nits") attached to hair shafts. Care must be taken in diagnosis of egg casings, since specks of dried epidermis may resemble them. Firm adherence of the egg casings to the hair shaft will usually make differentiation possible. In case of doubt a simple hand-lens will usually settle the issue. Differing rates of lice infestation between various groups may reflect degrees of social contact. Scalp hair inspection also frequently uncovers hair adornments – metal, bone or textile – and these also supply cultural information. Hair has been found to represent ideal sample sources for certain chemical analyses such as antemortem drug ingestion (Cartmell *et al.*, 1991) and stable isotope ratio measurements for dietary prediction (Tieszen *et al.*, 1995a). Its very high sulfur content, however, commonly binds contaminating exogenous heavy metals, negating the prediction of the antemortem body burden of these metals. At least part of the hair sample should be acquired with scalp attached to permit segmental studies that may reflect dietary changes over time (White, 1993).

Ears

Intentional deformation of external ear lobe soft tissues may supply cultural data useful to identify certain population subgroups, especially the elite (Inca nobility; shamans). If the lobes are perforated, the pendants (textile, wood, bone or metal) may generate even more specific information. Especially in groups near marine coasts, careful inspection of the external auditory canal should be carried out to identify the possible presence of benign osteomas (exostoses?) in those locations, since they are believed to be the product of diving or at least frequent exposure to cold marine water – a finding having important implications for subsistence strategy predictions. An otoscope can visualize the tympanic membrane and assist in extracting ear ossicles.

Eyes

The high water content of eyes commonly results in their collapse in a desiccated cadaver. Foreign bodies, often metal plates, may be found under or above the eyelids in some groups. Anthropogenic ("artificial") mummification practices may include the placement of foreign bodies in the orbits (clam shells with superimposed circular, black, centrally placed obsidian disks were placed into some of the *moais* or "stone heads" of Easter Island – Rapa Nui), or clay balls in the orbits of mummified Chinchorro bodies of coastal northern Chile as described by Allison *et al.* (1984).

Attempts have been made to predict eye color by the amount of pigment seen in microscopic sections of hydrated eyes (Andersen & Prause, 1989). One would expect that evaluation of vascular structure in eye sections might be able to predict the presence of hypertension or even diabetes in ancient bodies, but neither these nor other diseases seem to have been sought in ancient, mummified eyes to date.

Nose

Foreign bodies (vegetal, textile, clay) are often found in nasal passages. Their presence is usually interpreted as having the function of a tampon (in contrast to oral foreign bodies, see below). When the cultural group being studied may have indulged in nasal insufflation of drugs, it is useful to seek evidence of nasal septum ulceration or perforation if soft tissue preservation permits. Excessively high frequency of nasal bone fracture can reflect personal violence practices in some cultural groups (Kelley & Smeenk, 1995). Extensive nasal and skeletal soft tissue destruction can occur in syphilis, leprosy or leishmaniasis ("uta"). Egyptian embalmers created access to the cranial cavity by forcing a metal tube through the nostrils and ethmoid sinuses (transnasal craniotomy).

Mouth

Foreign bodies within the oral cavity are present quite frequently. While these may sometimes function as tampons, they are much more frequently interpreted by archaeologists as spiritual offerings. The range of objects found is remarkably wide, and varies with the cultural group being studied. In coastal Andean areas they are commonly of botanical nature – often leaves of the coca plant (Fig. 6.19). A protruding tongue (Fig. 6.20) may simply reflect postmortem effects of gravity, particularly if marked neck flexion is present in a well-preserved body in a sitting position, though sometimes

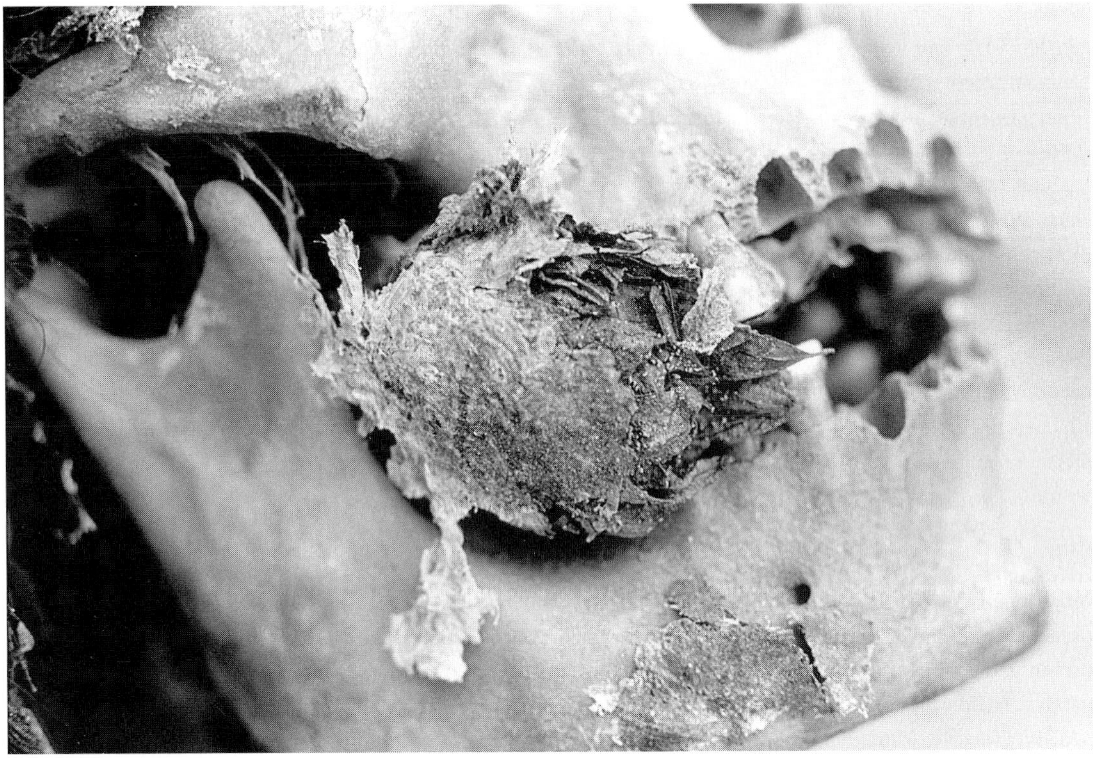

Fig. 6.19. Coca leaf quid in a mummy; SM, AD 1150–1300.
Removal of the cheek soft tissues exposes this quid whose leaves retain sufficient morphology to identify them as *Erythroxylum coca*. Chiribaya population at Algodonal cemetery near Ilo, Peru. Mummy no. 127. (Photo by ACA.)

it may have been pushed out by the internal gas formation of a postmortem fermentative process.

Neck

Soil adherent to the skin of the neck area needs to be removed with great care to avoid loss of beads or other ornamentation. The high water content of goiters frequently causes them to collapse to an unrecognizable degree unless extensive calcification and fibrosis has occurred (Gerszten *et al.*, 1976a).

Thorax

Inspection of the thoracic wall to evaluate its integrity is a useful effort at this point in the study. Defects in the soft tissues of the chest cage, whether by insects, scavengers or trauma are important. They can admit fungi, bacteria or rodents, and these can be destructive to the thoracic viscera. A completely intact thoracic wall is highly predictive of recognizable heart and lung structures within the thoracic cavity in a spontaneously mummified body.

A significant effort should be made to identify the presence of breasts (Fig. 6.21). Sex prediction from skeletal tissue alone is not invariably accurate and degenerative perineal changes often destroy the female external genitalia. Hence unequivocal identification of the presence of breast tissue can make an independent and valuable contribution to sex determination. While one might expect the presence of breasts to be one of the easier evaluations to carry out during a mummy

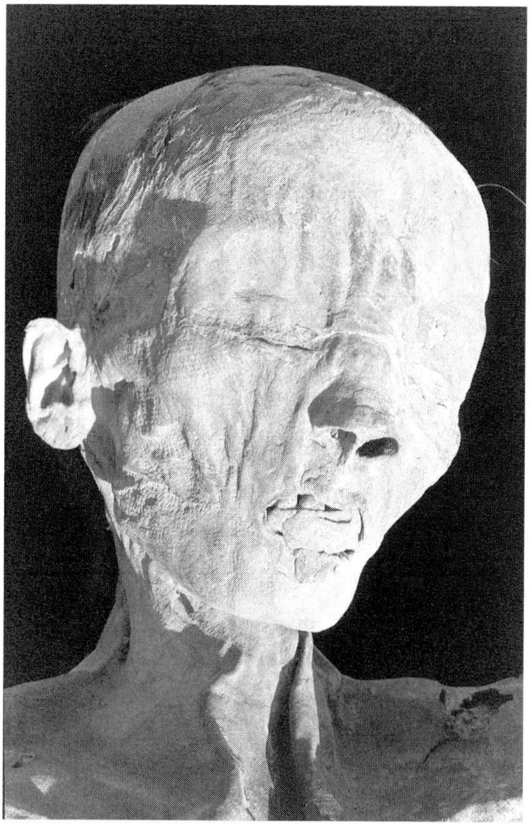

Fig. 6.20. Protruding tongue; SM, 30 BC to AD 395. A tongue apparently pinched between the teeth is usually the result of intra-abdominal, postmortem gas formation whose pressure forces oral cavity structures forward. Spontaneously mummified Roman Period Egyptian body from Kellis 1 cemetery at Dakhleh Oasis. Mummy no. 106. (Photo by ACA.)

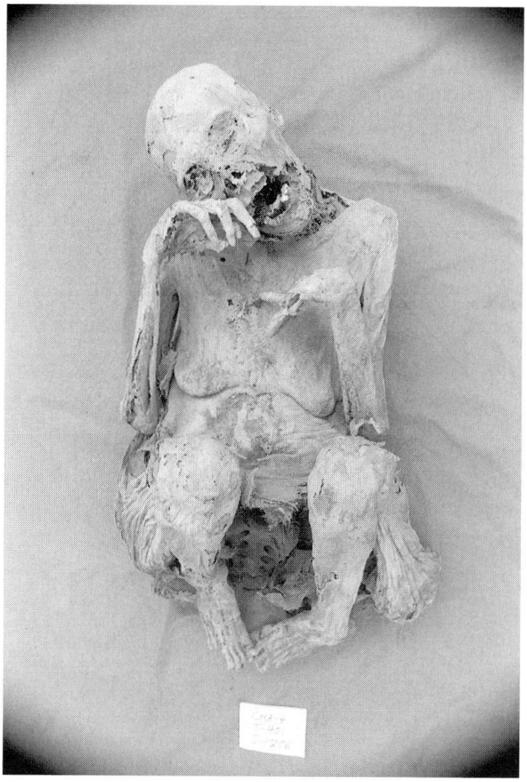

Fig. 6.21. Breast preservation; SM, AD 1150–1300. After death breast fat often hydrolyzes and the freed fatty acids seep away, reducing the size of the breast. The degree of collapse is quite variable, but may make the breasts unidentifiable. In other circumstances, as displayed here, they remain easily recognizable. Chirabaya population, Ilu, Peru. Adult female. Mummy no. 91. (Photo by ACA.)

study, it is often one of the most taxing. The range of variation in breast size (and especially its fat content) is substantial within a living population. It should not be surprising, then, that small breasts of women who have never been pregnant or whose breast content is composed primarily of fat would collapse to a point of imperceptibility in a spontaneously desiccating cadaver. They would be flattened further by tightly wound body wrappings. In such bodies the breasts may be indistinguishable on the basis of form, being recognizable only by their nipples; occasionally breasts of female mummies can not be identified at all. Nipples are difficult to identify among spontaneously mummified males when the body is well preserved, and may not be detectable at all in many. Thus the absence of apparent breast tissue does not exclude the female sex, while the presence of identifiable nipples without detectable underlying breast tissue is not proof of the male gender.

Abdomen

Integrity of the abdominal wall (and that of the dorsal skin as well) has the same implications as does that of the thorax, as discussed above. In one memorable

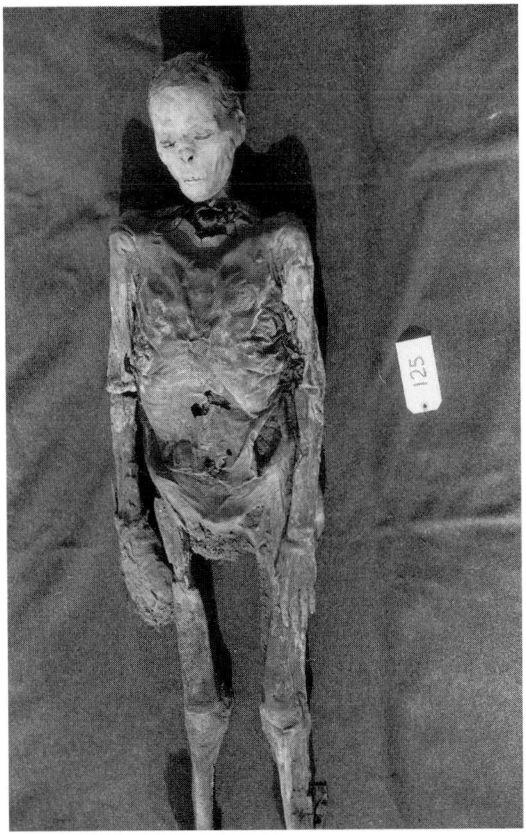

Fig. 6.22. Abdominal evisceration incision; SM, 30 BC to AD 395.
An open evisceration incision is evident in the left side of the anterior abdominal wall in this anthropogenically mummified Roman Period mummy from Kellis 1 cemetery in Egypt's Dakhleh Oasis. Mummy no. 125. (Photo by ACA.)

dissection, gnaw marks at the edge of an insect perforation of the anterior abdominal wall reflected an entry site into the abdominal cavity for a rodent. Removal of the anterior abdominal wall revealed a mouse's nest within the abdomen, laboriously composed of hairs from a piece of camelid hide in which the body was wrapped. Absence of visceral organs suggested the source of the animal's food. Within the nest were adult mouse scats and the skeletons of two infant mice. Additional observations should include a search for evidence of an evisceration incision (Fig. 6.22). If one appears to be present, this may need to be differentiated from postmortem dehiscence of the anterior abdominal wall. This latter term is applied to a condition occurring in a body that is the target of advanced putrefactive fermentation with softening of the abdominal wall soft tissues. Subsequent intra-abdominal gas formation may create more tension than the softened, distended abdominal wall can withstand, with consequent rupture. This is more apt to occur at the site of an abdominal wall scar (McGee & Coe, 1981). This phenomenon, well known in modern forensic medicine, has not been reported in paleopathology literature and I have never encountered it – probably because a body with such advanced decay would only rarely progress to spontaneous mummification.

External genitalia and perineum

Several postmortem factors operate to reduce the possibility of sex prediction on the basis of external genitalia identification. Postmortem body position is usually supine or sitting, resulting in near or actual contact of genitalia with the surface underlying the body in many cases. Contact retards the drying process, providing time for soft tissue obliteration by decay. In addition, these areas are among the most dependent in such body positions, thus becoming the ultimate locale of draining, gravity-directed, enzyme-laden abdominal fluids. In addition the cavernous penile structure may collapse after its blood content dries so that it blends with underlying scrotal skin. The penis is most obviously recognized if its burial position is pointed cranially, draped across the symphysis pubis. I have dissected several spontaneously mummified bodies in which the penis could not be distinguished easily but appeared to lie in the midline, pointing caudally, blending with the underlying scrotum and depressing it sufficiently so that the scrotal bulges on each side resembled female labia. In females the labia frequently blend with surrounding perineal skin so effectively that they can not be distinguished. In such cases only identification of both anal and vaginal orifices can provide positive sex identification.

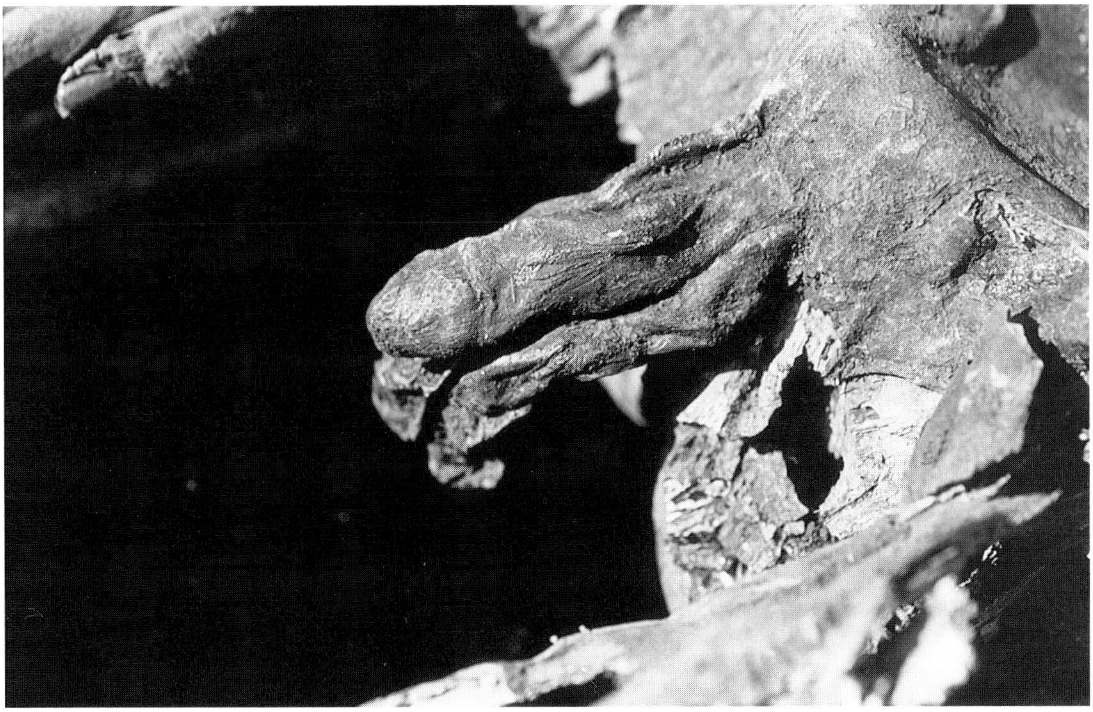

Fig. 6.23. Male external genitalia; SM, 30 BC to AD 395.
Preservation of external genitalia is quite variable. Sometimes, as displayed here, soft tissue preservation is of sufficient quality to identify that circumcision had been carried out. Anthropogenic mummy of Roman Period at Kellis 1 cemetery in Egypt's Dakhleh Oasis. Mummy no. 110. (Photo by ACA.)

I have rarely been able to identify testes with certainty in spontaneously mummified male cadavers. Soft tissue penile preservation quality occasionally is sufficient to evaluate the presence of circumcision (Fig. 6.23).

Both anal and vaginal orifices commonly contain tampons of either vegetal or textile composition (Fig. 6.24). Additionally, postmortem intra-abdominal gas formation may result in sufficient pressure to herniate bowel through the anus as well as the vagina. In one of my cases segments of the female internal genitalia including even ovaries protruded from the vaginal orifice.

Extremities

Their position should be noted in detail because this may be a culturally related feature. This is also true of the details of their enveloping wrappings, including whether the fingers are wrapped individually. Nails are tied into position in some groups. The possibility of gravity-directed postmortem body position change should also be considered if bodies were buried in chambers. Clues to such an event may lie in focal areas of epidermal loss where skin-to-skin contact at time of burial slowed desiccation. For example, in a significantly flexed, "sitting" body the chin may come to rest on the extensor surface of the hands placed on the upper chest, with resulting loss of skin at those sites (Fig. 6.21). Relaxation of the arms in a chamber burial may cause them to sag downward until they come to rest on the tomb floor. The position at time of burial in such a case could be reconstructed by observation of the location of the sites of focal skin loss.

After these observations have been made, nails can be collected as specimens, since further body manipulation often results in their detachment and loss.

Fig. 6.24. Vaginal tampon; SM, 2000 BC.
A breech "cloth" made of reeds is pictured. Perineal tissues in this mummy were poorly preserved, but the labia majora, though detached, remained recognizable, seen here lying on the breech cloth. The labia remain adherent to a mass of parallel reed segments, doubled over and inserted as a tampon into the vaginal orifice. Chinchorro culture. Mummy no. Mo 1–6, T-53. (Photo by ACA.)

Finally the hands need to be inspected carefully, for they are commonly found to be grasping an artifact that may even be secured with a cord.

Anatomical study methods: internal dissection

Extremity and head disarticulation

Following external examination, if the dissection is to include study of the viscera, access to the thoracic and abdominal cavities is normally achieved by removal of their ventral walls. If the burial position was extended, the operator can proceed directly with that step. If, however, the body position has resulted in placement of one or more extremities over the ventral aspect of the trunk, these must first be removed. This discussion will assume that the entire body will eventually be dissected and that all extremities and the neck are maximally flexed (see also Modifications of anatomical dissection techniques, below).

The most superficial (most ventral) extremity (usually leg) is removed first (Fig. 6.25A,B). By means of a rigid-bladed knife with a 15 cm long blade, an incision is made through the skin and all underlying soft tissues, beginning in the groin and encircling the upper tissues of the femur neck region. In spontaneously desiccated bodies dissection of these soft tissue structures will be performed after body cavity examination. The goal at this point is simply to disarticulate the leg at the

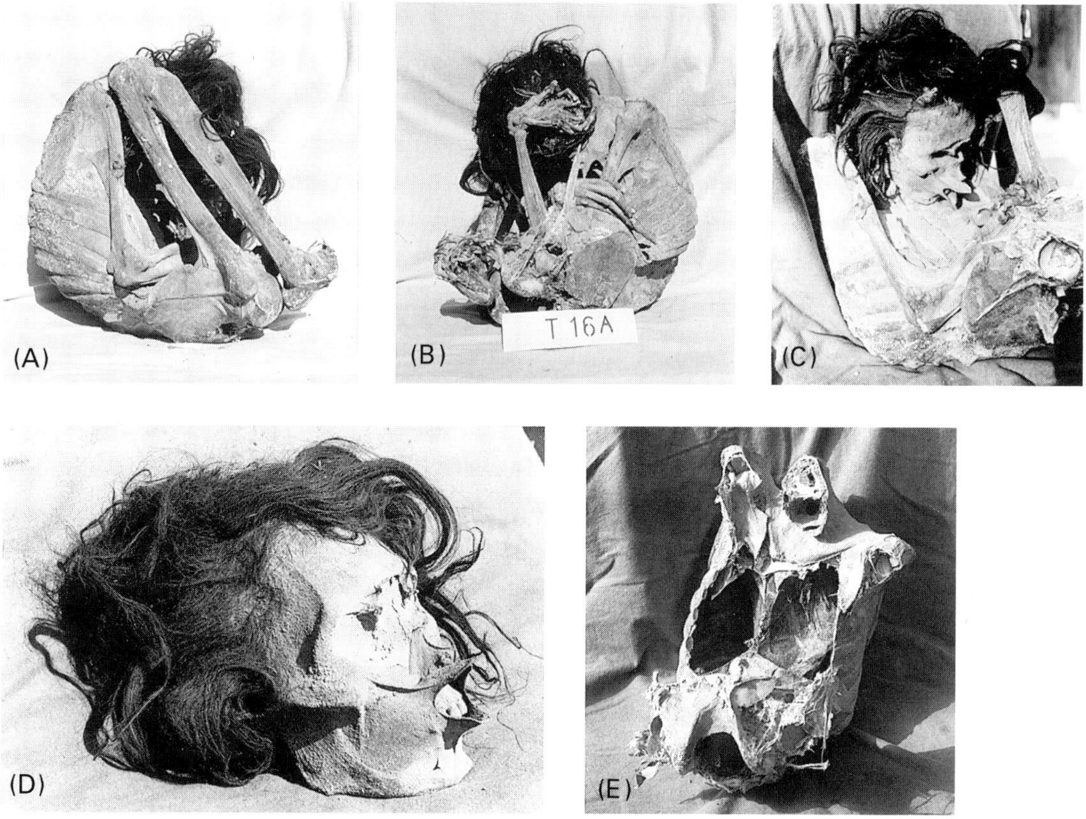

Fig. 6.25. Mummy disarticulation sequence; SM, 1000 BC to AD 400.
This tightly flexed, spontaneously mummified body (A) was subjected to complete dissection. Initially lower extremities were disarticulated (B) followed by disarticulation of arms (C) and cranium (D). After examination of viscera in the torso (E), extensive soft tissue sampling was followed by a complete osteological examination. Alto Ramirez population, near Arica, Chile. Mummy no. AZ-115, T-16 (Photo by ACA.)

hip. This can be done most simply by transecting the rigid, brittle, dry soft tissues as indicated. The extremity is then rotated, usually in a counterclockwise direction. If enough soft tissues have been severed, the remainder commonly split along anatomical planes when the extremity is rotated, and only a few, local, ligamentous attachments need to be severed to permit the femur head to be rotated out of the acetabulum. If attempts to rotate the extremity are met with marked resistance, further soft tissue sectioning should be carried out, since forcing rotation efforts with insufficient soft tissue transection commonly results in fracture of the femur neck.

After disarticulation of both legs has been achieved, the arms may be removed in a similar manner. Here two options may be exercised. An incision may be carried out around the scapula and a flat-bladed knife inserted through the incision to separate the scapula from the thoracic cage. The soft tissues around the humerus head can then be transected. Subsequent rotation similar to that of the leg can then separate the arm with attached scapula. The second option involves a cut through the soft tissues in the plane of the acromion and just under it; this is followed by the usual soft tissue section incision around the humerus head, with subsequent rotation of the arm. This will disarticulate the arm but leave the scapula intact (Fig. 6.25C).

The head is mostly easily disarticulated by an incision beginning at the lower edge of the mandibular symphysis (tip of the chin) and carrying it dorsally in a sawing motion to transect soft tissue structures smoothly along the inferior mandibular edge until the knife encounters the cervical spine. Depending on the degree of neck flexion, this may occur at the C1–C4 level. The remaining muscle and soft tissue structure over the lateral and posterior parts of the neck in the plane of section are then severed and a stiff-bladed knife or saw is used to transect the intervertebral space in the plane of section to complete head disarticulation (Fig. 6.25D).

Thoracic cavity dissection

The initial step involves removal of the "anterior chest plate". Several options are available to achieve this. If the clavicle is to be salvaged intact it must first be dissected out and removed. In dry, brittle tissue of a desiccated mummy, this can be quite time-consuming. If the particular study will not be compromised by clavicular transection, this may be performed and one may proceed to the next step (Fig. 6.25E).

The sternum is then removed. The decision here involves either transecting at the costochondral junction of the ribs with the sternum, or transecting the individual ribs more laterally, perhaps at the midclavicular line or even more laterally. The first option will produce intact ribs for later study, but creates a relatively narrow opening (the width of the sternum) through which all dissection inside the thoracic cavity must be carried out – an awkward activity that also usually precludes photography of intrathoracic pathology *in situ*. Sectioning each rib (and, of course, the soft tissue in the intercostal spaces) in a more lateral location will produce wide access to the thoracic contents, provide an unimpeded view (Fig. 6.26) for dissection and photography, but transect each rib into two segments. Rib (and clavicle?) transection can be carried out with a small hand saw or even scroll saw, but the autopsy saw (Stryker ™ or equivalent) is the most appropriate instrument for this purpose if electricity is available.

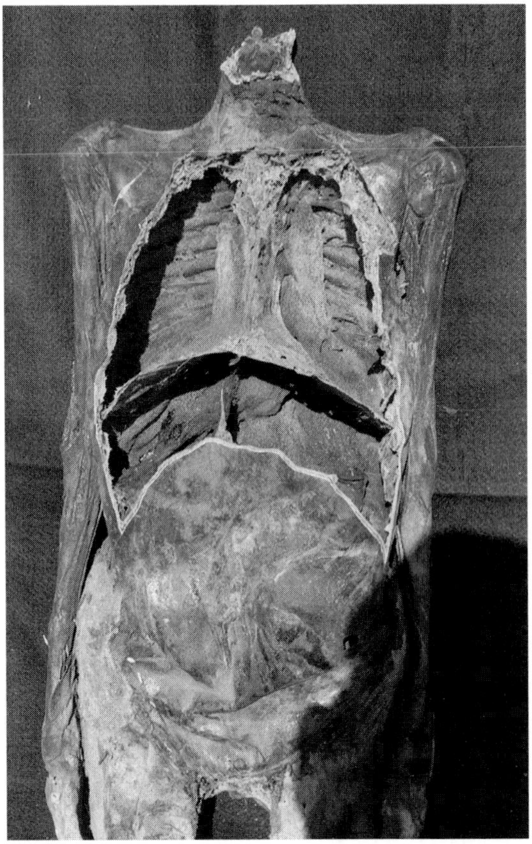

Fig. 6.26. Viscera in a spontaneously mummified, desiccated body; SM, 30 BC to AD 395.
The unopened pericardial sac that encloses the heart is in the chest's midline, and is flanked by two lungs that appear as flat structures extending from the midline about half the distance to the lateral chest wall. The diaphragm is intact and the liver appears as a black structure immediately under it on the body's right side. Roman Period, Kellis 1 cemetery, Egypt's Dakhleh Oasis. 14-year-old male. Mummy no. 1. (Photo by ACA.)

After all ribs (or costochondral junctions) and clavicles have been transected and intercostal space soft tissues cut as well, one other step remains. Mediastinal tissues, including the anterior midline portion of the pericardial sac, will be adherent to the intrathoracic surface of the sternum. If these are not severed but the sternum (with attached ribs) is simply pulled off, these structures will frequently be torn off with the sternum. Thus unpredictable parts of the heart, trachea and other structures may be fractured and torn from their *in situ* positions. These are not only important structures in themselves but may be needed as anatomical landmarks in subsequent dissection. Destruction can be avoided by gently lifting the sternal edge or a transected rib no more than necessary to insert a long-bladed knife with the flat blade surface in a plane parallel to the inner sternal surface, moving it in such a way as to sever all soft tissue connections between the sternum and mediastinum. The blade must be kept closely applied to the sternum to avoid lacerating a preserved, underlying heart. Transection of the soft tissues along the lower rib margin should then make it possible to remove the triangular-shaped "anterior chest plate".

Pleural cavity

Normally the parietal pleural surface is smooth and may even be shiny. If internal soft tissue preservation is good, it will be easy to identify the intact diaphragms. Mediastinal tissues extend from the anterior vertebral bodies to the sternum. Salt-like crystals scattered over visceral and parietal pleural surfaces is evidence of pleural effusion at time of death. If these are accompanied by larger, white, "softer" crystal masses of a powdery consistency (protein, usually predominantly albumin), the effusion was probably of inflammatory origin.

Heart

In spite of care used to separate sternum from mediastinal tissues, inadvertent laceration of the anterior pericardial sac is common. Heart preservation patterns emphasize the dual structure of this organ. The myocardium, rich in proteolytic enzyme content, and slow to desiccate (due to its liquid milieu of pericardial fluid outside and blood inside the heart) is far more apt to liquefy after death than are the lungs. The heart often is represented by little more than a black (hemoglobin breakdown), amorphic stain on the inner surfaces of the pericardial sac. This frustrates detection of antemortem myocardial infarction, as predicted by Zimmerman's (1972, 1978, 1993) simulation studies. The more fibrous or elastic tissue components of the heart valves, chordae tendineae and major arterial trunks, however, are more resistant to proteolysis (Fig. 5.7). Antemortem deformations or diseases of these structures are more apt to retain sufficient gross morphology to permit recognition (Zimmerman, 1993).

Lungs

Except for the major arterial and bronchial tubular structures near the hilum, lung is composed predominantly of capillaries and air. After death, air is absorbed and the high water content of blood disappears during the desiccating process of spontaneous mummification. The result of these changes is postmortem pulmonary collapse, with lung tissue being anchored to mediastinal tissues at the hilum. In bodies desiccated in a supine position the collapsed lung is represented by a 1–2 mm thin, black or brown (hemoglobin breakdown products) membrane. This flat structure is draped over the lateral aspects of the mid thoracic vertebral bodies and over the posteromedial half of their associated ribs, conforming to the rib cage curvature of the inner posterior portion of the thoracic cage in bodies desiccated in a supine position. The right lower lobe of the lung often covers the liver's "bare area". Over this area the lung's visceral pleura may display a fine, granular surface that reflects the irregular, underlying hepatic parenchymal lobularity. This must not be misinterpreted as disease. A previous episode of pneumonia may result in fibrous adhesions between the lung surface and the chest wall, preventing the lung from collapsing after death (Fig. 6.27). The trachea can normally be identified in the neck and can be dissected. Bronchi can be opened distally to about the third bronchial ramifications on each side.

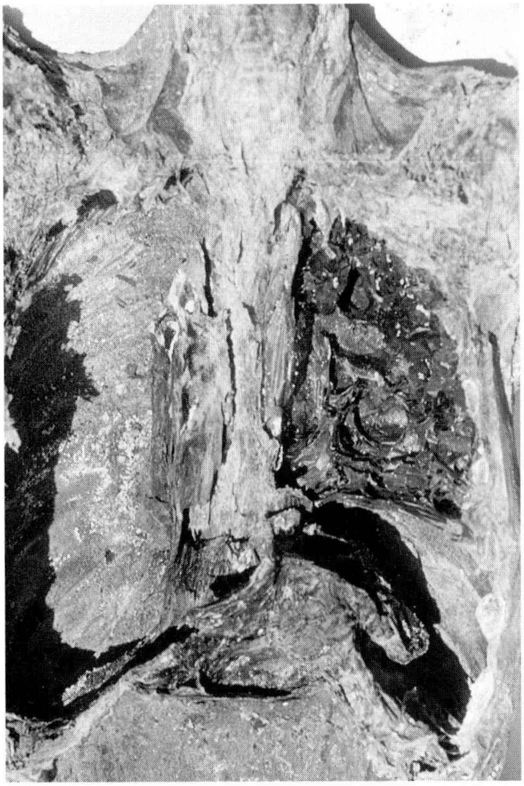

Fig. 6.27. Pulmonary adhesions; SM, 2000 BC. This body's right lung has collapsed into its usual position as a flat plate lying on the inner aspect of the posterior chest. The left lung is firmly bound to the chest wall by adhesions resulting from a past episode of pneumonia, preventing its collapse after death. Chinchorro population, northern Chile. Mummy no. Mo 1–6, T-22. (Photo by ACA.)

Esophagus

Although the esophagus is usually collapsed, it can generally be identified quite easily in the neck at the plane of the transection performed to disarticulate the head. Here it lies immediately behind the trachea. From this point it can be dissected distally. If its junction with the stomach can be identified, it is useful to identify it with a clamp or other marker that can serve as a reference point during the later dissection of the left upper abdominal quadrant.

Thymus

Postmortem blood cell lysis effectively deletes this organ from the field of paleopathology in most mummies.

Neck

If the head disarticulation transection has been carried out as suggested earlier, then the plane of section will probably pass through the upper end of, or above, the larynx. The vocal cords are usually easily noted, but the thyroid is commonly unidentifiably collapsed unless it is diseased. While the postmortem lysis of lymphocytes generally makes recognition of normal cervical and hilar lymph nodes a hopeless goal, these may be quite easily identified if they contain significant amounts of scar-embedded anthracotic pigment or are otherwise abnormal. The esophagus can be identified and traced by its location (see above). Parathyroid glands routinely escape attention in desiccated bodies. Major neck arteries and even veins can be isolated with a regularity encouraging their study.

Abdominal dissection

The abdominal viscera are most easily studied after the thoracic cavity has been eviscerated. The anterior abdominal wall is removed in a manner quite analogous to removal of the sternum except that no bone needs to be severed. An incision can begin in either the left or right upper quadrant laterally, since the removal of the anterior chest wall has already included a horizontal incision along the lower edge of the rib cage. This is then extended around the periphery of the anterior abdominal wall. If this incision is made too far laterally it will result in failure to enter the peritoneal cavity. Similarly the lower, horizontal portion should pass several centimeters above (cranial to) the symphysis pubis to avoid the space of Retzius. Because bowel loops are commonly adherent to the peritoneal surface of the anterior abdominal wall, the latter should not be removed forcibly. The same long-bladed knife used to separate the sternum can be employed in a similar manner to separate adherent structures prior to removal of the anterior abdominal wall.

Dissection of the abdominal organs is commonly dictated by the state of their preservation. In the following discussion it will be assumed that they are all recognizably present. However, the dissector is encouraged to individualize the dissecting sequence as the anatomical circumstances suggest, since postmortem changes and specific pathologic changes may alter anatomical relationships substantially. Because the liver is useful as a "navigational landmark" I usually make it one of the last of the abdominal viscera to be excised.

Gastrointestinal tract

The various loops of small intestine commonly present themselves first to the dissector following removal of the anterior abdominal wall. Their condition varies enormously. Most often the superficial loops are not easily recognized as intestine because they have been flattened against the anterior abdominal wall postmortem and therefore often have no identifiable lumen. As these are peeled off manually the dissector usually encounters at a deeper level loops that retain a lumen and present their usual tubular structure. These are often so thin walled as to be translucent, without recognizable mucosa or other individual layers of bowel wall.

It is useful at this point to focus the dissection on the area expected to be occupied by the colon. To understand what is frequently encountered there, the dissector must be aware that the cecum will commonly migrate during the postmortem interval into the region of the cul-de-sac, where it can easily be mistaken for the rectum. This is particularly apt to happen when the cecum is filled with feces at the time of death. The ascending colon is identified without too much difficulty in most cases, although the inflexibility of desiccated postmortem structures makes dissection at the hepatic flexure difficult. Demonstration of the colon's continuity in this region may fail. If, however, the liver is still *in situ* at this point, a segment of the transverse colon often can be identified and retrograde dissection may solve the "continuity" problem in the left upper quadrant. Throughout the pursuit of the colon, its exposure is achieved by manually peeling away overlying small bowel loops. In the pelvic area care must be exercised to avoid inadvertent destruction of the urinary bladder or other pelvic organs.

Before completing dissection of the left half of the transverse colon it is my practice to shift attention next to the left lower quadrant, where exposure of the descending colon is carried out in a manner not dissimilar to that on the right side. After exposure of the left colon at any point the dissection is carried out in a retrograde direction. This is done because the left upper quadrant contains several organs of interest: the stomach, splenic flexure of the colon, spleen and (retroperitoneally) the left kidney. With imperfect preservation of soft tissues, the spleen and left kidney can be mistaken for each other. Gaseous distention of the stomach or colon may displace the organs of this area. Their movement often is enhanced further by postmortem partial decay processes and gravity effects secondary to body position. Dissection of this area, therefore, benefits from the firm identification of as many anatomical structures as possible. Of these the colon is often the most superficial, and effort expended to identify it first is usually rewarding. The splenic flexure, however, is commonly so high that spleen or stomach may be encountered before the colon dissection can be completed to the level of the splenic flexure. Further dissection in this area then needs to be completed slowly and meticulously, with frequent references to demonstrated landmarks.

The stomach may be surprisingly difficult to identify. If the esophageal dissection has identified its junction with the stomach, the problem is solved. If not, then dissection of this area must expose it by location, confirming its identification by opening it and finding the esophageal – gastric junction from the gastric side. Only occasionally can the duodenum be found, probably the victim of postmortem pancreatic digestion.

After completion of dissection in the left upper

Fig. 6.28. Coprolites (desiccated feces); SM, AD 1150–1300.
Coprolites removed from mummy colon can provide valuable data regarding diet, season of death, parasites and other information. Chiribaya population, Ilo, Peru. Mummy no. 80. (Photo by ACA.)

quadrant, the left colon can then be followed into the rectum. However, I usually delay that dissection until most other steps in abdominal dissection have been completed because of the hazard of destroying delicate pelvic structures (urinary bladder, internal female genital organs) during colorectal dissection. Careful collection (at all stages of dissection) of discrete coprolites or any gastrointestinal content, no matter what its form, will supply specimens that have great analytical value (Fig. 6.28).

Spleen

This is sought during dissection of the left upper quadrant as described above. Its cellular structure renders a normal spleen very susceptible to postmortem lysis. In many spontaneously mummified bodies it does not survive at all in a recognizable form, while in others it is reduced to varying sized lumps of amorphous black material, all less than the size of a normal spleen. Most frustrating of all is the displacement it enjoys because of its mobility. In many cases only the subsequently demonstrated microscopic pattern of its fibrous trabeculae will confirm its identity. On the other hand, spleens affected by disease may be obvious.

Kidneys

The left kidney is profitably pursued during dissection of the left upper quadrant, because its identification can supply another anatomical reference point. Renal parenchymal degeneration is initiated promptly after death – even before bacteria disseminate sufficiently to reach it (Aturaliya *et al.*, 1995). As a result, on numerous occasions in desiccated bodies I have found the renal fossa to be occupied solely by a reniform cavity containing no solid tissue. In most cases some kidney tissue remains, even if it can only be identified by anatomical relationships. Among these the most valuable is demonstration of an attached ureter. Because of this, the preceding left and right colon dissection should be carried out with care to avoid destroying a possible preserved ureter. Pathological kidneys are more easily recognized. Radiological study before dissection of removed kidneys may reveal an unexpected calculus and assist in finding it.

The dissection is then shifted back to the right abdominal quadrants, and the right kidney is sought in a manner similar to that on the left. Here the problem often is to separate it from the liver, and is the principal reason for delaying liver removal until after the right kidney dissection has been completed. Again, ureteral identification is initially attempted because of its value if successful. If this fails, direct but careful dissection of the right renal fossa area is pursued, and the same criteria are used on the right as for the left side. Differentiation from the nearby liver is essential.

Adrenals

I have never identified adrenal glands in spontaneously desiccated bodies, probably because of their size and well-known predisposition to rapid postmortem autolysis.

Pancreas

The same can be said about the pancreas as is stated for the adrenal glands above.

Liver

Perhaps because of its size, at least some liver tissue survives postmortem effects more regularly than any other organ. For the same reason, however, postmortem displacement of the liver can be substantial. On several occasions I have found identifiable liver tissue as an isolated mass external to the body intermingled with soil. This usually occurred in a desiccated corpse so badly decayed that no other viscera could be identified and the body itself approached skeletonization. In other cases I have found an unattached liver in the left upper abdominal quadrant, where its presence could be confused easily with spleen or left kidney. In another instance it was lying loose in the left pleural cavity. In spite of these extremes, its usual position in desiccated bodies is its normal antemortem location (Fig. 6.26).

In desiccated bodies the liver is usually a black, hard, discrete, brittle mass. Postmortem desiccation shrinkage tends to obscure the normal, living liver's gross morphology, although the thinner left lobe can still be detected in many desiccated mummies. The normal gall bladder is only occasionally identifiable, while the extrahepatic bile ducts can usually be recognized only if they are dilated. It is quite probable that postmortem pancreatic autolysis also destroys the distal biliary drainage tract in many bodies.

Pelvic organs

If it has survived, the urinary bladder is commonly in its usual location, although its wall may be only a thin, translucent membrane. Failure to find this organ is common, probably because it empties and dries in the collapsed state. It is probably often inadvertently removed during the stripping of the small bowel loops. A normal prostate is rarely found, probably for the same reason. Nonpregnant uteri may be reduced to a walnut-sized, hard, fibrous-appearing nodule. The fallopian tubes and ovaries have a very fibrous structure, and special care is necessary for their identification. In spite of such precautions, however, it is a common experience to find no identifiable structures remaining in the pelvis of spontaneously desiccated bodies. The action of enzyme-laden peritoneal fluids augmented by colon bacteria (especially *Clostridia* sp.), directed by gravity with accumulation in the pelvis, are probably responsible for the frequent postmortem destruction of these structures.

Head

Following the external examination of the head's surface structures, further dissection is initiated by removal of the eyes from the orbits. A circular incision through the soft tissues around the edges of the orbit will detach all of the desiccated ocular structures from the skull except for the optic nerve. By gently pushing aside these structures the optic nerve is exposed and can then be transected, most conveniently with scissors bearing a curved blade. The dissector needs to be alert for the presence of foreign bodies, often metal plates, especially beneath the eyelids.

An incision through the scalp's midline, initiating posteriorly at the nape and carried up to about the junction of the sagittal with the lambdoid sutures, will assist in scalp removal. Peeling the scalp forward, it can usually be separated from the skull at the level of the orbits and laterally about at the external auditory meati.

Removal of the masseter and other facial muscles and soft tissues will expose the temporomandibular joint area. Further bilateral dissection of this region will permit detachment of the mandible. Completion of soft tissue removal from the malar and other areas then prepares the skull for osteological examination. Preparatory to this the cranial content can be visualized through the foramen magnum. The condition of the brain will vary as dictated by immediate postmortem conditions and body position. If desiccation is slow, the brain may have liquefied and seeped from the cranial cavity. In such circumstances the skull will be empty except for persistence of the dura mater. This can be extracted via the foramen magnum using a Kelly™ clamp. In other circumstances the largely liquefied brain does not seep out of the skull, especially in extended bodies. Instead, it first liquefies, then desiccates, the solutes taking the form of multiple, varying sized granules resembling sand or fine gravel. In still other conditions brain degeneration does not reach the stage of liquefaction. Instead it may persist as multiple, often soft (but sometimes hard), brown, irregular masses 5 cm or more in diameter (adipocere?). Since they rarely contain preserved, intact anatomicalal structure, their size can be reduced by transection and also extracted through the foramen magnum. In the very unusual circumstance that the brain remains as a reduced, but still cerebriform, single structure, it is probably useful to remove the calvarium of the skull and take out the brain intact. Calvarial removal also permits inspection of the pituitary fossa and evaluation of cranial suture closure. In Egyptian mummies the presence of transnasal craniotomy can be established.

Extremities

Samples from tissues never exposed to human contact either in antiquity or during and after excavation and that are designed for biochemical (especially molecular biological) study can be obtained best in many bodies from extremity muscles using an aseptic technique. At least the major joints can also be examined easily for evidence of arthritides. Major arteries can be evaluated for evidence of calcification or atherosclerosis. Following such studies, soft tissue (including nails) can be removed and the disarticulated bones prepared for osteological study.

Mandible and dentition

Removal of the soft tissues composing the floor of the oral cavity will permit identification of the tongue and usually the epiglottis. Careful dissection of the gingiva will expose the mandibular and maxillary teeth to inspection. While study of these belongs properly to the osteological examination, paleodentition specialists frequently are not available in the field. Therefore a brief outline for field collection of paleodentition data is presented. Individual paleodentition specialists differ widely (and often intensely) in the methods that they use to transcribe such data and in the nature of the items recorded. For that reason a form, designed by Odin Langsjoen, D.D.S., in my paleobiology laboratory (Fig. 6.29), is reprinted here and offers the method he devised and uses in my laboratory. Experience has indicated that this simple system is sufficiently flexible to accommodate most interests (Fig. 6.30). Such records can easily be transferred into a computerized version by workers with minimal specialized skills. Furthermore, the printed format is easily "read", yet requires minimal record space.

Modifications of anatomical dissection technique

Preservation of internal soft tissues in spontaneously desiccated mummies has been my principal interest (and therefore experience). Nevertheless, research opportunities can be exploited even with a more restricted type of anatomical dissection. In many circumstances concerns regarding destructive dissection may preclude some of the above-described methods of study. This is particularly true if exhibition of the studied body is planned. Furthermore, alternative methods of mummification will require different study methods, though some of these also permit the application of other instrumentation that will generate unique types of information. For these reasons the following pages will identify a variety of study modifications that can be employed to accommodate such varying circumstances, together with the loss or gain of information related to these alternative methods.

Desiccated mummies

Spontaneous desiccation

Unopened bundles or unwrapped mummies.

Inspection and palpation

Simply inspecting and palpating an unopened bundle can provide strongly suggestive or definite information about the following:

1. Whether the body is articulated or disarticulated (the latter implies minimal or absent soft tissue preservation).
2. Presence of possible artifacts and their general nature (soft=textiles? hard=bone, wood, shell, metal, etc.)
3. Textile nature (wool, cotton, vegetal, etc.) weave patterns, designs.
4. Radiocarbon dating sample of the bundle's textiles.

Radiological studies

Plain X-rays can increase the specificity of the state of articulation data estimated by palpation alone, and greatly enhance the identification of nontextile artifacts. Major fractures can also be identified. CT studies of unopened bundles can increase the efficiency of demonstrating paleopathological changes and better define the morphology of artifacts. Using densitometric techniques, CT radiology may even identify their composition (Notman, 1995). Application of three-dimensional (3D) and other software programs can make contributions unique to these methods. In some cases sex estimation may be possible and certain radiological features (epiphyseal closure pattern, vertebral osteophytosis, etc.) may even suggest at least broad age-at-death estimates.

Opened bundles (or unwrapped mummies)

Inspection

This step can confirm, alter or add further data on the presence and nature of artifacts as well as the presence and degree of soft tissue preservation. Additionally, it adds the very substantial mass of information derived by simple inspection of the body's external surface as described above. This database is enhanced significantly by analytical studies of specimens acquired from biopsies with no or minimal further dissection.

Dissection modification when extremities can not be disarticulated

Endoscopy

This method has distinct but limited applications in mummy studies. The technique consists of the introduction of a flexible optical instrument into a body

Museum ______________ Culture ______________

Registration # ______________ Site ______________

Site# ______________

Access# ______________

M.N.I. ______________

Object ______________

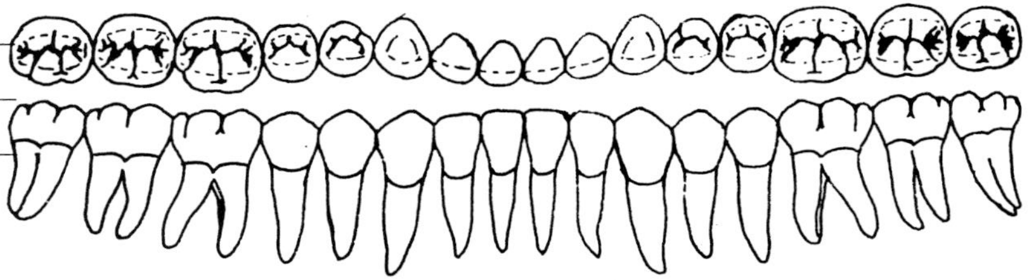

ANNOTATION	32	31	30	29	28	27	26	25	24	23	22	21	20	19	18	17	COMMENTS
1. Present in Arch																	
2. Absent from Arch																	
3. Malpositioned																	
4. Anomalous Form																	
5. Hypoplasia																	
6. Crown Traits																	
7. Attrition																	
8. Calculus																	
9. Caries																	
10. Alv. Abscess																	
11. Alv. Bone Level																	
12. Hypercementosis																	
13. m.-o. dia Crown																	
14. Fa.-Li. dia Crown																	
15. Length of Crown																	
16. Length Overall																	
17. Absent from Arch - Primary																	
18. Present in Arch - Primary																	

PRIMARY ANNOTATION ☛	T	S	R	Q	P	O	N	M	L	K	

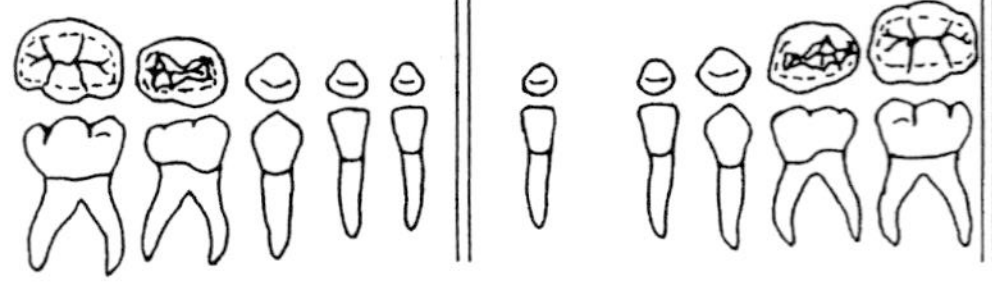

Estimate of Age ______ Sex ______

Photographs ______________

Radiographs ______________

Examiner ______ Date ______

Fig. 6.29. Dental field record for mandibular teeth.

This form was designed and used by Odin Langsjoen, D.D.S., for recording dental observations on mandibular teeth in the field. Alv., alveolar; M, mesial; Fa.–Li., facial–lingual; d, distal; Dia, diameter.

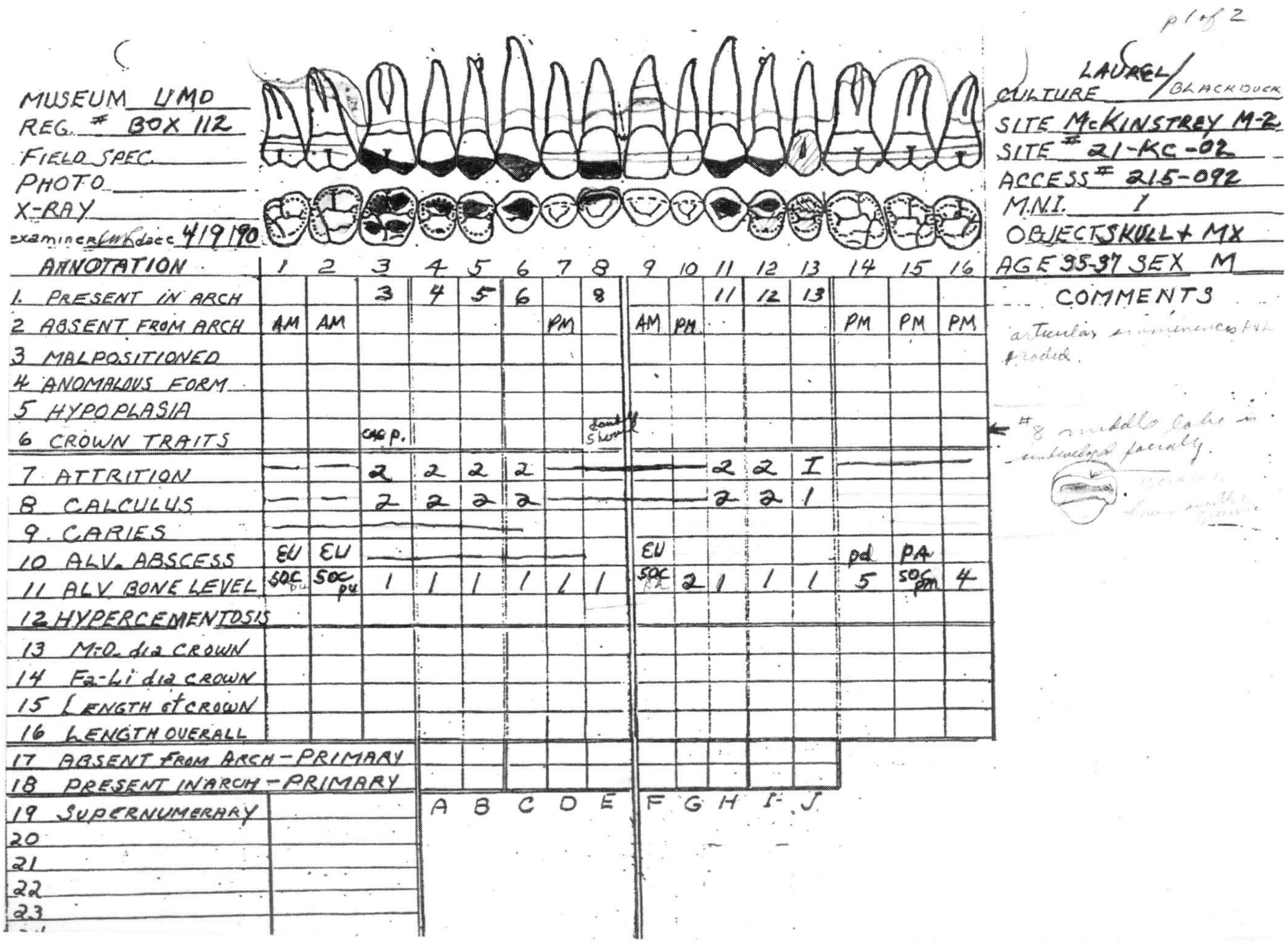

Fig. 6.30. Dental field record for maxillary teeth.
Designed and used by Odin Langsjoen, D.D.S., for maxillary teeth observations in the field.

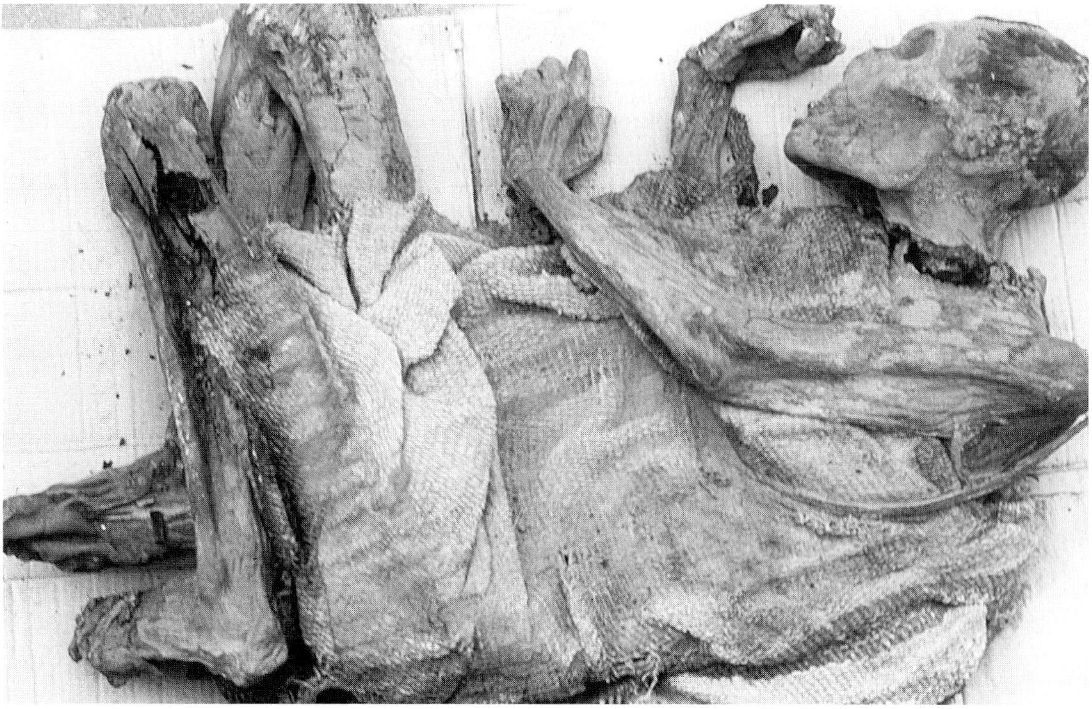

Fig. 6.31. Intact soft tissues on left side; SM, 1000 BC to AD 400.
The spontaneously mummified body lay on its right side in the tomb. Soft tissues on the left side dried sufficiently rapidly after death to result in preservation. Alto Ramirez population at Pisagua site, northern Chile (see Fig. 6.32). Mummy no. PSG-7, T-723. (Photo by ACA.)

cavity (abdominal, thoracic, cranial) in order to visualize and biopsy organs and tissues. The method is described in detail in the section Physical methods: Endoscopy, below.

Visceral dissection without extremity disarticulation or ventral dissection

Occasionally body wall defects due to natural decay can be employed when any dissection of the mummy's ventral surface is precluded because the body is going to be displayed after the study is completed. Contact of the mummy's body surface with any other surface commonly retards those contact areas' desiccation and subsequent decay follows. This is very common in the perineal area of a body in a sitting position after death and is especially frequent in the dorsal or lateral areas of bodies in dorsal or lateral supine positions. Decay may destroy an entire lateral chest and abdominal wall sufficiently to provide access to those viscera without entry dissection, permitting display of the opposite, well-preserved side (Figs. 6.31 and 6.32).

In circumstances under which both thoracic and abdominal walls are intact, yet no ventral dissection is desired because of intended exhibition of the mummy, a dorsal (posterior) approach can be employed. The position of the body during dissection will be dictated by its fixed, postmortem position. If that is an extended position, the body can be turned on its side, or placed in a prone position with its rigid extremities and head appropriately supported and protected (Fig. 6.33). If, as is more common, the upper and lower extremities are maximally flexed, the body is usually more easily dissected if it is placed in a vertical, sitting position, again with extremity and general support.

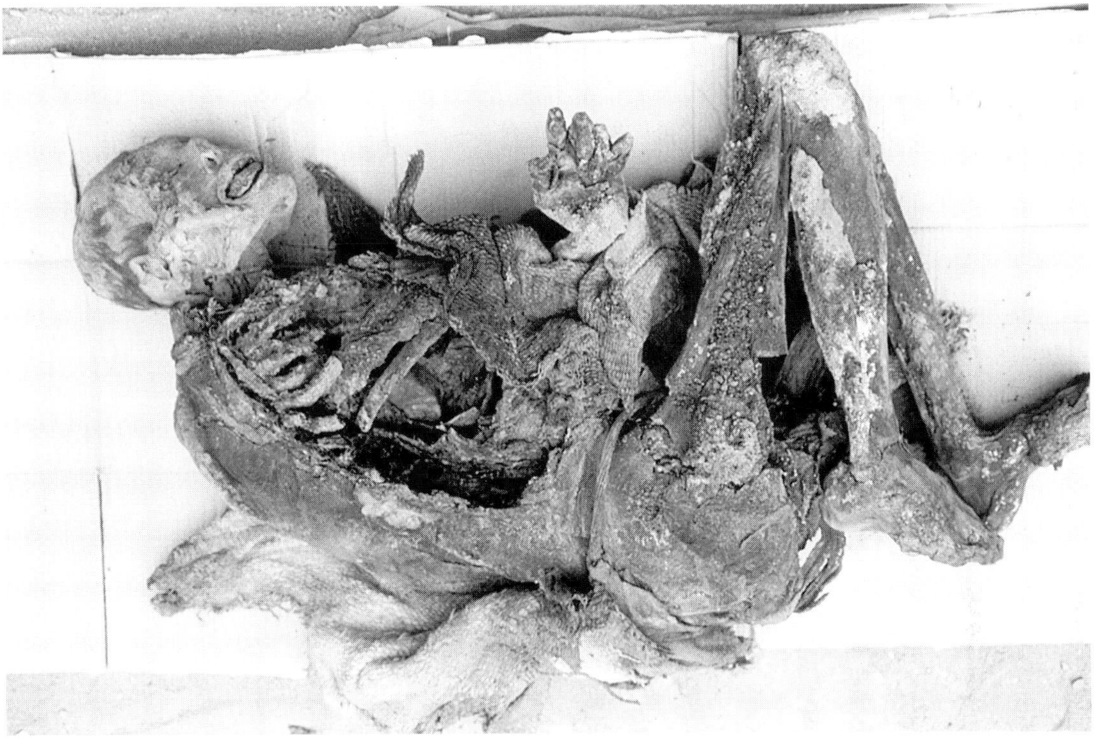

Fig. 6.32. Decayed soft tissue on right side; SM, 1000 BC to AD 400.
Same mummy as Fig. 6.31. but opposite side. Lying on its right side in the tomb, the mummy's soft tissues on that side did not desiccate fast enough to prevent decay there. This provided an opportunity to sample the viscera without dissection, yet permitting the mummy to be exhibited with its left side exposed. (Photo by ACA.)

The most useful area to create access through a posterior chest wall defect is usually the left lower thoracic area, though in certain individuals the pathology revealed by radiological study might indicate otherwise. Using a Stryker™ saw the fourth to the tenth ribs can be transected just lateral to their attachment to the spine medially. Location of the lateral line of transection of these same ribs will depend on what the exhibit plans will permit. The more lateral this is made, the wider the aperture and therefore the more effective subsequent dissection will be. Cutting of the rectangular portion of chest wall to be removed is completed by horizontal transection of soft tissues between these two lines in the intercostal spaces (Fig. 6.34). Before removing this chest wall section the operator should use a knife with a long, flat blade to separate soft tissue possibly adherent to the internal surface of the chest wall.

This approach provides relatively good access to the heart, lungs and usually also liver, the right colon and often right kidney. Recognition and dissection of the stomach, spleen, left colon, left kidney and pelvic organs are far less predictable, while the larynx and esophagus are usually not reached through this aperture. The virtue of this approach is that considerable information can be derived from the viscera that can be recovered. Its disadvantage lies in the loss of potential information that the unrecoverable viscera might have supplied with a conventional approach. It should also be recognized that it will be extremely difficult to evaluate an area of pathology extensive enough to alter substantially the normal gross anatomy, as might occur, for example, in pneumonia with a lung abscess, and associated empyema perforating the diaphragm to form a subdiaphragmatic abscess. Either the removed spine

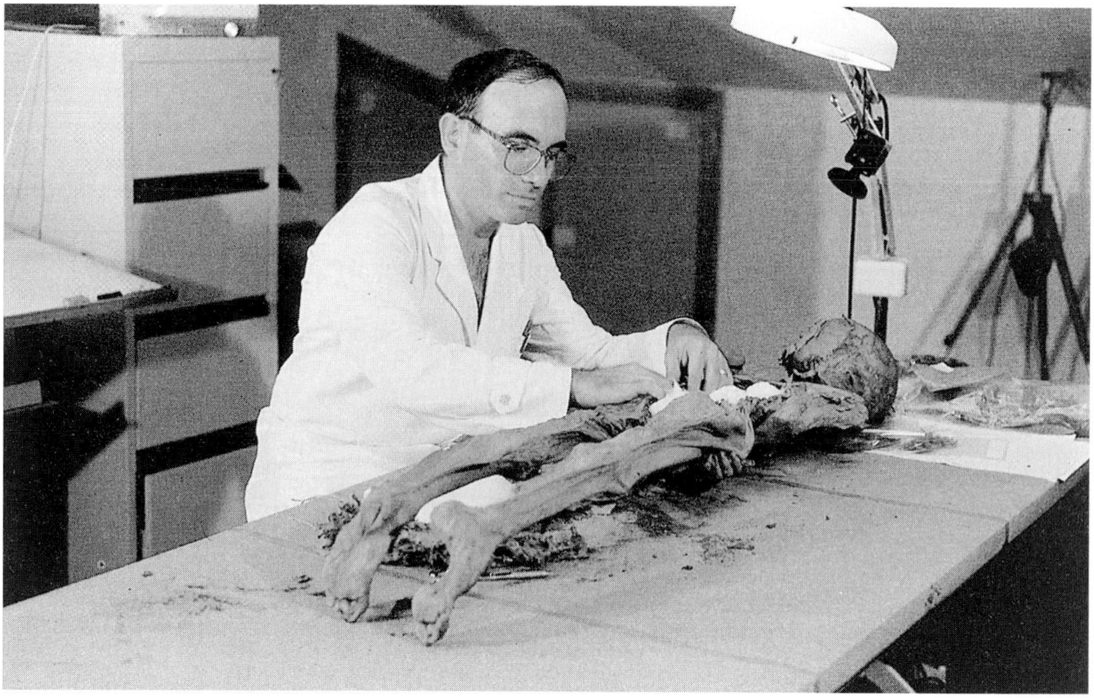

Fig. 6.33. Mummy dissection via the back; SM, ca. AD 1050.
The back of the thoracic and abdominal cavities of this spontaneously mummified child have been opened to gain access to the viscera for their study by Dr Gino Fornaciari of the University of Pisa (Gieseppina-Banchieri, 1993). This approach was selected to permit the body to be exhibited in a manner exposing the body's ventral side (see Fig. 6.34). (Photo by and courtesy of Dr Fornaciari and by the Director, Musei Varese, Italy.)

segment needs to be reinserted following the dissection, or a wood splint can be employed as a stabilizing prop.

Anthropogenic mummy desiccation

General principles

The questions to be answered quite routinely in this group of mummies involve the treatment of the wrappings, body cavities, extremities and head.

Wrappings

Their source (human or nonhuman), nature (vegetal, animal, textile), preparation (raw or tanned hides, weaving patterns), and application (specific wrapping of extremities, trunk, etc.) are routinely required observations. Equally important is a search for artifacts commonly included within wrappings. Examples include the goat hides in which the Guanches on Tenerife (Canary Islands) wrapped their mummies, the burial mats made of split reeds by northern Chile's ancient Chinchorro people and, of course, the elaborate linen wrapping ritual of Egypt's New Kingdom and Late Period mummies.

Body cavity

The principal piece of information here is whether or not evisceration was carried out. It is important to determine the presence or absence of an abdominal wall and, if present, whether or not it displays an evisceration incision (Fig. 6.35). If evisceration has occurred, the operator must determine whether

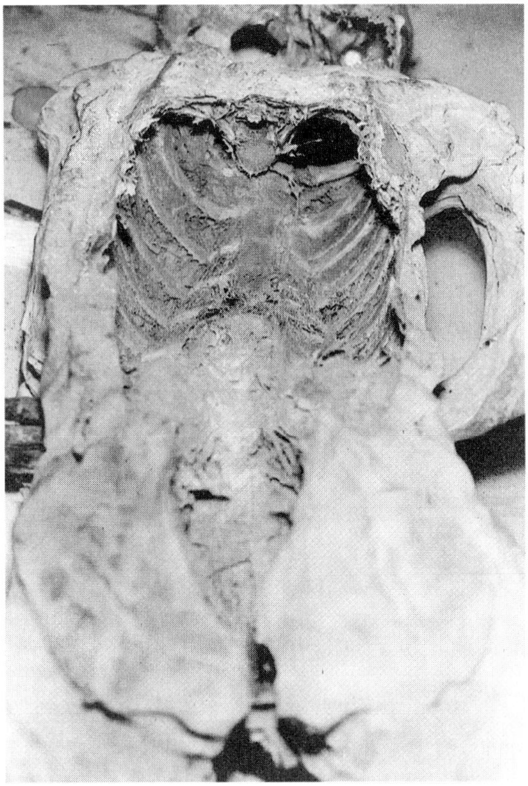

Fig. 6.34. Dorsal view of body eviscerated from the rear; SM, ca. AD 1050.
The body seen undergoing dissection from the dorsal aspect in Fig. 6.33 is here seen after completion of the removal of the posterior body wall and spine followed by evacuation of the viscera for study. (Photo courtesy of Dr Gino Fornaciari and by the Director, Musei Varese, Italy.)

foreign items (hides, soil, artifacts, etc.) were inserted to replace the viscera (Fig. 6.36). If neither an evisceration incision nor visceral organs can be found it is important to evaluate whether the organ absence is consistent with other evidence of natural decay or whether the perineal area reveals evidence of evisceration through that route.

Extremities

In addition to noting details of position, the hands and feet need to be inspected for evidence of defleshing, tying together of digits, tying fingernails into position and presence of artifacts in hands. Evidence of defleshing arms and legs should be sought. If present, the nature of materials, if any, used to replace the removed soft tissues needs to be determined. These may be quite varied, complex and multilayered (Allison *et al.*, 1984)

Head

The presence of excerebration attempts, their nature and their success must be evaluated, as well as replacement with foreign materials. Foreign objects (clay, shell, etc.) in the orbits are common in anthropogenic mummies and should always be sought. The nature and structure of facial masks are data useful for cultural interpretation. The operator also should be alert to the fact that many forms of anthropogenic mummification result in hair loss; hence, the use of adult hair as replacement in children, or wigs composed of vegetal, fur or wool in all age groups is not unusual (Fig. 6.37).

Methods

In addition to radiological studies, the principal methods of ascertaining the above conditions is simple dissection. If, as is common, museum exhibition of the studied body is planned, dissection of the posterior surfaces only or solely unilateral dissection of extremities is a practical compromise. An approach to the body cavities other than ventral is less apt to be successful in supplying the information sought in anthropogenically mummified bodies. Much, but usually not all, of the information desired about head treatment can be obtained by limiting dissection to the posterior aspect of the head.

The operator must remember that the various mummifying procedures and materials may alter the tissues sufficiently that analytical results from biopsy specimens (especially in radiocarbon dating specimens) may result in artifactual effects (Nissenbaum, 1992).

Applications

There is probably no better studied group of anthropogenically desiccated mummies outside of ancient

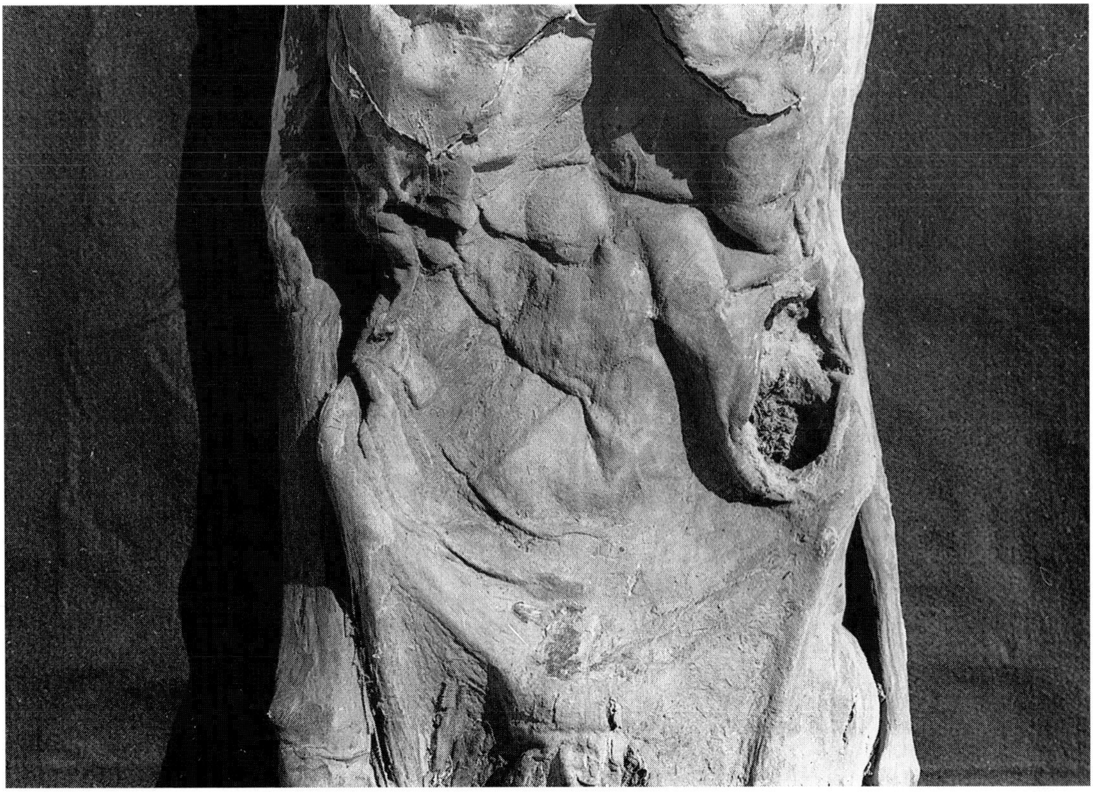

Fig. 6.35. Evisceration incision in abdominal wall; AM, 30 BC to AD 395.
The incision in the left upper quadrant of the anterior abdominal wall is the usual location prior to Egypt's New Kingdom in the Nile Valley. Roman Period, Kellis 1 cemetery, Dakhleh Oasis. Mummy no. 3. (Photo by ACA.)

Egypt than those of the Chinchorro people. This maritime, Preceramic group of sea hunters appeared at the coastal sites of northern Chile about 9000 years ago, probably shortly after the more major Andean glaciers had retreated from the Andean peaks. While the remainder of their cultural traits differed little from the other maritime coastal populations that ultimately extended northward as far as Panama, the ritualistic mummification of at least part of their population was unique. The details of this are recorded elsewhere in this volume (Chapter 4: Coastal and low valley sites near Arica, Chile), but the descriptions of these mummies can serve as useful models for the dissection of anthropogenically mummified bodies. Beginning at least about 5000 years ago, these bodies demonstrate a highly complicated alteration of body tissues at time of death. After removal of the skin, the trunk was eviscerated and dried, apparently by inserting glowing embers and hot ashes into the body cavities; these were then filled with any of several foreign materials including clay, soil, hides, vegetals or other items. Extremities were defleshed, the bones splinted with sticks and the form of the removed muscle and other soft tissues was simulated with material similar to that placed in the body cavities. The vertebrae were individually isolated, cleaned of soft tissue and then splinted with one or more sticks (sometimes placed into the spinal canal). The disarticulated head was excerebrated via a calvarial defect, filled with clay and impaled on one of the splinting sticks projecting from the neck. A clay mask

Fig. 6.36. Body cavity filling; AM, 30 BC to AD 395.
Rolls of resin-saturated linen were employed to replace the viscera in the mummy of Fig. 6.35. (Photo by ACA.)

commonly covered the face (Fig. 6.37) and the entire body was painted black (magnesium) or red (ocher). The body's own skin or that of an animal was then used to replace the originally removed skin. Later the process was simplified, involving external coating of the body with a cement-like sand mixture until all intentional mummification practices ceased about 4000 years ago.

Now that Allison *et al.* (1984) have established how these bodies were preserved, it is clear that application of the above principles would make it unnecessary to dissect the entire body in all subsequently excavated Chinchorros. The body cavities would need to be examined sufficiently to identify evisceration and the replacing materials. Only parts of one arm and one leg would require evaluation of how extremities were treated. Focal areas of dissection of the head could establish how it was treated. Perhaps X-rays could clarify the method used for spine support. Careful reading of the descriptions by Allison *et al.* (1984) can teach a scientist who is anticipating examination of this type of mummified body how to plan the dissection (Bittman & Munizaga, 1976; Allison *et al.*, 1984; Aufderheide *et al.*, 1993).

The mummification of Egyptian bodies, at least after the Old Kingdom, is also discussed elsewhere in this volume (Chapter 4: Mummies from Egypt and elsewhere in Africa). While the details differ considerably from those in the Chinchorro people, application of the general principles as was suggested above will be equally effective for Egyptian mummies. It is obvious that knowledge about mummification details,

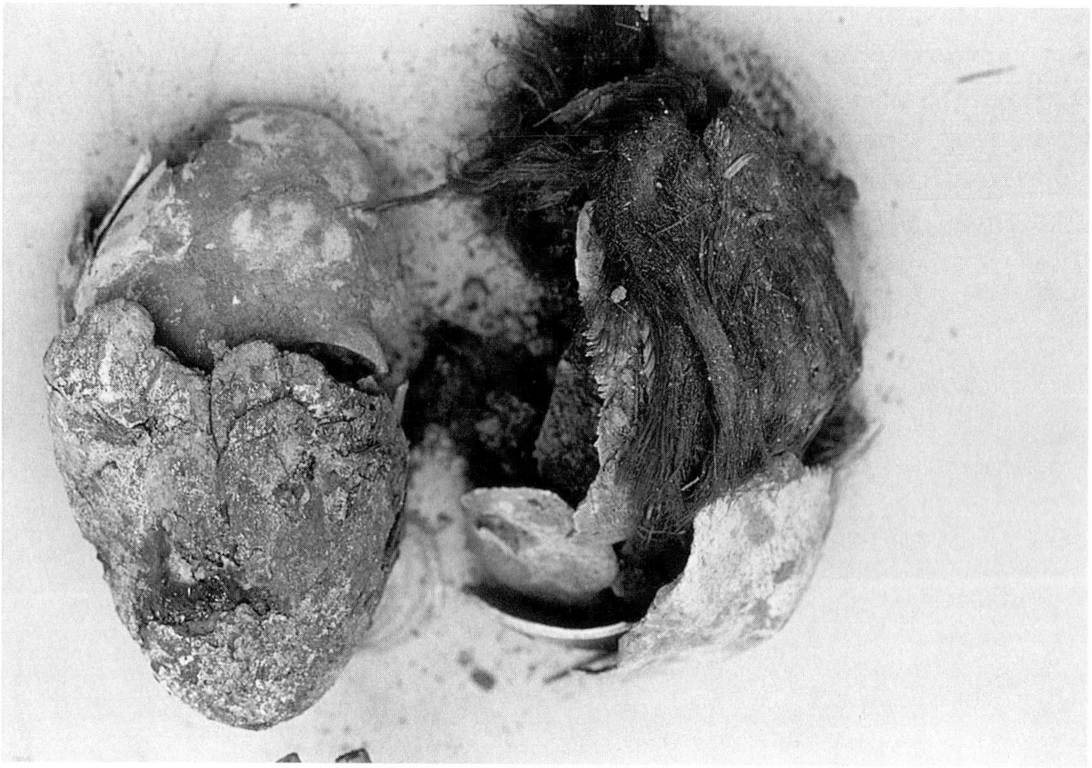

Fig. 6.37. Mummy's face mask; AM, 4020 BC.
The facial area of the Chinchorro population's anthropogenic mummies (5000 BC) was usually covered by a clay mask with a wig of hair or wool. From Cam-17 site at the mouth of the Camarones River, northern Chile. Mummy no. Cam-17, T-1, C4. (Photo by ACA.)

if available, will be helpful in planning the dissection procedure.

Chemically preserved mummies

General principles

I use "chemically preserved" to include those bodies in whom the primary preservation mechanism involved the action of chemicals other than those used to induce desiccation (e.g., natron) or to minimize subsequent rehydration (e.g., resins). The best examples we have are those ancient bodies found in peat bogs of northern Europe, a few from the medieval and Renaissance periods in Italy and the modern bodies of the nineteenth and twentieth centuries in America.

Spontaneous preservation of the bog bodies of northern Europe

Hundreds of bodies, most from about 400 BC to AD 400 have been exhumed from the peat bogs of northern Europe during the past century. The remarkable preservation of some of these bodies' tissues has been attributed to the action of sphagnan (see Chapter 4: Peat bogs) in the bog waters. Unfortunately most of these have had no or only superficial scientific study and the tissues of no more than perhaps a dozen or so are avail-

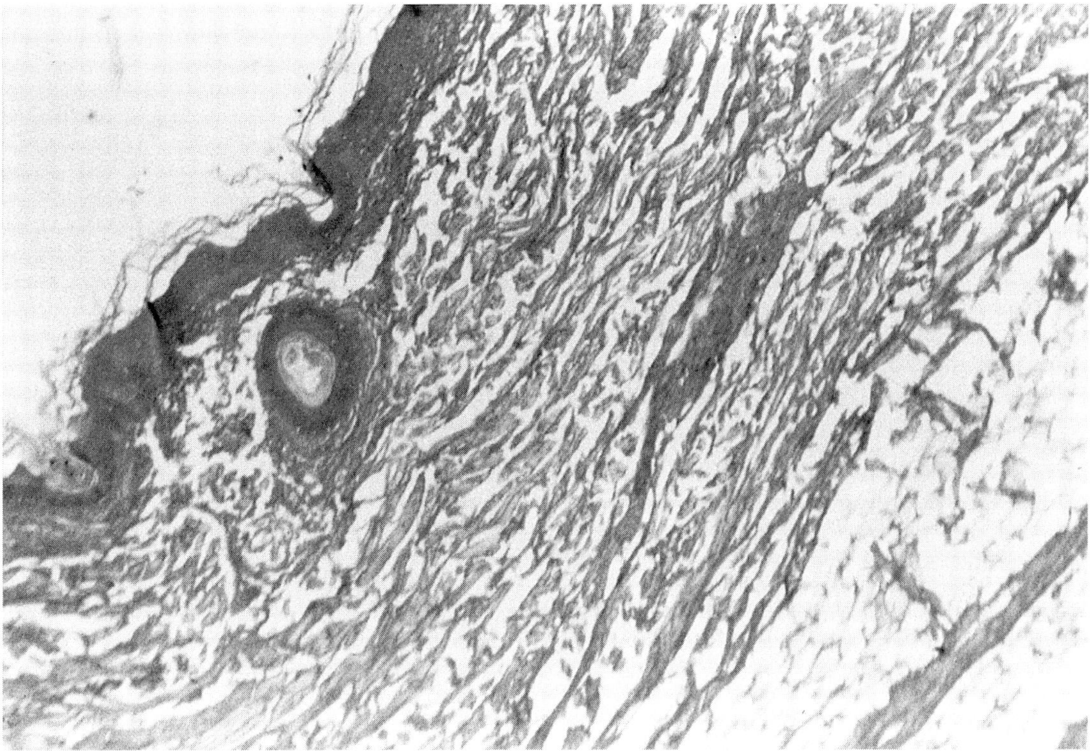

Fig. 6.38. Photomicrograph of mummy skin; AM, AD 1911.
The epidermis, dermis with its vascular structures and subcutaneous fat are all easily recognized in this tissue section of a skin sample from a body preserved by intra-arterial injection of an arsenic solution (Elmer McCurdy; by Snow & Reyman, 1977). (Photo courtesy of T.A. Reyman.)

able for examination today. In contrast to past practices, those found recently have been studied exhaustively, such as Lindow Man (Stead *et al.*, 1986). While these are too few to establish a protocol certain to be useful in all future exhumations of such mummies, several unique features found in these deserve attention and anticipatory planning. Discovery conditions will probably always vary but rapid change in air-exposed tissues can be a serious problem. Skeletal decalcification secondary to the acidic peat waters can be anticipated and have a profound effect on stability of body form, requiring special handling methods in the field. Since these are hydrated bodies, if they fit within the available space of the magnetic resonance imaging instruments, we can expect good anatomical visualization of the soft tissues, though CT will probably also contribute useful skeletal details (if bone mineral is retained). The limited number of published studies on these bodies prevent further generalizations. Those mummy investigators working in parts of the world where they would expect to be called in the event of unexpected exhumation of such mummies would be best prepared to respond effectively if they were completely familiar with the studies published to date (Brothwell, 1986).

A variation of such preservation of brain tissue has been encountered in a Florida "sinkhole". Since those waters have a pH near neutrality, the preserving mechanism is not at all clear, although anaerobiasis and the high mineral content of the water have been suggested (but not tested) as probable mechanisms (Stephenson, 1985; Doran *et al.*, 1986). Dissection modifications are similar to those of the bog bodies. These have permitted recovery, amplification and study of their 8000-year-old DNA content.

Intentional chemical preservation

Fornaciari (1985) has reported several examples of Renaissance mummies in which efforts were made at least to preserve human soft tissues after death by the deliberate application of chemicals. In his most obvious example an eviscerated child demonstrated numerous radiopaque blotches within an externally applied paste. Chemical analysis of this paste demonstrated a mercury compound (Fig. 3.11). The profound alkalinizing effect of lime was employed to preserve the bodies of some slain soldiers during America's civil war in the nineteenth century (Johnson *et al.*, 1990). Intra-arterial injection of heavy metals, especially arsenic (Fig. 6.38), became popular in Europe about the same time and a little later also in the USA (Johnson *et al.*, 1990:23–57; Quigley, 1998:60–70). None of these requires any special modification of study methods other than suspicion of their use, appropriate sampling and subsequent analysis.

Cryopreserved mummies

General principles

To date we have no evidence of human presence on the Antarctic continent during antiquity, and arctic temperatures rise above the freezing point of water at least briefly every year. Consequently there is little precedent for study protocols of ancient frozen bodies. From those that have been studied and reported (Rudenko, 1970; Zimmerman & Smith, 1975; Zimmerman & Aufderheide, 1984; Beattie & Geiger, 1987; Hart Hansen *et al.*, 1991; Spindler, 1994) the following items are of interest. Discovery circumstances are rarely under the mummy specialist's control, but preservation of protein, microbiological study options and many other research opportunities are dependent on maintaining the frozen condition of the body tissues. If thawing has occurred it may be wise to refreeze the body while a research protocol can be established (Lippert & Spindler, 1995). Ice is quite radiopaque, requiring postponement of radiological studies until thawing has occurred (Notman *et al.*, 1987). While thawed bodies can be expected to be good candidates for soft tissue study with magnetic resonance imaging, none has been reported to date; thus this research opportunity should be kept in mind when one is designing a study protocol. While freezing has been demonstrated to reduce the viability of bacteria (Aturaliya *et al.*, 1995), it usually does not sterilize the tissue. Hence microbiological studies may also be included in such protocols. The publication by Hart Hansen *et al.* (1991) of studies on the Greenland mummies is a dramatic example of an interdisciplinary study on frozen mummies and can be read with profit when designing such study protocols.

Table 6.3. *Classification of cost estimates*

Class no.	$US
1	0–25
2	26–100
3	101–500
4	>500

Morphology-based methods: paleohistology

For those who want only an overview of these methods, an abstract is presented here. Following the abstract the individual steps of the procedures are detailed.

The estimated cost of procedures described in this chapter is indicated by the classification listed in Table 6.3.

Abstract of paleohistology

Principle

For light microscopic study, desiccated tissue specimens are first rehydrated, then fixed, followed by dehydration with organic solvents and replacement of the removed water with paraffin. After the paraffin has hardened, the tissue may be sectioned into slices about 6 μm thick by a microtome designed for that purpose. These slices are transferred to a glass slide, the paraffin removed chemically and the tissue stained. A drop of mounting fluid is placed on the tissue and covered with a glass slide to form a permanent mount. The slide is then examined in a light microscope.

To be examined with up to 1000× greater magnification the tissue is similarly prepared but embedded in a polymerized resin and sectioned to produce a much thinner slice for examination by an electron beam in a transmission electron microscope (TEM). For examination of tissue specimens' surface features, a similar electron beam is used to energize the surface and the low energy electrons emitted from the surface area translated into a two-dimensional image (scanning electron microscopy or SEM).

Applications

Light microscopy usually is capable of visualizing tissue alterations characteristic of biological changes affecting the studied tissues. This includes pneumonia, liver cirrhosis, lactating breasts, some parasites or their ova and many infectious agents. Viruses can sometimes be visualized by electron microscopy (EM). Additional measuring methods may be coupled to EM to identify the chemical or crystalline nature of structures found within tissues.

Limitations

The postmortem autolytic loss of many soft tissue elements may delete anatomical changes necessary for the recognition of many diseases, especially those in which the altered cells are epithelial. This limits to varying degrees the ability to demonstrate the diagnostic changes of hepatitis, acute myocardial infarction and most carcinomas.

Cost

For light microscopy the cost is in Class 1 range once the equipment has been acquired. The EM instrumentation is in the Class 4 range but operational cost for a specific specimen is in the Class 2 range (see Table 6.3). The various steps in the procedure are detailed below.

Although gross examination of mummified tissues provides the basis for paleopathological studies, paleohistology can extend such observations substantially in some circumstances. The examining investigator considering the microscopic study of mummified tissues, however, must be aware that autolysis by intracytoplasmic hydrolytic enzymes begins promptly after death. Cells with the greatest concentration of lysosomes will be most vulnerable to such destructive postmortem changes. These are also commonly the most metabolically active body cells. Thus renal tubules, hepatic parenchymal cells, white blood cells and myocardial muscle fibers are usually the least well preserved structures in human mummies, though often all epithelial cells have undergone autolysis in mummified bodies. Since most diseases are recognizable histologically by changes in epithelial cells, the postmortem disintegration of these cells imposes serious constraints upon the information that can be recovered by paleohistological studies (Figs. 6.39 and 6.40). Nevertheless, in many cases alterations of connective, skeletal muscular and elastic tissues can be of diagnostic value. Parasite ova and cysts commonly retain recognizable morphology (Fig. 6.41) as do the nonbronchial pulmonary tissues.

In spite of these limitations the paleopathology literature includes ample examples in which histological studies enabled useful information (Fig. 6.42). Hence I will here review methods of preparing sections of mummified tissue for microscopic examination.

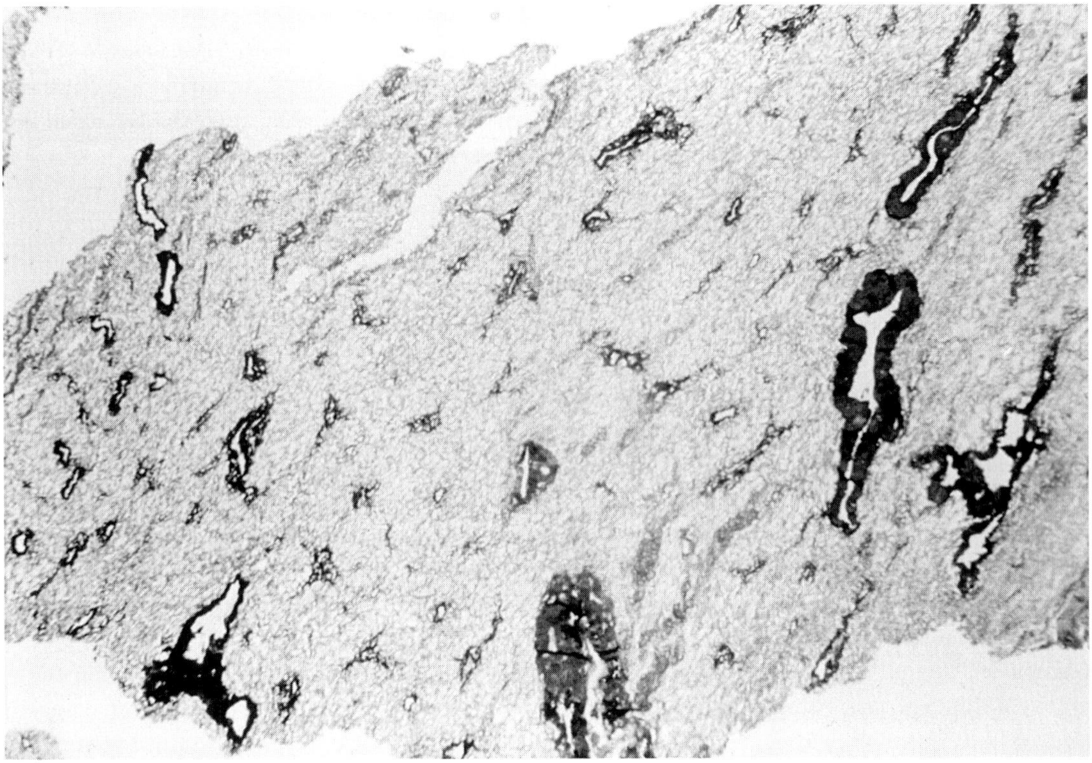

Fig. 6.39. Photomicrograph of mummy liver; SM, AD 400–1000.
The epithelial cells composing most of the living liver have disappeared in this liver tissue from a 1500-year-old spontaneous mummy of the Cabuza culture in northern Chile. The structure of the portal areas, however, survived well. Mummy no. AZ-141, T-36. (Photo by ACA.)

Rehydration of desiccated tissues

Before desiccated tissues can be infiltrated with paraffin prior to sectioning, they must be rehydrated. Among methods employed by nineteenth and early twentieth century workers, Wilder (1904) used potassium hydroxide but the technique proposed by Sir Marc Armand Ruffer in his studies on Egyptian mummy tissues in Cairo has remained most popular (Ruffer, 1921b:64). This employs sodium carbonate to enable penetration of water into the hard, dry tissue, and alcohol to discourage rehydrated proteases in the tissue from wreaking their destructive action (see Table 6.4 for composition). Unfortunately the sodium carbonate achieves its effect by breaking bonds between collagen fibers, thus reducing the physical integrity of the tissue. Since collagen content varies substantially among the different body tissues, the concentration of this component must either be individualized or the time interval of tissue immersion in the solution must be varied according to the nature of the tissue. Often overlooked is the fact that Ruffer cautioned readers, pointing out that he did both, and advised monitoring the individual tissues during their immersion in order to prevent their disintegration by excessive action of the sodium carbonate.

Subsequently various authors have made adjustments to Ruffer's basic solution to deal with this limitation. Sandison (1955b) substituted 1% formalin in place of the water component that Ruffer used, while

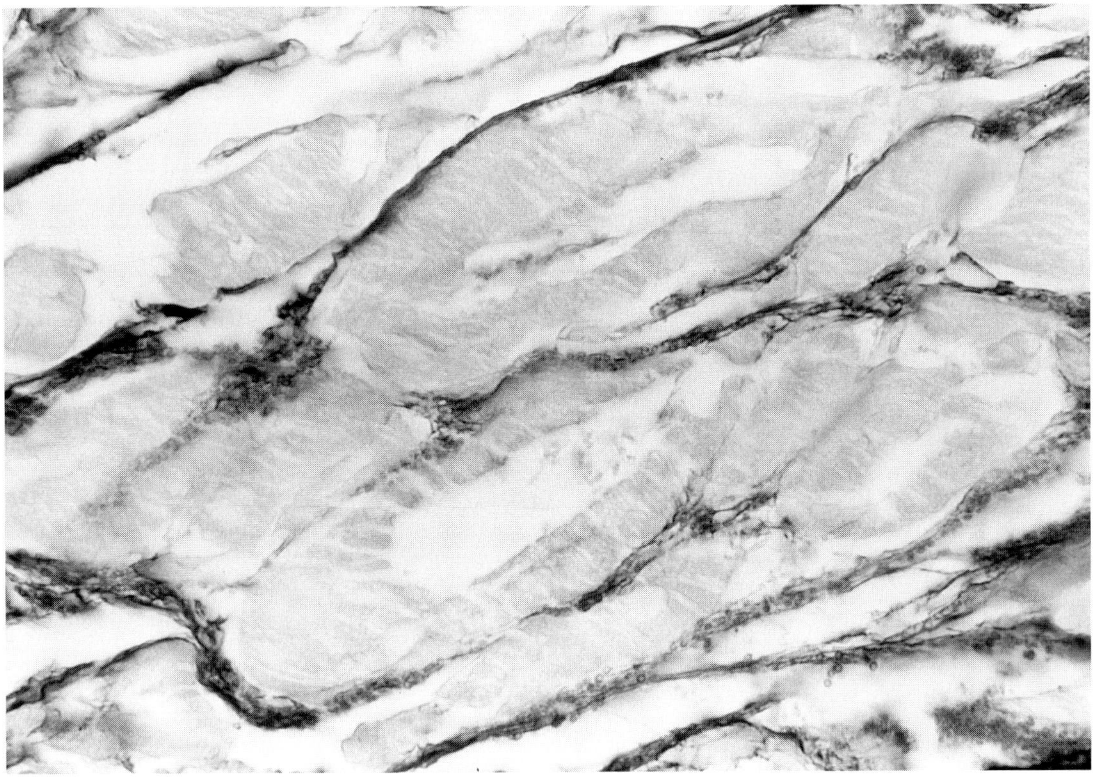

Fig. 6.40. Kidney histology; SM, AD 400–1000.
Metabolically active cells of renal proximal convoluted tubules degenerate postmortem even before bacteria arrive, but the tubular shapes remain recognizable much longer. Andean mummy no. AZ-141, T-36. Cabuza culture. (Photo by ACA.)

Graf (1949) used saline followed by formalin. Zimmerman (1972) employed a succession of solutions of increasing alcohol concentration. More recently, workers have attempted to standardize rehydration to avoid the frustrating need to individualize the solution or immersion time to each tissue. Herrman (1982) proposes desiccated tissue immersion into human serum (whose enzymatic action had been inactivated by exposure to 56 °C for 30 minutes), followed by immersion in 75% ethanol. Fulcheri *et al.* (1985) modified this procedure by exposing the tissue to formaldehyde vapors in an airtight chamber for three hours at 60 °C before immersion into the human serum. Several commercially available fabric softeners have been proposed, including one labeled "Comfort" produced by Lever Bros. (Turner & Holtom, 1981) and another called "Coleo", a detergent marketed by Colgate Co. (Zugibe & Costello, 1985). Recently a comparative study of these and modifications by the authors was reported (Mekota *et al.*, 2000). Table 6.4 details the composition of several of the most popular of these.

Although one might expect that a succession of modifications would result in ever better results, improvement using these techniques has been slow. The detergent and serum methods appear to have overcome the all too often tissue-destructive effects of the procedures that incorporate sodium carbonate, but the varying needs of differing tissues have constrained standardization of rehydration methodology, and some individualization still seems to be necessary for best results.

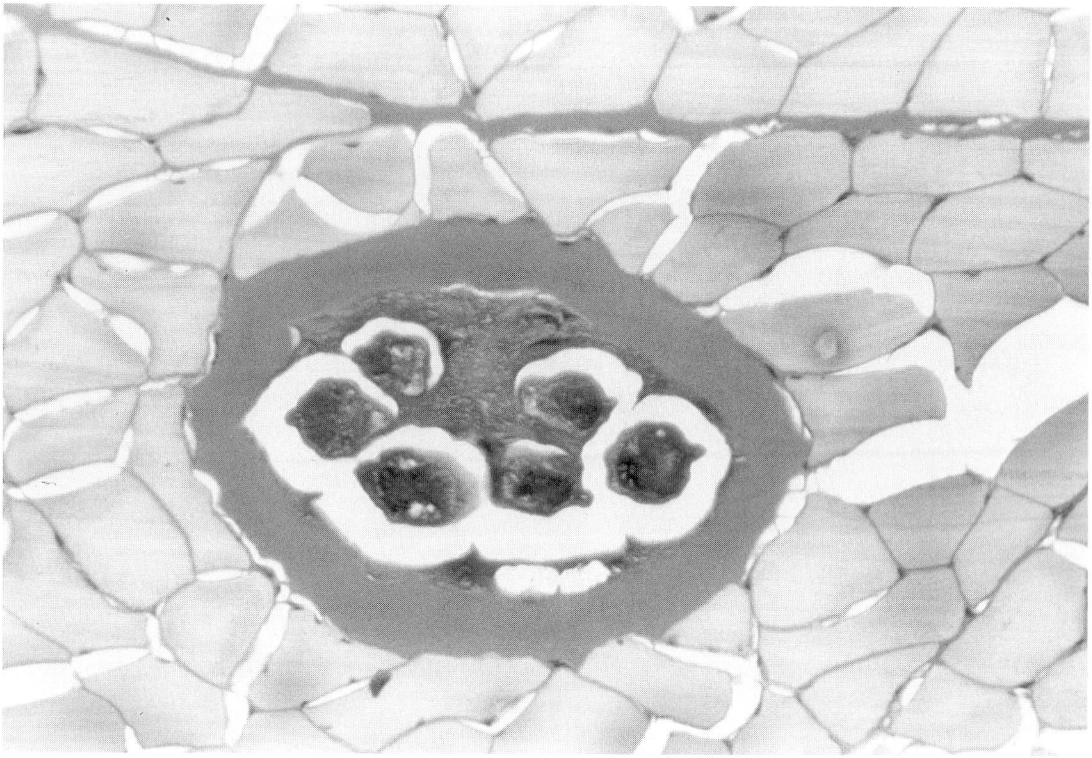

Fig. 6.41. Mummified muscle with encysted larvae of *Trichinella spiralis*; AM, AD 1911.
This tissue sample was acquired from the mummy of Elmer McCurdy, a nineteenth century outlaw in the USA described in Chapter 4. (Photo courtesy of T.A. Reyman.)

Embedding and processing of mummified tissues

Following rehydration by whatever method is chosen, a 24–48 hour period of fixation in 4% formalin is desirable. Thereafter the fixed, rehydrated tissue may be processed in a manner no different from that employed by hospital pathology departments for their routine, fixed tissues excised surgically daily from living patients (for one example, see Sandison, 1955b). This usually involves removal of water by organic solvents and replacement with paraffin, though methacrylate may be used for special needs (Troyer & Bablich, 1981). Neophytes should know that stock formalin solutions are marketed in a 40% concentration. Working solutions for tissue fixation are usually prepared as 1:9 dilution of this stock. That dilution is commonly called "10% formalin" solution, but because the stock solution is only 40% (not 100%) concentration, the actual formalin concentration of the diluted solution is 4%.

Sectioning and staining mummified tissues

Conventional microtomes used by hospital pathology departments will permit sections cut to 6 μm thickness, floated in hot water baths onto glass slides. The usual hematoxylin and eosin stains used for tissues obtained from living people are designed for maximal detail in preserved epithelium. Since mummified tissues usually lack intact epithelium, stains selected for connective and muscular tissue such as Masson's

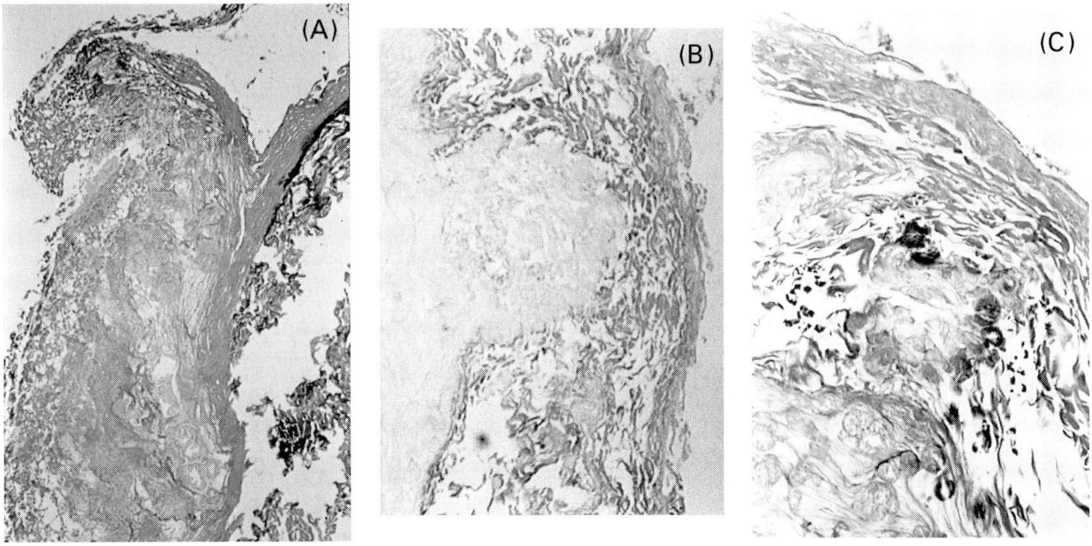

Fig. 6.42. Histology of rectal carcinoma; SM, 30 BC to AD 395.
Epithelial cells of the tumor have degenerated but its noninvasive, adenomatous structure is recognizable in the left panel, invasion into submucosa is evident in the central panel, while the infiltrating, glandular, component can be seen on the right. Spontaneous Egyptian adult male mummy, Roman Period, Dakhleh Oasis. (Photo courtesy of Michael Zimmerman.)

trichrome stain (Masson, 1929; Luna, 1968:94) are more informative for slides of mummified tissues. Gomori's aldehyde fuchsin stain is very useful for elastic tissue (Gomori, 1946; Luna, 1968:78). My own choice has been a modification of Movat's pentachrome stain (Russell, 1972) because it stains nuclei, elastic tissue, collagen, mucin, fibrin, muscle and nerves differentially. The text edited by Luna (1968) is a highly utilitarian histology staining resource.

Applications

The microscopic examination of Egyptian mummy skin by a Viennese physician, Johan Czermak (1852), is probably the first reported histological study of any human mummified tissue. Wilder (1904) repeated such an investigation on skin of North and South American spontaneously mummified bodies, and five years later Shattock (1909) demonstrated microscopically atherosclerosis in an Egyptian pharaoh's aorta. Ruffer's (1921a:64) contributions to paleohistology have already been noted above. In the late 1920s Simandl (1928) studied skin and muscle of an Egyptian mummy and Williams (1927) examined viscera of two spontaneously mummified Peruvian bodies. Williams identified atherosclerosis in a posterior tibial artery and what he thought were red blood cells (probably actually fungal spores). By 1938 Shaw had described the histological appearance of many different viscera found in a canopic jar in the tomb of an Egyptian mummy, noting evidence of anthracosis, pulmonary emphysema and pulmonary adhesions. Most of these workers called attention to the absence of epithelium and cell nuclei. More extensive tissue sampling and careful gross evaluations subsequently led to a broad range of diagnoses in mummified human remains. These include silicate pneumoconiosis (el-Najjar *et al.*, 1985), anthracosis (Zimmerman & Aufderheide, 1984), colon adenocarcinoma (Marchetti *et al.*, 1996; Zimmerman, 2001), pneumonia and pulmonary edema (Walker *et al.*, 1987), cirrhosis (Zimmerman, 1990) and many others. In addition, bone has also been

Table 6.4. *Method for rehydration of desiccated soft tissues*

Ruffer, 1921b	
Alcohol	30 parts
Water	50 parts
5% Sodium carbonate	20 parts
Sandison, 1955b	
Ethyl alcohol	30 parts
Formalin (1% aqueous)	50 parts
Sodium carbonate, 5% aqueous	20 parts
Zimmerman, 1972	
Initially rehydrated as in Ruffer (1921b)	
Subsequently: replace 1/3 of volume daily with absolute ethyl alcohol.	
Herrmann, 1982	
Inactivate human serum at 56 °C for 30 minutes	
Store at 4 °C	
Immerse specimen in inactivated serum at 4 °C, 24 hours	
Wash and transfer to 75% ethyl alcohol	
Fulcheri et al., 1985	
Expose desiccated tissue in paraformaldehyde vapors in airtight container chamber at 60 °C, 3 hours	
Then immerse tissue in inactivated serum as in Herrmann (1982)	

Notes:
The degree of preservation of each listed item is assigned an estimated number of "units" varying from 0 (absence of the structure) to a maximum of 10 (complete preservation). The visceral index is the sum of such assigned values.

studied microscopically by grinding a slab of cortical femur diaphysis to a thickness of about 100 μm The ratio of intact to restored osteons has been found to correlate with age-at-death (Stout, 1986), and a bone sample of the similarly prepared femur head demonstrated a histological pattern of "tunneling resorption" typically seen in persons with hyperparathyroidism (Cook *et al.*, 1988). Basset *et al.* (1980) also identified fluorescence characteristic of tetracycline deposition in Nubian bones as did Cook *et al.* (1988) in Roman Period Egyptian bones, probably reflecting ingestion of grain contaminated with fungi.

Readers interested in more detailed reviews of this field may wish to consult those by Sandison (1963), Tapp (1992) and Schultz (1993).

Electron microscopy (EM)

Resolution refers to the human eyes' ability to discriminate two separate points of an image. The nature of visible light waves limits optical light microscope resolution and, therefore, maximal magnification. Light microscopy resolution is about 0.2 μm (achieving about 1000× magnification). The shorter wave length of a beam of electrons can permit the construction of transmission electron microscopes with resolution potential of about 0.2 nanometers (nm, 1000 times smaller than light microscopy can reach). TE microscopes create "lenses" by manipulating an electron beam with electromagnets. The principle is similar to that of an X-ray machine in that a beam of electrons is allowed to pass through a specimen. The differing densities of the various structures in the specimen will absorb varying amounts of energy from the penetrating electron beam. Detectors below the specimen allow manipulation to translate these differences into a viewable image. This image can be magnified almost 1000 times more than the maximum achievable by light microscopy. TE microscopes became available commercially in the late 1930s (Postek *et al.*, 1980:1–11).

Although TEM has and probably will continue to be useful in paleopathology studies, it is SEM that appears to be making more regular contributions to the field of mummy studies. In this procedure an electron beam scans the specimen's *surface*, inducing the release of low energy electrons from the surface of the specimen. The latter are collected by detectors and translated into an image of the specimen's surface. If the TE microscope is analogous to an X-ray that images the *internal* structure of a specimen, then the SE microscope is conceptually similar to a camera in that it produces an image of the specimen's surface only. Its resolution is a little less than that of TEM, but the high depth of field that SEM makes possible is a valuable feature.

Specimen preparation and processing for EM

Although Desanti (1977) had demonstrated that certain results can be obtained by sectioning without rehydration, most investigators working with desiccated, mummified human remains prefer to rehydrate the tissue in a manner similar to that described above. Fixation for TEM is commonly achieved by immersion in glutaraldehyde. Because this fixative penetrates only about 0.25 mm into hydrated tissue, the specimen diameter should not exceed 0.50 mm. Following fixation the TEM specimen is dehydrated by organic solvents in a manner similar to preparation for study by light microscopy. However, the water is not replaced with paraffin (as in light microscopy), but instead the tissue is placed within a resin (commonly methacrylate) that hardens by polymerization. This enables the cutting of ultrathin tissue slices of about 30 nm with glass knives and 10 nm with diamond knives (200–600 times thinner than sections for light microscopy) (Hayat, 1989).

Specimens designed for SEM are commonly fixed by exposure to osmium tetroxide vapors. Completely desiccated mummified tissue may be examined by SEM without prior rehydration or fixation, since sectioning may not be required. If, however, the goal is to restore the tissue structures as much as possible to their living appearance, then rehydration is desirable. If carried out, the tissue must be fixed and again dehydrated, usually with organic solvents or by a gentler process termed critical point drying. The specimen is then attached to a carbon stub for handling. In addition, SEM also requires the tissue to be rendered conductive to electricity to avoid the building up of electric charge on the tissue when it is exposed to the electron beam. This is accomplished by placing it in a special chamber and exposing it to a vapor of conductive material. The latter should be nonreactive with the tissue surface components and slow to oxidize. Probably the most commonly employed coating material is gold or gold-palladium (Postek *et al.*, 1980:151). The specimen is then ready for entry into the SEM viewing instrument.

Electron microscopy: applications

The earliest application of TEM to ancient specimens was Leeson's (1959) report in which he studied a rehydrated specimen of skin from the desiccated, spontaneously mummified remains of a Native American from British Columbia, Canada, about 480 kilometers north of Vancouver. Leeson noted that cell wall and nuclear membranes could be distinguished but common organelles such as mitochondria, endoplasmic reticulum and Golgi apparatus were not recognizable. Lewin's (1968) efforts 9 years later were equally fruitless in providing intracellular detail. Most of the efforts in this area then shifted to the more promising SEM form (see below). However, later work found several applications of TEM that generated useful information. Mikhailov *et al.*, (1984) studied the intestinal lining of the 10 000-year-old mummy of the Yuribey mammoth and found spherical and linear structures suggestive of virus structure. This approach was refined further when Fornaciari & Marchetti (1986) found a structure consistent with smallpox virus in a skin biopsy of an Italian medieval child whose facial rash was characteristic of this disease; these authors confirmed their diagnosis with immunohistochemical methods. Fornaciari *et al.*, (1992a) also demonstrated the amastigote form of *Trypanosoma cruzi* in a section of a dilated esophagus spontaneously mummified from a Peruvian Inca and also confirmed this as Chagas' disease by immunohistochemical methods. An excellent example of additional information that can be acquired when TEM is coupled to other measurement methods is the investigation of tissue acquired by needle biopsy from the lung of the Tyrolean Iceman. Electron energy loss spectroscopy (EELS) enabled identification of lighter elements. Energy dispersive X-ray analysis (EDXA) can recognize elements from carbon to uranium. Energy-filtering transmission electron microscopy (EFTEM) permitted two-dimensional spatial distribution of chemical elements (Pabst & Hofer, 1998). An intact lymphocyte was visualized by TEM in a brain tissue sample of a spontaneously mummified body of an adult Andean male dated to AD 1500 (Gerszten *et al.*, 1997).

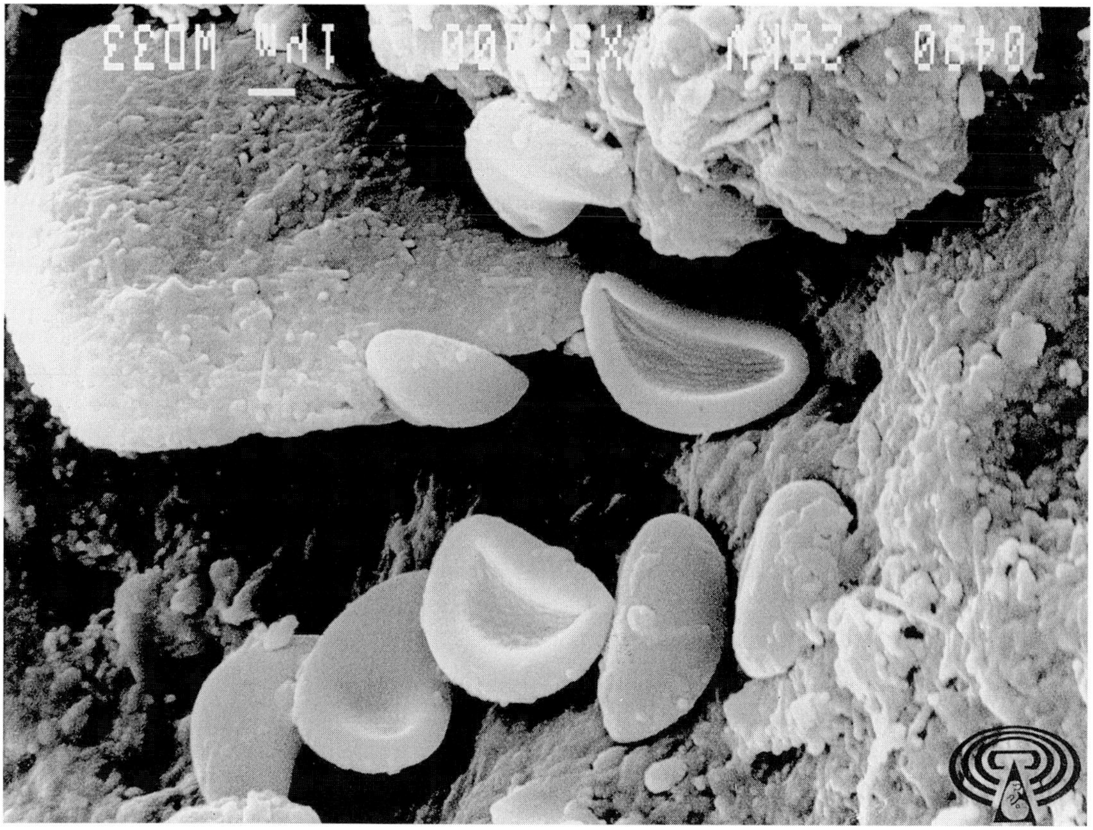

Fig. 6.43. Red blood cells showing sickle-cell structure; SM, 180 BC.
Scanning electron microscopy reveals distortion of red blood cells characteristic of sickle cell anemia in an ancient bone from the Middle East. (Photo courtesy of G. Maat and M.S. Baig (1990).)

The depth of field and magnification features of SEM were recognized to be of value to paleopathology studies quite early. Lamendin (1974) found it possible to view dentinal tubuli and recognized odontoblastic processes by SEM. Undecalcified sections of vertebrae visualized by SEM revealed excellent histomorphometric features (Wakely *et al.*, 1989), as did the investigation of mummy hair morphology (Conti-Fuhrman & Rabino-Massa, 1972). The supplementation of other methods of measurement to which SEM lends itself quite readily has proved to be of inestimable value. Thus SEM permitted recognition of crystalline deposits in the lung sections of southwestern U.S.A. mummies. When SEM was coupled with EDXA, the crystalline structures were confirmed as mica and feldspar (el-Najjar *et al.*, 1985). These studies established that an ancient occupational hazard (agricultural dust inhalation by these individuals while alive) resulted in pulmonary fibrosis. SEM also enabled Maat & Baig (1990; 1996) to recognize sickled red blood cells in the bone marrow (orbit, vertebra) of an Arabian adult male dated to 2130 years BP (Fig. 6.43). Great care must be employed to differentiate between ancient red blood cells and fungi and other artifacts (Maat, 1991; Horne & Jarzen, 1993).

Physical methods: radiographical methods

For readers who want only an overview of the field, two abstracts are presented here. Following the abstracts the individual steps of the procedures are detailed below.

Abstract of radiographical methods: plain radiography

Purpose

Display interior of body or structure.

Principle

Short-wave energy X-rays are passed through the body and detected by photographic film as they emerge on the other side. Internal body structures absorb these X-rays to differing degrees, depending on their differing tissue density. The resulting differences in the intensity of the emerging X-rays impacting on the different areas of the film reflect the pattern of the internal structures. The developed film is a permanent record of this image.

Paleopathology application

Demonstrate skeletal structures of mummified body, associated artifacts.

Limitations

(1) In desiccated bodies soft tissue densities are commonly too similar to visualize them. (2) Overlapping structures and/or soil and artifacts obscure structures of interest. (3) Body or body part must be transported to the location of the X-ray machine, though portable devices are available for field use.

Cost/test

Class 1

References

Gray, 1967; Harris & Wente, 1980.

Abstract of radiographical methods: computed tomography

Purpose

Image internal body structure in detail greater than plain X-ray examination provides.

Principle

Similar to plain radiography but an arm rotates the X-ray source around the body in a plane tangential to the long axis of the body, and the energy of the X-rays at their multiple emerging sites is detected and stored electronically instead of by a photographic film. The electronically reconstructed cross-sectional image is equivalent to a single, radiographical image of a body slice 1–10 mm thick. The result eliminates overlapping structures, and electronic manipulation can enhance detail.

Paleopathology application

(1) Visualize body interior structure in uncluttered detail. (2) Because electronic information is stored, multiple slice images can be reconstructed in different planes, permitting their presentation in more familiar, 3D pattern. (3) Density of a structure of interest (bone, artifact) can be quantified and compared with values of known standards for identification. (4) Can be combined with other methods to generate sculpted morphological structures. (5) Wrapped, mummified body within a coffin can be visualized without unwrapping. (6) Details of textiles and coffin structure can be visualized. (7) CT can be employed to guide a biopsy device introduced into the body through a small aperture.

Limitations

(1) Body must be transported to the location of the CT machine. (2) Many mummified bodies are too large and rigid to fit within the chamber of the CT

machine especially if they are contained in an unopened coffin. (3) Measurements are relatively slow, whole body scans requiring several hours. (4) Interpretations usually need radiologist consultation. (5) Visceral desiccation distortion may confuse radiographical interpretation.

Cost/test

Class 2 (3 if professional interpretive services require payment; see Table 6.3).

References

Wong, 1981; Vahey & Brown, 1984 (CT of Egyptian mummy); Notman, 1995 (general review); Notman & Lupton, 1995 (three-dimensional radiography); Davis, 1997 (review of CT applications in bioanthropological studies).

History

W.K. Roentgen (1895) announced his astounding discovery of X-rays as an oral presentation at a scientific assembly in December 1895. Within a year W. Koenig (1896) had employed the method to examine a mummified child and cat (Fig. 1.10). Later, in 1896, the British worker C.T. Holland (1937) used X-rays to demonstrate that an Egyptian bundle contained a mummified bird. Although apparently the work was never published, Fiori & Grazia-Nunzi (1995) note that C.L. Leonard used an X-ray machine constructed at the University of Pennsylvania to radiograph a Peruvian Mochica mummy in 1897. By 1898 W.M. Flinders Petrie had examined mummy extremities with the then still novel technique (Fig. 6.44) and, in 1912, Grafton Elliott Smith (1912b) reported that in 1904 he had been aided in his age-at-death estimate of Tuthmosis IV by visualizing his bone epiphyses with X-rays. However, bringing the mummies to the fixed X-ray installations common to this early period tended to discourage its use (Smith used a Cairo taxi). Large-scale radiographic mummy surveys followed the development of portable equipment and in 1931 R.L. Moodie studied the collection at the Field Museum in Chicago, while P.H.K. Gray much later did the same for several hundred mummies in London's British Museum of Natural History and in other United Kingdom and continental museums (Gray, 1973). Such equipment also permitted similar study of the pharaohs in the Cairo museum (Harris & Wente, 1980). When CT methodology became available, Lewin & Harwood-Nash applied it successfully to a mummy in 1977. This remains the most widely used technique for mummy studies, since Notman (1983) demonstrated that the still more recent imaging technique – magnetic resonance imaging (MRI) – can not be used in desiccated mummies. One could anticipate that MRI's

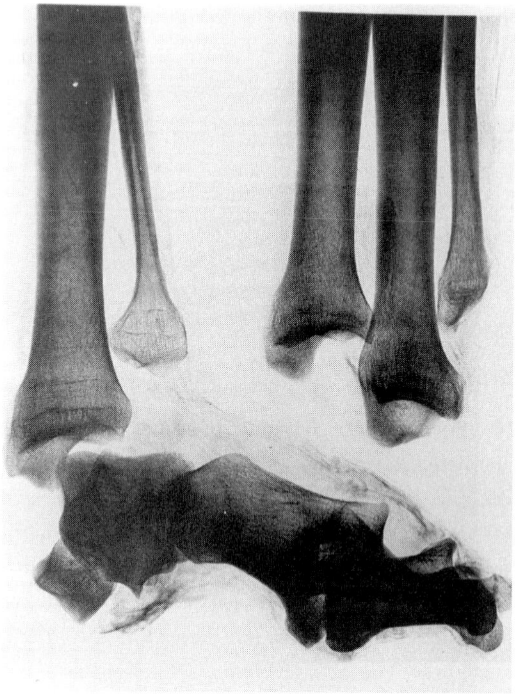

Fig. 6.44. Early X-ray of a mummy; AM, 1549–1069 BC.
In 1897 the British Egyptologist W.M. Flinders Petrie examined mummy extremities to determine age on the basis of epiphyseal status. (From Fiori & Grazia-Nunzi, 1995.)

exquisite resolution of soft tissue density differences could be exploited in bog mummies. This has now been tested and its utility confirmed during the study of the bog body Lindow Man in the United Kingdom (Bourke, 1986).

Equipment

Plain radiography

In this form a beam of X-rays is passed through the mummy and then allowed to impact on the silver-containing emulsion film within an appropriately placed film-containing cassette. As they pass through the body organs, some of the X-rays are absorbed and never reach the film. The degree of absorption depends on the tissue density (the number and type of atoms per unit volume of tissue). Thus muscle may absorb only a small fraction of the X-rays passing through it but bone might "block" most of them, permitting easily recognizable separation ("resolution") of these two structures on the developed film. The significant density differences of various living viscera, however, decrease substantially following postmortem desiccation, and this results in greater difficulty in visualizing individual organs *in situ*. Nevertheless, plain films commonly generate considerable basic information and can be prepared for only a little more than the cost of the film.

Their clinical value means that such equipment will be found in virtually every health care facility. Most hospitals also have mobile units used to prepare X-rays on nonambulatory patients within their hospital rooms. Such units can be wheeled into the autopsy or other conveniently available room when exposures on mummies are desired without seriously interrupting the hospital patient care delivery routine. This still, however, requires the mummy to be brought to the health facility. In many instances it is desirable to have X-ray services available in the field, especially when extensive mummy dissections are planned. For such circumstances "suit-case" type portable X-ray equipment is available that can be powered by a small, equally portable, gasoline generator. A darkroom tent is a necessity because it can provide valuable immediate-developing capacity, either in open trays or relatively simple electrically operated units. Such an arrangement was employed successfully by Notman *et al.* (1987) during a Canadian Arctic excavation, as well as a Himalyan expedition to Mt Hotse-Shar (Fig. 6.45).

Computed tomography

Resolution can be enhanced enormously and the obscuring effect of superimposed structures can be minimized by use of CT. This bulky, fixed and expensive equipment places the body in a position that surrounds it with a set of X-ray detectors (rather than film) and then rotates a beam from an X-ray source in a circular fashion, exposing a *segment* ("slice") one or more millimeters thick of the mummy's body. In essence, each detector records the pattern of penetrating X-rays as a separate exposure. This information from each of the detectors for each of the 360° exposure sets is then passed to, and stored in, a computer whose software will resolve each anatomical unit within the field, and present it to the viewer as a cross-sectional slice of the body of a thickness that can be varied over a range of 1–10mm or more. Electronic controls can be adjusted for a variety of desired effects, such as density range, enhancement of edge differences (improving resolution) and others. Information accumulated from such successive slices at predetermined intervals can be used with software that will "fill in" the spaces between the scanned slices and join them in a continuous line image in sagittal, cross-sectional or any other desired plane (three-dimensional imaging: 3D-CT). Other potential uses are outlined in the section Applications, below. The time required to scan an entire adult human mummy will depend on the number and thicknesses of the slices selected and can vary from two to six hours with traditional equipment. Recent fast helical CT scanning can accomplish this in minutes. The cost and amortization requirements for this sophisticated equipment, the substantial

Fig. 6.45. Portable X-ray unit.
The utility of a portable X-ray unit is apparent in this photo of its use on a Himalayan expedition to Mt Hotse-Shar at 4836 meters' altitude. Applications to field study of mummies are valuable. (Photo courtesy of Derek Notman.)

technician time and the cost of interpretation by professional radiologists, provide a powerful incentive to employ this potentially valuable diagnostic tool in a highly selective manner when circumstances require compensation for these services.

Magnetic resonance imaging (MRI)

This imaging method does not use X-rays. When a living body is placed within a powerful electromagnetic field, the magnetic fields surrounding each individual hydrogen and certain other atoms in the body tissues become aligned with the externally imposed field. When the latter is reduced or interrupted, the miniature fields surrounding the individual atoms resume their original alignments. Energy differences between these two states can be detected, located and quantified. Computer software can translate them into images (similar to CT) whose resolution for soft tissues is unsurpassed. Exquisite as such anatomical displays are, the above magnetic movements can occur only in a liquid aqueous or neutral fat matrix. Hence, Notman (1995) was not surprised when MRI failed to generate images in desiccated Egyptian mummies, though he found the method useful to monitor the desiccation process during an experimental mummification study (Notman & Aufderheide, 1995). Piepenbrink *et al.* (1986) produced successful MRI images on rehydrated, naturally mummified Peruvian extremities. The potential value of MRI studies in mummies with hydrated preserved soft tissues was established unequivocally by Bourke (1986) in his study of Lindow Man. The bulk of this equipment and its other limiting operational features are similar to (even greater than) those of CT and therefore impose similar limitations.

Applications

Plain radiography applications

This is the most utilitarian of radiographical methodology for mummy studies. It is simple, applicable in the field and economical. Its principal contribution is a noninvasive method to answer the question: What is inside (the coffin, cartonnage, textile bundle, other container or even inside the body itself)? Possibilities of interest to the investigator include:

1. A body? If so, further questions commonly are:
   (a) One or more bodies?
   (b) Human or animal?
   (c) If human, are sex, stature and age-at-death predictable?
   (d) Are lines of growth arrest visible in long bones?
   (e) Are fractures or other pathological lesions evident?
   (f) Is soft tissue present on the body's surface (if so, bones can be expected to be in an articulated array even if soft tissue resolution on the X-ray film is insufficient to identify it)?
2. If an unclothed body is radiographed, further questions include:
   (a) Are viscera present? If not, further dissection may not be of interest.
   (b) Is soil present within the body cavities?
3. Artifacts?
   (a) If present, is their material composition predictable? Are their shapes sufficiently distinctive to identify their nature?

For example, a very common situation encountered during the study of museum-curated mummies deals with the exhibition of a spectacularly decorated Egyptian cartonnage. It is the container itself that usually has special display value. Further information derivable from its content is also of interest but commonly not of a degree justifying the risk of destruction of the fragile container that would be involved with efforts to open it and remove its content (if any). Standard simple radiography may expand the field of knowledge relating to a specific container substantially by answering at least some of the more basic of the above-listed questions.

Egyptian mummies wrapped to suggest that they contain a child have been especially rewarding when radiographed. When Diener (1980, 1981) X-rayed all the Egyptian mummies in Stockholm's Museum of Mediterranean Near Eastern Antiquities he found that one bundle enclosed only an adult lower leg. Diener lists five other reported examples similar to the latter, to which I can add another that was shown to me during a visit to the Fine Arts Museum in Budapest in 1988.

Identification of pathological lesions is usually limited to the more obvious displaced fractures or equivalent structural distortion when mummies still inside containers are radiographed, because the detail and resolution required to identify the more subtle changes reflecting soft tissue abnormalities are often obliterated by overlying materials. However, if an *unclothed* body is studied with plain X-rays in this way, useful information is derived so regularly as to justify its routine inclusion in a mummy study protocol. Figure 6.46 demonstrates calcified thoracic lymph nodes probably secondary to tuberculosis in the mummy of a pre-Columbian adult in northern Chile. Their size and location would have carried a significant probability that they would be overlooked during subsequent dissection without the forewarning provided by the X-ray. Gray's (1967) X-ray survey of museum-curated mummies revealed not only numerous fractures but also arterial calcifications as well as gallstones, bone infarcts and congenital skeletal abnormalities.

While plain X-rays are valuable orienting aids, they do have significant limitations. Loss of the epithelial components of viscera and water often reduce the density of desiccated abdominal organs to the point where differentiation from muscle and other tissues becomes difficult. Separation of kidney density from the underlying psoas muscle may become impossible. Coprolites overlying muscle or liver may not be

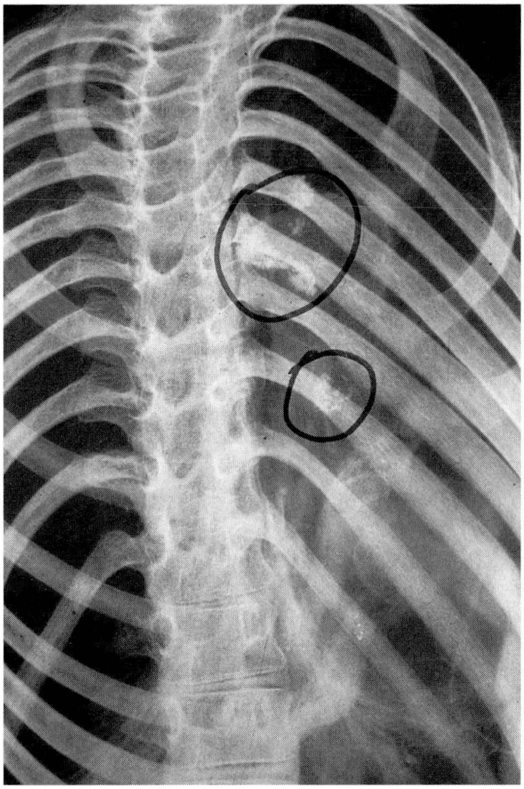

Fig. 6.46. Calcified lesions revealed by X-ray; SM, AD 400–1000.
The plain X-ray of this mummy prior to dissection demonstrates several small calcifications that might have been overlooked during dissection if the researcher had not been alerted by these findings. Cabuza population, northern Chile. Mummy no. AZ-141, T-38. (Photo by ACA.)

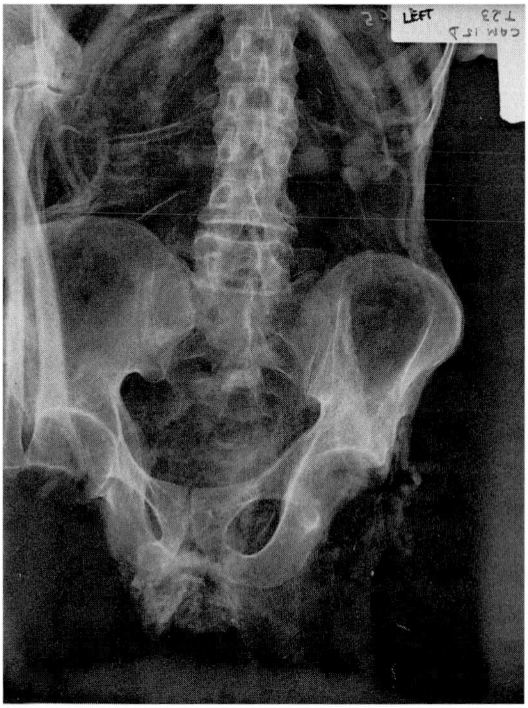

Fig. 6.47. Coprolites revealed by plain radiography; SM, 3000 BC.
The lobulated masses in the abdomen's left upper quadrant are coprolites in the mummy's colon. From Cam 15D site at the mouth of Camarones River in northern Chile. Mummy no. Cam 15D, T-23. (Photo by ACA.)

visualized, although they can usually be recognized easily when located in a sufficiently lateral position so they are not superimposed on other structures (Fig. 6.47). The poor discrimination between normal and pathological lung has been particularly dismaying (Fig. 6.48). Of greatest concern, however, is the fact that everything within the field of exposure contributes to the X-ray image. In flexed mummies (Fig. 6.49) or in mummies within containers, this can produce such a clutter of bones, textiles and artifacts superimposed on the body image that major visceral pathology becomes obscured. If available, CT usually can resolve the competing, overlapping images (Davis, 1997).

Plain radiography can be used to evaluate microscopic bone changes if a thin slice of bone is laid directly on radiographic film, exposed to low-energy X-rays and the film examined under the microscope (microradiography) (Rowland & Farnham, 1959; Stenn *et al.*, 1975).

Computed tomography applications

The ability of CT to limit the visualized body portion to a "slice" thickness as small as 1 mm and present the image in cross-sectional view (Fig. 6.50) is without

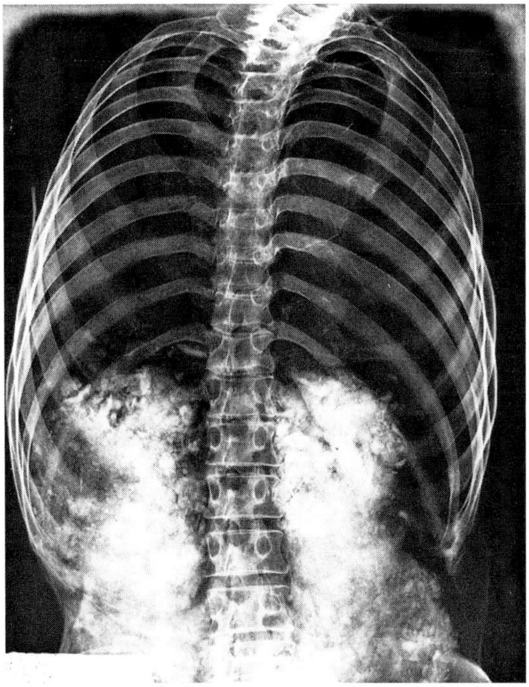

Fig. 6.48. Failure of plain radiology to delineate viscera.
While plain radiography demonstrated much radiopaque foreign material (sand) in the abdomen, it failed to demonstrate the mediastinal structures and lungs that were present above the intact diaphragm in dissection. Mummy no. AZ-75, T-57. (Photo by ACA.)

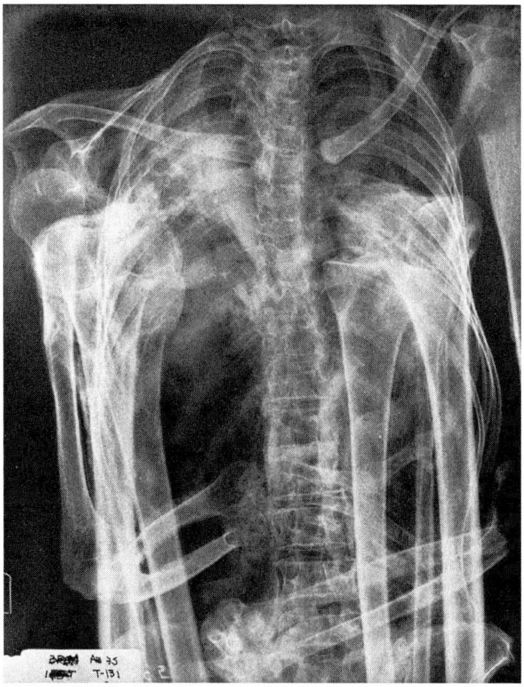

Fig. 6.49. Superimposition of extremities in X-rays of a flexed mummy; SM, ca. 50 BC.
X-rays of flexed bodies usually result in a clutter of superimposed extremities that obliterate the content of the body cavities. Mummy no. AZ-75, T-131 (from same site as mummy in Fig. 6.48). (Photo by ACA.)

question the most useful contribution this method makes to paleoradiological studies (Wong, 1981). However, the profound anatomical distortion and sometimes even dislocation that visceral organs undergo during desiccation can tax the interpretive skill of a radiologist accustomed to dealing only with hydrated, living tissue (Appelboom & Struyven, 1999). Kolta (1995) used CT to delineate more clearly structures that had been initially observed as poorly defined irregularities on plain X-ray films. Similarly, by using CT to eliminate obscuring effects of overlying items, Notman (1986) could define precisely that a structure lying between the lower extremities of a young adult female Egyptian mummy in a cartonnage container was the isolated head of an adult male. Vahey & Brown (1984) used CT to determine whether the mummification method revealed by radiography was consistent with that suggested by the ornamentation on the cartonnage.

The ability to make soft tissue structures more easily visible by electronically exaggerating small density differences contributes also to the power of this imaging method to produce new information. Lupton (1995) was able to exploit this feature to identify the location of the abdominal evisceration incision and the nasal and ethmoidal bone defects reflecting cerebrectomy via transnasal craniotomy in an unwrapped Egyptian mummy as part of a study of mummification methods.

The fact that all of the scanned data are stored

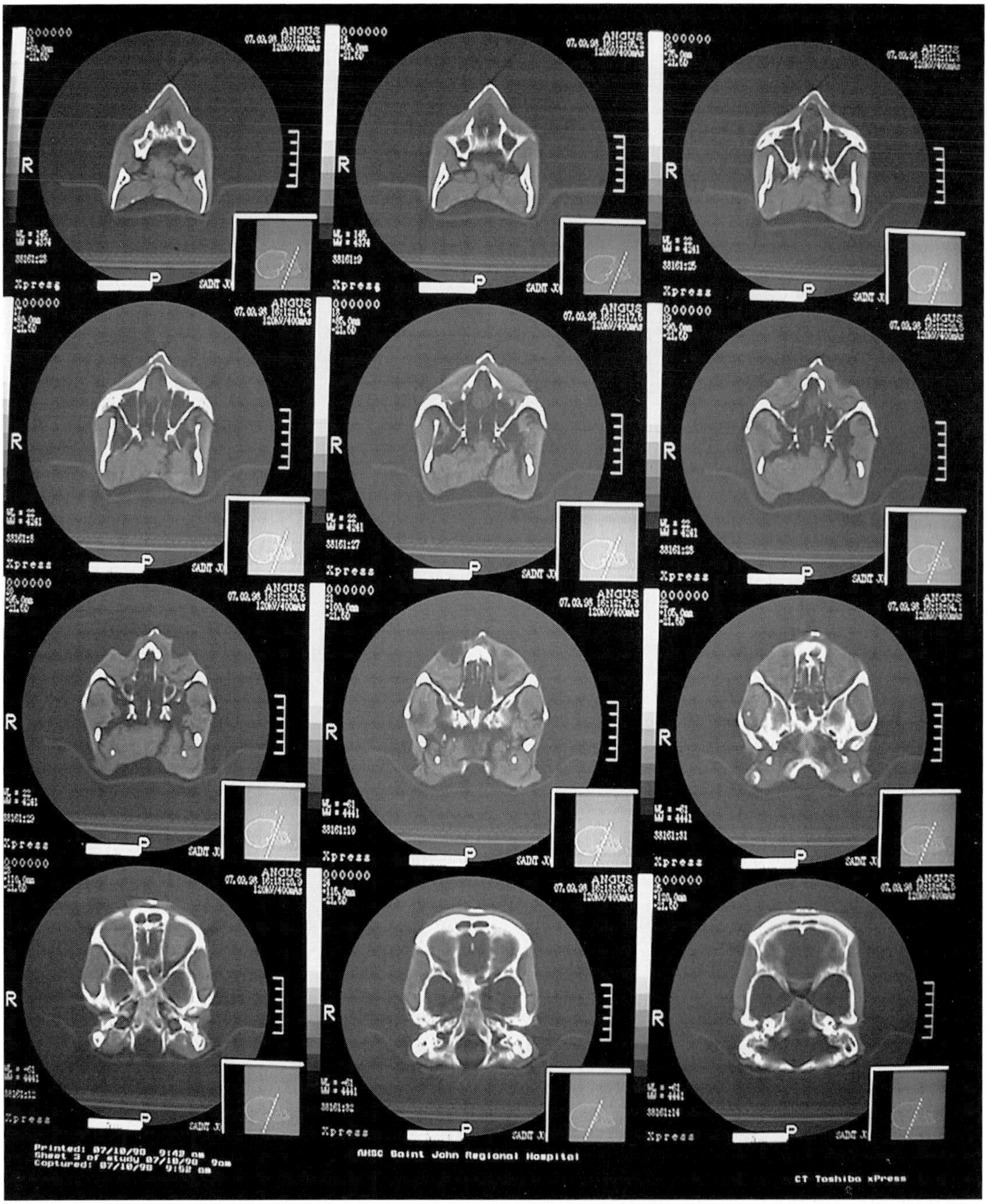

Fig. 6.50. Head scan (SM, ca. AD 1800) visualized by computed tomography (CT scan).
The scan data are made visible in the form of "slices" (planes) whose thickness can be controlled. Images are progressive levels of the examined area. Overmodeled skulls curated in New Brunswick Museum in St John, New Brunswick. (CT scans courtesy of Ian Anderson, M.D., and Andrea Kirkpatrick.)

permanently in the computer unit has led to the development of software methods to present the anatomical information in more easily comprehended forms. 3D-CT programs permit rearrangement of the scanned images in such a manner as to make them appear as they would be if viewed from different perspectives (Fig. 6.51). This makes it possible, for example, to "view" the cranial suture closure status from the inside of the skull of a never-unwrapped Egyptian mummy. Other programs permit "subtraction" or "addition" of parts of the image. A still-evolving application is reconstruction of facial soft tissue features by combining CT skull scans with computer graphic programs. Such features have obvious forensic applications, especially in cases of missing persons.

CT has also made some unanticipated contributions. It is capable of providing exquisite detail in imaging of textiles, making it possible to reconstruct at least some aspects of wrapping techniques noninvasively; details of coffin construction are visualized with similar ease (Lupton, 1995). Furthermore, in an experimental study, Notman & Lupton (1995) demonstrated the ability of CT to identify correctly seven different types of wood on the basis of their density differences. After the development of an appropriate database one can expect application of this approach to identify noninvasively the material composing visualized artifacts within containers or textiles. Since CT is already used to measure bone density in osteoporosis studies of living populations, its routine application to mummies for that purpose can be anticipated soon. This has already been pioneered by Pahl *et al.* (1988) using a specialized (double photon) radiation device.

Magnetic resonance imaging applications

MRI has no applications in desiccated mummies. It has been used in rehydrated, spontaneously mummified extremities (Piepenbrink *et al.*, 1986), where this application might be indicated if CT resolution is inadequate to answer a special question and the rehydration process does not invalidate other desired studies. In living individuals soft tissue resolution of MRI often exceeds that of CT. Notman & Aufderheide (1995) used it to monitor taphonomic changes during experimental desiccating mummification of a dog because of its ability to separate signals originating in an aqueous medium from those emanating from fat. Its use in bog body studies has been established (Bourke, 1986).

Physical methods: endoscopy

For those who want only an overview of the method, an abstract is presented here. Following the abstract, the individual steps of the procedure are detailed below.

Abstract of endoscopy

Procedure

A flexible fiberoptic tube is inserted into an enclosed area such as body cavity or coffin interior.

Purpose

To visualize and/or biopsy the interior of such enclosed areas: brain, heart, viscera or whole body.

Limitations

The visualized structure must be located within a space to which the endoscope can be given access. Often the desiccated mummy tissue is too brittle to be grasped effectively by the biopsy jaws that are designed for use in hydrated, living tissue on this medical instrument.

Cost

Class 1 (if the cost and maintenance of the instrument and the operator's time is not included; see Table 6.3).

References

Tapp *et al.*, 1984 (hydatid cysts); Bonfils *et al.*, 1986–1987 (methodology); Brothwell, 1986:63 (brain); Notman *et al.*, 1986 (heart identification).

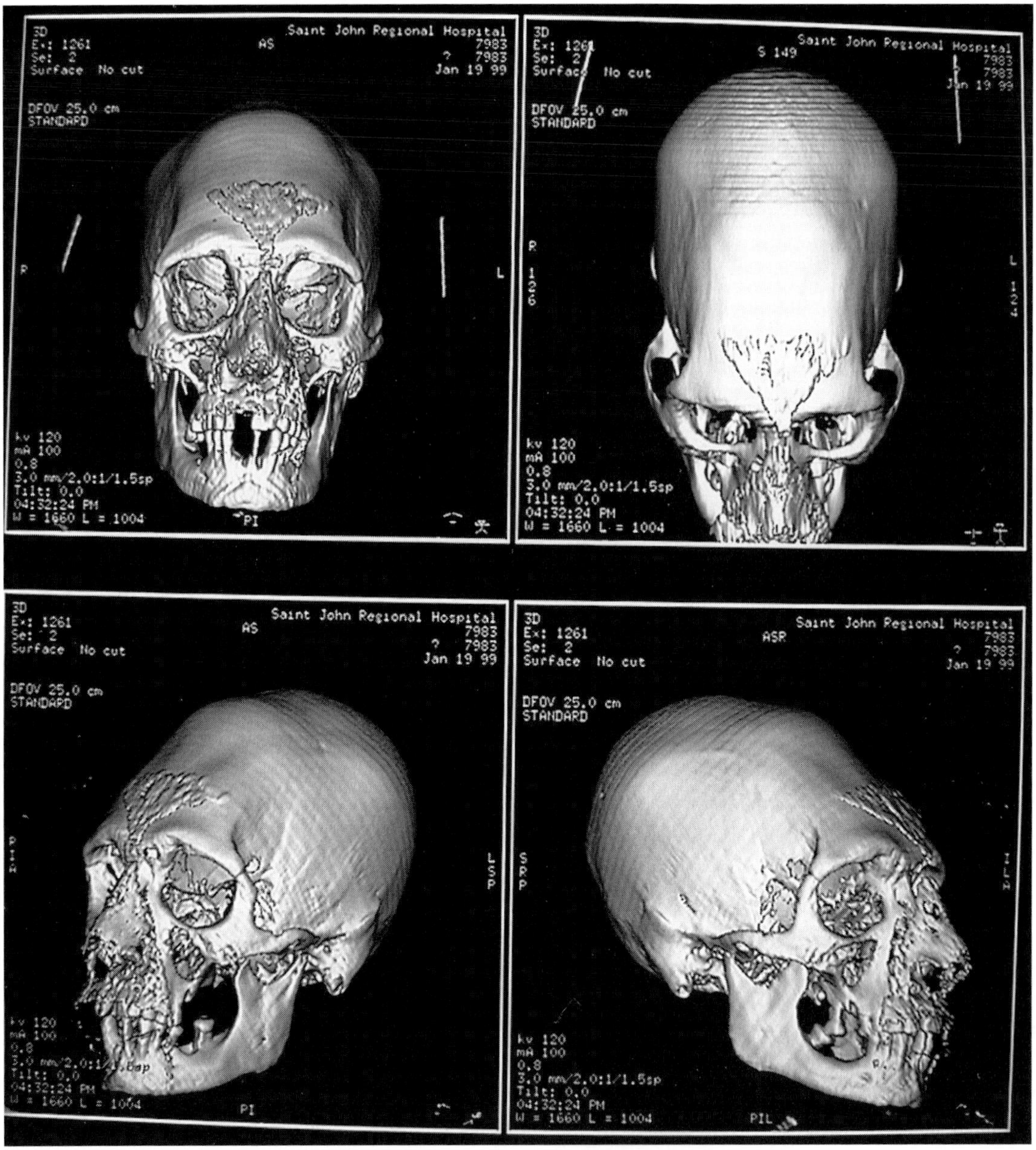

Fig. 6.51. Three-dimensional reconstruction of CT-scanned skull; SM, AD 1800.
The CT data of successive scanned skull "slices" have been reconstructed by a three-dimensional program to present the skull in multiple, three-dimensional views (compare with Fig. 6.50). (Photos courtesy of Ian Anderson, M.D., and Andrea Kirkpatrick.)

Endoscopic technique is a method borrowed from the field of medicine and, to a lesser extent, from industry. An endoscope is an optical, usually tubular, device that can be inserted through a small (2 cm or less) aperture into various body cavities in order to view and/or biopsy their contents. The earlier models were simple, straight, commonly metallic, rigid tubes equipped with an ocular at the operator's end as well as a light source and manipulatable biopsy forceps at the distal end. Those employed today have a lens at the distal end whose image is transmitted by bundles of optical fibers to the operator's eye. The fiberoptic core permits the endoscope's body to be flexible, and the distal end can be manipulated through wide arcs and planes, maximizing the viewable area. Most modern endoscopes are also fitted with beam-splitters that permit simultaneous video recording and display of the image on a video screen. Endoscopes designed for pediatric use are usually of less than 2 cm in diameter.

Endoscopes have been applied in mummy studies primarily when information about the contents of body cavities is desired in mummies on whom destructive dissection is considered to be prohibitive. For example, plain radiography revealed an irregular area of radiopacity in the left upper chest of the nineteenth dynasty pharaoh Merneptah. Endoscopy of the thorax revealed a white, calcareous mass, biopsy of which was analyzed chemically and found to be composed of a collection of natron salts (Manialawi *et al.*, 1978). Tapp *et al.* (1984) evaluated a similar radiographic appearance in a Late Period Egyptian mummy by endoscopic biopsy that established it as a hydatid cyst, produced by infection with the dog tapeworm *Echinococcus granulosus*, and identified a similar cyst in the brain of another. These authors also used endoscopy to determine that the dense lung of another Egyptian mummy was due to silicate pneumoconiosis, secondary to lifelong inhalation of Egyptian desert dust. Again, Notman *et al.* (1986) proved that such a radioopacity was the displaced but retained heart covered in part by resin in an Egyptian mummy, while Brothwell (1986:63) demonstrated endoscopically that an irregular mass in the cranial cavity of the British peat bog body Lindow Man was a shrunken brain.

Other useful but less common applications of endoscopy in the study of mummies include the retrieval of radiographically visualized coprolites for study from the abdomen of a spontaneously mummified 8-year-old child, using a rigid, straight endoscope passed through the abdominal wall and manually manipulated forceps (Fernand Filh *et al.*, 1988; Ferriera *et al.*, 1983). Fornaciari & Torino (1995) passed an endoscope through the intact lid of the sarcophagus believed to contain the medieval prince of Fano in a central Italy church, in order to evaluate the presence and state of preservation of the body inside. The study value of prints made from selected frames of the video record of endoscopic studies carried out on Peruvian and Egyptian mummies was demonstrated quite dramatically by Schaefer *et al.* (1995). Ferriera *et al.* (1983) retrieved a coprolite specimen from a spontaneously mummified Brazilian child by use of a straight, tubular rectosigmoidoscope.

Limitations of endoscopy for such purposes are several and include the requirement that the visualized structures must be located in an air-containing space that is accessible to the endoscope. Sometimes such access can exploit an already existing aperture such as an evisceration wound or the previously created transnasal craniotomy passage. However, modern endoscopes are of such small diameter that an aperture created by dissection can usually be prepared in a location easily disguised subsequently if the body is to be exhibited. Using a thin-bladed saw wheel on an electric drill, I created a 1.5 × 1.5 cm defect with beveled edges on the left lateral aspect of a rib during a study on the St Paul Science Museum Egyptian mummy (Fig. 6.52). This easily accommodated a pediatric endoscope, and after completion of the endoscopy, replacement of the excised rib plug produced a fit so snug that it was virtually undetectable. In many mummies the anterior abdominal wall has collapsed so completely upon the underlying viscera that insufficient space exists for the introduction of the endoscope. In others, pleural adhesions obliterate the thoracic cavity. The

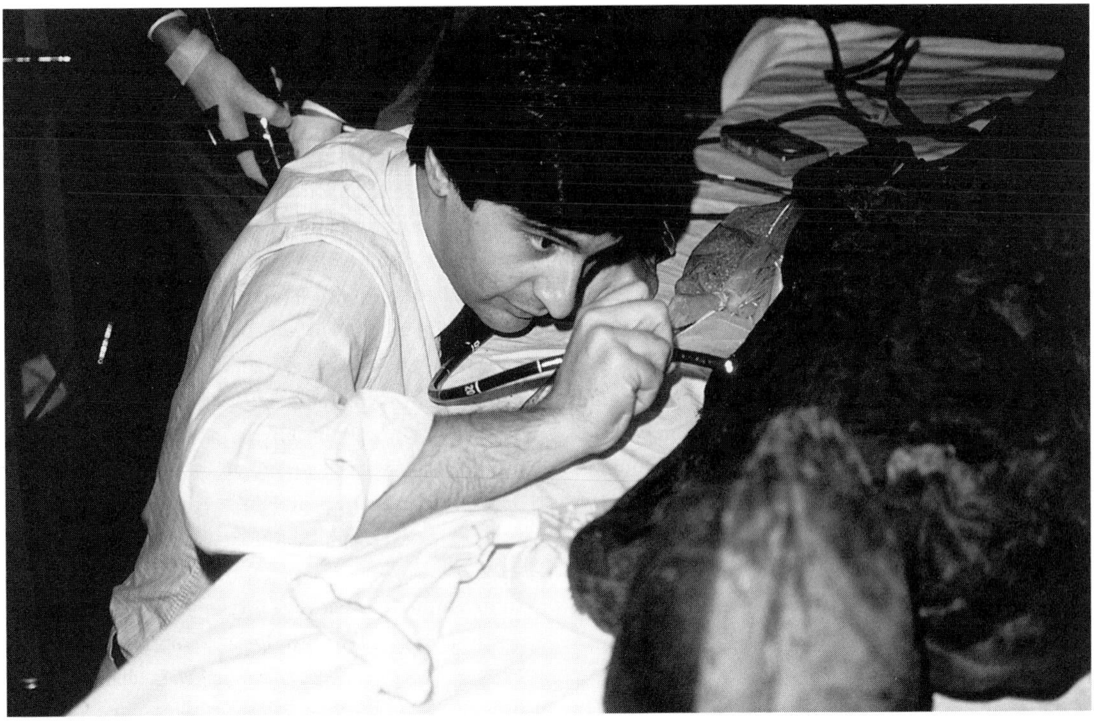

Fig. 6.52. Endoscopy of a mummy; AM, ca. 1549–1069 BC.
The endoscopist is inserting a pediatric endoscope into the St Paul Science Museum mummy via a small aperture I created in a rib. (Photo by ACA.)

black inner walls of the abdomen or thorax that have been painted with resin in Egyptian mummies may require more illumination than is available for ideal viewing conditions (Bonfils *et al.*, 1986–1987). Most frustrating, however, is the fact that the metal jaws of the endoscope's biopsy device (which is designed to grasp hydrated soft tissue in living people) often fails to grasp desiccated, brittle, mummified tissue or fragments it beyond recovery.

Biochemical methods

Molecular (DNA) chemistry

What is measured

A segment of DNA.

Principle

DNA is extracted from the sample and the two strands separated by heat. One strand is then amplified enzymatically, employing the polymerase chain reaction (PCR), by flanking the segment of interest with specific chemically synthesized primers (Saiki *et al.*, 1985, 1988). The amplified segment of predicted size is then separated by electrophoresis, and probed or sequenced to confirm its identity. An early application of PCR to ancient DNA was carried out on brain tissue of paleo-Indians by Pääbo *et al.* (1988).

Technique

Genes in the DNA molecules within a diploid eukaryotic cell nucleus are present as single copies within each of the two sets of chromosomes. After death the cells are normally exposed to the devastating effect of destructive lysosomal enzymes within the cell's cytoplasm, followed by the action of powerful enzymes of putrefactive bacteria and later those of necrophilic insects such as the dermestids. As a result, most mummified human remains contain less than 1% (often as little as 0.1%) of the cellular DNA present at time of death. Hence it is not surprising that nuclear DNA (which contains most of the information useful for research) is less frequently and consistently recoverable, though numerous, successful, individual efforts have been reported (Thuesen *et al.*, 1995). In addition to nuclear DNA, however, human cells also contain a relatively short, circular chain of DNA in the cytoplasm's mitchondria (mtDNA), inherited only from the mother. Since cells contain an average of nearly 1000 mitochondria, the residuum of even 1% of this DNA is often sufficient to be detectable.

Even this small quantity is normally fragmented into segments less than 200 nucleotides long (bases), many of which are rendered chemically unresponsive as a consequence of alterations by oxidation or deletion of the bases. The laboratory methodology of DNA chemistry has been designed for amounts of DNA normally found in living tissues. Hence such small amounts of ancient DNA (aDNA) must be amplified before they can be manipulated chemically. This can be achieved less regularly by direct cloning techniques, but enzymatic amplification is successful in most human mummy tissues (Pääbo *et al.*, 1988; Pääbo, 1989).

Amplification is achieved by heating the specimen, which uncoils the DNA and separates it into two single strands. The segment of interest is then "marked" by adding previously prepared (synthesized) short DNA chains (primers) specifically designed to bind to each end of the segment to be amplified. This is followed by the addition of the four base units: adenine (A), guanine (G), cytosine (C) and thymine (T). With the action of a heat-stable polymerase enzyme, appropriate bases are added to each DNA strand downstream from the primer binding sites. The result is the creation of a double-stranded DNA molecule from each of the two single strands. Repeating this procedure 20 times results in a theoretical million-fold amplification of the original trace amounts of residual DNA in the tissue. This amplification method is called the polymerase chain reaction (PCR). While in practice the reaction efficiency is less than 100%, the amplification is usually sufficient to yield the amounts needed for subsequent chemical study. Problems preventing amplification come from protein (especially hemoglobin) breakdown products that inhibit the reaction, other unknown inhibitors or arrest of enzyme action when the polymerase enzyme encounters a site of deletion or chemically altered bases. Following amplification, the amplified segment is isolated by electrophoresis. If that procedure demonstrates a band of the expected length, the amplified segments can be excised from the gel, amplified again and then the sequence of bases can be determined to establish their ultimate identity.

It is not uncommon to encounter problems at every step of these procedures. The most frequent is contamination by human DNA – of ancient handlers of the corpse, of modern excavators or of laboratory workers. Ancient contamination sources can often be minimized by appropriate selection of the sample (e.g., dental dentine), but laboratory contamination avoidance demands both physical isolation of the various steps and meticulous bench technique.

Bioanthropological applications

Using ancient mitochondrial DNA (amtDNA) techniques, applications have taken the form of attempts to determine biological distance by comparing variable DNA segments in samples from multiple ancient skeletal populations or blood (Merriwether *et al.*, 1994, 1995; Merriwether, 1999; Moraga *et al.*, 2000). Identification of an infectious agent in ancient tissue has also been achieved for cases of tuberculosis (Spigelman & Lemma, 1993; Salo *et al.*, 1994; Baron *et al.*, 1996; Taylor *et al.*, 1996; Donoghue *et al.*, 1998;

Fig. 6.53. Spitzbergen burial site of influenza victims; SM, AD 1918.
(Photo by and courtesy of Peter Lewin.)

Nerlich *et al.*, 1997), leprosy (Rafi *et al.*, 1994), virus (Li *et al.*, 1999; Sonoda *et al.*, 2000) HTLV-1 (Taubenberger *et al.*, 1997; Ham, 2000), 1918 "Spanish" influenza, and American trypanosomiasis or Chagas' disease (Guhl *et al.*, 1999; Madden *et al.*, 2001). Attempts to identify ancient intestinal bacteria by similar techniques have been at least partially successful (Cano & Borucki, 1995; Spigelman *et al.*, 1995; Ubaldi *et al.*, 1998), and similar coprolite studies may eventually identify the source (specific animal) of ingested meat. An expedition to the arctic island of Spitzbergen in Svalbard, supported in part by the National Institute of Allergy and Infectious Diseases, has excavated and sampled bodies of seven sailors that died there during the 1918 epidemic of the "Spanish 'flu" in an effort to recover and sequence that virus (Figs. 6.53 and 6.54). To date their results have not been reported (Davis *et al.*, 2000). Workers in this field are eagerly awaiting development of predictably successful methods of amplifying ancient *nuclear* DNA with the same ease as ancient mitochondrial DNA.

Limitations

Ancient DNA may not survive the postmortem environment at all. Ancient nuclear DNA, present in only two copies per cell, is less frequently recoverable. Mitochondrial DNA (present in hundreds of copies per cell) can be recovered from the majority of ancient human mummified tissue samples. Postmortem decay

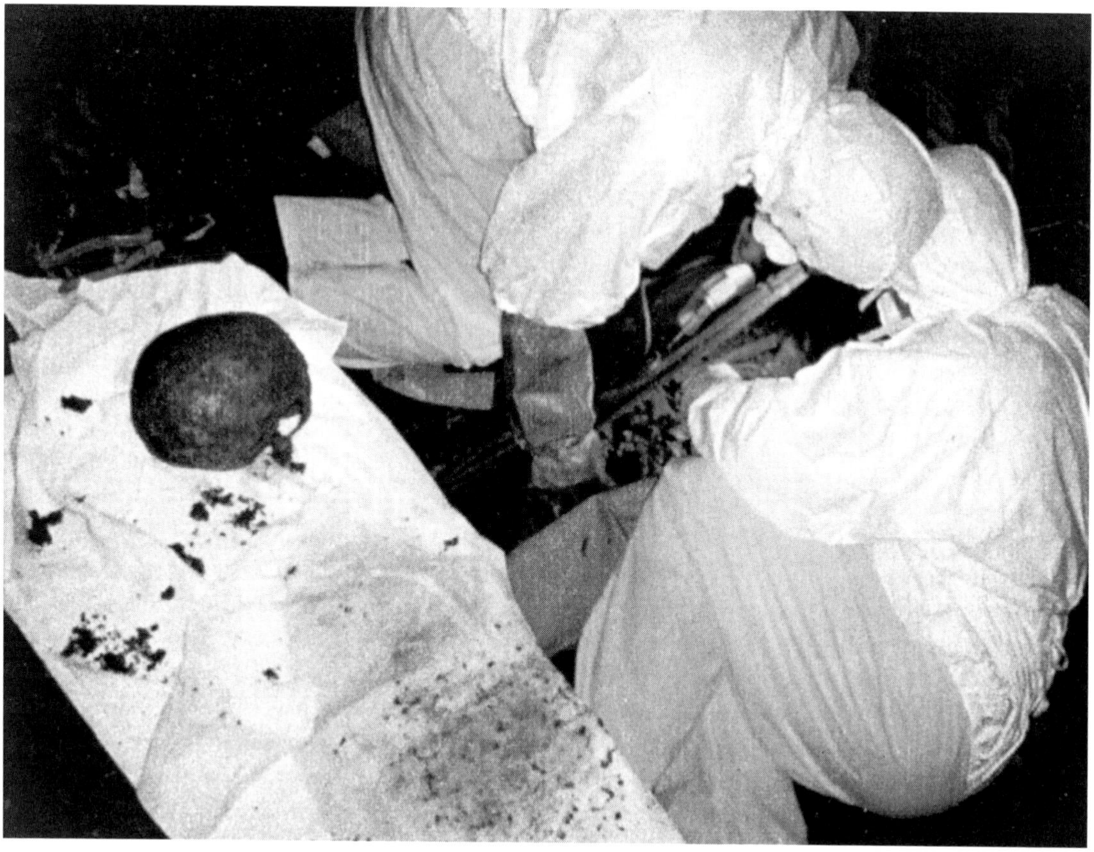

Fig. 6.54. Recovery of samples from influenza victims; SM, AD 1918.
Members of the Spitzbergen, Svalbard, expedition team dressed in protective suits recover tissue samples of 1918 influenza victims in an effort to reconstitute the virus' genetic structure. (Photo by and courtesy of Peter Lewin.)

products as well as base oxidation or deletion may inhibit the PCR. Most recoverable ancient DNA segments are fewer than 200 base-pairs (bp) in length.

Cost

Commercial laboratories commonly have a separate charge for each major step of the procedure.

References

Merriwether *et al.*, 1994 (biological distance); Bagasra *et al.*, 1995 (*in situ* PCR for EM and histochemical applications); Cano & Borucki, 1995 (aDNA of insect embedded in amber); Nielsen and Thuesen, 1998:355–362 (method).

Amino acid racemization (AAR)

What is measured

Using a simple polarimeter, measurement is made of the change in degree and direction of rotation (right or left) of plane-polarized light that has passed through a solution of an amino acid (usually aspartic acid, Asp) that has been extracted from a sample of protein-containing tissue by hydrolysis and separated by ion exchange chromatography.

The instrument (polarimeter)

Light of a known wavelength (sodium or mercury vapor lamp) is passed through a set of prisms, causing it to become polarized, i.e., vibrating in a single plane, spiraling in a circular manner clockwise (toward the left, L). When a solution of optically active amino acid is placed in such a beam's path, the plane of rotation is displaced in a direction and to a degree that is determined by the ratio of D (toward the right) and L forms in the amino acid solution (Willard *et al.*, 1981:412–429).

Polarimetry is simple and inexpensive, but requires a substantial quantity of specimen. Alternative methods have been developed to harvest the benefits of modern micromethodology, using the principles of either chromatography (discussed above) or nuclear magnetic resonance (NMR) spectroscopy. The initial steps of specimen hydrolysis to free the protein-bound amino acids are the same.

The Asp D/L mixture can then be separated with considerable sensitivity by the methods of chromatography discussed. The Asp D/L mixture, however, cannot be directly separated by chromatographic techniques. The mixture is made separable by reacting it with ("deriving") one of certain pure optically active compounds. This process of deriving a different product is called derivitization. Such derived D and L products are not enantiomers ("mirror images") and so are easily, rapidly and sensitively separated chromatographically. Alternatively NMR spectroscopy can also be employed for their separation. This technique identifies molecular structure by imposing a huge magnetic field upon a solution of the substances of interest. This interacts with magnetic fields of certain atomic nuclei (especially hydrogen) and changes in the atomic magnetic moment in response to changes in the external field are dependent on molecular structure and are measurable. These methods' sensitivity permits measurements on very small samples.

Precision

Depends partly on the purity of the aspartic acid solution as separated from the other amino acids; this is usually not a major problem.

Applications

Amino acids (AAs) are the building blocks of proteins, analogous to the role of pearls on a necklace. AAs (except glycine) exist in two molecular forms that are mirror images (enantiomers) of each other and are not separable by the usual chromatography techniques. They are "optically active"; i.e., when polarized light passes through a solution containing AAs, one of the molecular forms rotates the beam toward the left (L form) and the other to the right (D form). Since the rotation values for pure solutions of D and L forms are known, the concentration of each form can be calculated from the measured value of the mixed forms in a sample. Animal biosynthetic enzymes employ L forms exclusively. Hence, living animals and human proteins are composed almost exclusively of L forms. As long as such AAs are involved in metabolically active processes, they retain their L mode. When such metabolic activity ceases, as it does at some sites during life (ocular lens, dental dentine) and in all proteins after death, the L forms gradually convert to D forms spontaneously at a rate unique to each AA until they reach equal concentrations. This conversion is termed racemization. While Pasteur recognized the optical activity of AAs as early as the mid nineteenth century, serious study of these properties in archaeological samples has been carried out principally since the early 1970s (Bada, 1982).

Aspartic acid (Asp) has received the most study because its racemization rate (Asp r) is the most rapid: 43×10^4 years at 0°C. These rates are temperature-sensitive: Asp $r = 0.30\times10^4$ years at 25°C (Bada, 1982). Early studies assumed no other significant variables were operational, and Bada *et al.* (1974) used the Asp *r* value to date Californian human fossil bones to as much as 48 000 years BP. These authors "calibrated" their method by assuming that the temperature to which the sampled bones had been exposed was similar to that of a skull ("Laguna skull") from the same site whose radiocarbon date was 17150 ± 1470 years. Establishing the Asp *r* rate at that site (assuming that both the Laguna skull and the other fossil bones had experienced the same temperatures) on the basis of this age they then applied it to a Los Angeles human skull whose earlier reported radiocarbon date of >23600 (later retested value was 3560 years: Taylor *et al.*, 1985) seemed to compare favorably with its calculated Asp age of 26000 years. Similar applications to several human bones in a San Diego museum, however, provided stunningly early dates for humans in the New World, ranging from 28000 to 48000 years.

Such unprecedented early New World human remains stimulated multiple studies, searching for other variables affecting the racemization process. We now know that, in addition to mean temperature, such rates are influenced by the water content (slowed by desiccation: Bada & Schroeder, 1975), metal ions (hastened, if bound: Bada, 1985), collagen conformation integrity (delayed if helical form is retained: Collins *et al.*, 1999), radiation (expedites: Lubec *et al.*, 1994), pH (significant deviations from neutrality accelerate: Bada, 1985), the state of the AA (faster at elevated temperatures if bound to protein than if free: Bada, 1982) and the state of bone protein preservation (delayed if poor: Masters, 1986). Subsequent accelerator mass spectrometry (AMS) ^{14}C dates of the California bones producing such unexpectedly early AAR dates proved them to be less than 6000 years old (Taylor, 1983; Taylor *et al.*, 1985; Bada, 1985). This discrepancy led to a high level of distrust for the use of AAR. It is only now, after definition of many of the variables noted above that AAR is gradually finding a niche (unstable as that may still be) in our scientific methodological armamentarium for dating ancient specimens.

It is self-evident that the same formula employed to predict a sample's age can predict paleotemperature if the sample's age is known (Bada & Schroeder, 1975).

Certain tissues within the body have metabolically inert proteins (dentine collagen, some ocular lens proteins). During life these are maintained at body temperature that accelerates the racemization rate sufficiently to be detectable (Helfman & Bada, 1975; Masters *et al.*, 1978). Under circumstances such as bodies found in glaciers in which the body is maintained in a frozen state after death (which arrests racemization), the age-at-death can be calculated, assuming an Asp *r* of 0.1% per year during life (Masters & Zimmerman, 1978).

Since AAR rates and DNA postmortem integrity are affected by the same variables (especially presence of water, temperature and chelation of certain metal ions to proteins), a potentially very useful application of AAR is based on the observation that AAR and spontaneous postmortem DNA degradation correlate well enough that samples selected for DNA study can first be screened by the relatively simple AAR technique to identify those in which DNA recovery is unlikely (those in which the D/L Asp ratio exceeds 0.08) (Poinar *et al.*, 1996). Some authors, however, still note variables that can confound such measurements (Collins *et al.*, 1999). These include the tissue source of the sample (insoluble portion of dentine is best), contamination from bacterial walls (they may contain D amino acids but, more importantly, are made up of L amino acids), size of the sample (other contaminating proteins if sample is too small) and amino acid selected for measurement (alanine and leucine are excessively temperature-sensitive) (Rodríguez-Albarrán *et al.*, 1999).

Limitations

Most of these have been identified in the initial section above. They may be summarized here as follows:

1. The requirement for the knowledge or assumption of the mean temperature to which the sample was exposed.
2. Collagen integrity may deteriorate with time, altering the racemization rate.

3. Sample site within the specimen (bone or mummy) needs to be standardized because collagen protein turnover rates vary with its location (the less metabolically active sites have higher rates).
4. Desiccation retards the rate (an untested factor in mummies) (Aitken, 1990:308).
5. The above demand calibration of the rate at each site.
6. Heavy modern bacterial or fungal contamination can contribute diagenetic modern amino acids.
7. Cost (operating): Class 2 (see Table 6.3).

References

1. Bada, 1982, 1985. Overview of AAR.
2. Bada & Protsch, 1973; Rae *et al.*, 1987; Csapo *et al.*, 1994. Prediction of a sample's age beyond the radiocarbon range
3. Poinar *et al.*, 1996; Collins *et al.*, 1999; Rodríguez-Albarrán *et al.*, 1999. Prediction of DNA integrity in an ancient sample
4. Bada & Schroeder, 1975. Prediction of paleotemperatures (if the specimen's age is known).
5. Helfman & Bada, 1975; Masters & Zimmerman, 1978. Prediction of age-at-death.

Other methods

The above methods were selected because they have been employed quite frequently to extract information from mummified soft tissue remains. Other techniques including Fourier transform, dating by eggshell protein diagenesis, laser lithography, infrared spectrophotometry, gas chromatography, mass spectrometry, Raman spectroscopy etc. have also found their way into bioanthropological studies, but do so more exceptionally. Undoubtedly both the number and the range of laboratory manipulations will increase dramatically during the next several decades. However, the field of bioanthropology is already replete with examples of the hazards incurred when one is interpreting results from ancient mummy tissues analyzed by methods designed for study of fresh or living tissues. I expect that heads of bioanthropology training programs in the near future will find it desirable or even necessary to include in their curriculum a course dealing with an introduction to instrumental analysis. This, in turn, will generate a need for an appropriate student text with special emphasis on the effects of postmortem and diagenetic processes on such analyses, and the necessary interpretational adjustments.

Chapter 7. Mummification of animals

Introduction

Just as is true with human bodies, the soft tissues of an animal's corpse may be preserved as a result of deliberate human effort (anthropogenic or "artificial" mummification) or as a consequence of environmental conditions (spontaneous or "natural" mummification). Since the cultural significance of each of these two methods is different, they are considered separately here.

Anthropogenic animal mummification

Origin of the practice

In Chapter 2 we pursued the purpose of human mummification under the assumption that such a dramatic and expensive act would survive as a group's cultural feature only if the mummifier derived great benefit from it. We concluded that among the Incas, Egyptians and probably others, human mummification originated as an effort to enhance the power and security of the self-declared deification of the nation's political leader. Among other, less technologically developed groups mummification was commonly involved with appeasement or exploitation of the deceased's spirit. Because we have the most information about animal mummification from ancient Egyptian practices, we will focus our quest for the purpose and origin of animal mummification among them. It will be recalled that mummification among Egyptians was initially restricted to the pharaoh. As the administrative structure of unified Egypt expanded, it became necessary to delegate authority to selected nobles, who then demanded mummification opportunities, probably recognizing its political value. Still later mummification privileges were extended also to the clergy that carried out the religious rituals, a benefit gained during periodic struggles for power between the clergy and the pharaohs. The political model seems applicable to all these groups. Not until the latter portion of Egypt's national history, especially during the New Kingdom, was mummification democratized and became available to Egyptians of nearly all socioeconomic classes. While the rulers remained conscious of mummification's political value (it was shrewdly embraced even by the Ptolemies: Grant, 1972:22–23), millennia of mummification rituals had generated a religious motivation to participate in an eternal afterlife among the general population.

Efforts to understand Egyptian mummification of animals can also be directed by the assumption of the mummifiers' expectations of personal gain from the practice, but one must acknowledge some substantial differences in the practices of mummifying between human bodies and those of animals. Principal among these is that mummification of human bodies originated among the earliest conquerors of Lower Egypt, with subsequent national "unification", while the recognition of an intimate relationship between certain deities and distinctly identified animals may be a concept far older than the ancient Egyptian state. Linen-wrapped animal bodies were included in Upper Egyptian tombs as early as the Badarian Predynastic Period (5500–4000 BC), and ceramic designs employing animals during the Naqada II (3500–3150 BC) interval can be interpreted as an effort to display them as deity symbols (David, 1982:24–26). Though the limitations of rupestrian art, archaeology and mythology constrain our elaboration of its nature, these sources imply the presence of such a concept in the earliest of predynastic records and at least suggest some of the possible motivating factors responsible for its existence. Note, however, that D'Auria *et al.* (1988:230) suggest the possibility that Predynastic animal images may have served merely as regional symbols (totems), becoming attached to deities at the onset of the Dynastic Period. In either case, it is in these early periods, therefore, that we must focus our efforts to understand such relationships. Animal symbols have also been interpreted as images that reflect a population's social stratification. Selection of such animal symbols may have been motivated by the need for a form sufficiently flexible to respond to social changes over time (Levy, 1995). Certainly Egyptians were comfortable with the practice of assigning a given animal symbol to different gods having different functions at different times.

A prominent principle of modern behavioral science is that the fear of death underlies much of human behavior. Ancient humans, encountering natural phenomena beyond their understanding, can be expected to have assigned the cause of such events to acts by one or more supernatural forces (deities) (Hamilton-Paterson & Andrews, 1979:13). Since these early humans were hunters, their dependence upon availability of animals would have been self-evident. Fluctuations in ready accessibility to animals were a frequent threat to life. Because the causes of varying hunting success were not always obvious, we can expect such hunters to assume that the barren periods were acts of these supernatural forces. Propitiation of such deities would appear to be the only avenue available to unsuccessful hunters for influencing the gods to act in their favor (Howey, 1972:148). Such perceptions are commonly believed to be responsible for the evolution of deity-appeasing taboos among early societies and for sacrifices practiced by the more structured populations. Ethnological studies confirm such relationships among modern, isolated hunting–gathering groups.

Without denying that some groups may have or do idolize the animal itself, the term "animal worship", so often applied to any relationship between humans and animals involving supernatural concepts, is in many cases a misnomer (Gandhi, 1954:4). To understand Egyptian animal mummification it is essential to be aware of the precise nature of such a relationship. In the majority of cases the human perceives the animal as responsive to a supernatural force (a deity). At most, the deity may be transiently occupying the animal's body to manifest itself (incarnation) and to direct the animal's behavior. Since certain animals are vital to a hunter's survival, specific deities become associated with specific animals. Thus god-appeasing acts by the hunter, though carried out physically for the animal itself, are actually directed at the deity represented by the animal. Such relationships may be elaborate and can be individual and personal, or extend to a group or clan, a condition I shall include conveniently in the term "totemism".

Modern ethnological examples are abundant. Until recently almost every animal hunted by arctic Inuit required some form of conciliation to appease the animal's spirit offended by the slaying. Examples range from such minor taboos as arranging fish (caught through a hole in the ice) in a circular manner around the hole with heads pointed toward the hole, to major festivals following capture of a whale, designed to honor the spirits of all whales and encourage them to return the following year. Frazer (1976:600) relates that Kamchatkans, before killing sea mammals, begged their forgiveness, and postslaughter feasts honored the animal, trying to convince its spirit that their enemies, not the Kamchatkans, were responsible for the death and requesting its spirit to entice its kinfolk in the sea to come and be caught.

With this concept of "totemism" we can now approach the border of prehistory and history that affected the Predynastic–Dynastic junction in Egyptian history.

Pastoralism and animal cults

In the Middle East, North Africa, Andean area and other locations the subsistence chronology was hunting–gathering → pastoralism → agriculture. In the Fertile Crescent sheep-herding can be detected as early as 9000 BC and cattle domestication appears by 6000 BC (Herre, 1963). Thereafter involvement of cattle with religious ritual is easily demonstrable. Indeed, if our reconstruction of the development of totemism among hunters is essentially correct, then replacement of hunted prey by cattle in the new, pastoral subsistence mode would lead one to expect that cattle would also replace the prey animals in spiritual concepts and rituals.

Schwabe (1978:15–29) notes the following observations: aurochs (from which domesticated cattle are derived) were central to the rituals of early Sumerians, and the earlier name of the principal Babylonian god Marduk meant "young bull of the sun", while that of the Akkadian god from Uruk meant "bull of heaven". A shrine at Catal Hüyük (6000 BC) is rich in bull symbols. The legend of Theseus and Ariadne relating to Crete's Minoan civilization (3000–1500 BC) dealt with the beast Minotaur, born to King Minos' wife as a consequence of bull god–human copulation. The palace at Knossus in Crete is dominated by bull-with-horns statuary and murals involving bulls. Bull worship is the basis from which the later belief system of Zoroastrianism is derived. Evidence of cattle domestication in China is identifiable as early as 4000 BC, and several of the legendary dynasties of the following periods such as the Hsia had deities with cattle features. Furthermore, in some of these areas the cattle-related deities were also linked to the societies' principal political leaders. Thus, from the Middle East to eastern Asia, cattle pastoralism is established in many areas between 6000 and 4000 BC, and frequently such sites provide archaeological or rupestrian evidence of incorporating cattle into rituals.

Hoffman (1979:242ff.) shows that North Africa, too, demonstrates a similar trend. Paleolithic hunters of the Western (Libyan) Desert evolved into a mixed Neolithic agropastoral economy about 6000 BC, in which cattle herding was dominant, and contact with the people of the Upper Nile Valley *may* have been made during the subsequent millennium. By 5500 BC agricultural villages appear in Upper Egypt, and rock art near Aqaba from about 4500 BC reveals a predominance of cattle. Pre-Egyptian Nilotes (Badarians) about 5000 BC inhabited the entire length of the Nile Valley, had domesticated cattle and may have worshipped the bull (Schwabe, 1978:30 ff.). Schwabe also believes that the modern Dinka of the Nile Valley in southern Sudan may be their direct descendants, as may be the Dinka's relatives – the Masai. Both are cattle-centered cultures and incorporate cattle into their rituals. Archaeology had demonstrated evidence of cattle burials in Egypt during the Predynastic Badarian and Naqada periods (Schwabe, 1978:54). These Nilotes may have been eventually displaced by, or more likely evolved into, cattle pastoralists whose social structure consisted of totemic clans also practicing some agriculture. Thus the development of pastoralism about 6000 BC that accompanied the repopulation of the Sahara at that time included domesticated cattle and this practice spread eastward until, by the onset of the Dynastic Period, a totemic culture occupied Upper and Lower Egypt, and a cult of the bull had developed in the Nile Valley that included cattle features in religious ritual including cattle burial.

Dynastic Egypt and animal cults

The unification of Egypt found its Nile Valley to be occupied by large numbers of individual populations, each of which had its own gods. These were portrayed in the forms of local animals, implying a totemic role.

As part of the nationalization process of Upper and Lower Egypt, these local gods were merged and some of them (e.g., Horus-falcon, Isis-ibis) became national deities. As time passed and the intellectual concepts of the new state became more complex, the representation of these deities gradually assumed more human morphology until the entire body was human except for the animal head. The latter probably was retained for the purpose of instant and universal recognition of the identity of the portrayed deity. The fact that the original associations between local gods and the local animal totem were simply added to the national pantheon had resulted in the now often confusing situation in which the same god may be represented by several different animals, or the same animal may be a totem of several different deities. Furthermore, persistence of local practices also created a potential for conflict of interest when a local animal totem is not respected by a neighboring village that uses such an animal as a food source.

The earliest representations of apparent Egyptian deities is seen in the fourth millennium BC, long before the first hieroglyphs (Guirand, 1968:9) and some of these Predynastic symbols persisted to the end of Egyptian civilization (James, 1957:235). Diodorus Siculus (ca. 50 BC/1967, I:82–90) describes some of these practices, contributing his personal speculations about the ancient Egyptians' motives. Herodotus (450 BC/1981, II:65–76) describes the animals in great detail but says "were I to declare the reason why they are dedicated, I should be brought to speak of matters of divinity, of which I am especially unwilling to treat . . .". A list of totem animals associated with national Egyptian deities is itemized in Table 7.1, below. The rituals relating to some of the more frequently mummified animals are discussed below.

The Apis bull cult

This is the earliest of the totem cults. It focused on the bull as a symbol of strength and fertility, and incorporated mummification of the totem animal routinely. The discovery of bull burials in the late Predynastic Period documents its antiquity. Pharaohs of the first dynasty, probably attracted by its image of power and fertility, incorporated some of the bull's features by appropriating its name, initiating the bull god-king concept that persisted throughout the Dynastic Period (Conrad, 1957:72). The cult center was in Memphis, where the bull was considered to be the incarnation of the creator gods Ptah and Osiris. Only one bull in all of Egypt was recognized as a living totem at any one time. The animal was selected on the basis of certain symbolic skin markings and was installed in a series of grand ceremonies. This bull was treated with the utmost care while housed in his special quarters at a temple built for that purpose, and his movements (believed to have divination significance) were observed closely.

The Apis bull was usually allowed to die a natural death, though long-term survivors appear to have been sacrificed in their twenty-eighth year. This number may be related to the 28 phases of the moon, since the legendary origin of Apis identifies him as born from a moonbeam (Mackenzie, 1978:70). Death of an Apis bull was cause for profound national mourning, with elaborate funeral ceremonies, including embalming. In a Memphis temple a special room for that purpose contained an alabaster table large enough to incorporate its body. This embalming table was adorned with leonine features and equipped with fluid drainage channels. Spencer (1982:200) notes that embalming methods included desiccation but usually not evisceration. He speculates that organ destruction was probably by intra-anal oil injection as described by Herodotus. D'Auria *et al.* (1988:231), however, describe evisceration including cerebrectomy. Artificial eyes and a gilded plaster head enhanced the bull's appearance (Spencer, 1982:201). Mummified in the crouched position, the process was completed with elaborate bandaging (el Mahdy, 1989:166). Before the New Kingdom, individual burials, often in wooden coffins were the rule, accompanied by valuable gems and charms. The funeral rites were identical to those of humans. These were carried out in a special temple called the Serapeum at nearby Saqqara. During the

New Kingdom the mummified bull's body was interred in a massive (up to nearly 60 tonnes) sarcophagus placed in a special subterranean chamber beneath the funeral temple. At Saqqara a sequence of such chambers, each once accommodating a mummified Apis bull was excavated by Auguste Mariette in 1851 (Mariette, 1882). Sacred bulls were also kept at Armant near Luxor (Buchis bull) associated with the god Mont who was linked to the sun god Re. Their catacomb was called the Bucheum and the bulls were buried stapled to a large board (Spencer, 1982:203). A Mneves bull associated with the sun god at Heliopolis is also known. Two of these have been found in Heliopolis (relating to the reign of Ramses II and III) as individual burials (el Mahdy, 1989:168).

If the Apis bull was so highly revered, it is not surprising that his mother, the cow, would enjoy similar veneration. The sacred cow was identified with Hathor, Isis, Nut and Nepthys (Mackenzie, 1978:72), and commonly was buried in a separate cemetery. The earliest of these, from about 350 BC, was near the Buchis bull cemetery (Bucheum) and that near the Serapeum is about 50 years later, although texts suggest the practice is several centuries older (Spencer, 1982:205). Embalming methods were similar to those for the bulls.

The Apis bull was the first animal cult for which we have archaeological documentation (Paton, 1925:7) and also survived the restricting influence of Christianity longest. Destruction of the Serapeum followed an edict of the Roman Emperor Honorius, who banned sanctuaries for pagan use in the fourth century AD (D'Auria *et al.*, 1988).

The ibis cult

Though textual references suggest ibis cult practices in the early New Kingdom Period, archaeological evidence focuses primarily on the Ptolemaic and especially Roman periods. Dedicated to the Egyptian god of wisdom, Thoth, the cult's rituals were centered in his widely distributed temples (Mackenzie, 1978:75). This ritual characterized the impact of the democratization of personal participation in religious rites among the general Egyptian population. In contrast to the ancient bull cult, in which only a single living animal at any one time symbolized the god's incarnation, that of the much later ibis cult extended veneration to *all* members of the species. Members of the general public visiting the temples purchased mummified ibises from the priests and presented them as votive offerings. To meet this demand, ibises were reared on the temple premises, killed (frequently as juveniles), mummified (often by dipping them in tar) and wrapped (Moodie, 1931:58). After serving their ritual purposes the mummified bodies were placed in stoppered ceramic pots, occasionally in coffins or sarcophagi. When sufficient numbers had accumulated, priests held special ceremonies for their interment in subterranean chambers. The number of mummified ibises in some of such burial sites is phenomenal. The site at Saqqara has been estimated to contain hundreds of thousands (Spencer, 1982:205) or half a million, a rate of about 10000 per year (D'Auria *et al.*, 1988:232), while el Mahdy (1989:162) states that more than four million were found at Tuna-el- Gerbel, a Thoth city. Veneration of the ibis reached a point at which the killing of an ibis in an act unrelated to ritual was punishable by death (Diodorus of Sicily, ca. 50 BC/1947:108). In contrast, Chapter 10 (Mummy as a drug) describes how seventeenth century Egyptian entrepreneurs pulverized the contents of some of such cemeteries and marketed the "mummy powder" as a medicant in Europe (Dawson, 1927c).

An ibis cemetery with 1500 mummified birds was found at Abydos (Leonard & Loat, 1914) (Figs. 7.1 and 7.2). The mummification technique there included evisceration and desiccation. The head and neck were commonly doubled back on the body and the corpse thinly wrapped (Lupton, 1986; D'Auria *et al.*, 1988:232) (Fig. 7.3). The high demand for these bodies may have taxed the production system periodically, because X-rays of apparent ibis mummy bundles quite frequently contain only a part of a body. Even greater liberties appear to have been taken by the embalmers of falcons, killed and mummified for a Horus cult ritual,

Fig. 7.1. The ibis cemetery at Abydos; AM, ca. AD 150.
Leonard & Loat reported their discovery of this huge ibis cemetery at Abydos in 1914. (Photo from Leonard & Loat, 1914.)

for many of these prove to include not only other birds of prey but virtually every bird found in Egypt (Diener, 1981; el Mahdy, 1989:162).

In addition to the two contrasting cult examples given above, it must be noted that, at one time or another in various locations, animals representing most of the species found in ancient Egypt were buried. Among them were the following, more commonly mummified ones.

Baboons

These represent another totem of Thoth, are found in relation to his temples and ibis cemeteries, and were important enough to appear on canopic jars (together with jackals and falcons). They were reared at these temple sites; 400 were found buried at Saqqara (Spencer, 1982:205). Some of these were within wooden chests packed in gypsum plaster (Emery, 1970). In addition to their ritual function, some pharaohs appear to have used them as pets. Nerlich *et al.* (1993) added considerable information to our understanding of the living conditions provided to these animals at their temple quarters when they examined the remains of 82 baboons buried in sarcophagi and coffins in an animal burying ground in central Egypt. They found evidence of malnutrition, fractures, osteomyelitis, vitamin D deficiency (rickets) and death before or at their prime of life in a large fraction of the total.

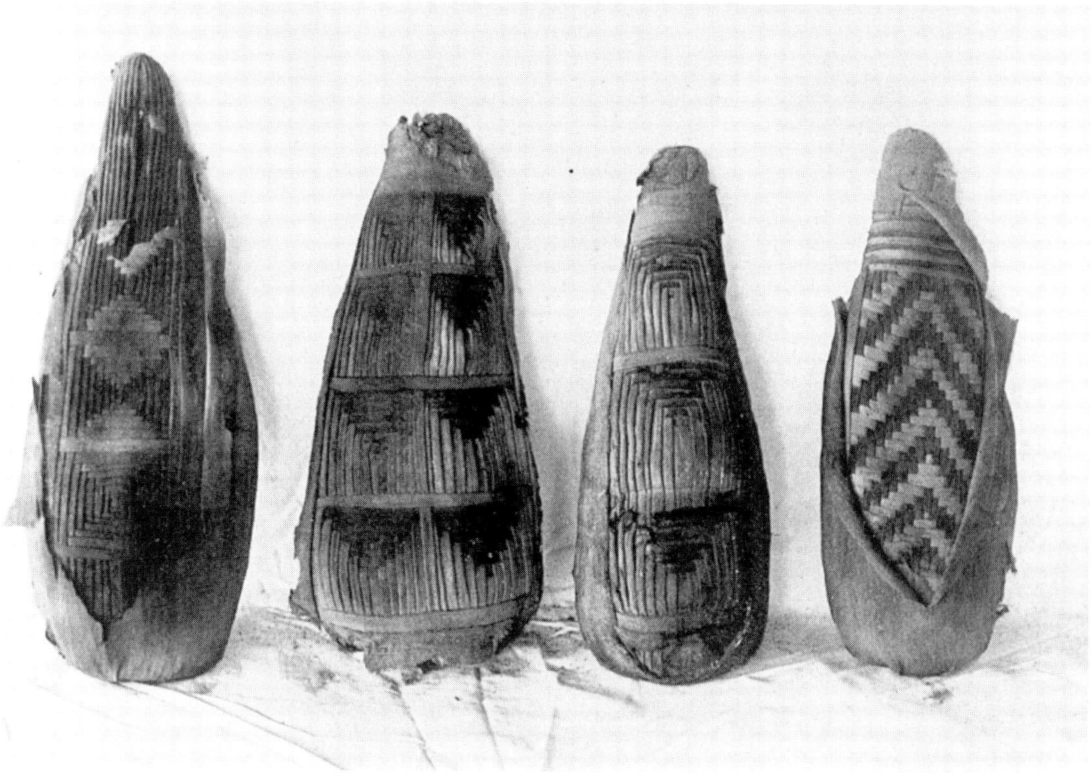

Fig. 7.2. Ibis mummies; AM, ca. AD 150.
Individual wrapped ibis mummies from urns in cemetery shown in Fig. 7.1. (Photo from Leonard & Loat, 1914.)

Cats

Cats were common pets among Egyptians. Their ritual function relates to the goddess Bast or Bastet. Most examples of this cult are from the Ptolemaic Period (Malek, 1993). The most commonly mummified cat in Egypt corresponds closely to the feral *Felis libyca*, while less frequently the larger *Felis chaus* was used (Morrison-Scott, 1952). Evidence of temple rearing of these animals has been found. The cult was centered at Bubastis (Tell Basta) in the Nile delta, and Herodotus (450 BC/1981, II:67) describes the embalming and burial of dead cats in sacred buildings of this city. The popularity of cat mummies as votive offerings rivals that of ibises (el Mahdy, 1989:165). At Beni Hasan, a cemetery of mummified cats contained so many that it was exploited commercially as a source of fertilizer in modern times (Guirand, 1968:48). Herdman (1889–1890) examined a few of these and felt that they represented *Felis caligata*. Mummification was accomplished by evisceration and desiccation, some without natron and some by wrapping the eviscerated body in natron-soaked linen bandages. Body cavities of some were filled with soil. The head was sometimes set at right angles to the neck to provide a cylindrical package. Forelimbs were stretched back on the trunk's ventral surface, and hindlegs were flexed against the pelvis (Armitage & Clutton-Brock, 1981). These workers also found two separate peaks with respect to age-at-death: one to four months and more than 2 years. Occipital bone displacement suggested death by strangulation. Linen bandaging was often elaborate,

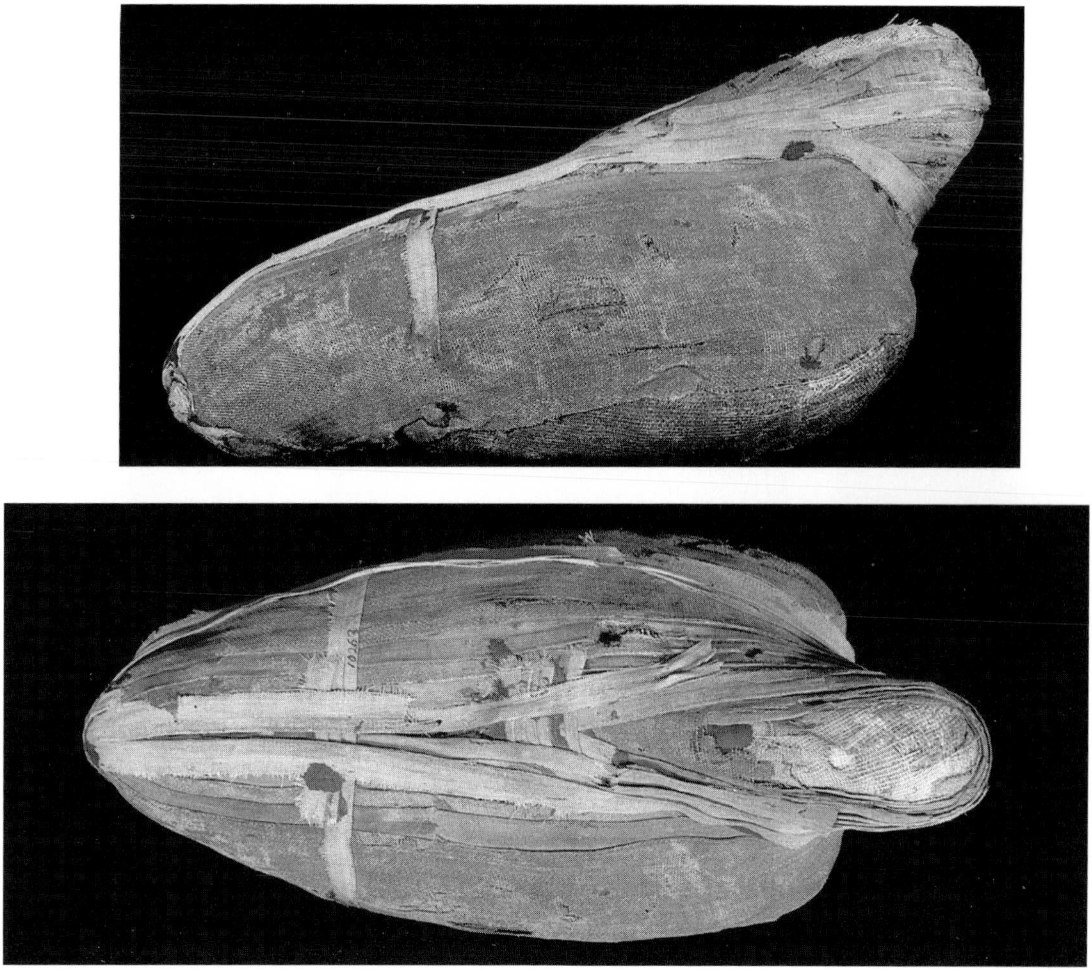

Fig. 7.3. Wrapped ibis mummy; AM, 30 BC to AD 395.
Huge numbers of such wrapped mummified ibis bodies represent votive offerings from the later periods of Egyptian history. *Upper*, lateral view; *lower*, superior view. (Photo by ACA, courtesy Carter Lupkin and the Milwaukee Public Museum.)

including complex, two-color, geometrical patterns. However, as with the ibis cult, X-rays and unwrappings have indicated that many apparent cat mummy bundles actually enclose not only merely parts of cat bodies, but even miscellaneous human bones (Pahl, 1986). Furthermore, Burleigh (1980) dated two of the British Museum's mummified cats by radiocarbon studies and found them to be from the Ptolemaic Period. Similarly, the "cat mummy" examined by Pahl was also of Ptolemaic date. Clearly, while many "false mummies" in European or American collections were prepared for the modern tourist trade, at least *some* of them can be attributed to ancient embalmers.

Crocodiles

It has been suggested that this fierce animal was selected early so that its use as a battle standard dis-

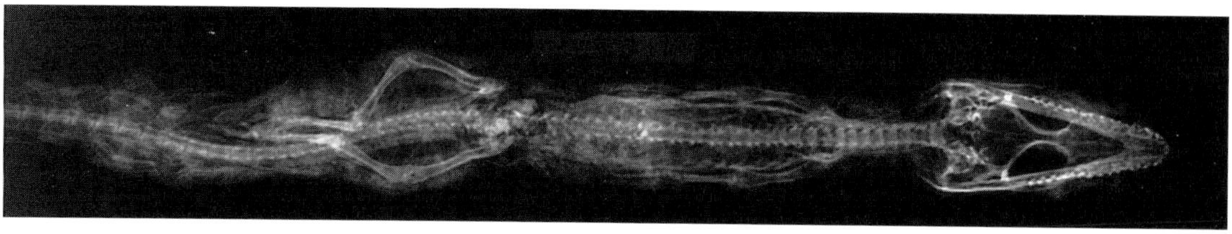

Fig.7.4. X-ray of Egyptian mummified crocodile bundle; AM, 30 BC to AD 395.
Most crocodile votive offerings in Egypt were preserved simply by dipping them in resin or bitumen. Principally juveniles were selected for this purpose. (Photo and radiography by Patrick Horne; specimen housed in Royal Ontario Museum, Toronto.)

played during wars would terrify the enemy; alternatively, its selection may have been an attempt at pacification of its spirit. While interesting, such musings are speculative. Dedicated to the fertility god Sebek (later a creator god associated with the sun god Re) it represents one of the earlier customs. Diodorus Siculus (50 BC/1967) suggests that Egypt's unifying conqueror and founder of the first dynasty, Menes, established the cult in the Fayum. Archaeological evidence has confirmed the city of Crocodilopus in the Fayum as the center of this cult, and identified a crocodile necropolis there. Other necropolises were at Kom Ombo and Tebtunis. Schwabe (1978:61) cites Herodotus and Strabo as saying that it spread from the Fayum throughout Egypt during Dynasty 12, when considerable contact was established at the Fayum between Egyptians and the Upper Nile's pastoralists. At Tebtunis, temple visitors buried about 2000 crocodile mummies in shallow sand graves (D'Auria *et al.*, 1988:234). Few large animals are represented; most are juveniles or small lizards (Fig. 7.4). Little preservation effort other than the application of black resins appears to have been made. Linen wrappings, however, are abundant, with complex geometrical patterns. Many of the smaller bundles allegedly representing juvenile crocodiles contain only reeds or random body parts (Spencer, 1982:212). Herodotus (450 BC/1981, II:90) comments dryly that drowned bodies and those killed by a crocodile are expected to be embalmed by their finders, being considered "more than human" and to be touched only by priests of the Nile.

Rams

This cult resembles the bull cult in that the animal symbolized fertility and only one living member at any one time represented the deities Amen and Khnum (the latter a creator god). The cult was focused at Elephantine at the first cataract of the Nile, though Mendes was also a center for rams. Linen wrappings were elaborate, with burials in coffins or sarcophagi.

Other animals

In addition to the above, Egyptian animal mummification included the Egyptian mongoose (ichneumon) and shrew (Horus), dogs and jackals (Anubis), birds of all types (Horus), fish (oxyrhynchus: Seth), serpent and eel (Atum) and even beetles (*Cheperá*). A sacred animal necropolis at north Saqqara was investigated in 1992. It consisted of a catacomb with dozens of branches and short halls leading to rooms filled to capacity with various embalmed animals, some wrapped in cloth and some in ceramic jars. Animals represented there included baboons, "mothers of apis, hawks, ibis, rodents . . .". Mummification involved desiccation and generous application of resin (Nicholson, 1994). The scope of animals mummified reflects the inclusion of local deities following nationalization of the Egyptian state, though in a few cases no deity has been found to be related to the mummified animal. At the Fine Arts Museum on Hero's Square (Budapest, Hungary), in 1988 we found a fine collection of such mummified Egyptian animals including several cats, an ichneumon, a crocodile (juvenile), a snake, a falcon, several ibis as well as several late New Kingdom human mummies. The most complete list of animals I could identify that were mummified by Egyptians is itemized in Table 7.1.

Summary of anthropogenic animal mummification in Egypt

Anthropogenic animal mummification in Egypt is an expression of totemism, a common cosmic view in antiquity that acknowledges the existence of supernatural forces ("deities") who are symbolized by the various local animals. The deities may even occasionally manifest themselves in the animals of the region occupied by the cultural group as, for example, when priests used the actions of the Apis bull for prophetic purposes or oracular answers. It is important to recognize that Egyptians did not worship the animals themselves. They worshipped the invisible deity, employing the animal principally as a symbol of that deity, or occasionally as a deity's vehicle to communicate with humans. In this text I am including this definition within the range of meanings embraced by the term "totemism". The frequency of totemism in antiquity is probably a reflection of ancient humans' recognition of their dependence on animals (which ones would vary with the region and their subsistence strategy), with the accompanying need to assure such animals' availability through propitiation of the deity that is symbolized by the animal and that, at times, controls the animals' behavior. The scope of animals mummified in Egypt reflects incorporation of the many local deities embraced by individual Nilotic tribes at the time of Egyptian unification, with subsequent nationalization of certain of these deities, and integration, not suppression, of the remainder. The presence of the many necropolises suggests that the mummified animal bodies were votive offerings and not animals mummified for personal utilitarian employment in the afterlife. The oldest cult – Apis bull – was derived from the cattle pastoralism that characterized early Predynastic tribal subsistence strategy, as did that of the ram. In these cults a single animal was maintained as a representative of the deity. Most of the other cults flourished during the later dynasties, becoming almost explosive during the Ptolemaic Period. Under these foreign rulers Egyptians undoubtedly experienced rising personal tension, with associated increased perception of need for divine assistance. Since the votive offerings (mummified animals) were expected to enhance the success for their wish-fulfillment, self-interest as motivation for animal mummification is again evident. These practices declined under the influence of Christianity in Egypt, disappearing during the fourth century AD. The presence of massive numbers of mummified cats, birds and other animals in some necropolises provides unparalleled opportunities for bioanthropological research that is just beginning to be exploited (Nerlich *et al.*, 1993). Special efforts are being made to preserve the large collection of mummified animals at the Cairo Museum and research on samples from these has been initiated through a program at the American University in Cairo.

Non-Egyptian mummified animals

Environmental conditions may result in preservation of a cadaver's soft tissues without any human intervention to bring it about. As pointed out in Chapter 3 (Principles of mummification), this may result from the action of thermal extremes, chemical environment, aridity and other factors. Animal tissues occasionally have been preserved in this manner. Because these often present rare research opportunities little exploited to date, some of the more well known are discussed below, although I recognize that most remain both unstudied and unreported.

Horses

Central Altai burials

The domesticated horse appears to be derived from *Equus przewalskii*, domestication occurring during the third millennium BC. Initially it was a beast of burden

Table 7.1. *Mummified animals in ancient Egypt*

Common name	Scientific name	Related god(s)	Reference
Mammal			
Cow	*Bos* sp.	Hathor, Isis	B, L
Mouse (striped)	*Acomys cahirinus*	Uncertain	M,E
Wild sheep	*Ammotragus tragelaphus*	Uncertain	M
Bull	*Bos africanus*	Ptah, Osiris, Montu (Buchis)	A,D,E,M,S
Antelope	*Bubalis buselaphus*	Satis	M
Jackal	*Canis aureus*	Anubis (Anfu), Apu-, Duamutef	M,B,L
Dog	*Canis familiaris*	Seth, Khentamentiu	M,E
Monkey	*Cercopithecus* sp.	Thoth, Chensu	M,B
Shrew	*Crocidura gigantea*	Horus	M
Shrew	*Crocidura religiosa*	Horus	M,B
Wild cat	*Felis maniculata*	Re, Horus, Bast	M,L
Cat, domestic	*Felis maniculata* var. *domestica*	Bast	M,B
Gazelle	*Gazella dorcas*	Anuket	M,E,L
Gazelle	*Gazella isabella*	Anuket	M,L
Mongoose	*Herpestes ichneumon*	Horus, Re, Wadjet	D
Goat	*Hircus mambricus*	Ba-Neb Djedet	M,B
Goat	*Hircus reversus*, L.	Ba-Neb Djedet	L
Goat	*Hircus thebaicus*	Ba-Neb Djedet	M
Rat	*Mus alexandrinus*	Uncertain	M,E
Sheep	*Ovis longpipes*, var. *paleoaegyptica*	Amun, Knum, Herischef, Kherti	M,L
Sheep	*O. platyura*, var. *aegyptica*	Amun, Osiris, Shu, Geb	M,L
Baboon	*Papio anubis*	Thoth, Chensu	M,B
Baboon	*Papio hamadryas*	Thoth, Chensu	M,B
Hippopotamus	*Hippopotamus amphibius*	Tweret	L
Lion	*Felis leo*	Re, Ruty, Shu, Tefnut, Aker	L
Lioness	*Felis leo*	Re, Sekhmet, Mut ,Bastet	L
Avifauna			
Sparrow hawk	*Accipter nisus*	Uncertain	M,L
Goose	*Anser albifrons*	Amun, Horus, Setk	M
Eagle	*Aquila imperialis*	Uncertain	M,L
Eagle	*Aquila maculata*	Uncertain	M,L
Eagle	*Aquila pennata*	Uncertain	M,L
Owl	*Asio brachyotus*	Uncertain	M,L
Owl	*Asio otus*	Uncertain	M,L
Owl	*Bubo ascalaphus*	Uncertain	M,L
Hawk	*Buteo desertorum*	Uncertain	M,L
Hawk	*Buteo ferox*	Horus	M,L
Hawk	*Buteo vulgaris*	Horus	M,L
Kestrel	*Cerchneis tinnunculus*	Horus	M,L
Kestrel	*Cerchneis cenchris*	Uncertain	M,L
Eagle	*Circaetus gallicus*	Uncertain	M,L
Hawk	*Circus aeruginosus*	Horus	M,L
Hawk	*Circus cyaneus*	Horus	M,L
Hawk	*Circus macrourus*	Horus	M,L
Hawk	*Circus pygargus*	Horus	M,L

Table 7.1 (*cont.*)

Common name	Scientific name	Related god(s)	Reference
Grackle	*Coracias garrula*	Uncertain	M,L
Cuckoo	*Cuculus canorus*	Uncertain	M,L
Kite	*Elanus caeruleus*	Uncertain	M,L
Falcon	*Falco babylonicus*	Horus, Re	A,B,D,M,L
Falcon	*Falco barbarus*	Horus, Re	A,B,D,M,L
Falcon	*Falco feldeggi*	Montu, Sokar, Horus, Re-Hodakhty	A,B,D,M,L
Falcon	*Falco subbuteo*	Horus, Re	A,B,D,M,L
Eagle	*Haliaeetus albicillus*	Uncertain	M,L
Gyrfalcon	*Hierofalco saker*	Horus	M,L
Swallow	*Hirundo rustica*	Uncertain	M,L
Ibis	*Ibis aethiopica*	Thoth	B,E,M,L
Goshawk	*Melierax gabar*	Thoth	M,L
Kite	*Milvus aegyptius*	Thoth	M,L
Kite	*Milvus regalis*	Thoth	M,L
Vulture	*Neophron percnopterus*	Nekhebet, Mut, Neit, Ptah	B,L
Plover	*Oedicnemus oedicnemus*	Uncertain	M,L
Osprey	*Pandion haliaetus*	Uncertain	M,L
Buzzard	*Pernis aviporus*	Uncertain	M,L
Ibis	*Plegadis falcinellus*	Thoth	M,L
Grouse	*Pteroclurus senegallus*	Thoth	M,L
Bird (unknown)	*Querquedula crecca*	Thoth	L
Bird (unknown)	*Querquedula circia*	Thoth	L
Owl	*Scops aldrovandi*	Thoth	M,L
Owl	*Strix flammea*	Thoth	M,L
Reptiles			M,L
Crocodile	*Crocodilus niloticus*	Sebek	A,B,E,M,S
Lizard	Not determined	Sebek	L
Snake	*Mabuia quinquetaeniata*	Mehen, Merseker, Buto, Renenet	L
Cobra	*Naja haje*	Mehen, Merseker, Buto, Renenet	B
Fish			
Eel	*Anguilla* sp.	Atum	D
Nile perch	*Lates niloticus*	Neith	M,B
Nile perch	*Latus* sp.	Neith	B
Nile carp	*Lepidotus* sp.	Seth	B
Oxyrhyncus sp.	*Oxyrhyncus* sp.	Hathor, Isis, Mut, Osiris	B,L
Bream	*Phagrus* sp.	Seth	B
Catfish	*Silurus* sp.	Hathor, Isis, Mut	B

Notes:
This table was constructed from articles or books published by Budge, 1894 (B), Moodie, 1931 (M), Guirand, 1968 (A), Spencer, 1982 (S), D'Auria *et al.*, 1988 (D), Lurker, 1974 (L), and el Mahdy, 1989 (E) – see reference list.

but nomads of the Asian steppe converted it to a riding animal about 1500–1000 BC (Herre, 1963). The Turano–Uyuk Valley high in the Altai mountains in central Asia (about 52 °N and 92 °E) on the border between Siberia and Outer Mongolia contains several chieftain burials of the horse-centered nomadic people of the grasslands. They were buried about the middle of the last millennium BC. Here the latitude and altitude combine to create an area of permafrost. Burials were carried out by excavating a pit approximately 8 meters square and 5 meters deep, constructing a wood-enclosed burial chamber within it, roofing it with logs and covering it all with a mound of boulders and gravel several meters high (Rudenko, 1970). One of these barrows (*"khargans"*) contained the permanently frozen body of a tribal chief, his wife or concubine and enough material goods to have proved irresistible to ancient grave looters. Although human bodies were usually "embalmed" (eviscerated and sometimes defleshed), the permanently frozen ground surrounding the tomb together with its insulating cover is principally responsible for the superb preservation of both bodies and artifacts in these graves (Marsadolov, 1993).

These people were skilled horsemen. Judging from their tomb contents, a chieftain's horse must have ranked in importance very near to that of his family members. In the burials near Pazyryk at least several horses (some of which were eviscerated) were commonly included in the grave, sometimes within the wooden tomb chamber and sometimes between the chamber wall and the pit wall (Artamonov, 1965). Their accompanying saddles are handsomely inscribed with animal images and their leather harnesses are laden with artistically crafted faunal, floral, patterned or surrealistic images made of leather, wood and horn. Some bear leather masks for horses, one of which is even fitted with a set of horns to resemble a deer. A few barrows include wooden carts.

The largest of these barrows is that located near the community of Arzhan. Its mound is 100 meters in diameter. In addition to its central wooden tomb housing the human bodies and artifacts, multiple wooden chambers (each designed to house a horse) are arranged in galleries radiating from the central tomb in a stellate manner resembling a "sunburst" arrangement. A total of 180 such chambers for horses was identified (Marsadolov, 1993). Alignments of the galleries with stone mounds in the surrounding area strongly implies that this burial structure has astronomical and probably ritual relationships. Because of the central importance of the horse in Scythian culture, its inclusion in a setting laden with religious symbolism justifies viewing it as a votive offering. We can only speculate about the benefit anticipated by those "mummifying" (eviscerating and then freezing) the horses' corpses, but most probably it was spiritual pacification, born out of apprehension when the governing leader died.

Whereas the artifacts relating to these horses have been described in detail, I am not aware of publications regarding bioanthropological studies of their body tissues. Most of the archaeological items, including some samples of equine tissues, are curated at the Hermitage Museum in St Petersburg, Russia.

In Book IV:70–72 Herodotus (450 BC/1981) describes a chieftan's burial. Modern archaeological findings confirm his description of the barrow construction. He also describes evisceration, saying they filled the abdomen with marsh plants, frankincense, parsley and anise seed. These items have not been confirmed archaeologically, nor has his statement that the body was covered with wax. He also describes a dramatic ceremony 1 year after the death of the chief in which 50 of the chief's squires and their horses were killed, the horses mounted on posts and the riders fixed on the horses' backs with stakes. This, too, has not been confirmed archaeologically.

Miscellaneous horse mummies

Skeletal horse tissues have, of course, been found in North America (Quackenbush, 1909) but without associated soft tissues. However, what could be interpreted as a modern animal "mummy" is the taxidermist's product of the famous Australian racehorse Phar Lap, whose tanned hide was stuffed and exhibited in a

Melbourne museum, though its skeleton may be viewed in the Wellington Museum in New Zealand where he was born.

A military horse has been commemorated in a similar manner. When General George Custer of the 7th United States Cavalry marched his men into the Little Bighorn Valley in Montana (USA) in May of 1876 he was riding a horse that had already been wounded repeatedly in prior battles. Indeed, when the horse received an arrow wound in the flank during a battle with Comanche Indians, according to soldiers he "screamed like a Comanche . . .", so Comanche became his name. As is well known, Custer encountered an unexpectedly large number of American Indian warriors in that valley. When the battle ("Custer's Last Stand") was finished Custer and every one of his soldiers was dead (Dary, 1976:4). Two days later an army detachment arrived at the site and found Comanche was the only thing alive – barely! Lying on his side with seven wounds (a bullet had traversed his neck), he was unable to stand. He was carried to the river, transported by boat to Fort Lincoln where a year passed before he recovered his health. Here his valiant service and his status as the only survivor of Custer's unit was acknowledged by his retirement from service at 14 years of age. He remained a respected symbol of military valor until his death in 1891 at age 29 years. The 7th Cavalry officers contracted with a taxidermist, who built a wooden frame with excelsior to mold it to form and then stretched Comanche's tanned hide over it (Fig. 7.5). The taxidermist donated the preserved horse to the University of Kansas Natural History Museum in Lawrence, Kansas, where it may be viewed today in Dyche Hall.

Another military horse that was similarly honored was one of General Thomas J. (Stonewall) Jackson's favorite battle horses. Called "Little Sorrel" because of his size and color, this undistinguished-looking animal proved to be ideal in battle. He was used in many violent encounters by Jackson during the height of the American Civil War until Jackson's death in 1863 (he was wounded while riding Little Sorrel). Mrs Jackson and various friends stabled the horse for some years after Jackson's death. Little Sorrel survived both his master and Mrs Jackson until 1886 at age 36 years. Following Mrs Jackson's expressed desire before she died, the horse's tanned hide was mounted over a plaster mold and placed on exhibit at the Virginia Military Institute in Lexington, Virginia, where he can still be admired today (Dooley, 1975:34–39; Weil, 1992:296–304).

Undoubtedly the most unique mummified horse is that prepared by Honore Fragonard (1732–1799). Surgeon, anatomist and director of the École Royal Véterinaire in Lyons, France, this man prepared anatomical specimens for study in a distinctly unusual manner. He flayed his human and animal cadavers in a mixture of alcohol, pepper and herbs, injecting some arteries with this mixture also. After removing their skin he injected colored wax and turpentine into the arteries and veins, dissecting the specimens to reveal the vascular structures and ligaments, then posed them for study. His "horse and rider" (Fig. 7.6) is most well known, often called "The Apocalypse" by critics. Though eventually incarcerated as insane, his school today displays his work in a museum, and his specimens themselves are a declaration of perfection in soft tissue preservation (Brier, 1998:57).

Dogs

Because of difficulty differentiating dog bones from those of the jackal or wolf, it is not certain just when the dog was first domesticated. However, Schwabe (1978:10) estimates that it occurred about 12000 BC, and believes that dogs helped humans to determine where their sheep and goats grazed by 9000 BC in the Fertile Crescent. This would be consistent also with the observation that the dog was a prominent totem in ancient Babylonia (Howey, 1972:150). Idaho cave sites suggest an "earliest" North American date of about 9500–8400 BC, and at the Cayönü site in Turkey about 7000 BC (Clutton-Brock, 1963). Brothwell *et al.* (1979) studied 42 dog skulls dating from 1030 BC to AD 1324, in Lima museums, and found two distinct forms represented. Eaton (1912) found a dog tibia among human

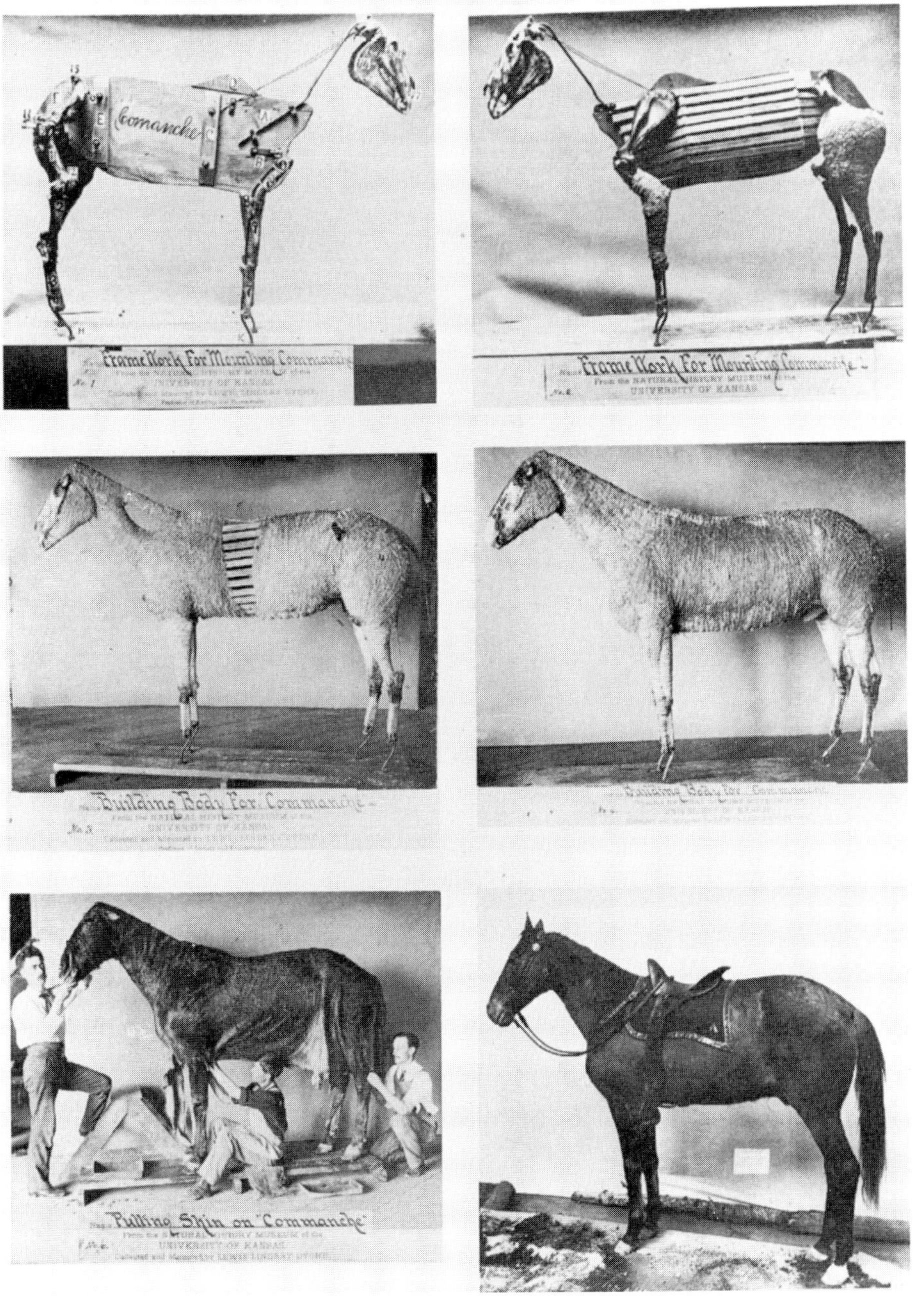

Fig. 7.5. The preparation of Comanche's remains; AM, AD 1891.
The only survivor of the US Seventh Cavalry Unit at the Battle of Little Bighorn was General George Custer's horse called Comanche. The wounded horse's valor was celebrated by preserving his tanned hide after his death years later and mounting it. It is displayed currently at Dyche Hall at the University of Kansas Natural History Museum in Lawrence, Kansas (Dary, 1976). (Photo courtesy of the Kansas State Historical Society, Topeka.)

Fig. 7.6. Fragonard's mummified horse and rider; AM, AD 1785.
Honoré Fragonard (1732–1799) was an anatomist, surgeon and director of France's École Royal Vétérinaire in Lyons. He prepared anatomical demonstrations of human and animal bodies by soaking them in a mixture of alcohol, pepper and herbs, then injected the mixture intra-arterially and flayed them. After dissecting them to reveal ligaments and vascular structure and injecting turpentine and colored wax, he posed them in their exhibit position. This pictured example is frequently dubbed "The Apocalypse" (Brier, 1998:57). (Photo: Homme et Cheval Anatomisés. Pièce Naturelle de Henri Fragonard. Collection du Musée de l'École Vétérinaire de Maisons-Alfort. Ph. © Gilles Capée.)

bones collected by Hiram Bingham during the Yale Peruvian expedition of 1911 and notes that the Incas had three types of dog resembling the modern bulldog, dachshund and wolf.

Mummified dog tissues, however, are not common. A dog cemetery has been found at Abydos in Egypt, probably relating to the area's necropolis-god Khentamentiu (Spencer, 1982:211), though these, of course, were anthropogenically mummified. Allison *et al.* (1982a) described eight spontaneously mummified dogs from South America's Andean area at low valley-site tombs near Chile's northern coast at Arica in the Atacama Desert. These dated from 500 BC to AD 1500. All were from partly or primarily agricultural societies. One of these (from the Inca Period) contained coprolites; no parasite ova were found within them. Cultural identification was based on accompanying grave goods and/or burial within a culturally identified human cemetery. One was a simple burial without grave goods, and another was found together with a human burial. The remainder were ceremonial burials with associated grave goods. These commonly contained food (marine, avian and agricultural items), cloth, lithics and everyday domestic items (baskets, blankets, rope, ceramics and tools). Allison *et al.* (1982a) also point out that in the Andean area dogs were used as food as well as pets in prehistoric times and their skin was employed in handicrafts. They also identified dogs in rupestrian art of the region's valleys (Azapa, Codpa, Camarones) and cite other studies to identify the antiquity of dog skeletal tissues in the Andean highlands as early as 9160 BC. Other than coprolite analysis, no bioanthropological studies were performed. They conclude that dogs were brought from the highlands to the low valley and coastal sites at this latitude by highland shepherd migrants (locally known as Alto Ramirez) with their arrival at the coast 1000–500 BC. Moodie (1931) found a carefully wrapped, spontaneously mummified dog buried in a basket at Pachacamac, Peru. Reiss & Stuebel (1880–1887) excavated a spontaneously mummified dog (Nasca Period?) from the huge cemetery at Ancón, Peru.

A nondescript, 17 kg North American mongrel dog gained fame in 1985 when, after his death in that year, his body was anthropogenically mummified successfully using an Egyptian eighteenth dynasty technique with natron. This was one of the largest animal bodies employed to study that technique at that time (Notman & Aufderheide, 1995) and set the stage for a similar later study on a human cadaver (Brier, 1997).

Pet mummification for sentimental reasons by a freeze-dry process followed by taxidermy methodology (Helser, 1994) and also by embedding in resin (Ra, 1998) (Fig. 7.7) is a late twentieth century phenomenon, detailed in Chapters 3 and 10. Together with several spontaneously mummified bodies of Native

Fig. 7.7. Modern bronze container for mummified pet Doberman Pincher; AM, ca. AD 1985.
This custom-made container was shaped for its occupant, a pet dog. Mummification and container (© Mummiform) was carried out by Summum, described in Chapter 3 and in Fig. 3.24. (Photo courtesy of Corky Ra, Summum.)

Americans in an Arkansas Ozark Bluff rock shelter, Riddick (1995) also reports the spontaneously mummified body of a dog. Similarly, Strong & Evans (1952) identified the spontaneously mummified remains of two Peruvian dogs in a midden at Castillo de Tomaval (Chavín) site, one with offerings, the other not. Another, found in 1982 in an Andean grave in the Azapa Valley of northern Chile dated to about 1000 BC, may have been a pet of the child whose spontaneously mummified body was also found within that tomb (Fig. 7.8). Chemical reconstruction of ancient domesticated dogs' diets indicate that their principal

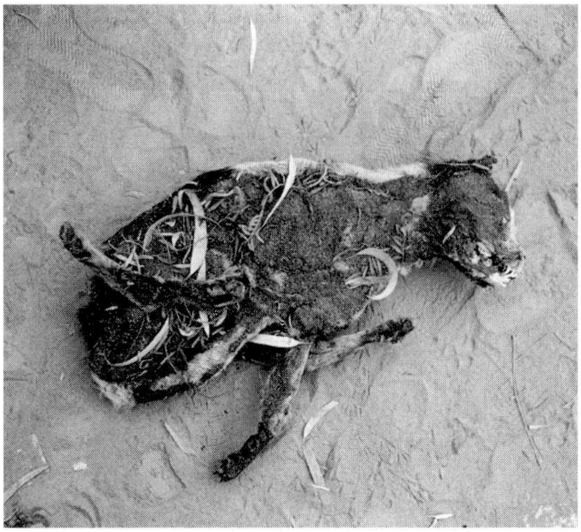

Fig. 7.8. Andean dog; SM, 1000 BC to AD 400. Mummified dogs when found as isolated burials are commonly interpreted as grave offerings, but, when present in a child's tomb, it may have been a pet. This one was found in a cemetery occupied by bodies of highland migrants (Alto Ramirez) who arrived at the Pacific coast of northern Chile 1000 BC–AD 400 (Allison *et al.*, 1982a).(Photo by ACA.)

dietary components commonly are similar to those of their human masters (Aufderheide *et al.*, 1988).

Mammoths

Some have speculated that this large-tusked relative (*Mammuthus primigenius* (Blumbach)) of the modern elephant may have been the principal target that lured the northeast Siberian natives across Beringia before this animal became extinct about 10000 years ago. Its remains have been recovered primarily from remote arctic sites in bog or postglacial environments. Only a few of these have demonstrated soft tissue preservation of significant degree. However, the discovery conditions of most of them have been such that extensive deterioration of the soft tissues occurred before the body could be removed from these usually isolated sites and brought to locations where conservation practices could be applied. Of the 40 examples of frozen Siberian mammoth remains, only five were complete or nearly so. These, and the plants found in the stomach of one from the Berezovka River are characterized by Farrand (1961). Farrand also feels that the bodies had deteriorated in part and show predator effects, indicating that they had not frozen to death.

Probably the most well-known is the frozen baby mammoth ("Dima") from the Magadan province of Siberia. There a bulldozer operator in June, 1977, uncovered a baby mammoth mummy. It was recovered sufficiently promptly that a good deal of soft tissue was still intact. Radiocarbon study indicated death about 44000 years previously. Goodman *et al.* (1979) reported studies on a desiccated 1 g sample of muscle and 0.2 ml sample of dried blood. Scanning electron microscopy studies demonstrated intact red and white blood cells within muscle blood vessels, and cross-striations in skeletal muscle cells. However, the muscle protein had deteriorated considerably. Only one specimen of mammoth muscle extract demonstrated a very faint precipitin line of partial identity with an antibody produced in chickens that had been immunized with a homogenate of modern elephant (*Elephas maximus*) muscle. The antibody may be directed against albumin. This reaction was no stronger than that from several close relatives of Proboscidea. Thus little of the extracted protein's antigenic structure may remain intact (Goodman *et al.*, 1979). Cockburn (1980) makes an ardent plea for the meticulous conservation of frozen tissue beginning with its discovery. He notes that a broad range of studies could be carried out on frozen cells that have not been thawed after freezing. Recent interest in causes of megafauna extinction, including mammoths, has led to speculation about the role of diseases, as well as climatic changes that impact negatively dietary sources (Stevens, 1997). Biological studies of well-preserved tissues could shed light on these possible effects.

In addition to the Magadan mammoth from the Taimyr Peninsula (Siberia), tissues of two other mammoths and a prehistoric horse from Siberia were also made available by Russian scientists. One mammoth (53000 years BP) was an adult, also from the Taimyr Peninsula, and the other (10000 years BP), also an adult, was from the Yuribei River on the Gydan Peninsula. The ancient horse provenience was not provided but it was dated to 37000 years BP. Histological studies revealed good preservation only in the Yuribei mammoth tissues (Reyman & Goodman, 1981). Johnson *et al.* (1981) extracted samples from the Magadan, Taimir and Yurigei mammoths and were able to identify DNA fragments in the latter two. Nucleic acid hydridization techniques revealed only 2–5% sequence identity with that of modern Asian elephants. Using immunological techniques, however, Prager *et al.* (1980) demonstrated a strong reactivity of modern Asian elephant albumin with antibodies induced in rabbits by injection of mammoth muscle tissue.

In the New World as of 1909, Quackenbush lists seven mammoth finds as well as one horse, two bison and a walrus in Alaskan and Klondike areas but this number has been greatly expanded since then. More recent finds have been identified in Utah and even more southern areas (Schaedler *et al.*, 1992). One of the mammoths found by Quackenbush at Eschscholtz Bay had muscle, skin, hair and wool attached to the femur. Much later these were studied by Ezra & Cook (1959), who found the femur diaphysis to have normal histology similar to that of a modern elephant and normal bone collagen and carbonate content. Several intact and seven fragmented mammoths in northern Idaho have been reported but none had preserved soft tissue (Miller, 1998). In 1998 Noro *et al.* successfully sequenced the mitochondrial cytochrome *b* and 12 S ribosomal genes of a woolly mammoth and compared them with those of modern African and Asian elephants. They found that, in contrast to results from earlier morphologically based comparisons, the mammoth was more closely related to the Asian than the African elephant, as judged by the molecular studies. These basic studies indicate that soft tissues attached to mammoth bones retain sufficient chemical integrity to make a broad array of studies possible. This provides support for Cockburn's plea to exert prompt efforts at time of discovery to preserve such tissues appropriately. An 11000-year-old juvenile mammoth that was excavated at the Hebior site (Kramer, 1996) in southeastern Wisconsin was studied in my laboratory. Chronic osteomyelitis of a metatarsal joint was identified with central abscesses, cloacae and secondary degenerative arthritis (Fig. 7.9). Several associated interphalangeal joints were also involved. No soft tissues were found with this animal's remains, although cut-marks on skeletal structures together with the presence of bifaced lithics imply that the animal was butchered by paleo-Indians. It is conceivable that pain resulting from weight-bearing on the foot afflicted with osteomyelitis may have impaired the mammoth's defenses against its hunters. More recently Stone (1999) describes the find of a Siberian mammoth mummy from the Taimyr Peninsula that was found completely frozen except for part of the head. It was recovered by helicopter and delivered to the nearby city of Khatanga in the frozen state. Suggestions that researchers might attempt to clone this mammoth were negated by Tschentscher (1999), who noted the improbability of the DNA whole-molecule-preservation required for cloning. As of March, 2002, no results of scientific studies have been announced or reported. This find fulfills the desired recovery conditions and promises well-preserved tissue for scientific study.

Insects

Various insects are often encountered in mummified remains. Probably the most common necrophilous insect encountered is the dermestid beetle. Maggots (Diptera larvae) are common but their high water content rarely permits them to survive the long-term postmortem milieu. Identification of insects in Andean mummy bundles by examination using scanning electron microscopy provided some insight into Andean burial practices. The various identified stages enabled estimation of the time interval between death and burial. Insects were commonly found in bundles with skeletons, but rarely in those in which soft tissues were preserved. Detailed examination of the preserved bodies led to evidence suggesting that they had been

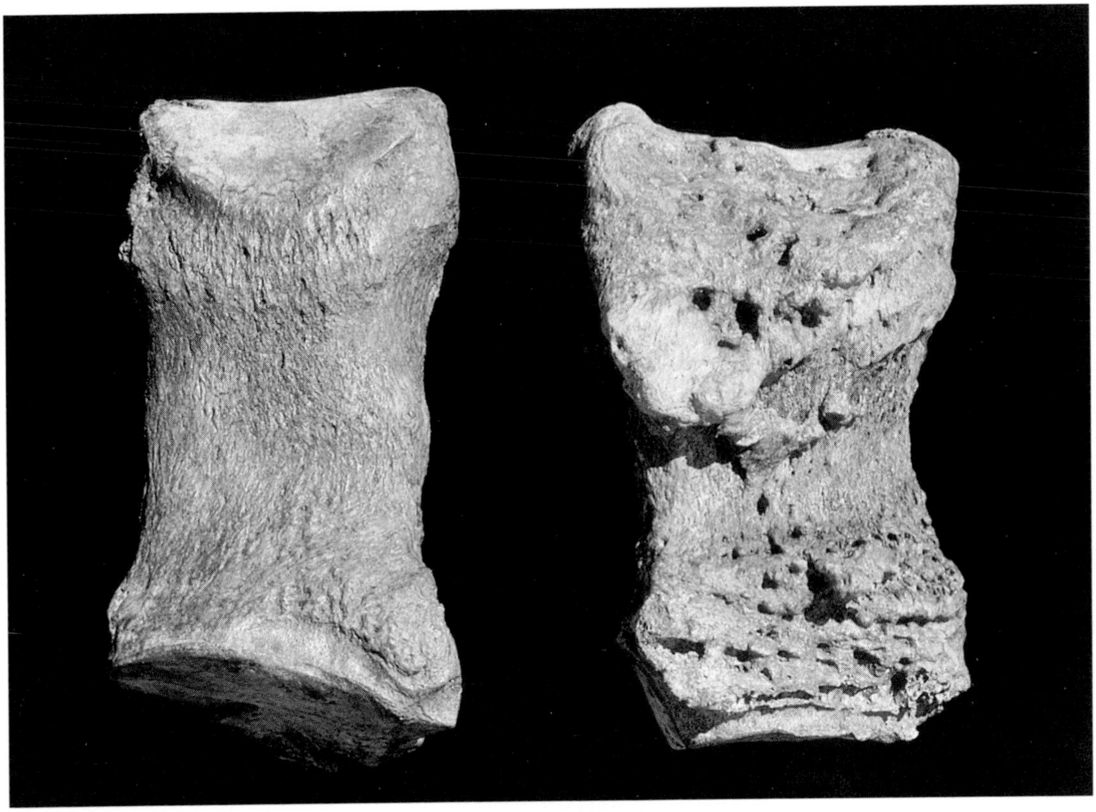

Fig. 7.9. Osteomyelitis of mammoth metatarsal; SM, ca. 7000 BC.
The bone on the left is a normal metatarsal of a mammoth while that on the right is a metatarsal of the same animal affected by osteomyelitis. X-rays revealed large medullary abscesses. Extensive involucrum formation and cloacae together with secondary osteoarthritis are apparent externally. (Photo by ACA, courtesy of Jon Kramer of Potomac Museum Group.)

exposed to heat before burial (Riddle & Vreeland, 1982). Fulcheri *et al.* (1994) identified larvae within an ancient Egyptian head initially by radiography followed by anatomical dissection. Insects trapped in liquid amber that subsequently hardens upon drying can preserve not only anatomical insect structure but even its molecular integrity. Cano & Borucki (1995) isolated a segment of DNA unique to *Bacillus* spp. from the abdomen of a stingless bee trapped in amber (that was 23 million years old) from the Dominican Republic in the Gulf of Mexico, and subsequently cultured it. No one, however, has duplicated this feat subsequently, though a staphylococcus has been so identified from similar amber (Lambert *et al.*, 1998). Hino *et al.* (1982) were able to demonstrate the ultrastructure of bacterial spores in the skin of an Egyptian mummy anthropogenically embalmed in AD 100–300. Pediculosis seems to have been a common ancient affliction. Reinhard (cited in Aufderheide, 1990a) was able to use the frequency of this condition in several Peruvian groups to estimate the degree of social contact between groups. Both egg casings adherent to hair shafts (nits) and the louse itself (Horne, 1979) have been demonstrated in Aleutian mummy hair by scanning electron microscopy. The water environment of peat bog mummies often results in a host of ancient insects accompanying the mummy. More than 35 species were found with the body of the Iron Age

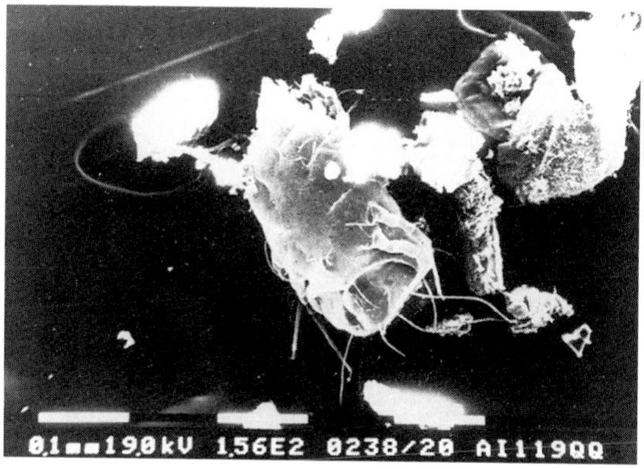

Fig. 7.10. Mites from mummies; SM, ca. AD 1550. Mites found on a medieval mummy in Italy are believed to have been responsible for extensive disfiguration of the mummy's skin. (Scanning electron microscope photo by and courtesy of Gino Fornaciari.)

Lindow Man in England. Midges (chironomids) and water fleas (Cladocera) were especially abundant (Dayton, 1986; Girling, 1986). Occasional archaeological sites provide a chemical milieu that enhances calcium salts deposition in soft tissues. This fossilizes the specimen sufficiently faithfully to permit identification, as demonstrated among arthropods found at a number of British sites (Girling, 1986). Mites were found to have contributed to soft tissue degeneration in a medieval Italian mummy (Fig. 7.10).

Miscellaneous animal mummies

In Peru animal mummies are frequently found in association with human mummies, and are commonly spontaneously mummified. They include macaws, parrots, dogs, cats, viscachas, guinea pigs, herons, flamingos, doves, llamas, vicuñas and alpacas. A spontaneously mummified monkey was found in a shaft tomb with a similarly spontaneously mummified child's body dating to about AD 900 (Late Tiwanaku Period) near Arica, Chile (Fig. 7.11). This was particularly instructive since the burial site is on the coastal area of the Atacama Desert and the pet monkey's jungle origin was separated from the coast by the Andean mountain range whose highlands are at about 4000 meters' altitude.

Fig. 7.11. Andean desiccated monkey; SM, ca. AD 900. This spontaneously mummified (desiccated) monkey was found in an AD 900 tomb on the coast of northern Chile's Atacama Desert that also contained the spontaneously mummified body of a child. The monkey may have been the child's pet. (Photo by ACA.)

Spontaneously mummified llamas are particularly common. Ancient stone structures (ritual, wall, etc.) in Peru and northern Chile frequently are found to have a llama buried beneath a cornerstone (Fig. 7.12). A cache of 23 spontaneously mummified alpacas and llamas about 1000 years old were found near Moquegua, southern Peru, at a site occupied by Chiribaya people. Comparison of these animals with modern equivalents suggested that a breakdown in specialized breeding occurred after the Spanish Conquest in 1532 (Wheeler *et al.*, 1998).

The subsistence value of an animal is not always consistent with a population's selection of animals for religious sacrifice. It is true that early Saxons, for example, sacrificed animals (horses and dogs) not normally employed in their daily life (Crabtree, 1995). On the other hand, the Middle East Bedouin sacrifice the very animals (sheep and goats) that form a staple of their existence (Klenck, 1995). Ancient Andean indigenes also selected animals that formed their principal meat source (camelids and guinea pigs) for their religious rituals. Thus guinea pigs were bred domestically

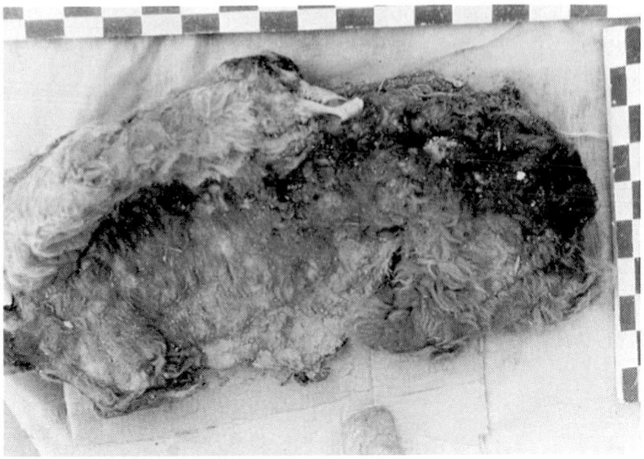

Fig. 7.12. Sacrificial llama; SM, AD 1150–1300.
The Andean llama was found under the cornerstone of structure 70 at the El Yaral site in southern Peru's Osmore Valley near Moquegua. It was one of a cache of 23 at that site. (Photo by and courtesy of Jane Wheeler, San Marcos University, Lima.)

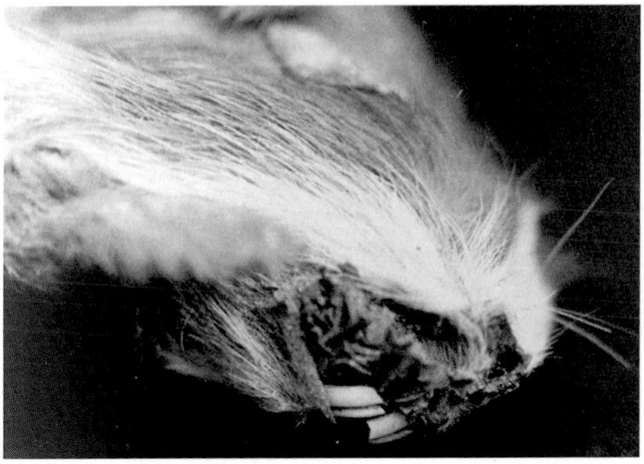

Fig. 7.13. Sacrificed Andean guinea pig; SM, AD 1150–1300.
Guinea pigs in the Andes were raised not only for food consumption but also for sacrifice. Note the coca leaves in the mouth of this one. (Photo by and courtesy of Juan Rofes.)

specifically for food consumption, a practice that has survived the Spanish Conquest into modern times. Evidence for this can be found in accumulation of their excrements not only in breeding areas but also in middens. Their religious function is easily established archaeologically. For example at the Yaral site near Moquegua in southern Peru's Osmore Valley, in a Chiribaya site from about AD 1150, a cluster of 112 spontaneously mummified guinea pigs were found beneath the floor of a house's living quarters. Many of these had been decapitated. Leaves of the coca plant, which were also employed commonly as a religious offering, were found stuffed into the oral cavity of many of the guinea pigs (Rofes, 2000) (Fig. 7.13). Similar evidence has been found at other sites including Cahuachi (Silverman, 1988) and Los Demás (Sandweiss & Wing, 1997). Coca leaves were also present in the oral cavity of many spontaneously mummified human bodies from the Yaral population (Aufderheide, 1990a). A large group of similarly spontaneously mummified guinea pigs were found beneath the temple foundation at the Early Horizon Peruvian site of Janabarriu, where they appear to represent a propitiatory offering (Burger, 1995b:43, 169). Andean chroniclers confirm their use for sacrifices (Gade, 1967), and their entrail patterns were employed for divining purposes and for healing rituals (Bolton, 1979).

What could be described as a dinosaur mummy is the fossilized soft tissue present on an articulated skeleton of a duck-billed dinosaur in Wyoming found in 1908. Currently it is curated at the American Museum of Natural History (2002).

The frozen bodies of crabeater seals (*Lobodon carcinophagus*) have been found in abundance in the ice-free areas of Antarctica (Fig. 7.14). These are commonly a long distance (2–60 kilometers) from water. Scattered ice-fed lakes are believed to have attracted them and the seals died when they were unable to find food in those lakes. Because that region does not enjoy thawing temperatures, the bodies have probably remained frozen since death. Radiocarbon dates on some of these well-preserved seal mummies were estimated at 1600–2500 years old (Péwé *et al.*, 1959). An equivalent phenomenon occurs in East Greenland where the intact, frozen bodies of caribou, swans and musk oxen up to 4000 years old have been identified (Meldgaard & Hart Hansen, 1998).

A spontaneously mummified cat was found in a glaciated area at an altitude of more than 2000 meters in

Fig. 7.14. Antarctic crabeater seal; SM, ca. 100 BC. This *Lobodon carcinophagus* (crabeater seal) was found in the Wright Valley near the Bull Pass intersect within the Antarctic's McMurdo dry valleys. Radiocarbon dating of one of these indicated that it was 2000 ± 500 years old (Péwé *et al.*, 1959). (This photo by Tony K. Meunier, US Geological Survey.)

the Tyrolean Alps in 1994. Radiocarbon dates suggested an age of about 30–35 years. Each year glaciers of that region "deliver" human victims of glacial accidents. Such bodies usually are from 30 to several hundred years old. Considerable soft tissue of this cat survived the postmortem interval. This was principally the product of freeze-drying conditions and partly as the result of adipocere formation in several subcutaneous areas as well as within the abdomen. Most viscera, however, were unidentifiable (Weisgram *et al.*, 1996).

At the huge burial site of Ancón, Peru, numerous spontaneously mummified small animals, wrapped in cloth, often in baskets, were found within human tombs. These included dogs (*Canis ingae*), a parrot (*Chrysotis farinosa*) (Fig. 7.15), a Peruvian rat (*Hespermys* sp.) and guinea pigs (*Cavia cobaya*). Whether these were offerings or pets was not obvious. Several large animals including large dogs and llamas were found in separate pits; these were not wrapped (Reiss & Stuebel, 1880–1887).

Fig. 7.15. Mummified parrot; SM, ca. AD 100. This spontaneously mummified parrot was among one of many animals excavated from the large cemetery at Ancon, Peru, 1880–1887, by Reiss & Stuebel. Grave goods suggested contemporaneity with the Nasca Period. (From Reiss & Stuebel, 1880–1887.)

Chapter 8. Soft tissue paleopathology: diseases of the viscera*

Introduction

This chapter deals with paleopathological changes in the body's soft tissues with special emphasis on the viscera. The diseases arising within these organs only rarely affect bone, and even when they do so, osseous changes are secondary to the effects on the primary soft tissue organ in which the condition arose.

Postmortem morphological alterations are usually much more marked in soft tissues than in bone. While space does not permit a detailed discussion of postmortem taphonomic processes, the reader's interpretive skills may be usefully enhanced by a review of a few basic principles operative to achieve preservation of soft tissues.

Factors influencing preservation

Postmortem decay is an enzymatic process. Enzymes require an aqueous medium. If tissue dries before it decays, it remains so until it is rehydrated. This simple truth is exploited by those who bring us beef jerky and dried fruit. Spontaneous ("natural") mummification occurs by this mechanism in hot, dry environments, such as deserts, that lead to rapid desiccation. However, the race between decay and dehydration is a highly competitive one and, like most races, small effects can determine the difference between a win or a loss. Thus differences in delays in the interval between death and interment, presence of bacteremia at time of death, weather fluctuations and similar features are responsible for the broad range of variation in degree of soft tissue preservation usually observed in bodies from the same burial site. Other approaches to slowing or blocking enzymatic decay can make use of heavy metals to "poison" the enzyme. For example, numerous reports document the localized conservation of skin and muscle of the wrist underlying a copper bracelet in an otherwise skeletonized body. Finally, an enzyme's action is usually highly specific, and, if the chemical form of the tissue is altered (denaturation), the enzyme can not "recognize" it and so does not act on it. Oxidation of fatty acids can so alter body fat (production of adipocere), rendering it immune to enzymatic decay. Anthropogenic ("artificial") forms of mummification have employed (sometimes even combined) all of these features.

Variable response of body tissues

Not all body tissues decay at the same rate. Epithelial cells deteriorate much more rapidly than do the supporting tissues such as collagen, elastic tissue and cartilage or skeletal muscle. Epithelial cells contain destructive enzymes normally carefully controlled to destroy undesired particles or organisms that gain

* Most of this chapter is taken directly from Chapter 8 of Aufderheide & Rodríguez (1998).

access to the living cell's cytoplasm. After death this control is lost, and these same enzymes then destroy the cell itself. In general, the more metabolically active the living cell is, the more rapidly it will be destroyed after death. Thus the vital nature of the heart's muscle fibers, the kidney's tubular cells and the adrenal secretory cells predict their rapid postmortem dissolution, while the body's large mass of skeletal muscle deteriorates so slowly that often much of it is still intact when tissue dehydration arrests decay. These selective decay rate differences will dictate the amount of information soft tissue paleopathologists can derive from the various organs.

Recognition of morphological alterations by disease

From the above it will be clear that morphological changes produced by taphonomic processes are much more profound in soft tissues than in bone. Loss of color and texture impair a soft tissue paleopathologist's ability to evaluate or even recognize an organ. Even anatomical location is not a sacrosanct feature, since an organ's attachment to a site may be lost, with subsequent organ migration. We have found a stone-containing gall bladder lying free in the pelvis and the entire liver in the thorax. Thus in spontaneously mummified bodies adrenal pathology has only rarely been described, and the necrotic lesion of myocardium secondary to acute myocardial infarction, has not been reported. Indeed, in some cases the heart is represented by only a black stain on the inner lining of the pericardial sac. The kidney may be identifiable in some only by a reniform space at the appropriate retroperitoneal location. Compared with hospital autopsies, paleopathological dissections of mummified human remains can be viewed as "salvage pathology". The above factors seriously constrain recoverable antemortem information and paleopathologists need to keep their expectations realistic. Nevertheless, as in other fields, thorough knowledge of fresh tissue pathological changes combined with experience with advanced postmortem alterations can avoid the pitfalls of pseudopathology and enhance recovery of data. The experienced forensic pathologist brings highly applicable specialized skills to the field of soft tissue paleopathology.

Topics covered

It is difficult to recognize a disease of whose existence one is not aware. Many modern diseases have not appeared in the paleopathological literature. This may be due to a difference in the frequency of a disease in antiquity versus the modern period. In many cases, however, it may be possible that it is simply being overlooked. Hence I have chosen to discuss the most common diseases affecting the visceral organs today, even if they have never been reported in archaeological bodies. Illustrations for such, of course, will need to employ modern examples, the paleopathological appearance of which the reader will need to extrapolate from the illustration. By employing all of one's knowledge, meticulous dissection and reasonably conservative reconstruction of findings, it is often surprising how much useful information about the individual's antemortem health status can be predicted from such careful dissections by properly prepared operators.

Soft tissue lesions of the head and neck

The brain and meninges

Brain preservation

Together with the adrenal and kidney, the brain is among the first organs to undergo postmortem autolysis. Every experienced medical examiner has seen repeated examples of near or actual brain liquefaction in corpses exposed to summer temperatures for even as little time as a week. Insects gaining access to the cranial cavity can accelerate the destructive process. Hope (1834) examined a series of mummified Egyptian heads in the British Museum and found residua of the genera *Necrobia*, *Dermestis* and even *Diptera* within the cranial cavity; the numbers and stages present led him to believe that they entered the skull at the time of embalming. I examined the victim of a summer stabbing

18 days after the murder in which several wounds had penetrated the skull. The cranial cavity was devoid of brain or meningeal tissue but was filled with teeming blowfly larvae, indicating how rapidly the brain can be not only liquified but actually consumed after death.

Yet, in spite of the brain's tendency toward rapid postmortem dissolution, examples of its unexpected preservation abound. In many of these, adipocere formation from the brain's lipid content is responsible for the preservation. In others an excessively arid environment may prevent complete cerebral dissolution, while in still others the mechanism is not at all clear.

Probably the oldest preserved brain was reported from a site near Moscow (Walter, 1929) in which two cerebriform fossilized human brains (without skulls) were preserved together with parts of a woolly mammoth. In all probability, adipocere formation preserved the gross brain form initially and during the period that acid groundwater gradually dissolved the skull, eventually leading to mineral replacement of all the organic matter. Mineral springs have also preserved human brains by a means, probably chemical, not yet well understood but capable of retaining gross cerebral morphology for 7000–8000 years (Royal & Clark, 1960; Doran *et al.*, 1986; Pääbo *et al.*, 1988).

Adipocere formation was clearly the mechanism that preserved a clay-buried, 5000-year-old Swiss brain (Oakley, 1960) as well as that of a Bronze Age bog body (Powers, 1960), although in other bog bodies in which some brain tissue remained, adipocere was not described (Brothwell, 1986:63). Klohn *et al.* (1988) described a cerebral concretion that appears to have originated by a combination of calcium with fatty acids hydrolyzed from the brain's neutral fat to form calcium soaps, and adipocere was also the effective agent in preserving brain tissue in 56 of 74 medieval skulls excavated at a Danish monastery (Tkocz *et al.*, 1979). Finding retained brain tissue (some cerebriform) within the cranial cavities of persons who died 45 years prior to exhumation of cranial gunshot wounds, Radanov *et al.* (1992) suggest that the microclimate within the cranial cavity (resulting from the crania's gunshot-induced perforations) permitted rapid evaporation of cerebral fluids and consequent brain preservation without adipocere formation.

The incredibly arid and rain-free climate of northern Chile's Atacama Desert results in a high rate of soft tissue preservation of biological material interred in its nitrate-rich soil. Gerszten & Martínez (1995) examined the desiccated cerebral tissue from 15 of the many spontaneously mummified human remains excavated from sites in this area. Most preserved brains were shapeless, shrunken brown masses but at least one retained its gyriform surface morphology (see Intracranial hemorrhage, below) I have examined many mummies from this area and found the brown-stained dura mater to be present very frequently when abundant soft tissue of the rest of the body was also retained. In one 4000-year-old population of 16 adults, all of whose heads were present, brain tissue was found in 8. This was usually present as an amorphous, brown mass of pasty consistency (Fig. 8.1). In several, however, about one-fourth of the cranial cavity was filled with granular, pebble-like, brown material, the individual granules averaging only a millimeter or two in diameter. The taphonomic process that generated this appearance was probably initiated by liquefaction of the brain, with subsequent precipitation of the solutes as the body desiccated. In one body the particles were of sand-like quality.

Intracranial hemorrhage

In spite of all these observations of retained brain tissue, remarkably few paleoneuropathological reports have appeared. Evidence of epidural hematoma was commonly found among 45 infants from three infant populations at a site in Germany (Teegen & Schultz, 1994), which the authors identified not on the basis of soft tissue changes, but of stained skull bones and attributed them to birth injuries or child abuse. Gerszten & Martínez (1995), examining 15 spontaneously mummified human remains at sites ranging from 100 BC to AD 800 in the Atacama Desert identified several with retained brain tissue showing subdural hematomas related to head injuries. They also

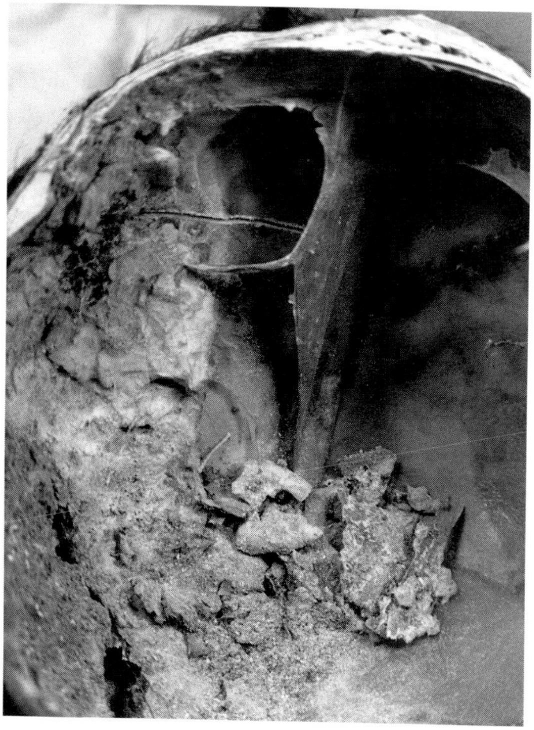

Fig. 8.1. Brain and meninges preservation; SM, ca. AD 100.
The calvarium (top of the skull vault) has been removed to demonstrate preservation of parts of the dura mater (falx cerebri). The brain has been reduced to several irregular, brown, amorphous masses of pasty material. Adult male (PBL.)

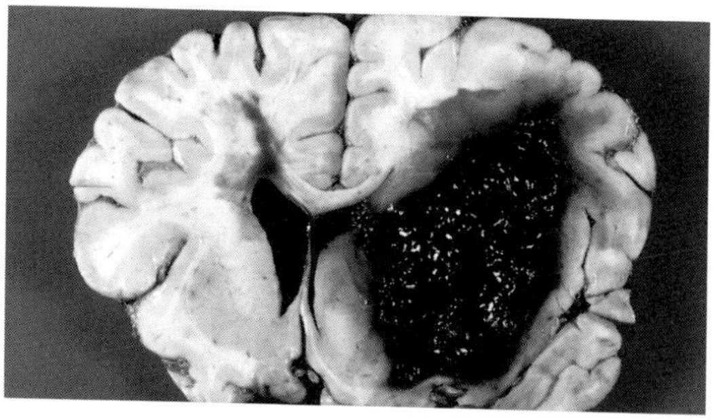

Fig. 8.2. Intracerebral hemorrhage.
Section of gross brain demonstrates that a large area of brain tissue has been disrupted by hemorrhage from a ruptured cerebral artery. Modern adult male. (PBL.)

found one with subarachnoid hemorrhage over the cerebral convexity. In another, an example of intracerebral hemorrhage was obvious in a 45-year-old male; severe renal atherosclerosis in that body suggested that hypertension, often related to this kidney condition, may have caused the cerebral hemorrhage (though the illustrated lesion is also consistent with a ruptured "berry" aneurysm of a cerebral artery). Figure 8.2 is an example of this type of hemorrhage.

Infectious diseases

Dalton *et al.* (1976) describe a 2-year-old child mummy from southern Peru with resolving pneumonia and a "thick layer of dried pus on the spinal cord". The brain had disintegrated. Histology revealed small bacilli consistent with *Haemophilus influenzae* (the most common cause of meningitis in infants today), but immunostaining efforts produced no reactive changes. Although osseous changes may sometimes reflect chronic meningitis, we have not been able to find other examples of pus-coated meninges in the paleopathological literature.

Neoplasms

Primary malignant tumors of cerebral tissue constitute about 10% of human primary cancers. Their frequency age distribution curve is bimodal with the first peak in children's age group and a second in that of adults. Because of the generally poor postmortem preservation of cerebral soft tissue, we must keep our expectations of encountering these in ancient mummified bodies low. Clues to their presence could include focal areas of hemorrhage and necrosis, especially in the common high grade astrocytic tumors. Low grade astrocytic tumors of the cerebellum in children may be recognizable because of their cystic nature. Small foci of calcification may also betray the presence of an antemortem oligodendroglioma. Many of these tumors will block the flow of cerebrospinal fluid and produce obvious hydrocephalus. The possibility of a primary

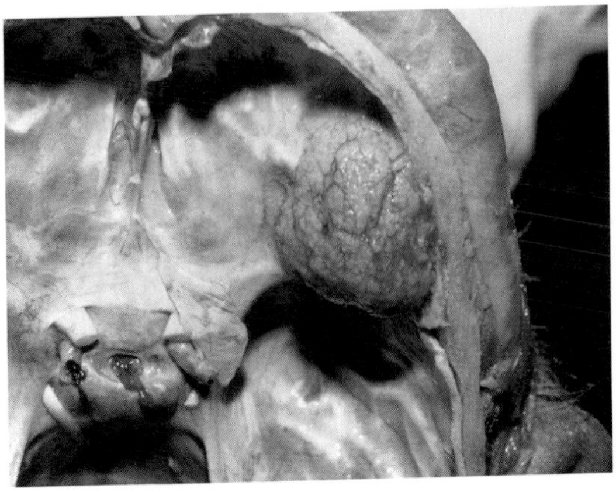

Fig. 8.3. Meningioma.
Removal of the calvarium demonstrates a fibrous-appearing 6 cm × 5 cm × 4 cm ovate tumor that arose from the meninges in the lateral aspect of the right anterior fossa of the cranial cavity. Its benign histology is consistent with its discrete, sharply demarcated and noninvasive periphery. Modern adult male. (PBL.)

cerebral neoplasm should be suspected when finding hydrocephalus in the brain of an adult archaeological body. Most of these tumors in antiquity would have progressed to produce eventual increased intracranial pressure with convulsions, coma and death.

Meningiomas arise from the meninges in older adults and can be located anywhere within the cranial cavity. They have a dense, fibrous matrix that would be expected to survive the postmortem changes quite well. About one-fifth of intracranial tumors are meningiomas and, except for the occasional meningiosarcoma that is capable of brain invasion, meningiomas are generally benign. They present a mass of hard or rubbery, lobulated, sharply demarcated lump of tissue intimately attached to the meninges (Fig. 8.3). Although benign, their slow but progressive enlargement may compress the brain sufficiently to cause increased intracranial pressure and even death unless removed. In antiquity this tumor may well have produced symptoms of headache, personality change, focal motor loss and/or convulsions progressively over a period of many years eventually leading to death. The clinical symptoms and signs vary with location and size.

While we have no record of identification of the soft tissues of a meningioma, the tumor frequently has left an indication of its presence in the form of localized marked hyperostosis that meningiomas may induce in the bone overlying it. While the tumor may erode the skull's inner table, the most dramatic examples of induced hyperostosis we have are present on the outer table. Campillo (1994–1995:95) demonstrates what he suspects is a meningiosarcoma (or possibly even osteosarcoma) in a lesion presenting an enormous mass over the skull's frontal lobe, similar to the well-known Chavín culture skull from Peru (MacCurdy, 1923). He also cites five other cases his group has published demonstrating meningioma-induced hyperostosis, the oldest of which is of Neolithic age, and notes that the oldest example, published by de Lumley (1962), was found in a 200 000-year-old skull of *Homo erectus* from a French cave. Rogers (1949) also found such characteristic cranial hyperostosis in a first dynasty Egyptian body and in another from the twentieth dynasty.

Soft tissue lesions of the eye

In spontaneously desiccated mummies the globe of the eye loses its fluid content, the muscles shrivel and the orbital soft tissues collapse into a small, irregular and inconspicuous mass at the rear of the orbit, often still anchored by the optic nerve. Among those Egyptian mummies in whom these dried tissues have not been replaced with artificial eyes, the appearance is not very different from that of naturally mummified bodies. The common conditions of conjunctivitis, trichiasis and others were well known in antiquity as judged from the recommendations for their treatment in old medical texts such as the Ebers papyrus, the Vedic literature and others. However, without tissue rehydration most of these did not produce a lesion clearly visible to an examining paleopathologist. Consequently in this text I acknowledge the ancients' awareness of many ophthalmological conditions, but I will limit the discussion principally to those few that would be recognizable without further tissue manipulation by a paleopathologist.

Trachoma

The most common cause of blindness in the world today is trachoma. It is an infection most frequent in arid, sandy areas, caused by *Chlamydia trachomatis* and spread largely by person-to-person contact, though transmission by flies has also been suspected. It begins as an infection of the upper eyelid conjunctiva, penetrating to the cartilaginous tarsal plate where destructive abscesses may occur. Eventual spread to the lower lid is common, but more important is extension to the cornea. Here the inflammation stimulates extensive granulation tissue. Healing occurs by organization of the granulation tissue that involves fibrous tissue replacement, which impairs vision. Paleopathologists are most apt to encounter the end stages of this disease, with eyelids often grotesquely deformed by the advanced destructive lesions. Defective closure of such deformed eyelids invites drying of the conjunctiva, with secondary bacterial infection, especially by the bacterium *Hemophilus aegypticus*. The widespread scarring of the cornea usually renders it opaque and the eye becomes sightless (Strano & Font, 1976). The paleopathologist's attention will probably be attracted by the lid deformation, and subsequent rehydration of the eye will render the fibrotic cornea apparent.

Trachoma is, in all likelihood, an ancient disease. Egypt's Ebers papyrus, written ca. 1500 BC but containing portions written probably as early as 3000 BC, includes specific remedies (application of antimony, ocher and natron) for trachoma (Ebbell, 1937: 69, column 57). The Ayurvedic texts of India were probably organized about the sixth century BC but assumed their present form about the sixth century AD. They describe an eye condition that most probably is trachoma (Chakravorty, 1993). Mi disease, described in the second century BC book *Shan Hai Ching*, is also most likely to be trachoma (Gwei-Djen & Needham, 1993). Trachoma ("*aspritudo*") was common in ancient Rome (Birely, 1992; Andersen, 1994:17). While the Middle Ages and early Renaissance texts of Europe and the Middle and Far East commonly describe trachoma, Crosby (1986) doubts it was a pre-Columbian New World disease. In spite of this frequency, the condition has not been reported in mummified bodies. Webb (1990) feels that a localized, lytic lesion of the orbital roof is the osseous "marker" for trachoma among prehistoric Australian aboriginal skulls. If so, the condition there can be traced back to the late Pleistocene Period. The relationship of the osteological lesion to the soft tissue infection by *Chlamydia trachomatis*, however, is speculative and circumstantial. Karasch (1993) has prepared a superb history of trachoma under the title "Ophthalmia".

Blindness

In Egyptian art, blindness is indicated by closed eyelids. Fuchs (1964) notes that, in addition, blind individuals are also characterized in tomb reliefs as having puffy eyelids and rolls of abdominal fat. These are features that he feels may have a metabolic and endocrine basis, resulting in focal fatty degeneration and water retention. Blind individuals are commonly pictured as harpists in ancient Egypt. Jantzen (1968) points out that one of these, from the eighteenth dynasty tomb of Paätenemheb, demonstrates a remarkably prominent, tortuous temporal artery, suggesting that his blindness was the consequence of thrombosed ocular arteries secondary to their involvement by temporal (granulomatous, giant-cell) arteritis. He attributes this to the then current artistic fashion of literal rendition of models. The earliest evidence of recognition of this entity appears to be a report by an Arabian physician about AD 1000 (Andersen, 1994:81). One of the well-known ceramics of Peru's Moche culture (AD 100–800) demonstrates the drooping corner of the mouth characteristic of left facial paralysis and a shrunken, empty right eye area (blindness) that might represent this same etiology.

In medieval Europe blinding was a common form of punishment as a substitute for capital punishment (hanging). It began with the support of the church.

Crimes punished by blinding included rebellion and theft quite commonly, but the same penalty was also inflicted for arson, poaching, perjury, rape and counterfeiting. Blinding methods were violent and commonly carried out surgically. In some instances it may have been carried out with hot iron instruments (Bruce, 1941).

Symbol

The eye has an age-old history as a symbol. Ancient Egypt's principal Old Kingdom sun god was Horus, symbolized as a health-bringing, stylized "*wedjat*" eye (related to his loss of an eye in a struggle for power with his uncle, the god Seth and which was restored by the god Thoth). Nordic mythology relates how their principal deity, Odin, traded one of his eyes for access to a well of knowledge (Guirand, 1978:257). Aztec iconography frequently depicts the gouging of an eye (which commonly dangles on the cheek, suspended by its optic nerve), but it is unclear whether the depictions are literal or symbolic. Arrington (1959) comments that eye gouging was a common ritual element among many ancient groups. All of these examples appear to treat the eye, or the loss of it, as being related to the acquisition of special knowledge. Donington (1974:122) cites the blindness of Samson and Oedipus as a "turning of their sight inwards", i.e., gaining unusual insight of the meaning of their actions dictated by their subconscious self.

Cataract

The eye's lens is normally transparent. Cataract is a term employed for the condition characterized by vision loss secondary to opacification of the lens. Such a process is the consequence of the precipitation of proteins within the lens. The most common form of lens opacification is due to aging. Lens proteins have virtually zero turnover (Berman, 1991:202) so that interactions with sugar (glycosylation) and changes in intermolecular linkages that produce protein precipitation will result in progressive accumulation of protein debris within the lens, leading to loss of its transparency and eventual blindness. While this process is age-related, certain conditions such as diabetes, ultraviolet radiation, drugs and many inherited enzyme deficiencies can accelerate it to a marked degree.

Cataract is surely an ancient disease (Andersen, 1994:49). The surgical approach called "couching" (insertion of a needle into the eye and pushing the hardened, opaque lens downward, out of the path of vision) began in India certainly before the fifth century AD and was popularized by Arab physicians in the medieval period (Hirschberg, 1982). Reflection of light from the opacified lens in a living individual makes the presence of a cataract obvious to even a casual observer and its treatment today is simply surgical removal with replacement by a plastic lens. It is, however, not simply recognizable within the shrivelled, desiccated remains of a mummy's eye, so that little has been written about it in the paleopathology literature. One would expect rehydration to make the lens identifiable, though whether the precipitated proteins of a cataract would become recognizable is not known. Now that chemical methods have become available to detect and quantify the insoluble lens' proteins we can expect the frequency and antiquity of cataract to be defined in the near future.

Artificial eyes

The use of artificial eyes in statuary has been known since ancient times. The "*wedjat*" eye was one of the early symbols of dynastic Egypt's initial principal god – Horus (Sandison, 1957; Andersen, 1994:40). Highly stylized eyes composed of copper or brass outlines filled in with white plaster, shell, glass or porcelain and with obsidian for the iris were commonly attached to Egyptian coffins, statuary (Tonkelaar et al., 1991) and (beginning with the New Kingdom) sometimes even in mummies (Wilson, 1972). Maegraith et al. (1991) feel the eye symbol of the Viking-type ship that appears on the Memoir II cover of the Liverpool School of Tropical Medicine, February, 1990, issue (*Report of the*

Malaria Expedition to West Africa, August, 1889) is the eye of Horus. The *moais* ("stone heads") of Rapa Nui (Easter Island), used by indigenes as spiritual containers for both spirits of ancestors and those of deities, were fitted with artificial eyes of coral and red scoria (van Tilburg, 1994: plate 25). Thin silver and gold wafers were found beneath the eyelids of mummies representing the Chiribaya people of coastal southern Peru from about AD 1200 (Aufderheide, 1990a). Rising (1866) found the crystalline lens of large, Pacific cuttlefish placed in the orbits of mummies from Arica, a northern Chile port city. However, in spite of extensive employment of the eye as a symbol, for decorative purpose, for statuary and even for mummies, there is no convincing evidence that artificial eyes were ever used in antiquity for cosmetic purposes in living persons.

Neoplasms

The two principal eye neoplasms are malignant melanoma and retinoblastoma. Melanomas arise from pigment-forming cells in the eye and most (but not all) of these tumors retain their melaninogenic capacity, the gross tumor presenting a black appearance. Growth is slow but the neoplasm eventually destroys vision and both invades locally and metastasizes hematogenously. It affects primarily adults.

Retinoblastoma is a cancer of childhood arising in the eye's retina. More than 90% are of sporadic origin; in the remainder a family history of these tumors is prominent. Such children inherit a mutation of the Rb gene (a tumor suppressor) on one of the two chromosomes for that site and thus are vulnerable for the development of this tumor if they suffer a mutation of the other also after birth since homozygous nonfunctional mutations of the Rb gene usually lead to growth of this neoplasm. It is a white or cream-colored tumor. It may spread into the brain by retrograde growth along the optic nerve, or it may metastasize hematogenously.

Thus an eye-destroying tumor may well remain recognizable in mummified human remains. The black pigmented appearance of a melanoma usually will be sufficient to separate it from neuroblastoma.

Soft tissue lesions of the ear

The only externally visible soft tissue lesions of the ear of interest to paleopathologists are the mutilations designed to support earrings. Several examples of ear drum perforations have been reported. Both Egyptian mummies labelled PUM II and PUM III were found to have perforations of the tympanic membrane (Benítez, 1988) as was the 2100-year-old body of the Marquis of Tai (Cockburn, 1974).

Otosclerosis

Sound is transferred from the vibrating tympanic membrane ("ear drum") to several articulated ossicles through the middle ear to the cochlea (hearing organ) in the inner ear. These ossicles must retain their normal articulation to transmit the sound vibrations. For reasons not yet well understood (though it behaves much like an autosomal dominant mutation), bony proliferation (otosclerosis) may occur at sites of articulation with subsequent fixation, resulting in decreased or absent sound transmission and consequent loss of hearing. Particularly common is fixation of the stapes footplate to the oval window. Using a surgical microscope, Birkby & Gregg (1975) were the first to report otosclerosis in the body of one of 11 Spaniards buried in a New Mexico (USA) cemetery about two centuries earlier. They found the stapes footplate fused to the round window by hard bone. Previously they had examined 4064 temporal bones harvested from 2600 Native American Indian skeletons without finding such a lesion. This was consistent with the virtual absence of this condition in the modern USA Indian populations in contrast to the known clinical prevalence of 1% (and histological evidence in about 9%) among modern Caucasions (Dalby *et al.*, 1993). In 1164 temporal bones from various British sites between the fourth and seventeenth centuries, Dalby *et al.* (1993) found a prevalence of about 0.9%. This study used endoscopy to visualize the ear ossicles in the intact skull. The simplicity of this noninvasive procedure can be expected to encourage its application in the future.

Otitis media

Infection of the middle ear is primarily a disease of children; more than half acquire it under the age of 4 years (Rudberg, 1954: Table 25). A wide range of bacteria can cause this disease, though pneumococcus (*Streptococcus pneumoniae*) and *Haemophilus influenzae* are responsible for about half of the cases. Acute pharyngitis causes sufficient swelling to occlude the eustachian tube, providing an opportunity for bacteria, trapped in the now undrained space, to proliferate. First fluid, then pus can fill the space, causing the tympanic membrane (ear drum) to become reddened and bulge into the external auditory canal. Eventually the exudate may digest a portion of the tympanic membrane, allowing the pus to drain by this route. Subsequent healing is the rule, though recurrences are common. The tympanic membrane defect may persist into adulthood, serving as an entry point for bacterial access with recurrent or chronic infections. In addition, the squamous epithelium lining the external auditory canal may extend through the defect into the middle ear where its desquamated cells (keratin) accumulate, blocking the lumen as an irregular expanding mass of material referred to clinically as a cholesteatoma. In spite of its name, this is not a neoplasm – merely a localized accumulation of exfoliated, dead, benign squamous epithelium.

Complications of otitis media include not only recurrent or chronic infection but also extension of the bacteria to the blood, resulting in a life-threatening septicemia. In addition, local destruction of the ear ossicles and/or cochlear branch of auditory nerve can lead to deafness, while extension to the cranial cavity can produce fatal meningitis or posterior extension can reach the air cells of the mastoid process. Since the air cells of the mastoid develop during the same period of infant growth that is coincident with a high prevalence of otitis media, extension of the infection to the mastoid may break down the thin walls between the cells and impair further air cell production. In some cases the mastoid abscess may perforate the cortical surface and drain exteriorly.

Paleopathology

The paleopathologist can expect to identify the perforated tympanic membrane by direct observation, though a simple, hand-held otoscope will provide a desirably illuminated view and, if required, a magnifying lens. Evaluation of middle ear ossicles for either destruction by otitis media or otosclerotic fixation is most simply and nondestructively achieved with an endoscope (Dalby *et al.*, 1993). Benítez (1985) extirpated the temporal bone and prepared histological sections, identifying an antemortem tympanic membrane perforation found in a young adult Egyptian mummy. Smith & Wood Jones (1910:284) found a draining sinus in the posterosuperior part of the mastoid in a Nubian mummy (subject no. 23:60:J) from the New Kingdom and Leek (1986:191) identified an even earlier example of such a sinus in the lower mastoid that exposed underlying air cells in an aged male excavated from Cheops' western necropolis of the Old Kingdom era. Gregg *et al.* (1965) and Gregg & Steele (1982) used radiological studies to evaluate the presence of mastoid air cell patterns demonstrating changes produced by otitis media in both living and ancient Native American skulls. They found that about half of the mastoids revealed evidence of alteration by otitis media with mastoiditis, and that pre-Columbian frequency of such changes was lower than that for those dating to the post-1492 period. In a large Swedish study of modern patients, Rudberg (1954) found clinical evidence of mastoiditis in 17.2% of children with otitis media not treated with antibiotics while chemotherapy reduced this to less than 1.5%. Comparison of the Gregg *et al.* study with that by Rudberg suggests that radiological evaluation may be the most sensitive method for detection of mastoiditis. Certainly persistence into adulthood of a draining sinus via the mastoid occurs so infrequently that it would serve as an insufficiently sensitive "marker" for mastoiditis prevalence study.

Exostoses of the external auditory canal

Definition

This benign lesion is composed of a skin-covered, circumscribed mass of dense bone located at the meatus or within the external auditory canal. (Fig. 8.4).

Nature of the lesion

Whether this lesion is a reactive change or a benign neoplasm is not at all clear. The very early literature refers to it as a benign neoplasm (osteoma) while more recent articles favor the term "exostosis". Mingled within these are reports by several authors who feel that both occur and can be differentiated.

The floor, all of the anterior wall and about half of the posterior wall of the external auditory canal is formed by the tympanic portion of the temporal bone, while the roof and upper half of the posterior wall are created by the squamous portion (Sobotta & McMurrich, 1939: 38–45). Collectively this tubular structure is termed the tympanic ring, while the posterior junction of the tympanic bone with the squamous roof (the latter forming part of the anterior mastoid structure) is named the tympanomastoid suture while the homologous junction on the anterior wall is termed the tympanosquamous suture. Graham (1979) defines what he felt were true benign, neoplastic (osteoma) lesions as roughly spherical but pedunculated nodules of dense osseous tissue arising from either of the two sutures (i.e. from either the anterior or posterior wall) that were solitary and tended to increase sufficiently so as to occlude the canal, with a histological structure of fibrovascular channels embedded in lamellar bone. He admits that any given lesion might lack at least one of these features. He characterizes reactive exostoses as being of sessile, broad-based forms, commonly bilateral and often multiple, invariably involving the tympanic bone and that its histological structure was composed of concentric layers of subperiosteal bone with abundant osteocytes and far fewer fibrovascular channels. Gregg & Bass (1970), however, describe four different forms, later described simply as spongy, knob-like, linear and spicular (Gregg & Gregg, 1987:90). Of these only the linear type fits Graham's (1979) description of exostoses, while a single, additional example of the nodular type was described that might conform to Graham's osteoma classification (but pedunculation is not noted though it was unilateral). Gregg & Bass depict their single, nodular, fifth example as "tumor type", but seem to equate the terms exostosis and osteoma for all their other examples. While DiBartolomeo (1979) cites Graham's (1979) classification, he presents all of his own 50 cases as exostoses even though some are solitary and pedunculated. The custom of referring to these lesions as osteomas in the older literature appears to have evolved into terming them exostoses in more recent reports. There appears to be little sense of

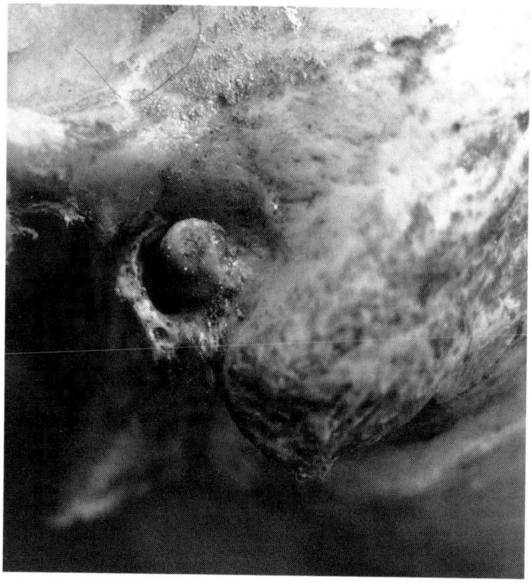

Fig. 8.4. Exostosis at external auditory canal meatus; SM, AD 1000.
A white, smooth-surfaced osseous nodule located on the anterior wall of the aural canal meatus was 4 mm in diameter. Adult male, coastal Peru. (PBL.)

Table 8.1. *Proposed etiology of external auditory canal exostoses*

Proposed etiology	References
Alcoholism	Toynbee,1849
Genetic	*Seligman, 1864; *Turner, 1879; Blake, 1880; *Dalby, 1889; *Hartmann, 1893; Hrdlicka, 1935; *Adis-Castro & Neumann, 1947; *Berry & Berry, 1967
Gout and/or rheumatism	Toynbee, 1849; *Field, 1882; *Virchow, 1893; *Vail, 1928
Ear piercing	von Luschan, 1896a,b
Cranial deformation	Jackson, 1909; *Oetteking, 1930; *Ostman, 1930
Bathing	*Tod, 1910
Chronic infection	*St. John Roosa, 1866; Cassels, 1877; *Field, 1878
Canal form	*Stewart, 1933
Mastication stress	*Burton, 1923
Chronic irritation	Hrdlička, 1935; Sheehy, 1958; DiBartolomeo, 1979
Swimming	*Wyman, 1874; *Field, 1878;* Kelson, 1900; *Moore, 1900; Jackson, 1909; *McKenzie, 1920; *Belgraver, 1938; *Harrison, 1951; Roche, 1964; *Starachowicz & Koterba, 1977
Cold water	van Gilse, 1938; Fowler & Osmon, 1942; *Adams, 1951; *Ascenzi & Balistreri, 1975; Kennedy, 1986.

Notes:
This table was created from the narrative portions of the above-mentioned references listed in Kennedy (1986) together with additions.
*For references not present in the bibliography of this volume, see Kennedy, (1986).

urgency to distinguish benign neoplasm from reactive exostosis today.

Antiquity

In Japan these lesions have been found in Neolithic bones from about 5000 BC. Among early marine coastal populations of northern Chile, these structures were common as early as 7000 BC (Aufderheide *et al.*, 1993), but were absent in nearby highlanders (Standen *et al.*, 1995). A few Egyptian examples beginning with the twelfth dynasty (roughly 1800 BC) have been reported (Hrdlička, 1935). Seligman (1870) found them in Inca skulls. Such lesions were also found in the fossil hominids of Shanidar 1 (Trinkaus, 1983) and La Chapelle aux Saints (Trinkaus, 1985).

Location of the lesions

About 70% of the lesions are found on the posterior wall, beginning near the tympanomastoid suture; most of the remainder are on the anterior wall (near that suture) with a minority on the floor near the tympanic membrane (Blake, 1880; Roche, 1964; Gregg & Bass, 1970; DiBartolomeo, 1979; Gregg & Gregg, 1987). Most report a high frequency of bilateral lesions (DiBartolomeo, 1979:62%), but Gregg & Gregg (1987:90, Table 4.7) found that only 58% of their upper Missouri River archaeological skulls of Native Americans were bilateral. Of the unilateral lesions, 75% were on the left side. All of DiBartolomeo's clinical cases revealed more than one lesion/meatus.

Age and sex

All series identify a high male:female ratio varying from 6:1 (Gregg & Gregg, 1987) to 14:1 (Roche, 1964). Both archaeological studies (Hrdlička, 1935; DiBartolomeo, 1979; Gregg & Gregg, 1987; Manzi *et al.*, 1991) and clinical studies (DiBartolomeo, 1979; Gregg & Gregg, 1987) identified either no or only an occasional such structure in individuals under age 20 years, strongly implying that this is an acquired condition. In 177 skulls from Peru, Capasso (1988) found 15 with ear canal exostoses; all but one were in males and all were adults. Similarly, none could be found in examination of about 10 000 schoolchildren (Bezold, 1885, cited in Hrdlička, 1935).

Frequency

Archaeological population frequencies reveal some geographical differences. North American upper mid-western archaeological populations averaged about 4.9% (Gregg & Gregg, 1987), Hrdlička's (1935) multinational collection revealed about twice that many (10.8%), while a large Australian collection demonstrated the highest recorded frequency of 27.9%. Furthermore, Gregg & Gregg (1987:90) report that the frequency in their archaeological specimens (4.9%) approximates that found in surveys of those groups' living descendants from the same region, suggesting that the soft tissue covering these lesions probably is not obscuring underlying lesions when examined in current populations. For that reason DiBartolomeo's (1979) finding of only 0.6% in about 11 000 patients seen in a otolaryngologist's clinical practice is the product of patient selection, reflecting the asymptomatic nature of a large fraction of these lesions.

Etiology

The older literature contains a generous quantity of speculation about the cause of these lesions. DiBartolomeo (1979) and Kennedy (1986) have collected these and Table 8.1 has been constructed from their articles and some other sources. Some of these, such as gout or alcoholism, do not warrant serious consideration, while others (cranial deformation or mastication stress) appear conceivable but not probable. Racial affinity and "constitutional predisposition" are general terms that assume the effect of some type of genetic population difference leading to this condition, but its virtual absence under age 20 years suggests an environment mechanism instead.

Because chronic infections involving the periosteum are known to provide a powerful stimulus to new bone formation, the possibility of chronic external auditory canal infection has been postulated (Toynbee, 1849; Cassels, 1877). In spite of the plausibility of individual reports, however, DiBartolomeo (1979) found that while as many as 30% of his patients admitted to a history of an acute ear infection in childhood, none had chronic infections.

The turn of the century produced a variety of reports dealing with individuals exposed to cold water immersion to a far greater degree than the general population (Kelson, 1900; Jackson, 1909). Thus, DiBartolomeo (1979) found that many of his patients from California's southern coast were ocean surfers, swimmers or divers. Kennedy (1986) notes that populations in whom this lesion is common are those living between latitudes 30–45° N and 30–45° S, and who exploit marine or freshwater resources. A study of British military recruits identified a frequency of about 5% among swimmers (same among fresh- and sea-water swimmers) compared with fewer than 1% of nonswimmers (Harrison, 1962). Recently a high frequency of exostoses found among wealthy, male, imperial Romans (in contrast to none in nearby slaves and laborers) was attributed to their use of daily baths and the absence of lesions in their women was associated with the men's traditional use of cold water baths in the frigidarium (Manzi *et al.*, 1991).

Only a few efforts have been made to reproduce these lesions experimentally. Van Gilse (1938), cited by DiBartolomeo (1979), observed that Italian swimmers in the warm Adriatic waters did not develop aural exostoses, while the Dutch bathing in the cold milieu of the North Sea were vulnerable to these lesions. He attempted to test the role of the thermal element. Irrigating human external auditory canals with both warm and cold water he created no exostoses but found prolonged (45 minutes) hyperemia of the canal's epidermal lining following a 30 second cold water irrigation in contrast to only 1 minute of hyperemia after using warm water. A more aggressive attempt to reproduce these lesions employed cold water irrigation of the external auditory canal in guinea pigs (Fowler & Osmun, 1942). This produced no lesions in the external canal but did induce a diffuse proliferative skeletal response in the middle ear. Twenty years later

Harrison (1962) also irrigated anesthetized guinea pig external aural canals at weekly intervals for a total of up to 11 hours. Histological sections demonstrated vasodilatation in the underlying bone and in five animals diffuse histological evidence of new bone formation was found deep in the meatus adjacent to the tympanic membrane and in one animal a "seedling" osteoma was found at that site. Emphasizing the thin layer of soft tissue overlying the external auditory canal's bony structure, he postulated that the increased blood flow stimulated by repeated cold water exposure may be sufficient to evoke a proliferative response in the underlying periosteum.

A critical summary of the evidence for the etiological role of cold water exposure in the development of external auditory canal exostoses and/or osteomas reveals that cold water exposure initiates a prolonged vasodilatation of the canal structures involving not only the soft tissues in humans, but also the superficial aspects of the underlying bone in guinea pigs. The implication that the demonstrated proliferative skeletal changes in the studied guinea pigs would have proceeded and eventually formed gross lesions characteristic of the exostoses under consideration in humans is provocative but speculative. While current authors frequently declare that these experimental studies have "proved" the etiology of this condition, this body of evidence is fragile indeed. The simple fact remains that the gross lesions have not yet been reproduced experimentally. In addition, the cold water theory is not a plausible explanation for the presence of these bony nodules among inland populations that have a frequency rivalling that of coastal groups (Blake, 1880; Gregg & Gregg, 1987). I consider the etiology of external auditory canal exostoses and/or osteomas as an unresolved issue.

Recognition in archaeological remains

A defined, benign outgrowth of bony tissue of any form located anywhere from the meatus to the tympanic membrane is eligible for the assignation of "exostosis". Until the separation of benign neoplasm (osteoma) from exostosis is better defined, the term "exostosis" is probably acceptable (though Kennedy's (1986) suggestion of the descriptive term "auditory hyperostosis" could be a practical solution in the interim). More explicit description of the location, shape, base and size of the lesions as well as their histology in future reports could help to resolve the reactive or neoplastic nature of these structures.

Soft tissue lesions of the nose

Nasal polyps

Nasal polyps have a considerable antiquity. An Egyptian remedy for "fetid nose" (*ozaena*?) probably deals with an infected, ulcerated, foul-smelling nasal polyp (Ebbell, 1937:105 column 20). The Hippocratic writers about 400 BC described five types of polyps, recommending removal with strings (Pahor, 1992). Medieval Arabian physicians preferred to remove them with scissors while Fabricius (Renaissance) used copper forceps, tongs and scissors (Pahor & Kimura, 1990).

The nasal "polyp" is not a neoplasm but an area of nasal mucosa in which a localized expression of allergy has resulted in so much fluid extrusion that its sheer weight causes the mucosa to descend to near (or actually protrude from) the nares in polypoid form. Drying with secondary infection results in ulceration and often bleeding. Their high moisture content would result in a dramatic size decrease in desiccated bodies, though fibrosis secondary to ulcerative infection can be expected to survive the desiccating process sufficiently for gross recognition. In spite of this, Allison (1992) found a nasal polyp large enough to fill the nasal space in a skull otherwise devoid of soft tissue from the Atacama Desert in northern Chile, dated to about AD 1000.

Syphilis

Nasal soft tissue destruction may be extensive. The earlier forms destroy the cartilaginous support of the nasal bridge and the resulting depression of the overlying skin is known as "saddle nose". Later destructive changes include the skin and bone, exposing the nasal chamber.

Mucocutaneous leishmaniasis

This is a parasitic infection of the skin and mucous membranes. An isolated, often self-healing form of cutaneous leishmaniasis called "oriental sore" is found in the Old World and is caused by a variety of species. The mucocutaneous form of the New World type is caused largely by *Leishmania braziliensis* and is present in Central and South America. The natural reservoir lies in many forest mammals and the organism is transmitted by blood-sucking sandflies such as those of the genus *Lutzomyia* (Connor & Neafie, 1976). Initial lesions appear on skin surfaces exposed to fly bites. Years later the nasal mucocutaneous areas and even associated bone may become involved and can be devastatingly destructive with extensive disfigurement. This form is called espundia, and ultimate death by bacterial infection is common (Locksley, 1991). A primary, localized facial lesion found in the Peruvian area is known as "uta". Diagnosis is established by demonstrating the organism histologically. A skull of a 45–50-year-old female of a population occupying a low valley (Azapa) site near the modern port of Arica, Chile, about AD 1100 showed extensive skeletal destruction of the nasal areas as well as the right orbit and paranasal sinus walls, strongly suggestive of this disease (Allison *et al.*, 1982b), though the absence of soft tissue prevented histological search for the organism.

Sinusitis

Chronic infection of the paranasal sinuses with characteristic acute exacerbations provides recurring stimulation to the periosteum immediately underlying the mucosa lining. The eventual result is irregular first fibrotic and later calcific thickening of these structures. In an intact skull or head, these are usually visualized in skull radiographs. Frequently seen in modern individuals, examples have also been reported in the paleopathology literature. Lewis *et al.* (1994) report multiple cases from the collection at the University of Bradford characterized by maxillary antrum bone deposits and cite dental abscesses, poor ventilation, air pollution and allergies as etiological conditions leading to the infection.

Neoplasms

Nasopharyngeal carcinoma is the principal neoplasm of concern in this area. It arises from the soft tissues of the posterior nasopharynx, invades aggressively into the surrounding structures (including the bone) of the palate, sinuses, orbit and eventually even reaches the cranial cavities. Regional lymph node involvement is common and late stages include distant hematogenous metastases. Histological type is most commonly squamous cell carcinoma of varying degrees of differentiation including one type exhibiting extensive lymphocytic infiltrate (lymphoepithelioma). Causes remain unknown. While it has a global distribution, the populations in southern China are particularly susceptible, where it constitutes almost 20% of all cancers in contrast to its world-wide estimate of 0.25% (Balogh, 1994). The tumor is also found in arctic natives, whose practice of chewing raw hides has been suggested as being an etiological contributor, but only on an anecdotal basis. The Epstein–Barr virus is frequently present in the cells of this tumor, but its causal status remains undefined.

Archaeological skulls demonstrating lytic lesions consistent with nasopharyngeal carcinoma have been reported frequently. In addition to an Egyptian case he reported himself (Strouhal, 1978), nasopharyngeal carcinoma was the diagnosis made by half of the authors of individual articles reporting primary soft tissue carcinomas collected and summarized by Strouhal (1994). Most were found in Egyptian mummies. A 1000-year-old skull from a 45–50-year-old female from northern Chile demonstrates a widely destructive process in the right facial bones consistent with either leishmaniasis (uta) or nasopharyngeal carcinoma (Allison & Gerszten, 1981). Brothwell (1967a) cites the description by Derry (1909) of the skull of a pre-Christian Nubian adult of the sixth century AD that had an extensive lytic process of the skull base including the nasopharyngeal area that Derry thought most probably began in the nasopharynx. Brothwell also cites a similar case reported by Wells (1963) in an early Old Kingdom mummy, as well as an example published by Krogman (1940) from an Iranian site more than 5000 years old. Unfortunately, in none of these was it possible to confirm the suspected nasopharyngeal carcinoma diagnosis by soft tissue histology. Compared with the relatively low frequency of this lesion today, the cases diagnosed as nasopharyngeal carcinoma in ancient bodies are proportionately very high, suggesting diagnostic imprecision.

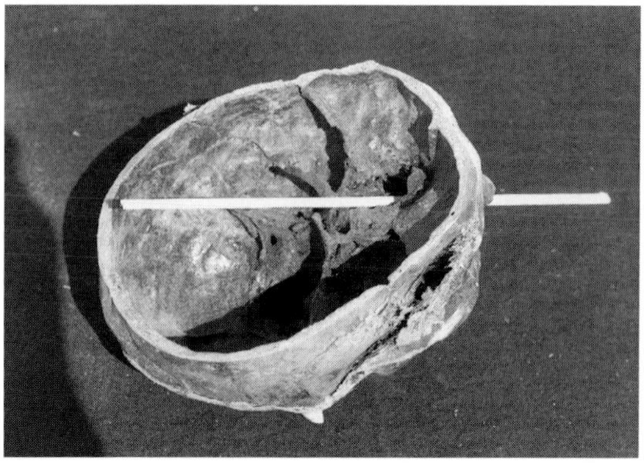

Fig. 8.5. Transnasal craniotomy; 30 BC to AD 395, AM. Egyptian mummification commonly involved the creation of a channel from the nasal cavity to the cranial cavity via the ethmoid sinuses. The white rod in this illustration indicates the traverse of this defect (titled transnasal craniotomy here) in the skull of an Egyptian mummy from Dakhleh Oasis from the Ptolemaic Period. (PBL.)

Transnasal craniotomy (TNC)

During the Egyptian mummification process a metal rod was inserted by the embalmers into one of the cadaver's nostrils (most often the left) and forced upward through the skull bones, penetrating the ethmoid sinuses and reaching the cranial cavity (Fig. 8.5). Allegedly this was done to enable subsequent brain extraction through the defect. This can be and was performed without visible external deformation. The modern paleopathologist can identify this defect easily by inserting a straight probe into the nostril. When the inner end has reached the cranial cavity, movement of the outer end in a circular manner can provide an impression of the defect's size at the level of the cribriform plate. Visual inspection of the nasal chamber permits a view of the extent of nasal structure destruction and whether resin or linen cloth are visible. Radiology of the skull can detail these changes. Sometimes the defect will be large enough to accommodate the passage of a pediatric endoscope, permitting an internal view of the cranial cavity to determine the possible presence of residual brain or meningeal tissue as well as resin or linen.

The early Greek historian Herodotus (ca. 450 BC/1981 II: 86) described the creation of the defect by embalmers using a hooked rod with which they also removed the brain through the channel they had produced. That this could have been performed was demonstrated by Karl Sudhoff (1911), who collected instruments that were believed to have been used by ancient Egyptians for that purpose, and used them to create the TNC defect in human cadavers, and also to remove some brain tissue. The latter was achieved by plunging the metal rod through the defect into the brain then withdrawing it and wiping off the adherent brain tissue. Repeated efforts of this nature resulted in the removal of a significant amount of cerebral tissue. Subsequent brain maceration was achieved by vigorous rotation of the metal rod, mixing the macerated tissue with the cerebrospinal fluid from the brain ventricles. Turning the body prone caused this semiliquid suspension to drain out through the defect. These procedures were later repeated on both sheep and human bodies with similar results (Leek, 1969).

Earlier Smith (1902), Smith & Dawson (1924) and others (Macalister, 1894) had noted that TNC first appeared in the eighteenth dynasty body of Ahmose I

(Brier, 1994:62), leading to the conclusion that this was a procedure limited to the latter half of Egyptian history. However, Strouhal (1986) found TNC in four Middle Kingdom (twelfth dynasty) mummies and noted that one of Manchester Museum's "Two Brothers" from the Middle Kingdom also revealed such a defect, as does the mummy from the same period currently curated at the Boston Museum of Fine Arts. He also cites the report of a body with TNC found in Imhotep's tomb (third dynasty) though this would be more convincing if it were radiocarbon dated. Thus the initiation of this practice can be traced credibly to at least the Middle Kingdom.

Yet certain incongruities remain in our reconstruction of this practice. Among these the most obvious is that, while all tangible parts of an Egyptian ruler's body were considered sacred and all his extirpated viscera were preserved, no one has ever found brain included with those organs (Smith & Dawson, 1924:68). This is most easily explained if the suggested method of excerebration (repeated insertion of a metal rod into the brain followed by wiping off the soft, cerebral tissue adhering to the withdrawn rod) is accurate, since no cerebral structure would remain in recognizable form. Such preserved brain-smeared bandages would not be distinguishable from other masses of used natron-containing packages often interred with or near the body. Strouhal (1992b:260) expresses doubt that the brain was removed routinely in the manner described by Herodotus.

The most disturbing, however, is the presence of TNC in bodies in which no other effort was made to preserve soft tissues. For example, all 10 of the anthropogenically mummified Ptolemaic Period bodies examined by me at Egypt's Dakhleh Oasis revealed not only the usual evisceration and other obvious soft tissue conservation procedures, but also the expected TNC. The TNC defect, however, was also present in 18 of 21 additional bodies that were removed from the same tombs and clearly spontaneously mummified without evisceration or other efforts to preserve soft tissue (Aufderheide *et al.*, 1999b). If the purpose of TNC were to remove brain in order to prevent its putrefaction, then it would appear illogical to remove it without also eviscerating the abdomen, since the latter organs are obvious sources of offensive putrefaction and are easily removed. Furthermore, Egyptian belief system identified the heart, not the brain, as the site of emotions and features of character. It is conceivable that the practice of TNC was initiated as part of the body soft tissue conservation effort and in later periods survived the curtailment of the other elements, such as evisceration, perpetuated only as a ritual feature. We have only the comment by Herodotus (not an eye witness) that the purpose of TNC was brain removal. Alternatively, therefore, creation of this defect could have been introduced initially solely for ritual purposes. That would be consistent with the religious nature of the entire mummification process. Many cultures have beliefs involving the postmortem exit or entry of souls (Frazer, 1976:208); for the Egyptians it was the ka and ba that would move out of and into the tomb and its contained body after death. The TNC defect thus could be conceived as an enabling device for such passage, similar to that of the door often painted upon the tomb wall. Such a viewpoint could accommodate the variations and inconsistencies in its application as well as the related variable cranial cavity content. Indeed, Smith & Dawson (1924:100) cite Maspero as explaining that a hole in the cranial vault of the skull of Meneptah (as well as Sety II and Ramses IV, V, and VI) was created by embalmers "in order to allow evil spirits a free exit from the head . . .".

I am aware of no other cultural group that practiced TNC.

Soft tissue lesions of the neck

Cysts

These are usually embryonic remnants. Thyroglossal duct cysts, as their name implies, are products of that structure. During fetal life the thyroid migrates from a position equivalent to the base of the tongue, down to its final locale, resting on the trachea just below the larynx. Its "wake" consists of the thyroglossal duct that normally atrophies and essentially disappears. Persistence of that duct's lumen in the central portion, with obliteration at both ends, results in entrapment of the fluid secreted by its epithelial lining, generating a cyst varying from a few millimeters to several centimeters in diameter. These are harmless and always in the midline.

Similarly, during the embryonic stage of "gill clefts", the pharynx communicates with the skin surface of the side of the neck. Subsequent failure of obliteration of these structures may produce a cyst at any of the several branchial cleft levels. Fusion of the inner, pharyngeal level with lumen persistence of the central and outer portions may produce a skin surface defect. Complete persistence of the entire structure results in a skin–pharynx fistula. Other branchial cleft structures such as cartilaginous fragments are also often present. The cyst size range is similar to that of the thyroglossal duct origin, but one feature (location) easily differentiates the two: the thyroglossal duct cysts are in the midline while those of branchial cleft origin are lateral.

After desiccation such cysts lose their liquid content and often collapse to the point of difficult recognition. In others the globoid shape is retained in spite of its fluid loss, often with a thin, peripheral rim of calcification. We know of only one paleopathology report of a young adult male from a Frankish thirteenth century cemetery in Greece that demonstrated a thyroglossal cyst (Barnes, 1995), though their frequency in most modern populations today suggests they are being overlooked in mummified, ancient bodies.

Torticollis

"Wry neck" is characterized by involuntary turning of the head secondary to dystonic muscle contraction. Initially it may be only transient, but it commonly progresses to a nearly fixed deformity. An Egyptian mummy from the Birmingham Collection demonstrates this condition (Pahor, 1992: 864).

Thyroid lesions

See Chapter 6 (Mummy study methodology).

Heart and pericardial diseases

Examination of the heart in desiccated mummies is very difficult because of the brittle nature of the tissue. However, if the entire heart is immersed in rehydration fluid, the various structures of interest can be dissected in a useful manner. Parts of the heart intended for biochemical, immunological or other study methods can first be removed before rehydration from areas not expected to show anatomical changes and these samples can be retained for such other investigations.

The heart is a mass of muscle (myocardium) whose almost only purpose is to keep blood in motion so that oxygen and nutrients can be delivered to tissues while metabolic waste products are transported to excretory organs (lungs, kidney, gut) for elimination. The muscle mass is so thick that the heart has a set of its own vessels (coronary arteries) that nourish the heart muscle. The most common modern diseases of the heart involve conditions (especially atherosclerosis, discussed in Aufderheide & Rodríguez-Martín (1998: Chapter 5) that decrease or arrest blood flow through these arteries. The result of this is necrosis (infarction) of the myocardium. A weakened heart may not be able to pump blood at a normal rate ("heart failure"), resulting in accumulation of blood (congestion) in the lungs. Untreated this can displace air space for gas exchange to a degree sufficient to be lethal. Other diseases can affect the heart valves, whose purpose is to regulate direction of blood flow. Diseased valves can either obstruct the flow of blood or permit reflux, either of which reduces its pumping efficiency and often leads to heart failure. Finally, these and many other heart conditions may interfere with the normal conductivity of the heart's electrical impulse that is necessary for muscle contraction, and this may result in sudden cardiac arrest. Those conditions that may result in a grossly recognizable lesion are discussed below.

Fig. 8.6. Heart: fibrinous pericarditis; SM, AD 350–1000.
The light gray coating over the visceral pericardium of this heart represents fibrin that leaked from pericardial and epicardial vessels. Cause of death was end-stage renal failure with uremia. Young adult male from northern Chile, Cabuza culture. (PBL.)

Coronary thrombosis and infarction

Stenosing (narrowing) atheromas (fibrofatty plaques) in the coronary arteries are commonly accompanied by pain (angina) originating in the ischemic (oxygen-deprived) myocardium. Eventually ulceration of such an atheroma may be followed by acute coronary artery thrombosis ("heart attack") and the ischemic myocardium often becomes necrotic (infarcted). The heart muscle so weakened may not be able to carry out its pumping action efficiently, and death by cardiac arrest, heart failure, or rupture of the infarcted ventricular wall may occur. The latter will fill the pericardial sac (that envelops the heart) with blood, compressing the heart and producing cardiac arrest. If the affected individual survives, the infarcted myocardium is eventually replaced by fibrous scar. These complications can be expected to be recognizable in mummified remains, though postmortem autolysis often obliterates the evidence of antemortem myocardial infarction occurring shortly before death (Zimmerman, 1978).

Infections

Myocarditis

Various species of viruses, bacteria, fungi and protozoa can all infect the myocardium. Immunocompromised persons are particularly susceptible to myocarditis by organisms normally not virulent enough to produce human infection. Human hosts with a normal immune system, however, may suffer infection by many bacteria, especially if these have established an infective focus within the bloodstream, such as a scarred heart valve deformed by a previous episode of rheumatic fever. In endemic areas of South America, *Trypanosoma cruzi* myocarditis (Chagas' disease) is a common and often lethal infection, and cardiomegaly in residents of such areas suggests the possibility of this condition. In mummified human remains all forms of myocarditis may be recognizable by the presence of an enlarged, dilated heart sometimes covered with fibrin (Fig. 8.6). Bacterial and fungal infective agents may be demonstrable with appropriate stains.

Valvular heart disease

The mitral and aortic heart valves (rarely those of the right heart) may be damaged by rheumatic fever, and the aortic may also be involved by atherosclerosis or by the dynamic stress to which a congenital bicuspid aortic valve deformity is subjected. The result of these conditions is commonly extensive fibrotic thickening and even calcification. The additional workload imposed on the heart by these deformities often causes

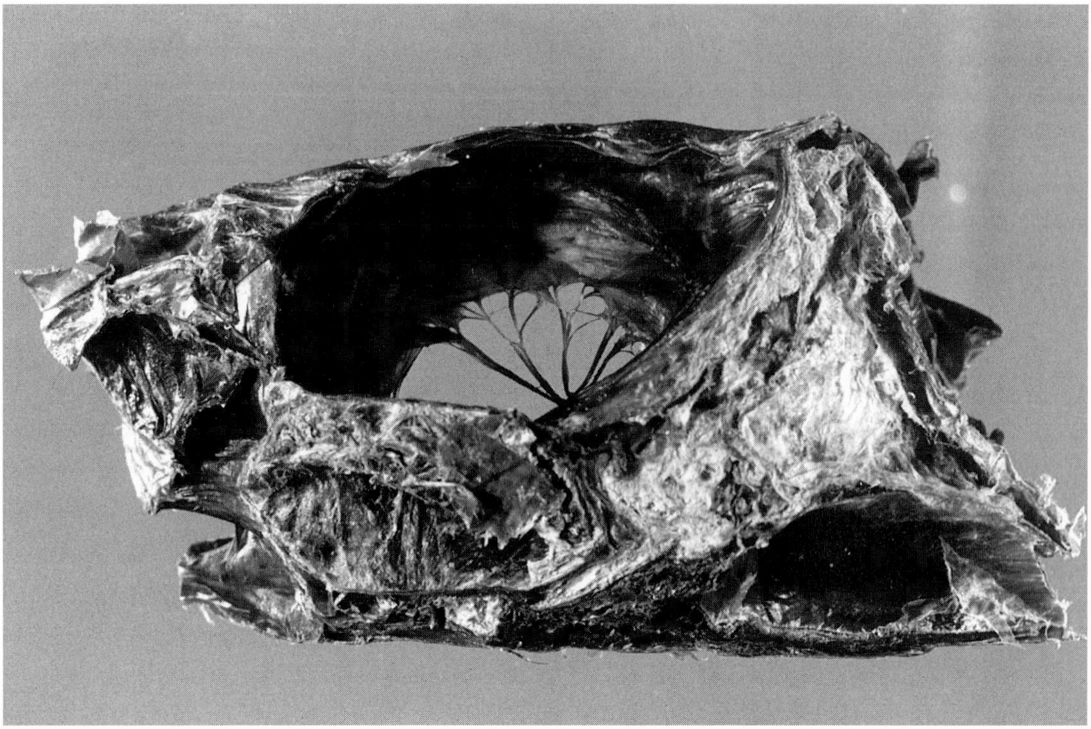

Fig. 8.7. Heart: normal mitral valve; SM, AD 500–800.
The actively metabolizing myocardial fibers autolyze rapidly after death but the fibrotic and elastic tissues are more resistant. In this infant's heart, most of the muscle was lost, but the delicate fibroelastic structures of the normal mitral valve remain virtually unchanged. Two-year-old infant, northern Chile. (PBL.)

heart failure that, untreated, may be fatal by early to middle adulthood (mitral) or 10–20 years later (aortic). Iron-containing macrophages in lung alveolae may testify to antemortem heart failure in human mummies. Such valves are also subject to local infection (endocarditis) with production of a clot upon the infected valve; its untreated course is often fatal over a period of months. Calcification is frequently most extensive in the aortic valve, which can become so encrusted with calcium as to retain its diagnostic appearance even in skeletonized bodies. The larger mitral valve is often more irregularly and incompletely calcified. Their fibroelastic structure often permits the valves to survive postmortem changes far better than the muscle does, permitting their evaluation for pathology even when much of the myocardium has decayed (Fig. 8.7).

Heart disease due to miscellaneous causes

Enormous heart enlargement due to indeterminate reasons has been termed idiopathic cardiomyopathy. It may present as a heart with markedly thickened walls (hypertrophic form) or one with increased musculature but whose striking change lies in enlargement of its chambers (dilated cardiomyopathy). Occasionally metabolic disturbances can result in inappropriate deposits of substances (glycogen, amyloid) into the heart muscle that weaken the myocardial fibers, causing heart failure. Nutritional deficiencies (thiamine: common in chronic alcoholics) can have a similar effect. While the enormously enlarged heart of dilated or hypertrophic cardiomyopathy can be expected to be recognizable in mummified tissue, only infrequently will its cause be demonstrable because of the myocardium's rapid postmortem autolysis.

Paleopathology of cardiac pathology

Column 37 of the Ebers papyrus from Egypt provides a graphic description of angina pectoris, including the pain radiating to the arms (Ebbell, 1937:48): "If thou examinest a man for illness in his cardia, and he has pains in his arms, in his breasts, and one side of his cardia . . . it is death threatening him'. Mortuary inscriptions have been interpreted to describe acute, sudden death (Bruetsch, 1959). Peng (1995) reported a Chinese noblewoman buried about 200 BC who suffered coronary atherosclerosis with severe stenosis. That the process was not limited to highly structured civilizations is evident by observations of atheromas in aortic and coronary arteries of Alaskan spontaneously mummified bodies (Zimmerman *et al.*, 1971, 1981b; Zimmerman & Aufderheide, 1984; Zimmerman, 1985). Experimental studies carried out by Zimmerman (1978) suggested that acute myocardial infarction could probably not be distinguished from postmortem autolysis of normal myocardium in spontaneously mummified bodies, but that the fibrotic scar of a healed infarct could be. These predictions are consistent with the observations by Long (1931) of fibrous foci in the myocardium of a twenty-first dynasty Egyptian mummy whose coronary arteries (as well as her aorta and renal arteries) were atherosclerotic.

Localized calcifications in otherwise undeformed mitral valves were interpreted by Long (1931) and by Zimmerman (1978) as healed endocarditis. No such reports have been described from the many spontaneously mummified bodies examined at the University of Tarapaca in Arica, Chile. *Trypanosoma cruzi* amastigote organisms have been demonstrated to be responsible for the enlarged heart of a Peruvian Inca mummy (Fornaciari *et al.*, 1992a,b). Another example of infectious myocarditis was the finding of multiple myocardial abscesses containing Gram-negative bacilli in a late nineteenth or early twentieth century Aleutian mummy (Zimmerman, 1993). The heart of an 8-year-old child (from northern Chile, about AD 300) encased in pale exudate, was reported by Allison (1981a). The etiology probably was related to the pneumonia found in the lung. Extension of suppurative, bacillary lung infection to the pericardium, producing acute, purulent pericarditis is a well-known and often lethal complication of pneumonia. Blackman *et al.* (1991) found a similar-appearing fibrinous pericarditis enveloping the heart of an 18-year-old male from the Cabuza culture (AD 350–1000) from northern Chile. In this case, however, it was not an infectious process but rather a metabolic complication of uremia due to renal failure with diffuse renal calcification, a product of secondary hyperparathyroidism.

The entire array of cardiac shunts, valvular deformities and malpositions that characterize congenital lesions may, of course, be encountered by the paleopathologist, but in most of these cases postmortem alterations can be expected to obscure and complicate their identification. In spite of myocardial preservation difficulties, Robertson (1988) reported an example of a congenital heart lesion (common atrioventricular canal) in a spontaneously mummified body of an infant member of the Anizazi culture from Canyon de Chelly, dating to about AD 600. Cardiomegaly was obvious and right heart failure was manifested by ascites. Lung fragments from tissue retrieved from an isolated canopic jar demonstrated microscopic presence of intra-alveolar fluid and protein that might have been the result of pulmonary edema secondary to left ventricular failure, though an inflammatory origin could not be excluded (Walker *et al.*, 1987).

Pulmonary diseases

Of all the viscera, the lungs (or at least part of them) are most frequently preserved in mummified remains.

This is most probably due to the small fraction of their mass represented by epithelial cells, because epithelium is richly endowed with lysosomes filled with hydrolases that initiate the autolytic decay process. The rate of desiccation is also enhanced by the enormous collective surface area of the pulmonary capillary bed and the relative paucity of chest wall soft tissues.

Congenital conditions

Bronchogenic cyst

Derived from primitive foregut, these cysts are usually found near the tracheal bifurcation. The wall is composed of bronchial elements, though cartilage may not be prominent. Accumulated mucus secretion distends the lumen. This lesion has not been reported in archaeological remains.

Bronchopulmonary sequestration

This malformation is composed of pulmonary tissue located either within normal lung or adjacent to it, but whose bronchi and vasculature usually are separate from the normal lung. In such cases the bronchial secretions accumulate, since they cannot drain via the normal bronchial system. The mucus-distended bronchi expand until they dominate the gross structure, compressing the softer intervening structure into a mass of scar. The anomaly has not been identified in mummified tissue.

Pulmonary infections

Pneumonia

I generate a working diagnosis of pneumonia upon finding a thickened or consolidated area in a desiccated, mummified lung. Validation of that approach was achieved by reviewing more than 250 autopsy records in a community hospital for the years 1925 through 1928, a period prior to the availability of effective pneumonia therapy. Only 5% of pulmonary areas of consolidation or masses proved to be noninfectious lesions, most commonly pulmonary infarcts and metastatic carcinoma (A.C. Aufderheide, unpublished results). Thus, a consolidated area of lung bears a 95% probability of representing pneumonia. A diffuse area of pulmonary adhesions may prevent postmortem collapse of the lung, but will not result in consolidation.

Tuberculosis

Tuberculous changes may also be recognizable. The characteristic appearance of a primary tuberculous Ghon complex was obvious in the lung of an adult female Andean mummy from about AD 1000 (Salo *et al.*, 1994). A pulmonary hollow, peripheral, calcified nodule was associated with two enlarged hilar lymph nodes (Fig. 8.8). Zimmerman (1979) identified pulmonary adhesions with recent pulmonary hemorrhage and found acid-fast bacilli in vertebral bone from an Egyptian mummy. Allison *et al.* (1973) were able to demonstrate acid-fast bacilli in soft tissues (including miliary tubercles) of several Andean mummies, as did García-Friás (1940), though some question the pre-Columbian context of the latter. Thus it appears that ancient tuberculous lungs manifest the fibrotic and calcifying changes with which pathologists examining modern lungs are familiar, but without the inflammatory cellular infiltrate. More recently Salo *et al.* (1994) isolated a segment of DNA unique to the *Mycobacterium tuberculosis* complex from the partly calcified hilar lymph node of a 1000-year-old Andean mummy. In future studies the methods of molecular biology can be expected to expand greatly the number of etiological agents responsible for deaths in antiquity. In the New World, soft tissue tuberculous infections date back to only about AD 300, but in the Old World it has been identified in Egyptian New Kingdom mummies. Excellent summaries of this disease's antiquity and history can be found in Morse (1961), Buikstra (1977, 1981), Manchester (1983:39–41), Buikstra & Williams (1991), Strouhal (1991) and Johnston (1993). Walker *et al.* (1987) identified foci of calcification in tissues removed from Egyptian canopic

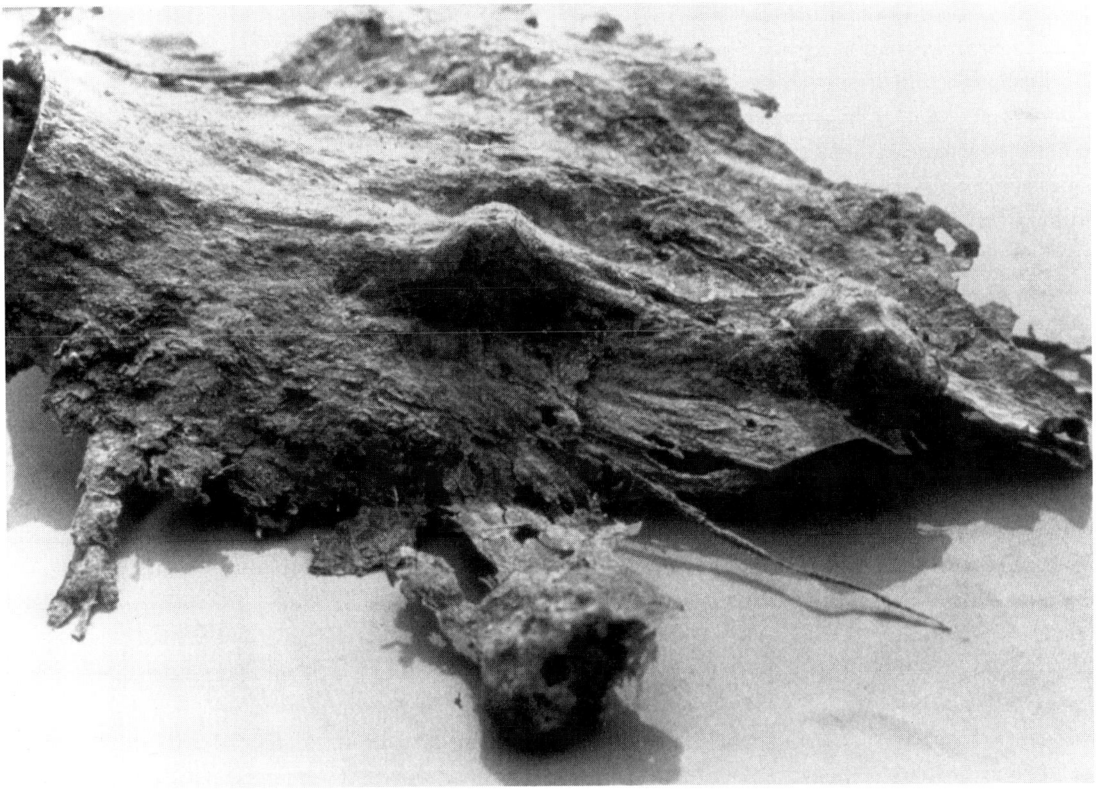

Fig. 8.8. Primary tuberculosis (Ghon complex); SM, AD 1000.
The nodule at the bottom of the illustration is a 1.2 cm calcified shell located in the subpleural area of pulmonary parenchyma and represents the residuum of the initial (now arrested) pulmonary tuberculous infection. The two mounds present in the mediastinal tissues (upper half of the photograph) are partly calcified hilar lymph nodes, enlarged because tubercle bacilli had drained into them from the primary focus in the lung and infected these nodes. From an extract of one of these nodes, DNA characteristic of the tuberculosis-causing bacterial complex was isolated and sequenced. Adult 30–35-year-old female, Peru. (PBL.)

jars, and felt that they probably represented earlier tuberculous infection.

Fungal infections

Since at least several common fungal agents such as *Blastomyces dermatitidis* and *Coccidioides immitis* would be expected to retain their diagnostic structure in tissues, one would anticipate reports of their presence in the lungs of occasional ancient bodies. Separation of the common saprophytes contaminating postmortem archaeological tissues is not usually difficult. However, few published cases of pulmonary infections by pathogenic fungi can be identified.

Aspergillus sp. can colonize lung cysts, abscesses or bronchial mucosa, particularly in previously diseased pulmonary parenchyma. Horne (1995) reported findings in an Egyptian mummy with *Aspergillus* sp. lining a lung cavity. A newspaper article (*Vancouver Sun*, December 20, 1995) stated that *Aspergillus* sp. was recovered from the lung of the "Iceman" (Tyrolean glacier mummy examined at Innsbruck, Austria) but at the time of this writing this information has not appeared in a scientific journal. Harrison *et al.* (1991)

found *Coccidioides* sp. in the bone of an archaeological skeleton from the southwestern USA demonstrating the lesion of Pott's disease. In view of the current hazard of infection by this organism following exposure to dust inhalation in this geographical region, the lungs of mummies from this area should be examined with special focus on fungal infection. Allison *et al.* (1979) reported a case of pulmonary paracoccidioidomycosis in a northern Chile mummy and demonstrated the unique, "spokewheel" structure of the organism in tissue sections.

Fornaciari & Marchetti (1986) and Fornaciari *et al.* (1989b) have identified smallpox virus in tissues of a medieval body in Italy, confirmed by immunostaining, and Dalton *et al.* (1976) have published an electron micrograph of a mummy kidney tissue section resembling a virus structure, but to date no convincing case of virus pneumonia in an ancient body had been reported.

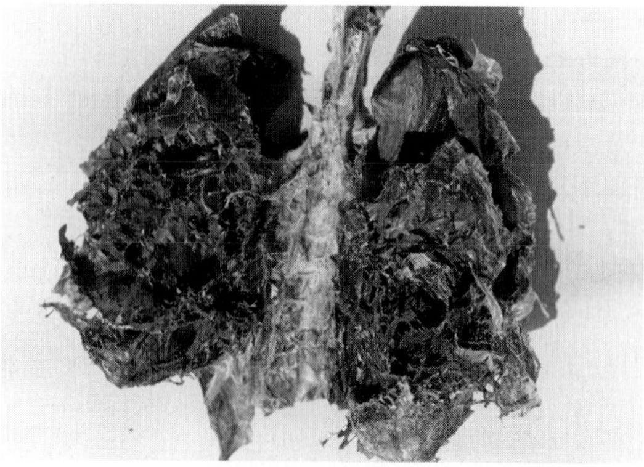

Fig. 8.9. Pulmonary emphysema; SM, AD 450–800. Bullous (cystic) changes of varying degree are scattered throughout both lungs. Adult male from northern Chile. (PBL.)

Degenerative conditions

Emphysema

The lung is masterfully constructed to effect gas exchange with the blood efficiently. As the bronchi extend into the lung periphery they bifurcate into rami of ever-diminishing diameter, terminating as a series of blind-ended air sacs (alveoli). The walls of these alveoli are thin and fragile, occupied principally by thin-walled blood vessels and capillaries. It is here that gas exchange between inspired air and circulating blood occurs. Indeed, functionally a lung can be conceptualized as a mass of capillaries suspended in air, maximizing the surface area of such gaseous movement.

Emphysema is characterized by destruction of such alveolar walls, together with the capillary normally present in that wall. The two alveoli adjacent to the destroyed wall thus become a common chamber at the expense of reduced vascular surface area secondary to the loss of the capillary in the destroyed wall. Slow progression of this diffuse process eventually reduces the pulmonary vascular bed to the point of insufficient gaseous exchange to sustain life (pulmonary insufficiency) or places an intolerable strain on the heart's right ventricle, which is obligated to perfuse the damaged lung (right heart failure).

While destruction of shared alveolar walls begins at the microscopic level, progression of the process eventually can produce easily recognizable, balloon-like cystic, air-filled structures called "bullae" that may measure 10 or more centimeters in diameter (Fig. 8.9). The end result, however, varies in structure ranging from the bullous form to that composed principally of multiple diminutive cystic areas that are quite subtle visually and not easily defined without magnification.

The etiology and mechanism of the process is not understood satisfactorily. Lung structure integrity is viewed as the result of a balance between a certain tissue-destructive enzyme (elastase) released by inflammatory cells and the presence of a compound in the blood (alpha-1 antitrypsin) that normally inhibits this enzyme's action. Imbalance resulting from either an increase in elastase production or a decrease in the amount of inhibitor can result in destruction of the alveolar wall by the elastase.

The most common form of emphysema is related to chronic smoke inhalation. The irritative smoke effect on tissues attracts a surplus of inflammatory cells

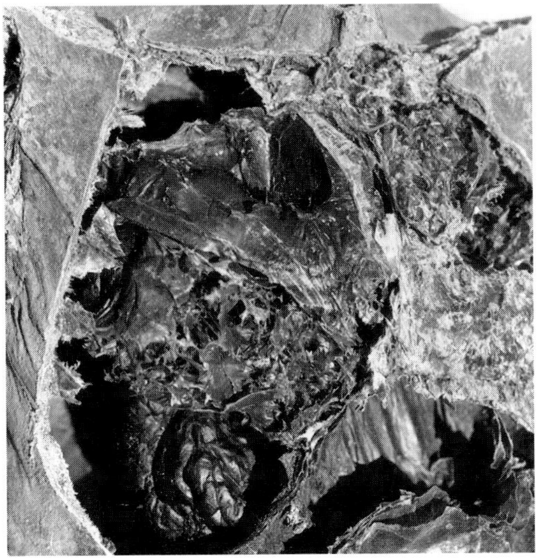

Fig. 8.10. Pulmonary emphysema; SM, AD 500–800. Almost the entire right lower lobe (arrow) has been converted into a huge, single, bloodless translucent bulla. The middle and upper lobes show similar but smaller cysts, exposed in this photograph by removing the overlying pleura that had made the lung adherent to the chest wall. Adult male from northern Chile. (PBL.)

whose high content of elastase overwhelms the inhibitor. Another, less common mechanism involves the mutation of the gene responsible for the production of the inhibitor. Individuals homozygous for such defective alleles may produce little effective alpha-1 antitrypsin, permitting the unopposed elastase to carry out its destructive action.

The gross appearance of the bullous form of emphysema can be quite dramatic. For example, the entire right lower lobe in Fig. 8.10 has been converted into a single, large bulla without any internal structure. Differentiation of the two types of emphysema can be effected by determining the presence of soot particles in a microscopic section of lung tissue. This can not be evaluated with certainty grossly because hemoglobin breakdown products can impart a black color to the tissue that is just as intense as that due to extensive carbon deposition from chronic smoke inhalation. Emphysema in the absence of substantial carbon deposition suggests the possibility of alpha-1 antitrypsin deficiency. In such an instance, liver histology may be helpful. The inhibitor is formed in the liver but not released from the liver cell to the blood. Liver cells become engorged with the retained inhibitor where it may be demonstrable with immunostaining methods. In later stages, the injured liver cells may die and be replaced by scar, leading to cirrhosis of the liver.

In addition to these two principal forms of emphysema, any chronic, scarring lung disease can distort bronchial air flow sufficiently to result in multiple areas of hyperinflation of pulmonary tissue. Focal areas of localized emphysema are also common at the periphery of areas of pneumonia or calcification. These only rarely create health problems.

Primary emphysema has no lengthy historical record. It was not described until the publication of René Laennec's 1826 edition of *Treatise on Diseases of the Chest* (Knudson, 1993), and our understanding of the pathophysiological processes is still evolving. It is clear, however, that the disease is much older than its historical documentation. Allison 1984a:522, (Table 20.7) lists emphysema as a common finding in spontaneously mummified bodies from northern Chile, but it is not clear what fraction of these represent the focal, relatively benign form. Their illustration of the lungs of an 11-year-old spontaneously mummified child from a northern Chile member of the Cabuza culture (AD 450–1000), however is unmistakably an advanced, bullous form of emphysema, most probably (based on his age) a severe form of alpha-antitrypsin deficiency (Allison, 1984b). A moderate degree of emphysema was identified in an Egyptian early New Kingdom mummy (Shaw, 1938). Tissue from Egyptian canopic jars also demonstrated emphysema (Walker *et al.*, 1987).

Pneumoconioses: anthracosis

With every breath we inhale dust grains, germ-laden moisture droplets and other particles, many of which may be potentially harmful. The lung has several defense mechanisms that are so effective they maintain a microbiologically sterile pulmonary environment. The most important of these is the mucosal lining of the bronchi. These cells not only secrete a mucus blanket but, using a surface-located, whip-like organelle (cilium), produce a constant upward, cleansing flow of mucus. Most inhaled particles adhere to this sticky mucus and are effectively removed. The process can, however, be overwhelmed. Frequent smoke inhalation introduces so many particles that some of them reach the walls of the lung alveoli. Smoke particles vary enormously in their composition and size, but most are carbon-containing compounds. While inflammatory cells can ingest, liquefy and remove many other foreign inhaled particles, the body has no effective method for disposal of most smoke particles. Many of those brown or black particles that manage to reach the alveolar walls are retained there; others are carried to lymph nodes in the hilum where they become trapped permanently. Continued smoke inhalation thus results in a cumulative lung burden of smoke particles that eventually discolor the lung. The quantity of accumulated smoke particles (and, therefore, the degree of dark discoloration) is a measure of past smoke exposure. This process is termed "anthracosis" (Fig. 8.11). An alternative source of black discoloration of lung tissue is coal dust, usually inhaled by miners or as a result of other industrial coal exposure.

Although wood smoke includes an enormous range of chemically heterogeneous particles, only a minority invoke locally toxic effects. Some coal miners, for example, may have completely blackened lungs that, however, function normally. The term "anthracosis" is limited to deposition of carbon-containing pigment in the lung, and does not necessarily imply limitation of pulmonary function.

In other circumstances anthracotic pigment deposition is accompanied by reactive and destructive scar formation. Whether this is due to the deposited black pigment or whether another, different and toxic substance was inhaled and deposited simultaneously is not always obvious. Some coal dust, for example, contains quartz, some forms of which are known lung toxins

Fig. 8.11. Anthracosis: pleural adhesion; SM, 2000 BC. The left lung is adhesive to the chest wall over a broad area, preventing its postmortem collapse. The dark color proved to be anthracosis histologically. Twenty-two-year-old male from northern Chile. (PBL.)

(see Pneumoconioses: Silicosis, below). Slatkin *et al.* (1978) found the stable carbon isotope ratio of pulmonary anthracotic pigment to be distinctly lower than that of endogenous tissues, indicating its exogenous origin (Fig. 8.10).

Simple anthracosis is a common finding in many mummies (Brothwell *et al.*, 1969; Zimmerman, 1977; Walker *et al.*, 1987). This is probably a reflection of poor ventilation in rooms with open hearths. Occasionally, however, additional smoke exposure may be related to rituals. Furthermore, scientists at the General Electric Environmental Science Laboratory found no dioxins or dibenzylforans in mummy lung "soot", suggesting that it was not derived from burning wood (*Wall Street Journal*, 24–25 October, 1989). The seal oil commonly used by arctic natives for both light and heat is also notorious for generating a smoky flame, and all such native lungs examined by me have been profoundly anthracotic (Zimmerman & Aufderheide, 1984), some of which have been accompanied by pulmonary fibrosis (Zimmerman, 1986a). Shaw (1938) also found it in a New Kingdom mummy.

Pneumoconioses: silicosis

Sandy soils contain abundant silicon, most of which is in the silicon salt form (silicate), but some commonly is combined with oxygen as silicon dioxide (silica) When inhaled and deposited in the lung, silica has a local toxic effect causing tissue necrosis that heals by fibrosis. Lesser exposure can riddle the lung with isolated, fibrotic nodules, but as these enlarge with further silica accumulation they coalesce to produce broad areas of fibrous replacement of lung tissue. Though the lung is more tolerant of silicate than of silica, even the silicate form can produce a diffusely fibrotic effect if the deposit is of major degree (Sherwin *et al.*, 1979).

Given the inevitable dust exposure suffered by most agriculturalists, it is not surprising that varying degrees of pulmonary fibrosis secondary to silica or even silicate deposition have been reported in mummy tissues. Since such aboriginals commonly are also exposed to smoke, the combined effect may be a dark and fibrotic lung (anthracosilicosis). An example of such a seriously diseased lung is that of a pre-Columbian adult male from the southwestern USA in whom it was considered to be the cause of death by el-Najjar *et al.* (1985). Zimmerman (1977) has observed this condition in an Egyptian mummy and also in Peruvian (Zimmerman *et al.*, 1981b) and arctic mummies (Zimmerman, 1995: and personal communication). Tapp *et al.* (1975) described it in an Egyptian mummy, calling it sand pneumoconiosis. Similar findings were noted in 26 mummified lungs from northern Chile (Abraham *et al.*, 1990). The lesser degree of reactive fibrosis to the silicate inhaled from the Negev Desert dust has been attributed to changes in the crystal surface resulting from weathering, rendering it less tissue-toxic (Pollicard & Collet, 1952; Bar-Ziv & Golderg, 1974) and leading these workers to suggest the term "simple siliceous pneumoconiosis". However, Sherwin *et al.* (1979) found a broad range of fibrosis, including severely destructive forms, in southern California farmers with pulmonary silicate deposition. Whether these different observations reflect differing soil composition (perhaps with varying silica content) or some other unidentified factor is not yet obvious. Munizaga *et al.* (1975) also described the condition in sixteenth century Chilean natives whom the Spanish compelled to work in dusty mines.

The diagnosis is not too difficult to establish in mummified lungs. The black color itself is suggestive of anthracosis, though it should be verified histologically because hemoglobin breakdown products can also blacken a lung. While silica itself (silicon dioxide) is not refractile when viewed by polarized light, silicate crystals are. Silicates are invariably inhaled and deposited with silica, and they are therefore of "marker" value for the etiology of the fibrotic effects. One important interpretive caveat is essential: bodies buried in sandy soils commonly accommodate the postmortem aspiration of sand into the bronchi. This can extend down into remarkably small bronchi. Such a postmortem artifact must be differentiated from dust silicates inhaled and deposited in lungs antemortem. This can usually be achieved by evaluating two criteria: silicates penetrating the lungs of a buried body postmortem will be characteristic of the burial soil (i.e., the range of crystal sizes will be great), in contrast, only the finer particles (usually 5 μm or less) of silicates will find their way into the tissues during life. The other criterion is location: postmortem sand contamination is found within bronchial lumens (though in histological sections the microtome blade may drag some of this into the adjacent tissues), while antemortem silicate deposits are primarily distributed within the pulmonary fibrotic areas. This source of contamination has also frustrated my effort to quantify the silica in fibrotic, mummified lungs. On the other hand, Abraham *et al.* (1990) describe a correlation between silica content and degree of pulmonary fibrosis in lungs of mummies from the Atacama Desert. These differing results may be related to selection of mummies for study and the anatomical portion of lung sampled.

Shaw (1981) cites Collis (1915) as suggesting that the modern curse of "knapper's rot" (silicotic pneumoconiosis occurring in contemporary industrial knappers who shaped flint fragments into gunflints) also must have affected prehistoric populations who found, mined and knapped flint from these same mines making this the world's oldest occupational disease. However, Shaw (1981) points out that modern miners and knappers only created enough dust to be fibrogenic if they employed modern metal tools in unventilated enclosures, conditions not existing in ancient times.

Pneumoconioses: asbestosis

The protein and iron-containing, coated, asbestos fibers called "ferruginous bodies" could be expected to retain their appearance in mummified lungs, as would the often prominently thickened pleura found in asbestos-exposed tissues. No such findings have been reported, including one study in which the investigators specifically sought them in vain (Abraham *et al.*, 1990). These observations emphasize the broad range of asbestos contamination of soils in different geographical areas.

Vascular conditions

Pulmonary embolism and infarction

The process of tissue necrosis following abrupt interruption of its blood supply (ischemia) is called infarction. If the individual survives the ischemic event, the infarcted tissue undergoes autolytic liquefaction, collapse, at least partial resorption and is replaced by a linear scar. Lung tissue, however, has two blood sources: the principal one is composed of pulmonary artery branches while the other consists of multiple branches from the aorta directed to the bronchi themselves. Thus, if a pulmonary artery branch is suddenly occluded, the lung tissue it normally supplies may become infarcted. However, the bronchial arteries supplying the bronchi of the infarcted lung area remain open and continue to pump blood into the area. Within 48 hours the necrotic vessels in the infarcted lung tissues will leak the blood from the bronchial artery supply, converting the dying lung tissue into a hemorrhagic mass. As stated above, up to 5% of modern pulmonary consolidation areas are infarcts (and, except for neoplasms, almost all the others represent pneumonia). Such infarcts in mummies can expect to be discrete masses without demonstrable integrity of internal structure. The blood would make the affected area dark or even black. Red blood cells might or might not persist in recognizable form, but bacteria or fungi should be absent. In spite of its modern frequency and the favorable rate of lung preservation in mummies, to date no pulmonary infarct in mummified human lung has been reported. The most likely reason for this is that such lesions are being misinterpreted as pneumonia.

The usual cause of a lung infarct is an embolus. This is a clot, commonly formed in a leg vein, that becomes detached, is swept "downstream" to the right atrium, then the right ventricle and finally enters the pulmonary artery where it is passed into successively smaller branches until it lodges in one slightly smaller than the clot itself. The clot is called a thrombus while still attached at its original site of formation in the leg vein, but called an embolus after it becomes detached and transported to the lung. The lung may survive occlusion of a diminutive arterial branch but become infarcted if a larger branch is blocked. On the other hand, if the embolus is huge (several centimeters or more in diameter) it may block the main pulmonary artery, causing sudden death. Given the brittle nature of the main, desiccated arterial trunks in mummified bodies, it would be difficult to demonstrate the presence of such a fatal embolus in the pulmonary artery, particularly since desiccation and at least partial autolysis would shrink it substantially. No such reports of paleopathological infarcts are identifiable.

Atherosclerosis

Atherosclerosis of the main pulmonary artery occurs only when the pulmonary systolic blood pressure (normally about 20 mm Hg) rises above 60 for a prolonged period. If present the atheromas can be identified easily – they are identical to the much more common aortic atheromas. Their presence is proof of severe, chronic pulmonary hypertension, such as occurs either in chronic fibrotic pulmonary diseases or is of an idiopathic nature. Pulmonary artery atheromas rarely ulcerate or calcify, perhaps because the right ventricle cannot withstand pressures of this magnitude long enough. The accompanying thick-walled small pulmonary arteries and arterioles may be demonstrable histologically. Pulmonary artery atherosclerosis has not been reported in mummified bodies.

Neoplasms

Hamartoma

Few benign tumors occur in the lung. The most common of these is a hamartoma. By definition, a hamartoma is a benign tumor composed of cells normally present in the organ in which it is found but arranged in architectural disarray. Thus we could expect to find such a tumor made up of cells of bronchial mucosa, smooth muscle, cartilage and blood vessels. These are, indeed, found in lung hamartomas but cartilage is usually the dominant tissue and is easily recognizable. Since cartilage survives the postmortem environment better than epithelium, lung hamartomas could be encountered at mummy dissection. It usually presents in adults as spherical, solid masses up to 5–6 cm in diameter in any part of the lung. Its gross appearance depends on its structural elements but most commonly resembles a chondroma. It rarely impairs the lung function in any manner. It is, however, a quite uncommon tumor.

Brochogenic carcinoma

Most other lung tumors arise from bronchi (bronchogenic carcinoma). While these cancers are reaching epidemic proportions in modern developed countries, the bulk of them have been linked statistically to cigarette smoking. Hence their frequency in antiquity would be expected to differ little from that in modern communities where cigarette smoking is uncommon. Such populations rarely suffer from bronchogenic carcinoma. The occasional cancers of this type in such groups are probably the result of inhaling other carcinogens (radon gas, heavy metal mining dust, aerosolized asbestos and many unknown substances). Bronchogenic carcinoma tends to grow into the bronchial lumen, blocking it and causing pneumonia in the portion of the affected lung. Simultaneously and/or alternatively invasion through the bronchial wall into the pulmonary parenchyma is also common. In these, tumor cells enter the lymphatics and spread to hilar lymph nodes and later to cervical nodes. They also penetrate bronchial veins, where the detached cells are swept and transported by the blood to any organ of the body. The brain, adrenals and bones are particularly common metastatic sites, though no area or structure is immune to spread from bronchogenic carcinoma.

Mesothelioma

An extremely rare malignant tumor can arise from the lung pleura: mesothelioma. This usually quite fibrous tumor spreads diffusely and envelops the lung in a thick rind of white, usually quite fibrous, tumor as much as 6–8 cm thick. It is found almost exclusively in persons working with asbestos and is viewed as a malignant response to the inhalation of asbestos fibers. Hence encountering this tumor in ancient bodies would be expected to be exceptional.

Most of these tumors occur in the older age group after many years of exposure to the offending carcinogen. Since the malignant lung cancers today have a low rate of cure, the natural history we associate with them in modern times is probably quite similar to the occasional lung cancer that occurred in antiquity.

If a lung tumor is encountered in mummified human remains and it is not a hamartoma, it is more apt to be a metastatic carcinoma to the lung from an extrathoracic primary source. Osteosarcomas and other soft tissue sarcomas commonly metastasize to the lung, as do primary visceral cancers late in their course. If more than one tumor is found in the lung, the probability of its metastatic nature is high. In contrast to primary bronchogenic carcinomas, metastases are more apt to be spherical and multiple, with reasonably sharply defined periphery unless extremely large.

Foreign bodies

Probably the most common material aspirated in modern times is food. Precipitating factors include malfunction of the swallowing musculature secondary to cerebral disease (hemorrhage, infarction, neoplasm, demyelinating diseases, etc.) or alcoholic intoxication. Local factors, such as dentures, may play a role. In antiquity some of these factors may have operated as well. The bronchi of a spontaneously mummified 22-year-old male from northern Chile were found by me to be filled with masticated food items; the absence of pneumonia implied this was a terminal event. Fatal complications probably abbreviated the lifespan of individuals with chronic conditions. Nevertheless several reports document aspiration of nonfood items in prehistoric bodies. Zimmerman & Smith (1975) found the trachea and major bronchi obstructed by aspirated moss in a 1600-year-old arctic female body. She had apparently been buried in a mud-slide, a fairly common arctic phenomenon in terrain of the type present on St Lawrence Island, where the body was found. Allison *et al.* (1974b) found that pneumonia in the mummified remains of an Andean adult male from AD 950 was due to bronchial obstruction by an aspirated tooth. Allison (personal communication) also found a lithic point embedded within the lung of a 40-year-old northern Chile mummy from about AD 1350 without evidence of an entry wound, suggesting the injury had been incurred a long time previously. El-Najjar *et al.* (1985) also reported a probable tooth aspiration in a child mummy from the southwestern USA. The reconstructed trajectory of an arrowhead found in the Tyrolean Iceman indicates that it had penetrated the lung, probably with fatal consequences (Richburg, 2001).

Liver and biliary tract diseases

Preservation and appearance

The liver is essentially an epithelial organ. After death the prompt autolytic changes in hepatic cells often delete antemortem patterns of primarily hepatic epithelial cells, such as hepatitis. The liver at time of death weighs about 1500 g in an adult, but in desiccated cadavers weights of 100–200 g are common. While the liver normally is quite well anchored to the diaphragm and lower mediastinal tissues, degenerative postmortem changes abetted by gravity secondary to burial position can result in detachment of the liver from these structures. This can result in displacement of the liver into the pelvis, the left thoracic cavity or elsewhere.

While the hepatic cells usually degenerate rapidly, the fibrous portal skeleton of the liver commonly persists. Thus antemortem conditions characterized by portal area alterations may remain recognizable. These include the various forms of cirrhosis. Evaluation of portal patterns in mummy livers, however, requires awareness of a possible common misinterpretation. In most mummies much of the degenerated hepatic cell debris remains (sometimes even the profile of the former living cells is still apparent), and in these circumstances the relationship of the portal areas to the hepatic cellular tissue can be judged reliably. In some bodies, however, the hepatic cell degeneration proceeded to liquefaction before desiccation occurred. Accordingly, the surviving fibrous portal areas collapse, closely approximating each other sufficiently to create the illusion of a pattern of interconnected, proliferative portal areas suggestive of cirrhosis. Simple awareness of this form of pseudopathology is usually sufficient to allow its recognition. Indeed, the gross appearance of a light brown, spongy structure (instead of the usual firm, black, often brittle appearance of a normal, desiccated liver) can anticipate the microscopic appearance of this phenomenon.

Hepatitis

For reasons given above it is not likely that a diagnosis of hepatitis can be made on the liver of a desiccated mummy. A possible exception could be an Egyptian mummy that was promptly eviscerated and the liver placed in a canopic jar with sufficient natron and/or resin to effect rapid desiccation. In chemical or thermal forms of mummification, sufficient liver cell survival to allow diagnosis is also conceivable. However, I am not aware of any reports of hepatitis identification in the paleopathological literature.

Cirrhosis

Bile formed by liver cells is drained by a network of bile ducts at the periphery of the individual lobules of liver cells. Normally these are enveloped by a thin layer of fibrous tissue also containing blood vessels and collectively these are termed a "portal area". Chronic liver cell injury and destruction due to toxins (alcohol), microbiological agents (hepatitis viruses) or even autoimmune mechanisms is often followed by a proliferative, fibrous, scarring response in the portal structures surrounding the lobules of affected liver cells. Given enough time the portal areas enlarge substantially, intercommunicate freely and finally expand to an extent that virtually surrounds such liver lobules in a fibrous envelope (cirrhosis). The resulting antemortem gross appearance of such a cirrhotic liver is striking: a web of sunken, retracted fibrous bands with bulging masses of liver cellular tissue (representing compensatory hyperplasia) between the fibrous cords (Fig. 8.12). While every clinical pathologist is familiar with this picture, it must be emphasized that the bulging hepatic cell masses so prominent before death can be expected to have collapsed in the mummified body because of the postmortem epithelial cell degeneration. The gross cirrhotic changes may thus be too subtle for certain diagnosis. However, the persistence of fibrous tissue in desiccated cadavers usually permits ready recognition upon histological examination. Pathologists examining the liver of an Egyptian mummy, Nakht, noted some surface irregularity but were not certain of the diagnosis of cirrhosis (in that case secondary to schistosomiasis) until it was studied in microscopic sections (Reyman, 1977).

The antiquity of cirrhosis is considerable. Chen & Chen (1993) note that Erasistratus (300 BC) associated abdominal fluid (ascites) with liver disease and that Aretaeus the Cappadocian in the third century AD recognized that cirrhosis could follow hepatitis and that carcinoma could arise in a cirrhotic liver. They also point out that the British physician Matthew Baille in 1793 recognized the role of alcohol in the production of this condition and that René Laennec assigned it the term "cirrhosis". In addition to the Egyptian mummy reported by Reyman (1977), cirrhosis has also been reported in two medieval mummies from Naples (Fornaciari *et al.*, 1988b).

It is unlikely that tissue preservation in mummified livers will permit the separation of alcoholic or post-viral cirrhosis from some of its less frequent forms (primary biliary cirrhosis or that following primary sclerosing cholangitis), since these differentiations are usually dependent on the demonstration of various patterns of inflammatory cells that only rarely survive the mummification processes. However, the cirrhosis representing a late complication of common bile duct obstruction (see below) should be recognizable in ancient bodies, but remains unreported to date.

Fatty degeneration of the liver

Accumulation of fat droplets within the cytoplasm of liver cells is frequently seen in alcoholics, certain types of toxin exposure and especially in circumstances of severe malnutrition. The fat per se is not injurious but the conditions producing it frequently are. It is a readily reversible phenomenon if the primary condition is treated appropriately. Its recognition in ancient

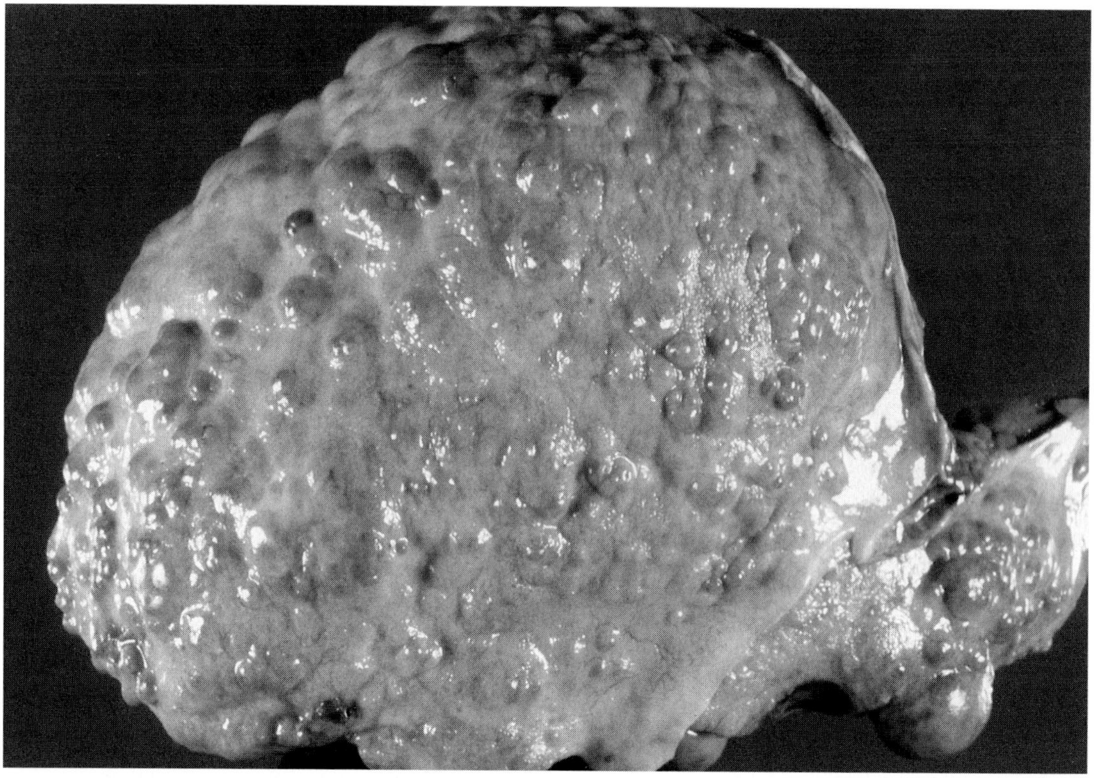

Fig. 8.12. Cirrhosis of the liver.
The knobby surface of this liver is composed of regenerative nodules of liver tissue enveloped by dense scar. Modern adult male. (PBL.)

bodies could serve usefully as a guide to the probability of a serious state of malnutrition before death. If present in a substantial fraction of a studied population, it might signal famine conditions. Since this condition is so frequent in malnutrition, including the late circumstances accompanying a fatal chronic illness, the probability of its presence in at least a few of the mummies that have been autopsied is great. Yet this lesion has not appeared in the paleopathology literature. The most probable reason for this failure of recognition lies in the postmortem taphonomic processes. This fat accumulates as triglycerides, and these are normally split into fatty acids by the lysosomal enzymes during postmortem autolytic action. This would render the lipid invisible on histological examination. Thus this otherwise valuable clue is probably lost to us, at least in spontaneously desiccated bodies.

Congenital lesions

Both single cysts and a polycystic form have been reported in modern cases. These ought to be identifiable if occurring in mummified livers (Fig. 8.13). Their flattened epithelial lining, however, would probably not survive. The cysts could probably be separated from artifacts of gas bubbles produced by postmortem growth of bacteria by simply identifying the bacterial presence with appropriate histological stains. Cysts in the region of the extrahepatic bile ducts also occur.

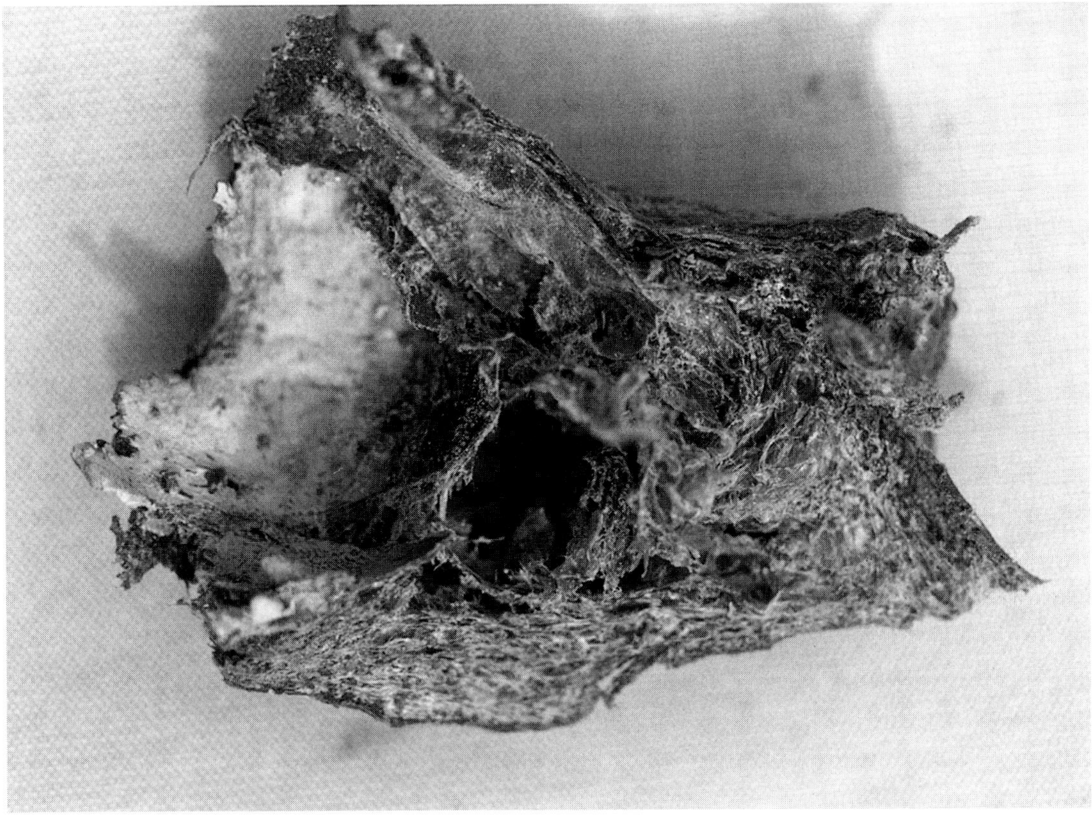

Fig. 8.13. Congenital cystic disease of the liver (juvenile form); SM, AD 450–800.
The larger cyst demonstrated in this sectioned liver is 4 cm in diameter. Multiple others ranging from 1 to 2 cm were present in other areas. Histologically few normal portal areas were identified. Infant (2 years old) from northern Chile. (PBL.)

Gas gangrene septicemia

Clostridium perfringens, a gas-producing Gram-positive bacillus common in soil and even occasionally in normal human feces, flourishes in the anaerobic milieu of the necrotic muscle of a contaminated crushing injury and usually kills the affected individual swiftly by invading the blood stream. Before death the gas produced in the wound by the bacilli will accumulate locally there, but that produced by the organisms in the blood does not accumulate because during the blood's passage through the lungs the gas diffuses into the lung air and is exhaled. However, after death the blood no longer circulates. Bacterial growth for a few hours after death before the body temperature drops too low is common. At the moment of death when circulation ceases, hundred or thousands of bacilli in such an affected person will be present in the blood vessels of an organ as vascular as the liver. The continued production of gas postmortem by these bacteria in the liver will result in its accumulation in the form of trapped bubbles within the liver. Hence the transected liver of such an individual will produce the dramatic and easily recognizable spongy appearance seen in Fig. 8.14.

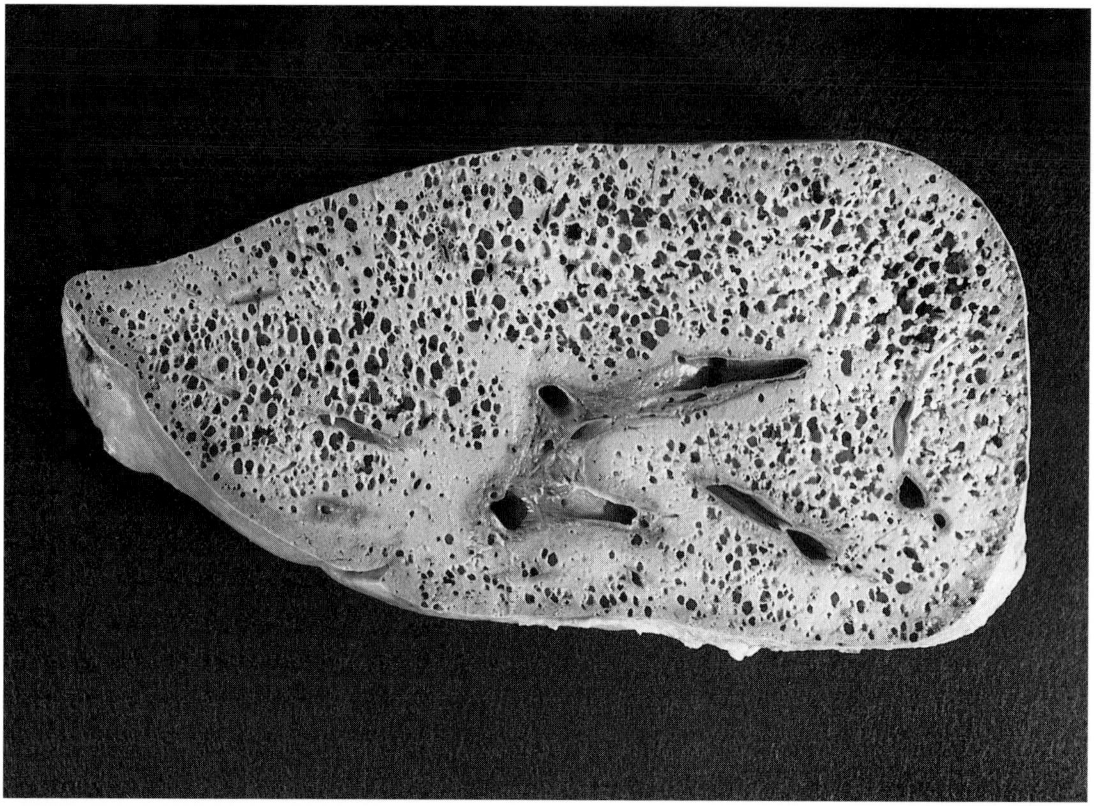

Fig. 8.14. Liver: gas bacillus septicemia.
This individual developed *Clostridium perfringens* septicemia following surgical bone tumor removal from the femur. Gas bubbles formed by postmortem growth of bacteria present in the liver at the time of death produced this spongy texture. Modern adult female. (PBL.)

Abscesses

Single or multiple liver abscesses can occur as a consequence of septicemia by a virulent, pyogenic organism, but more common is the consequence of bowel organisms gaining access to the extrahepatic biliary tract. The most frequent mechanism for the latter is a gallstone impacted at the ampulla of Vater, where the common bile duct drains into the intestine. Such an impaction disturbs the effectiveness of the sphincter muscle there that usually effectively prevents bacterial entry. With this protective mechanism impaired by the gallstone, bacteria may enter the bile duct and ascend rapidly into the smaller biliary ducts within the liver (ascending cholangitis) giving rise to multiple abscesses there. These, in turn, may "feed" their bacteria into the bloodstream, generating fatal septicemia. While this condition was not rare prior to effective antibiotic and surgical treatment, and while such abscesses should not present diagnostic difficulties at a mummy autopsy, I have not identified a published case of this kind in an archaeological body. I have had personal experience in the case of a spontaneously mummified body from a northern Chile coastal population of about AD 1300 in which a lobar pneumonia of the right lung was complicated by an empyema of the right lower thoracic cavity and a fistula which had penetrated the diaphragm near the "bare area" to produce

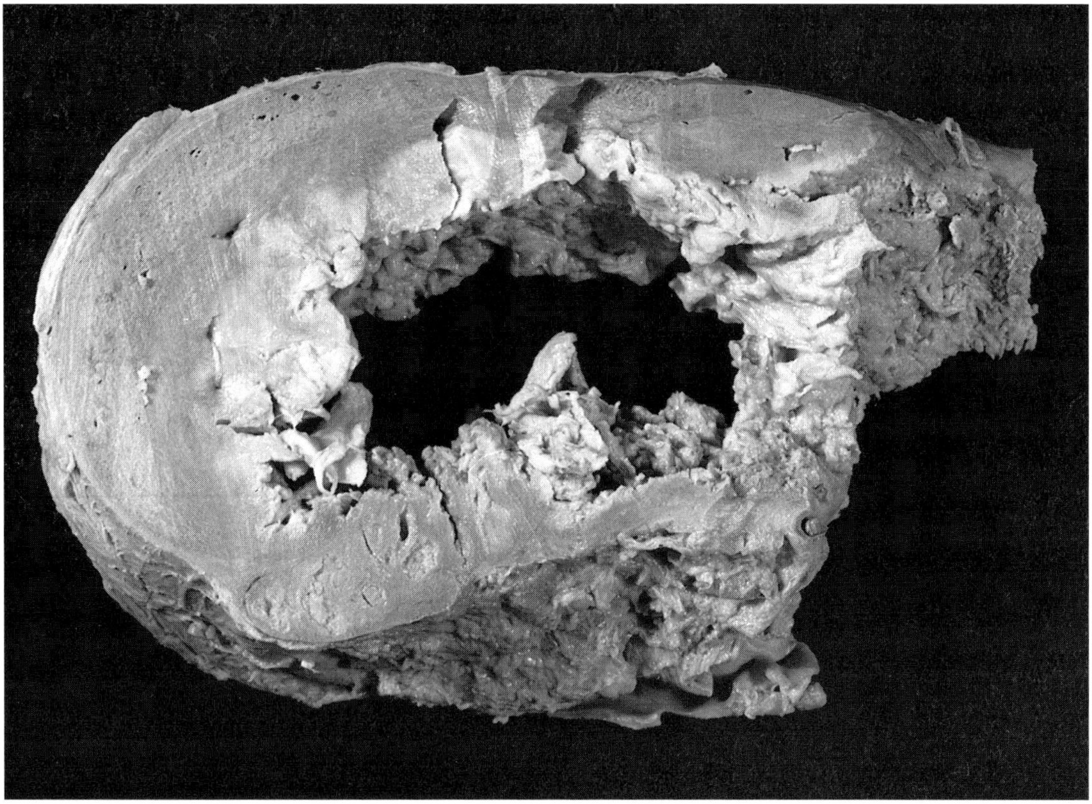

Fig. 8.15. Liver: amebic abscess.
The central cavity represents an area of liquid pus that has drained from this liver. The pale area surrounding it is a mixture of amebae-containing pus infiltrating liver tissue. Amebae infected the colon, entered mesenteric veins and were transported by these and the portal vein into the liver where they flourished and produced this abscess. Modern adult female. (PBL.)

a right subdiaphragmatic abscess penetrating into the liver. Like abscesses elsewhere, such an abscess may or may not retain grossly recognizable pus whose appearance can vary from a thin, light-colored membrane lining the abscess wall to a granular, sand-like material. Bacterial stains usually easily demonstrate the offending organism histologically.

Additional causes of abscesses include parasites. These include amebae (*Amoeba histolytica*) as seen in Fig. 8.15, dog tapeworm (*Echinococcus granulosus*) and the blood fluke, *Schistosoma* sp. The latter does not produce actual abscesses but its ova, deposited in the liver in large numbers initiate a destructive, scarring inflammatory response that results in a cirrhosis-like structure. All of these are discussed in more detail in Rubin & Farber (1994:765–766).

Neoplasms

Primary hepatoma

An occasional benign adenoma composed of liver cells can be found in the liver. The malignant tumor of importance is hepatocellular carcinoma arising from liver cells. This tumor today is of relatively low incidence in the developed world, where most cases arise in cirrhotic liver. In the Mediterranean countries, Africa and especially the Orient it is often the most common cancer in those populations, parallelling the frequency of hepatitis B virus (HBV). It has also been linked to the mycotoxin aflatoxin (Patten, 1981) in India. We do not know the frequency of these agencies in antiquity, but this tumor has not been described in mummified remains.

Metastatic carcinoma

Much more common than any of these today is carcinoma metastatic to the liver from a primary tumor elsewhere, usually the gastrointestinal tract, bronchus or breast. The cancers that are primary in the liver usually present as a large, irregular, knobby mass, while the metastatic lesions commonly are multiple, smaller and often spherical nodules. While bone changes suggestive of metastases have been reported in ancient skeletons, no definite soft tissue liver metastases have been found, although Zimmerman's (1990) experimental work suggests that such tumor cells would probably survive better than benign liver cells.

Biliary tract diseases

Bile serves as an excretory mechanism for cholesterol, bilirubin (a hemoglobin breakdown product derived from destroyed red blood cells) and bile salts (useful for fat digestion in the intestine). In addition, a stabilizing compound, lecithin, is also present. Since cholesterol is not water soluble, the proportions of these four substances in the bile becomes critical in order to keep them in solution. Circumstances substantially altering normal conditions may upset the equilibrium sufficiently to result in supersaturation of one of these substances with the consequence of its crystalization and ultimate formation of a precipitate in the form of a stone (Greek: *lithos*), gallstone or cholelith.

Bile is formed by liver cells, excreted into the smallest of the bile ducts ("canaliculi") that then join others to form ever larger ducts within the liver. These finally emerge from the liver as two hepatic ducts that unite into a single channel – the common bile duct. After traversing the head of the pancreas (where it is joined by the pancreatic duct) the common bile duct empties into the intestine through a sphincter-guarded nipple termed the ampulla of Vater. In about the middle of its course, the gallbladder joins the common bile duct via a smaller one called the cystic duct. Thus, when the sphincter at the ampulla of Vater is closed, bile can "back up" via the cystic duct into the gallbladder, where it can be stored until the next meal is ingested.

Two problems of interest to paleopathology can arise within this system: the formation of calculi (bile precipitates: gallstones or choleliths) and infection of the biliary tract.

Gallstones (Cholelithiasis)

While mixtures are common, gallstones can be classified into two major groupings based on their predominant chemical element: cholesterol or pigment (bilirubin) stones. The frequency of gallstones is age dependent. In the Western, developed, countries cholesterol stones are most common while in the remainder of the world pigment stones predominate. Their frequency at autopsy in aged USA Caucasians is about 25% (Fraire *et al.*, 1988), though some Native American groups are thrice that. Rural, developing countries often have much lower rates. Both environmental and hereditary ("thrifty gene?") factors probably are influential (Neel, 1982).

While gallstones can form within the common bile duct, most occur in the gallbladder, probably at least partly because bile stagnation favors precipitation of bile constituents. Initially these crystals formed are so diminutive that bile merely becomes more viscid ("sludge"). Larger agglutinations may form calculi 0.5–1.0mm in diameter ("gravel") small enough to pass from the gallbladder through the cystic duct and out via the common bile duct. Physicochemical and mechanical factors determine the appearance of larger stones. Occasionally a single stone may be formed, assuming the egg-shape and size of the gallbladder lumen. More commonly multiple calculi averaging

Fig. 8.16. Gallstones (cholelithiasis) and chronic cholecystitis; SM, 2000 BC.
Gallstones with flattened (faceted) surface are present within a gall bladder whose dense 1 cm thick wall has been thickened by fibrotic scar. Sixty-year-old male from northern Chile (Chinchorro culture). (PBL.)

about 1 cm in diameter occur. Forced together by gallbladder contraction, such stones abrade each other, each gradually acquiring multiple flat (faceted) surfaces. Pure cholesterol stones are pale yellow and translucent while pigment (bilirubin) stones vary from brown to black. About 15% of cholesterol stones and 50% of pigment stones contain sufficient calcium salts to be detectable radiologically.

The majority of gallstones within the gallbladder are silent, producing no symptoms. A stone approximating to the size of the cystic duct may be passed with painful difficulty. If a stone becomes impacted in the cystic duct and blocks it, bacteria may reach the gallbladder lumen and initiate a violent, acute infection, turning the gallbladder into a pus-filled bag (empyema) that may perforate, initiating a lethal acute peritonitis. Alternatively, erosion of the gallbladder mucosa by stones within its lumen may generate chronic inflammation (cholecystitis) that may be symptomatic intermittently. A calculus that reaches the common bile duct but is not immediately expelled may continue to expand by further crystal deposition, then pass to the ampulla of Vater, blocking the main flow of bile at this constricted location. Unable to enter the intestine, the bile pressure rises and the small intrahepatic ducts rupture. The spilled bile is absorbed by the blood, where the increase in reabsorbed bilirubin pigment imparts a yellowish-orange color to all body tissues, including the skin (jaundice). Intestinal bacteria commonly also make their way into the biliary tract, causing acute ascending cholangitis.

Aufderheide & Allison (1994) found a thick-walled (1 cm) gallbladder filled with brown and black, faceted gallstones lying loose in the pelvis of a spontaneously mummified body of a north Chile Preceramic coastal population (Chinchorro) of about 2000 BC (Fig. 8.16).

Angel (1973) describes two brown, faceted gallstones found between the ribs and iliac crest of an arthritic male skeleton from Grave circle B at Mycenae, dating to about 1600 BC. Several Egyptian mummies are known to harbor gallstones. Smith & Dawson (1924: 156, Figure 60) demonstrate a gallstone-filled gallbladder adherent to the liver of a priestess of Amen from the twenty-first dynasty, and Gray (1966) identified radiopaque gallstones in X-rays of an Egyptian mummy of unknown provenience curated at the University of Leiden. The body of a noblewoman from the Han Dynasty (about 200 BC) in the Human province of China had multiple gallstones (Wei, 1973). Munizaga *et al.* (1978) found multiple gallstones present in 2 out of 75 spontaneously mummified bodies from a coastal northern Chile population who occupied that area about AD 100–200. A north Colombian mummy (desiccated by exposure to heat from a fire) of the Muisca culture (about AD 1200) was X-rayed by Cárdenas (1994) and demonstrated multiple gallstones. Studying a group of medieval mummies from a cathedral in Naples, Italy, Fornaciari *et al.* (1988b) found cholelithiasis and cholecystitis in a mummy that also demonstrated cirrhosis. In North America, Steinbock (1990) cites a personal communication from C.O. Lovejoy that 6 of more than 1000 burials at the Ohio Libben Woodland (AD 1000–1200) site were associated with gallstones.

Steinbock (1990) has also summarized the earlier

literature in an excellent and well-documented presentation in which he noted that Hippocrates and Aristotle discuss jaundice but do not mention gallstones. Nevertheless, animal gallstones from that general period were not only known but used as a source of yellow pigment. Interestingly Galen (AD 130–201, cited in Glenn, 1971), describing obstructive jaundice, may have mistaken a small gallstone for pomegranate or grain seeds. It was at those early universities at Padua and Bologna that the first accurate and detailed autopsy was performed by a graduate (Gentile da Foligno) of the University of Bologna who carried out the postmortem dissection at Padua in 1341. Not surprisingly, the report includes a description of a gallstone imbedded in the gallbladder's cystic duct (Steinbock, 1990, 1993b). Later contributions by A. Beniveni, A. Vesalius and others finally led to G.P. Morgagni's (1682–1771) classical study of all clinical and pathological aspects of cholelithiasis (Glenn, 1971).

Miscellaneous conditions

Hypervitaminosis A

If enough vitamin A is ingested, carnivore liver contains sufficient of the vitamin to produce hepatotoxicity as well as periosteal stimulation. A pattern of coarsely woven, periosteal new bone formation of the long bones and skull, with randomly placed lacunae found on fossil skeletal parts of a *Homo erectus* fossil from Kenya, East Africa, was felt to conform to the expected pattern for this condition (Walker *et al.*, 1982; Zimmerman, 1990).

Gastrointestinal tract diseases

The organ occupying the most space in the abdomen (gastrointestinal tract) is distinctly underrepresented in the paleopathological literature. Examination of a mummy's gut provides the obvious reason. Most intestinal lesions are a product of the mucosal epithelium, but the mucosa rarely survives the postmortem process. The smooth muscle, collagenous and vascular elements of the wall fare much better, but the frequency of lesions arising primarily in these is low. The mummified intestine is thin, translucent and friable. Tubular segments are common but perimortem compression and fracture from postexcavation handling often prevents detailed examination of much of it. Yet, in spite of all the factors, including its bacterial content, that act to generate postmortem loss of colon tissue for study, recovery of desiccated feces (coprolites) for laboratory study is common. Search for these can be facilitated by awareness that the lower rectum in both living and dead bodies is usually devoid of feces, but that gravitational forces commonly result in prolapse of the often coprolite-containing cecum into the lower pelvis, where it frequently lies on the perineal floor overlying the anus or adjacent to it. Because the types of lesions affecting the various segments of the gastrointestinal tract are quite similar, we will review them by a mechanism of pathological process rather than the more usual practice of dividing the tract into segmental levels.

Congenital lesions

Omphalocele

This is a rare lesion. During fetal development much of the bowel transiently enters the umbilical cord during a rotational process and subsequently re-enters the abdomen. Interference with this process may result in its arrest while the gut is enclosed within the cord's membranes. At birth the infant is born with not only most of the intestine but often also other abdominal organs, most commonly the liver, extruded from the abdominal cavity, covered by thin, translucent membranes from the center of which the umbilical cord emerges. This is not to be confused with a superficially similar-appearing lesion (gastroschisis) that consists of an anterior abdominal wall defect *adjacent* to the umbilicus through which much of the gut herniates, uncovered by a membrane. This can be differentiated by identifying a normal, intact umbilical cord arising from the skin of the defect's edge. While prompt modern surgical care can salvage some of these infants, the mortality is high.

Esophageal atresia

This is also a rare lesion. It usually involves the proximal half. While variations are common, the most frequent form results in the upper portion of the esophagus ending in a blind pouch while the lower esophageal segment originates from the trachea and communicates with its lumen (tracheoesophageal fistula). Pneumonia due to aspiration of food is the principal cause of death in unoperated infants.

Imperforate anus

This is another segment of the gastrointestinal tract that can fail to develop a lumen – in this case that of the cloacal membrane. The complete anal closure causes prompt intestinal obstruction that can be relieved surgically today by anal incision. Atresia of any other level of the gastrointestinal tract may occur rarely, all of which cause intestinal obstruction.

Pyloric stenosis

This is a consequence of deficiency of an enzyme involving the production of nitric oxide. The absence of this endogenous smooth-muscle relaxant results in spasm of the pyloric muscle at the stomach's outlet, preventing the emptying of the stomach contents. It is characterized by the appearance of projectile vomiting, abdominal colic often with a palpable mass (the hypertrophic, spasmodic pyloric muscle) and sometimes visible reverse peristalsis all appearing during about the second week of life. The Ramstedt operation that incises and transects part of the pyloric muscle can cure the condition.

Meckel's diverticulum

A blind-ended outpouching of the gut is called a diverticulum. That described by Meckel is located near the end of the ileal part of the small intestine (Fig. 8.17). In most cases it produces no symptoms and is related to the intestinal movements described for the omphalocele lesion, therefore it may not be surprising that its wall often contains ectopic gastrointestinal tissues. The most common of these include pancreatic tissue and gastric mucosa. The former of these rarely may produce a lump large enough to generate the complication called intussusception (described under Intestinal obstruction, below) but much more common is ulceration of the diverticulum's mucosa from the hydrochloric acid produced by the ectopic gastric mucosa. Erosion of the ulcer into a submucosal artery can produce hemorrhage into the intestinal lumen serious enough to demand surgical excision.

Megacolon (Hirschsprung's disease)

Normal intestinal contractions need to be precisely coordinated in order to produce a progressive, peristaltic wave that effectively propels ingested food along the length of the gut. Such integration is achieved by nerves and their related ganglion cells located in the gut wall. In about 1 in 5000 births (Cotran *et al.*, 1994:786) such ganglion cells fail to develop, most commonly in the rectosigmoid area. The aganglionic segment is therefore incapable of propelling food, and the ingested food accumulates in the colon just proximal to this defective segment, with progressive dilatation of the otherwise anatomically normal part of the colon. Eventually complete intestinal obstruction may occur; in some the dilated colon may become so thinned that it ruptures with resulting (usually lethal) peritonitis. Surgical excision of the aganglionic (not the dilated) segment is usually curative.

Diaphragmatic hernia

Normally the diaphragm effectively separates the abdominal from the thoracic cavities. Developmental failure of one side (usually the left) of the diaphragm permits the stomach, a variable amount of the intestine, sometimes the liver or other abdominal organs to

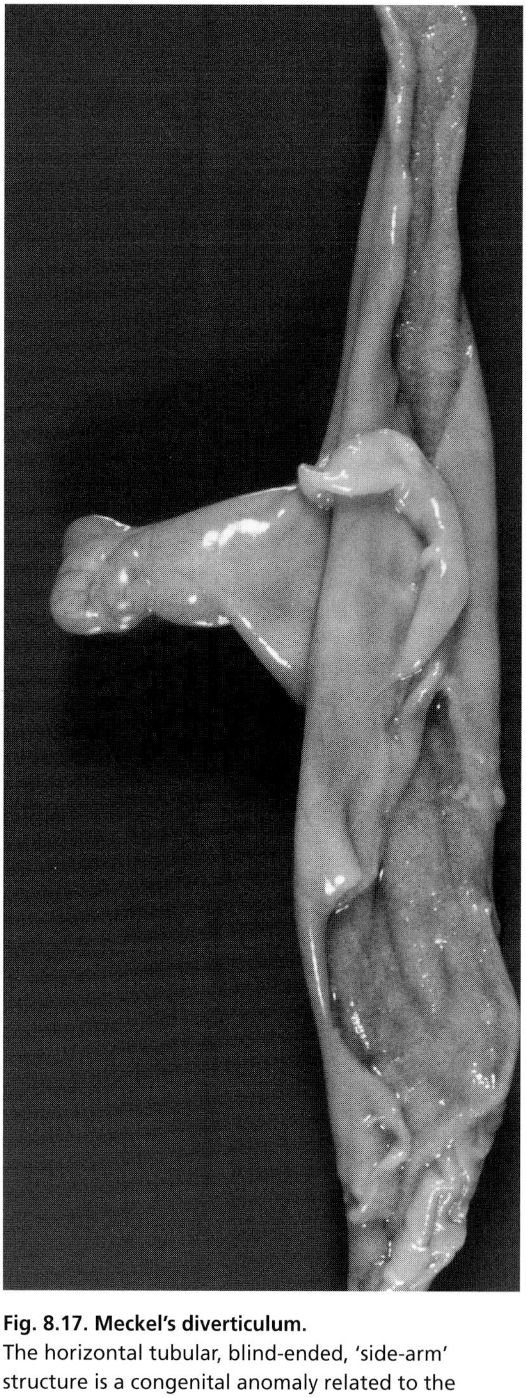

Fig. 8.17. Meckel's diverticulum.
The horizontal tubular, blind-ended, 'side-arm' structure is a congenital anomaly related to the umbilicus. It is of clinical significance because some of these include ectopic gastric mucosa or ectopic pancreatic tissue. Modern adult female. (PBL.)

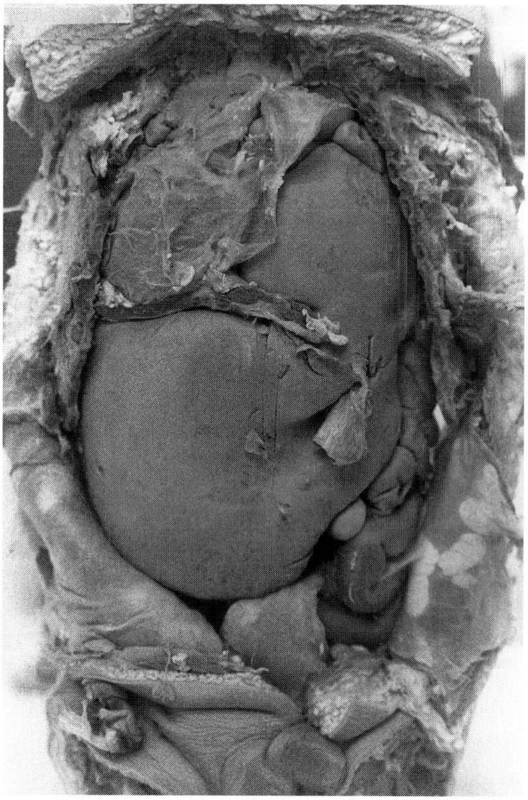

Fig. 8.18. Diaphragmatic hernia.
Congenital absence of the diaphragm permits abdominal viscera to become displaced into the thoracic cavity, inhibiting growth of the lung. Modern infant. (PBL.)

move upward and occupy the thoracic cavity on the affected side (Fig. 8.18). The diaphragmatic defect is believed to be related to premature return of the intestine from the umbilical area during its development. The resulting lung compression may inhibit development of that organ to a degree that can not support life at birth. In some cases the diaphragmatic defect is not total. In these the smaller defect may produce a much lesser mass of herniated structures, producing no serious lung growth interference and these may be compatible with life, sometimes even if unrepaired.

This condition should not be confused with an acquired form called hiatal hernia. The esophagus, of course, must traverse the diaphragm to reach the stomach. As it does so, the diaphragmatic muscle fibers and ligaments are structured to support the esophagus in such a manner as to prevent reflux of the acid stomach secretions into the esophagus. In some individuals the esophagus-enveloping tissues are too loosely structured. In the simple "sliding" form, this may permit an upward sliding movement of a portion of stomach into the thoracic cavity. The resulting reflux of acid gastric juice produces unpleasant but not life-threatening symptoms. In a small minority of cases the esophagogastric junction remains intact, but the tissues adjacent to it permit a portion of the upper (cardiac) end of the stomach to squeeze upward into the thorax adjacent to the esophagus. Such a space is usually small, and this paresophageal hernia can be more serious because the blood supply of the herniated portion may be so compressed that the herniated segment of stomach may become infarcted. In such circumstances only prompt, corrective surgery before it ruptures can avert a fatal outcome. Both sliding and paraesophageal forms of hiatal hernia are acquired and occur commonly in adults. The mechanism is believed to be due to abdominal pressure generated while straining at stool to expel a relatively desiccated mass of feces. It is thought to be rare in undeveloped countries because of the large fecal content of dietary fiber in the latter (Burkitt, 1981:143).

Paleopathology reports

In antiquity both omphalocele and gastroschisis would have been fatal uniformly and promptly, the infants succumbing to infection. Furthermore, the cord membranes enveloping the omphalocele's content would rarely survive the mummifying process, increasing the difficulty of differentiating them in a mummified body. I have found no case reports of either of these in mummified human remains. Prompt recognition with immediate surgical repair of tracheoesophageal fistula cases can save many of these infants, but in antiquity they all must have perished within the first 10 days as a consequence of aspiration pneumonia. The lesion, however, is unreported in the paleopathological literature, perhaps because mummification changes make it difficult to carry out the necessary delicate dissection, and attention may be diverted to the more obvious changes of pneumonia. Imperforate anus could probably have been relieved in some cases in antiquity by blind incision of the area if it were an entity known to them, but its low frequency would have prevented common knowledge of the condition. Atresia at other intestinal levels would all have ended with irreparable intestinal obstruction, as would congenital pyloric obstruction. Most cases of Meckel's diverticulum do not develop complications and one would expect them to be identifiable. The absence of any reports is probably simple testimony of postmortem preservation problems and the gut's fragility even if this lesion were present. Consequently I was unable to find reports of any of the above-mentioned conditions in archaeological bodies.

However, Munizaga *et al.* (1978) dissected an adult female mummy from coastal northern Chile who died about AD 250. They found two loops of jejunum that had herniated through a defect in the posterior part of the left diaphragm into the thoracic cavity. The blood supply of these jejunal loops had been sufficiently compressed to render the loops ischemic with subsequent infarction. The herniated bowel loops were thickened and black, reflecting the changes of hemorrhagic infarction. Gerszten *et al.* (1976b) reported a coastal Andean adult female from the Colonial period that died of infarction of a part of the stomach that had formed a paraesophageal hernia.

The only case of congenital megacolon in an archaeological body that I could find was reported in a Roman Period Egyptian mummy by Ruffer (1911a).

Ulcerative gastrointestinal lesions

Gastric and duodenal ulcers

In modern Western populations gastric or duodenal ulcers are found in about 10% of males and about 4% of females. Their direct cause is controversial but it is clear that gastric colonization with *Helicobacter pylori*, hyperacidity and ingestion of nonsteroidal anti-inflammatory agents play a major role. They begin as superficial mucosal erosions that extend to the submucosa, where an intense chronic inflammatory response results in a thickened, fibrotic wall. Complications include gastric obstruction from the fibrosing base of a pyloric ulcer, a surgically remediable lesion today. Perforation of the gastric or duodenal wall by a penetrating ulcer occurs in about 5% of the cases and may be a fatal complication. Gastric juices and food spill out of the bowel wall defect, commonly causing widespread peritonitis, though a slowly perforating ulcer may localize the infection in the form of an abscess. A more frequent complication is hemorrhage: about 20% of ulcers bleed. Extension of the ulcer base may erode a capillary or venule that results in a constant trickle of blood. Such individuals may present with iron-deficiency anemia. Erosion of an artery in the ulcer base, however, results in a blood-spurting lesion that can produce a hemorrhage of massive proportions and prompt exsanguination.

Esophageal varices

Another cause of such massive upper gastrointestinal hemorrhage is ruptured esophageal varicose veins. Such persons usually have a liver so densely scarred by cirrhosis that the portal vein blood flow is impeded and the pressure within that vein rises substantially, forcing much of its blood into collateral vessels that bypass the liver. The esophageal submucosal veins form part of that collateral venous system and become engorged with blood under increased pressure. They are covered only by relatively thin and soft mucosa, erosion of which causes the underlying vein to bleed directly into the esophageal lumen, a complication with a 50% mortality and almost inevitable recurrence in those surviving the first bleeding episode.

Mallory–Weiss syndrome

Yet another cause of upper gastrointestinal hemorrhage is the Mallory–Weiss syndrome, in which violent retching (usually in an inebriated person) can cause a laceration of the mucosa at the esophagogastric junction. Life-threatening hemorrhage can result when the laceration includes a large submucosal artery or vein, or when the laceration penetrates the entire wall, spilling stomach contents into the left thoracic cavity, with resulting massive infection there.

Angiodysplasia

Lower gastrointestinal hemorrhage in the elderly can result from mucosal erosion of a focal, degenerative, tangled mass of blood vessels (angiodysplasia), most commonly found in the cecum. They, too, can bleed massively into the colon.

Paleopathological reports of ulcerations

In mummified bodies the mucosa is absent, making the mucosal part of the mucosal defect unrecognizable. However, the dense, fibrotic zone in the submucosa that extends even into the muscular wall underlying the ulcer can be expected to survive the postmortem atmosphere quite well in an otherwise mummified body and should be easily detectable. In a perforated ulcer either the presence of a local abscess or exudate reflecting generalized peritonitis should stimulate a search for the origin – a search that should include not only appendicitis and other causes but also ulcer perforation. Similarly the blood-filled stomach, esophagus and upper small intestine from a bleeding ulcer, Mallory–Weiss syndrome laceration, or esophageal varix rupture (or even a colon distended with blood from a bleeding angiodysplasia lesion) ought to be obvious. While there is no reason to believe that many risk factors for gastric and duodenal ulcers and other bleeding lesions were not present in antiquity, the described lesions have not been reported in archaeological bodies. Sievers & Fisher (1981:212), however, report that today gastric ulcers are less common and duodenal ulcers rare among full-blooded Native Americans from the southwestern USA. Rowling (1967) notes that, while no peptic ulcers have been

Fig. 8.19. Megaesophagus; SM, AD 1100.
The esophagus is dilated down to its junction with the stomach. In this case the dilatation resulted from destruction of esophageal ganglion cells due to infection with *Trypanosoma cruzi*. Adult female, northern Chile. (PBL.)

found in Egyptian mummies, the Ebers papyrus (Ebbell, 1937:30, column 39) describes diagnostic findings strongly suggestive of upper gastrointestinal hemorrhage: "if thou examinest his obstacle in his cardia, and thou findest that he has . . . turned deathly pale, his mind goes away, and his cardia becomes dry [thirst?], then thou shalt say of him: it is a blood-nest which has not yet attached itself . . .". Nevertheless, the characteristic, food-relieved, epigastric pain of ulcers is not mentioned in any Egyptian papyrus. However, while the Hippocratic authors did not mention ulcers, both Celsus and Pliny were familiar with peptic ulcers (Majno, 1975:360), testifying to an antiquity of at least two millennia for this condition.

Intestinal obstruction

Achalasia

This is a serious complication of a wide variety of bowel lesions and can occur at any level from the esophagus to the anus. Lower esophageal sphincter spasm results in retention of food in the esophagus, with distention of that segment (achalasia). The distended esophagus (megaesophagus) can also be the result of destruction of its ganglion cells by infection with *Trypanosoma cruzi* (see Fig. 8.19). Obstruction of the gastric outlet by a scarring pyloric ulcer has already been mentioned above.

Adhesions

Abdominal adhesions, often from prior abdominal infections or surgery may provide a rigid, fixed structure that can impede an intestinal wave of peristalsis, permitting the intestine to twist around it and compress its own blood supply. A segment of bowel (most often the sigmoid colon) containing a heavy fecal load may even undergo such a twisting contortion (volvulus) without prior adhesions. The resulting compression of the blood supply in its mesentery can lead to infarction of the twisted segment with hemorrhage into the infarcted portion, rendering it thickened and black.

Neoplasms

Neoplasms arising in the bowel mucosa (adenocarcinoma) or muscular layer (leiomyoma) of the wall can also obstruct the intestinal lumen, especially in the colon.

Intussusception

Intussusception is a process in which a usually upper small bowel segment "telescopes" into the lumen of a lower segment causing bowel obstruction and (if untreated) also infarction. Imperforate anus has been discussed in the section on Congenital lesions, above.

Hernia

Perhaps the most common cause of intestinal obstruction is a hernia, which represents protrusion of a segment of intestine through a defect in the fascial layer of the abdominal wall, most commonly in the inguinal area or scrotum, producing a subcutaneous bulge in the affected area. If the fascial defect is large, pressure on the bulge can push ("reduce") the bowel back into the abdominal cavity. Sometimes inflammation will cause the bowel to become permanently adherent to the hernia sac, preventing reduction by manual pressure ("incarcerated" hernia). The herniated bowel, of course, also drags its mesentery into the hernia sac, and, if the fascial defect is small, the mesentery (containing its blood supply) may become compressed and render the herniating bowel ischemic ("strangulated" hernia) – a surgical emergency.

In these various conditions the afflicted person suffers abdominal distention, pain and vomiting with loss of fluids that produces major metabolic changes. The distention of the bowel can impair its blood circulation, leading to gangrene and perforation with peritonitis. The inguinal fascial defects are responsible for the most common form of hernia and in the male occur there as the result of incomplete fusion of the canal along which the embryonic testis migrates from the intra-abdominal position to the extra-abdominal scrotal location. Another common hernia site is the umbilicus, though obstruction at this site is uncommon because of the usually wide mouth of the hernia sac.

Paleopathological reports of intestinal obstruction

Except for the paraesophageal and diaphragmatic hernias described under Congenital lesions in this chapter, I could find no examples of intestinal obstruction or any of the other discussed causes in mummies. Rowling (1967) suggests that this may be the case in Egyptian mummies because embalmers would have removed the bowel from the hernia sac during evisceration. An enlarged scrotum found in the mummy of Ramses V has been suggested to have been caused by a hernia (Rowling, 1967), as was the large scrotum found by Dzierzykray-Rogalski (1988) in a baby. In neither case was bowel demonstrated in the scrotum. Pharaoh Merneptah's scrotum was removed by the embalmer though his penis remained intact (Smith & Dawson, 1924:100). Harris & Went (1980) suggest its removal may have been performed by the embalmers because of a hernia. In view of the fact that scrotal enlargement could be due to a number of other causes such as hydrocele or, more commonly, postmortem putrefactive changes, assignation of hernia as the cause of an expanded scrotum rests on slender evidence indeed. Textual and other artistic evidence, however, more strongly suggests that the presence and mechanism of hernia production were known in antiquity at least as ancient as the Egyptian civilization. The Ebers papyrus (Ebbell, 1937:123, column 106) describes a hernia quite convincingly in its instructions to a physician concerning "a swelling of the covering on his belly's horns [i.e. inguinal areas] . . . above his pudenda, then thou shalt place thy finger on it . . . and knock on thy fingers; if . . . that has come out and has arisen by his cough . . . thou shalt heat it [i.e. the swelling] in order to shut (it) up in his belly . . ."

Umbilical hernias are clearly delineated in Egyptian funerary art, both paintings and reliefs, some of which (reliefs from mastabas of Ptah-Hetep and of Mehu, Old Kingdom, Saqqara) are of a degree of clarity that they leave no diagnostic doubt (Ghalioungui & Dawakhly, 1965: Figure 43 and 44), but the diagnosis of inguinal hernias descending into the scrotum are of far more marginal diagnostic specificity and dependent on only modest degrees of scrotal enlargement (Ghalioungui & Dawakhly, 1965: Figure 42 and 45 – reliefs from mastabas of Ankh-Ma-Hor and of Menu, Saqqara, Old Kingdom). However, a statuette, thought to be that of the imported god Bes commonly represented in Lower Egypt, is more convincing. This stone figure was found in an ancient Phoenician cemetery at Sousa dated to about 900 BC by Gouvet but first reported by F. Poncet (1895) in a French journal *Le Progrès Medical* (June 1, 1895). An English translation appeared in the Lancet (October 5, 1895). It demonstrates a seated male figure wearing a tough, fascial cloth skillfully bound to support what appear to be bilateral inguinal hernia masses – probably the earliest known truss. While perhaps not unequivocally diagnostic, the suggested diagnosis is certainly the most plausible.

No case report of intestinal obstruction by volvulus in ancient bodies could be identified by me but again the Egyptian papyrus Ebers (Ebbell, 1937:39, column 25) suggests its presence at that time: "if thou examinest one who suffers . . . with colicky pains, and whose belly is stiff through it, and who has pain in his cardia . . . nor is there any way it can come out, then it shall rot in his belly . . . it grows into a twist in the bowel . . ."

Megaesophagus has been reported in ancient bodies with Chagas' disease, but I searched in vain for cases of intestinal obstruction in mummified bodies due to adhesions, intussusception or neoplasms.

Infectious conditions

Salmonellosis

The gastrointestinal tract is the target of viruses, bacteria and parasites so frequently that most modern people suffer at least one mild infection annually. Most of these, however, produce either no grossly recognizable lesion at all, or one that is limited to the mucosa and hence is lost in the process of postmortem mucosal autolysis. Among these are cholera, almost all the viral gastroenteritides, salmonellosis and others. Probably the only recognizable enteric lesion of typhoid fever in a mummy would be that of its complication: perforation with acute peritonitis. In view of the substantial fraction of infant mortality attributed to such infections in the developing countries of the modern world, the inability to recognize these anatomically in ancient mummies is a major impediment in the study of paleo-epidemiology. More sophisticated laboratory methodology will need to be developed for their recognition. Sawicki *et al.* (1976) have identified one such possible path by inoculating a rabbit with solutions of powdered Peruvian mummies' coprolites and demonstrating a substantial rise in *Salmonella* antibody titer (method of antibody induction). Molecular biology methodology has also shown promise of its ability to demonstrate infectious agents' DNA in ancient tissue (Salo *et al.*, 1994).

Appendicitis

There are, however, a few specific forms that can produce identifiable lesions. One of these is appendicitis. The usual cause of appendiceal infection is occlusion of its orifice, commonly by a fecalith (pebble of dried fecal material). The gas forming, normal colon bacilli within the luman continue to form gas, the escape of which the occluded orifice will now prevent. The pressure within the appendix rises, compresses the mucosal blood vessels and those in the wall. The bacteria then invade the ischemic, necrotic wall, which eventually perforates, producing either a localized abscess or, more commonly, generalized peritonitis.

Crohn's disease

An ulcerating enteritis that may be evident in a mummy is Crohn's disease. About two-thirds of the cases involve the small intestine and the remainder affect the colon. Its cause is unknown. It begins as a diffusely ulcerative mucosal process, but is accompanied by a very extensively granulomatous, scarring process that greatly thickens the wall. Extension to adjacent structures such as other bowel loops or the bladder results in fistulae between such structures. The disease is chronic and has no known cure.

Diverticulosis

While there are several forms of congenital diverticulum formation (jejunal; Meckel's) that bulge from the

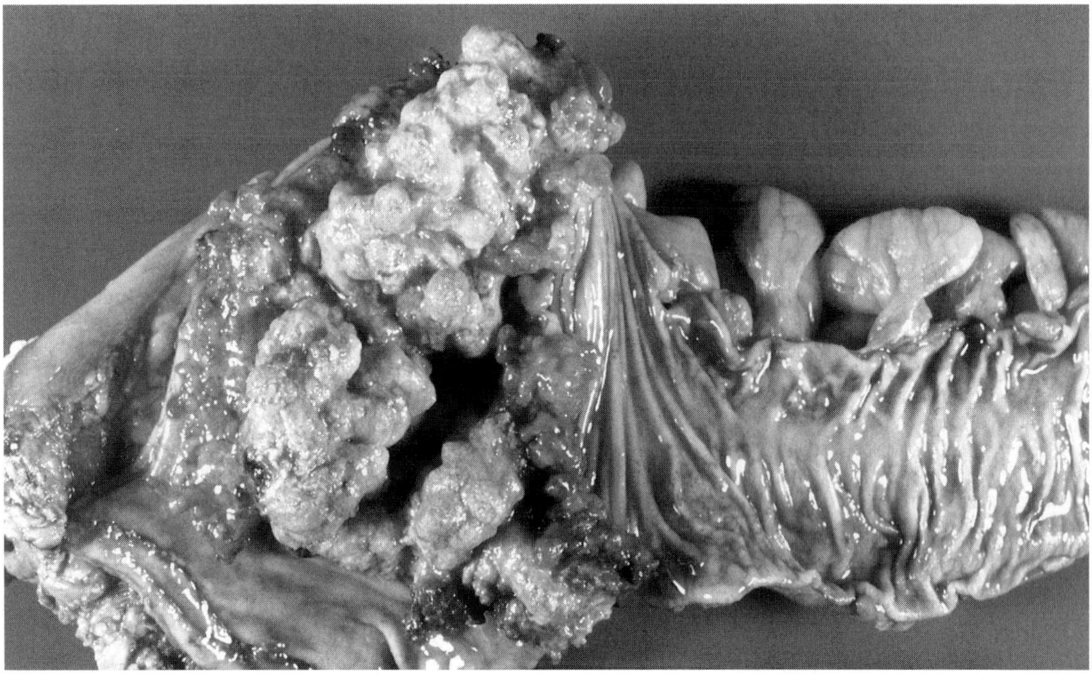

Fig. 8.20. Colon carcinoma.
Because of the colon's large diameter, cancers can attain a substantial size before calling attention to themselves by obstructing the intestine. Modern adult female. (PBL.)

antimesenteric border of the bowel, acquired forms develop on the mesenteric border where blood vessels penetrate the wall. Bulging through such a wall defect, the soft mucosa is covered only by the thin serosal layer. Not surprisingly this fragile diverticular wall often ruptures when it becomes infected secondary to erosion of the mucosal lining by hard fecal fragments. Believed to be produced by years of straining at stool, their development becomes a function of age, found most commonly in the descending colon of elderly people. Perforation of diverticula below the peritoneal line of reflection commonly results in a localized abscess that may fistulize, frequently to the perineum. Perforation at higher levels usually spills the pus directly into the peritoneal cavity, causing generalized peritonitis.

Paleopathological reports

The description of right abdominal pain in Egypt's Ebers papyrus (Ebbell, 1937:52, column 41) is too nonspecific to serve as evidence of appendicitis in that period, but a single case of healed appendicitis in a Byzantine Nubian, uneviscerated mummy was identified by Smith & Wood Jones (1910:268). The appendix was found in the pelvis of this young woman, attached to the opposite (left) pelvis wall by a thick adhesive band, suggesting that she had survived appendiceal rupture with abscess formation. This finding is highly suggestive, but is not absolute proof of a primary appendiceal infection, because the appendiceal serosa can become adhesive to a nearby inflammatory process such as acute salpingitis. No other case was identified in the literature. It is conceivable that the highly fibrous, bulky diet of ancient people produced sufficiently rapid food transit to prevent fecal desiccation in the cecum, though the incidence of acute appendicitis has not been studied formally in human vegetarians versus those on a general diet. Modern epidemiological observations, however, note that acute appendicitis is rare in tribal societies (Burkitt, 1981:142).

Crohn's disease was first reported in 1932 in the USA. It is believed to be of recent development. Its world prevalence pattern reveals a predominance in Western developed countries (Hutt & Burkitt, 1986:22). One would expect that the thickly scarred wall would be recognized easily even in ancient bodies whose mucosa has dissolved and the muscle layer thinned, but it has not been seen in mummies.

Diverticulosis coli has not been reported in mummies. Being an acquired disease primarily afflicting the elderly, the small fraction of mummies reaching that age may explain the absence of this complication.

Neoplasms

Adenocarcinoma

The most common intestinal tumor today in Western countries is colon adenocarcinoma. A similar tumor also affects the small intestine and even the stomach. It is believed to arise from a pre-existing adenomatous polyp. Most cases are sporadic but a small fraction occur within families demonstrating a heritable syndrome of one of several types. Multiple specific mutations involving both oncogene (*ras*) and suppressor genes have been shown to occur prior to the growth of invasive cancer. The cause of these mutations is unknown but epidemiological studies show significant correlations with increased dietary animal fat and decreased dietary fiber (and subsequent slow fecal transit time permitting longer exposure of colon mucosa to intestinal content: Hutt & Burkitt, 1986:27–30). The capacious lumen of the cecum and ascending colon can accommodate a very large tumor mass before the latter interferes with stool passage (Fig. 8.20). For this reason tumors of the proximal colon are usually of considerable size before detection. The surface of these large cancers commonly erode, causing enough blood oozing from its ulcerated areas to result in anemia, a common initial symptom. In the smaller diameter of the more distal colon, similar tumors are more apt to call attention to themselves earlier by blocking the lumen and causing intestinal obstruction. The frequency of small intestine tumors is only about one-tenth that in the colon. The large volume of the stomach also permits the growth of very large tumors unless they arise at the relatively narrow opening of the esophageal junction with the stomach or at its pyloric outlet into the duodenum. All of these intestinal cancers tend to metastasize into the mesenteric lymph nodes and, later, invade veins that carry the metastatic cancer cells into the liver where they usually flourish.

Paleopathology

Because of their substantial size, one would normally expect to recognize gastrointestinal cancers readily in mummified human remains. Perhaps postmortem liquefaction and destruction by bacteria is responsible for the paucity of these paleopathological reports describing them, yet the colon is quite commonly recognizable in spontaneously mummified bodies. Fornaciari *et al.* (1993) have reported such a case in the mummified remains of a fifteenth century king from Naples, and Zimmerman (1995, and see Fig. 4.62) found a similar one in a Ptolemaic Period body from Dakhleh Oasis in Egypt's Western Desert. Rowling (1961) suggests the possibility of carcinoma of the rectum in an Egyptian body, but the diagnosis is based entirely on pelvic bone changes, unfortunately without associated soft tissues. Leiomyoma, composed of smooth muscle cells, is a benign and common gastrointestinal tumor in modern times, and would be expected to survive postmortem effects even better than cancer, yet we can find no report of this lesion in archaeological bodies.

Miscellaneous lesions

Hemorrhoids

Hemorrhoids may be the lesion for which an ointment is described in the Eber papyrus (Ebbell, 1937:44, column 33): "to relieve the vessels of the hinder part . . .". In addition the same text (p. 42, column 31) may describe either pruritus ani or fistula-in-ano: "to expel

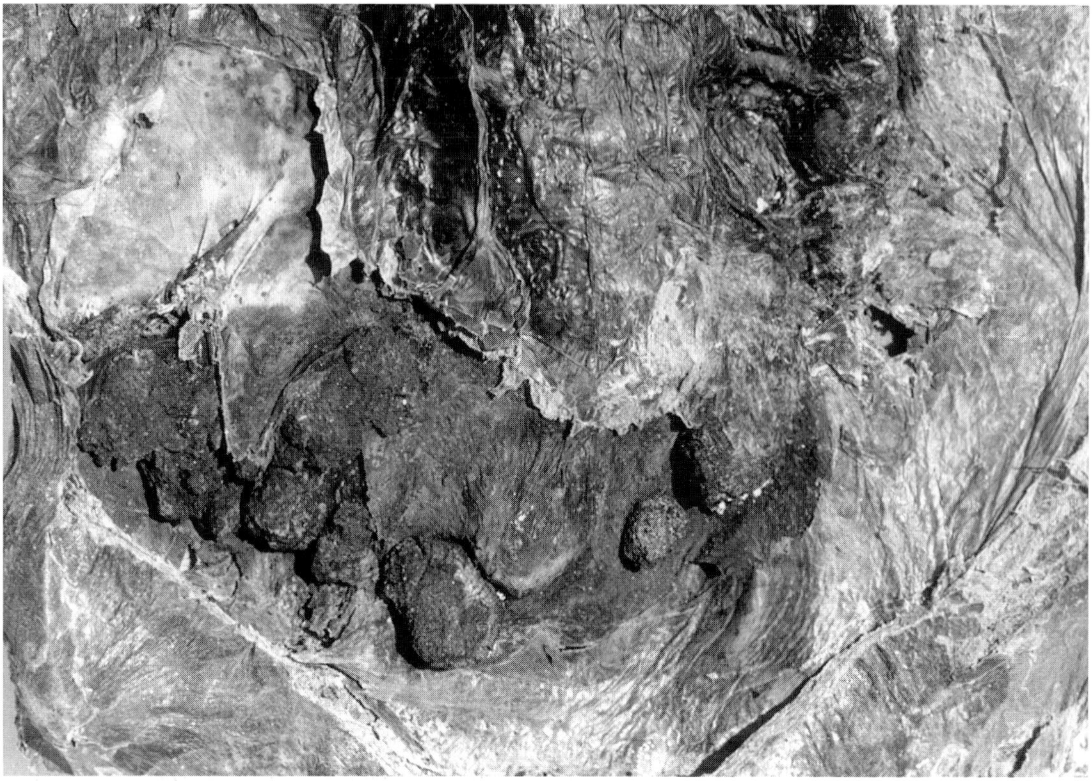

Fig. 8.21. Coprolites (dehydrated feces); SM, 2000 BC.
After death the coprolite-filled cecum has become displaced by gravity from the right lower quadrant of the abdomen, sliding downward so that its medial aspect lies just above the anus. A 15-year-old male, northern Chile. (PBL.)

burning in the anus . . . that is accompanied by many flatus without his perceiving it . . . to expel purlency in the anus . . .".

Rectal prolapse

Ruffer (1913) describes a case of rectal prolapse in a Coptic mummy. The Ebers papyrus (Ebbell, 1937:43, column 32) suggests this may have been a problem among living Egyptians when it recommends application of a mixture of compounds as a hot pack for "dislocation in the hinder part . . .". In spontaneously mummified bodies, however, we have observed numerous examples of rectal prolapse with features that suggest these were postmortem processes of fermentation with gas formation and increased abdominal pressure (Zumwalt & Fierro, 1990).

Gastric contents and coprolites

Finally, gastric contents and colon feces (coprolites) (Fig. 8.21) contain an enormous mass of data about ancient diets and parasites, and these can be employed to predict season of death, method of subsistence and other features valuable for reconstruction of an ancient population's lifestyle. Interested readers can begin their search with references to Ruffer (1920), Halbaek (1958), Horne (1985), Scaife (1986), Aaronson (1989), Reinhard (1990), and Miller et al. (1993).

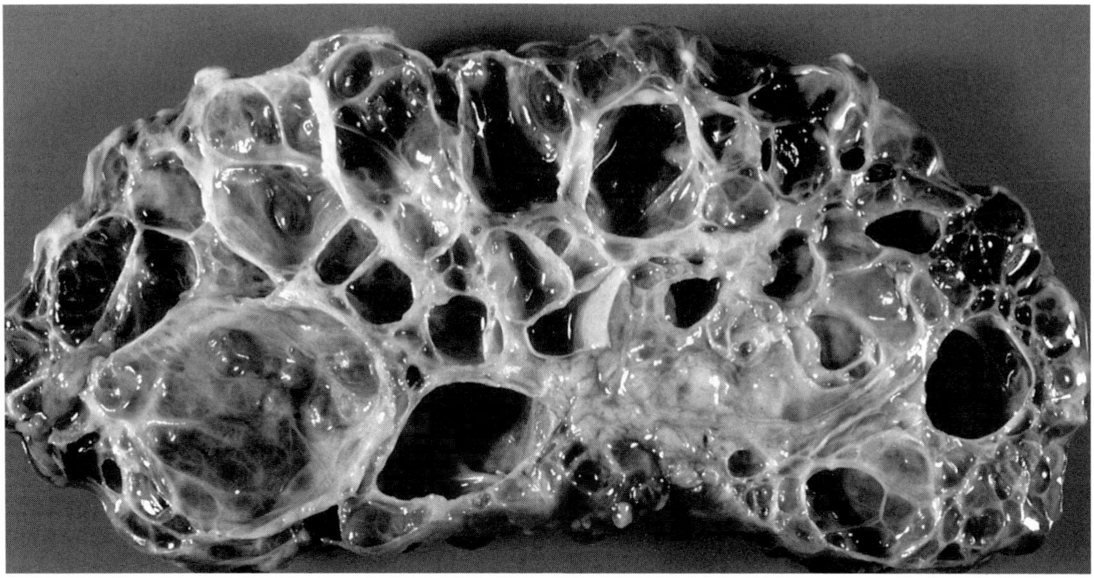

Fig. 8.22. Polycystic kidney disease.
Both kidneys demonstrated enormous enlargement of the organ due to expansion of the many cysts with atrophy of the renal parenchyma between the cysts. Modern adult male aged 50 years. (PBL.)

Urinary diseases

Congenital lesions

Polycystic disease

While normal kidneys frequently have occasional, harmless cysts, the most common form of polycystic kidney disease (PKD) is the product of an autosomal dominant gene located on chromosome 16 (European Polycystic Kidney Disease Consortium, 1994) or another on chromosome 4 (Kimberling *et al.*, 1993). PKD is characterized by the presence of a large number of cysts scattered throughout the renal parenchyma (Fig. 8.22). These cysts somehow induce self-destruction (apoptosis) of the intervening renal elements until insufficient functioning renal parenchyma remains to sustain life, and renal failure (uremia) ensues (Woo, 1995). The rate of progression, however, varies enormously (depending on the nature of the gene mutation?) and may lead to death in early childhood or as late as the ninth decade. The condition is bilateral and end-stage PKD commonly reveals kidneys of enormous size (10× increase) composed principally of fluid-filled cysts varying from microscopic size to some that are 4–8 cm in diameter. Treatment today consists of dialysis and transplantation. PKD patients constitute about 10% of renal transplantations in most centers. In antiquity, of course, no effective therapy was available, and the common adult form onset was late enough to permit child-bearing and thus gene transmission. In addition to the renal pathology, nearly half of these individuals also have a few cysts in the liver (not enough to affect hepatic function) and a few demonstrate cerebral artery aneurysms whose rupture may be fatal. About one-third of PKD deaths are due to renal failure, another third due to a complication of the commonly accompanying hypertension and the remainder to unrelated causes (Cotran *et al.*, 1994:935). Other, much less common forms of kidney cystic disease usually have small cysts; there are several in which the cysts are limited to the renal medulla, and that are inherited in a variety of patterns.

Horseshoe kidney

The other congenital renal lesion of a frequency about that of polycystic kidneys (about 1 in 750 in clinical, modern autopsy series) is fusion of the (usually) lower pole of both kidneys. This results in a horseshoe-shaped single structure, but commonly two ureters emerge. It normally does not interfere with function, although the renal arteries in some cases may compress one or both ureters sufficiently to cause ureteral obstruction.

Paleopathology of congenital renal lesions

I have been able to find the report of only one case of polycystic renal disease in an archaeological body. Allison, (1981b) describes this condition in a 17–19-year-old spontaneously mummified body excavated from a site in northern Chile dating to about AD 3000 and illustrated it in Case no. 7 in the Paleopathology Club series. The horseshoe malformation has not appeared in the paleopathology literature.

Infectious conditions

Pyelonephritis

Both the renal pelvis and the kidney parenchyma are so commonly involved in kidney infections as to justify use of the term pyelonephritis. While bacilli may reach the kidney via a septicemic episode, they usually arrive there as a result of infections in any part of the lower urinary tract (urethra, bladder, and ureters). The entrance of the ureters into the bladder is normally guarded by a valve-like structure that permits urine to enter the bladder, but prevents reflux of urine from the bladder back up into the ureter. In cases of bladder infection, dysfunction of these valves may permit reflux of bacteria-laden urine into the ureters that subsequently can reach the level of the kidney. Other obstructions may involve urethral constriction by an enlarging prostate, tumors encroaching upon any part of the urine-conducting system or impaction of ureteral calculi. Because of the proximity of the external urethral orifice to the anus, particularly in females, it is not surprising that the most common organism involved is the one most prevalent in the human colon: *Escherichia coli*. In diabetics staphylococcal infections are common, often reaching the kidney through the bloodstream.

In acute pyelonephritis the onset may be quite abrupt, with chills, fever, malaise and burning urination. Subsequent course is dependent on the nature of the initiating factor, especially in those in whom obstruction played a role. Untreated multiple abscesses may develop in both kidneys (pyonephrosis); in some (especially diabetics), bacilli may reach beyond the renal capsule, producing a perinephric abscess. Fatal septicemia by the infecting bacillus may terminate the process. Alternatively, a combination of the immune system's response plus at least temporary or partial relief of obstructive effects may allow recovery. Without correction of the obstructing effect (e.g., passage of a ureteral stone), however, recurrences are common and lead to a continuing condition of chronic pyelonephritis with acute exacerbations. Since each episode results in some amount of renal parenchymal destruction, the end result often is bilateral renal fibrosis and atrophy manifested as chronic renal failure with uremia (Fig. 8.23). A chronic xanthogranulomatous form often extends beyond the kidney, involving retroperitoneal tissues with dense scar formation (Scully *et al.*, 1995).

Renal tuberculosis

Renal tuberculosis usually is the result of tuberculous septicemia (miliary tuberculosis). Subsequently caseous necrosis of a progressively expanding lesion occurs until it coalesces with a neighboring lesion and/or erodes into a hollow viscus through which the nectoric caseous material is discharged. That hollow viscus in the lung is, of course, a bronchus, while in the kidney it is commonly one of the urine-filled calyces of the renal pelvis. The bacilli-laden urine then passes down the ureters to the bladder and urethra.

Fig. 8.23. Kidney: chronic pyelonephritis; SM, AD 450–800.
This kidney measured 7 cm × 4 cm × 2 cm and its mate was about two-thirds as large. In addition to the bilateral size reduction, they demonstrated both finely and coarsely granular surfaces reflecting profound, diffuse irregular parenchymal atrophy characteristic of chronic pyelonephritis or other causes of diffuse parenchymal damage (renal arteriolosclerosis, chronic glomerulonephritis). A small calculus found in the renal pelvis is pictured lying free beneath the right side of the kidney. An 18-year-old male, northern Chile. (PBL.)

Secondary infection of these structures as well as others draining into them (prostate, seminal vesicles, epididymis and even testis) is a common complication. In a significant percentage of people suffering miliary spread of the tubercle bacillus, the body's immune system may eliminate organisms in all organs except one. Examples of such processes include those persons with active urinary tract tuberculosis but no evidence of active disease in the lung or elsewhere. Occasionally penetration of the renal capsule will result in a perinephric tuberculous abscess. The tuberculous process may reach the vertebrae via lymphatics and veins that drain into Batson's plexus of veins. The progression of renal tuberculosis is slow, commonly involving many years before renal failure or a complication of the disease leads to death in untreated persons.

Paleopathology of infectious renal lesions

Chronic pyelonephritis and perirenal abscess have both been reported in South American mummies by Gerszten *et al.* (1983). Ruffer (1921a) identified multiple, bilateral kidney abscesses with stainable Gram-negative bacteria in an Egyptian mummy. He also found schistosome ova in two kidneys. I have been unable to identify tuberculous kidneys in the paleopathological literature, though psoas muscle abscesses in relation to tuberculous vertebritis are dramatically demonstrated in an Egyptian mummy reported by Ruffer & Ferguson (1911).

Hydronephrosis

This term applies to kidneys with greatly dilated calyces and renal pelvis due to urinary flow obstruction. However, the obstruction may occur at any point in the urinary tract with dilatation of structures proximal to the obstructed site. This would include the ureter (Fig. 8.24) (calculus, tumor, compression by gravid uterus of ureter at pelvic brim, defective valve at ureterovesical junction) or the bladder (urethral stricture, tumor, prostatic hypertrophy (benign or malignant), calculus). Obstruction distal to the bladder obviously would result in bilateral hydronephrosis. If the obstruction persists unrelieved, the elevated urinary pressure will result in atrophy of functional renal tissue and precipitate renal failure with uremia if bilateral. Such an affected kidney is enlarged due to the distended urine-filled calyces that eventually replace most of the renal parenchyma as the latter undergoes progressive atrophy (Fig. 8.25). Modern treatment involves endoscopic or open surgical relief of the obstruction. A serious complication of unrelieved obstruction occurs when bacteria gain access to the urine proximal to the obstruction, initiating an acute, violent pyelonephritis that can be lethal.

Paleopathology of hydronephrosis

I am aware of only one reported case of this lesion in antiquity. Allison & Gerszten (1983) demonstrated a

Fig. 8.24. Hydroureter; SM, AD 450–800.
The distal half of this ureter is dilated to about three times its normal diameter. Two small calculi (renoliths) were present in the renal pelvis. Although no calculus was present within the ureter at the time of death, in all probability a small calculus had passed from the renal pelvis into the ureter, obstructing the ureter at about its midcourse long enough to dilate it, then was expelled into the bladder and excreted. An 18-year-old male, northern Chile. (PBL.)

hydronephrotic kidney in their Case no. 13 of the Paleopathology Club series. The right renal pelvis and upper quarter of associated ureter were clearly distended, though no calculi could be found. In modern clinical cases this circumstance is often due to ureteral compression by an aberrant renal artery or ureteral stricture.

Glomerulonephritis

The purpose of the kidney is to eliminate the body's metabolic waste products (MWP) and toxins. It achieves this in a two-step process.

1. The kidney first separates the plasma from the blood cells and proteins, returning the latter to the blood. This separation (filtration) is carried out within thousands of microscopic-sized, spherical masses of tangled capillaries called glomeruli. As blood flows through these capillaries the plasma leaks through the specialized capillary membrane into the space around it.
2. The microscopic, cell-lined tube (tubule) surrounding the glomerulus collects the plasma filtrate and eventually empties into the renal pelvis to join that from other tubules; the filtrate is then called urine. However, during its traverse along the tubule, 95% of the water and all of the sugar and other nutrients are reabsorbed into the blood, but the MWP remain within the tubule and form the principal solute of urine.

This process responds to many physiological stimuli to maintain a constant blood level of appropriate nutrients and other vital substances in spite of a wide range of continually changing food and beverage intake and varying metabolic expenditure. The integrity of the glomerular basement membrane is essential to this control. A vast array of conditions, however, can affect this membrane's structure negatively. The most common are the infectious or autoimmune diseases in which antigen–antibody protein complexes are deposited on or in the glomerular basement membrane and which tend to "clog the filter". Among the most frequent of these are infections by certain strains of beta

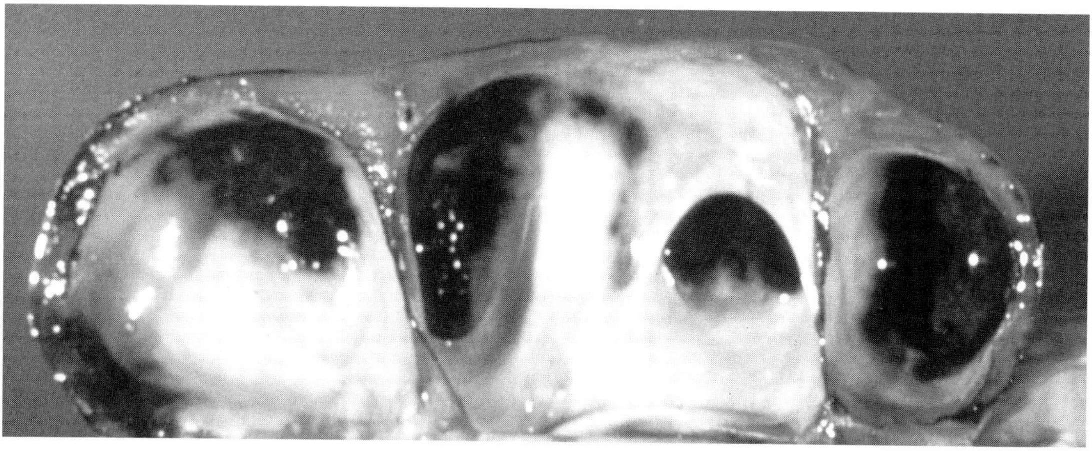

Fig. 8.25. Kidney: hydronephrosis.
A bladder tumor obstructed the ureter, causing unilateral hydronephrosis. The increased pressure caused the renal pelvis and calyces to dilate enormously, resulting in atrophy of the renal parenchyma. Modern adult male. (PBL.)

hemolytic streptococci whose antibodies not only obstruct the filtering membrane, but also stimulate swelling of the cells lining the capillaries to a degree that obstructs their lumen. Without plasma flow through the capillaries or through a membrane partially obstructed by deposition of such protein complexes, no urine can be formed. Untreated, the MPW accumulate in the blood (uremia) and the individual may die within a few weeks. Modern dialysis therapy supports the patient through this often self-limited, acute phase. Still other antigen–antibody complex-producing conditions are often of autoimmune nature, the classic model of which is lupus erythematosus. A variety of other bacilli, viral or even protozoal infections occasionally produce a similar effect. Collectively these conditions are termed acute glomerulonephritis.

Other conditions may injure the membrane in such a way that it leaks proteins. Since renal tubules are not designed to reabsorb proteins, such proteins are lost to the body when excreted in the urine. Massive urinary protein loss lowers the serum albumin level dramatically so that it can not carry out its function of osmotic fluid balance (nephrosis). The result is accumulation of water in tissues, creating a bloated, edematous appearance.

If the acute forms of the nephritides are survived, the condition may progress into an often years-long chronic glomerulonephritis phase that may eventuate in the uremic state of end-stage chronic renal disease.

Death in the acute phase often presents a kidney of up to twice normal size. The progressive loss of glomeruli in the chronic phase is accompanied by the loss of their associated tubules. The result is usually diminutive, atrophic, granular-surfaced kidneys.

Paleopathology

Estes (1989:46) refers to reports of glomerulosclerosis in Egyptian mummies, but he does not cite them. End-stage renal disease due to chronic glomerulonephritis often demonstrates abundant fibrotic glomeruli, but this change also occurs with other chronic renal diseases. No other diagnoses of glomerulonephritis in archaeological kidneys could be found by me.

Metabolic conditions

Urinary calculi

The most common of these that confront the paleopathologist is the formation of urinary calculi (urolithiasis). Stones arise by precipitation of the urine's solutes. The precise circumstances leading to such a state of supersaturation are not well understood, but the chemical and physical state of the crystallized solute (urolith) varies in relation to location, secretion of hypernormal levels of the solute, age, gender, urine pH, crystal inhibitors, urinary infection and the presence of minute particles that serve as nucleating agents to initiate precipitation. The cause of the increased urinary solute secretion is clearly defined in some cases: hyperparathyroidism leaches calcium from bones and excretes it into urine in large amounts; metabolic derangements in gout patients cause elevated levels of uric acid in both blood and urine. In many persons, however, the cause is not obvious.

Calculi in the urinary tract create two types of problem: they may obstruct the flow of urine and they may erode the urinary tract's lining mucosa resulting in hemorrhage (hematuria) or rendering it susceptible to infection. Because the nature of the stones differ, it is customary to divide them into those that precipitate in the upper tract (renal or kidney stones) and those that form in the urinary bladder (vesical stones).

Renal calculi are most commonly formed by the precipitation of calcium in the form of calcium oxalate or calcium phosphate. If the mucosal areas abraded by the calculi become infected it is often by bacilli that split urea, producing ammonia. The resulting more alkaline urine will then often enlarge the calculus by precipitating magnesium ammonium phosphate ("struvite") onto it. Progressive crystal growth may produce a single calculus that fills the entire renal pelvis and its calyces ("staghorn calculus"). Alternatively, the renal pelvis may expel the stone into the ureter while it is still no more than 2 or 3 mm in diameter. The smaller stones may represent a greater hazard because they may enter the ureter, impact there and obstruct urine flow. The ureter's peristaltic contractions in its effort to propel the stone through the ureter into the bladder can produce notoriously agonizing pain. Permanent arrest of the stone in the ureter may cause total obstruction with subsequent hydronephrosis and loss of that kidney's function if the stone is not removed. Renal calculi occur largely in adults, with little difference between sexes.

Urinary bladder calculi may occur in adults but are most common in children, and are 10-fold more frequent in boys. They may reach enormous size in the bladder; diameters of 8–10 cm and occasionally greater have been reported. While calcium oxalate composition is also common in these stones, uric acid calculi are frequent as well. Schistosome ova have been found in Egyptian uroliths, suggesting they acted as a nucleation particle for stone formation. In certain body positions, gravity-directed movement of the stone may obstruct the urethral opening, preventing micturition and causing painful bladder distention. Infection can also generate a struvite stone in the bladder.

The epidemiology of urolithiasis is unique. Bladder stones are common among the world's poor and largely vegetarian people (e.g., Thailand). During the past several centuries it has become clear that, as such populations develop modern economies and concordantly increase their dietary meat component, bladder stones disappear while the frequency of renal calculi increases. Certain obvious exceptions to this generalization, however, make it clear that factors other than meat protein are operating. Another major difference between renal and bladder stones is their response to surgical excision: renal calculi commonly recur while bladder stones do so only rarely.

Paleopathology

The interpretation of the history of this disease is challenging and the interested reader is referred to its superb documentation by Steinbock (1985, 1989, 1993a). Space restrictions compel me to extract from his account only a few of the essential features. Because uroliths are discrete, tangible structures, the diagnostic certainty of ancient literature is often reassuring. The Vedas of India include bladder stone as a proper prayer subject as early as 1500 BC. Hippocratic writers recognized both renal and bladder stones. About 300 years later Ammonius of Alexandria created a stone-crushing instrument for use within the bladder. Rufus of

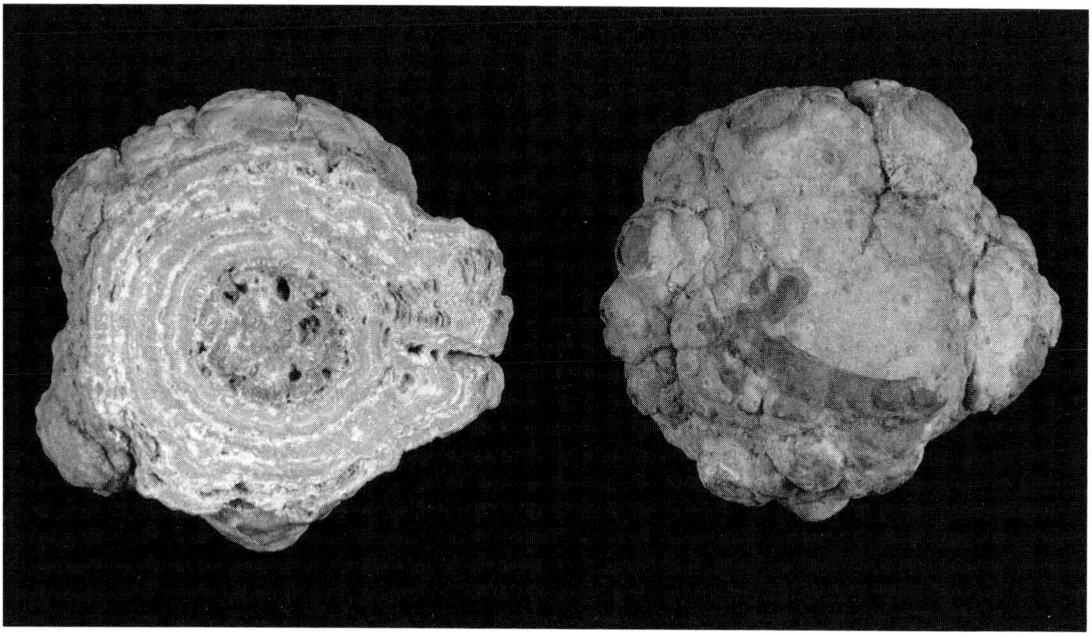

Fig. 8.26. Vesical calculus (bladder stone); SM, AD 500.
This 3 cm × 2.7 cm × 2.5 cm calcium oxalate stone was found within an anatomically normal bladder. It has been bisected to reveal both its knobby surface and the concentric layers of urine solutes that precipitated around what was probably organic matter (protein?, pus?) now represented by the small central defects in the sectioned plane. Male, 30–40 years old, southwestern USA. (PBL.)

Ephesus (first century AD) described a transverse perineal incision to remove bladder stones as did Celsus (Rome) about the same time (cited in Steinbock, 1993a). In the eighteenth century, hospitals at Norwich, England, and at Paris became centers for bladder stone removal, averaging about one case per month for more than a century.

The actual antiquity of urolithiasis, however, exceeds its literary record by a considerable margin. A calcium oxalate stone was found in the urinary bladder of a 1500-year-old spontaneously mummified body of a native North American (Streitz *et al.*, 1981: see Fig. 8.26). Paleopathological findings of both renal and bladder stones have been documented carefully by Steinbock (1985) and his table is reproduced in Table 8.2 by permission. The Mesolithic bladder calculus from Sicily (Piperno, 1976) testifies to this condition's antiquity there of more than eight millennia. Though Smith & Wood-Jones (1910:56) were able to identify only a single Egyptian case among about 10 000 bodies (not all mummies), excavation methods and endemic geographical differences may be responsible for the larger number of finds subsequent to their studies. Both Europe and the Americas are well represented in Steinbock's table, to which I can add a renal calcium oxalate calculus from the skeleton of an adult Maori female in New Zealand from 150–200 years ago (Houghton, 1975). Steinbock's accompanying text annotates each of the reports.

Several further examples have appeared since that 1985 article. Di Tota *et al.* (1992) reported at the ninth session of the Paleopathology Association in Barcelona, identifying a lamellated, concentric pattern of a 1.3 cm probable bladder stone found in the pelvic area of an adult female skeleton from Alfadena (Aquila, central Italy). Sánchez Sánchez & Gómez Bellard

(1988) found renal and biliary calculi in a seventeenth century young adult female excavated in the Basque country in Spain and also found similar calculus forms in necropoles from two different archaeological sites from the seventh and sixteenth centuries. Morris & Rogers (1989) also report two large calcareous bodies, composed of hydroxyapatite (calcium phosphate), each up to 4 cm in diameter, lying adjacent to the lumbar vertebrae of an adult woman's skeleton from a South African site. Fulcheri & Grilletto (1988) provided observations of a probable renal calculus in an Old Kingdom Egyptian body. Blackman *et al.* (1991) detected two small calcium oxalate stones in the left kidney of an 18-year-old male spontaneously mummified body of a member of northern Chile's coastal Cabuza culture (AD 450–1000). Diffuse, punctuate bilateral calcification of the renal parenchyma, bilateral hydroureter, osteopenia and fibrinous pericarditis suggested end-stage renal disease with uremia and secondary hyperparathyroidism leading to bilateral nephrocalcinosis. More recently Catalano & Passarello (1988) documented the presence of 24 left renal calculi in an adult female from a fifth to sixth century BC site near Rome.

Individual case reports remain justifiable because the number of stones reported to date is not yet large enough to spell out specific patterns, though what I do have available supports the historical documents and the epidemiological concepts noted above.

Hypertension and diabetes mellitus

Atherosclerosis of the larger renal arteries is accelerated in individuals with supranormal blood pressure levels (hypertension). One of the effects of this process is stenosis (lumen size reduction) and complete lumen obstruction by thrombosis. The kidney parenchyma nourished by the now occluded artery commonly becomes necrotic (infarcted) and subsequently undergoes liquefaction, absorption and replacement by a localized and grossly visible scar. In addition, microscopic-sized arteries may suffer from stenosis due to smooth muscle proliferation of the wall at the expense of lumen size, or deposition of protein deposits in the wall of arterioles. These changes are exaggerated in diabetes in which further restriction of blood flow occurs by extension of this process to the glomerular capillaries. The consequence of this is atrophy of the individual, diminutive areas supplied by each of these arterioles, resulting in atrophy of the kidney as a whole, and finely granular surface secondary to sunken atrophic microareas of renal parenchyma of pinhead size juxtaposed to surviving parenchymal portions. Thus these conditions may result in a collapsed, sharply defined fibrotic area in an otherwise grossly normal kidney produced by infarction secondary to occlusion of a large renal artery branch; or arteriolosclerosis and/or glomerulosclerosis can cause a granular-surfaced, small atrophic kidney. Most hypertensive individuals, however, do not have sufficient involvement to make these alterations recognizable in mummified tissue.

Paleopathology

Except for the occasional example of hand and/or foot joint destruction due to diabetic neuropathy, I have no reliable "marker" for the presence of diabetes in antiquity. Glomerulosclerosis and prominent arteriolosclerosis in renal arterioles could at least conceivably be suggestive of diabetes, but it has not been reported to date. Indeed, Long's (1931) histological observation of renal medium-sized arterial atherosclerosis in an Egyptian adult female (Teye), consistent with hypertension (though not necessarily diabetes) is an exception, and was accompanied by coronary atherosclerosis with myocardial fibrosis. Gerszten *et al.* (1983) identified both medium and small artery lesions in South American spontaneously mummified bodies.

Neoplasms

Hypernephroma

Hypernephroma (adenocarcinoma of the renal cortex) is the most common adult renal neoplasm. It often

Table 8.2. *Paleopathology reports of renal and bladder calculi*

Date	Location	B/K*	Reference
8500 BC	Sicily	B	Piperno, 1976
3500 BC	Egypt	B	*Shattock, 1905
3500 BC	Egypt	B	Ruffer, 1910
3300 BC	Kentucky, USA	B,B,K	*Smith, 1948
3100 BC	Egypt	K	*Bitschai, 1952
2800 BC	Egypt	K	*Shattock, 1905
2100 BC	France	B	*Duday, 1980
2000–700 BC	England	B	*Mortimer, 1905
1500 BC	Illinois, USA	K	*Beck & Mulvaney, 1966
1000 BC	Egypt	B	Smith & Dawson, 1924
1000 BC	Egypt	K	Gray, 1966
1000 BC	Sudan	B	Brothwell, 1967b
550 BC	Italy	K	Catalano & Passarello, 1988
100 BC	Italy	B	Di Tota *et al.*, 1992
100 BC to AD 500	Arizona, USA	B	*Williams, 1926
0–AD 200	Sinai	K	*Basset, 1982 (personal communication)
AD 450–1000	England	B	Brothwell, 1967b
AD 500–750	Arizona, USA	B	Streitz *et al.*, 1981
AD 1000	Chile	K	*Allison & Gerszten, 1982 (E. Gerszten, personal communication)
AD 1350–1500	Denmark	K	Møller-Christensen, 1958
AD 1500	Indiana, USA	K	*Beck & Mulvaney, 1966
AD 1650	W. Virginia, USA	K	*Metress, 1982 (personal communication)
AD 1650	Spain	K	*Sánchez Sánchez & Góméz Bellard, 1990
AD 1700	South Africa	K	Morris & Rogers, 1989
AD 1800	New Zealand	K	Houghton, 1975

Notes:
B, bladder; K, kidney.
The content of this Table was abstracted from Steinbock (1985) with his permission and expanded with references post-1985.
Source: *For references not listed in the bibliography of this volume see Steinbock (1989).

reaches an enormous size (Fig. 8.27), and invades the renal vein, via which metastases commonly reach the lung. However, the paleopathological literature does not record any malignant kidney tumor of any kind. Today this tumor represents about 2% of all visceral cancers (Cotran *et al.*, 1994:986). If it had been this frequent 5000 years ago, one would surely have expected G. Elliot Smith or F. Wood Jones to have encountered it because Egyptian bodies commonly retained the kidneys, and renal cancers of this type are often so large they would be difficult to overlook. Even though most adult Egyptians died before age 60, a significant fraction reached an age in which this tumor commonly occurs. Papillary cancers of the renal pelvis are much less common and less obvious. In the bladder, identical papillary carcinomas are more easily recognized, though other types, such as squamous cell carcinomas, are more apt to be flat, and obviously invasive into a thickened bladder wall.

Genital diseases

Preservation of external genitalia

In spontaneously mummified bodies the enzyme-laden fluids (especially those from the abdominal cavity) tend

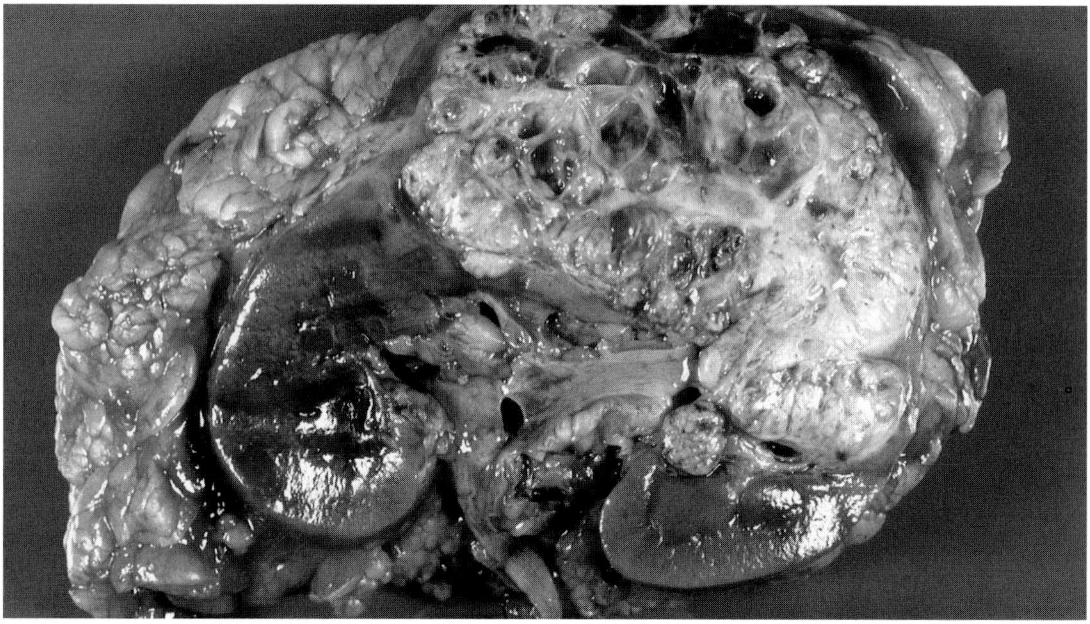

Fig. 8.27. Kidney: hypernephroma.
The uninvolved upper and lower ends of the kidney are visible at the left and right lower edges of the mass. The remainder of the kidney has been replaced by a huge, primary adenocarcinoma (hypernephroma) of the kidney that has extended considerably beyond the kidney itself and invaded surrounding tissue. Modern older female. (PBL.)

to gravitate to the most dependent portion of the body's burial position. In many cases this is a sitting position, and in such burials the perineum is the most dependent area. Postburial desiccation in such mummies is frequently slow enough to permit the enzymatic decay process sufficient time to destroy the external genitalia beyond recognition in both males and females. Because of their prominence and contours, those of the male may remain identifiable more frequently than those of the female (Fig. 8.28). The supine position offers a higher frequency of identification. In anthropogenic forms of mummification, recognition of these structures is heavily dependent upon mummification practice traditions. Egyptian embalmers, for example, frequently made considerable effort to preserve external genitalia, at least among males of royal or noble classes. Unfortunately, during the later periods of extensive resin use, their common custom of wrapping the penis in hot, resin-soaked linen often destroyed the soft tissues; in some, such wrappings were used to model a penis whose soft tissues had not survived the desiccating action of natron. The presence or absence of circumcision can be judged quite frequently in Egyptian male bodies, even in many preserved by spontaneous mummification. Among Egyptian boys it was part of a puberty ritual even in predynastic times (Brewer & Teeter, 1999:98 Fig. 8.29). In those, however, that were eviscerated via the perineum and the perineal defect closed with resin-soaked linen, dissection usually fails to identify the external genitalia. Evidence of the more mutilating types of females "circumcision" has not been reported in Egyptian mummies and those of lesser degree probably would not be easy to detect. In spontaneously mummified bodies, genitalia are preserved less frequently. Aufderheide & Allison (1994) found 71% of 21 males but only 29% of 22 females had identifiable external genitalia in a Preceramic group

Fig. 8.28. Male external genitalia; AM, 140 BC. Male genitalia are preserved in recognizable form more frequently than are those of females in spontaneously mummified bodies and both are preserved better with the anthropogenic mummification methods used in this example. A 23–27-year-old male Egyptian (Ptolemaic), Dakhleh Oasis. (PBL.)

from northern Chile that had been buried in a supine position. Recognition of female breasts is discussed below in Paleopathological aspects of the breasts.

Sexually transmitted diseases (STD)

Three of these diseases common today are syphilis, gonorrhea and anogenital warts (condyloma acuminata). Acute gonorrheal urethritis produces a white discharge and intense burning on urination. Retrograde extension to the prostate and epididymis causes abscesses there that are exquisitely tender and can be serious by producing septicemia. In women the spread is from the cervix to the "adnexae" (fallopian tubes and ovaries), producing serious abscesses there called pelvic inflammatory disease (Fig. 8.30) or leading to a chronic state of adnexal infection with extensive adhesion formation. Today chlamydial infections have replaced gonorrhea as the most common cause of such pathology in Western countries. Syphilis has been described by Arrizabalaga (1993) and will not be repeated here. Anogenital warts are virus-induced papillomas of the external genitalia and perianal areas of both sexes,

Fig. 8.29. Male circumcision; AM, 30 BC to AD 395. This appeared to be a puberty rite during Egypt's Dynastic Period, but was probably practiced even in predynastic times. Young adult Egyptian male from Dakhleh Oasis, Ptolemaic Period. (PBL.)

caused by certain subtypes of the human papilloma virus (HPV). They are not tissue-destructive, though some strains have been linked to cervix cancer in women and to anal cancer in immunosuppressed males.

Paleopathology

Whether or not the Hippocratic references to painful urination represent gonorrhea is controversial, since they do not associate it with venereal transmission. Some (Rothenberg, 1993) regard Egyptian remedies for painful urination as supportive of the diagnosis of gonorrhea, and label this as the oldest of venereal diseases. Oriel (1973a,b), however, feels that even the

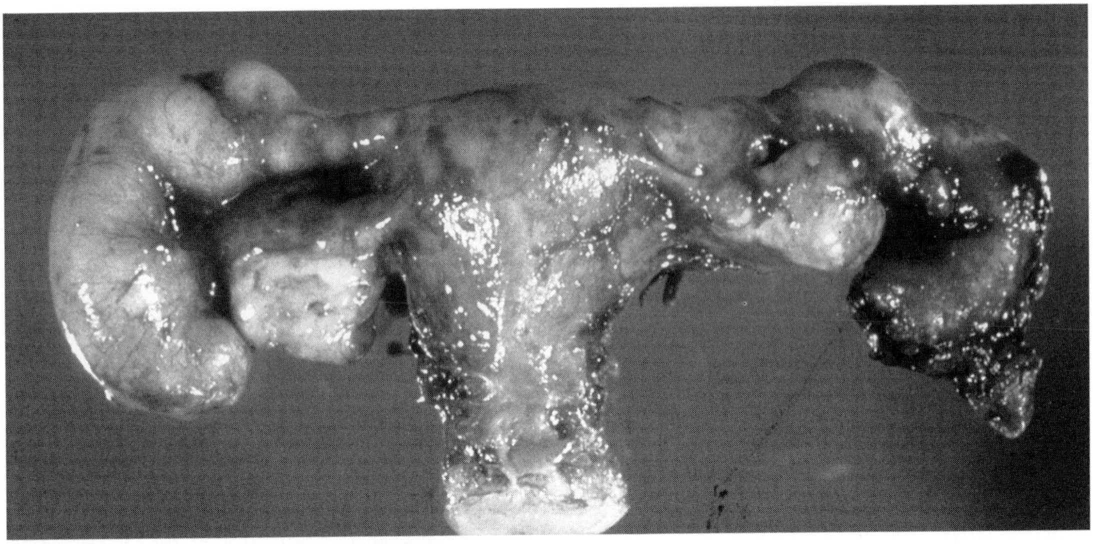

Fig. 8.30. Acute pelvic inflammatory disease.
Both fallopian tubes are very distended with liquefied pus. Ovaries are enlarged due to diffuse infection without abscesses. Modern young adult female. (PBL.)

citations of Galen and Celsus (who describe urethral discharge) are not acceptable evidence of gonorrhea because they describe neither pain nor the association with sexual intercourse. While James Fernel (in France) challenged the prevailing opinion that gonorrhea and syphilis were manifestations of the same condition as early as the sixteenth century, it was not until the mid nineteenth century that Phillippe Record clearly separated these two venereal diseases, and shortly thereafter (1879), Albert Neisser identified the gonorrheal organism that still bears his name (Rothenberg, 1993). The disease has, however, not been identified in mummified tissue, and is not likely ever to be except in testicular, joint or other complicating abscesses. The absence of reference to such a disease by Roman authors also led Oriel (1973b) to doubt that gonorrhea existed in the Western world at that time. However, the opposite is true about anogenital warts, which were called condyloma or thymion by the Greeks (slang: ficus or fig by the Romans). Celsus (first century AD) described penile warts and clearly associated this condition with intercourse. Lay writers viewed anogenital warts with amused contempt and commonly equated their presence on the male perineum with sodomy. We are, however, unable to find any reports of a paleopathological lesion of such papillomas. Thus, while the literary record of gonorrhea prior to the Middle Ages is not universally convincing, that of venereal warts certainly extends at least two millennia backward in time.

Testis

Though the scrotum is often identifiable, the testis rarely survives spontaneous mummification and does so only infrequently in anthropogenic forms of soft tissue preservation. Scrotal enlargements in mummies have been attributed to hernia (see Gastrointestinal tract diseases, above) but are most probably a postmortem artifact of bacterial fermentation with gas formation. One would expect that the fibrotic reactions of gonorrheal or tubercular testicular infections would remain detectable, but none has been reported, nor have testicular neoplasms, even though the most common of these (seminoma; embryonal carcinoma) today demonstrate frequencies that peak between 20

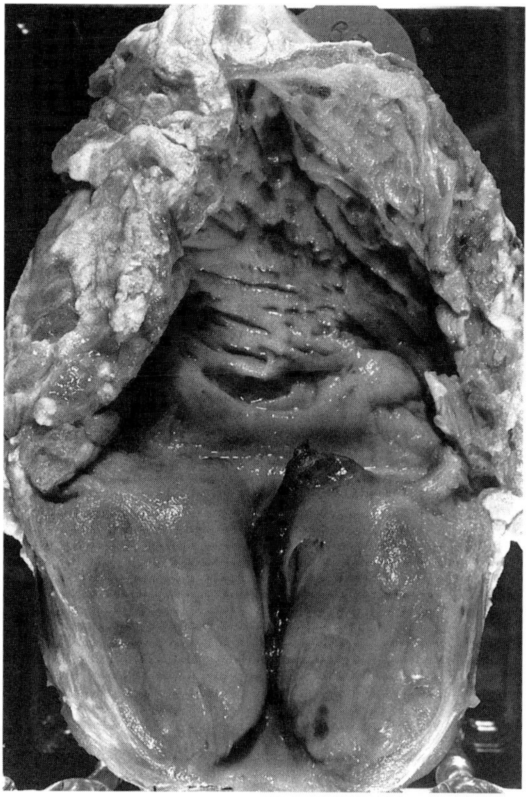

Fig. 8.31. Benign prostatic hypertrophy. The greatly enlarged prostate in the lower part of the picture surrounds and compresses the urethra, interfering with the outflow of urine from the bladder above it. The increased work required to force urine through the narrowed urethra has resulted in hypertrophy of the muscle in the bladder wall, reflected in the folds evident in its inner surface. Modern, aged male. (PBL.)

and 30 years – ages commonly represented in archaeological human remains.

Prostate

In males the urethra that drains the urinary bladder traverses the normal, walnut-sized prostate gland, after which it enters the penis. It is, therefore, not surprising that enlargement of the prostate gland will constrict the urethra, impairing emptying of the urinary bladder (Fig. 8.31). The most common reason for such increase in size is benign prostatic hypertrophy (BPH), an age-related change probably secondary to a not-well-understood hormonal action. The expanding prostate, however, may be due to a malignant change in the organ, that invades surrounding tissue, metastasizes to pelvic lymph nodes and to the spine where the vertebral response is frequently both lytic and proliferative ("osteoblastic").

Paleopathology

Since remedies for a distended bladder mentioned in the Ebers papyrus (Ebbell, 1937:60, column 48) are not always linked to the presence of bladder stones, it is probable that at least some cases of vesical distention may have been due to prostatic enlargement. The failure to identify even normal-sized prostate tissue at autopsy in spontaneously as well as in many anthropogenically mummified bodies is probably the result of the same processes as those related to the testis (see above). Consequently, neither benign nor malignant forms of prostatic enlargement have appeared in the paleopathology literature.

Abnormalities of female internal genitalia

Normal ovaries respond at roughly monthly intervals to pituitary hormones by "developing" one of the many ova present in adult ovaries. The uterine lining (endometrium) responds (by a rapid increase in thickness) to estrogen formed by the follicle of the developing ovum. About mid cycle the ovarian-developed follicle ruptures, releasing the ovum, which is drawn into the fallopian tube (salpinx) and transported to the uterine cavity on a layer of mucus propelled by the whip-like action of cilia on the surface of the tube's lining cells. Following follicular rupture the endometrium responds to progesterone now formed by the ovary and develops substantial and specialized mucus secretion, in anticipation of the needs of a fertilized ovum. If the ovum is not fertilized, ovarian secretions cease, the arterial supply of the now thickened, secreting endometrium undergoes clotting, the ischemic endometrium becomes necrotic and drains out of the uterus via the vagina as a trickle of bloody, necrotic debris (menstruation) after which the cycle is repeated.

Endometriosis

Endometriosis represents ectopic endometrium implanted outside the uterus, most commonly on the adnexal structures. Many of these respond to normal hormones and bleed. In their ectopic position, such recurrent hemorrhage into tissues of the tube and ovary commonly produces focal accumulation of old blood surrounded by a fibrous wall ("chocolate cyst").

Neoplasms

Abnormalities of menstrual chronology are commonly secondary to hormone production, while changes in the amount of menstruation frequently are due to local organ pathology. Those of interest to the paleopathologist are tumors affecting the endometrium. The endometrial mucosa covering the thumb-sized endometrial polyps often demonstrate surface mucosal erosion with subsequent bleeding unrelated to menstruation. The thick mass of smooth muscle constituting the uterus often produces benign smooth muscle tumors (leiomyomas: Fig. 8.32). Their twisted bundles of muscle frequently compress the tumor's blood vessels and occasional arteries undergo thrombosis, causing a local area of infarction in the tumor. Such necrotic tissue may liquefy and be absorbed, healing by scar formation, or it may be replaced by calcification. If the tumor is located under the endometrium the mucosa may be stretched, become ischemic and bleed, also resulting in vaginal bleeding not related to menstruation. Endometrial carcinoma usually ulcerates and causes bloody discharge. Untreated it invades the uterine wall and adjacent tissues including the bladder, adnexae (tubes and ovaries) and bowel, metastasizing to pelvic lymph nodes. These painful masses may destroy the tissues they occupy and produce dense adhesions. Carcinoma may also begin in the cervix, causing ulceration and a bloody discharge. Cervical carcinoma is probably related to strains of the human papilloma virus (which also causes anogenital warts), and spreads in a manner similar to that of endometrial carcinoma.

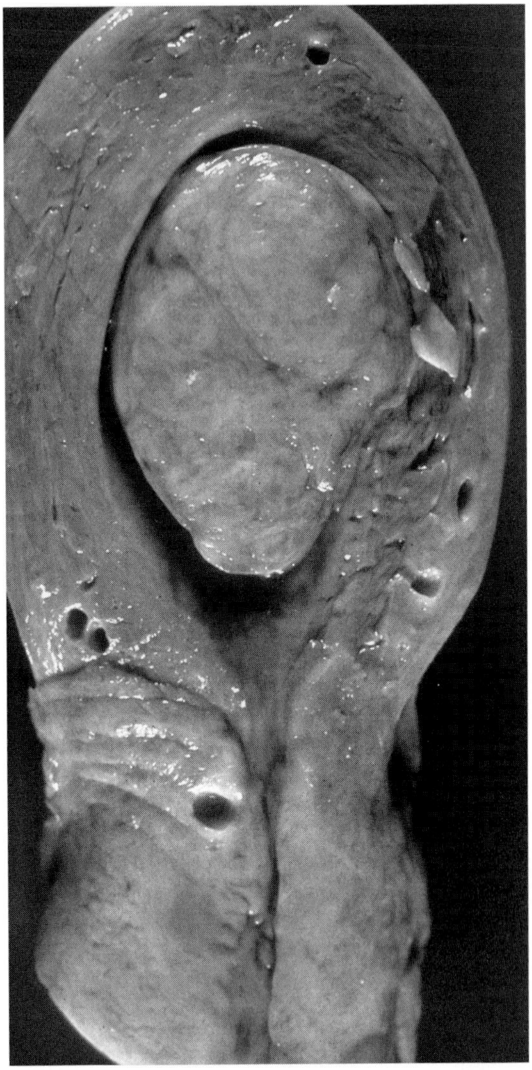

Fig. 8.32. Leiomyoma.
A 3.0 cm × 2.0 cm benign tumor of smooth muscle cells, arising in the smooth muscle of the uterine wall, is growing into the uterine lumen. In other examples it may be found entirely within the wall or projecting from the surface. Infarction with calcification is common in these benign tumors. Modern 43-year-old female. (PBL.)

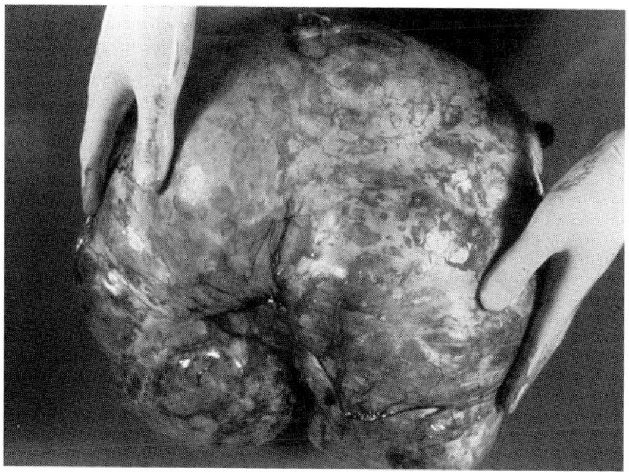

Fig. 8.33. Benign ovarian cystadenoma.
The huge size and weight of this tumor is due to the fluid secreted by its lining cells and which collects in the smaller cysts within the main tumor. Modern middle-aged adult female. (PBL.)

While pathology texts commonly present a bewildering complex of ovarian neoplasms, the postmortem loss of epithelium prevents the paleopathologist from identifying most of the specific histological types. A pragmatic paleopathological compromise, therefore, is to view ovarian tumors as either solid or cystic, each of which can display benign or malignant features. The lining cells of cystic tumors secrete a watery (serous) or a slimy (mucinous) fluid. Their size ranges from a few centimeters to some that have been reported to weigh more than 25 kg, almost all of it due to the contained fluid (Figs. 8.33 and 8.34). About one-third of serous cystadenomas are malignant, while cancerous forms constitute only about one-tenth of the mucinous forms. A general, though not invariable rule is that the greater the mass of solid tissue within a cystic ovarian tumor, the more likely is its malignant nature (Fig. 8.35). The malignant forms tend to implant tumor cells at many sites on the peritoneal surface, each of which may grow into an independent tumor mass, interfering with bowel and other abdominal organ function, leading to death. A special form of cystic tumor unrelated to the above is a dermoid cyst, a cystic form of teratoma. Teratomas are tumors containing structures composed of all three germ layers (endoderm, mesoderm and ectoderm). When found in extraovarian locations most of these are solid, malignant tumors. Such may also occur in the ovary, but peculiarly, most ovarian teratomas are cystic and benign. Their cystic fluid is usually an oily secretion produced by glands of the skin tissue lining the cyst wall. The dermal elements in the tumor can also produce abundant hair. Especially striking is the production of calcified masses (in about 25–30% of dermoid cysts) the majority of which prove to be malformed teeth, most prominently misshapen crowns that are radiologically and visually obvious (Figs. 8.36:Thornton, 1988).

Uterine prolapse

Prolapse of the vaginal mucosa or uterus occurs most commonly as a consequence of weakened pelvic ligaments and especially musculature, frequently following childbirth. Pressure of a urine-filled bladder or feces-laden rectum can push the vaginal mucosa toward and even through the vaginal orifice. The suspensory ligaments may permit the uterus to descend or actually protrude out beyond the labia.

Paleopathology of gynecological lesions

Apparent prolapse of the vagina, uterus or rectum are common forms of pseudopathology in mummified human remains. These are usually the consequence of postmortem fermentative processes with gas formation that result in intra-abdominal pressure increases, forcing viscera out of the available orifices (vagina, rectum or possibly even the paraesophageal area). I have seen one case in a spontaneously mummified body of the Chinchorro culture of northern Chile in which not only the uterus but both fallopian tubes and ovaries had prolapsed externally – undoubtedly a postmortem artifact. Smith & Wood Jones (1910:268) feel, however, that a fibrous vaginal polyp may have been responsible for an antemortem vaginal mucosal prolapse in an adult Nubian female body from the Byzantine Period, and in another from the same excavation area, a vaginal cyst had dragged the vaginal mucosa to the introitus. That uterine prolapse in living persons was familiar to

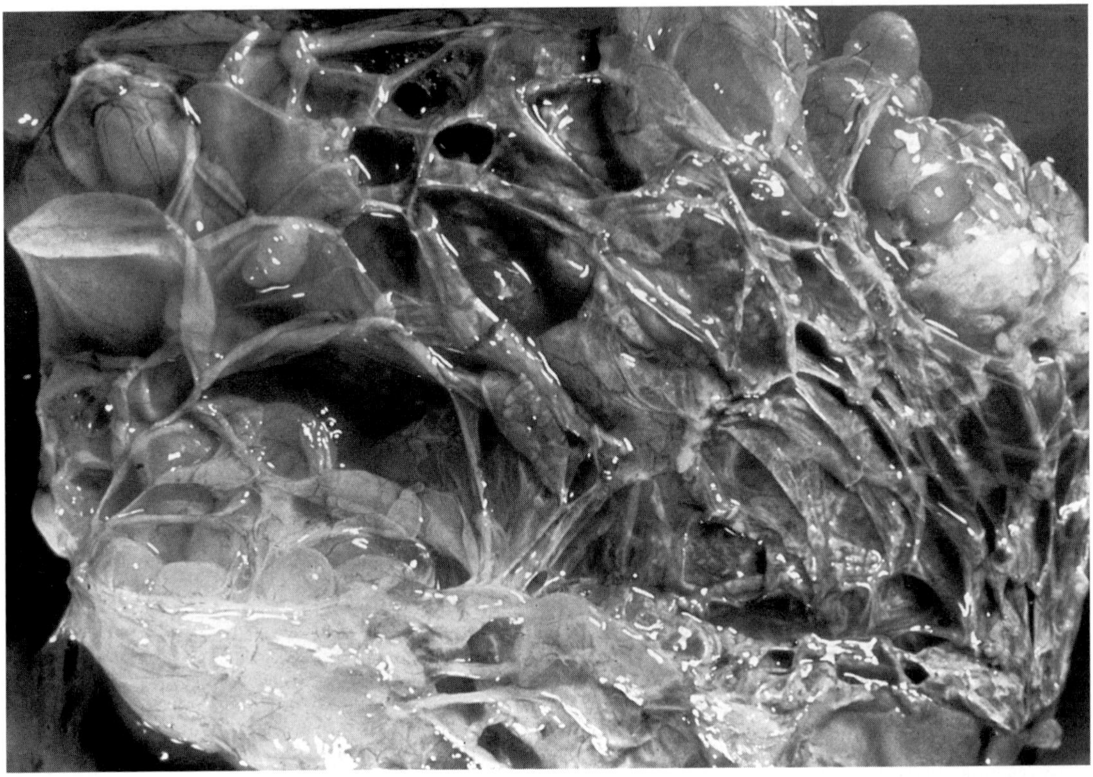

Fig. 8.34. Benign ovarian cystadenoma.
Sectioned surface of tumor shown in Fig. 8.33. Note the multilocular internal structure (numerous smaller cysts) virtually devoid of solid tissue. Modern middle-aged adult female. (PBL.)

Egyptian physicians is evidenced by the Ebers papyrus (Ebbell, 1937:109, column 43): "remedy to cause a woman's womb [uterus] to go to its place . . .". Janssens (1970:118) describes a vesicovaginal fistula in the mummified body of the Egyptian princess Hehenit of the eleventh dynasty, speculating that it may have been the product of a protracted delivery.

At least two benign neoplasms of the uterine wall have been reported in archaeological bodies. These are believed to be smooth muscle tumors (leiomyomas, often called fibroids), but usually their survival of the postmortem decay process is enhanced by calcification of infarcted areas. Strouhal & Jungwirth (1977) found a 123 mm long nodular structure in the pelvis of an adult Nubian (Sayala) female skeleton dating to the third to fourth century AD. It was extensively calcified but also demonstrated an organic matrix. Similarly, Kramar *et al.* (1983) also employed radiological, chemical, histological and ultrastructural techniques on a 50 mm calcified (hydroxyapatite) mass that was also found to contain collagen; this mass was associated with a group of human skeletons from a 5000-year-old Neolithic burial ground in Switzerland. In both cases uterine leiomyoma seems the most probable diagnosis.

One would expect that the malformed teeth found in about one-third of ovarian benign teratomas (dermoid cysts) would have been identified even in skeletonized bodies but we have not been able to find such a report. Their small size and atypical appearance may escape attention at excavation and dissection of mummies, but in skeletons only awareness of the significance of

Fig. 8.35. Ovary: cystadenocarcinoma.
Note the partly cystic and partly solid nature of this malignant cystic tumor. Modern adult female. (PBL.)

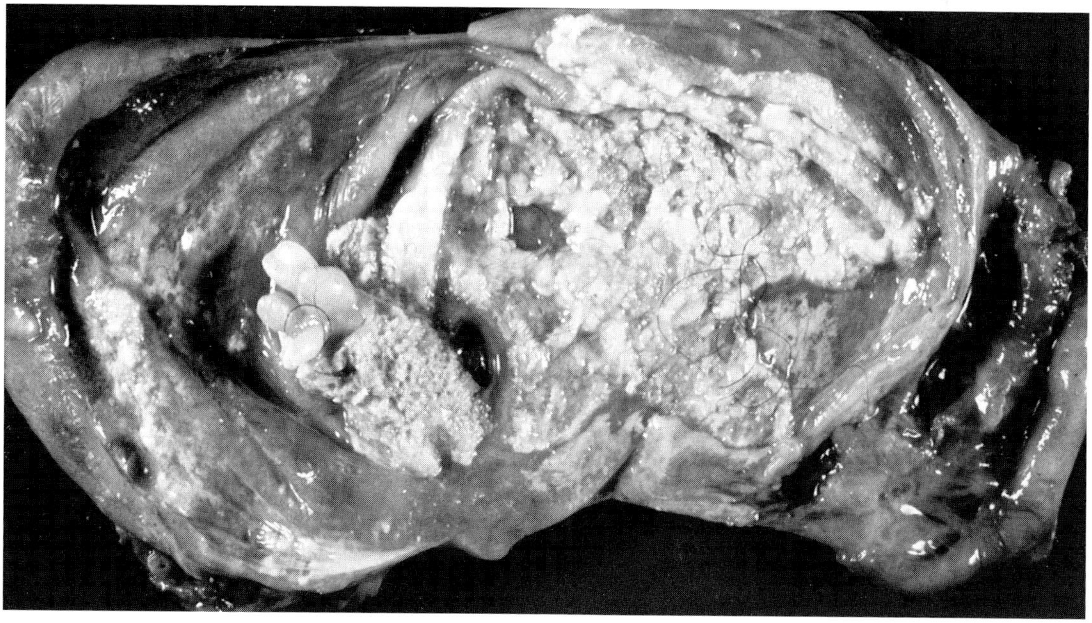

Fig. 8.36. Benign ovarian teratoma ("dermoid cyst").
While structures of all three germ layers are usually present, in benign ovarian teratomas the ectodermal elements predominate. Epidermis lines the cyst, filling it with oily sebum from the sebaceous glands and hair from the skin's hair follicles. In addition, malformed tooth crowns are commonly prominent in the cyst wall. The latter can be recognized by their malformed morphology when these are the only surviving tissues in skeletons. Modern adult female. (PBL.)

finding isolated, misshapen teeth would lead to the diagnosis (Thornton, 1988). While no malignant tumors of female internal genitalia have been identified in mummified bodies, the Ebers papyrus suggests that Egyptian physicians may have been aware of them (Ebbell, 1937:111, column 45): "Remedy against 'eating' [cancer?] in the womb which produces phagadena in her vagina."

In addition to the above we have a description of an adult Egyptian mummy dissected by Granville (1825:298) in which he states that "the disease which appears to have destroyed her was ovarian dropsy attended with structural derangement of the uterine system general . . .". He also notes that the uterus was larger than that expected in a 50–55-year-old woman, that there must have been "a large sac connected with the left ovarium . . ." and that "the ovarium and broad ligament of the right side are enveloped in a mass of diseased structure . . ." and that the right fallopian tube is normal. Figure 1 of Plate XXII (Granville, 1825:316) contains a drawing purporting to give evidence of these changes. The drawing reveals a normal-appearing right tube with the right ovary and parametrial tissues in a tangle of structures continuous with, and overlying much of, the uterine body. The upper part of what appears to be a bulbous uterus emerges from behind these tangled structures. The left tube area is shredded into three linear, strap-like structures terminating in two, rounded folds. Granville's interpretation of what today would probably be called bilateral ovarian cystadenomas or cystadenocarcinomas is consistent with this drawing. The uterus would not be expected to show a substantial enlargement without distortion of its form if it were invaded. The cystic nature of the lesions is not obvious in the drawing. If the lesions were not cystic, other causes need to be considered. Alternatively, malignancy of the uterus (either endometrial or cervical) with parametrial invasion is also possible. Even a simple cystadenoma of the left ovary and late stage pelvic inflammatory disease of the right ovary and parametrium cannot be excluded (though the uninvolved right tube would make that unlikely).

Pregnancy-related conditions

Ectopic pregnancy and abortion

We can expect women in antiquity to have been vulnerable to all of the pathophysiological, metabolic and hemorrhagic hazards of pregnancy as well as those relating to the placenta, to parturition and to postpartum complications.

Normally the ovum meets the sperm and is fertilized during its traverse through the fallopian tube, though sometimes this encounter occurs prior to entry into the salpinx. If the fertilized ovum never enters the tube, it may implant on the ovary (rarely the peritoneal surface). If its passage through the fallopian tube is delayed or arrested (often by mucosal scarring from previous infection, with associated lack of mucus propulsion due to loss of ciliated mucosal cells), implantation in the tube (ectopic pregnancy) may occur. The latter is by far the most frequent ectopic site. The hazard at an ectopic site is one of the hemorrhage, since the placenta invading the structure supporting the ectopic implantation commonly penetrates the full thickness of the supporting tissue. In the tube this occurs about a month after implantation (6 weeks after the last menstrual period). The external surface, with its exposed, highly vascular placental base, may result in serious and even lethal, sudden intra-abdominal hemorrhage (Fig. 8.37). Ovarian implantations may survive a few weeks longer but suffer the same hemorrhagic hazard. The muscle underlying peritoneal implantation sites, however, may support fetal growth into the third trimester.

A substantial fraction of pregnancies today terminate in spontaneous abortion – most in the first trimester. Death of the developing embryo is usually followed by placental necrosis as well. The entire necrotic mass may be sloughed with little greater bleeding than that anticipated in a normal menstrual period. Incomplete placental necrosis with continuing circulation of blood into the nonnecrotic portion may result in more abundant bleeding and occasionally this can be life-threatening without some form of surgical

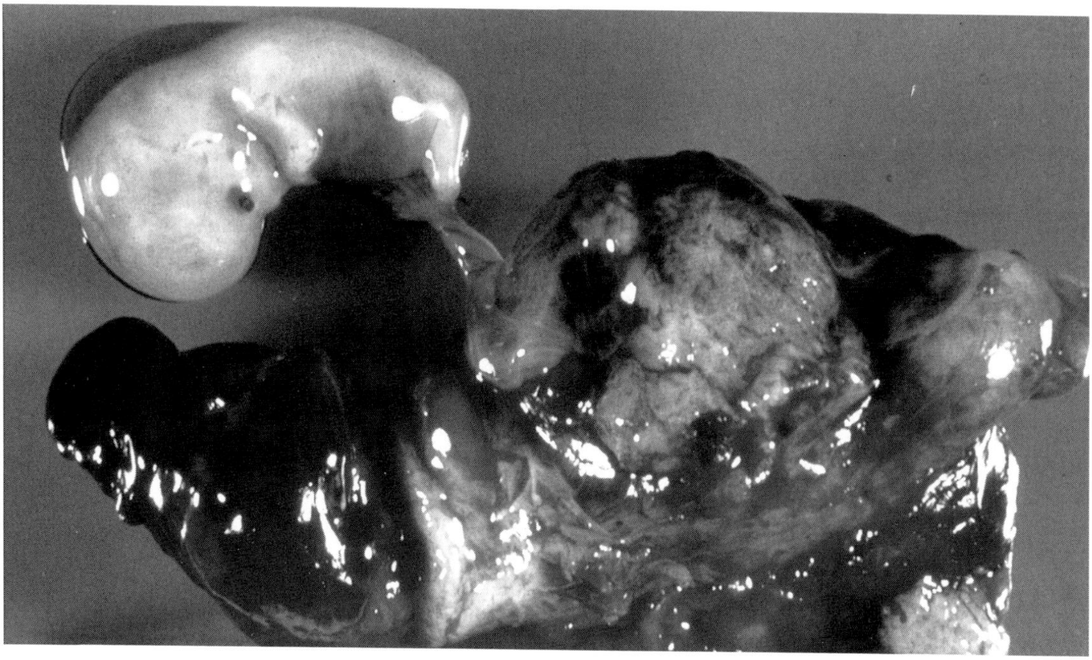

Fig. 8.37. Fallopian tube: ruptured ectopic tubal pregnancy.
The placental growth has expanded the diameter of the tube. Complete placental penetration of the tube wall resulted in hemorrhage, rupture of the tube and expulsion of the fetus. Modern young adult female. (PBL.)

intervention. The more advanced the pregnancy, the more probable is serious hemorrhage when spontaneous abortion occurs.

Paleopathology of ectopic pregnancy and abortion

Since almost 1% of pregnancies implant at ectopic sites today, and since 90% of these are in the tube, the soft tissue paleopathologist has a reasonable expectation of encountering this complication. In spite of these statistics, the condition has not been reported. If it has not been overlooked, the most likely reason for its absence is that the principal risk factor (pelvic inflammatory disease complicating gonorrheal or chlamydial venereal infections) was rare or did not exist in antiquity. While the lesion in the tube might not be identified with ease, the exsanguinating abdominal hemorrhage would be obvious.

A large number of abortions today are secondary to nonviable chromosomal damage or gross DNA alterations, so it is probable that this problem also plagued the women of ancient populations. Lethal hemorrhage from this source may not be obvious in a mummified body. Residual blood may still be found in the uterus or even the vagina, though the latter area may be cleaned by those preparing the body for burial. Furthermore, abortion may occur so early that the uterine enlargement may not be overt either. Control of family size may have included induced abortion by amniotic sac perforation in some ancient groups. This would surely have involved a greater hazard of hemorrhage because it most probably would have been performed at a somewhat later stage of pregnancy and the lack of aseptic technique can be expected to have been followed by a substantial increase in uterine infections. Whether or not abortion was induced (usually by a probe or chemicals as some were: Sandison & Wells,1967:499) would not be readily detectable but the exudate in the uterus and its thickened wall together with parametrial tissue involvement might well betray the infectious complication.

Lithopedion

In exceptional circumstances (about 0.0005% of all pregnancies) the dead fetus and pregnancy products become necrotic but are not expelled by the uterus. A few rare examples of this have been reported in cases of normal, intrauterine implantation, but the overwhelming majority are the product of ectopic, intra-abdominal, extrauterine implantation of the fertilized ovum. Christianson (1992) describes the earlier post-mortem stages in swine. Autolysis of the fetus is initiated promptly after death, with resulting edema and swelling of the body. This swelling may be equivalent to the human form of hydrops fetalis: an immediately premortem event in the human fetus. Uterine absorption of the fluid converts the body into a desiccated, shriveled "mummy". Such necrotic tissue will occasionally undergo dystrophic calcification. The form of the fetus is retained and often remains easily recognizable. More than 100 years ago Keuchenmeister (1881) described the three forms such calcification may take, involving (1) only the membranes, lithokelyphos; (2) both membranes and fetus, lithokelyphopedion (Fig. 8.38); and (3) only the fetus, lithopedion. This useful classification is still employed today. The three forms of lithopedion may remain *in situ* for years and such women may be otherwise asymptomatic, although erosion of the bones into the colon may result in their eventual expulsion. More than 300 cases of such retained, calcified feti and/or pregnancy products have been reported (Spirtos *et al.*, 1987). Rothschild *et al.* (1994) found vertebrae and calcified membranes of a lithokelyphopedion in the level corresponding to 3100 years BC of a Texas sinkhole used as a burial site in Texas' Archaic Period. Their diagnosis rests entirely on the assumption that the concave calcified mass represents calcified placental membranes. This is a reasonable assumption, but weakened by its burial context.

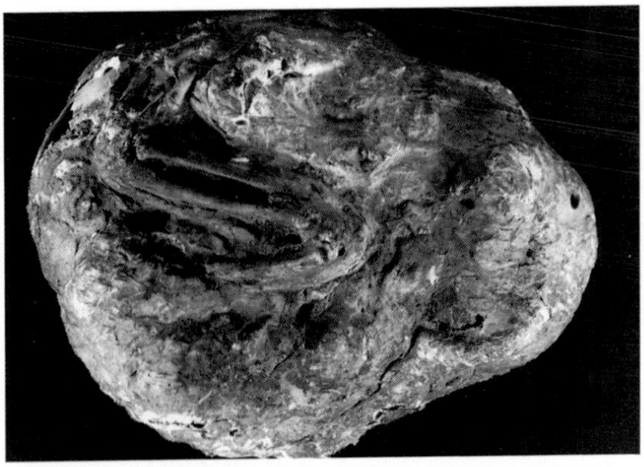

Fig. 8.38. Lithokelyphopedion.
Both the calcified placental membranes (lithokelyphos) and calcified parts of the infant's lower extremities (lithopedion) are evident. Provenience data unavailable but specimen is from modern adult female. (NMHM.)

Had it been found in an isolated, adult burial, certainty would have been enhanced. The next-oldest case was reported by Caspar Bauhin (1586) followed by several other sixteenth century physicians cited by King (1954) and D'Aunoy & King (1922).

Eclampsia

During the later stages of pregnancy, certain poorly understood physiological changes may occur that are grouped under the term eclampsia. This condition is characterized by a constellation of findings, the principal ones of which include hypertension, acute renal failure, convulsions and multifocal areas of hemorrhage. The kidney and brain lesions may be fatal. The glomerular changes are too subtle to expect them to be identifiable in a mummified body, but the uremic pericarditis with precipitated exudate on the visceral pericardium can be recognized grossly. The focal hemorrhages in the brain, liver or other organs are usually large enough to be apparent. Thus the finding of a fetus consistent with the third trimester in a mummified body showing either pericardial exudate and/or brain or other hemorrhages can suggest this diagnosis. To date it has not been reported in the paleopathology literature.

Placenta previa

During labor two major maternal problems may occur: hemorrhage and obstruction to fetal passage; lesser but still major concerns include perineal lacerations. The lacerations themselves may result in vesicovaginal or rectovaginal fistulae. The past Arabian custom of packing the vagina with salt may initiate enough inflammation to generate such fistulae even without lacerations (Naim & Fahmy, 1965). Hemorrhage can be life-threatening in cases of placenta previa. In this circumstance the ovum implants in the lower portion of the uterus and part of its placenta may lie across the internal cervical os. Thus during labor that opening expands to allow the fetus exitus from the uterus. As it opens (usually to about 10cm) the portion of the placenta covering it is torn. Resulting hemorrhage may be minor, especially if the amniotic sac ruptures and permits the fetal head to descend to the cervix and compress the placenta there. If the placenta completely covers the cervical exit area, however, the bleeding may be so major as to be fatal. Since, in antiquity, the infant might well be allowed to remain *in situ* at burial, this condition could be recognizable in a mummy. There is, however, no such report as yet.

Obstruction to fetal passage

This commonly is due to abnormal pelvic skeletal structure or an atypical fetal position. Skeletal alterations can be secondary to nutritional deficiency (rickets), fracture deformations, infections (Micozzi, 1982) or skeletal neoplasms. While such problems today are dealt with surgically (cesarean section), in antiquity they could represent a survival threat to both mother and child. In marginal situations a good deal of soft tissue molding can permit eventual passage, but cranial dimensions dictate the ultimate parameters. On the other hand, certain fetal positions are incompatible with delivery, but conceivably in antiquity efforts (sometimes successful?) might be made by those attending the mother to alter the position.

Paleopathology of birth obstructions

The paleopathology literature provides examples of both types of problem. A pelvis deformed by the presence of multiple exostoses provided so much obstruction to fetal passage that it completely blocked delivery of a mature fetus in a medieval woman. The mother's skeleton with its restrictive pelvis was found with that of a full term child in the birth position (Sjovold *et al.*, 1974). In another case (Joseph G. Bohlen, personal communication) the mother's skeletonized body was found with that of the fetus in a transverse position but lacking an arm. The missing arm may represent an unsuccessful effort to extract the fetus, though it could of course have been amputated postmortem (Fig. 8.39). Williams (1929b) also cites a description by Derry of a negress (in a Coptic cemetery) whose congenitally absent right sacroiliac joint and small right innominate bone narrowed the birth canal sufficiently to result in impaction of the fetal head at the constricted site. More recently van Gerven (1981) identified the skeleton of a Nubian woman from the Christian period who died with a term fetus in a breech position in the birth canal. A similar case has been reported by Regöly- Mérei (1961), cited in Sandison & Wells (1967).

Postpartum complications

These include infection (puerperal sepsis), placental hemorrhage and uterine prolapse. Infection of a recently pregnant uterus often produces septicemia quite abruptly. Occasionally maternal feces contain *Clostridium perfringens* and if the uterus becomes contaminated by such feces postpartum, gas gangrene septicemia may occur, with fatal results. Such contamination may occur via manipulation by midwives (e.g., efforts to assist expulsion of the placenta). Indeed, recognition of the cause of puerperal sepsis through contamination by attending physicians did not occur until 1795 when Gordon called attention to it and Semmelweis popularized the value of aseptic obstetrical techniques about 40 years later (Lowis, 1993).

Fig. 8.39. Obstructed birth canal; SM AD 1100.
Maternal female skeleton with near term fetal bones in pelvic area arranged in transverse (undeliverable) position. Note the right humerus present in the birth canal. All fetal bones were present except right forearm and hand. Young adult female. (Personal communication, Joseph Bohlen, MD., Springfield, IL.)

In mummies puerperal fever may be so fulminant that only the indications of pregnancy may be apparent. If the process was less fulminant, intrauterine exudate and involvement of parametrial tissues may suggest the diagnosis. A "foamy liver" (Fig. 8.14) in a mummy with signs of recent birth can reflect puerperal fever due to *Clostrium perfringens*.

Hemorrhage secondary to abruptio placenta may occur at this stage of pregnancy. In this condition the placenta is only partially separated from the uterine lining by uterine contractions. The opened veins may then suck in sufficient amniotic fluid to cause intravascular coagulation that depletes the blood's coagulation proteins, producing major or lethal hemorrhage. In a mummified body, evidence of the multiple areas of hemorrhage in various organs and tissues of a body also showing signs of term pregnancy or recent delivery could imply this diagnosis. However, I could find no report of such a case in antiquity. In placenta accreta, the placenta can not be separated from the uterus after delivery due to fibrotic changes. This can also result in serious hemorrhage, and in mummies at least part of the placenta would be found still attached within the uterus.

At least one case of perineal laceration with resulting fistula between the bladder and vagina has been reported (Williams, 1929b). Rabino-Massa (1982) has also demonstrated that a prolapsed uterus in an Egyptian naturally mummified body buried with a newborn child was an antemortem finding by presenting histological evidence that the uterus showed complete inversion.

Maternal mortality in antiquity is difficult to evaluate. The expression of mortality rates in many ancient populations often demonstrates a distinct increase in female deaths in the 20–25 year age group (see Manchester, 1983:9, Figure 5). This is commonly interpreted as pregnancy-related deaths, though little supportive, specific physical evidence of such a cause is available. Mafart (1994), however, suggests that the maternal mortality in two French skeletal series from the fifth/sixth and the eleventh/twelfth centuries AD did identify childbearing risks, but that these were quantitatively insufficient to contribute significantly to the general female mortality. Interpretation of this demographic feature requires further analysis involving demonstrated causes of death. Increasing mummy studies may render this possible in the future.

A highly informative study of maternal mortality was carried out by Arriaza *et al.* (1988) on 18 examples of childbirth-complicated deaths extracted from 128 female, spontaneously mummified bodies of childbearing age (12–45 years) who lived in a single northern Chile valley over an interval of 2700 years.This would suggest a maternal mortality rate of about 14%. This is in sharp contrast to the approximately 1% value in rural parts of undeveloped countries today (Harker, 1993). The large difference in these two figures probably lies in the small sample of the study involving ancient populations. These workers found that 3 of their 18 bodies had failed to complete delivery (one was breech presentation, one had failed to deliver the second twin and the third died of unknown causes with a 7–8-month fetus in the uterus – eclampsia?). The cause of death in the remainder was not obvious, but occurred in the puerperium, though in several the placenta was still present. Of diagnostic value for the recognition of death during or a short period before or after parturition are the intrauterine presence of a fetus, presence of placenta or umbilical cord, enlarged uterus, many redundant abdominal skin folds, lactating breasts as well as separation of the pubic symphysis and/or sacroiliac joint.

Breast diseases

Normal

The normal human breast consists of 5–12 clusters of potentially milk-forming, glandular tissue separated by loose fibrous tissue interwoven with a variable amount of fat. Each cluster is drained by a duct leading to the nipple. Substantial differences in fat and fibrous content lead to a significant normal variation in size, texture and morphology not necessarily dependent on somatic size or total body fat. In addition the cells composing the ductular and glandular tissue contain estrogen and progesterone receptors, and thus respond to these blood hormone level fluctuations, both those involved in the normal menstrual cycle as well as in pregnancy. Thus the newborn breast of either sex will demonstrate a transient increase in the neonatal period in response to the maternal estrogen stimulus at term, and again at puberty (regularly in the female and transiently in some males). The prolonged estrogen elevations of pregnancy provide time for substantial increase in glandular tissue at the expense of the fat, with a further response to other hormones during the later pregnancy stages, leading to production of milk. The retardation of ovarian function after menopause drops the blood estrogen level to the low quantities contributed primarily by the adrenal cortex, and the breast glandular tissue undergoes substantial atrophy, though only in the advanced elderly state does the glandular tissue virtually disappear, merely the ducts remaining embedded in fibrous stroma.

Prior to the domestication of cattle, the importance of the human breast was obvious: it was essential for the survival of the infant and, therefore, for that of the population. Human milk supplies all of the infant's essential nutrients except vitamin D and iron. The maternal antibodies of the IgG class present in the milk cross the placenta and provide at least a useful degree of protection until the infant's immune system can supply its own. Even the transient decline in body iron content in the purely milk-fed infant may have had some beneficial inhibitory effect against infection because most pathological bacteria require an abundant, readily available iron supply from their environment (Stuart-Macadam & Kent, 1992:5). Bottle-feeding with baby food "formulas" has been associated with a striking increase in infant mortality in developing countries, due principally to lack of aseptic techniques, with resulting infectious infant diarrhea (Lithell, 1981). Weaning time also influences infant mortality. Premature withdrawal of breast feeding may introduce pathogenic bacteria into the infant's intestine before its immune system is prepared to deal effectively with them, resulting in similar infections (Rothschild & Chapman, 1981:586). Furthermore, lactation prolongs the postnatal period of amenorrhea with its associated lowered fertility level (Ellison, 1987).

Infectious lesions

In spite of every precautionary effort, even modern breastfeeding practices place the breast at increased risk for infection. Staphylococci, streptococci or other organisms may gain access to the dermal areas through small skin fissures. The resulting infection may spread widely and rapidly in the form of a cellulitis (commonly caused by streptococci) or, more frequently, produce a localized abscess (usually staphylococci). Untreated either of these, but especially the former, may spread to the bloodstream and represent a threat to the mother. The localized lesions often drain spontaneously. The nursing infant is at risk in either form of infection. The localized abscess may reach substantial proportions, and usually heals by scar formation that may distort the breast.

Fibrocystic disease of the breast

The cause of the condition is unknown but manifests as either the formation of one or more cysts within the breast, as foci of dense fibrosis or, most commonly, as both. The cysts may be diminutive but can reach 5 or 6 cm in diameter (Fig. 8.40). They are lined by epithelium similar to that lining normal breast ducts. Hence they frequently contain estrogen receptors and enlarge and become tender at certain stages of the menstrual cycle. This disease is very common in Western countries today and usually appears during the late second or third decade of life. Except for a small subgroup that also demonstrates marked epithelial hyperplasia with dysplasia microscopically, the condition is benign.

Neoplasms of the breast

Fibroadenoma

The most common breast tumor is a benign, sharply circumscribed lesion called fibroadenoma (Fig. 8.41). Its size varies from a nearly microscopic dimension to 4 or 5 cm. It is usually discovered and removed surgically when it is about 2 or 3 cm in diameter. It afflicts the same age group as fibrocystic disease, with which it occasionally coexists. Its histology is a simple mixture of breast ducts entwined within and compressed by lobules of fibrous tissue.

Papilloma

A far less common tumor of the breast is an intraductal papilloma. This is a benign, branching tumor with a tree-like structure. It arises as a solitary lesion in one

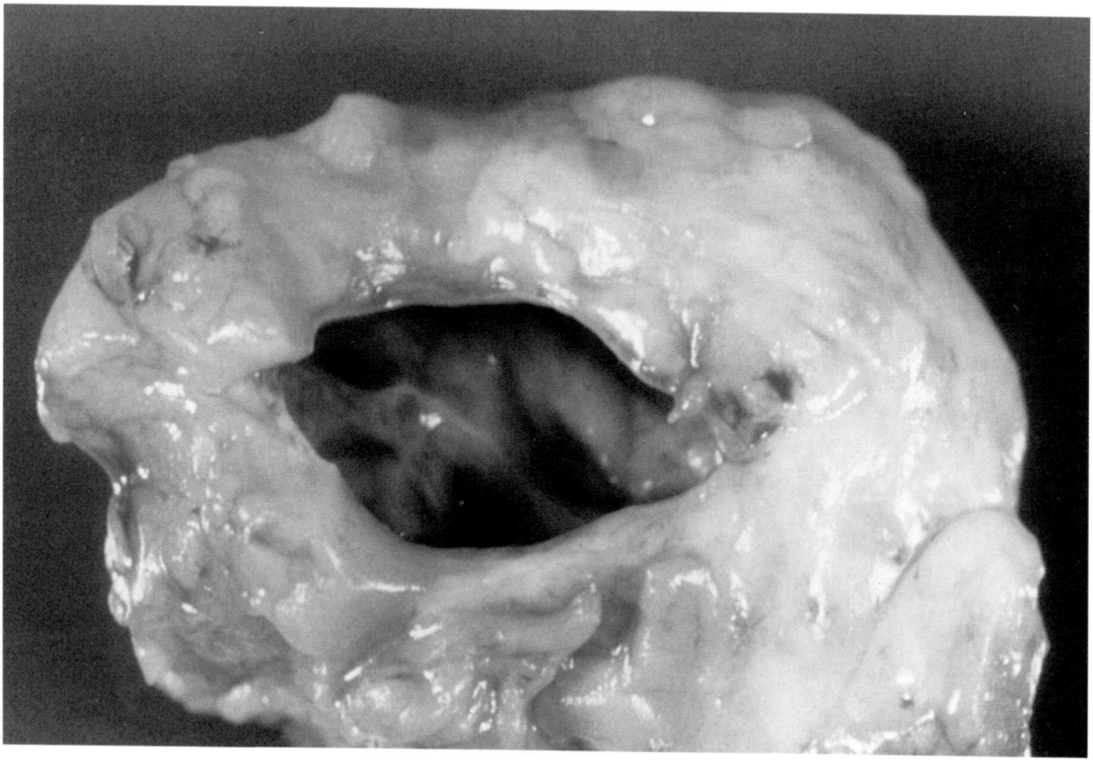

Fig. 8.40. Breast: benign cyst.
A smooth-lined, simple 2 cm cyst is embedded in a fibrofatty area of the breast. These are very common in Western countries today, and their size varies from microscopic to several centimeters in diameter. If they were present in antiquity, they would be expected to be recognizable in mummified human remains. Modern young adult female. (PBL.)

of the main ducts along its course to the nipple. It seldom exceeds 2 cm in diameter and is so fragile that normal pressures on the breast may break off an "arm" or "branch", resulting in expulsion of a few drops of blood from the nipple (which is the usual initial symptom). With few exceptions this, too, is a benign lesion.

Carcinoma

Carcinoma arising in the breast is the most feared neoplasm. It occurs infrequently in women less than 30 years of age, its frequency then rising subsequently with progressive age. Most cases occur after menopause. Variations in gross and microscopic features as well as location are enormous. Nevertheless, the most common forms are characterized by tissue hardness due to an abundant fibrous stroma that often shows focally calcified areas ("scirrhous" carcinoma). The tumor causes necrosis of normal structures, replacing them with tumor growth until a mass of about 1 cm or more is reached. Quite commonly tumor cells then infiltrate along local structures, resulting in a central tumor "body" with numerous, linear extensions radiating from it into surrounding tissues in every plane (Fig. 8.42). This gross appearance of a central mass with peripheral extensions suggested the structure of a crab to the early pathologists, leading to the term "cancer" (Latin term for crab) for this and any other malignant tumor.

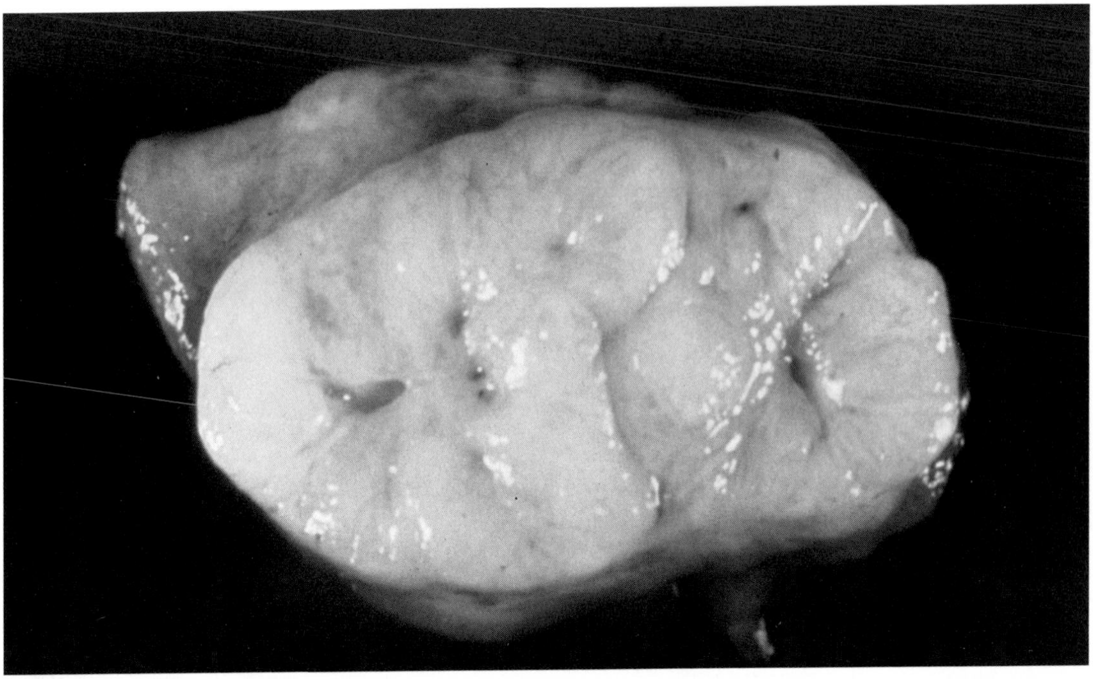

Fig. 8.41. Breast: benign fibroadenoma.
A 3.5 cm × 2 cm × 2 cm white, predominantly fibrous breast nodule has been excised surgically. Its sharply demarcated periphery suggests benignity. Because of its almost exclusively fibrous tissue component this lesion has a high probability of remaining intact in a mummified body. Modern young adult female. (PBL.)

Untreated the tumor commonly extends to the breast's skin surface (Fig. 8.43), where it ulcerates and usually becomes infected (a stage one would expect in archaeological bodies if breast cancer occurred in antiquity). It also may grow into the chest wall, penetrate it and induce pleural effusion or even invade the lung. In later stages it spreads to the nodes draining lymph from the area. The most evident involved nodes are usually those in the axilla on the same side, though those above the clavicle or in the retrosternal regions are often involved as well. Finally, spread (metastasis) may occur through the bloodstream, which can carry tumor cells to any organ in the body. Though no patterns of metastatic areas are unique to breast carcinoma, soft tissue sites that are particularly common include the lung and liver. In end-stage disease bones are often involved by metastases that are usually of lytic nature, though a minority induce distinct and even obvious osteoblastic features. (Bullough, 1992).

Gynecomastia

Enlargement of the male breast is called gynecomastia (literally: female breast). At birth and during childhood the histology of both male and female breasts is identical: simple ducts surrounded by fibrous tissue. Throughout life, in fact, male breast lining cells remain sensitive to estrogen, as manifested by their identical response at birth and initially at puberty as outlined above. In females the rising and sustained estrogen and progesterone production from the ovary stimulates continued cellular proliferation, with the final development of the glandular breast elements. In the male,

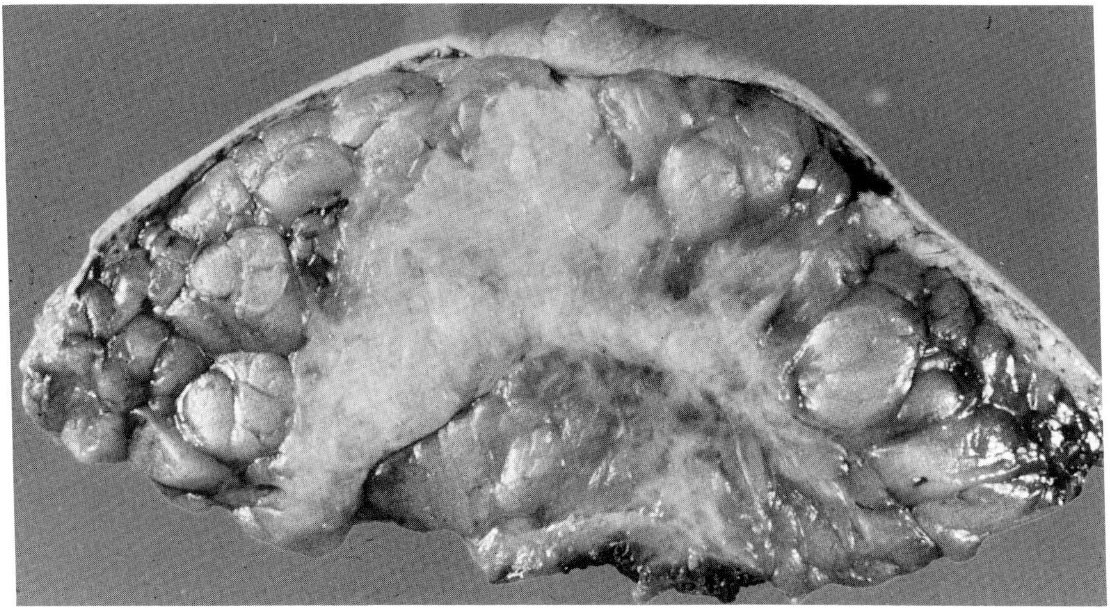

Fig. 8.42. Breast: scirrhous carcinoma.
The breast has been bisected through the tumor. The nipple-containing skin is at the top of the photograph. The underlying breast tissue demonstrates the lobulated, fatty appearance of a non-lactating breast, much of which has been replaced centrally by an irregularly shaped mass of white cancer tissue. This was hard (scirrhous) to palpation and at its periphery poorly defined, finger-like extensions blend with the fatty tissue where the cancer is invading it. Modern adult female. (PBL.).

normally testicular androgen production causes the stimulated cellular proliferation to cease and finally to subside, returning it to the prepubertal state. In addition, as in the female, small amounts of estrogen are also made by the male adrenal cortex (the testis also generates small quantities of estrogen precursors that can be converted into estrogen by enzymes in fat and muscle). Thus maintenance of the appropriate ratio of (large amounts of testicular-formed) testosterone to (the adrenal and extratesticular sources of) estrogen will maintain a male breast in its usual atrophic, adult form. When the male ages and develops testicular atrophy, the testosterone level drops, and his breast elements may once again respond to the proliferative stimulus of endogenous estrogen that continues to be secreted from the extragonadal sources.

This discussion makes it clear that the male breast is free to respond to an estrogenic stimulus if the testosterone:estrogen ratio (about 300:1) is altered by either an increase in estrogen or a decrease in testosterone. The latter is characteristic of the elderly male with testicular atrophy or the genetic XXY abnormality known as Klinefelter's syndrome in which no functioning testicular tissue is formed. Increased estrogen in males has several potential mechanisms that include estrogen-secreting testicular tumors or a variety of genetic mutations that affect the function of extragonadal enzymes that convert testicular precursors into estrogen, or that can block the action of testosterone on the breast cells. Some drugs (digitoxin, marijuana) can also duplicate these effects (Wilson, 1991; Braunstein, 1993). Cirrhosis or other chronic liver injury (drugs) can reduce the liver's ability to metabolize and destroy estrogen, raising the effective estrogen level.

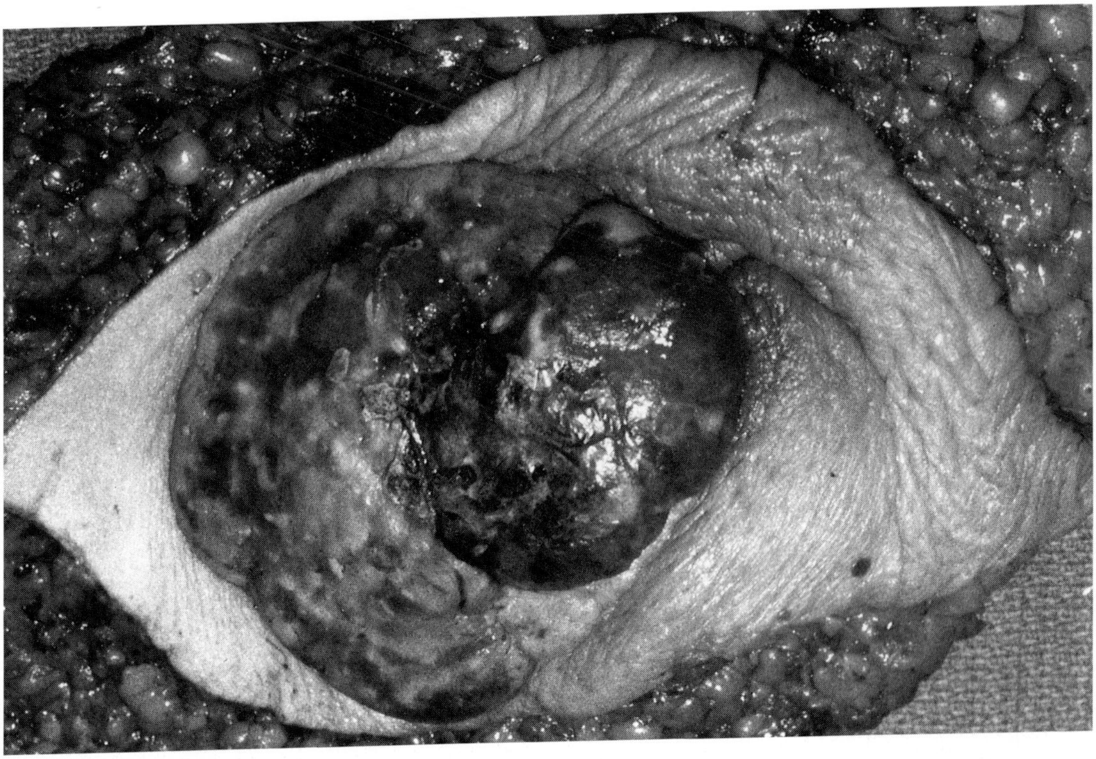

Fig. 8.43. Breast: late stage carcinoma.
The cancer has extended through the skin. It continuing expansion had caused it to project as an irregular mass above the level of the surrounding skin. If breast cancer occurred in antiquity one would anticipate that its untreated course would bring it to this easily recognized stage at the time of death frequently. Modern, aged adult female. (PBL.)

With this admittedly unpretentious background, the ongoing controversy of the presentation of apparent gynecomastia in certain royal Egyptian families presented below can be understood.

Paleopathological aspects of the breast

Appearance of the mummified breast

In addition to the variations in the size of the normal, resting (i.e. nonlactating) breast in living women, further variations are added by differing ratios of glandular:fat tissue (Fig. 8.44). Postmortem hydrolysis of neutral fat by endogenous and bacterial enzymes generates fatty acids which, in an alkaline environment become ionized and removable by groundwater (or seep diffusely into surrounding tissues). In women with a low glandular:fat breast tissue ration, removal of breast fat by the processes mentioned above can collapse the breast of a spontaneously mummified body to a degree that minimizes the breast morphology beyond recognition. In an unpublished study (A. Aufderheide) the sex of only about half of spontaneously mummified adult, nonpregnant female bodies was identifiable with certainty on the basis of a grossly recognizable breast. Transection of nonpregnant and nonlactating breasts reveals what one would expect, since epithelium does not normally survive the postmortem environment:

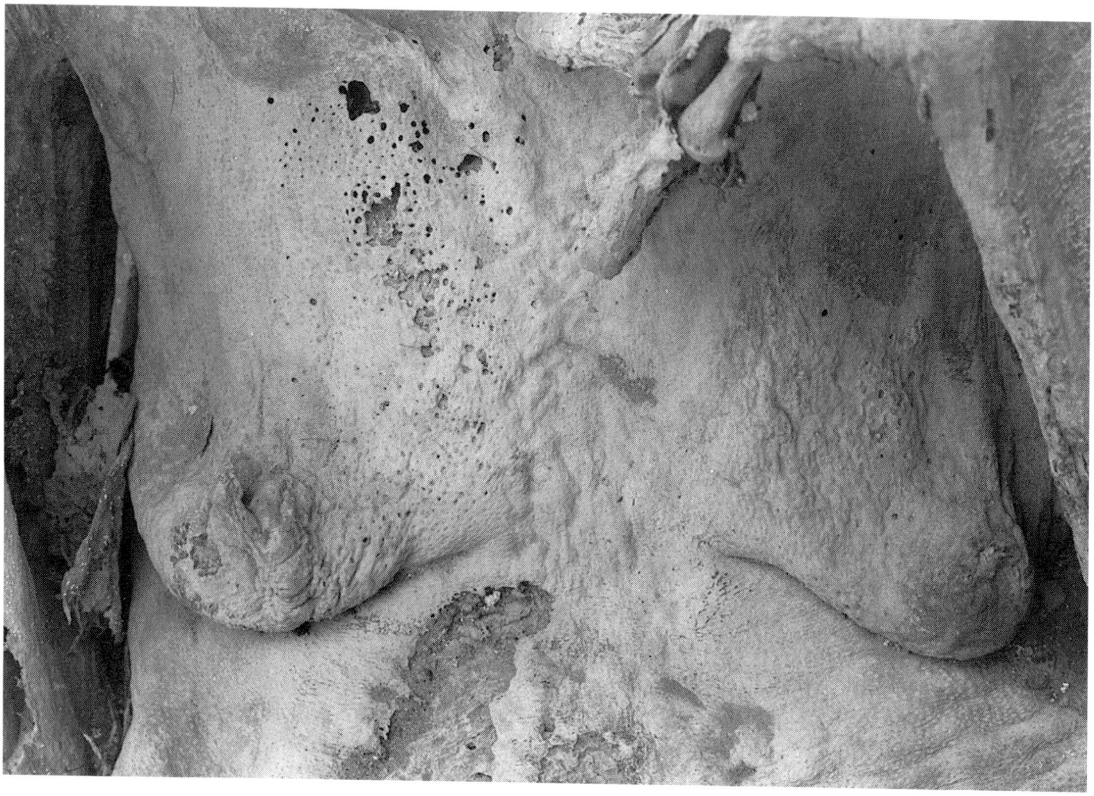

Fig. 8.44. Breast: normal; SM, AD 1100.
Varying rates and degrees of postmortem breast fat hydrolysis results in a wide range of breast prominence in spontaneously mummified bodies. Some, like this example, are obvious, but I have found sufficiently well-preserved breast tissue to permit sex identification in only about half of spontaneously mummified female human remains. Female, 55–60 years old, southern Peru. (PBL.)

the "fibrous skeleton" of the breast remains in the form of interconnecting, fibrous "leaves" separated by empty space. In such a breast, scars from previous abscesses would not be easy to detect, but most invasive carcinomas would be overt, demonstrating a hard, fibrous mass at least several centimeters in size, often ulcerating, with irregular peripheral "fingers" reaching the skin surface and/or chest wall, and with secondary axillary masses. To date, no such lesion has been described in a mummy.

Lactating breasts, or those in later stages of pregnancy, however, are commonly obvious grossly. Histologically the glandular pattern can even be recognized after the epithelium has degenerated (Zimmerman & Aufderheide, 1984).

The breast as symbol

One would expect to find diseases affecting an external organ so visible, accessible and vital to the ancients as was the breast, to be well represented in ancient lore, literature and art. The fact that they are not, suggests that the conditions of principal concern affecting the breast in modern populations (benign fibrocystic disease and carcinoma) may not have been common problems in antiquity. It is true that the female breast is a central feature in many items of ancient art, but it is employed there primarily in a symbolic manner, usually to represent fertility. The so-called "Venus figures" (Lyons & Petrucelli, 1987:24, figure 16) of paleolithic and mesolithic times demonstrate grotesque exaggeration of breast size, almost surely intended as symbolism (though at least one author has viewed them as literally representations of the pathological condition called "massive" or "virginal" breast hypertrophy: Harding, 1976). During the Hellenistic period we have no better example than the statue of Artemis in her temple at Ephesus, where her nurturing character is portrayed as a human female body with 16 breasts. The importance of the breast in antiquity can be appreciated by noting that this organ was elevated to the level of divinity in Egypt (Hathor, the cow goddess) and reliefs demonstrate pharaohs and even commoners sucking at the teats of Hathor's exaggerated udder (Strouhal, 1992b:25). The later Romans repeated the icon, substituting a wolf suckling Romulus and Remus, Rome's legendary founders. The breast is also occasionally used to symbolize close emotional attachment, as in the Chinese statuette demonstrating a woman breastfeeding her aged grandfather (Lyons & Petrucelli, 1987:122, figure 192). A similar principle undoubtedly led to the former occasional Inuit practice of a mother greeting an adult child (returning from a prolonged absence) by exposing her breast to which her child responded by kissing it. The nutritive image of the breast is also evident in an antipodal mode among the Amazons – a legendary tribe composed entirely of bellicose females whose armies beheaded or subjugated males. Whether or not the tale is based upon a historical group, as Schecter (1962) seems to feel, is less important for our consideration than that they were believed to have amputated their right breast in order to enhance the use of their weaponry.

Paleopathological aspects of breastfeeding

Most of these relate to symbolism and have been dealt with in the first part of this section. In addition, we may note that weaning times were probably much later than at present. Aufderheide & Allison (1994) have used bone strontium levels to predict weaning in pre-Columbian, Preceramic natives of northern Chile, estimating the weaning time to have occurred between 2 and 3 years. Nitrogen stable isotope ratios may achieve similar predictions (Fogel & Tuross, 1988–1989). Sandison & Wells (1967) identify the weaning time in Mesopotamia and ancient Egypt at about 3 years, presumably based on literary sources. The vulnerability of the recently weaned infant to enteric infections was also pointed out by Strouhal (1992b:23), who found that childhood mortality peaked in that age group. The Egyptian papyri contain only a few references to breast problems, including recommendation for application of a lotion to "a breast that is ill" (Ebbell, 1937:110, column 95), though Sandison & Wells (1967) cite the Edwin Smith papyrus with inflammatory (probably lactation-related) breast conditions including abscess. They also cite Soranus (Alexandrian physician practicing in Rome about AD 100) with the observation that sudden regression of the breast was predictive of impending spontaneous abortion. Strouhal (1992b:22) describes an exotic lotion believed to have properties that would increase the flow in a mother's breast, and notes that most aristocratic Egyptian mothers contracted with wet-nurses who often came to play an important family role. Seventh century Mesopotamian texts include symptoms of, and treatment for, breast diseases (Saggs, 1962:462). Because of breastfeeding's importance, the adequacy of milk flow appears always to have been an item of concern to mothers. Even as late as the Aztec Period, Ortiz de Montellano (1990:178) cites Hernández (1577/1959) as stating that Aztec mothers employed sympathetic magic in order to stimulate their flow of milk by rubbing their breasts with plants that exude milky latex. Lactation may have been employed deliberately as a method of family planning in antiquity. Although viewed as a notoriously unreliable contraceptive regimen today, Ellison (1987) notes that breastfeeding an infant "on demand" (i.e., about every 15 minutes as is done today in many remote groups and was probably so performed in antiquity) enormously enhances its antiovulatory effect. Such an effect has been documented by the annual pregnancies of upper-class Elizabethan women and the much lower fertility of the poorer women to whom breastfeeding of the wealthy women's children was commonly delegated (McLaren, 1979).

Thus, while there is clear literary and graphic evidence that mothers have concerned themselves with the adequacy of their breast milk productivity for many millennia, and that they at least occasionally suffered inflammatory complications of lactation including abscess formation, such lesions have not been reported in ancient human bodies.

Fibrocystic disease

With its densely fibrous areas and gross cysts it would be expected that fibrocystic disease would be recognized in ancient mummified bodies, yet to date I could find only one reported case. The explanation may lie in its modern, worldwide epidemiological features. Although its etiology remains undefined, today it is a condition found with significant frequency primarily in Western developed countries (Hutt & Burkitt, 1986:115). Similarly, the fibroadenoma, composed primarily of fibrous tissue, would be easily identifiable, but only one of these (in an adult, Egyptian female) appears in the paleopathological literature (Reyman & Peck, 1980:93). The intraductal papilloma, however, is almost entirely epithelial and is rare, so the absence of this tumor among paleopathological reports is not surprising.

Neoplasms

Historical sources and artistic representations, however, can help us to trace the antiquity of breast cancer back several millennia. A reference in the Edmund Smith papyrus has been interpreted to refer to breast cancer (Breasted, 1930:403–406; Baum, 1993), but is now believed to be too insubstantial for such a diagnosis. The Syrian surgeon Archigenes wrote profusely about cancer of both breast and uterus about AD 125. Benedek & Kiple (1993) provide a valuable compilation of ancient sources relevant to breast cancer. The following items are abbreviated from their listing. The Alexandrian surgeon Leonides not only recognized but incised and cauterized malignant breast lesions, and Galen (ca. AD 129–201) suggested the cancer lesion represented a coagulum of black bile. The ninth century Arabic physician Rhazes (AD 850–932) warned that operating on breast cancer only made it worse unless excision was complete and accompanied by cautery. By the seventeenth century, operations for breast cancer included the axillary nodes. Sandison & Wells (1967) also feel that Celsus (25 BC to AD 50) was familiar with breast cancer.

Reading breast cancer into artistic displays, however, is less convincing. Few examples of graphic art better exemplify the hazards of making medical diagnoses of artistic representations than that of Michelangelo's *La Notte* (Night). On opposite walls of the Florentine Medici chapel lie the bodies of two, politically important Medici dukes, encased within sarcophagi. Above each the artist has created an exquisitely carved likeness of the entombed individuals. Two additional nude figures (one male and one female) are placed on the lid of each sarcophagus. The males are labeled Day and Dusk while the females are Dawn and Night. Michelangelo considered this a major work and devoted a significant fraction of his working years to it.

Interest is directed at the breasts of Night (*La Notte*). These are deformed by multiple surface depressions up to about 5 cm in diameter, without evidence of ulceration. These "lesions" have led to varying interpretations by several different professionals. Medically trained persons are apt to agree with Rosenzweig (1983) that they represent the characteristic appearance of skin retraction produced by incorporation of the breast's Cooper's ligaments into a deep-lying scirrhous carcinoma. After viewing the statue, I felt that was certainly a serious consideration, but that other scarring processes such as healed fat necrosis or breast abscesses could duplicate such an appearance. Some poetic-minded viewers note that the physical features of the entire bodies of all four figures would be consistent with an interpretation that they represent the four stages of life: birth (Dawn), maturity (Day), decline (Dusk), and death (Night), and that the breast deformation of the latter simply indicates organ disintegration leading to death. On the other hand, viewers knowledgeable about all other aspects of an artist's life and works (such as art historians) may incorporate this additional information into their judgments. Levy (1965:70) feels that "the breast deformities reflect the deliberate expression of Michelangelo's hostility toward women". I feel that the sculptor intended *La Notte*'s deformations to symbolize the physical and – by extension – probably also mental and emotional deterioration and exhaustion of the pre-death state and may even have deliberately selected a model possessing these features. Whether or not he was aware of the specific nature and appearance of breast malignancy is an interesting intellectual diversion but makes no contribution to the history and antiquity of breast cancer, since abundant evidence of greater certainty (some of which is listed above) testifies to its recognition many centuries prior to the Renaissance. Probably viewers from other backgrounds have generated still other interpretations, but those listed above are sufficient to remind us of the uncertainties inherent in many medical interpretations of iconography. An ancient Greek statuette illustrated by Long (1965:6, plate 1) stated to be of votive nature demonstrates a similar deformity in the left breast, which Long also interprets as carcinoma, though this judgment is subject to the same constraints as those of Michelangelo's sculpture. Thus literary sources suggest recognition of beast carcinoma about 2000 years ago, though iconography contributes little unequivocally supportive evidence.

Carcinoma of the breast has been suggested as causing metastatic bone lesions in a skeletonized body of a female from about AD 750 in northern Chile, representing a Tiwanaku-related culture. Both lytic and sclerotic lesions were identified in the skull, pelvis, spine and femur (Allison *et al.*, 1980). As Allison *et al.* point out, it is not possible to indict the breast unequivocally as the primary source of the metastases, since the metastatic pattern can be duplicated by other tumors. The presence of a lytic as well as an osteoblastic feature in the metastases increases the probability of the breast as the primary tumor but, like other suggested cases of such a primary source, it requires demonstration of the soft tissue lesion to be certain of that specific diagnosis. Furthermore, Zimmerman & Kelley (1982:207) remind us that "there is not one single instance of a tissue diagnosis in mummies of cancer of the breast . . .'. Nevertheless, if breast cancer had been as frequent during antiquity as it is in the USA today, it would still require examination of a huge number of mummified bodies to establish a statistically valid assessment of its prevalence – so many, in fact, that the absence of any reported malignant tumor in breast soft tissues to date can not be used to imply its absence in antiquity.

Gynecomastia

As explained above, enlargement of the male breast can occur under certain normal, physiological as well as pathological circumstances. The abundance of Egyptian graphic art and statuary includes some apparent examples of the finding. Undoubtedly, the most well-known example is the breast prominence of Tutankhamun, his two brothers (Smenkhare and Amenophis IV, also known as Akhenaten) and his father, Amenophis III. While the exact relationships between those four is not known with certainty, the above listing seems to be a most popular one currently (Harrison *et al.*, 1969). Their mothers, however, have not been established. The appropriate interpretation of these enlarged breasts in paintings, relief and statuary has been the object of intense controversy. Arguments have been made to support a long list of possibilities that include artistic license (Drioton, 1950:84–89; Harrison, 1973; Hawkes, 1973:438; Swales, 1973; Taitz, 1973), mytho-theological significance (Fazzini, 1975:65), political significance (Fairman, 1972), acquired medical condition (Aldred, 1963; Ghaliounguli & el Dawakhly, 1965: 18, figure 37, schistosomal liver cirrhosis; Weller, 1972, adrenal feminizing tumor) and pathological medical genetic anomalies (Walsche, 1973, Klinefelter's syndrome (KS) or Wilson's disease (WD); Paulshock, 1980, familial gynecomastia). Many of the arguments supportive of or conflicting with these diagnoses are reviewed in Paulshock (1980). There are, however, other examples of enlarged breasts in Egyptian iconography: two Nile gods with obvious breasts are depicted on a ostracon from the Ramesside Period (Peck & Ross, 1978:48, figure 48); in a Ramesside Period erotic drawing, the alleged male is pictured with a breast of a size equal to that of his apparent female partner (Peck & Ross, 1978:85, figure 84). Probably one of the best examples is that of the superbly crafted painted limestone statue of "The Louvre Scribe" originally in an Egyptian tomb chapel at Saqqara from the late fourth or early fifth dynasty and now in Paris (Strouhal, 1992b:214). Unfortunately, no corporeal support for any of the medical suggestions has appeared for the Tutankhamun controversy (Harrison & Abdalla, 1972; Harrison, 1973) nor have I been able to identify a report of gynecomastia in any other mummy. I consider the question unresolved and await future reports of gynecomastia in mummified male human remains.

Dermatological diseases

The skin competes with muscle and bone for the distinction of being the largest body organ. It forms the interface between the body and its environment. Its protective function makes it vulnerable to the effects of its interaction with infectious and toxic agents as well as thermal and mechanical assaults. The openings of its sweat and sebaceous glands may become obstructed and the skin areas exposed to sunlight are subject to the effects of ultraviolet irradiation.

It is not surprising that skin should be one of the first mummified tissues studied. Some of these early efforts at rehydration and microscopic examination include those by Czermack (1852), Fouquet (1886), and especially by Wilder in 1904 as cited by Simandl (1928). Wilder examined skin from various areas including the head, thumb and limbs of several spontaneously mummified bodies from the southwestern USA (Basketmakers; Utah cliff dwellers) as well as several Peruvian bodies housed in New York museums.

The many references to dermatological conditions in the Bible, the Babylonian literature and others are almost irresistible to those who enjoy reconstructing diagnoses from ancient narrative clues. Sandison (1967) lists some of the many diseases that have been suggested from such sources. However, except for a few obvious ones such as pediculosis, most descriptions are not sufficiently detailed to establish specific, etiological entities with any great level of certitude. Listed below are a few examples of conditions most of which were made possible by their identification in mummified human remains.

Preservation

The epidermis is often absent over much of a mummy's skin surface. During the early days of mummy studies in Egypt, Herodotus' (ca. 450 BC/1981) description of Egyptian embalming was interpreted as involving immersion of the body in liquid natron ("the body is then immersed in natron for the proscribed number of days . . ."). Perhaps suggested by their experience with modern examples of drownings, early Egyptologists attributed absence of much of the epidermis in Egyptian mummies to such postulated immersion (Simandl, 1928). In fact, by somewhat circular reasoning, epidermal absence was considered supportive evidence of such an interpretation, and a good deal of nonsense was postulated on the basis of the use of absent epidermis as an immersion marker.

The current concept of Egyptian embalming methods does not include immersion. Natron is believed to have been employed as a pulverized, finely granular, dry salt applied directly to the skin and the surfaces of eviscerated body cavities. Efforts to reproduce this procedure demonstrated that the epidermis becomes so adherent to the encrusted natron that much of the outer skin layers are removed as the natron is peeled from the body (Notman & Aufderheide, 1995). Perhaps even more important, it must be remembered that all the moisture of the internal tissues (except for the eviscerated body cavities) must traverse the skin during the process of dehydrating the body and that such moisture is laden with the endogenous proteases of destroyed cells by the time it reaches the skin surface. Thus it is not surprising that the skin (having been exposed to destructive enzyme action) is the last of the tissues to dry, and that it would disintegrate and peel off readily upon removal of the natron. For these reasons, areas containing minimal quantities of underlying soft tissues (ear, finger, scalp) commonly retain epidermis (Post & Daniels, 1969, 1973) even when it is lost over the abdomen or thigh of the same body (personal observation; Daniels & Post, 1970). This can be true even in spontaneously mummified bodies in whom soil is often adherent to skin surfaces, though Giacometti & Chiarelli (1968) found more frequent epidermal preservation in Predynastic mummies. Thus, absent epidermis in a skin section from a mummy is more apt to reflect the quantity of soft tissue beneath the point of biopsy than to be predictive of immersion in a liquid.

Traumatic dermatological lesions

Lacerations, mutilations such as scarifications, tattoos and perforations are discussed by Hambly (1925) and Bietak & Strouhal (1974).

Infectious dermatological lesions

An unmistakable example of smallpox in a seventeenth century priest found in a Mexico City church has been reported by Allison & Gerszten (1989). The face was literally replaced by hundreds if not thousands of pustules. In a medieval child mummy from Naples, Fornaciari & Marchetti (1986) demonstrated that the virus may retain its morphological appearance in tissues and even its antigenic structure, though unsuccessful culture efforts indicated that it was no longer viable (Marennikova *et al.*, 1990).The initial impression of smallpox in an Egyptian mummy by Ruffer & Ferguson (1911) proved to be a bacterial lesion histologically, and the facial lesions of Ramses V have not been verified histologically.

Other infectious conditions described in the paleopathological literature include infectious dermatosis (Wilder, 1904), subcorneal pustular dermatosis in an Egyptian mummy (Zimmerman & Clark, 1976), and the angiomatous eruptive lesions of the second stage (verruga peruana) of bartonellosis (Carrion's disease) reported by Allison *et al.* (1974b). Lice infestation (Aufderheide, 1990a) is common in mummies from South America's Pacific coastal areas (Aufderheide, 1990a). An ulcerating, destructive facial lesion ("noma") has been seen by Pahl & Undeutsch (1988) in an Egyptian mummy, and Saskin (1984) reports acute vasculitis of unknown cause in a spontaneously mummified adult female body from about AD 350 in extreme northern Chile. Ruffer (1921a) identified a decubitus ulcer on the heel of Amenhotep II. The lesion known as histiocytoma found in an Egyptian mummy (Zimmerman, 1981) might also represent the residual scar of a local infection instead of a benign tumor. Rashes of unknown cause are displayed on clay Peruvian figurines in the Weisman (1965) collection.

Metabolic dermatological lesions

Obesity is not a commonly reported condition in mummified human remains. The exaggerated folds of abdominal skin seen in the royal mummies of Ramses III, Merenptah and Thutmose II by Ruffer (1921b) indicate it can be obvious when present, so the paucity of paleopathological reports imply that it was infrequent in ancient times. Baldness was not only recognized in Egyptian mummies by Ruffer (1921b), but Ebbell (1937:79, column 66) found a suggested remedy for it in the Ebers papyrus; Ruffer also describes comedones in royal mummies. Since melanin is quite resistant to enzymatic decay, Frost (1988) suggests its use to predict skin color. Wilder (1904) has also described the dermatological changes of infantile eczema.

Dermatological neoplasms

Considering the amount of ultraviolet radiation that the exposed skin of many ancient groups probably endured, it is surprising that to date I have not been able to find reports of a verified, primary skin cancer in the paleopathological literature. In fact, only a few benign skin tumors have been reported. They include angiokeratoma circumscripta from the leg of an Inca child that had been left as a sacrificial offering at a ritual site on a high Andean peak (Horne, 1986a,b) and a common wart (verruca vulgaris, Fulcheri, 1987c) in an Egyptian mummy. A simple lipoma on an adolescent male's chest was reported by Allison & Gerszten (1983), though observers must be careful not to confuse skin folds resulting from clothing compression for the true neoplasm these authors describe.

Miscellaneous conditions

Examination of the hairs enabled the identification of human skin that had been nailed to medieval church doors in England; these are believed to represent punishment for sacrilege (Quekett, 1848). Chest and leg papules found on a 2100-year-old Chinese noblewoman's body proved to be an example of pseudopathology when Kuo-liang *et al.* (1982) showed them to be foci of adipocere formation. Frost (1988) has also found that male skin has a darker color than that of females and that lighter skin color is generally perceived as a feminine feature. Hino *et al.* (1982) even identified a bacterial spore with intact morphological structure using electron microscopy of the skin of an Egyptian mummy from the Roman Period. In past centuries dermatological problems were common enough in Ethiopia to develop a therapeutic industry in thermal baths (Pankhurst, 1990).

Chapter 9. The museology of mummies

Introduction

The directors of most large museums today would agree that their mission includes collection of objects, their conservation and preservation, exhibition and interpretation (education) as well as research on the curated, collected material (Tyson, 1995). This is an impressive list of quite varied responsibilities. When mummies are included among the collected items, each of these features merits scrutiny to determine their application to the mummies housed in a museum. To these the San Diego Museum of Man has added a sixth concern: viewer sensitivity (Tyson, 1995).

Collection

Museum acquisition of mummy collections varies considerably. The Egyptomania that swept the Western countries in the nineteenth and twentieth centuries led many a wealthy American or European to purchase a mummy from an Egyptian street hawker during a visit to the pyramids of Giza. When the novelty of such a possession eventually eroded, the mummy frequently was donated to a local museum. Devoid of provenience data, it was commonly displayed as a curiosity for some time, following which it was relegated to the museum storage area, where it often deteriorated slowly in an uncontrolled environment. Small museums sometimes acquired a few non-Egyptian mummies that were excavated regionally, usually the climatic product of a spontaneous mummification process. Only a few, large museums received a large number of mummified bodies, generally the consequence of a major excavation carried out by members of the museum's professional staff. Some, such as Chicago's Field Museum, were the recipient of a collector's assemblage. Few museum boards defined the collection of mummies as a prospective goal within their mission. Once acquired, however, the proper preservation and conservation of a mummy makes serious demands on the museum's budget. Upon acquisition it must be prepared for either exhibition or storage, a potentially elaborate process often involving extensive cleaning. The recent preparation of Egyptian mummies for exhibition after more than a century of storage at the Niagara Falls Museum is such an example (Gibson, 2000). Thereafter, whether exhibited or stored, it must be housed under climate-controlled conditions. Subsequently it should be maintained with regular monitoring, retarding and repairing processes of deterioration. Lacking such a plan and commitment, the mummy is quite likely to undergo slow but progressive deterioration.

Exhibition

Documentation and interpretation are museological functions closely linked with exhibition. In the past it was common for museums simply to exhibit their collected items for whatever satisfaction their viewing brought to the visiting public. Today, however, collec-

Fig. 9.1. San Diego Museum of Man mummy exhibit.
This spectacularly successful exhibit owed its public acceptance to its educational focus. CAT, computer-aided tomography. (Photo courtesy of Rose Tyson.)

tions are exhibited for specific purposes, usually to expand the knowledge of the viewing public about the items on display. In short, the goal of exhibits is usually an educational one. This becomes particularly important in the public exhibition of human remains. Several generations ago American funeral customs, probably derived from Irish and Scots traditions, commonly included a wake in which the often unembalmed body was kept in the home for several days to be visited and viewed by friends until the burial day. The increasing tendency toward denial of death's realities have now so sheltered Americans that many of them mature without ever having seen a dead human body. It is not surprising, therefore, if the first view of a human corpse in the form of a museum-exhibited mummy becomes a disquieting experience. If the mummy is displayed merely as a curiosity, there is little redeeming value for such a viewer. If, however, the mummy is part of a carefully constructed, highly educational exhibition in which the mummy's absence would depreciate the exhibition's ability to be as instructive as possible, then it seems inappropriate to deprive the visitors of that opportunity.

In *Manual of Curatorship*, Thompson *et al.* (1984:535) warn that curators of human remains should "avoid giving rise to public outrage or offense in management of such material . . .". That this can be achieved successfully has been demonstrated repeatedly, most recently by the popularity of several sensitively prepared and conducted mummy exhibitions presented by the San Diego Museum of Man (Fig. 9.1). Viewers' comments were uniformly positive. In addition, the superb educational program employing the mummified body of an Egyptian mummy following its extensive scientific study at Toronto's Royal Ontario Museum is a model for this type of exhibit (Pope, 1995). The museum's visiting public responded enthusiastically, but when the exhibit was moved outside the museum to a public area (subway station) it received sufficient criticism to result in its removal (Laytner, 1975). Probably the lesson to be learned from this experience is that it is unwise to deprive the general public of its option whether or not to view a mummy. A visitor reaction survey at an exhibit of indigenous Canary Islanders' mummies at the Tenerife Archaeological Museum in 1990–1991 drew only a 6% unfavorable response (Fariña-Trujillo *et al.*, 1995).

Some museums have responded to public concerns about human body displays by refusing to exhibit human remains under any circumstances. For example, Seipel (1996) documents the extreme example of Angelo Soliman, an Arab prince, abducted into slavery, who rose to the level of consultant to kings. On orders of the emperor, at death his body was flayed, the skin mounted on a wooden model and displayed publicly. Seipel seems to view this not as an effort to honor a public servant but rather as callous indifference to dignity (which it certainly would be in modern Western culture). However, his expressed view appears too confining for universal application when he states that: "In my opinion the public display of mummies or parts of mummies or of human preparations is intrinsically inhuman and quite inappropriate to our cultural consciousness – regardless of their historical and scientific importance, their state of conservation or public interest." Such an absolute view ignores the profound importance (often religious) of the corpse in public mortuary rituals of some globally distributed cultural groups even today. For museums in Western countries, his citation of the more moderate view of the International Council of Museums' Code of Ethics (Edson, 1997:251) appears more workable: "it is occasionally necessary to use human remains and other sensitive material in interpretive exhibits[;] this must be done with tact and with respect for the feelings for human dignity held by all peoples . . .". A valuable bibliography on the topic of visitor response to museum exhibitions is available (Elliott & Loomis, 1975).

Conservation and preservation

Primary (direct-acting) agents causing destruction of mummy tissues

Microbiological agents (bacteria and fungi)

Here we address the potential destruction of already mummified tissue by bacteria and fungi. Most of the bacteria present on and in mummified tissue are not the residuum of bacteria present at time of death. Instead, they are commonly acquired from the museum or storage site's environment. The one possible exception to this generalization is the category of spore-forming bacteria, especially those of the genera *Clostridium* and *Bacillus.* The spores of these organisms are so durable that they can often survive the postmortem environment that leads to mummification. Since these bacteria are often present in small numbers in normal colonic content, they can occur in mummified remains. In fact, Cano & Borucki (1995) report recovering such viable spores from the abdomen of a stingless bee embedded in Dominican amber for 40 million years. In addition, Ubaldi *et al.* (1998) have demonstrated the presence (but not viability) of *Clostridium* sp. in the dried feces of a 1000-year-old Andean mummy. Such anaerobes, however, usually require a relative humidity greater than 70% as well as a virtual absence of oxygen for growth (Valentin *et al.*, 1995). These conditions could be met in mummies removed from peat bogs, but are unlikely to be encountered in desiccated mummified bodies. Nevertheless, in British bog mummy Lindow Man, only the bacterium *Pseudomonas* sp. could be recovered from the skin surface up to 6 months after recovery of the body, and even that organism does not appear to have caused significant skin deterioration (Daniels, 1996; Ridgway *et al.*, 1986). Microbiological destruction of mummy tissue deep to the skin surface, however, is more apt to be due to bacteria than to fungi.

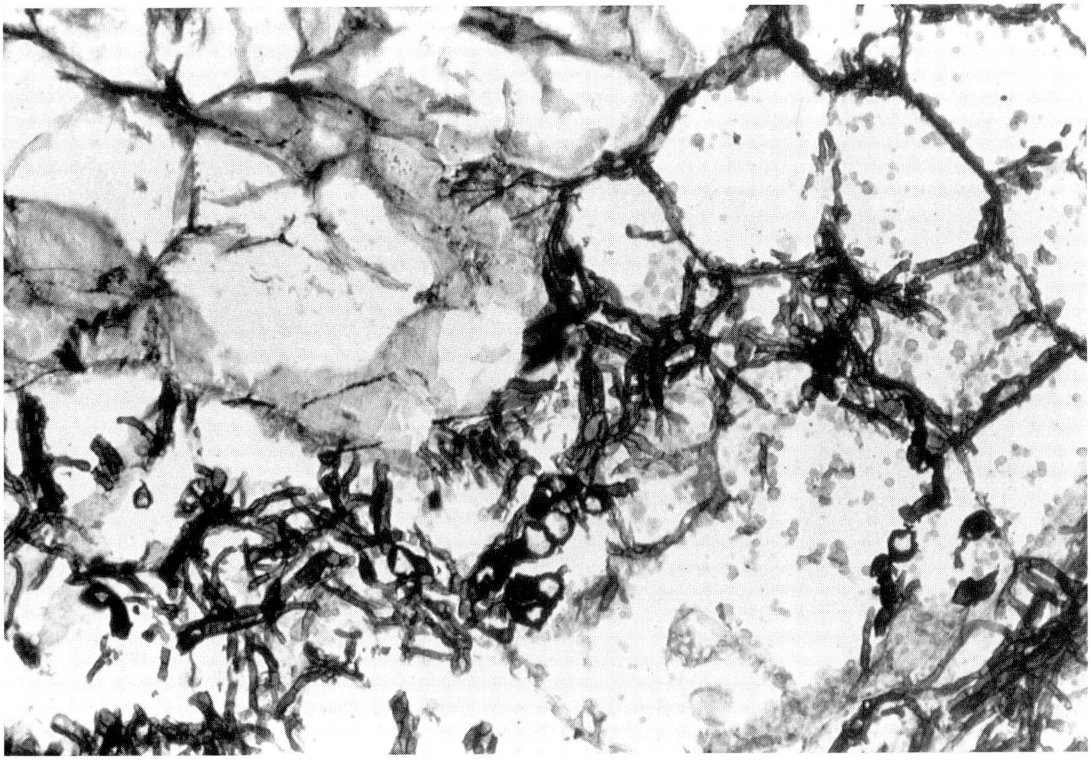

Fig. 9.2. Fungal invasion of Egyptian mummy soft tissue; AM, 1549–1069 BC.
The Grocot silver stain of a necrotic area of muscle in an Egyptian mummy from Turin demonstrates active fungus growth (hyphae). From (Fulcheri *et al.*, 1999. Photo courtesy of Ezio Fulcheri and Rosa Boano.)

Fungi on mummified bodies are virtually always derived from the environment. They are rarely found in the deeper tissues but colonize skin and other accessible surfaces when conditions are favorable for their growth. The two most important features that will enable their proliferation are high humidity and a low volume of air flow. Under these conditions they tend to grow at the edge of clothing – protected skin and locations of opposing skin such as the groin or axilla. Other sites include protected facial areas such as the oral cavity and the eyes, spreading from these to the open facial surfaces. These organisms are capable of digesting the epidermis. In areas of exposed muscle they can digest the muscle fibers but often leave the collagenous portions of muscle structures intact (Fig. 9.2). They may even use the Haversian canals of exposed bone to reach and destroy bone protein, rendering the bone mineral more vulnerable to groundwater resorption (Perotti *et al.*, 1998).

Aspergillus nigra is commonly represented, even on coffins (Subert *et al.*, 1993), and has been demonstrated to coat the bronchial surfaces, occasionally growing into a spherical mass of hyphae known as a "fungus ball" (Horne, 1996). Six months after Lindow Man's recovery a variety of fungi were found growing on his body skin surface, including species of the genera *Penicillium, Mucor, Vertissilium* (Daniels, 1996) and, quite consistently, *Candida* (Ridgway *et al.*, 1986). Unopposed, their slow destructive effect can, over the course of years, produce a serious degree of tissue destruction, though the conditions influencing this need to be defined much more specifically by further research. Histological demonstration of fungi in necrotic, deep-lying tissue may merely be reflecting secondary fungal colonization of tissue already rendered necrotic by other mechanisms.

Insects

While bacteria and even fungi are capable of destroying mummy soft tissues under some circumstances, insects are both more aggressive and effective. Insect activity is partly temperature related. Their breeding is substantial at room temperature, accelerated in warmer environments and essentially ceases below 10 °C. While a broad array of insects plague museum curators, it is principally members of the family Dermestidae (carpet beetles) that attack mummified human tissue. The most common genera include species of *Anthrenus, Attagenus and Dermestes* (Pinniger, 1994). Other insects recovered from 42 corpses removed from crypts in mausolea in the Czech Republic by Subert *et al.* (1993) included species of *Tineidas*, Ptinidae, Cryptophagidae and Psoidae. Beetles are particularly destructive because, in addition to the tissues they (and, among some, their larvae) destroy, they can create defects in the chest and abdominal walls. These then provide entrance into the thoracic and abdominal cavities that can be utilized not only by other insects but enlarged by gnawing and admit rodents as well, resulting in the destruction of the desiccated viscera inside (Fig. 9.3). If a recent biological acquisition can tolerate it, Pinniger & Harmon (1999) suggest consideration of subjecting it to freezing at −18 °C (2 weeks) or −40 °C (3 days) to kill possible infesting insects.

Fig. 9.3. Rodent invasion of mummy soft tissue; SM, ca. 50 BC.
Gnaw marks at the edge of an abdominal wall defect indicate that a field mouse has gained access to the mummy's abdominal cavity. Spontaneous mummification of a warehouse-stored mummy in the Azapa Valley. Mummy no. AZ-75, T-57. (Photo by ACA.)

Secondary (indirect-acting) agents

Light

Direct sunlight on mummies can be destructive. Part of this effect is the elevation of temperature that can soften skin, melt lipids and bleach colors, as well as degrade collagen. On one occasion while photographing a recently excavated mummy in a subtropical area, I left the mummy exposed to the sun during the lunch hour. Upon returning, the sun had bleached the bright colors painted on the enveloping straw mat, caused sticky fat to ooze from skin lacerations and softened facial skin sufficiently to distort it to a degree that made me suspect that my colleague had substituted another mummy for the one I had left. Incandescent indoor lights can have similar but milder effects. In addition, ultraviolet light included in both sunlight and neon light, if unfiltered, can degrade collagen. For biological collections the United Kingdom's Museum and Galleries Commission (Paine, 1992) recommends a 50 lux light limit and a maximum ultraviolet light of 75 microwatts/lumen (Table 9.1).

Chemical changes

If mummified tissue becomes rehydrated, destructive enzymes within the tissues can be reactivated.

Temperature

Elevated temperature can, as noted, stimulate insect activity and breeding, and accelerate bacterial prolifer-

Table 9.1. *Recommended environment for biological collections*

Reference	Temperature range (°C)	RH range (%)	Light W	Light UV
Carter & Walker, 1999	14 ± 1	45 ± 5	—	—
Paine, 1992	18	50	50	75
Brown, 1999	<30	—	—	—
Pearce, 1990	7–21 ± 5	45–60 ± 7.5	—	—
Rae, 1996	21 ± 3	50 ± 5	<200	70

Notes:
RH, Relative humidity (%); W, white (lux); UV, ultraviolet (microwatts/lumen).

ation, while temperatures below usual room levels can suppress these effects. Unfortunately, prevention of temperature-related impact occurs at levels too low for full-time workers' comfort. One approach to a solution lies in keeping storage areas at temperatures (ca. 15 °C) so low as to require additional clothing for workers that need to carry out prolonged tasks in such rooms. The United Kingdom's Museum and Galleries Commission (Paine, 1992) even advises temperatures as low as 10 °C when possible for biological collections. Such temperatures are too low for full-time personnel in such areas.

Humidity

Most museum professionals feel that relative humidity is a critical feature contributing to long-term preservation. Elevated humidity levels encourage microbial and insect activity. Considerable disagreement, however, prevails on the actual desirable levels. Fungi tend to cease proliferation at relative humidity levels below about 70%, as do many (but far from all) bacteria. Creases in mummy textiles tend to disappear at higher humidity levels, thus deleting potential wear information (Cooke, 1988), while values less than 30–40% will cause textile fibers to become brittle to the point of fracturing when handled. Glover (1984) also stresses the importance of keeping the relative humidity constant, since fluctuations result in contraction and expansion of textile fibers that can produce abrasion sufficient to cause fiber fractures.

Readers will note that some of these suggestions incorporate a conflict of interest. Very low storage area temperatures are desirable to retard (though not necessarily eliminate all) insect and microbial activity, but larger museums that require workers to be exposed to these levels for lengthy periods can expect to encounter employee resistance. Specimens being moved from such storage areas require special handling to avoid condensation. Relative humidity levels that are appropriate for textiles (>40% but <70%) are probably still higher than ideal for soft tissue preservation in mummies, since all microbial action in such a range is not necessarily arrested at such levels.

Dust

This can form a hygroscopic layer on tissue that will tend to retain water. In addition, dust includes particulates that can serve as food for some destructive insects such as silverfish (Rae, 1996).

Oxygen

Abundant available oxygen will stimulate the growth of aerobic bacteria such as *Escherichia coli* (normal colon flora), while its absence will encourage proliferation of anaerobes (*Clostridium*, *Bacillus*).

Types of bodies requiring conservation

For purposes of preservation, mummies can be classified as either "wet" (those having normal or near-normal living water content) or "dry" (those with substantially less than normal living water content). Preservation methods for these two groups are discussed below. As is apparent in some of these categories, the many different methods employed for long-term preservation reflect the fact that insufficient data exist to make a well-informed selection.

Consultation with curators who have employed methods under consideration is highly advisable.

"Wet" mummies

Bog mummies

A variety of conservation methods have been applied to these, including those of tanning (Grauballe Man: Glob, 1969:59); liquid fixatives, such as ethanol and formaldehyde (Tollund Man: Glob, 1969:36; Fischer, 1998:177); freeze-dry or lyophilization (Daniels, 1996) and simply frozen (Delany & O'Floinn, 1995).

Frozen mummies

These usually retain most of their water, though spontaneous desiccation can occur via the process of sublimation (transfer from solid to a gaseous state without an intermediate liquid state). These have been kept frozen (Bogucki, 1996; Reinhard, 1998), frozen-dried (Spindler, 1994) and preserved by gamma radiation (Johansson, 1989).

Modern bodies

These are most commonly mummified and preserved by intra-arterial injection of liquid fixatives such as formaldehyde or, during the nineteenth century, a heavy metal such as arsenic (Johnson *et al.*, 1990).

"Dry" bodies

The aim in preservation of most of these bodies is to maintain or increase the state of desiccation of these bodies. They include the following types of mummification.

Spontaneous desiccated mummification

These occur most commonly in the hyperarid deserts of the world including South America's Atacama, North Africa's Sahara, China's Taklimakan and Mongolia's Gobi. Tests in my laboratory indicate that 99.8% of the free water in such mummies' soft tissues has been removed.

Anthropogenic desiccated mummification

This may be carried out in some deserts (Egypt) but is more common in moist areas such as parts of Colombia and Torres Strait, where heat is applied to achieve desiccation.

Chemical changes

Heavy metals can be injected to paralyze enzymatic decay; spontaneous evaporation of water from such a body can achieve a desiccated state. By drastically raising the pH of the body to a level at which enzymes can not act, lime and lye treatment can produce a state of desiccation, though the caustic effects of these two agents obviate their use for museums' purposes.

Tanning and taxidermy

Most commonly applied to hides, these depend on chemical alteration of molecular structure, but the process usually involves also removal of a substantial part of the water content. General Custer's horse (Comanche) has been so preserved (Fig. 7.5) and remains today displayed in the University of Kansas Museum of Natural History (Dary, 1976).

Overmodeled skulls and trophy heads

Vegetal pastes are often employed in the production of overmodeled skulls, and trophy heads are desiccated with heat, both resulting in a desiccated state. Their smaller size (as compared with a whole human body) simplifies their preservation problems.

Relics

These medieval European items are usually small body parts such as a finger, heart or bone. They are easily dried and commonly kept in dry containers.

Adipocere formation

The fatty acids of which this material is composed have been modified so that they remain unrecognized by most enzymes resident in such tissues and are chemically stable. If kept dry they can be maintained for many years.

The general approach to all of these is to maintain them in a state of desiccation. The preservation methods to achieve this, as well as those for "wet" mummies are discussed in the next section.

Mummy preservation methods

General considerations

The following discussion is based on the assumption that the museum has just acquired an unclothed mummified human body. Initially it will also assume this body is in a desiccated state; exceptions to that rule and the handling of Egyptian desiccated bodies are presented at the end of this section. Responses to a questionnaire survey in the Canary Islands indicates that restoration, preparation and preservation methods vary enormously among museums (García & Herraez, 1995). This would be expected, given the paucity of standards, the limited database and the differing goals of the institutions. The information in this section is heavily dependent upon and abstracted from the following publications; Pinniger, 1994; Maekawa *et al.*, 1995; Melville, 1995; Pearlstein, 1995; Valentin *et al.*, 1995; Daniels, 1996; Rae, 1996; Fulcheri, 1997; and David, 1998.

Restoration and preparation

Cleaning

Following removal of all wrappings and clothing, the nude body usually is subjected to cleaning. After photographic recording of hair style and appearance, the hair is inspected for evidence of pediculosis. The hair can then be cleaned as described by Rae (1996), in which alcohol-soaked sawdust is gently worked into the hair and removed with a brush and low-flow vacuum cleaner after evaporation of the alcohol. Overt clumps of adherent soil are then removed manually and the skin surface of the entire body is covered with a brush and low-flow vacuum cleaner to remove dust. Some museums, e.g., the British Museum, may accompany or follow this by wiping the skin surface with alcohol-soaked cotton swabs.

Restoration

The degree and nature of restoration efforts is totally individualized, based upon the nature of the defects, the purpose for the acquisition (collection, display, research, etc.) and the interests of the conservator. These actions inevitably involve conflicts of interest. The bioanthropologist interested in chemical or immunological analysis of tissue samples may consider as undesirable the introduction of stabilizing substances and chemicals of any kind, while the body's curator may fear that an unstabilized long bone fracture will lacerate surrounding soft tissues when the body is handled, or uncontrolled insect invasion will threaten the body's integrity. While both views can usually be accommodated, the increasing view that, in most cases, restoration is not desirable (some even believe, unethical) has minimized this potential conflict. The following discussion assumes that the primary interest is preparation for exhibition. Omissions or modifications can be made to accommodate special interests.

Mechanical repair of principally postmortem bone fractures can involve metal or wood splints for major bone displacement, while rapidly polymerizing epoxy may be added or used alone for lesser breaks. During desiccation the skin often contracts sufficiently in some areas to become detached from the underlying subcutaneous tissue or muscle. Such a subdermal defect is very common in areas where the skin is "tented" over a bone such as the clavicle. The thin, brittle skin overlying such a defect is remarkably subject to splitting and crumbling into small fragments when pressure is applied to such areas. At the British museum, Rae (1996) has filled the defect with methacrylate polymer or plastic sheets that can fuse and solidify with heat application, thus supporting the overlying skin. Disfiguring skin cracks in the facial area can be filled with a suspension of $CaCO_3$ powder in polyvinyl acetate that, when dry, can be pigmented with acrylic paint. Orbits commonly require similar treatment. Chapped lips may be softened by local glycerol application, and a variety of waxes may be employed to build up structural defects such as the nose (Subert *et al.*, 1993).

Sterilization

Today's museum conservators are so conscious of the potentially disastrous effects that insects can wreak in a museum that they employ elaborate pest prevention programs. These minimize or eliminate the need for exercising chemical controls that were so popular in the past. Nevertheless, such prevention programs are not yet universal. Furthermore, new acquisitions may require treatment. The following is not intended as a literal guide, but rather to familiarize the reader with approaches that are available and have been used in the past. Decisions for their employment are best made with consultation.

Sterilization is sometimes desirable with new acquisitions or with poorly conserved bodies already acquired by the institution. Several different options are available to achieve this, all of which have some disadvantages. Probably the simplest is what David (1986:87–89) calls "wet sterilization". This involves the direct application by brush or spray of a disinfectant solution onto the skin surface. David has even applied this to the lining of body cavities by delivering the disinfectant through an endoscope. He recommends 1–5% of the sodium salt of pentochlorophenol in 100% ethanol. For specific insecticides, the reader is referred to an excellent tabulation and textual discussion by Pinniger (1994:40–50) in which he notes that some provide residual effects while others do not. The toxicity to insects and humans is documented for each insecticide as well as the details of their applications. Again, however, the reader is reminded that the application of chemicals to the body has become increasingly unpopular because they may interfere with future research studies.

An alternative to direct application is gas fumigation. This is a relatively simple process for small items (e.g., a mummified cat), but in the absence of a dedicated chamber specially constructed for this use, it can be cumbersome when applied to an entire human mummified body. Placing the body in a rigid-walled chamber permits evacuation of the air, which will enhance penetration of the sterilizing agent. Introduction of a gaseous toxic fumigant into a plastic bag enclosing a mummy is simpler but can not employ vacuum. Ethylene dioxide or sulfuryl fluoride (SO_2F_2) plus methyl bromide have been popularly employed for this purpose, as was phosphine (Pinniger, 1994). It is important to realize that such fumigation has no residual action and thus provides no defense against subsequent infestation (David, 1998). Since this is merely a different method of delivering a chemical, this, too, is rapidly disappearing from the list of museum practices.

Controlled drying

If an airtight vitrine (display case) can be constructed, the enclosed body can be exposed to an environment whose humidity can be decreased in a controlled manner over a period of several months until it reaches a point intolerable for insects or even bacteria. The key here is the vitrine construction, discussed below (David, 1986).

Radiation

I have already noted that the ultraviolet segment of sunlight and some forms of synthetic illumination can be harmful to mummified tissue (especially collagen) if allowed to act for long periods of time. Its sterilizing action, however, is usually limited to the surfaces directly exposed to the light. For purposes of sterilization, gamma radiation's ability to penetrate the entire body when delivered in adequate dosage can be a highly effective method. The radiation source commonly is in the form of radioactive cobalt (^{60}Co). While some of such instruments can be found in medical radiotherapy units, they are now also quite widely used in the food industry for both grains and meats. Radiation sterilization was employed for control of fungus growth on and in the body of Ramses II and it was transported from Cairo to Paris in 1976–1977 for such treatment. The radiotherapy delivered 1.8 megarads of gamma radiation over nearly 13 hours and achieved its sterilization goal (Balout & Roubet, 1985; Bucaille, 1990:194). Subsequently, the frozen bodies of Inuit natives found on Greenland's northwestern coast were sterilized using a similar dosage (Johansson, 1989). Just as was true of gas fumigation, it must also be pointed out here that, while sterilization is achieved at the time of treatment, this has no residual effect and provides no protection against subsequent exposure to infectious agents.

Prevention and maintenance

The control of a mummy's environmental atmosphere is absolutely essential to the maintenance of the integrity of a desiccated body. The potentially devastating effects of high temperature and humidity have been described above. Protection from such effects can be achieved by conserving/exhibiting the mummy (1) in a room all of whose space is so controlled, or (2) in a vitrine (glass display case) whose internal atmosphere may be controlled under conditions different from those in the rest of the room. Of these two options the vitrine is usually cheaper and easier to control, particularly if the mummy is in an exhibit area. The degree of control in an exhibition room is usually less. For example, most insects of concern for mummy conservation do not breed at temperatures less than 10–15 °C. This, however, is a temperature that would be unacceptable to the visiting public. Traffic through such a room can also strain the conditioning unit's capacity.

Independent control of the vitrine space would overcome these limitations, but until recently it has been difficult if not impossible to create a completely

Fig. 9.4. A vitrine prepared by the Getty Conservation Institute.
Prototype of the vitrine developed by the Getty Conservation Institute at Los Angeles. (Photo courtesy Shin Maekawa.)

airtight vitrine whose joints would remain sealed in spite of large variations in barometric pressure. This has led to several unique forms of vitrine control. The large collection of Egyptian mummies at Turin uses vitrines into whose space a chemically inert gas (nitrogen) is pumped. Because this gas gradually leaks out of the vitrines via cracks, continuous positive pressure and constant monitoring must be maintained. An alternative in England involves a vitrine into which room air is pumped after the air is first passed through a filter, a humidifying unit and a second sulfur dioxide (SO_2) filter. This air leaks out of the vitrine through deliberately created outlets and thus is returned to the room. Clean, filtered, humidified air under positive pressure is thus continuously filling the vitrine. However, the desired cool temperatures are not achieved unless the whole room temperature is controlled at that cool level (David, 1986).

During the past decade the Egyptian Department of Antiquities contacted the Conservation Unit of California's Getty Museum for assistance in conserving Egypt's ancient pharaohs' mummified bodies in the Cairo Museum. The staff there developed an airtight vitrine that could be filled with nitrogen gas and remain sealed for at least a decade (Fig. 9.4). Furthermore, it can function without electricity.

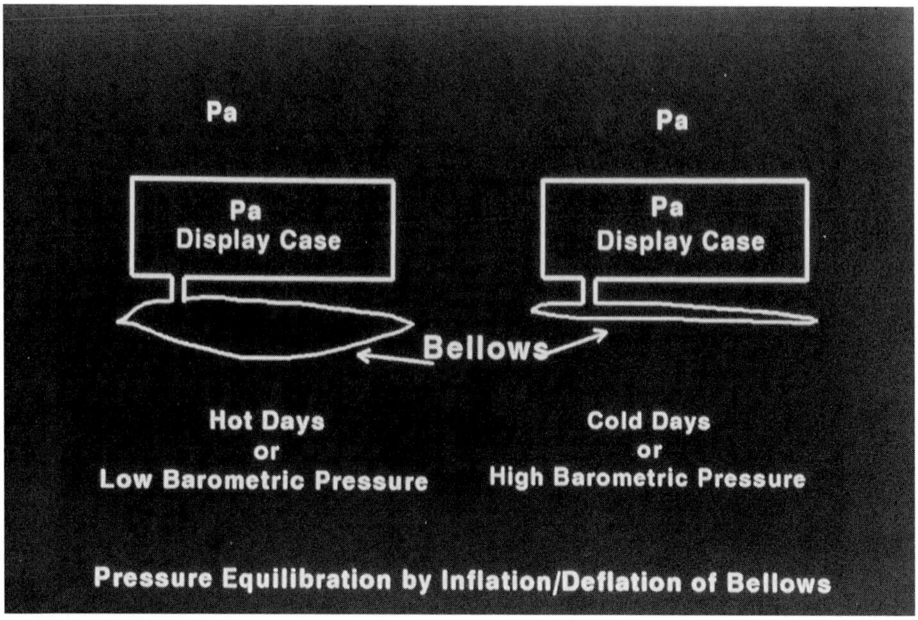

Fig. 9.5. Principle of bellows action in the Getty vitrine.
Passive expansion and contraction by the bellows that communicate with the vitrine chamber reduce stress and prevent leakage along the vitrine's seams due to barometric pressure changes. (Photo courtesy Shin Maekawa.)

Vitrines normally develop leaks as a consequence of the strain imposed on their joints (especially those relating to the glass viewing plate) by the large range of barometric pressure changes to which they are normally subjected. The Getty vitrine normalizes and stabilizes barometric pressure by installation of a passively responding bellows built into the vitrine's bottom (Figs. 9.5 and 9.6). Within the vitrine are trays containing chemicals that act as oxygen "scavengers" (FeO, KCl) that keep the oxygen level below 2%. Humidity control employs a silica gel buffer placed within the vitrine, keeping the relative humidity below 25%. If the Getty vitrine is filled with nitrogen at the desired humidity and then sealed, the permeation of oxygen into the vitrine is so slow that it is expected to require a change of gas no more than every 10 years, and perhaps as many as 20. Construction cost in the USA is approximately $10000 per vitrine. The principal pharaonic mummies on display in the Cairo Museum now are in such vitrines (Maekewa *et al.*, 1995; Maekewa & Valentin, 1996) (Fig. 9.7).

Humidity

It has already been observed above that low humidity levels discourage bacterial and fungal proliferation. Anaerobic growth requires high (>70%) humidity. Levels below 25% profoundly slow bacterial proliferation. Such low levels, however, render textiles brittle and subject to breakage when handled. Here two conflicting options arise and decisions must be made. Ideal preservation of mummy biological tissue requires a low humidity level; while we do not know the precise level required, it is most probably considerably below 25%. But textiles become brittle below 40–45%. A compromise between the two levels will probably ultimately damage both, though less than the extreme levels cited above. If humidity is controlled within a vitrine at all, in most museums that level lies between 40% and 55%. At this level microorganisms gaining access to the mummy in the vitrine can survive, even if not thrive. Ultimately the conservator must choose between conditions ideal for the mummy or those for the textiles.

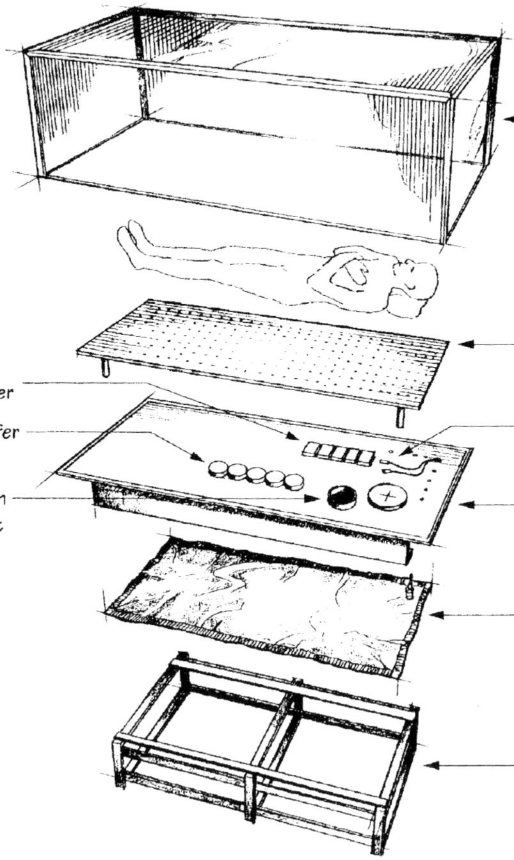

Fig. 9.6. "Exploded view" of the Getty vitrine.
The elements composing this vitrine are evident in this sketch. Key to the vitrine's success is the bellows action and the tight seams of the structure. RH, relative humidity. (Photo courtesy Shin Maekawa.)

Most select a compromise level at which deterioration of mummy tissues is quite slow but usually not totally arrested. An influencing factor can be the degree to which the textiles are expected to be handled. The ideal solution is the separation of the body from the textiles, and provision of storage conditions appropriate for each of these.

Air flow

If the vitrine is not as absolutely airtight as the Getty vitrine, then air flow considerations become relevant. We have observed that a continuous air flow of significant degree discourages fungal growth. Thus, if a vitrine can admit fungal or bacterial spore-containing air but limits air flow to low levels, conditions can occur that will admit microorganisms and encourage their growth. As an extreme example of the importance of air movement, it can be noted that it is probably only because of the vigorous air flow to which the Peruvian Chachapoya mummies in Peru's humid cloud forest (see Chapter 4: Mummies from Peru) were exposed that prevented their destruction under those otherwise humid conditions.

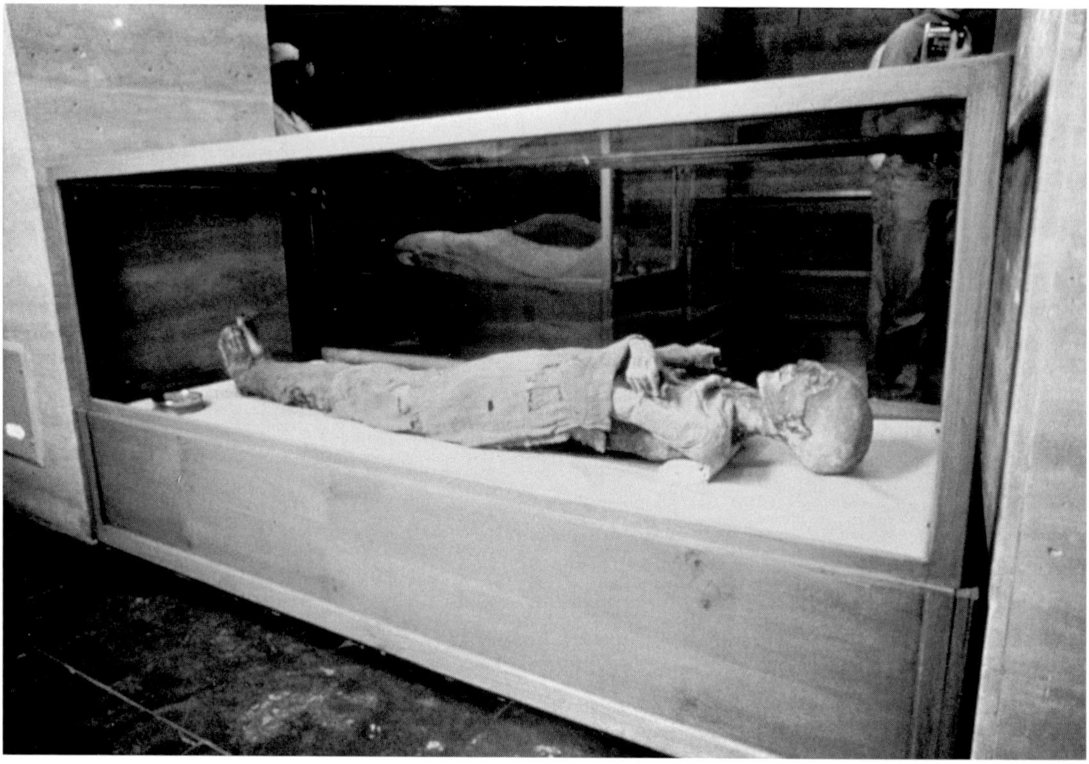

Fig. 9.7. Vitrine installed in Cairo Museum.
An Egyptian royal mummy (AM, 1549–1069 BC) is exhibited in a Getty-designed vitrine in the Cairo Museum. (Photo courtesy Shin Maekawa.)

Miscellaneous preservation methods

Freeze-drying (lyophilization)

Under these conditions the mummified body is first immersed for a month in a 15% solution of the consolidant polyethylene glycol (PEG). The body is then frozen at $>-20°C$ and placed in a container in which a strong vacuum is induced. This causes frozen water to turn into a gaseous vapor (sublimation) and this is removed from the chamber by the vacuum. After five days the frozen but now desiccated body is removed from the chamber, gradually brought to room temperature in a chamber where it can be viewed but in which the humidity is kept at such a low level that rehydration is prevented. This was the method chosen for conservation of the British bog body Lindow Man and the Irish bog body from Meenybraddan, Ireland (Omar *et al.*, 1989; Daniels, 1996). The freeze-dried and subsequently plastinated (see below) Lindow Man's body is currently on display in the British Museum.

Freeze

The Austrians and subsequently the Italians have chosen to maintain the frozen body of the >5000-year-old Iceman (Oetzi) found on a Tyrolean glacier in the frozen state. An elaborate metal container has been constructed for him as a combined storage and exhibit unit that maintains a temperature below the freezing point and a relative humidity probably in the region of about 40%. This body was at least partly spontaneously freeze-dried when found, and the described conditions will probably maintain him in that state (Hoepfel *et al.*, 1992:92; Spindler, 1994:155).

Egyptian mummy conservation

The conservation of Egyptian mummies presents special problems. Not the least of these occurs when radiological studies demonstrate the presence of a mummy within a strikingly beautiful and interestingly ornamented cartonnage, composed of fragile pasteboard with an extensively lacquered and painted surface. The risk of damage to the container during an attempted removal of the mummy is usually so high that many curators would rather sacrifice the mummy's integrity than risk injury to the cartonnage during an effort to extract and conserve it separately.

When a mummy separate from its container is acquired, it is usually hidden beneath a thick layer of linen wrapping. Again the question must be faced: how to conserve both the textiles and the body? If preservation of the body has priority, the chamber's relative humidity should be reduced to perhaps 25% or less at the risk of desiccating the textiles to the point of threatening their integrity.

If the wrappings are removed (rarely done today), the body will most likely be found to be covered to a varying extent and thickness with a coat of resin. A decision must be made whether to leave this resin layer intact or to remove it. Its continuing presence will enhance its desiccated state while its removal will improve the body's appearance. Resin removal may be difficult; layers two or more centimeters thick may require the employment of an approach as coarse as hammer and chisel. Thinner layers respond to swabs soaked in solvents. The display of the Egyptian mummy at the St Paul Science Museum in Minnesota represents a compromise to the question of wrapping removal. A linear incision in the ventral midline from the crown of the head to the pubis penetrates the wrappings to the level of the skin. The textiles on the left half of the body were then removed, leaving those on the right side intact. This had been carried out before the museum acquired that body; few curators would be comfortable with this approach today. Completely unwrapped bodies may be conserved in the same manner as that of any other desiccated mummy.

Plastinations

This process has been used primarily for the preservation of anatomical pathological structures for teaching purposes. While variations occur, the usual method involves subjecting the specimen initially to a freeze-drying (lyophilization) procedure followed by further acetone desiccation and subsequent impregnation under vacuum with a silicone polymer. After "curing", the specimen resists rehydration and retains a flexible structure. The authors describing this procedure suggest that the process ought to lend itself to intact whole bodies, though to my knowledge this has not been demonstrated (Wade & Lyons, 1995).

Summary: museological problems of mummy preservation

Acceptance of responsibility for the conservation and preservation of mummified human remains is a formidable undertaking. Section 3.1 of the International Council of Museum's code of ethics advises museums not to collect items that they are unlikely to be able to conserve appropriately. So little research on preservation of human mummies has been carried out that we need to use guidelines developed for biological specimen collections. As Carter & Walker (1999) note "the key to long-term preventive conservation in natural history collections is control of the collections' environment . . .". All authors seem to agree that the principal environmental factors influencing soft tissue decay are temperature and relative humidity, though the differing requirements of soft tissue and artifact preservation lead to differing choices of actual levels of control. While many museums have climate-controlled environments, a recent survey found that even some large museums curating human mummies lack facilities for such control, and the exhibited mummies sometimes demonstrated evidence of continuing soft tissue deterioration (Meier, 1997). Questionnaire surveys indicate a wide range of curation practices for preservation of mummified remains among museums. Several large museums recently have carried out a thorough assessment of their biological collections (some specifically targeting mummies), incorporated appropriate conservation of such collections into their mission statements and initiated the often expensive capital improvements to achieve those goals (Pearlstein, 1995: Brooklyn Museum, New York). One (Museum of Fine Arts, Boston, Massachusetts) even loaned part of its Egyptian curations for 10 years to another museum that had none, using the "loan fee" to fund a conservator position to assist with the museum's improvement process (Gänsicke, 1995). For those with only a few mummies, the Getty-designed vitrines could make a major contribution to the problem's solution. Those with much larger collections of animal and human remains face substantially greater challenges that can be expected to require individualized solutions, for it seems unlikely that the desired changes can be extracted from most existing budgets. Emphasis is also shifting from active intervention such as fumigation to preventive conservation that is designed to prevent pests from ever establishing themselves within a museum (Pinniger & Harmon, 1999). On the other hand, recognition that preservation and conservation treatments may impair subsequent research analysis and even erase evidence is becoming an influence that is in conflict with preservation goals (Brooks *et al.*, 1996; Paterakis, 1996). This is an issue that probably merits more emphasis than I have apportioned it in this presentation.

Research

Proper preservation of curated mummies is an appropriate part of a museum's mission, as is the sharing of extant knowledge about such an acquisition in educationally oriented exhibits. Research (the generation of new knowledge by scientific study of their collections) is a more recent addition to smaller museums' mission lists. Some large museums have carried out formal research studies for more than a century and have established full-time research positions on their staff. However, ambivalent views regarding the appropriate budget share for research activity is still sufficiently common among some major museums' boards of directors to have this topic become the focus of an international conference (Gee, 1990). Research may be directed at broadening the characterization of the mummies being curated ("applied" research), whose results will enhance the museum's interpretation and educational goals. Alternatively (or in addition) scholarly research can broaden the general database dealing with all mummies. Even the latter will contribute to the museum's immediate educational efforts by making it possible to place its locally curated mummies in a more global context (Tyson, 1995). A useful bibliography of studies on the educational evaluation and research in museums and public exhibits has been published by Screven (1984).

Chapter 10. Use and abuse of mummies

Introduction

Living humans embody two distinct aspects: the tangible and the intangible. The latter includes an individual's thoughts, words, emotions, actions . . .; in brief, all the items that shape a person's character and that is often called the essence or simply the spirit of the individual. The tangible aspect is the physical body that houses the spirit and enables its expression. The extent to which we can separate and relate to these two aspects independently in our mind is the principal determinant of how directly we identify with the physical body, especially after death. This degree of corporeal identification is enormously variable. Some do not differentiate these two aspects and thus insist that the corpse be treated precisely as it was while the spirit still resided within it in the living state. Others identify with the body to a lesser degree or not at all, basing their decisions on far more pragmatic concerns.

This chapter presents a variety of unusual ways in which mummies have been treated. The same variation in corporeal identification noted above can be expected among the readers of this volume. What may be viewed as *use* by one is *abuse* to another. The following ways that mummies have been handled are presented for the readers' interest without imposing my judgment upon them.

Mummy as a drug

It is difficult to believe today that translation errors by medieval scribes and the black color of Egyptian embalming materials could have led late medieval and Renaissance physicians to incorporate into their practice what many would call a form of human cannibalism for which the human body parts were made available, sometimes by illegal activities.

The Renaissance European medical practice of dispensing allegedly therapeutic powder composed of pulverized mummified human body tissues came about through a complex, circuitous and improbable but factual sequence of events. We initially hear about that black, tarry, asphalt-like material during the first century from such well-known historians as Pliny and Dioscorides (Gunther, 1959:52). They describe the spontaneous oozing of such material that we will hereafter call "bitumen", from the earth in several unique locales in Persia (now Iran), the floor of the Dead Sea (then called Lake Asphaltites), Babylonia (*liquidum*), Sidon (*terra*), and Apollonia (now Albania). The material from all of these the Greeks termed *pissasphaltum*. The Persian term for bitumen was derived from *mûm* (which means "wax") and its location in Persia was termed "mummy mountain" in English (Dawson, 1927d). Eventually the Persian product was called *mûmiyá* and the tenth century physician Rhazes is

Fig. 10.1. Bituminous masses on the Dead Sea shore.
Large masses of asphalt have been rising to the surface ashore since antiquity. In Ptolemaic times these were exported to Egypt for embalming purposes. (Photo courtesy of Arie Nissenbaum, Tel Aviv, and *Israel Journal of Earth Sciences* **48**: 302, 1999.)

credited with generating the Latin word *mumia* for bitumen (Dannenfeldt, 1985). How the English word *mummy* came to be applied to a desiccated, ancient Egyptian corpse is part of the saga detailed below.

Bitumen (asphalt) is a naturally occurring, black hydrocarbon of varying consistency. Modern medical pharmacopeias have reserved a narrow niche for derivatives of this material that have demonstrated usefulness in the treatment of certain dermatological diseases (Nissenbaum, 1999). Physicians of Pliny the Elder's time 2000 years ago, however, recommended it for both external applications and even internal use for a variety of conditions (Mendelsohn, 1952). By the end of the Middle Ages its applications had become as ubiquitous as those of aspirin today. It was recommended by the Arabian physician Avicenna as well as the Byzantine Rhazes for a list of conditions so long it almost would be easier to itemize the exceptions. However the very popularity of bitumen eventually strained its availability from the limited supply of its natural sources. Furthermore, by the eleventh century, translations of writings by such earlier authors as Serapion and even Rhazes were confusing the term *mumia* (bitumen), equating it with the black material found within and even externally in artificially preserved Egyptian bodies. Those embalmers had achieved soft tissue preservation primarily by desiccation with various salts (natron) and they then sheltered the dried body from subsequent rehydration by the application of botanical resins. These resins, applied in liquid, heated form, acquired a hard, brittle consistency upon cooling (see Chapter 4: Mummies from Egypt and the rest of Africa). Together with myrrh and spices these underwent chemical changes during the following centuries that commonly transformed their original brown or amber color to jet black, superficially resembling that of bitumen.

The course of subsequent events is predictable. Justified by the resemblance of bitumen to blackened resin, fortified by the above-mentioned mistranslations and operating in a milieu of high demand but limited supply, both the term *mumia* and the medicinal properties that originally had been reserved for bitumen were broadened to include the resins of ancient Egyptian bodies. Increasingly the expanding European medical market for *mumia* was met by the export of blackened resin surreptitiously stripped from the occupants of readily accessible, ancient Egyptian tombs. Dannenfeldt (1985) cites an early fifteenth century incident in which a band of Egyptians was apprehended for looting tombs, placing the ancient bodies in boiling water and skimming off the oil that floated to the surface, hardening it by cooling and marketing it as *mumia*. Ironically, such black material from those archaic human bodies may actually have been bitumen in part. Ancient Egyptian embalmers commonly employed expensive resins and spices. However, records from the Ptolemaic Period (the last three centuries or so before Christ) noted the increased trade with Palestine and recorded exportation of Dead Sea bitumen (Fig. 10.1) to Egypt for embalming purposes. Modern chemical studies have identified bitumen in occasional Egyptian mummies but only from these late periods (Nissenbaum, 1992) (Fig. 10.2).

At that point a further transformation took place. The hot, liquid resin poured into the eviscerated body cavities by ancient embalmers made contact with soft tissues, especially muscle, and incorporated these into the subsequently cooled and congealed resin. Eventually those resin blocks containing the visible brown, desiccated tissue fragments were thought by physicians to have exceptionally good therapeutic value. They attributed these effective characteristics to the inclusion of muscle tissue into the black material. Thus the medicinal value that was originally thought to be inherent in the naturally occurring bitumen oozing from the earth had been transferred first to the ancient resin in Egyptian bodies because of its similar appearance and erroneous translations, and now from the resin to the body tissues accidentally incorporated within the resin.

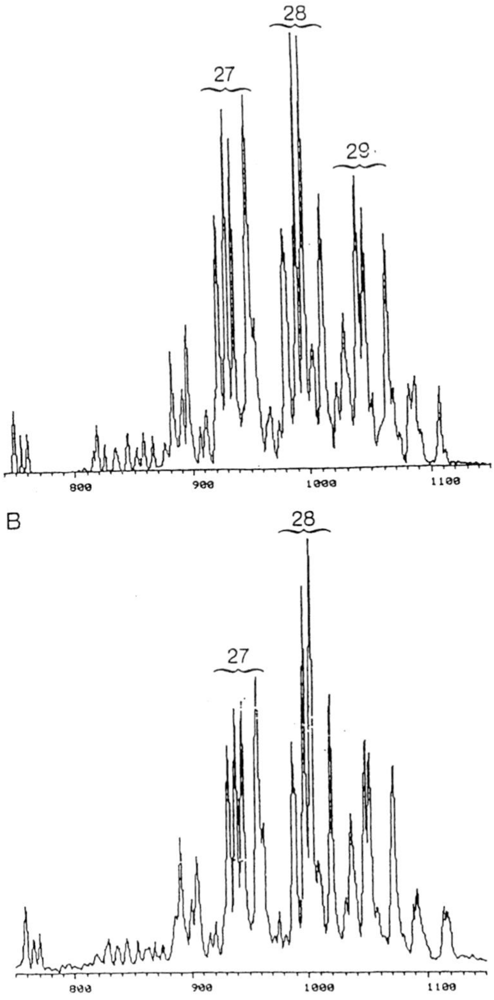

Fig. 10.2. Dead Sea bitumen as Egyptian resin component.
These two virtually identical patterns of gas chromatography studies of Dead Sea asphalt (top) and an extract of resin from a second century AD Egyptian mummy bottom document the use of Dead Sea asphalt incorporation into embalming resin during the Roman Period. (Photo courtesy of Arie Nissenbaum (1992) and *Archaeology* **19**: 1–6, 1992.)

This dramatically altered the nature of the mummy trade. The European medieval alchemists that had subsequently evolved into Renaissance apothecaries quickly accumulated assorted Egyptian mummy fingers, hands, feet, arms, legs and eventually even whole bodies, parts of which were ground into "mummy powder" as needed by these pharmacists and dispensed as medicine (Levey, 1966:40) (Plate 6). The demand for this new product soon exceeded even the supply of tomb-looted ancient Egyptian bodies. This led the "mummy" suppliers to the practice of looting Egyptian graves of the recently deceased. These modern bodies were oven-dried and marketed in various forms as *mumia*. Dawson (1927c) cites the experience of a French royal physician, Guy de la Fontaine, who visited Alexandria in 1564. There he found a *mumia* trader who showed him his stock of 40 bodies and who boasted that he had prepared them himself during the previous 5 years. Egyptian governmental prohibition of mummy exportation did exert some suppressive effect but that action seems principally to have shaped the practice into a more clandestine trade (Scott, 1837:28; Wilson, 1964:35).

The sixteenth century brought the first wave of reaction to this practice. The *Oxford English Dictionary* (Murray *et al.*, 1989) cites Richard Hakluyt (1599) as noting that "these dead bodies are the Mummie which the Phisitians and Apothecaries doe against our willes make vs to swallow . . .". Ambrose Paré, a French military surgeon, berated his colleagues for the use of this compound in whose medicinal value he had no confidence. He also deplored the fraudulent production of mummified bodies, which practice had extended to Europe by that time. Other voices joined his, and of these, probably that of the seventeenth century British physician Sir Thomas Browne (1658/1958) was expressed most poetically when he lamented that "Mummie is become mere merchandise, Mizraim [Egypt] cures wounds and Pharaoh is sold for balsoms . . .". Nevertheless, the practice subsided slowly, perhaps responding less to the persuasions of its critics than to the gradual decline of Galen's medical principles, the rise of rationalism and ultimately to the birth of bacteriology. While it was still listed in eighteenth century pharmaceutical catalogs, its clinical use declined substantially. Curiously a well-known German handbook of pharmaceutical practice still listed it in its 1972 edition (Dannenfeldt, 1985) and in 1973 a New York shop offered "witches' supplies" that included powdered Egyptian mummy (Harris & Weeks, 1973:92).

One might well ask how a product of such a bizarre source and of no known medical curative value could have attained this type of popularity for such a long time. In fact, it only becomes conceivable if we remember that germ theory had not yet evolved, and that medical thought was still dominated by the principles of balance and imbalance of body humours (fluids) popularized a millennium earlier by Galen. This concept almost precluded the idea that each disease state was the product of a specific, identifiable disease agent. Thus when recovery from an acute, self-limiting illness coincided with exposure to a certain substance, Galenic physicians were vulnerable to an explanation that attributed the clinical improvement to a restoration of humoral balance effected by that substance. That form of *weltschau* was supported by a popular view that every animal and plant was endowed by an intangible form of energy or life force (animus) and that this animus was transferable. The therapeutic effectiveness of a locally applied botanical product, ingested or inhaled, could be attributed to the transfer of the product's anima to the patient. Sixteenth century physicians, dispensing dried human muscle as medicine, would have been completely comfortable with our modern practice (as late as the 1970s) wherein we treated a child's growth hormone deficiency by injecting material extracted from a body part: crushed human pituitary glands that had been harvested from autopsied hospital corpses. Given the state of knowledge about human physiology at the time, the Renaissance physicians need to be judged less by their actions and more by their motivation for those acts.

Mummy as loot

The rewards of tomb-looting are not restricted to graves containing mummies, but the presence of mummified human remains in burial chambers has certainly enhanced their attractiveness in the eyes of the profes-

Fig. 10.3. Necklaces on a mummy; SM, 1000 BC.
Bejeweled necklaces were and are a common target for tomb robbers, even simple beaded ones as in this mummy. Spontaneous mummy of Alto Ramirez culture, from Pisagua in northern Chile, 1000 BC. (Photo by ACA.)

sional graverobber. Nor is tomb-looting a modern phenomenon. In Egypt it can be identified easily in tombs of the Middle Kingdom (2040–1640 BC) and it is obvious even in the impressive Old Kingdom mortuary edifices as early as that of first dynasty pharaoh Djer (Brier, 1994:81). In these older periods the universal target was the value of the grave goods entombed with the royal bodies. This has remained the principal attraction throughout the ages. As the value of these artifacts increased, so did the efforts of the thieves. Nearly successful efforts (fortunately aborted) to plunder the Dynasty 18 tomb of Tutankhamun seem to have been made about a decade after his death (Brier, 1994:128). The bribery and corruption that are essential components of royal tomb robbery were enabled particularly during periods of weakened central administration. A papyrus from the reign of Ramses IV (1134–1117 BC) records organized robbery of the Valley of the Kings tombs that involved participation not only by *fellahin* (peasants) and individual soldiers but even reached the king's vizier himself. The mummies were commonly disarticulated: heads removed to obtain necklaces (Fig. 10.3); hands and feet detached to gain access to wristlets and anklets; arms broken off to strip bejeweled armlets; wrappings torn or slashed to extract enfolded amulets. The pressure of possible discovery led to maximal haste in recovery of such items, leaving the tomb a chaos of mummy parts. Rowling (1961) refers to a Ptolemaic papyrus lamenting the destruction of many mummies when, in their precipitous departure, robbers failed to close the tomb entrance.

The Egyptian clerics' response to this blasphemous treatment of royal mummies often led to the equivalent of a cat and mouse relationship. Learning of the impending plunder of a specific royal tomb, priests would move the mummy occupant to a different tomb – sometimes even placing it in a different sarcophagus. The mummy of Ramses III was moved three times (Ceram, 1964:162). Eventually pressed into desperation, the priests collected groups of royal mummies, concentrating them into a single pit tomb near that of the imposing structure created by Queen Hatshepsut at Deir-el-Bahri. This one remained unmolested from about 1120 BC to AD 1875, almost exactly three millennia, when it was discovered by an entire village of tomb robbers that had practiced grave-robbing as a profession for generations (Ceram, 1964:163–167). The thieves were apprehended when a flood of the distinctive, royal jewelry appeared in the art market and was traced to them. No fewer than 40 mummified Egyptian monarchs and royal family members have been found in a common sepulchre. These included the mummies of Ramses the Great (II) (Smith & Dawson, 1924:171–183) as well as those of Sekenenre, Ahmose I, Amenophis I, Tuthmose I, II, III, Sety I and Ramses III. A second cache of royal and elite mummies was found in the tomb of Amenophis II that housed the bodies of Amenophis II, III, Tuthmose IV, Meneptah, Siptah, Sety II, Ramses IV, V, VI and many queens and their children (Smith & Dawson, 1924:171–183).

We have already commented above on the looting of mummies for sale in the medical mummy powder trade with Europe. These activities, together with those detailed below, had reached an intensity that caused that venerable patriarch of mummy studies, Sir Marc Armand Ruffer, and A. Rietti to plead for stricter control of access to mummies, lamenting that "unless this is done soon the present waste of scientific material will continue, only to be bitterly regretted before long" (Ruffer & Rietti, 1912). Their message, unfortunately, came too late for those hundreds (probably thousands) of recovered and examined royal and aristocratic mummies excavated principally during the early nineteenth century prior to those by Petrie and colleagues. With a moderate number of exceptions the results of those examinations were embarrassingly poorly recorded and the few tissue samples collected were not recoverably registered.

Napoleon's brief invasion of Egypt in 1798 had many repercussions. Accompanying his army were about a hundred members of the French Academy of Sciences whose publications during the two decades following that disastrous military defeat electrified the world, initiating what can be most accurately called Egyptomania. The passion for possession of even the most trivial ancient Egyptian memorabilia created and inflated an insatiable market. The 1800s witnessed a virtual invasion of Egypt by archaeological treasure hunters. Exceptional among these were the relatively few serious excavators, often sponsored by major museums, who virtually invented archaeology as we know it today. They included Sir Flinders Petrie (British School of Archaeology), George Reisner (Harvard and Boston Museum of Fine Arts), Herbert Winlock (Metropolitan Museum of Art), James Breasted (University of Chicago Oriental Institute), Auguste Mariette (The Louvre), Gaston Maspero (French Academy of Sciences) and others.

Many others, however, were less responsible and some were pure adventurers. Giovanni Belzoni was, at various times, all of the above. In 1816 he constructed the necessary machinery and mobilized the labor necessary to retrieve the massive stone head from the Ramesseum and float it down to Cairo (Fig. 10.4). From there Henry Salt, then the British Consul-General in Egypt, dispatched it to England. Belzoni then moved to the Valley of the Kings and, with the help of a battering ram, looted four tombs including that of Sety I (Wilson, 1964:28).

Tourists were an additional market to which the mummy trade catered. Both Europe and America were caught up in the wave of Egyptomania. The methods of the Industrial Revolution that were applied to the almost unlimited and readily accessible resources of the USA catapulted many from the lower socioeconomic classes into well-to-do circumstances. Many of

Fig. 10.4. Belzoni's removal of the stone head from the Ramesseum.
Giovanne Belzoni used aggressive measures to harvest ancient Egyptian artifacts. (From Mayes, 1961 with permission.)

these *noveaux riches* celebrated their new status in the form of a European voyage. As interest in Egyptian prehistory grew, it became ever more popular to include a stop in Egypt to acquire an appropriate memento (Chamberlin, 1983:40). As early as 1837 C.R. Scott deplored "the eagerness with which every sort of trash is purchased by travelers . . ." (Scott 1837:29). Intact mummies became an item readily available. After relatives and friends of these returning travelers had seen and appropriately admired such mummies, the bodies often found their way into the vitrines of the local museums. Many of the major American and European museums accumulated their modern inventory of Egyptian mummies in this manner. Unfortunately most of these mummies were acquired from Cairo's street hawkers without provenience data, and thus are of limited educational value unless studied. Furthermore, modern radiography of many such mummy bundles has revealed that some are modern frauds.

Looting mummies from ancient tombs is, however, not only an ancient practice. In remote areas discoveries of ancient burial grounds often come to the attention of archaeologists when the discarded, fragmented mummies of newly looted tombs are found and delivered by a passer-by to the local university or museum. If archaeologists do not immediately excavate the area scientifically it is common for the entire ancient cemetery to be looted in the subsequent months. On one occasion in South America a peddler came to my hotel room to offer me local ancient artifacts for sale. When I recognized these items, it became clear that our archaeologist group was excavating that burial area by day while the peddler was looting it by night!

Mummy as display

Perhaps the best-researched example of the use of a mummy as a display item is the case of Elmer McCurdy. As told by Snow & Reyman (1977), Elmer was Oklahoma's bad boy at the turn of the century. As a young man he had already graduated from the

Oklahoma territorial penitentiary and by 1911 had turned to train-robbing. Frustrated by the lack of money aboard a train he and companions had raided, Elmer grabbed two demijohns of whiskey and fled into the hills. Showing the effects of the first bottle, he arrived at a ranch where he was granted the privilege of sleeping in a hay loft; there he turned to consuming the second. This was where the posse found him and where he died of a chest wound during the subsequent shoot-out. His body was taken to Pawhuska, Oklahoma, where the local funeral director embalmed him. Months later, claiming he was his brother, a carnival operator made off with Elmer and launched him on a decades-long career as a traveling side show attraction. But the constant travel was wearing on Elmer, so he was sold to a wax museum as a mannequin. There a motion picture film crew rented him for a horror film production. While moving the body between scenes they were appalled when his arm broke off, exposing a bone and revealing that this "wax" body was actually a human corpse. The Los Angeles medical examiner autopsied Elmer, identified the gunshot wound, and was astonished at the high quality of microscopic tissue preservation. Chemical analysis of Elmer's tissues revealed high quantities of arsenic, characteristic of some contemporary embalmings (though this arsenic source has been questioned: Warrick, 1998). Retracing Elmer's wanderings, the department's investigation identified his origin. The Oklahoma Historical Society gratefully welcomed the return of their prodigal son and buried him, unequivocally terminating Elmer's wanderings by sealing his grave with two cubic yards of cement.

Wilson (1964:37) describes an early nineteenth century entrepreneur who purchased several mummies in Egypt, and sold them to Peale's Museum and Gallery of Fine Arts in New York. Later they became star attractions in one of P.T. Barnum's sideshows for 16 years, only to end their career as victims of the great Chicago fire in 1871. Another mummy gazed benignly upon the surgical activity in the operating theater at Massachusetts General Hospital for many years. The eighteenth century physician William Hunter initiated a school of anatomy, and when the mistress of his assistant lecturer (John Sheldon) succumbed to tuberculosis, Sheldon suggested Hunter attempt to embalm her. Sheldon was so pleased with the result that he kept her in his bedroom until the woman he subsequently married complained. Later Hunter also embalmed the Duke of Hamilton (at his request) as well as the wife of his friend Martin van Butchell, who exhibited the body in a glass case at his home until he died (Kobler, 1960:138–139; Broaddus & Harmony, 1984). More recently a funeral director exhibited (without a viewing fee) the unclaimed body of a 1928 drowning victim (Henry "Speedy" Atkins) in Paducah, Kentucky, for many years. After the funeral director died his wife continued the practice, washing and disinfecting the body three times a year until she tired of this effort 66 years after Atkin's death (Quigley, 1998:94–96).

Early nineteenth century "mummy unrollings", though commonly advertised in a manner suggestive of a dignified, scientific endeavor, in actual practice frequently included the sale of tickets and other features more characteristic of a carnival sideshow than an educational study. It was Pettigrew (Smith, 1910:194–195; Dawson, 1934), Granville (1825) and similar pioneers of soft tissue pathology (see Chapter 1) who rescued the practice of mummy dissections from the hands of the hucksters. Putnam (1993:41) cites the nineteenth century museum practice of collecting Egyptian heads and displaying them under glass. Common to all these described practices is the exhibition of mummies simply as items of curiosity.

Public interest in viewing mummies persists and they are on exhibit even today in museums ranging from the British Museum of Natural History to regional county historical museums. In most of these, however, the curators take their mission very seriously and present the exhibits as part of an educational program. An excellent example is the Science Museum in St Paul, Minnesota, where their star attraction is an Egyptian mummy. Considerable anatomical, radiographical, biochemical and bioanthropological research has been carried out on this body. The results of these, together with historical data are presented as a constructive, educational experience, comparable with instructive presentations on other topics elsewhere in that museum. The isolated, noninformative display of a mummy primarily as an attendance lure in museums is a disappearing practice.

Mummy as paper

Transforming mummies into paper must rank among the more bizarre forms of mummy commerce. Not the bodies themselves, of course, but their wrappings. Following the body desiccation stage of the embalming process, Egyptian mummies were completely encapsulated in a linen cocoon. In its most elaborated form the extremities were separately wrapped and then both were often incorporated into those of the torso. Many yards of cloth strips varying from a few centimeters to 30 cm wide were employed on the various body parts, sometimes covered by broad sheets. The cloth was virtually always linen, and for royal bodies was specifically prepared later for that purpose. As was true for other stages of the embalming process, this too was a ritual act. Appropriate prayers were not only spoken at specified intervals, but for the aristocratic class often were also inscribed on the linens themselves (see Chapter 4: Mummies from Egypt and the rest of Africa). The weaving quality varied, but those near the final surface were commonly composed of fine, densely woven material. In addition to amulets enclosed between linen layers, liquid resin was daubed at intervals onto the cloth to hold the strips in position after the resin cooled. In the Late Period vast quantities of resin were employed. In some of these mummies an attempt appears to have been made to apply sufficient resin to create a protective shell that would impair spontaneous rehydration in perpetuity. In later periods the embalming efforts were more casually performed, but the wrappings were applied in a manner resulting in a very attractive, complex geometric design. Since it commonly required as much as 16 kg of linen to achieve this, the cost of the wrappings must have constituted a significant fraction of the funeral costs.

In 1855 an article relating to the local papermaking industry was allegedly written by a local academician, Dr Isaiah Deck, speculating on the favorable economic aspects of using Egyptian mummy wrappings as a substitute for the locally purchased rags in papermaking. Dr Deck noted that mummy wrappings could be bought in Egypt and shipped to New York for about three cents per pound (0.5 kg), approximately half the price of local rags. Furthermore, recovery of the aromatic oils and resins might find a secondary market for use as incense in church liturgical services (Hunter, 1943:290). The article made no comment regarding any precedence for the conversion of mummy wrappings into paper, so we remain ignorant of what led Dr Deck to think in these terms. A year later, however, the newspaper *Daily Standard* (Syracuse, NY) reprinted an article from the *Albany Journal* (New York) that an unnamed individual from Onondaga County, New York, was preparing paper from Egpytian mummy wrappings. No further details were supplied. I have been unable to confirm a persistent rumor about paper manufacture from mummy wrappings in Canada.

The most reliable account of American paper production from mummy wrappings is detailed within a section of a book on the history and technique of papermaking by Dard Hunter (1943:287–291). I. Augustus Stanwood and William Tower began a papermaking company in Gardiner, Maine, in 1863. Hunter states that he obtained the following information from Stanwood's son, a retired professor of International Law. Because of a shortage of rags during the American Civil War, his father (I. Augustus Stanwood) imported several shiploads of Egyptian mummies to his mill, stripped the linen wrappings from the bodies and threw both the wrappings and the papyri found among the layers of cloth into the paper mill's beaters. The resulting coarse, brown paper found its way into the regional grocery and other stores. He also commented that the only competitor his father had for the purchase of mummies was the Egyptian railroad that used the very combustible, resin-soaked mummies as a sole source of fuel for a 10-year interval during the mid 1850s (see below). When Stanwood died, on March 6, 1914, many newspapers commented on his use of mummies for paper manufacture, providing validity to the above account. The introduction of ground wood as a rag substitute about 1863, however, suggests that the mummy wrappings experiment may not have been long lived. The fate of the mummy bodies, stripped of their wrappings is not identified. Considering shipping costs it is probable that the linen wrappings may have been stripped from the bodies in Egypt before they were shipped.

One incident recalled by Professor Stanwood is that the ragpickers stripping the wrappings from the mummy suffered from an outbreak of cholera. *Vibrio cholerae*, the organism causing cholera, however, is not a spore-former and in an environment other than water, is a rather fragile organism. It is conceivable, however, that the ship delivering the mummies may have had water in the hold contaminated with cholera organisms that gained access to the mummies in transit from Egypt to New York. If so and if the mummies or their wrappings were handled while still wet, the incident is conceivable. More likely, however, is the ingestion by the ragpickers of contaminated drinking water aboard the ship while unloading. It is surprising that no one has yet invoked "The Mummy's Curse" as an explanation! However, Dane (1995) has been unable to identify any of the original documents relating to the manufacture of paper from mummy wrappings and regards all the allegations as of an unverified anecdotal nature.

Mummy as fuel

Spontaneously or anthropogenically mummified bodies commonly are nearly devoid of water (Aturaliya & Lukasewycz, 1999). In addition, those from Egypt are enveloped in up to 16kg or more of linen wrappings. Furthermore a variable quantity of resins (and occasionally, in the Egyptian Late and Ptolemaic Periods, also bitumen) is incorporated within the linens and often the body cavities and tissues as well. The consequence of this preparation is the production of a highly inflammable product weighing 24kg or more. It was inevitable that someone would exploit this vulnerable feature of mummies.

We have already alluded above to the possible use of Egyptian mummies as fuel to power the first Egyptian railroad. This account was popularized by Mark Twain (1903:421) in his tale *Innocents Abroad* (as part of a facetious recital of other pseudoevents). I have been unable to document this practice from any reliable, other source and hence consider this a product of "poetic license". On a lesser scale we can gain some sense of the degree of reverence for antiquity from the incident in which Egyptian *fellahin* of the late eighteenth century found a group of papyri from the Ptolemaic Period. Unable to find an interested buyer in Cairo they used them as kindling in their stoves. Similarly mummy wrappings and even mummy containers were burned for heat in nineteenth century Thebes (Wilson, 1964:11). Even the unembalmed, spontaneously desiccated bodies of Predynastic Egyptians buried in the desert's sands sometimes were employed as fuel for Bedouin campfires (Giacometti & Chiarelli, 1968).

Mummy as a commercial product

The items included under this heading differ from those immediately above only in degree, but the "retail" nature of these activities seems to justify grouping them separately. In 1968 a publication (Ettinger, 1968) appeared regarding the possibility of long-term mummification by rapid freezing for the purpose of future thawing with successful "revival" after effective therapy had been developed for a currently incurable disease. Three years later the first customer for such a commercially available service was treated by a shock-freezing technique by immersion in liquid nitrogen (Vandenberg, 1975:122). It is probably only the expense of the method that limits consumers for this product. In terms of classification of mortuary services, this procedure differs from the almost routine formaldehyde-based embalming carried out on most American deaths. Differing materials are used for rapid freezing and long-term maintenance is required.

The modern treatment of a mummy as a commodity is more obvious in the case of some caches of mummified Egyptian animals. During the latter part of the Pharaonic Period a tradition arose of using mummified bodies of animals symbolizing specific gods as votive offerings. These were mummified by the temples' priests according to a prescribed ritual, and sold to the public, who brought them to the temple. There an offering ritual was carried out and the mummified animal was then buried in underground chambers. Animals as large as bulls or small as birds were mummified. For the smaller animals, tens of thousands of such prepared, desiccated mummies accumulated in the disposal sites. In the early Christian zeal to extinguish what they regarded as blasphemous and pagan practices, some of these animal cemeteries were destroyed. However, in 1890 an entrepreneur excavated and shipped to England more than 200 000 mummified cat bodies from the community of Beni-Hassan that had housed a temple to the cat goddess Bast. In Liverpool these votive offerings were pulverized and sold as fertilizer (Herdman, 1889–1890).

More recently dried vegetal remains from the bottom of an Egyptian container were analyzed, after which similar modern botanicals were employed to create a brew that allegedly was very similar to the ancient Egyptian concoction. It was marketed as "Tutankhamun Ale" (Reuter News Agency, 1996:131). Without claims to simulation of ancient Egyptian confectionery, a modern company offered "Yummy Mummy", a cigar-shaped candy item whose wrapper was emblazoned with mummy cartoons.

Egyptian mummies themselves, of course, have been a commodity for 2000 years, but recently their marketing has been modernized. On March 3, 2000, an Egyptian human mummy was offered for sale for $US40 000 on the Internet at http://www.thestrange.com.mummy.html.

A unique exploitation of mummies resulted when eighteenth century European artists discovered that pulverized, mummified animal tissues from Egyptian tombs mixed with resin and suspended in appropriate solvents functioned as a highly effective brown pigment for oil painting. It owed its popularity to the fact that it did not crack after drying when applied to the canvas (Wilson, 1964).

An oft-repeated anecdote deals with the manner in which sailships, supplying the Spanish colonialists in Peru following the Conquest, sometimes solved the problem of the empty holds after unloading their cargo. Even by the nineteenth century they had few products made in south central Peru's remote areas with which to fill these holds. Hence, at locations such as Ancón near Lima they simply used boatloads of the abundantly available mummified bodies from the local necropolis as ballast (McHargue, 1972: 94 – but no source offered). However, other accounts indicate that the ballast actually was sand from the site, and that mummies were transported (for commercial purposes?) stored in the sand to maintain their arid status.

Another unusual form of mummy as a commodity came to my attention during an afternoon of shopping in a South American city. Passing a store that advertised the availability of pre-Hispanic jewelry replicas, I paused to admire both the gems and the finely crafted, realistic-appearing mannequins that displayed them. Examining them closely, however, I was surprised to find that the manequins owed their realistic appearance to the fact that they were, indeed, the mummified bodies of excavated, desiccated human remains.

Mummy as curse

The fear of supernatural retribution for those who violate the sanctity and isolation of an ancient Egyptian tomb is almost as old as the tombs themselves. El Mahdy (1989:170) describes how 500-year-old Arab texts reflect this concern. She notes that the strange hieroglyphics together with tomb murals suggesting revivified mummies ("opening of the mouth" ritual) may have inspired such concerns, augmented by beliefs in magic commonly noted among contemporary Egyptians when Arabs dealt with Egypt. Hamilton-Paterson & Andrews (1979:192) point out that after Champollion deciphered the Rosetta Stone texts, pseudoscholars reveled in their eagerness to mystify these ancient texts for the purpose of sheer sensationalism, to which the public seemed to respond with enthusiasm.

It was, however, the discovery (November, 1922) and opening (February, 1923) of Pharaoh Tutankhamun's tomb that brought this type of interest into a specific focus (Carter & Mace, 1963:106). At the time of the tomb's opening the American novelist Marie Corelli wrote a letter to the *New York Times* stating that she owned a rare Arabic book dealing with ancient Egyptian burials. This included a "curse" against tomb robbers that "death comes on wings to he who enters the tomb of a pharaoh . . ." (El Mahdy, 1989:172). She predicted that all involved in the opening of Tutankhamun's tomb would die as a result of that act. Nine days later Lord Carnarvon (George E.S.M. Herbert, Earl of Carnarvon) was dead. He had provided the funding for the archaeologist Howard Carter and had been among those present when the tomb was opened. At this point various circumstances combined to sensationalize this incident. Hordes of reporters had been sent to Egypt to "cover" the excavation. Their limited access to the tomb and excavators, together with the slow progress of the tomb contents' removal caused them to seize upon otherwise minor or peripheral features of the affair. When they queried Sir Arthur Conan Doyle (physician and author of the Sherlock Holmes stories) about Corelli's letter and Carnarvon's death, he replied that he regarded the fatality as a product of some "elements" with which the ancient priests had imbued the tomb. Desperate for a story, the reporters noted that Doyle had "validated" Corelli's warnings. This notoriety enshrined the concept of "the mummy's curse" and even invented the subsequently accepted allegation that a "winged death" curse notice had been found within Tutankhamun's tomb. The "curse" was embraced by the Press and lecturing entertainers, citing it upon the occasion of every death, no matter how remote in time or geography, of anyone even most peripherally related to that pharaoh's handling (Wilson, 1964:165). Allegations that pathogenic fungi in dried bat droppings were responsible for the death have not been substantiated.

As the years passed "curse stories" were bred, nourished, expanded, dramatized and further falsified. One unusually long-lived example began when the British Museum acquired a mummy with an unusually attractive coffin lid. When this lid allegedly caused the museum much stress it was sold to an American, shipped on the liner *Titanic*, but rescued from it as it sank. After the disasters caused to its American owners, they sold it to a Canadian whose subsequent misfortunes then led him to ship it back to England, only to have that ship sink in the St Lawrence River. As the keepers of the British Museum in charge of this lid proclaim, however, none of this ever happened (Hamilton-Paterson & Andrews, 1979:195). The lid was never sold nor did it travel and could still be seen in one of the museum's Egyptian rooms at the time in 1979; its file number is 22542.

Skeptics, nevertheless, point out that half of those attending Tutankhamun's tomb opening died within a short time (Vandenberg, 1975:28). In addition to Carter, 20 others attended that opening. These are listed in Table 10.1. Of these I have been able to identify the date of death exactly in 11 and approximately in 2 additional ones. The mean age-at-death of these 13 persons is 68.1 (±11.7: one standard deviation) years and the median age is 66 (range 56–90). In the 1920s and 1930s national American and British statistics would indicate that death at 66 years would not be considered a truncated lifespan. Carter, who spent more time in the tomb than any of the others, died at 66 years. If we calculate the interval between the date of the tomb opening and the date of death we find the mean is 19.7 (±13.4) years and the median is 19 years. Inclusion of the two Carter assistants, whose age-at-death is unknown but time of death is known to be no later than 1929, lowers these values by only 3–4 years and does not alter the conclusions. If these values are typical also for those on whom I have no information, then we can conclude that a curse inflicts punishment only an average of two decades following an alleged insult, and that the usual age-at-death is not altered significantly. Vandenberg's (1975:28) claim that "by 1929 twenty-two people who had been directly or *indi-*

Table 10.1. *Persons in attendance at opening of tomb of Pharaoh Tutankhamun*

Name	Date of death	Age-at-death (years)	Years after 1923	Status
Carter, Howard	1939	66	16	Archaeologist
Carnarvon, Lord (G.E.S.M. Herbert)	1923	57	0	Carter's patron
Herbert, Lady Evelyn	1929	63	6	Carnarvon's wife
Suleman, H.E. Abd de Haum Pasha				
Lacau, Pierre	1963	90	40	Director, Service of Antiquities
Garstin, Sir Williams				
Cist, Sir Charles				
Lythgoe, Albert M.	1934	56	11	Curator Egyptian Dept. MOMA[a]
Breasted, James Henry	1935	70	12	Archaeologist
Gardiner, Sir Alan	1963	84	40	Egyptologist
Winlock, Herbert Eustis	1950	66	27	Archaeologist, Egyptologist
Herbert, Mervyn				
Bethell, Richard				Secretary to Carter
Engelbach, Reginald	1946	58	23	Inspector, Department of Antiquities
Fahmy, Bey				
Sirdah, Sir Lee Stack				
Lucas, Alfred	1945	78	22	Chemist, Department of Antiquities
Mace, Arthur Cruttenden	1928	54	5	
Astor	"by" 1929		0–6	
Bruere				
Callender	"by" 1929			

Notes:
[a] Metropolitan Museum of Modern Art.

rectly involved [italics are mine] with Tutankhamun and his tomb had died prematurely" is true only if we broaden the studied sample by including others who died (*indirectly involved*) but exclude others similarly exposed who did not die. I could only identify four of those involved in opening the tomb who had died by 1929. A statistically valid comparison would require each of those somehow involved with this pharaoh to be matched with an individual of the same cultural group of identical age, gender, degree of exposure to soil, climate and other variables, but without involvement with Tutankhamun's (or other Egyptian?) tomb. The formidable problems involved in identifying such individuals will most likely discourage any serious statistician from undertaking the problem. This is particularly true because most who consider the curse a real possibility are not apt to be swayed by statistical findings (Wilson, 1964:165).

El Mahdy (1989:174) insists that no true inscribed curse of this alleged type has ever been found in any ancient Egyptian tomb. The occasional apparent exception to this can be explained as forceful pleas by the deceased's relatives urging priests to persist in prescribed mortuary practices, or sometimes represent grievances against the deceased by a survivor. To the contrary, the fate of the deceased's spirit ("ka") depended on the living priest and relatives to carry out the appropriate rituals. Clearly, it is time to rebury the myth of the curse.

Mummy as deception

Deliberate deception in the manufacture, sale or use of a mummy is an ancient practice (Moodie, 1931:55). Judging from the quality of the mummies produced in pharaonic times, most of those embalmers clearly were committed to high professional principles. Yet the occasional scoundrel among them can be identified today by his product. Shoddy work may be reflected by focal areas of soft tissue decay reflecting insufficient desiccation, reuse of linen wrappings, and omission of some of the time-consuming details hidden beneath the wrappings. More overt are mutilating practices such as foreshortening the legs in order to fit the mummy into a smaller, prepared container. Nevertheless, the ancient embalmers of Egypt are probably innocent of many charges of preparing hoaxes. As early as 1792 Blumenbach opened several mummy bundles held in private British collections. Three of these were very attractive packages, each designed to resemble a human infant mummy. Opening them, however, revealed that one contained a resin-soaked mass of wrappings from an adult mummy placed inside a wooden "sarcophagus" fastened with modern, iron nails; the second contained disarticulated animal bones (principally ibis); and the third contained only a human subadult humerus. It had been wrapped in modern paper and cotton balls were tied around it to create an anthropoid form. In several others the wooden coffin container had obviously been "restored" (repainted). Several adult-size bodies were merely skeletons, some with painted bones and one with a mask of obviously modern preparation. At the Budapest Museum of Fine Arts I saw a very meticulously wrapped, anthropoid-shaped bundle, one meter long, fitted with a painted child's face; however, its X-ray image revealed that an adult lower leg was contained within the bundle. When Moodie (1931) reported his radiographical findings of Chicago's Field Museum's mummy collection he had similar experiences in which the bundle content was seen to be only wood, wire, clay or other nonbiological material. In a surprising number the "child mummy" bundle contained a mummified bird (ibis). Similar findings were also noted by Gray & Dawson (1968) in British Museum Egyptian collections, though in a few cases Gray (1966) felt the findings reflected sincere efforts to replace missing body parts.

The common feature of the hoaxes cited above is their presence in Europe or America. These are the very countries that engaged in active trade of *mummy* for medicinal purposes in the early Renaissance, or whose travelers purchased mummies from Cairo street hawkers and brought them home. As detailed in the above section (Mummy as a drug), the demand for mummies generated a clandestine operation involving the manufacture of mummies that was lamented as early as 1837 (Scott, 1837). Many of the hoaxes presented here are exactly what one would expect under these circumstances. A recent, non-Western hoax in Pakistan offered what was alleged to be an ancient mummy of the daughter of the Persian ruler Xerxes (519–465 BC). Suspicious potential buyers launched an

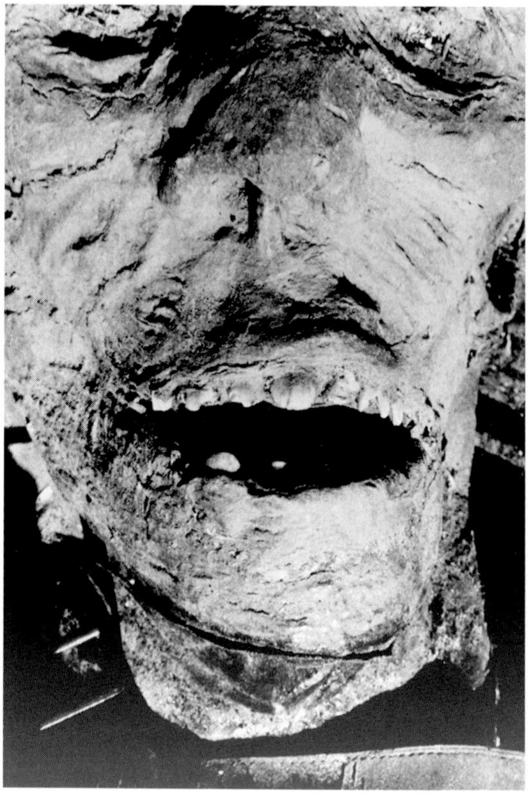

Fig. 10.5. "Dentition" in mummy hoax.
The teeth of mummy hoaxes often are overtly inappropriate. Frequently, as in this example (Martindale mummy) principally animal teeth are employed. (Photo courtesy of Larry Cartmell.)

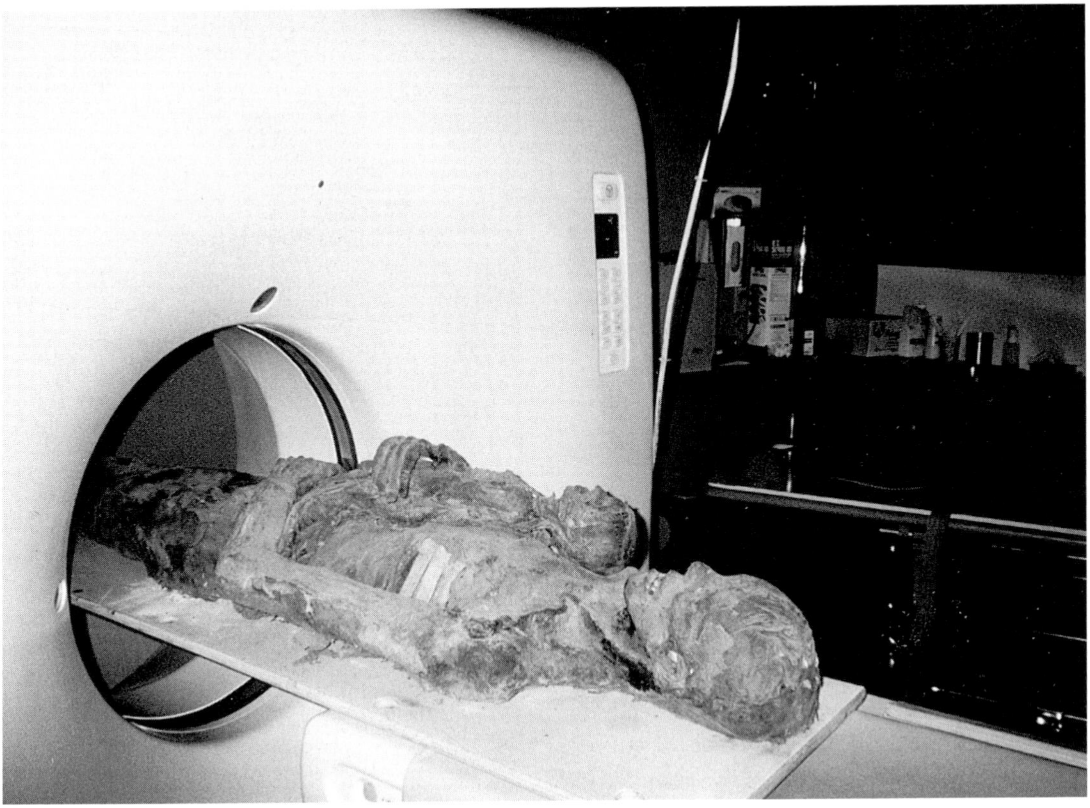

Fig. 10.6. Radiography for hoax detection.
Though this mummy hoax (Martindale mummy) is undergoing CT scanning, simple X-rays are usually sufficient to identify a hoax. (Photo courtesy of Larry Cartmell.)

investigation that identified the body as a recent death with features suggestive of homicide (Koenig, 2001).

Although on a much more diminutive scale, the manufacture of mummies continues today. I have consulted on several cases, both secular and forensic, to validate the authenticity of an alleged mummy that was being displayed for a viewing fee. In most cases this is easily established by simple inspection. Teeth are commonly made visible to the viewer in such hoaxes, probably to enhance the illusion of a human body. These, however, are usually the clearest evidence of the deception (Fig. 10.5). Some have been animal teeth; in others they were human deciduous teeth (probably more commonly available to the hoaxers) but placed in nonanatomical positions (Macon *et al.*, 1983). In a few instances exposed body parts have been so skillfully synthesized that one can not reach a satisfactory degree of certainty merely by viewing them in a glass-covered container. Simple "flat plate" radiography virtually always settles the issue (Figs. 10.6 and 10.7). For obvious reasons, it is common practice to envelop false mummies in cloaks or clothing, only exposing small areas of "skin". The well-known and widely displayed Martindale mummy from the central USA incorporated most of such features (Fig. 10.8). In its advertising, this mummy, together with several others, was alleged to have been discovered in a cave in California's Yosemite Valley in June 1891 by G.F. Martindale and three other gold prospectors. For nearly a century thereafter the mummy survived the

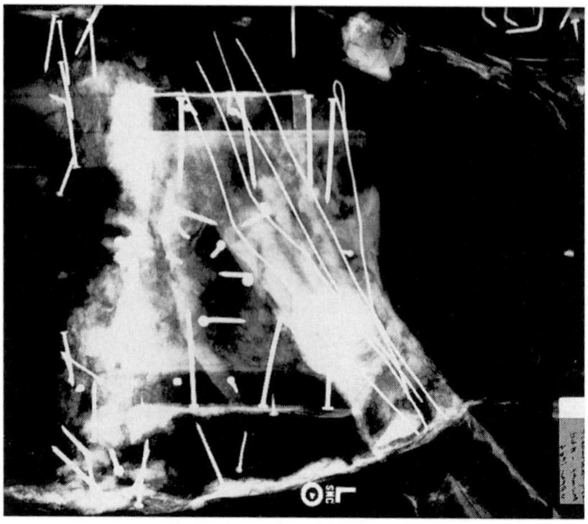

Fig. 10.7. Radiographical evidence of mummy hoax. The wire and nails employed as a structure around which to mold clay to resemble a human hand is made evident by simple X-rays in the Martindale mummy. (Photo courtesy of Larry Cartmell.)

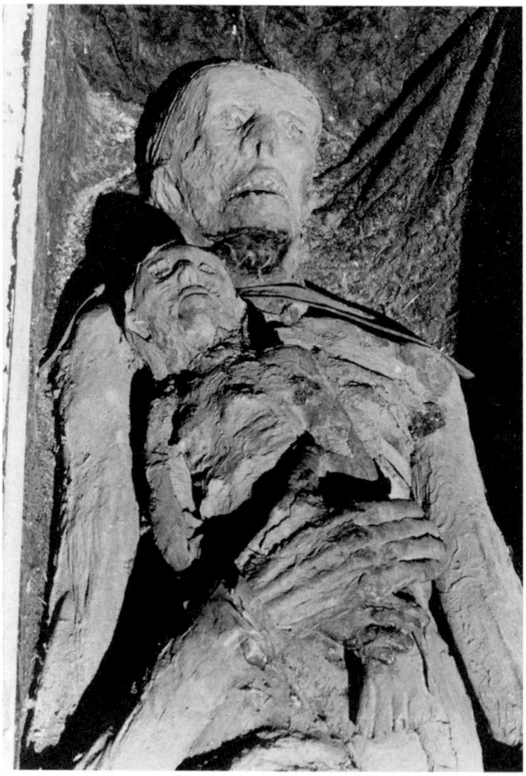

Fig. 10.8. Martindale mummy hoax. The apparent adult mummy (Martindale) holding a "child" is an obvious hoax. The metacarpals of the adult hand are overtly of nonanatomical structure. (Photo courtesy of Larry Cartmell.)

carnival sideshows, county fairs, and other occasions at which it was displayed by a succession of owners. Eventually it was acquired by a physician and paleopathologist from Ada, Oklahoma, who used radiological study to demonstrate that the "mummy" was a manufactured hoax (Cartmell, 1996). Some of these synthetic mummies are so obviously false that one can only assume they remain marketable because viewers are amused by their detection of the fraud.

Falling within the definition of deception involving human remains are newspaper reports of the very differently motivated attempts to smuggle aboard embalmed, recently deceased bodies in garment containers in the passenger compartment of an aircraft (Heller, 1990:J1). This has commonly involved an immigrant to America that is returning the body of a recently deceased family member for interment in the deceased's city of origin, but who can not afford the air freight costs for shipping such bodies.

Pettigrew (1834:228) records the description of a British traveler to Egypt who found a native Egyptian living in an ancient tomb, manufacturing mummies for sale.

Mummy as neglect

In contrast to the active exploitation of mummies for commercial or other purposes, the loss of a mummy's potential contribution can occur also through simple neglect. The most obvious example would include the institution possessing a mummified body whose partial loss of preserved soft tissue has rendered it ineligible as an exhibit item. Since proper conservation of mummified human remains can be an expensive and time-consuming procedure, such undisplayable items may be relegated to storage under conditions that often do not arrest the decay process. While such remains could make a valuable contribution if subjected to scientific study, they have only infrequently been used for research purposes in the past. It is encouraging to note that the increasing recognition of their value is resulting in a higher frequency of interdisciplinary investigations of such specimens in the past several decades.

The Aswan Dam, constructed in Upper Egypt to generate electricity for this nation with limited natural energy resources, has been responsible for destruction of untold numbers of mummies. It is true that Reisner's 1907–1908 survey of the future flooded area prior to raising the lower dam provided a wealth of new archaeological data from that area. In addition the international effort prior to building the even higher dam recently rescued several invaluable archaeological monuments. It is unfortunate, however, that these were essentially salvage operations, and that Lake Nasser created by the dam may have inundated a hundredfold greater number of mummies than were salvaged.

Mummy as language, literature and film

One of Egypt's greatest gifts to posterity has been a written language. Its earliest form appeared at about the beginning of the Pharaonic Period and is composed of both ideograms and phonograms (hieroglyphics; "picture-writing") (Fig. 10.9). This form was employed for the first 2000 years of Egyptian history by specially trained persons (scribes), usually members of the royal court or related units. Near the end of the New Kingdom an abbreviated version that included a cursive feature appeared (hieratic script). About the eighth century BC a simplified form of hieratic (demotic script) evolved and was used thereafter for nonliturgical and administrative purposes, though the scribes continued to use the hieratic form for religious purposes until the time of the Ptolemies.

Scribes wrote verses from the Book of the Dead on papyri and placed them in royal tombs. They also inscribed the walls of the tombs, the coffins ("coffin text") and even the bandages of the mummy wrappings with prayers, supplications, divine praises and other messages that have helped enormously to shape our

Fig. 10.9. Egyptian hieroglyphs.
The inscription on this obelisk uses one of the early forms of written languages, employing both ideograms and phonograms. (Photo by ACA.)

understanding of Egyptian life and especially the Egyptian vision of the afterlife.

The secular Egyptian literature that has survived helps us to bridge the millennia between us more effectively than any artifact can. Interested readers will enjoy the anthology of stories and inspirational poems translated into English made available by Simpson (1972). Even the Egyptians themselves enjoyed reading stories about mummies, and a few of these have survived since they were written about 200 BC (Brier, 1994:300).

If the mummies themselves are of sufficient interest to be the focus of personal collections, museum exhibits, scientific study and even motion pictures, it should come as no surprise they have been equally attractive to authors. While the passion for mummies has fluctuated in intensity since Napoleon's invasion, it has never subsided completely. Indeed a host of English language books dealing with mummies and targeting a juvenile readership have appeared during the past decade or so, while many more have appealed to the mature interest.

An exceptionally moving poem reflecting the poet's response to viewing the Guanajuato mummies (Chapter 4: Mummies from Mexico) is reproduced here with the kind permission of the author, Phyllis Janik, professor at Moraine Valley Community College, Paleo Hills, Illinois.

From Mexico
the mummies of Guanajuato
by Phyllis Janik

Outside, this Sunday morning, there's nothing
happening, An old postcard on the desk
the loudest thing here. It's the conversation
these people in the card are engaged in
and their green bodies
underground light and the camera have captured.
The unearthed and standing dead of Guanajuato
who are not wholly gone
but have only shrivelled in the dry, special dirt
are saying something.
The skin pulled tight and
– their delights have vanished! Necks crane,
they have turned their bald heads
toward one another. Arms folded hard
around stomachs, closed eyes,
gaping mouths: it's the look
of the passionate desperate
who went upstairs maybe for thread
and came down without it.

Indeed, the bard himself included "witches' mummy" as a cauldron ingredient (*Macbeth*, Act IV, Scene 1). Describing the handkerchief given to him by his mother, Othello tells Desdemona it was "dyed in mummy" (Act III, Scene 4). Falstaff was concerned that the swelling effect of water on his body would cause him to become a "mountain of mummy" in the *Merry Wives of Windsor*, Act III, Scene 5 (Dawson, 1927e). Later on, the term "mummy" passed into food preparation, "it must be thick and dry, and the rice not boiled to a mummy" (*Oxford English Dictionary*, 1989:X, 97). Table 10.2 lists some of the more popular prose publications dealing with mummies.

Mummies were also included (though minimally) in the volumes published by Napoleon's scientists who accompanied him during his invasion of Egypt in 1798 (Denon, 1821, reprinted 1997, Vol. 2: plates 49, 50). These remarkable volumes inflamed the imagination of the world (see Mummies as loot, above) and, while the passion for reading about mummies has fluctuated since that time, it, too, has never subsided completely.

In a serious essay deploring the difficulty of trying to recover the truth of past events from historical accounts, Poe revivifies an Egyptian mummy (calling him Allamistakeo) who points out how historical misinterpretations have misrepresented his times (Williams, 1983). In a distinctly lighter vein, however, the July 22, 1986, issue of the *Weekly World News* headlined: "Mummy Found With Artificial Heart". At the British Stock Exchange traders have incorporated the word mummy into their technical vocabulary and imbued it with an irreverent, geographical meaning by assigning to it the popular designation of stocks of Egyptian companies as "mummies" or "mummy stocks" (*Oxford English Dictionary*, 1989:X, 97).

As the motion picture industry also evolved and expanded during the "post-Tut" decades, it was probably inevitable that some of the novels would be rewritten as screenplays, some even transferring the novel's title to the film. Universal Studio's 1932 version of *The Mummy* starring Boris Karloff remains the model for mummy films (Figs. 10.10 and 10.11). The mummy was the star of the film, given a distinct personality and recently memorialized by the United States Postal Service in the form of a stamp bearing a scene from that film. Other mummies that have appeared on stamps in other countries include Tutankhamun (Egypt), the Andean mummy El Plomo (Chile) and Inuit mummies (Greenland). Unfortunately the successive mummy films that Universal produced during

Table 10.2. *Mummy literature*

Title	Film too?	Year	Author	Comments
Ring of Thoth	+		A. Conan Doyle	Reuniting of ancient lover (potion; immortal) with his mummified lover
Lot 249	+		A. Conan Doyle	Revivified mummy used by hero to murder his enemies
The Mummy! or, A Tale of the Twenty-Second Century		1827	Jane Webb Loudon	Revivified Cheops' mummy restores moral order to 2026 London
Romance of a Mummy		1857	Theophile Gautier	Romance in which story is flashback of Egyptian mummy found in tomb
She		1887	H. Rider Haggard	Reuniting of lovers through millennia
The Vanishing Man (later: *Eye of Osiris*)	+	1911	R.A. Freeman	First detective story to use mummy (prop)
Jewel of the Seven Stars	+	1912	Bram Stoker	Revivified mummy queen. Film: *The Awakening*
The Mummy Moves	+	1925	Mary Gaunt	Mummy turned murderer?
The Palgrave Mummy		1929	F.M. Pettee	Post Tut-tomb novel. Mummy and coffin with curse shipped to USA. Disaster
Literature of Ancient Egypt		1973	William Kelly Simpson (ed.)	Translations of ancient Egypt's stories, songs, poetry. Yale University Press
Egyptian Myth and Legend		1978	Donald A. Mackenzie	Rewritten translations of Egyptian myths and legends with anecdotes and explanations. Bell, New York
SUM VII		1979	T.W. Hard	Paleopathologist revives mummy. Methods = modern. Escapes. Adventures
The Mummy or Ramses the Damned		1989	Anne Rice	Lovers reunited throughout millennia. Mummy: handsome "Ramsey" → Cleopatra
The Scroll of Saqqara		1990	Pauline Gedge	Retelling of 200 BC story of Khwamas, magician and *Book of Thoth*
Voices from Ancient Egypt		1991	R.B. Parkinson	Anthology, Middle Kingdom writings translation. Secular, religious, every-day life. University of Oklahoma Press.
Murder in the Place of Anubis		1994	Lynda S. Robinson	Detective story set in Tut time regarding murder in embalmer's shop

Notes:
Much of the above was abstracted from Brier, 1994, augmented by my own collection.
+, film also produced using the same plot, often the same title. Tut, Tutankhamun.

the next 12 years did not live up to that standard. Further mummy films from Britain, Germany and France produced a few that were entertaining but most of them were of mediocre or worse quality. Several, such as *Abbot and Costello Meet the Mummy* were even comedies. Table 10.3 lists many, though certainly not all mummy films that have been made. Brier (1994:299–332; 1998) provides plot abstracts and many interesting production details about most of these. In the 1990s the Discovery television channel presented a succession of predominantly factual videos dealing with various aspects of mummies and mummification.

Mummy as a political or religious implement

In Chapter 2 of this volume I discussed the use of human mummification for the purpose of enhancing the validity of claims to divinity by chiefs of the Andean Inca empire as well as by the Egyptian pharaohs. The manipulation of their predecessor's

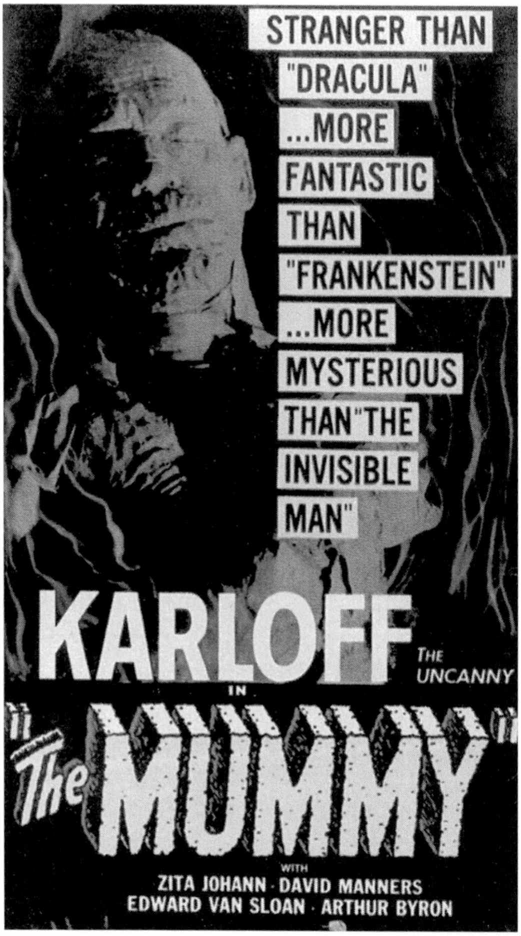

Fig. 10.10. Actor Boris Karloff as a mummy poster. Advertising poster for Universal's 1932 motion picture *The Mummy*, starring Boris Karloff. (Photo courtesy of Cinema Collectors, Hollywood, CA.)

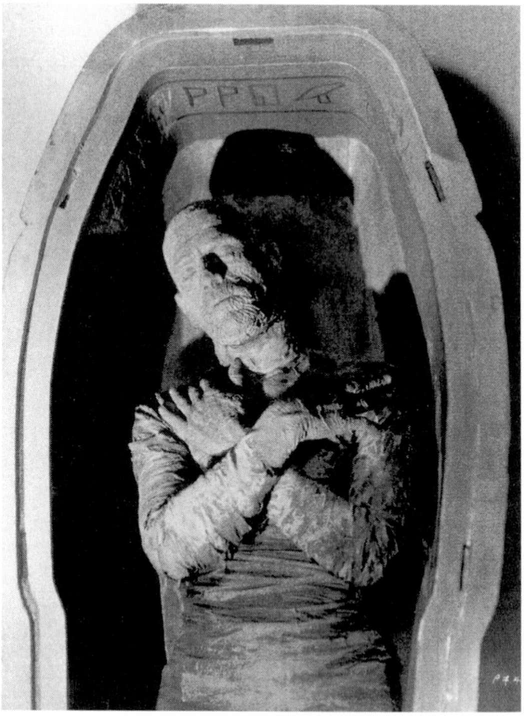

Fig. 10.11. Mummies in film. Boris Karloff made up as *The Mummy* in his coffin for Universal's well-known 1932 film. (Photo courtesy of Cinema Collectors, Hollywood, CA.)

mummy by these rulers was a tangible asset for the centralization of political power. Analogous to the theory underlying this postulate is the use of the remains of Alexander the Great. In little more than a decade this youthful Macedonian ruler conquered Greece, most of the Near East and even Egypt. This spectacular success against great odds led much of the populace to view him not only in heroic terms but as one who must have an unusual relationship with the gods. Although Alexander himself never made overt claims to divinity, he often consulted oracles such as those at Delphi (Greece) and the Western Desert oasis of Siwah (Egypt). Many of his comments suggested that he may have viewed himself as the direct son of Zeus; these were reinforced by some of his ritual acts. Even though he made no public stipulation of personal godliness, many among his army and the general populace so viewed him. Certainly after he died in Babylon at age 33 years, a ruler of some of Europe, all of ancient Persia, Arabia, Egypt and part of India, the concept of his divinity was literally thrust upon him and, by extension, on his body.

As his generals quarreled among themselves about the governing of this new empire after Alexander's

Table 10.3. *Mummy in film*

Film Title	Fictional mummy	Year	Studio	Mummy actor	Comments
The Mummy	Imhotep	1932	Universal	Boris Karloff	Director: Karl Freund. Script: John Balderston
The Ring of Thoth					From A. Conan Doyle's novel
Lot 249					From A. Conan Doyle's novel
The Palgrave Mummy	N	1932	Universal		
The Eye of Osiris					From novel *The Vanishing Man*
The Mummy Moves					From Gaunt novel
The Awakening		1980		Charlton Heston	Revivified mummy queen From Stoker's *Jewel of Seven Seas*
The Mummy of King Ramses		1909			French film
The Mummy		1911			French film
The Mummy		1911			British film
The Mummy		1911			American film
Eyes of the Mummy		1918			German film
Mummy Love		1926			German film
I was a Teenage Mummy		1992			
The Mummy's Hand	Ramses III	1940	Universal	Tom Tyler	Fast action (vs. brooding Karloff)
The Mummy's Tomb	Ramses III	1942	Universal	Lon Chaney	
The Mummy's Ghost	Ramses III	1944	Universal	Lon Chaney	
The Mummy's Curse	Ramses III	1945	Universal	Lon Chaney	
Abbot & Costello Meet the Mummy		1955	Universal	Eddie Parker	
The Mummy		1959	Hammer (UK)		
Curse of the Mummy's Tomb		1964	Hammer (UK)		
The Mummy's Shroud		1967	Hammer (UK)		
Blood from the Mummy's Tomb		1971	Hammer (UK)		
Curse of the Mayan Mummy					
Robot vs the Aztec Mummy					
Rock 'n' Roll Wrestling Woman vs Aztec Mummy					
Wrestling Woman vs. the Aztec Mummy					
Dawn of the Mummy					
Mummy's Dummies					
Mystery! Poirot, V: The Adventure of the Egyptian Tomb					Made for television. Detective story.
The Mummy Lives		1993		Tony Curtis	Remake of Karloff's *The Mummy*

death, one of Alexander's principal generals, Ptolemy, returned with his military division to Egypt. Ptolemy recognized that the virtual deification of Alexander was so ardent that possession of Alexander's body would impart enormous esteem and, consequently, power to its owner. When the cortège returning Alexander's body to Macedonia had left Babylon, Ptolemy dispatched a military unit to intercept it and bring the embalmed body to Egypt. He installed it first in a Memphis temple and later in a special shrine in Alexandria, appropriate for a son of Zeus. He exploited what might be called the "Cult of Alexander" and the political advantage with which this body empowered him. Alexander's mummy has not been seen since the

Roman general Octavian (later called Caesar Augustus) visited him.

In similar principle, though to a lesser degree, the legendary feats of the eleventh century Spanish hero El Cid were employed to rally his troops and terrify the enemy. After he was slain, his body is said to have been strapped to his horse and led into a battle leading to a victory (Harris, 1995:32). This alleged incident forms part of Spanish folklore. Even Joseph Smith, founder of the Church of Latter Day Saints (Mormons), may have been influenced, perhaps subconsciously, by this concept when he acquired an Egyptian papyrus from a traveling exhibit. Claiming the ability to translate the inscription, he declared it to be "The Book of Abraham" written in his own hand. Modern analysis by Egyptology scholars, however, has revealed it to have been written in late Ptolemaic Period script and that it represents an abstract from the Book of the Dead (Hamilton-Paterson & Andrews, 1979:188).

The principle involved in the above narration has been employed repeatedly. Probably the most obvious example in modern times is the enshrined body of V. Ilyich Lenin, the ideological author and leader of the Russian Communist movement. At his death, the ruling party's officials had Lenin's body embalmed by a mechanism not yet revealed to date, enclosed in a glass case and placed on public display in a building in Moscow's Central Plaza. For the subsequent seven decades the daily long line of Russian pilgrims to view the body was a fixture of Red Square. To only a slightly lesser extent, the apparently identically treated body of Josef Stalin was similarly revered. Their bodies were a daily reminder of their ideology and commitment to the Communist cause, support for which the contemporary government was exacting from the populace.

It is not obvious which of several possible interpretations correctly describes the items incorporated in the "Great Seal of the United States of America" that is also imprinted on the back of the United States $1 bill. A 13-step pyramid is inscribed with the Roman numerals for 1776 (year of the American Declaration of Independence). The Latin words *Annuit coeptus* (freely translated as "he who has favored our undertakings") are above it and the image of a sun-centered eye (that of God or Horus?) is at its apex. The Latin words below the pyramid can be translated as "A new order of the ages". A Giza pyramid, complete with camel, appeared on the first American $50 bill (Torodash, 1989:332; MacArthur, 2000).

On Kodiak Island at the time of European contact local Aleut fishermen embarking on a fishing trip sometimes would carry at least a part (often a finger) of a mummified human elder both as personal protection and for guarantee of fishing success (see Chapter 2).

All these can be viewed as examples of using mummified human remains in some manner for personal advantage by the living.

Mummy as science

The history of the uses of mummies for scientific purposes has been detailed in Chapter 1, especially the studies carried out since 1970. It will suffice here to call attention to the fact that information extracted from mummified bodies can and has been very useful for the reconstruction of both the cultural and the health status of past populations. The anthropological information about prehistoric peoples helps us to understand ourselves today, while the biomedical data these mummies supply tells us how the diseases we presently suffer have evolved. The former can help us in the shaping of our present and future behavior, while the recognition of how environment and human behavior influenced specific diseases in the past may provide clues to control of present and future afflictions. In summary, such knowledge is unique and relevant.

References

Aaronson, S. (1989) Fungal parasites of grasses and cereals: Their role as food or medicine, now and in the past. *Antiquity* **63**:247–257.

Abelson, P.H. (1957) Some aspects of paleobiochemistry. *Annals of the New York Academy of Sciences* **69**:276–285.

Abi-'Aoun, B., Baroudi, F., Ghaouche, A., Maroun, A., Maroun, R. & Karam, R.P. (1994) *Momies du Liban.* Antelias-Liban: Gersl (Groupe d'Études et de Recherches Souterraines du Liban).

Abraham, J.L., Hunt, A., Burnett, B., Allison, M. & Gerstzen, E. (1990) Light microscopic and microanalytic study of lungs from pre-Colombian Chilean mummies: Correlations with cultural data and comparison with contemporary pneumoconiosis. (Department of Pathology, Medical College of Virginia, Richmond, VA) *Paleopathology Club Newsletter* **41**:3–5.

Ackerknecht, E.H. (1943) Primitive autopsies and the history of medicine. *Bulletin of the History of Medicine* **13**:334–339.

Ackerknecht, E.H. (1944) Origin and Distribution of Skull Cults. *Ciba Symposia*, Vol. 5.

Aitken, M.J. (1990) *Science-based Dating in Archaeology.* London & New York: Longman.

Aldred, C. (1963) *Akhenaten and Nefertiti.* New York: Viking Press.

Alexander, C. (1994) Little Men. *Outside*, April, pp. 98–101, 174–179.

Allen, C. (1981) The Nasca creatures: Some problems of iconography. *Anthropology* **5**:43–70.

Allison, M.J. (1981a) Case no. 8: Fibrinous pericarditis. (Department of Pathology, Medical College of Virginia, Richmond, VA) *Paleopathology Club Newsletter* **10**:1.

Allison, M.J. (1981b) Case no. 7: Polycystic kidney disease. (Department of Pathology, Medical School of Virginia, Richmond, VA) *Paleopathology Club Newsletter* **9**:2.

Allison, M.J. (1984a) Case no. 16: Emphysema. (Department of Pathology, Medical College of Virginia, Richmond, VA) *Paleopathology Club Newsletter* **19**:1.

Allison, M.J. (1984b) Paleopathology in Peruvian and Chilean Populations. In *Paleopathology at the Origins of Agriculture,* ed. M.N. Cohen & G.J. Armelagos, pp. 515–529. New York: Academic Press.

Allison, M.J. (1985) Chile's ancient mummies. *Natural History* **94**:75–81.

Allison, M.J. (1992) Case no. 47: Nasal polyp. (Department of Pathology, Medical College of Virginia, Richmond, VA) *Paleopathology Club Newsletter* **51**:1.

Allison, M.J. & Gerszten, E. (1981) Case no. 9: Midline granuloma, probably uta. (Department of Pathology, Medical College of Virginia, Richmond, VA) *Paleopathology Club Newsletter* **12**:1.

Allison, M.J. & Gerszten, E. (1983) Case no 13: Hydronephrosis of the kidney, trauma, homicide, decapitation, abdominal laceration. (Department of Pathology, Medical College of Virginia, Richmond, VA) *Paleopathology Club Newsletter* **16**:1.

Allison, M.J. & Gerszten, E. (1984) Case no. 16: Pulmonary emphysema. (Department of Pathology, Medical College of Virginia, Richmond, VA) *Paleopathology Club Newsletter* **19**:1.

Allison, M.J. & Gerszten, E. (1989) Case no. 35: Smallpox. (Department of Pathology, Medical College of Virginia, Richmond, VA) *Paleopathology Club Newsletter* **39**:2.

Allison, M.J., Mendoza, D. & Pezzia, A. (1973)

Documentation of a case of tuberculosis in pre-Columbian America. *American Review of Respiratory Diseases* **107**:985–991.

Allison, M.J., Pezzia, A., Gerszten, E., Giffler, R.F. & Mendoza, D. (1974b) Aspiration pneumonia due to aspiration of teeth: A report ot two cases, 950 AD and 1973 AD. *Southern Medical Journal* **67**:479–483.

Allison, M.J., Pezzia, A., Gerszten, E. & Mendoza, D. (1974c) A case of Carrion's disease associated with human sacrifice from the Huari culture in southern Peru. *American Journal of Physical Anthropology* **2**:295–300.

Allison, M.J., Gerszten, E., Shadomy, J.H., Munizaga, J. & Gonzalez, M. (1979) Paracoccidioidomycosis in a northern Chilean mummy. *Bulletin of the New York Academy of Medicine* **55**:670–683.

Allison, M.J., Gerszten, E., Munizaga, J. & Santoro, C. (1980) Metastatic tumor of bone in a Tihuanako female. *Bulletin of the New York Academy of Medicine, Second Series* **56**:581–587.

Allison, M.J., Gerszten, E., Munizaga, J., Santoro, C. & Focacci, G. (1981) *La práctica de la deformación cránean entre los pueblos andinos precolombinos* [The practice of cranial deformation among pre-Columbian Andean populations]. *Chungará* (Arica, Chile) **7**:238–260.

Allison, M.J., Focacci, G. & Santoro, C. (1982a) The pre-Columbian dog from Arica, Chile. *American Journal of Physical Anthropology* **59**:299–304.

Allison, M.J., Gerszten, E. & Fouant, M. (1982b) Paleopathology: Today's laboratory investigates yesterday's diseases. *Diagnostic Medicine* **5** (Sept/Oct):28–44.

Allison, M.J., Focacci, G., Arriaza, B., Standen, V., Rivera, M. & Lowenstein, J.M. (1984) *Chinchorro, momias de preparación complicada: Métodos de momificación* [Chinchorro, mummies of complex mummification: Methods of mummification]. Chungará (Arica, Chile) **13**:155–173.

Allison, M.J., Figueroa, L., Razmilic, B. & González, M. (1995–1996) *Arsenicismo crónico en el norte grande* [Chronic arsenicism in the far north]. *Dialogo Andino* **15/16**:159–168.

Allison, P.A. & Briggs, D.E.G. (1991) *Taphonomy: Releasing the Data Locked in the Fossil Record.* New York: Plenum Press.

Alva, W. (1988) Discovering the New World's richest unlooted tomb. *National Geographic* **174**:570–555.

Alva, W. (1990) The Moche of Ancient Peru, New Tomb. *National Geographic* **177**(6):3–33.

Ambler, R.P. & Daniel, M. (1991) Proteins and molecular palaeontology. *Philosophical Transactions of the Royal Society of London: Series B* **333**:381–389.

American Museum of Natural History (2002) Dinosaur mummy. Website: http://www.amnh.org and "search" for Dinosaur.

Amy, R., Bhatnagar, R., Damkjar, E. & Beattie, O. (1986) The Last Franklin Expedition: Report of a postmortem examination of a crew member. *Canadian Medical Association Journal* **135**.

Andersen, S.R. (1994) *The Eye and its Diseases in Antiquity.* Copenhagen: Rhodos Publishers.

Andersen, S.R. & Prause, J.U. (1989) Histopathological examination of the eyes. In *Man and Society,* ed. J.P. Hart Hansen & H.C. Gullov, Vol. 12, *The Mummies From Qilakitsoq-Eskimos in the 15th Century,* 109–111. Copenhagen: Meddelser om Gronland.

Angel, L. (1973) Human skeletons from grave circles at Mycenae. In *The Grave Circle B of Mycenae*, ed. G.E. Mylonas, Appendix, pp. 379–397. Athens: Archaeological Society of Athens.

Anonymous (1991) *Las Momias de Guanajuato* [The Guanajuato Mummies]. Museum catalogue. Embajada, Mexico: Japon, Mexico.

Appelboom, T. & Struyven, J. (1999) Medical imaging of the Peruvian mummy Rascar Capac. *Lancet* **354**:2153–2155.

Appenzeller, O. & Aufderheide, A.C. (1999) Paleoneurobiology and the autonomic nervous system. In *Handbook of Clinical Neurology: The Autonomic Nervous System*, Part 1 *Normal Functions,* ed. O. Appenzeller, Vol. 74, part 30, pp. 181–197. Amsterdam: Elsevier.

Ara, P. (1996) *Eva Perón.* Buenos Aires: Editorial Sudamericana.

Araújo, A., Ferreira, F., Nuñez, L. & Oswaldo, F. (1983) Eggs of *Diphyllobothrium pacificum* in pre-Columbian human coprolites. *Paleopathology Newsletter* **41**:11–13.

Archaeological and Ethnographical Museum of Tenerife (1995) *Proceedings of the First World Congress on Mummy Studies*, February, 1992, vols. 1 and 2. Santa Cruz, Tenerife, Canary Islands: Archaeological and Ethnographic Museum of Tenerife.

Arené, E., Rivas, V., Padilla, A., Holguin, E., Rivera, H. & Otazo, R. (1995) Results of a softening technique for mummies. In *Proceedings of the First World Congress on Mummy Studies*, February, 1992, Vol. 1, pp. 237–240. Santa Cruz, Tenerife, Canary Islands: Archeological and Ethnographical Museum of Tenerife.

Armitage, P.L. & Clutton-Brock, J. (1981) A radiological and histological investigation into the mummification of cats from ancient Egypt. *Journal of Archaeological Science* **8**:185–196.

Arriaza, B.T. (1995a) Chile's Chinchorro mummies. *National Geographic* 187(3).

Arriaza, B.T. (1995b) Chinchorro bioarchaeology: Chronology and mummy seriation. *Latin American Antiquity* **6**(1):35–55.

Arriaza, B.T. (1995c) *Beyond Death, The Chinchorro Mummies of Ancient Chile.* Washington & London: Smithsonian Institution Press.

Arriaza, B.T., Allison, M.J., Standen, V., Focacci, G. & Chacama, J. (1986) Peinados Precolombinos en momias de Arica: Revista [Pre-Columbian hairstyles in mummies from Arica: Review]. (Arica, Chile) Chungará **16–17**:353–375.

Arriaza, B.T., Allison, M.J. & Gerszten, E. (1988) Maternal mortality in pre-Columbian Indians of Arica, Chile. *American Journal of Physical Anthropology* **77**:35–41.

Arrington, G.E. (1959) *A History of Ophthalmology.* New York: MD Publications, Inc.

Arrizabalaga, J. (1993) Syphilis. In *The Cambridge World History of Human Disease*, ed. K.F. Kiple, pp. 1025–1033. New York: Cambridge University Press.

Artamonov, M.I. (1965) *Frozen Tombs of the Scythians. Scientific American* **212**:101–109.

Ascenzi, A. (1998) The Uan Muhuggiag infant mummy. In *Mummies, Disease and Ancient Cultures,* ed. A. Cockburn, E. Cockburn & T.A. Reyman, pp. 281–283. Cambridge: Cambridge University Press.

Ascenzi, A., Bellelli, A., Brunori, M., Citro, G., Ippoliti, R., Lendaro, E. & Zito, R. (1991) Diagnosis of thalassemia in ancient bones: problems and prospects in pathology. In *Human Paleopathology: Current Syntheses and Future Options*, ed. D.J. Ortner & A.C. Aufderheide, pp. 73–75. Washington, DC: Smithsonian Institution Press.

Ascenzi, A., Binco, P., Nicoletti, R., Cerrarini, G., Fornaseri, M., Graziani, G. *et al.* (1996) The Roman mummy of Grottarossa. In *Human Mummies: A Global Survey of Their Status and the Techniques of Conservation*, ed. K. Spindler, H. Wilfing, E. Rastbichler-Zissernig, D. zur Nedden & H. Nothdurfter, pp. 205–218. New York: Springer-Verlag.

Ascher, R. & Clune, F. Jr. (1960) Waterfall Cave, southern Chihuahua, Mexico. *American Antiquity* **26**:270–274.

Aturaliya, S. & Lukasewycz, A. (1999) Experimental forensic and bioanthropological aspects of soft tissue taphonomy. 1. Factors influencing postmortem tissue desiccation rate. *Journal of Forensic Sciences* **44**:893–896.

Aturaliya, S., Wallgren, J. & Aufderheide, A.C. (1995) Studies in human taphonomy: An experimental animal model. In *Proceedings of the First World Congress on Mummy Studies*, February, 1992, Vol. 2, pp. 803–812. Santa Cruz, Tenerife, Canary Islands: Archaeological and Ethnographical Museum of Tenerife.

Aufderheide, A.C. (1982a) Mummies displayed at National Archaeological Museum in Lima, Peru. Unpublished manuscript on file in the Paleobiology Laboratory, Department of Pathology, University of Minnesota Duluth, School of Medicine, Duluth, MN 55812.

Aufderheide, A.C. (1982b) Description of mummies in Rafael Larco Herrara Museum (in Lima, Peru). Unpublished manuscript on file in the Paleobiology Laboratory, Department of Pathology, University of Minnesota Duluth, School of Medicine, Duluth, MN 55812.

Aufderheide, A.C. (1989a) Mummies viewed in China, 1989. Manuscript on file in the Paleobiology Laboratory in the Department of Pathology at the University of Minnesota Duluth, School of Medicine, Duluth, MN, 55812.

Aufderheide, A.C. (1989b) Chemical analysis of skeletal remains. In *Reconstruction of Life From the Skeleton,* ed. M.Y. İşcan & K.A.R. Kennedy, pp. 237–260. New York: Alan R. Liss.

Aufderheide, A.C. (1990a) Report of field study of mummified human remains Chiribaya Culture, Ilo, Peru. Manuscript on file in the Paleobiology Laboratory in the Department of Pathology at the University of Minnesota, Duluth, School of Medicine, Duluth, MN, 55812.

Aufderheide, A.C. (1990b) Anatomic observations on the Cambridge, England, Guanche mummy. Unpublished manuscript in the Paleobiology Laboratory in the Department of Pathology at the University of Minnesota Duluth, School of Medicine, Duluth, MN, 55812.

Aufderheide, A.C. (1996) Secondary applications of bioanthropological studies on South American Andean mummies. In *Human Mummies: A Global Survey of their Status and the Techniques of Conservation,* ed. K. Spindler, H. Wilfing, E. Rastbichler-Zissernig, D. zur Nedden & H. Nothdurfter pp. 141–151. New York: Springer-Verlag.

Aufderheide, A.C. & Aufderheide, M.L. (1991) Taphonomy of spontaneous ("natural") mummification with applications to the mummies of Venzone, Italy. In *Human Paleopathology: Current Syntheses and Future Options*, ed. D.J. Ortner & A.C. Aufderheide, 79–86. Washington & London: Smithsonian Institution Press.

Aufderheide, A.C. & Allison, M.J. (1994) Bioanthropological studies of spontaneously mummified bodies of a Late Phase Chinchorro site (Morro 1–6 in northern Chile). (Manuscript read at the 59th Annual Meeting of the Society for American Archaeologists. Anaheim, California, April 20–24.)

Aufderheide, A.C. & Allison, M.J. (1995a) Strontium patterns in infancy can validate retention of biogenic signal in human archaeological bone. In *Proceedings of the First World Congress on Mummy Studies,* February, 1992, Vol. 1, pp. 443–450. Santa Cruz, Tenerife, Canary Islands: Archaeological and Ethnographical Museum of Tenerife.

Aufderheide, A.C. & Allison, M.J. (1995b) Chemical dietary reconstruction of northern Chile prehistoric populations by trace mineral analysis. In *Proceedings of the First World Congress on Mummy Studies,* February, 1992, pp. 451–461. Santa Cruz, Tenerife, Canary Islands: Archaeological and Ethnographical Museum of Tenerife.

Aufderheide, A.C., Tieszen, L.L., Allison, M.J., Wallgren, J. & Rapp, G. (1988) Chemical reconstruction of components in complex diets: a pilot study. In *Diet and Subsistence: Current Archaeological Perspectives*, ed. B.V. Kennedy and G.M. Lemoine, pp. 301–306. Calgary: University of Calgary Archaeological Association.

Aufderheide, A.C. & Rodríguez-Martín, C. (1998) *The Cambridge Encyclopedia of Human Paleopathology.* Cambridge: Cambridge University Press.

Aufderheide, A.C., Muñoz, I. & Arriza, B. (1993) Seven Chinchorro mummies and the prehistory of northern Chile. *American Journal of Physical Anthropology* **91**:189–201.

Aufderheide, A.C., Rapp, G. Jr, Wittmers, L., Wallgren, J.E., Machiarelli, R., Fornaciari, G., Mallegni, F. & Corruccini, R.S. (1992) Lead exposure in Italy: 800 BC-700 AD. *International Journal of Anthropology* 7(2):9–15.

Aufderheide, A.C., Kelley, M.A., Rivera, M., Gray, L., Tieszen, L.L., Iversen, E. *et al.* (1994) Contributions of chemical dietary reconstruction to the assessment of adaptation by ancient highland immigrants (Alto Ramirez) to coastal conditions at Pisagua, north Chile. *Journal of Archaeological Science* **21**:515–524.

Aufderheide, A.C., Rodríguez-Martín, C., Estevez-González, F. & Torbenson, M. (1995a) Anatomic findings in studies of Guanche mummified human remains from Tenerife, Canary Islands. In *Proceedings of the First World Congress on Mummy Studies.* February, 1992, Vol. 1, pp. 113–124. Santa Cruz, Tenerife, Canary Islands: Archaeological and Ethnographical Museum of Tenerife.

Aufderheide, A.C., Rodríguez-Martín, C., Estevez-Gonzalez, F. & Torbenson, M. (1995b) Chemical dietary reconstruction of Tenerife's Guanche diet using skeletal trace element content. In *Proceedings of the First World Congress on Mummy Studies,* February, 1992, Vol. 1, pp. 33–40. Santa Cruz, Tenerife, Canary Islands: Archaeological and Ethnographical Museum of Tenerife.

Aufderheide, A.C., Ragsdale, B., Buikstra, J. & Ekberg, F. (1999a) An ancient osteosarcoma. *Journal of Paleopathology* **9**:101–106.

Aufderheide, A.C., Zlonis, M., Cartmell, L., Zimmerman, M.R., Sheldrick, P., Cook, M. & Molto, J.E. (1999b) Human mummification practices at Ismant El-Kharab. *Journal of Egyptian Archaeology* **85**:197–210, plus plate XXV.

Austin, J.J., Ross, A.J., Smith, A.B., Fortey, R.A. & Thomas, R.H. (1997) Problems of reproducibility – does geologically ancient DNA survive in amber-preserved insects? *Proceedings of the Royal Society of London: Series B* **264**:467–474.

Aveni, A. (ed.) (1990) *The Lines of Nazca*. Philadelphia: Memoirs of the American Philosophical Society.

Bada, J.L. (1982) Racemization of amino acids in nature. *Interdisciplinary Science Reviews* 7(1):30–46.

Bada, J.L. (1985) Racemization of amino acids. In *Chemistry and Biochemistry of the Amino Acids,* ed. G.C. Barrett, pp. 399–414. New York: Chapman & Hall.

Bada, J.L. & Protsch, R. (1973) Racemization reaction of aspartic acid and its use in dating fossil bones. *Proceedings of the National Academy of Sciences, USA* **70**:1331–1334.

Bada, J.L. & Schroeder, R.A. (1975) Amino acid racemization reactions and their geochemical implications. *Naturwissenschaften* **62**:71–79.

Bada, J.L., Schroeder, R.A. & Carter, G.F. (1974) New evidence for the antiquity of man in North America deduced from aspartic acid racemization. *Science* **184**:791–793.

Bada, J.L., Wang, X.S. & Hamilton, H. (1999) Preservation of key biomolecules in the fossil record: Current knowledge and future challenges. *Philosophical Transactions of the Royal Society of London: Series B* **354**:77–87.

Bagasra, O., Seshamma, T., Hansen, J., Bobroski, L., Saikumari, P. & Pomeranz, R.J. (1995) Application of in situ PCR methods in molecular biology. II. Special applications in electron microscopy, cytogenetics and immunohistochemistry. *Cell Vision* **2**(1):61–70.

Baggieri, G., di Giacomo, M. & Piazza, P. (2001) The mummies of Venzone. *Paleopathology Newsletter* **114**:6–8.

Bahn, P.G. (1996) The tragic end of the Romanovs. In *Tombs, Graves and Mummies*, ed. P.G. Bahn, pp. 56-59. London: Weidenfeld & Nicolson.

Baines, J. & Malek, J. (1983) *Atlas of Ancient Egypt*. Oxford: Phaidon Press, Ltd.

Ballinger, W.E. Jr (1988) Paleoneuropathology. (Department of Pathology, Medical College of Virginia, Richmond, VA) *Paleopathology Club Newsletter* **33**:6–9.

Balogh, K. (1994) The head and neck. In *Pathology*, ed. E. Rubin & J. Farber, 2nd edn, pp. 1239–1271. Philadelphia: Lippincott.

Balout, L. & Roubet, C. (1978) Operation Ramses II. A contribution from the laboratories of Egyptology of the National Museum of Natural History, Palais de Chaillot, Paris, France. *Bulletin de la Société Française d'Égyptologie* **83**:8–23.

Balout, L. & Roubet, C. (1985) *La Momie de Ramses II* [The Mummy of Ramses II)]. Paris: Éditions Recherche sur les Civilisations.

Banzer, D., Herrmann, B., Meyer, R.D., Precht, M. & Schutkowski, H. (1995) Evidence of intentional mummification in a pre-Colombian Peruvian mummy. In *Proceedings of the First World Congress on Mummy Studies*, February, 1992, Vol. 2, pp. 813–816. Santa Cruz, Tenerife, Canary Islands: Archaeological and Ethnographical Museum of Tenerife.

Bar-Ziv, J. & Goldberg, G.M. (1974) Simple siliceous pneumoconiosis in Negev Bedouins. *Archives of Environmental Health* **29**:121–126.

Baraybar, J.P. (1987) Cabezas trofeos Nasca: Nuevas evidencias [Nasca trophy heads: New evidence]. *Gaceta Arqueología Andina* **15**:6–11. Lima, Peru: Instituto Andino de Estudios Arqueológicos.

Barber, E. (1999) *The Mummies of Urumchi*. New York: W.W Norton & Co.

Barber, P.T. (1995) Mummification in the Tarim Basin. *Journal of Indo-European Studies* **23**:309–318.

Barbosa-Cánovas, G.V. & Vega-Mercado, H. (1996) *Dehydration of Foods*. New York: Chapman & Hall.

Barnes, C.L. (1896) *The Art and Science of Embalming*. Washington, DC.

Barnes, E. (1995) Case 59: Thyroglossal duct cyst. (Department of Pathology, Medical College of Virginia, Richmond, VA) *Paleopathology Club Newsletter* **63**:1.

Barnes, J. (1978) *Evita: First Lady. A Biography of Eva Perón*. New York: Grove Press.

Baron, H., Hummel, S. & Herrmann, B. (1996) *Mycobacterium tuberculosis* complex DNA in ancient human bones. *Journal of Archaeological Science* **23**:667–671.

Barraco, R.A. (1980) Paleobiochemistry. In *Mummies, Disease and Ancient Cultures*, ed. A. Cockburn & E. Cockburn, pp. 313–326. Cambridge: Cambridge University Press.

Barraco, R.A., Reyman, T.A. & Cockburn, T.A. (1977) Paleobiochemical analysis of an Egyptian mummy. *Journal of Human Evolution* **6**:533–546.

Barza, M. (1993) Case records of the Massachusetts General Hospital, ed. R.E. Scully. *New England Journal of Medicine* **328**:340–346.

Bassett, E., Keith, M.S., Armelagos, G.J., Martin, D.L. & Villanueva, A.R. (1980) Tetracycline-labelled human bone from ancient Sudanese Nubia (A.D. 350). *Science* **209**:1532–1534.

Batrawi, A. (1947) The pyramid studies: Anatomical reports. *Annales du Service de Antiquités de l'Égypte* **47**:97–111.

Battiscombe, C.F. (1956) *The Relics of St. Cuthbert*. Oxford: Oxford University Press.

Bauhin, C. (1586) *Gynaecorum sine de Mulierum Affectibus Commentari*. [Comments on Conditions Affecting Women]. Basel.

Baum, M. (1993) Breast cancer 2000 B.C. to 2000 A.D. – Time for a paradigm shift? *Acta Oncologica* **32**:3–8.

Baumann, B.B. (1960) The botanical aspects of ancient Egyptian embalming and burial. *Economic Botany* **14**:84–104.

Beaglehole, J.C. (ed.) (1955) *The Journals of Captain James Cook: I. The Voyage of the Endeavor* 1768–1771, Appendix IV, pp. 549–587. Cambridge: Cambridge University Press.

Beattie, O. (1995) The results of multidisciplinary research into preserved human tissues from the Franklin Arctic Expedition of 1845. In *Proceedings of the First World Congress on Mummy Studies,* February, 1992, Vol. 2, pp. 579–586. Santa Cruz, Tenerife, Canary Islands: Archaeological and Ethnographical Museum of Tenerife.

Beattie, O. & Geiger, J. (1987) *Frozen in Time.* London: Bloomsbury Publishing, Ltd.

Behrensmeyer, A.K. (1984) Taphonomy and the fossil record. *American Scientist* **72**:558–566.

Bell, Sir C.A. (1946) *Portrait of a Dalai Lama.* London: Wisdom Publications.

Benedek, T.G. & Kiple, K.F. (1993) Concepts of cancer. In *The Cambridge World History of Human Disease,* ed. K.F. Kiple, pp. 102–110. New York: Cambridge University Press.

Benedict, R. (1934) *Patterns of Culture.* New York: Hougton Millin Co., Cambridge: The Riverside Press.

Benítez, J.T. (1985) Temporal bone paleopathological studies from two members of the Barrow frozen family. Paper presented at the 12th Annual Meeting of the Paleopathology Association held in Knoxville, TN, April 9–10. *Paleopathology Newsletter* 50 (June) (Suppl.):5–6 (Abstract).

Benítez, J. T. (1988) Otopathology of Egyptian mummy PUM II: Final report. *Journal of Laryngology and Otology* **102**:485–490.

Bennike, P., Ebbesen, K. & Jorgensen, L.B. (1986) Early Neolithic skeletons from Bolkilde Bog, Denmark. *Antiquity* **60**:199–209.

Bereuter, T.L., Lorbeer, E., Reiter, C. & Seidler, H. (1996a) Post-mortem alterations of human lipids – Part I: Evaluation of adipocere formation and mummification by desiccation. In *Human Mummies: The Man in the Ice,* ed. K.Spindler, H. Wilfing, E. Rastbichler-Zissernig, D. zur Nedden & H. Nothdurfter, Vol. 3, pp. 275–278. New York: Springer-Verlag.

Bereuter, T.L., Reiter, C., Seidler, H. & Platzer, W. (1996b) Post-mortem alterations of human lipids – Part II: Lipid composition of a skin sample from the Iceman. In *Human Mummies; The Man in the Ice,* ed. K. Spindler, H. Wilfing, E. Rastbichler-Zissernig, D. zur Nedden, H. Nothdurfter, Vol. 3, pp. 275–278. New York: Springer-Verlag.

Berger, C. (1990) Sir Daniel Wilson. In *Dictionary of Canadian Biography*, Vol. 12, pp. 1109–1114. Toronto: University of Toronto Press.

Berger, J. (1999) The Fayum riddle. *Doubletake*, Spring 1999, pp. 141–143.

Berman, E.R. (1991) *Biochemistry of the Eye.* New York: Plenum Press.

Berman, S. (1995) Otitis media in children. *New England Journal of Medicine* **332**:1560–1565.

Berrizbeitia, E.L. (1991) Momias de Venezuela [Mummies of Venezuela]. *Natura* **92**:9–15.

Berrizbeitia, E.L. (1992) Estudio bioantropológico de la momia del Museo Arquidíocesano de Merida [Bioanthropological study of a mummy from the Arquidíocesano of Merida]. *Boletin del Archivo Arquidiocesano de Merida* **3**(10):6–18.

Berrizbeitia, E.L. (1995) Aporte al conocimiento de las momias indigenas de Venezuela [Contribution to the knowledge of indigenous mummies of Venezuela]. In *Proceedings of the First World Congress on Mummy Studies,* February, 1992, Vol 2, pp. 901–908. Santa Cruz, Tenerife, Canary Islands: Archaeological and Ethnographical Museum of Tenerife.

Bertoldi, F. & Fornaciari, G. (1997) A brief study of mummification techniques. *Paleopathology Newsletter* **99**:10–12.

Beverley, R. (1705/1947) *The History and Present State of Virginia.* Williamsburg, VA: Institute of Early American History and Culture/University of North Carolina Press.

Bezold, F. (1885) Schuluntersuchungen über das kindliche Gehörgang (School examination of the external auditory canal in childhood). *Ärtzliche Intelligenzblatt* **50**:2ff.

Biasoletto, B. (1829) *Cella Subterranea Venzonense ad Corpora Humana Emortua.* Exiccataque Lectus.

Bienert, H.D. (1991) Skull cult in the prehistoric Near East. *Journal of Prehistoric Religion* **5**:9–23.

Bietak, M. & Strouhal, E. (1974) Die Totesumstände des Pharaohs Seqenenre (17 Dynastie) (The circumstances of Pharaoh Seqenenre's death). *Annalen des Naturhistorischen Museums Wien* **78**:29–52.

Binford, L.R. (1965) Archaeological systematics and the study of cultural process. *American Antiquity* **31**:203–210.

Birch, S. (1850) Notes upon a mummy from the 26th Dynasty. *Archaelogical Journal* 8:273–280.

Bird, J. (1979) The "Copper Man": A prehistoric miner and his tools from northern Chile. In *Pre-Columbian Metallurgy of South America*, ed. E. Benson, pp. 106–131. Washington, DC: Dumbarton Oaks Research Library and Collections.

Birely, A.R. (1992) A case of eye disease (Lippitudo) on the Roman frontier in Britain. *Ophthalmologica* **81**:111–119.

Birkby, W.H. & Gregg, J.B. (1975) Otosclerotic stapedial footplate fixation in an eighteenth century burial. *American Journal of Physical Anthropology* **42**:81–84.

Birket-Smith, K. (1957) *Primitive Man and His Ways: Patterns of Life in Some Native Societies*, transl. R. Duffell. Cleveland & New York: The World Publishing Company.

Bittmann, B. & Munizaga, J. (1976) The earliest artificial mummification in the world? A study of the Chinchorro complex in northern Chile. *Folk* **18**:61–92.

Blackman, J., Allison, M.J., Aufderheide, A.C., Oldroyd, N. & Steinbock, T. (1991) Secondary hyperparathyroidism in an Andean mummy. In *Human Paleopathlogy*, ed. D.J. Ortner & A.C. Aufderheide, pp. 291–296. Washington, DC: Smithsonian Institution Press.

Blake, C. (1880) The occurrence of exostoses within the external auditory canal in prehistoric man. *American Journal of Otology* **2**:81–91.

Bloch, M. (1973) *The Royal Touch: Sacred Monarchy and Scrofula in England and France.* London: Routledge & Kegan Paul.

Blumenbach, J.F. (1794) Observations on some Egyptian mummies opened in London. *Philosophical Transactions of the Royal Society of London: Series B* **84**:177–195.

Bodenhoff, J., Geertinger, P. & Prause, J. (1979) Isolation of *Sporotrix fungorum* from a 500-year-old mummy found in Greenland. *Acta Pathologica et Microbiologica Scandinavica* **87**:201–203.

Bogucki, P. (1995) Hallstatt: An Early Iron Age mining centre. In *Great Archaeological Discoveries*, ed. P.G. Bahn, pp. 104–105. New York: Barnes & Noble.

Bogucki, P. (1996) Pazyryk and the Ukok Princess. In *Tombs, Graves and Mummies*, ed. P.G. Bahn, pp. 146–151. London: Weidenfeld & Nicolson.

Bolton, R. (1979) Guinea pigs, protein and ritual. *Ethnology* **18**:229–252.

Bonfils, P., Pirsig, W. & Ziegelmayer, P.F. (1986–1987) Flexible fiber endoscopy: New approaches. *OSSA* **13**:61–73.

Bonnichsen, R. (1982) Bone technology as a taphonomic factor: An introductory statement. *Canadian Journal of Anthropology* **2**:137–144.

Bonnichsen, R. & Schneider, A.L. (2000) Battle of the bones. *The Sciences* **40**:40–46.

Bourke, J.B. (1986) The Medical Investigation of Lindow Man. In *Lindow Man: The Body in the Bog*, ed. I.M Stead, J.B. Bourke & D. Brothwell, pp. 46–51. Ithaca, NY: Cornell University Press.

Bowen, C.D. (1939) *"Free Artist": The Story of Anton and Nicholas Rubinstein.* Atlantic Monthly Press Book. Boston: Little, Brown & Company.

Boyd, W.C. & Boyd, L.G. (1933) Blood grouping by means of preserved muscle. *Science* **78**:578–579.

Boyd, W.C. & Boyd, L.G. (1934) An attempt to determine the blood group of mummies. *Proceedings of the Society for Experimental Biology and Medicine* **31**:671–672.

Bradley, R. (1990) *The Passage of Arms.* Cambridge: Cambridge University Press.

Brady, J.E., Hasemann, G. & Fogarty, J.H. (1995) Harvest of skulls and bones. *Archaeology* **48** (May–June):36–40.

Braunstein, G.D. (1993) Gynecomastia. *New England Journal of Medicine* **328**:490–495.

Breasted, J.H. (1930) *The Edwin Smith Papyrus.* Chicago: University of Chicago Press.

Brewer, D.J. & Teeter, E. (1999) *Egypt and the Egyptians.* Cambridge: Cambridge University Press.

Brewster-Wray, C.C. (1982) Spatial patterning and the function of a Huari architectural compound. In *Investigations of the Andean Past*, ed. D. Sandweiss, pp. 122–127. Ithaca, NY: Cornell University.

Brier, B. (1994) *Egyptian Mummies.* New York: William Morrow.

Brier, B. (1997) The use of natron in human mummification: A modern experiment. *Zeitschrift fuer Aegyptische Sprache und Altertumskunde* **124**:89–100.

Brier, B. (1998) *The Encyclopedia of Mummies.* New York: Checkmark Books (Facts on File).

Briggs, C.S. (1995) Did they fall or were they pushed? Some unresolved questions about bog bodies. In *Bog Bodies*, ed. R.C. Turner & R.G. Scaife, pp. 168–182. London: British Museum Press.

Briggs, C.S. & Turner, R.C. (1986) A gazetteer of bog burials from Britain and Ireland. In *Lindow Man: The Body in the Bog*, ed. I.M. Stead, J.B. Bourke & D. Brothwell, pp. 181–195. Ithaca, NY: Cornell University Press.

Broaddus, B. & Harmony, S. (1984) William Hunter and the mummified mistress. *Medical Center Libraries Newsletter*, February 15 (University of Cincinnati).

Brooks, M., Lister, A., Eastop, D. & Bennett, T. (1996) Artifact or information? Articulating the conflicts in conserving archaeological textiles. In *Archaeological Conservation and Its Consequences*, ed. R. Ashok & P. Smith, pp. 16–21. London: International Institute for Conservation of Historic and Artistic Works.

Brothwell, D. (1967a) The evidence for neoplasms. In *Diseases in Antiquity*, ed. D. Brothwell & A.T. Sandison, pp. 320–345. Charles Thomas: Springfield.

Brothwell, D. (1967b) Evidence of endemic calculi in an early community. In *Diseases in Antiquity*, ed. D. Brothwell & A.T. Sandison, pp. 349–351. Springfield, OL: C.C. Thomas.

Brothwell, D. (1986) *The Bog Man and the Archaeology of People*. London: British Museum Press.

Brothwell, D. (1996) European bog bodies: Current state of research and preservation. In *Human Mummies: A Global Survey of Their Status and the Techniques of Conservation*, ed. K. Spindler, H. Wilfing, E. Rastbichler-Zissernig, D. zur Nedden & H. Nothdurfter, pp. 161–172. New York: Springer-Verlag.

Brothwell, D. & Sandison, A.T. (1967) *Diseases in Antiquity*. Springfield, IL: Charles Thomas.

Brothwell, D., Higgs, E. & Clark, G. (1963) *Science in Archaeology*. New York: Basic Books.

Brothwell, D., Sandison, A.T. & Gray, P.H.K. (1969) Human biological observations on a Guanche mummy with anthracosis. *American Journal of Physical Anthropology* **320**:333–347.

Brothwell, D., Malaga, A. & Burleigh, R. (1979) Studies on Amerindian dogs. 2: Variation in early Peruvian dogs. *Journal of Archaeological Science* **6**:139–161.

Brown, T.A. (1999) Genetic material. In *Care and Conservation of Natural History Collections*, ed. D. Carter & A.K. Walker, pp. 133–138. Oxford: Butterworth & Heinemann.

Browne, D.M., Silverman, H. & García, R. (1993) A cache of 48 nasca trophy heads from Cerro Carapo, Peru. *Latin American Antiquity* **4**:274–294.

Browne, T.D. (1658/1958) Hydriotaphia, Urne Buriall. In *Urn Burial and the Garden of Cyrus*, ed. J. Carter, pp. 2 (bookplate) & 48 (reference). Cambridge: Cambridge University Press.

Bruce, G. (1941) A note on penal blinding in the Middle Ages. *Annals of Medical History* **3**:369–371.

Bruetsch, W.L. (1959) The earliest record of sudden death possibly due to atherosclerotic coronary occlusion. Circulation **20**:438–441.

Bruhns, K.O. (1994) *Ancient South America*. Cambridge: Cambridge University Press.

Brunton, G. (1920) *Lahun* I: *The Treasure*. London: School of Archaeology in Egypt.

Bucaille, M. (1990) *Mummies of the Pharaohs: Modern Medical Examinations*, transl. A.D. Pannell. New York: St Martin's Press.

Buckley, S.A. & Evershed, R.P. (2001) Organic chemistry of embalming agents in pharaonic and Graeco-Roman mummies. *Nature* **413**:837–841.

Buckley, S.A., Stott, A.W. & Evershed, R.P. (1999) Studies of organic residues from ancient Egyptian mummies using high temperature–gas chromatography–mass spectrometry and sequential thermal desorption–gas chromatography–mass spectometry and pyrolysis–gas chromatography-mass spectrometry. *Analyst* **124**:443–452.

Budge, E.A.W. (1889) *The History of Alexander the Great*. Cambridge: Cambridge University Press.

Budge, E.A.W. (1894/1974) *The Mummy*. New York: Causeway Books.

Buikstra, J.E. (1977) Differential diagnosis: An epidemiological model. *Yearbook of Physical Anthropology* **20**:316–328.

Buikstra, J.E. (1981) *Prehistoric Tuberculosis in the Americas*. Evanston, IL: Northwestern University Archaeological Program.

Buikstra, J.E. & Williams, S. (1991) Tuberculosis in the Americas: Current perspectives. In *Human Paleopathology*, ed. D.J. Ortner & A.C. Aufderheide, pp. 161–172. Washington, DC: Smithsonian Institution Press.

Bullough, P.G. (1992) *Atlas of Orthopedic Pathology with Clinical and Radiologic Correlations*, 2nd edn, New York: Gower Medical Publishers.

Burger, H. (1978) HUM. *Harvard University Gazette*, October 6, p. 1.

Burger, R. (1995a) The early ceremonial center of Chavín de Huantar. In *Chavín and the Origins of Andean Civilization*, pp. 128–164. London: Thames & Hudson.

Burger, R. (1995b) *Chavín and the Origins of Andean Civilization*. London: Thames & Hudson.

Burkitt, D.P. (1981) Geography of disease: Purpose and possibilities from geographic medicine. In *Biocultural Aspects of Disease*, ed. Henry Rothschild, pp. 133–152. New York: Academic Press.

Burleigh, R. (1980) Radiocarbon dating of animal remains from Egypt. *MASCA Journal* **1**:188–189.

Burridge, A.L. (2000) Preliminary research report (Parts I and II): Marfan syndrome and the 18th dynasty family of ancient Egypt. *Paleopathology Newsletter* **110**:6–12 (Part I) and **111**:8–13 (Part II).

Campbell, M. (1994) Ice Maiden of the Steppes. *World Press Review*, June, p. 40.

Caminos, R.A. (1982) The Rendells mummy bandages. *Journal of Egyptian Archaeology* **68**:145–155.

Campillo, D. (1994/1995) *Paleopatología. Los Primeros Vestigios de la Enfermedad* [Paleopathology. The First Evidence of Disease]. Segunda Pare (Part 2). Barcelona: Fundación Uriach 1838.

Camporesi, P. (1988) *The Incorruptible Flesh*. Cambridge Studies in Oral and Literate Culture. Cambridge: Cambridge University Press.

Candela, P.B. (1939) Blood-group tests on stains, mummified tissues and cancellous bone. *American Journal of Physical Anthropology* **25**:187–213.

Canfora, L. (1987) *The Vanished Library*. Berkeley, CA: University of California Press.

Cano, R.J. (1998) The microbiology of amber. In *Digging for Pathogens*, ed. C.L. Greenblatt, pp. 71–95. Rehovot: Balaban Publishers.

Cano, R.J. & Borucki, M.K. (1995) Revival and identification of bacterial spores in 25- to 40-million-year-old Dominican amber. *Science* **268**:1060–1064.

Canziani, J.A. (1987) Analisis del complejo urbano Maranga-Chayavilka [Analysis of the urban site of Maranga-Chayavilka]. *Gaceta Arqueológica Andina* **4**(14):10–17. Lima, Peru: Instituto Andino de Estudios Arqueologicos.

Capasso, L. (1988) Exostoses of the auditory bony meatus in pre-Columbian Peruvians. *Journal of Paleopathology* **1**:113–116.

Capasso, L. & di Tota, G. (1991) The human mummies of Navelli: Natural mummification at new site in central Italy. *Paleopathology Newsletter* **68**:5–8.

Cárdenas, F. (1994) Form and function of original mummification in Colombia: A view from ethnohistory and modern technology. Paper presented at the 21st Annual Meeting of the Paleopathology Association Symposium, Modern Medical Research and Ancient Mummies, Denver, CO, March 29.

Cárdenas-Arroyo, F. (1989) La momificación indigena en Colombia [Indigenous mummification in Colombia]. *Boletin Museo del Oro* **25**:121–123.

Cárdenas-Arroyo, F. (1990) El emmochilado de Chiscas: Un caso de momificación en el siglo XVIII D.C., para la antropología física actual [The emmochilado of the Chiscas: a case of mummification in the eighteenth century A.D. to carry out a physical anthropological study]. *Memorias del V Congreso Nacional de Antropología*, pp. 235–249. Bogotá: Instituto Columbiano Antropología Nacional – Universidad de los Andes.

Cárdenas-Arroyo, F. (1993a) Cocaine and other hallucinogens in human archeological remains: Setting the cultural historical context for a growing field of enquiry in bioanthropology. (Abstract and unpublished manuscript.) *Paleopathology Newsletter* **82** (Suppl.):19.

Cárdenas-Arroyo, F. (1993b) La bioantropología De Maranga (Peru). Unpublished manuscript on file at the Paleobiology Laboratory Department, Department of Pathology, University of Minnesota Duluth, School of Medicine, Duluth, MN, 55812.

Cárdenas-Arroyo, F. (1993c) Paleodieta y paleodemográfico en poblaciones arqueológicas muiscas: sitios Las Delicias y Candelaria [Paleodiet and paleodemography of Muiscan archaeological populations at the sites of Las Delicias and Candelaria]. *Revista Colombiana de Antropología* **30**:131–148.

Cárdenas-Arroyo, F. (1994) Form and function of aboriginal mummification in Colombia: A view from ethnohistory and modern technology. Paper presented at the 21st Annual Meeting of the Paleopathology Association Symposium: Modern Medical Research and Ancient Mummies. Denver, Colorado, March 29.

Cárdenas-Arroyo, F. (1995) Momias, santuarios y ofrendas: el contexto ritual de la momificación en el Altiplano Central de los Andes Colombianos [Mummies, sanctuaries and offerings: The ritual context of mummification in the central Colombian Andean Highlands]. In *Proceedings of the First World Congress on Mummy Studies*, February, 1992, Vol. 2, pp. 633–642. Santa Cruz, Tenerife, Canary Islands: Archaeological and Ethnographical Museum of Tenerife.

Cárdenas-Arroyo, F. (1998) Mummies from Colombia and other South American areas: An archaeological context. In *Mummies, Disease and Ancient Culture*, ed. A. Cockburn, E. Cockburn & A. T. Reyman, pp. 197–210. Cambridge: Cambridge University Press.

Carey, J. (ed.) (1987) *Eyewitness to History.* New York: Avon Books.

Carmichael, P. H. (1988) Nasca mortuary customs: Death and society on the south coast of Peru. PhD dissertation, Department of Archaeology, University of Calgary (University Microfilms International, Ann Arbor, Michigan, no. 8918465.1990.)

Carmichael, P.H. (1994) The life from death continuum in Nasca imagery. *Andean Past* (Cornell) **4**:81–90.

Carter, D.J. & Walker, A.K. (1999) Collection environment. In *Care and Conservation of Natural History Collections*, ed. D. Carter & A.K. Walker, pp. 139–151. Oxford: Butterworth & Heinemann.

Carter, H. & Mace, A.C. (1963) *The Tomb of Tutankhamen*. New York: Cooper Square Publishers.

Cartmell, L.W. (1996) The Martindale mummy. Unpublished manuscript on file in the Paleobiology Laboratory in the Department of Pathology at the University of Minnesota, Duluth, School of Medicine in Duluth, MN, 55812.

Cartmell, L.W., Aufderheide, A.C., Springfield, A., Weems, C. & Arriaza, B. (1991) The frequency and antiquity of prehistoric coca leaf-chewing practices in northern Chile: Radioimmunoassay of a cocaine metabolite in human mummy hair. *Latin American Antiquity* **2**:260–268.

Carvajal, A., Martinez, A. & Campillo, D. (1995) Study of the internal structures by computerized tomography of a mummy dating from the beginning of the century. In *Proceedings of the IXth European Meeting of the Paleopathology Association*, pp. 115–122. Barcelona: Museo d'Arqueología de Cataluña.

Cassels, J.P. (1877) On the etiology of aural exostoses. *British Medical Journal* **2**:845.

Castillos, J.J. (1982) Analyses of Egyptian predynastic and early dynastic cemeteries: Final conclusions. *Journal for the Society for the Study of Egyptian Antiquities* **12**:29–53.

Catalano, P. & Passarello, P. (1988) Renal calculi from a skeleton of Gabii Iron Age necropolis (Rome). *Rivista di Antropologia (Roman)* **66** (Suppl.):221–230.

Ceram, C.W. (1964) *Gods, Graves and Scholars.* New York: Knopf.

Ceruti, M.C. (1998) La capacocha del Nevado de Chañi. Una appoximación preliminar desde la arqueología (The shrine on the snow-capped mountain of Chañi. A preliminary archaeological approximation). *Chungará* **33**:279–282.

Chakravorty, R.C. (1993) Diseases of antiquity in South Asia. In *The Cambridge World History of Human Disease,* ed, K.F. Kiple, pp. 408–413. New York: Cambridge University Press.

Chamberlin, R. (1983) *Loot! The Heritage of Plunder.* London: Thames & Hudson.

Chavez, R., Tellez-Giron, J.R., Alba, M., Marquez, L., Jiménez, J.C., Serrano, C. *et al.* (1995) An interdisciplinary study of mummies from northern Mexico. In *Proceedings of the Second World Congress on Mummy Studies*, pp. 8–9 (Abstract). Bogotá, Colombia: Department of Anthropology, Universidad de Andes.

Chen, T.S.N. & Chen, P.S.Y. (1993) Cirrhosis. In *The Cambridge World History of Human Disease,* ed. K.F. Kiple, pp. 649–653. New York: Cambridge University Press.

Cheng, T.O. (1984) Glimpses of the past from the recently unearthed ancient corpses in China. *Annals of Internal Medicine* **101**:714–715.

Chevreul, M.E. (1823) *Recherches Chimiques sur les Corpes gras d'Origine Animale* [Chemical Investigations on Fatty Cadavers of Animal Origin]. Paris: F.G. Levault.

Christianson, W.T. (1992) Stillbirths, mummies, abortions and early embryonic death. *Veterinary Clinics of North America* **8**:623–639.

Ciranni, R., Caramella, D., Iacconi, P. & Fornaciari, G. (1998) The "Lace Lady of Arezzo (central Italy): Imaging study and laparoscopy reveal a puerperal emergency of four centuries ago. *Paleopathology Newsletter* **103**:4–5.

Ciranni, R., Iacconi, P., Caramella, D. & Fornaciari, G. (1999) Imaging study and laparoscopy in a puerperal emergency of four centuries ago. *International Journal of Osteoarchaeology,* **9**:289–296.

Clausen, C. J., Cohen, A. D., Emiliani, C., Holman, J. A. & Stipp, J. J. (1979) Little Salt Spring, Florida: A unique underwater site. *Science* **203**:609–614.

Clutton-Brock, J. (1963) The origins of the dog. In *Science in*

Archaeology, ed. D. Brothwell, E. Higgs & G. Clark, pp. 269–308. New York: Basic Books.

Cobo, Father B. (1653/1990) *Inca Religions and Customs*, transl. R. Hamilton. Austin, TX: University of Texas,

Cockburn, A. (1978) Paleopathology and its association. *Journal of the American Medical Association* **240**(2):151–152.

Cockburn, A. (1980) Commentary by Aidan Cockburn. *Paleopathology Newsletter* **29**:11–14.

Cockburn, A. & Cockburn, E. (1977) Mummy – Peru. *Paleopathology Newsletter* **19**:3–6.

Cockburn, A., Barraco, R.A., Reyman, T.A. & Peck, W. (1975) Autopsy of an Egyptian mummy. *Science* **187**:1155–1160.

Cockburn, E. (1974) China. *Paleopathology Newsletter* **6**:3–4.

Cockburn, E. (1980) Chinese mummies. *Paleopathology Newsletter* **31**:9–10.

Coe, J.I. (1977) Postmortem chemistry of blood, cerebrospinal fluid and vitreous humor. In *Forensic Medicine*, ed. E. Tedeschi, C.G. Eckert, G. William & L.G. Tedeschi, pp. 1033–1060. Philadelphia: W.B. Saunders.

Cohen, S.G. (1996) John Paul Jones (1747–1792) American naval officer. *Allergy Asthma Proceedings* **17**:222–223.

Collins, M.J., Riley, M.S., Child, A.M. & Turner-Walker, G. (1995a) A basic mathematical simulation of the chemical degradation of ancient collagen. *Journal of Archaeological Science* **22**:175–183.

Collins, M.J., Riley, M.S. & Hedges, R.E.M. (1995b) Improving the analysis of ancient collagen: Establishing and testing a quantitative chemical model of collagen degradation. *ABI Newsletter* (Natural Environment Research Council, UK) no. 1 (November):31–32.

Collins, M.J., Waite, E.R. & van Duin, A.C.T. (1999) Predicting protein decomposition: The case of aspartic acid racemization kinetics. *Philosophical Transactions of the Royal Society of London, Series B* **354**(1379):51–64.

Collis, E.L. (1915) Milroy lectures 1915. Industrial pneumoconiosis with special reference to dust-phthisis. *Public Health* (London) **28**:252–264, 292–305; **29**:11–20, 37–44.

Colombini, M.P., Modugno, F., Silvano, F. & Onor, M. (2000) Characterization of the balm of an Egyptian mummy from the seventh century B.C. *Studies in Conservation* **45**:19–29.

Conlogue, G., Schlenk, M., Cerrone, F. & Ogden, J. (1989) Dr. Leidy's Soap Lady: Imaging the past. *Radiologic Technology* **60**:411–415.

Conlogue,G., Forcier, D., Airo, M., Kilosky, J., Gambardella, S., Mansfield, K. & Greenwood, L. (1997) Radiographic evaluation of the Soap Man mummy. *Radiologic Technology* **68**:391–398.

Connan J. (1991) Chemische Untersuchung altagyptiischer Mummien-Salbole [Chemical investigation of ancient Egyptian mummy oils]. In *Mumie und Computer*, ed. R. Drenkhahn & R. Germer, pp. 34–36. Hannover: Kesnter Museums.

Connan, J. & Dessort, D. (1991) De bitume dans des baumes de momies égyptiennes (1295 Av. J.C.–300 ap. J.C.): Détermination de son origine et évaluation de sa quantité [Bitumen in the balm of Egyptian mummies (1295 BC-AD 300): Determination of origin and quantification]. *Comptes Rendus del l'Académie des Sciences*, Series 2 **312**:1445–1452.

Connolly, R.C. & Harrison, R.G. (1969) Kinship of Smenkhare and Tutankhamen, confirmed by serologic micromethod. *Nature* **224**:325.

Connor, D.H. & Neafie, R.C. (1976) Cutaneous leishmaniasis. In *Tropical and Extraordinary Diseases*, ed. C.H. Binfords & D.H. Connor, pp. 258–264. Washington, DC: Armed Forces Institute of Pathology.

Conrad, J.R. (1957) *The Horn and the Sword: The History of the Bull as Symbol of Power and Fertility*. New York: E.P. Dutton and Co.

Conti-Fuhrman, A. & Rabino-Massa, E. (1972) Preliminary note on the ultrastructure of the hair from an Egyptian mummy using the scanning electron microscope. *Journal of Human Evolution* **1**:487–488.

Cook, M., Molto, E. & Anderson, C. (1988) Possible case of hyperparathyroidism in a Roman period skeleton from the Dakhleh Oasis, Egypt, diagnosed using bone histomorphometry. *American Journal of Physical Anthropology* **75**:23–30.

Cook, M., Molto, E. & Anderson, C. (1989) Fluorochrome labelling in Roman Period skeletons from Dakhleh Oasis, Egypt. *American Journal of Physical Anthropology* **80**:137–143.

Cooke, W. (1988) Creasing in ancient textiles. *Conservation News* **35**:27–30.

Cooper, R.A., Molan, P.C. & Harding, K.G. (1999) Antibacterial activity of honey against strains of *Staphylococcus aureus* from infected wounds. *Journal of the Royal Society of Medicine* **92**:283–285.

Correal-Urrego, G. & Florez, I. (1995) Estudio de las momias Guanes de la Mesa de los Santos, Departamento de

Santander, Colombia [Study of two Guanes mummies from the Highlands in the Department of Santander, Colombia]. In *Proceedings of the First World Congress on Mummy Studies*, February, 1992, pp. 307–313. Santa Cruz, Tenerife, Canary Islands: Archaeological and Ethnographical Museum of Tenerife.

Cortez, C. (1996) The tomb of Pacal at Palenque. In *Tombs, Graves and Mummies*, ed. P.G. Bahn, pp. 126–127. London: Weidenfeld & Nicolson.

Costa-Junqueira, M.A. (1988) Reconstrucción física y cultural de la población tardia cementerio de Quitor-6 (San Pedro de Atacama) [Physical and cultural reconstruction of the cemetery population of Quitor-6 from San Pedro de Atacama]. *Estudios Atacameños* **9**:99–126.

Cotran, R.S., Kumar, V. & Robbins, S.L. (1994) *Robbins' Pathologic Basis of Disease*, 5th edn. Philadelphia: Saunders.

Cotton, G.E., Aufderheide, A.C. & Goldschmidt, V. (1987) Preservation of human tissue immersed for five years in fresh water of known temperature. *Journal of Forensic Science* **32**:1125–1130.

Crabtree, P.J. (1995) The symbolic role of animals in Anglo-Saxon England: Evidence from burials and cremations. In *The Symbolic Role of Animals in Archaeology*, ed. K. Ryan & P.J. Crabtree, Vol. 5, pp. 20–37. Philadelphia: MASCA, University of Pennsylvania Museum of Archeology and Anthropology.

Crane, E. (1983) *The Archaeology of Beekeeping*. Ithaca, NY: Cornell University Press.

Crosby, A.W. (1986) *Ecological Imperialism: The Biological Expansion of Europe, 900–1900*. New York: Cambridge University Press.

Cruise, R.A. (1823) *Journal of a Ten Month's Residence in New Zealand*. Christchurch: Pegasus Press.

Csapo, J., Csapo-Kiss, Z., Nemethy, S., Folestad, S., Tivesten, A. & Martin, T.G. (1994) Age determination based on amino acid racemization: A new possibility. *Amino Acids* **7**:317–325.

Cuong, N.L. (1984) About the dead bodies of two Buddhist monks preserved in the form of statues at the Dau Pagoda. *OSSA* **9**(11):105–110.

Czapek, F. (1913) *Biochemie der Pflanzen* [The Biochemistry of Plants]. Jena: Verlag von Gustav Fischer.

Czermak, J. (1852) Beschreibung und mikroskopische Untersuchung zweier aegyptischen Mumien [Description and microscopic studies of two Egyptian mummies]. *Sonderberichte Akademie Wissenschaft Wien* **9**:427–469.

D'Aunoy, R. & King, E.L. (1922) Lithopedion formation in extrauterine fetal masses. *American Journal of Obstetrics and Gynecology* **3**:377–384.

D'Auria, S., Lacovara, P. & Roehrig, C.H. (1988) *Mummies and Magic: The Funerary Arts of Ancient Egypt*. Boston: Boston Museum of Fine Arts & Dallas Museum of Fine Arts.

d'Errico F., Villa, G. & Fornaciari, G. (1988) Dental esthetics of an Italian Renaissance noblewoman, Isabella d'Aragona: A case of chronic mercury intoxication. *OSSA* **13**:233–254.

Daggett, R.E. (1994) The Paracas mummy bundles of the great necropolis of Wari Kayan: A history. *Andean Past* **4**:53–75.

Dalby, G., Manchester, K. & Roberts, C. (1993) Otosclerosis and stapedial footplate fixation in archaeological material. *International Journal of Osteoarchaeology* **3**:207–212.

Dale, P.M. (1952) *John Paul Jones Medical Biographies: The Ailments of Thirty-Three Famous Persons*. Norman, OK: University of Oklahoma Press.

Dall, W.H. (1875) Alaskan mummies. *American Naturalist* **9**:433–438.

Dall, W.H. (1878) *On the Remains of Later Pre-Historic Man Obtained from Caves in the Catherina Archipelago, Alaska Territory, and Especially from the Caves of the Aleutian Islands: Smithsonian Contributions to Knowledge* **22**(318). Washington, DC: Smithsonian Institution Press.

Dalton, H.P., Allison, M.J. & Pezzia, A. (1976) The documentation of communicable disease in Peruvian mummies. *Medical College of Virginia (MCV) Quarterly* **12**(2):43–48.

Dane, J.A. (1995) The curse of the mummy paper. *Printing History* **17**(2):18–25.

Daniels, F. & Post, P. (1970) The histology and histochemistry of prehistoric mummy skin. In *Advances in the Biology of Skin*, Vol. X *The Dermis*, ed. W. Montagna, J.P. Bentley & R. L. Dobson, pp. 279–292. Publication no. 396 from the Oregon Regional Primate Research Center. New York: Appleton-Century-Crofts.

Daniels, V. (1996) Selection of a conservation process for Lindow Man. In *Human Mummies: A Global Survey of Their Status and the Techniques of Conservation*, ed. K.

Spindler, H. Wilfing, E. Rastbichler-Zissernig, D. zur Nedden & H. Nothdurfter, pp. 3–8. New York: Springer-Verlag.

Dannenfeldt, K.H. (1985) Egyptian mumia: The sixteenth century experience and debate. *Sixteenth Century Journal* **16**(2):163–180.

Dansie, A. (1997) Early Holocene Burials in Nevada. *Nevada* **40**(1):4–14.

Dary, D. (1976) *Comanche.* Lawrence, KS: University of Kansas.

David, A.E. (1986) Conservation of mummified Egyptian remains. In *Science in Egyptology*, ed. R.A. David, pp. 87–89. Manchester: Manchester University Press.

David, R.A. (1979) *Manchester Museum Mummy Project.* Manchester: Manchester University Press.

David, R.A. (1982) *The Ancient Egyptians: Religious Beliefs and Practices.* London: Routledge & Kegan Paul.

David, R.A. (ed.) (1986) *Science in Egyptology.* Manchester: Manchester University Press.

David, R.A. (1998) Benefits and disadvantages of some conservation treatments for Egyptian mummies. Manuscript read at the 3rd World Congress on Mummy Studies held at Arica, Chile, May 14–18.

David, R.A. (2000) Mummification. In *Ancient Egyptian Materials and Technology*, ed. P. Nicholson & I. Shaw, pp. 372–389. Cambridge: Cambridge University Press.

David, R.A. & David, A. (1992) *A Biographical Dictionary of Ancient Egypt.* London: Seaby.

Davies, W.V. (1987) *Egyptian Hieroglyphs.* London: University of California Press/British Museum.

Davis, E.N., Wallen, L.L., Goodwin, J.C., Rohwedder, W.K. & Rhodes, R.A. (1969) Microbial hydration of *cis*-9-alkenoic acids. *Lipids* **4**:356–362.

Davis, J.L., Hegin, J.A., Annan, A.P., Daniels, R.S., Berdal, B.P., Bergen, T. *et al.* (2000) Ground penetrating radar surveys to locate 1918 Spanish 'flu victims in permafrost. *Journal of Forensic Science* **45**:68–76.

Davis, R. (1997) Clinical radiography and archaeo-human remains. In *Radiography of Cultural Material*, ed. J. Lang & A. Middleton, pp. 117–135. Oxford: Butterworth/Heinemann.

Davy, J. (1819) Observations on the changes which the animal body undergoes in a hot climate soon after death. Copy of letter to Sir James Macgrigor, in the files of the Paleobiology Laboratory in the Department of Pathology, University of Minnesota Duluth, School of Medicine, Duluth, MN, 55812.

Dawson, W.R. (1924) A mummy from the Torres Straits. *Annals of Archaeology and Anthropology* **11**:87–96.

Dawson, W.R. (1927a) On two Egyptian mummies preserved in the museums of Edinburgh. *Proceedings of the Society of Antiquities of Scotland* **61**:290–296.

Dawson, W.R. (1927b) On two mummies formerly belonging to the Duke of Sutherland. *Journal of Egyptian Archaeology* **13**:155–161.

Dawson, W.R. (1927c) Mummy as a drug. *Proceedings of the Royal Society of Medicine* **21**:34–39.

Dawson, W.R. (1927d) A mummy of the Persian Period. *Journal of Egyptian Archaeology* **11**:76–77.

Dawson, W.R. (1927e) Contributions to the history of mummification. *Proceedings of the Royal Society of Medicine* **20**:832–854.

Dawson, W.R. (1927f) Making a mummy. *Journal of Egyptian Archaeology* **13**:40–49.

Dawson, W.R. (1928a) Mummification in Australia and in America. *Journal of the Royal Anthropological Institute of Great Britain and Ireland* **58**:115–138.

Dawson, W.R. (1928b) Two mummies from Colombia. *MAN* **28**:73–74.

Dawson, W.R. (1929) A bibliography of works relating to mummification in Egypt. *Memoires de l'Institute d'Égypte* **13**:1–49. Cairo: Imprimerie de L'Institut Français d'Archéologie Orientale.

Dawson, W.R. (1934) Pettigrew's demonstrations upon mummies. A chapter in the history of Egyptology. *Journal of Egyptian Archaeology* **20**:170–182.

Dawson, W.R. (ed.) (1938) *Sir Grafton Elliot Smith.* London: Jonathan Cape, Ltd.

Dayton, J.A. (1986) Animal remains – the Cladocera and Chironomidae. In *Lindow Man: The Body in the Bog*, ed. I.M. Stead, J.B. Bourke & D. Brothwell, pp. 93–98. Ithaca, NY: Cornell University Press.

de Acosta, J. (1590/1940) Historia natural y moral de las Indias [The natural and moral history of the Indies]. Citta del Messico.

de la Vega, G. (1609–1961) *The Incas.* ed A. Gheerbrant, transl. M. Jolas. New York: Avon Books.

de la Vega, G. (1609–1987) *Royal Commentaries of the Incas.* Austin, TX: University of Texas Press.

de Laguna, F. (1933) Mummified heads from Alaska. *American Anthropologist* **35**:742–744.

de Laguna, F. (1956) *Chugach Prehistory: The Archaeology of Prince William Sound, Alaska*. Seattle: University of Washington Press.

de Lumley M.A. (1962) Lésions osseuses de l'homme de Castellar [Osseous lesions of the man from Castellar]. *Bulletin du Musée d'Anthropologie Préhistorique de Monaco* **9**:191–205.

Deelder, A.M., De Jonge, N., Boerman, O.C., Fillie, Y.E., Hilberath, G.W., Rotmans, J.P. *et al.* (1989) Sensitive determination of circulating anodic antigen in *Schistosoma mansoni* infected individuals by an enzyme-linked immunosorbent assay using monoclonal antibodies. *American Journal of Tropical Medicine and Hygiene* **40**:268–272.

Deelder, A.M., Miller, R.L., De Jonge, N. & Krijger, F.W. (1990) Detection of schistosome antigen in mummies. *Lancet* **335**:724.

Delaney, M. & O'Floinn, R. (1995) A bog body from Meenybradden Bog, County Donegal, Ireland. In *Bog Bodies*, ed. R.C. Turner & R.G. Scaife, pp. 123–132. London: British Museum Press.

Delattre, R.P. (1888) Fouilles d'un cimetiere romain Revue [Excavation of a Roman cemetery]. *Archéologique* **12**(3):151–162.

Delattre, R.P. (1905) Sarcophage en pierre, orné de decors peints, trouvé à Carthage [An ornate, stone sarcophagus with well-chosen, artistic decoration, at Carthage]. Comptes Rendus de Séances de l'Académie des Inscriptions et Belles Lettres, pp. 482–487.

Delattre, R.P. (1929) Les fouilles de Bir-Ftouha [The excavations at Bir-Ftouha]. *Comptes Rendus des Seances de l'Académie des Inscriptions et Belles Lettres*, pp. 23–29.

Den Dooren de Jong, L.E. (1961) On the formation of adipocere from fats. *Antonie van Leuwenhoek Journal of Microbiology and Serology* **10**:337–361.

Denon, V. (1821) *Description de l'Égypte.* Published by order of Napoleon Bonaparte, reprinted 1994 by Benedikt Tasche: Koeln (Cologne).

Derry, D.E. (1909) Anatomical report. *Archives of the Survey of Nubia Bulletin* **3**:29–52.

Derry, D.E. (1939) Note of the remains of Shashanq. *Annales du Service des Antiquités de l'Égypte* **39**:549–551.

Derry, D.E. (1942) Mummification: Methods practiced at different periods. *Annales du Service des Antiquités de l'Égypte* **41**:240–265 plus two plates.

DeSalle, R., Gatesy, J., Wheeler, W. & Grimaldi, D. (1992) DNA sequences from a fossil termite in Oligo-Miocene amber and their phylogenetic implications. *Science* **257**:1933–1936.

Desanti, M.V. (1977) The structural reliability of mummy tissue in tracing disease ecology in prehistoric and recent human populations. *American Journal of Physical Anthropology* **47**(1):126 (Abstract).

Dexiang, W., Shenqi, H., Tiancheng, S., Wenxiu, H., Wenyuan, Y., Yunfang, L. *et al.* (1981) Parasitological investigation on the ancient corpse of the Western Han Dynasty unearthed from tomb no. 168 on Phoenix Hill in Jiangling County. *Acta Academiae Medicinae Wuhan* **1**:16–23.

Di Bartolomeo, J.R. (1979) Exostoses of the external auditory canal. *Annals of Otology, Rhinology and Laryngology* **88**(6) (Suppl. no. 61):2–20.

Di Tota, G., Capasso, L. & DiMuzio, M. (1992) A probable bladder stone from an Iron Age burial in central Italy. Abstract of manuscript read at the European meeting of the Paleopathology Association held in Barcelona, Spain, 1992, and reprinted in the *Paleopathology Newsletter*, September 1992, p. 10.

Dickinson, O.T.P.K. (1977) The origins of Mycenaean civilisation. *Studies in Mediterranean Archaeology* **49**:1–134. Goeteborg: Paul Astroems Foerlag.

Dieck, A. (1972) Stand und Aufgaben der Moorleichenforschung [The status and the problem of bog body research]. *Archaeologisches Korrespondenzblatt* **2**:365–368.

Dieck, A. (1986) Der Stand der europaischen Moorleichenforschung im Jahr 1986 sowie Materialvorlage von anthropologischen und medizinischen Sonderbefunden [The status of European bog body research in 1986 as well as presentation of material about anthropological and medical evidence]. *Telma* **16**:31–168.

Diener, L. (1980) Radiology of a "child-mummy" in Stockholm: Unexpected findings. *MASCA Journal* **1**:102–103.

Diener, L. (1981) Radiology of two "child-mummies" in Stockholm. *MASCA Journal* **1**:218–219.

Dio Cassius (ca. AD 200/1970) *Loeb Classical Library*. Cambridge, MA: Harvard University Press.

Diodorus of Sicily (ca. 50 BC/1947) *Historical Library*, transl. C.H. Oldfather. London: Heinemann.

Diodorus Siculus (cs. 50 BC/1947) *Loeb Classical Library*, Vol. 3. Cambridge, MA: Harvard University Press.

Doering, H.U. (1958) Bericht ueber archaeologische Feldarbeiten in Peru. I [Report of field work in Peru I]. *Ethnos* **23**(12–14):67–99.

Doering, H.U. (1959) Bericht ueber archaeologische Feldarbeiten in Peru. II [Report of field work in Peru II]. *Ethnos* **24**(1–2):1–32.

Doering, H.U. (1966) *On the Royal Highway of the Inca*. New York: Praeger.

Donington, R. (1974) *Wagner's Ring and Its Symbols*. Boston: Faber & Faber.

Donnan, C.B. (1988) Iconography of the Moche: Unravelling the mystery of the warrior-priest. *National Geographic* **174**:551–555.

Donnan, C.B. & Castillo, L.J. (1992) Finding the tomb of a Moche princess. *Archaeology* **45**(6):38–42.

Donnet, J. & Duclos, A. (1999) T*he Amazing Ibaloi Mummies of Kabayan*. Paris: Gamma News Agency, November 24.

Donoghue, H.D., Spigelman, M., Zias, J., Gernazy-Child, A.M. & Minnikin, D.E. (1998) *Mycobacterium tuberculosis* complex in calcified pleura from remains 1400 years old. *Letters in Applied Microbiology* **27**:265–269.

Dooley, L.K. (1975) A War-horse for Stonewall. *Army* (April), pp. 34–39.

Doran, G., Dickel, D.N., Basllinger, W.E. Jr, Agee, O.F., Laipis, P.J. & Hauswirth, W.W. (1986) Anatomic, cellular and molecular analysis of 8000 year old human brain tissue from the Windover site. *Nature* **323**:803–806.

Drioton, E. (1950) *Egyptian Art*. New York: Golden Griffin Books.

Drusini, A.G. & Baraybar, J.P. (1991) Anthropological study of Nasca trophy heads. *Homo* **41**:251–265.

Dunand, F. & Lichtenberg. R. (1991) *Un voyage dans l'éternité* [A Voyage to Eternity]. Paris: Découvertes Gallimard.

Dunand, F. & Lichtenberg, R. (1992) Douch, I. La nécropole de Douch: Oasis de Kharga [The Necropolis of Douch: Oasis of Kharga]. *Exploration Archéologique* (DFIFAO) (Cairo) **26**.

Dyson, S.L. (1979) The mummy of Middletown. *Archaeology* **32**:57–59.

Dzierzykray-Rogalski, T. (1988) Paleopathologic changes of the scrotum of a baby from the sixteenth century. Abstract of a manuscript presented at the Seventh European meeting of the Paleopathology Association held in Lyon, France, September 1–4, 1988 and reprinted in the *Paleopathology Association Newsletter* **20** (June) (Suppl.):4–5.

Eaton, G.F. (1912) Report of the remains of man and of lower animals from the vicinity of Cuzco, Peru. *American Journal of Science* **33**:325–333.

Ebbell, B. (transl.) (1937) *The Papyrus Ebers*. Copenhagen: Levin & Munksgaard.

Eckert, W.G. (1982) Identification of John Paul Jones. *American Journal of Forensic Medicine and Pathology* **33**: 143–152.

Eckert, W.G., James, S. & Katchis, S. (1988) Investigations of cremations and severely burned bodies. *American Journal of Medicine and Pathology* **9**:188–200.

Edgar, H.J.H. (1997) Paleopathology of the Wizards Beach Man (AHUR 2023) and the Spirit Cave Mummy (AHUR 2064). *Nevada* **40**(1):57–61.

Edson, G. (ed.) (1997) *Museum Ethics*. London: Routledge.

Efremov, J.A. (1940) Taphonomy: New branch of paleontology. *Pan-American Geologist* **74**(2):81–93.

Eglinton, G. (1998) The archaeological and geological fate of biomolecules. In *Digging for Pathogens*, ed. C.I. Greenblatt, pp. 299–327. Rehovot: Balaban.

Eglinton, G. & Logan, G. (1991) Molecular preservation. *Philosophical Transactions of the Royal Society of London, Series B* **333**:315–328.

Ehrenberg, K. (1927) Die Stellung der Palaeobiologie in der Biologie als Gesamtwissenschaft [The position of paleobiology among the general biological sciences]. *Biologia Generalis* **3**:213–244.

Eiselt, B.S. (1997) Fish remains from the Spirit Cave. *Paleofecal Material Nevada* **40**:117–139.

el Amir, M. (1948) The Shkos of Apis at Memphis. *Journal of Egyptian Archaeology* **34**:51–52.

el Mahdy, C. (1989) *Mummies, Myth and Magic in Ancient Egypt*. London: Thames & Hudson.

el-Najjar, M.Y., Benitez, J., Fry, G., Lynn, G.E., Ortner, D.J., Reyman, T.A. & Small, P.A. (1980) Autopsies on two Native American mummies. *American Journal of Physical Anthropology* **53**:197–202.

el-Najjar, M.Y., Aufderheide, A.C. & Ortner, D. (1985) Preserved human remains from the southern region of the North American continent: Report of autopsy findings. *Human Pathology* **16**:273–276.

el-Najjar, M.Y., Mulinski, T.M.J. & Reinhard, K. (1998) Mummies and mummification practices in the southern and southwestern United States. In *Mummies, Disease and Ancient Cultures*, ed. A. Cockburn, E. Cockburn & T.A. Reyman, pp. 121–139. Cambridge: Cambridge University Press.

Elder, J.R. (1934) *Marsden's Lieutenants.* Dunedin, New Zealand: Coulls Somerville Wilkie Ltd and A.R.H. Reed for the Otago University Council.

Elkin, A.P. (1948) *The Australian Aborigines: How to Understand Them.* Sydney & London: Angus & Robertson.

Elliott, P. & Loomis, R.J. (1975) *Studies of Visitor Behavior in Museums and Exhibitions: An Annotated Bibliography of Sources Primarily in the English Language.* Washington, DC: Office of Museum Programs, Smithsonian Institution Press.

Ellison, P. (1987) American Scientist interviews. *American Scientist* **75**:622–627.

Emery, W.B. (1970) Preliminary report on the excavations at North Saqqara, 1968–9. *Journal of Egyptian Archaeology* **56**:5–11.

Engel, F. (1963) A preceramic settlement on the central coast of Peru. *Transactions of the American Philosophical Society* **53**(4):3–119.

Engel, F. (1977) Early Holocene funeral bundles from the Central Andes. *Paleopathology Newsletter* **19**:7–9.

Esguerra, V. (1991) Las Momias de San Bernardo. *El Tiempo*, February 21, p. 28. Bogotá, Colombia.

Estes, J.W. (1989) *The Medical Skills of Ancient Egypt.* Canton, MA: Science History Publications.

Estes, J.W. (1993) *The Medical Skills of Ancient Egypt*, revised edition. Canton, MA: Science History Publications/ Watson Publishing International.

Ettinger, R.C.W. (1968) *The Prospect of Immortality.* New York: MacFadden-Bartell.

Etxeberria, F. & Herrasti, L. (1995) Ines Ruiz de Otálora, cuerpo momificado del siglo XVII [Ines Ruiz of Otálora, a mummified body from the seventeenth century]. *Proceedings of the First World Congress on Mummy Studies*, February, 1992, Vol. 2, p. 883. Santa Cruz, Tenerife, Canary Islands: Archaeological and Ethnographical Museum of Tenerife.

Etxeberria, F., Armendariz, A., Barrutiabengoa, J., Carnicero, H.L., Tamayo, G. & Ignacio-Vegas, J. (1994) Antropología, historia y creencias populares entorno a las momias conservadas en el País Vasco [Anthropology, history and popular beliefs relating to mummies conserved in the Basque country]. In *Cuadernos de Sección Ciencias Médicas*, ed. I.M. Barriola, pp. 11–51. Donostia: Sociedad de Estudios Vascos.

European Polycystic Kidney Disease Consortium (1994) The polycystic kidney disease 1 gene encodes a 14 kb transcript and lies within a duplicated region on chromosome 16. *Cell* **77**:881–894.

Evans, W.E.D. (1963) *The Chemistry of Death.* Springfield, IL: Charles Thomas.

Evershed, R.P. (1990) Lipids from samples of skin from seven Dutch bog bodies: preliminary report. *Archaeometry* **32**:139–153.

Evershed, R.P. (1992) Chemical composition of a bog body adipocere. *Archaeometry* **34**:253–265.

Evershed, R.P. (1993) Biomolecular archaeology and lipids. *World Archaeology* **25**:74–93.

Evershed, R.P., Turner-Walker, G., Hedges, R.E.M., Tuross, N. & Leyden, A. (1995) Preliminary results for the analysis of lipids in ancient bone. *Journal of Archaeological Science* **22**:277–290.

Evershed, R.P., Bland, H.A., van Bergen, P.F., Carter, J.F., Horton, M.C., Rowley-Conwy, P.A. (1997) Volatile compounds in archaeological plant remains and the Maillard reaction during decay of organic matter. *Science* **2787**:432–433.

Evershed, R.P., Dudd, S.N., Charters, S., Mottram, H., Stott, A.W., Raven, A. *et al.* (1999) Lipids as carriers of anthropogenic signals from prehistory. *Philosophical Transactions of the Royal Society of London* **354**(1379):19–32.

Ezra, H.C. & Cook, S.F. (1957) Amino acids in fossil bone. *Science* **126**(3263):80.

Ezra, H.C. & Cook, S.F. (1959) Histology of mammoth bone. *Science* **129**:465–466.

Faculdad de Filosofía y Letras (1992) Equipo De Rescate y Conservación de Colección de cuerpos Momificados del Museo Etnográfico de J.B. Ambrosetti [Rescue and conservation of the mummy collection at the Ethnographical Museum J.B. Ambrosetti]. Universidad De Buenos Aires, Argentina.

Fairman, H.W. (1972) Tutankhamun and the end of the 18th dynasty. *Antiquity* **46**:15–18.

Fariña-Trujillo, J.M., Iraola, H. & Ruiz, H. (1995) El publico ante las momias [Public attitudes toward mummies]. In *Proceedings of the First World Congress on Mummy Studies*, February, 1992, Vol. 1, pp. 187–195. Santa Cruz, Tenerife, Canary Islands: Archaeological and Ethnographical Museum of Tenerife.

Farrand, W.R. (1961) Frozen mammoths and modern geology. *Science* **133**(3455):729–735.

Favetti, C. & Pennacchi, A.M. (1992) *Le Mummie di Ferentillo* [The Ferentillo Mummies]. Perugia: Quattroemme.

Fazzini, R. (1975) *Images for Eternity.* New York: Brooklyn Museum.

Fernand-Filh, L., Araujo, A. & Confalonieri, U. (1988) The finding of helminth eggs in a Brazilian mummy. In *Paleoparasitologia no Brasil* [Paleoparasitology in Brazil], ed. L.F. Ferreira, A. Araujo & U. Confalonieri, pp. 41–45. Rio de Janeiro: PEC/ENSP.

Ferreira, L.F., Araujo, A.J.G. & Confalonieri, U.E.C. (1983) The finding of helminth eggs in a Brazilian mummy. *Transactions of the Royal Society of Tropical Medicine and Hygiene* **77**:65–67.

Figueroa-Tagle, L. (2000) *Arica Inserta en Una Region Arsenical: El Arsenico en el Ambiente Que Afecta y 45 Siglos de Arsenicismo Cronico* [Arica Is Located In An Arsenical Region: Ambient Arsenic Has Caused Chronic Arsenicism For 45 Centuries]. Arica, Chile: Universidad de Tarapaca.

Fiori, M.G. & Grazia-Nunzi, M. (1995) The earliest documented applications of X-rays to examination of mummified remains and archaeological materials. *Journal of the Royal Society of Medicine* **88**:67–69.

Firth, R. (1931) Maori material in the Vienna Museum. *Journal of Polynesian Society* **430**:95–102.

Fischer, C. (1951) *The Tollund Man and the Elling Woman Museum Catalog.* Silkeborg: Silkeborg Museum.

Fischer, C. (1998) Bog bodies of Denmark and northwestern Europe. In *Mummies, Disease and Ancient Cultures*, 2nd edn, ed. A. Cockburn, E. Cockburn & T.A. Reyman, pp. 237–262. Cambridge: Cambridge University Press.

Fisher, F. (1994) After 66 years, Speedy can finally rest in peace. *Duluth News-Tribune*, August 6, p. 5A.

Fleming, S.J., Miller, W.T. & Brahin, J.L. (1983) The mummies of Pachacamac, Peru: Max Uhle's early career. *MASCA Journal* **2**(5):139–141.

Flower, W.H. (1879) Illustrations of the Mode of Preserving the Dead in Darnley Island and in South Australia. *Journal of Royal Anthropological Institute of Great Britain and Ireland* **8**:389–395.

Fogel, M.L. & Tuross, N. (1988/1989) *Nitrogen Isotope Tracers of Human Lactation in Modern and Archaeological Populations.* Annual Report of the Director, Geophysical Laboratory, Carnegie Institute of Washington, pp. 111–116.

Fonda, D. (2000) Saving the dead. *Life*, April, pp. 69–71.

Force, R.W. (1984) Oceanian peoples and cultures. In *Encyclopaedia Britannica (Macropedia)*, 15th edn, ed. P.W. Goetz & M. Sutton, pp. 468–472. London: Helen Hemingway Benton.

Fornaciari, G. (1982) Natural mummies in Central Italy: A preliminary survey. *Paleopathology Newsletter* **40**:11–12.

Fornaciari, G. (1985) The mummies of the Abbey of Saint Domenico Maggiore in Naples: A preliminary report. *Archivio per l'Antropologia e la Etnologia* **115**:215–226.

Fornaciari, G. (1997) The mummies of the Basilica of San Francesco in Arezzo (Tuscany, Central Italy). *Paleopathology Newsletter* **97**:13–14.

Fornaciari, G., & Capasso, L. (1996) Natural and artificial 13th–19th century mummies in Italy. In *Human Mummies: A Global Survey of Their Status and the Techniques of Conservation*, ed. K. Spindler, H. Wilfing, E. Rastbichler-Zissernig, D. zur Nedden & H. Nothdurfter, pp. 3–8. New York: Springer-Verlag.

Fornaciari, G. & Marchetti, A. (1986) Intact smallpox virus particles in an Italian mummy of the XVI century: An immuno-electron microscopic study. *Paleopathology Newsletter* **56**:7–12.

Fornaciari, G. & Torino, M. (1995) Exploration of the tomb of Pandolfo III Malatesta (1370–1427), Prince of Fano (Central Italy). *Paleopathology Newsletter* **92** (December):7–12.

Fornaciari, G., Castagna, M., Tognetti, A. & Tornaboni, D. (1988a) Syphilis tertiaire dans une momie du XVIe siècle: Étude immunohistochimique et ultrastructurelle [Tertiary syphilis in a sixteenth century mummy: Immunohistochemical and ultrastructural study) Presented at the VIIth European Members Meeting in Lyons, France. September 1–4. *Paleopathology Newsletter* **48**(Suppl.):7 (Abstract).

Fornaciari, G., Tornaboni, D., Pollina, L. & Tognetti, A. (1988b) Pathologie pulmonaire et hépatique [Pulmonary and hepatic pathology]. In Papers on paleopathology presented at the VIIth European Members Meeting of the Paleopathology Association in Lyons, France on September 1–4. *Paleopathology Newsletter* **62** (Suppl.):9 (Abstract).

Fornaciari, G., Spremolla, G., Vergamini, P. & Benedetti, E. (1989a) Analysis of pulmonary tissue from a natural mummy of the XIII century (Saint Zita, Lucca, Tuscany, Italy) by FT-IR microspectroscopy. *Paleopathology Newsletter* **68**:5–8.

Fornaciari, G., Tornaboni, D., Castagna, M., Bevilacqua, G. & Tognetti, A. (1989b) Variole dans une momie du XIVe siècle de la basilique de S. Domenico Maggiore a

Naples: Étude immunohistochimique, ultrastructure et biologie moléculaire [Variola in a fourteenth century mummy from the Basilica of St Domenico Maggiore at Naples: immunological, ultrastructural and molecular biochemical study]. In *Advances in Paleopathology,* ed. C. Capasso, pp. 97–100. Chieti: Marino Solfanelli Editore.

Fornaciari, G., Castagna, M., Viacava, P., Tognetti, A., Bevilacqua, G. & Segura, E.L. (1992a) Chagas' disease in Peruvian Inca mummy. *Lancet* **339**:128–129.

Fornaciari, G., Castagna, M., Viacava, P., Tognetti, A., Esteva, M., Segura, E.L. & Bevilacqua, G. (1992b) Malattia di Chagas in una mummia peruviana Inca del Museo Nazionale di Antropología et Etnología di Firenze [Chagas' disease in a Peruvian Inca mummy in Florence's National Museum of Anthropology and Ethnology]. *Archivio per l'Antropología e la Etnología* **122**:369–376.

Fornaciari, G., Castagna, M., Naccarato, G., Collecci, P., Tognetti, A. & Bevilacqua, G. (1993) Adenocarcinoma in the mummy of Ferrante I of Aragon, King of Naples (AD 1431–1494). *Paleopathology Newsletter* **83**:5–8.

Fouquet, D. (1886) Observations relevées sur quelques mommies royales d'égypte. [Relevant observations on several royal mummies of Egypt]. *Bulletin de la Société d'Anthropologie de Paris* **3**:578–586.

Fouquet, D. (1897) Observations pathologiques. In *Recherches sur les Origines de l'Egypte,* ed. J. de Morgan, pp. 350–373. Paris: Ernest Leroux.

Fourcroy, A.F. de (1791) *Leçons Élémentaires d'Histoire naturelle et de Chimie.* Paris: Ernest Leroux.

Fowler, E. Jr & Osmon, P. (1942) New bone growth due to cold water in the ears. *Archives of Otolaryngology* **36**:455–466.

Fraire, A.E., Johnson, E.H., Kim H.S. & Titus, J.L. (1988) The relationship of obesity to gallstones – an autopsy study. Poster session at meeting of the United States and Canadian Academy of Pathology held at Washington, DC February 28 to March 4, 1988.

Francalacci, P. (1995) DNA analysis of ancient desiccated corpses from Xinjiang. *Journal of Indo-European Studies* **23**:385–397.

Frankfort, H. (1948) *Kingship and the Gods.* Chicago: Chicago University Press.

Frazer, Sir J.G. (1976) *The Golden Bough: A Study in Magic and Religion.* New York: Macmillan. (First published in 1922.)

Freese, B.L. (1996) Maronite mummies. *Archaeology* **49**(4):22.

Frost, P. (1988) Human skin color: A possible relationship between its sexual dimorphism and its social perception. *Perspectives in Biology and Medicine* **32**:38–58.

Fuchs, J. (1964) Physical alterations which occur in the blind and are illustrated on ancient Egyptian works of Art. *Annals of the New York Academy of Sciences* **117**:618–623.

Fulcheri, E. (1987a) Seconda relazione medica [Second Medical Report]. In *Cardinale Josyf Slipyj,* ed. G. Nolli, pp. 37–56. Rome: Editrice Elettongraf.

Fulcheri, E. (1987b) Seconda relazione medica: recognizioni effetuate sui reste S Chiara d'Assis. [Evaluation of the remains of Saint Chiara D'Assisi]. In *S. Chiara D'Assisi,* ed. G. Nolli, pp. 55–60. Rome: Editrice Elettrongraf.

Fulcheri, E. (1987c) Case no. 27: Verruca vulgaris. (Department of Pathology, Medical College of Virginia, Richmond, VA) *Paleopathology Club Newsletter* **31**:2.

Fulcheri, E. (1996) Mummies of Saints: A particular category of Italian mummies. In *Human Mummies: A Global Survey of Their Status and the Techniques of Conservation,* ed. K. Spindler, H. Wilfing, E. Rastbichler-Zissernig, D. zur Nedden & H. Nothdurfter, pp. 219–230. New York: Springer-Verlag.

Fulcheri, E. (1997) Esame istologico random nel controllo periodico dello stato di conservacione dei resti umani mummificati [Random histological sections as part of periodic control of the state of conservation of human mummified remains]. In *Nuove Tecnologie nella Tutela e nel Recupero delle Raccolte Naturalistiche, Museali e Scientifiche* [New Technology for the Protection and Repair of Scientific and Biological Museum Collections], ed. M. Girotti, pp. 81–89. Turin: Department of Biology, University of Turin.

Fulcheri, E. & Grilletto, R. (1988) Sur un probable calcul rénal daté ancien empire Égyptien. Manuscript read at the 7th European Meeting of the Paleopathology Association held in Lyons, France, 1988.

Fulcheri, E., Rabino-Massa, E. & Fenoglio, C. (1985) Improvement in the histological technique for mummified tissue. *Verhandlungen der Deutsche Gesellschaft fuer Pathologie* **68**:471 (only).

Fulcheri, E., Baracchni, P., Crestani, C., Drusini, A. & Rippa-Bonati, M. (1991) Studio preliminare delle mummie naturali di Ferentillo: Esame istologico ed immunoistochimico della cute [Preliminary study of the naturally mummified bodies from Ferentillo: An histological and immunohistochemical study of the skin]. *Rivista Italiana Medicina Legale* **13**:171–183.

Fulcheri, E., Baraccini, P., Doro Garetto, T., Pastorino, A. & Rabino Massa, E. (1994) Le mummie dell'antico Egitto custodite nei musesi Italiani [Ancient Egyptian mummy collections in Italian museums]. *Museological Science* **11**(1–2):1–11.

Fulcheri, E., Boanao, R., Grilletto, R., Savoia, D., Leospo, E. & Rabino-Massa, E. (1999) The preservation status of ancient Egyptian mummified remains estimated by histological analysis. *Paleopathology Newsletter* **108**:5–12.

Gade, D.W. (1967) The guinea pig in Andean folk culture. *Geographic Review* **57**:213–224.

Gandhi, M.K. (1954) *How to Serve the Cow.* Ahmedabad, India: Navajivan Publishing House.

Gannal, J.N. (1838/1840) *History of Embalming, and of Preparations in Anatomy, Pathology and Natural History, Including an Account of a New Process for Embalming,* transl. R. Harlan. Philadelphia: Judah Dobson.

Gänsicke, S. (1995) Collections care and management of the Egyptian collection at the Museum of Fine Arts, Boston. In *Conservation In Ancient Egyptian Collections*, ed. C.E. Brown, F. Macalister & M.M. Wright, pp. 109–117. London: Archetype Publications.

García, M. & Herraez, M.I. (1995) Conservación de momias. Un estudio comparado de los tratamientos de conservación preventivos y activos [Conservation of mummies: A survey on passive and active conservation treatments]. In *Studies on Ancient Mummies and Burial Archaeology,* ed. F. Cárdenas & C. Rodríguez-Martín, pp. 293–302. *Proceedings of the 2nd World Congress on Mummy Studies* in Cartagena, Colombia, February Bogotea, Colombia: Universidad de Los Andes.

García-Friás, J.E. (1940) La tuberculosis en los antiguous Peruanos [Tuberculosis in ancient Peruvians]. *Actualidad Medica Peruana* **5**:274–291.

Gardiner, Sir A. (1961) *Egypt of the Pharaohs.* London: Oxford University Press.

Garland, A.N. (1995) Worsley Man, England. In *Bog Bodies,* ed. R.C. Turner & R.G. Scaife, pp 104–107. London: British Museum Press.

Garner, R.C. (1979) Experimental mummification. In *The Manchester Museum Mummy Project,* ed. A.R. David, pp. 19–24. Manchester: Manchester University Press.

Garner, R.C. (1986) Insects and mummification. In *Science in Egyptology,* ed. R.A. David. pp. 97–100. Manchester: Manchester University Press.

Garraty, J.A. & Gay, P. (eds.) (1972) *The Columbia History of the World.* New York: Harper & Row.

Garrison, F.H. (1917) Memorial notice: Sir Marc Armand Ruffer. *Annals of Medical History* **1**:218–220.

Garrison, F.H. (1929) *An Introduction to the History of Medicine.* Philadelphia: Saunders.

Garstang, J. (1907) *The Burial Customs of Ancient Egypt.* London: Archibald Constable & Co. Ltd.

Gebuehr, M. (1979) Das Kindergrab von Windeby [The child's grave from Windeby]. *OFFA* **36**:75–107.

Gee, H. (1990) Directors agree on research. *Nature* **347**:506.

Gerszten, P. & Martínez, A.J. (1995) The neuropathology of South American mummies. *Neurosurgery* **36**:756–761.

Gerszten, E., Allison, M.J., Pezzia, A. & Klurfeld, D. (1976a) Thyroid disease in a Peruvian mummy. *MCV (Medical College of Virginia) Quarterly* **12**(2):52–53.

Gerszten, E., Munizaga, J., Allison, M. & Klurfeld, D.M. (1976b) Diaphragmatic hernia of the stomach in a Peruvian mummy. *Bulletin of the New York Academy of Medicine* **52**:601–604.

Gerszten, E., Allison, M.J. & Fouant, M.M. (1983) Renal diseases in South American mummies. *Laboratory Investigation* **48**(1):29A (Abstract).

Gerszten, P., Gerszten, E. & Allison, M.J. (1997) Ultrastructure of a well-preserved lymphocyte from a mummified human. *Journal of Electron Microscopy* **46**:443–445.

Gessler, W. (1998) Die gewheimnisvolle Lagune der Mumien [The mysterious lagoon of the mummies]. *PM*, March, pp. 8–14.

Ghalioungui, P. & el Dawakhly, Z. (1965) *Health and Healing in Ancient Egypt.* Cairo: The Egyptian Organization for Authorship and Translation.

Giacometti, L. & Chiarelli, B. (1968) The skin of Egyptian mummies. *Archives of Dermatology* **97**:712–716.

Gibbon, G. (1989) *Explanation in Archaeology*. Oxford: Basil Blackwell.

Gibbons, A. (2001) The peopling of the Pacific. *Science* **291**:1735–1737.

Gibbons, G. (1986) Oregon Trail: The itch to move West. *National Geographic* **170**:147–177.

Gibson, G. (2000) Names matter: The unfinished history of the Niagara Falls mummies. *KMT (Kemet)* **11**(4):18–39.

Gieseppina-Banchieri, D. (1993) La mummia dei Musei di

Varese. *Comune di Varese Assessorato alla Cultura Musei Civici.*

Gill, G.W. (1976) Two mummies from the Pitchfork Rock Shelter in northwestern Wyoming. *Plains Anthropologist, Journal of the Plains Anthropological Society* **21**(74):301–310

Gill, G.W. & Owsley, D.W. (1985) Electron microscopy of parasite remains on the Pitchfork Mummy and possible social implications. *Plains Anthropologist, Journal of the Plains Anthropological Society* **30**(107):45–50

Girling, M.A. (1986) The insects associated with Lindow Man. In *Lindow Man: The Body in the Bog,* ed. I.M. Stead, J.B. Bourke & D. Brothwell, pp. 90–91. Ithaca, NY: Cornell University Press.

Glenn, F. (1971) Biliary tract disease since antiquity. *Bulletin of the New York Academy of Medicine* **47**:329–350.

Glob, P.V. (1954) Lifelike man preserved 2000 years in peat. *National Geographic* **105**:419–430.

Glob, P.V. (1969) *The Bog People.* London: Faber & Faber.

Glover, J.M. (1984) Conservation and storage: Textiles. In *Manual of Curatorship,* ed. J.M.A. Thompson, pp. 333–355. London: Butterworths.

Gmelch, S.B. & Gmelch, G. (1980) Ireland's bountiful bogs. *Natural History* **89**(11):48–57.

Goetze, A. (1947) Old Babylonian omen texts. Yale Oriental Series. *Babylonian Texts,* **10**. New Haven, CT: Yale University Press.

Goff, M.L. (1991) Feast of clues. *The Sciences* **31**(4):30–35.

Goldstein, P. (1989) The Tiwanaku occupation of Moquegua. In Ecology, Settlement and History in the Osmore Drainage, Peru, ed. D.S. Rice, C. Stanish & P.R. Scarr, *British Archaeological Reports, International Series* **545**, Part 1, pp. 219–255.

Goldstein, P. (1995–1996) Tiwanaku settlement patterns of the Azapa valley, Chile: New data and the legacy of Percy Dauelsberg. *Dialogo Andino* (Universidad de Tarapaca, Arica, Chile) **14/15**:57–73.

Golenberg, E.M., Giannas, D.E., Clegg, M.T., Smiley, C.J., Durbin, M., Henderson, D. & Zurawski, G. (1990) Chloroplast DNA sequence from a Miocene Magnolia species. *Nature* **344**:656–658.

Gómara, F. López de (1522/1992) *General History of the Indies and the Conquest of Mexico*, transl. V. Hispania.

Gómez-Bellard, F. & Abel-Cortés, J.L. (1990) Trichinosis in the mummy of a young girl (Toledo, Spain). In papers on paleopathology presented at the 8th European Members Meeting of the Paleopathology Association, Cambridge, England. *Paleopathology Newsletter* **72**(Suppl.):11 (Abstract).

Gomori, G. (1946) A new histochemical test for glycogen and mucin. *Technical Bulletin for Registry of Medical Technologists* **7**:115–117.

Goodman, M., Shoshani, J. & Barnhart, M. (1979) Frozen mammoth muscle: Preliminary findings. *Paleopathology Newsletter* **25**:3–5.

Goodman, M.H. (1986) *The Last Dalai Lama: A Biography.* Boston: Shambhala.

Gorecki, P. (1996) The initial colonization of Vanuatu. In *Arts of Vanuatu,* ed. J. Bonnemaison, C. Kaufmann, K. Huffman & D. Tyron, pp. 63–65. Honolulu: University of Hawaii Press.

Gorman, J. (1999) The Iceman goeth. *Discover* **20** January:27.

Gotouda, H., Takatori, T., Terezawa, K., Nagao, M. & Tarao, H. (1988) The mechanism of experimental adipocere formation: Hydration and dehydrogenation in microbial synthesis of hydroxy and oxo fatty acids. *Forensic Science International* **37**:249–257.

Graf, W. (1949) Presence of a spasmogenic substance, presumably histamine, in extracts of mummy tissue. *Nature* **164**:701–702.

Graham, M. D. (1979) Osteomas and exostoses of the external auditory canal. *Annals of Otology, Rhinology and Laryngology* **88**:566–572.

Grant, M. (1972) *Cleopatra.* New York: Dorset Press.

Granville, A.B. (1825) An essay on Egyptian mummies; with observations on the art of embalming among the ancient Egyptians. *Philosophical Transactions of the Royal Society of London* **1**:269–316.

Gray, P.H.K. (1966) Embalmers' "Restorations". *Journal of Egyptian Archaeology* **52**:138–140.

Gray, P.H.K. (1967) Radiography of ancient Egyptian mummies. *Medical Radiography and Photography* **43**:34–44.

Gray, P.H.K. (1972) Notes concerning the position of arms and hands of mummies with a view to possible dating of the specimen. *Journal of Egyptian Archaeology* **58**:200–204.

Gray, P.H.K. (1973) The radiography of mummies of ancient Egyptians. *Journal of Human Evolution* **2**:51–53.

Gray, P.H.K. & Dawson, W.R. (1968) Catalogue of Egyptian antiquities in the British Museum. Vol. 1 *Mummies and Human Remains.* London: Oxford University Press.

Graves, R. (transl.) (1957) *Suetonius: The Twelve Caesars.* Harmondsworth, Middlesex: Penguin Books.

Green, C.J.S. (1977) *The Significance of Plaster Burials for the*

Recognition of Christian Cemeteries Burial in the Roman World. CBA Research Report **22**. London: Council for British Archaeology.

Green, C.J.S., Paterson, M. & Biek, L. (1981) A Roman coffin-burial from the crown buildings site, Dorchester: With particular reference to the head of well-preserved hair. *Proceedings of the Dorset Natural History and Archaeological Society* **103**:67–100.

Gregg, J.B. & Bass, W.H. (1970) Exostoses in the external auditory canals. *Annals of Otology, Rhinology and Laryngology* **79**:834–839.

Gregg, J.B. & Gregg, P.S. (1987) *Dry Bones: Dakota Territory Reflected*. Sioux Falls, SD: Sioux Printing, Inc.

Gregg, J.B. & Steele, J. P. (1982) Mastoid development in ancient and modern populations. *Journal of American Medical Association* **248**:459–464.

Gregg, J.B., Steele, J.P. & Sylvester, C. (1965) Ear disease in skulls from the Scully burial site. *Plains Anthropologist, Journal of the Plains Anthropological Society* **10**:233–239.

Group for Research of Japanese Mummies (1993) *Worship of Mummies in Japan and China* (in Japanese but with English chapter summaries). Tokyo: Heibonsha Limited, Publishers.

Grupe, G. & Hummel, S. (1991) Trace element studies on experimentally cremated bone. 1. Alteration of the chemical composition at high temperatures. *Journal of Archaeological Science* **18**:177–186.

Gruspier, K. (1997) The mummified remains from Qazone, Jordan (Poster). In papers on paleopathology, presented at the 24th Annual Meeting of the Paleopathology Association at St Louis, Missouri, April 1–2. *Paleopathology Newsletter* **31**(Suppl.):7

Gryaznov, M.P. (1950) *First Pazyryk Kurgan*. Leningrad: Hermitage Museum.

Guhl, F., Jaramillo, C., Chiriboga, M. & Cárdenas-Arroyo, F. (1992) Aislamiento y purificación de anticuerpos a partir de cuerpos momificados [Isolation and purification of antibodies from mummified bodies]. *Bioantropologia* (Universidad de Los Andes, Bogotá) **2**(1):6–7.

Guhl, F., Jaramillo, C., Vallejo, G.A., Yockteng, R., Cárdenas-Arroyo, F., Arriaza, B. & Aufderheide, A.C. (1999) Isolation of *Trypanosoma cruzi* DNA in 4000 year old mummified human tissue from northern Chile. *American Journal of Physical Anthropology* **108**:401–407.

Guillén, S. (1992) The Chinchorro culture: Mummies and crania in the reconstruction of preceramic coastal adaptation in the South Central Andes. Doctoral Dissertation on File at University of Michigan, Department of Anthropology.

Guillén-Oneeglio, S., Huertas, L., Boza, A.M., Cornejo, I. & Isla, E. (1995) Identificación y estudio de los restos del Virrey Conde de la Monclova en la Cripta Arzobispal de la Catedral de Lima, Peru [Identification and Study of the Remains of the Spanish Viceroy of Peru, the Count of Monclova, in the Archbishop's Crypt of Lima Cathedral]. In *Proceedings of the First World Congress on Mummy Studies*, February, 1992, Vol. 1, pp. 359–366. Santa Cruz, Tenerife, Canary Islands: Archaeological and Ethnographical Museum of Tenerife.

Guirand, F. (ed.) (1968) *New Larousse Encyclopedia of Mythology*, transl. R. Aldington & D. Ames. New York: Hamlyn.

Guirand, F. (ed.) (1978) *The New Larousse Encyclopedia of Mythology*, transl. R. Aldington & D. Ames. New York: Hamlyn.

Gülaçar, F.O., Susini, A. & Klohn, M. (1990) Preservation and postmortem transformations of lipids in samples from a 4000-year-old Nubian mummy. *Journal of Archaeological Science* **17**:691–705.

Gunther, R.T. (ed.) (1959) *The Greek Herbal of Discorides*. New York: Hafner Publishing Co.

Gurley, L.R., Valdez, J.G., Spall, W.B., Smith, B.F. & Gillette, D.D. (1991) Proteins in the fossil bone of the dinosaur, *Seismosaurus*. *Journal of Protein Chemistry* **10**:75–90.

Gwei-Djen, L. & Needham, J. (1993) Diseases of antiquity in China. In *The Cambridge World History of Human Disease*, ed. K.F. Kiple, pp. 345–354. New York: Cambridge University Press.

Habachi, L. (1939) A first dynasty cemetery at Abydos. *Annales du Service des Antiquités de l'Égypte* **39**:767–774 plus one plate.

Habenstein, R.W. & Lamers, W.M. (1981) *The History of American Funeral Directing*. Milwaukee, WI: National Funeral Directors Association of the United States, Inc.

Haddon, A.C. (1935) Reports of the Cambridge Anthropological Expedition to the Torres Straits, Vol. 1, *General Ethnography*. Cambridge: Cambridge University Press.

Hadingham, E. (1994) The mummies of Xinjiang. *Discover* **15**(4):68–77.

Hadley, J. (1764) An account of a mummy, inspected at London, 1763. *Philosophical Transactions of the Royal Society of London* **54**:1–2.

Hagelberg, E., Kayser, M., Nagy, M., Roewer, L., Zimdahl, H., Krawczak, M., Lio, P. & Schiefenhovel, W. (1999) Molecular genetic evidence for the human settlement of the Pacific: Analysis of mitochondrial DNA, Y chromosome and HLA markers. *Philosophical Transactions of the Royal Society of London* **354**:141–152.

Haglund, W.D. (1997) Dogs and coyotes. Postmortem involvement with human remains. In *Forensic Taphonomy*, ed. W.D. Haglund & M.H. Song, pp. 367–381. New York: CRC Press.

Haglund, W.D. & Sorg, M.H. (1997) *Forensic Taphonomy.* New York: CRC Press.

Hakluyt, R. (1599) *The Principall Navigations, Voiyages and Discouries of the English Nation. II.* London.

Halbaek, H. (1958) Grauballemandens sidste mueltid [The Grauballe man's last meal: An analysis of food remains in the stomach]. *KUML* 1958, pp. 83–116.

Haley, A. (1976) *Roots.* Garden City, NY: Doubleday.

Hall, A.J. (1974) A lady from China's Past. *National Geographic* **145**:661–681.

Hall, A.J. (ed.) (1981) Temperature and rainfall each month for selected places around the world. In *National Geographic Atlas of the World*, p. 236. Washington, DC: National Geographic Society.

Haller, J.S. Jr (1993) Ergotism. In *The Cambridge World History of Human Disease*, ed. K.F. Kiple, pp. 718–719. Cambridge: Cambridge University Press.

Ham, A. (2000) AFIP researchers complete second critical gene sequence on deadly 1918 Spanish flu. *AFIP Letter* **1598**(4):13.

Hamamsy, L.S.E. & Jones, M. (1984) The Arab Republic. In *The New Encyclopedia Britannica*, 15th edn, ed. P. Goetz (editor-in-chief), pp. 445–459. Chicago: Helen Hemingway, Publisher.

Hambly, W.D. (1925) *The History of Tattooing and its Significance.* London: H.F. & G. Witherby. (Republished in 1974 by Gale Research Co., Book Tower, Detroit.)

Hamilton-Paterson, J. & Andrews, C. (1979) *Mummies: Death and Life in Ancient Egypt.* New York: The Viking Press.

Hamlyn-Harris, R. (1912a) Mummification. *Memoirs of the Queensland Museum* **1**:7–23 with plates 4–11.

Hamlyn-Harris, R. (1912b) Papuan mummification: Queensland Museum. *Memoirs of the Queensland Museum* **1**:1–6 with plates 2–3.

Handler, J.S., Aufderheide, A.C., Corruccini, R.S., Brandon, E.M. & Wittmers, L.E. Jr (1986) Lead contact and poisoning in Barbados slaves: Historical, chemical and biological evidence. *Social Science History* **10**:399–425.

Harding, J.R. (1976) Certain Upper Paleolithic "Venus" statuettes considered in relation to the pathological condition known as massive hypertrophy of the breasts. *Man* (London) **11**:271–272.

Harker, W.B. (1993) Health in Pharaonic Egypt. In *Bioanthropology and the Study of Ancient Egypt*, ed. W.V. Davies & R. Walker, pp. 19–23. London: British Museum Press.

Harmer, R.M. (1984) Embalming, burial and cremation. In *Encyclopedia Britannica*, ed. P.W. Goetz, pp. 735–741. Chicago: Helen H. Benton.

Harner, M.J. (1972) *The Jivaro: People of the Sacred Waterfalls.* Berkeley, CA: University of California Press.

Harrington, S.P.M. (1996) Imperial jade shroud. *Archaeology* **49**(4):15.

Harris, J.E. & Weeks, K.R. (1973) *X-raying the Pharaohs.* New York: Scribner's and Sons.

Harris, J.E. & Wente, E.F. (1980) *An X-ray Atlas of the Royal Mummies.* Chicago: University of Chicago Press.

Harris, N. (1995) *Mummies: A Very Peculiar History.* New York: Franklin Watts.

Harrison, D.F.N. (1962) The relationship of osteomata of the external auditory meatus to swimming. *Annals of the Royal College of Surgeons in England* **31**:187–201.

Harrison, R.G. (1973) Tutankhamun postmortem. *Lancet* **1**:259.

Harrison, R.G. & Abdalla, A. (1972) The remains of Tutankhamun. *Antiquity* **46**:8–14.

Harrison, R.G., Connolly, R.C. & Abdalla, A. (1969) Kinship of Smenkhare and Tutankhamen affirmed by serological micromethod. *Nature* **224**:325–326.

Harrison, W.R., Merbs, C.F. & Leathers, C.R. (1991) Evidence of coccidioidomycosis in the skeleton of an ancient Arizona Indian. *Journal of Infectious Diseases* **164**:436–437.

Hart Hansen, J.P., Meldgaard, J. & Nordqvist, J. (eds.) (1991) *The Greenland Mummies Nuuk (Godthab), Greenland.* Greenland Museum & Copenhagen: Christion Ejlers' Forlag.

Hartleben, H. (1906) *Champollion: Sein Leben und sein Werk* [Champollion: His Life and Work]. Berlin: Weidmannscher Buchhandlung.

Haskell, N.H., Hall, R.D., Cervenka, V.J. & Clark, M.A.

(1997) On the body: Insects' life stage presence and their postmortem artifacts. In *Forensic Taphonomy*, ed. W.D. Haglund & M.H. Sorg, pp. 415–448. New York: CRC Press.

Hassan, F.A. (1988) The Predynastic of Egypt. *Journal of World Prehistory* **2**:135–185.

Hatfield, M.P. (1886) *The Funeral Directors' Official Text Book*. Chicago: Funeral Directors' Association of the United States and Canada (Donohue and Henneberry, printers).

Hatheway, C.L. (1990) Toxigenic *Clostridia*. *Clinical Microbiology Review* **3**:66–98.

Hawass, Z. (2000a) The valley of the gilded mummies. *KMT* (Kemet) **10**(4):72–85.

Hawass, Z. (2000b) *Valley of the Golden Mummies*. New York: Harry Abrams.

Hawkes, J. (1973) *The First Great Civilizations: Life in Mesopotamia, The Indus Valley and Egypt*. New York: Knopf.

Hayat, M.A. (1989) *Principles and Techniques of Electron Microscopy*. Boca Raton, FL: CRC Press.

Heinemann, S. (1974) Xeroradiography of a Pre-Columbian mummy. *Journal of the American Medical Association* **230**(3):1256.

Helfman, P.H. & Bada, J.L. (1975) Aspartic acid racemization in tooth enamel from living humans. *Proceedings of the National Academy of Sciences, USA* **72**:2891–2894.

Heller, J. (1990) Tales from the carry-on crypt. *St Petersburg Times*, Scripps Howard News Service, August 26, p. J1.

Helser, L. (1994) Freeze-dried Fluffy, Fido: A comfort to grieving owners. *Duluth News-Tribune*, October 23:1E.

Herdman, W.A. (1889–1890) Note on some mummy cats. *Egypt Proceedings of the Liverpool Biological Society* **4**:95–96.

Hernández, F. (1577/1959) *Historia Natural de la Nuera España*, 2 Vols. Medical City: Universidad Nacional Autónoma de Mexico.

Herodian of Antioch (250/1961) *History of the Roman Empire: From the Death of Marcus Aurelius to the Ascension of Gordian III*, transl. E.C. Echols. Berkeley, CA: University of California Press.

Herodotus (ca. 450 BC/1970) Burial in Honey (Babylonians). In *Herodotus*, p. 251. Loeb Classical Library. Cambridge, MA: Harvard University Press.

Herodotus (ca. 450 BC/1981) *Herodotus*, transl. A.D. Godley. Loeb Classical Library. Cambridge, MA: Harvard University Press.

Herre, W. (1963) The science and history of domestic animals. In *Science in Archaeology*, ed. D. Brothwell, E. Higgs & G. Clarke, pp. 235–249. New York: Basic Books.

Herrmann, B. (1982) A simple method for softening and rehydration of dried and mummified tissues. *Paleopathology Newsletter* **39**:10.

Herrmann, B. & Meyer, R.-D. (1993) *Sudamerikanische Mumien aus vorspanischer Zeit: Eine radiologische Untersuchung* [Pre-Hispanic South American Mummies: A Radiologic Investigation]. Berlin: Museums fuer Voelkerkunde.

Heyerdahl, T., Sandqeiss, D.H. & Narvaez, A. (1995) *Pyramids of Túcume: The Quest for Peru's Forgotten City*. London: Thames & Hudson.

Higuchi, R., Bowman, B., Freiberger, M., Ryder, O.A. & Wilson, A.C. (1984) DNA sequences from the quagga, an extinct member of the horse family. *Nature* **312**:282–284.

Hino, H., Ammitzbøll, T., Møller, R. & Asboe-Hansen, G. (1982) Ultrastructure of skin and hair of an Egyptian mummy. *Journal of Cutaneous Pathology* **9**:25–32.

Hirschberg, J. (1982) *The History of Ophthalmology*, 11 Vols, transl. F. Blodi. Bonn: J.P. Wayenborgh. (First published in German in 1899.)

Ho, T.Y. (1966) The isolation and amino acid composition of the bone collagen in Pleistocene mammals. *Comparative Biochemistry and Physiology* **18**:353–358.

Ho, T.Y. (1967) The amino acids of bone and dentine collagens in Pleistocene mammals. *Biochimica et Biophysica Acta* **133**:568–573.

Hodder, I. (1986) *Reading the Past*. New York: Cambridge University Press.

Hoepfel, F., Platzer, W. & Spindler, K. (1992) *Der Mann im Eis* [The Man in the Ice]. Innsbruck: Universitaet, Innsbruck.

Hoffman, M.A. (1979) *Egypt Before the Pharaohs*. New York: Alfred A. Knopf.

Holden, C. (1999a) Kennewick Man gets his day in the lab. *Science* **283**:1239–1240.

Holden, C. (1999b) Glacier man, a spring chicken. *Science* **286**:229.

Holden, C. (2000) Rediscovery of a pharaoh? *Science* **289**:539.

Holden, T.G. (1989) Preliminary work on South American mummies held at the British Museum. *Paleopathology Newsletter* **65**:5–9.

Holland, T. (1937) X-rays in 1896. *The Liverpool Medico-Chirurgical Journal* **45**:61.

Hongi, H. (1916) On mummification (A review by Mr. H. D. Skinner). *Journal of the Polynesian Society* **25**:169–172.

Hope, F.W. (1834) Several species of insects found in the heads of Egyptian mummies. *Journal of the Proceedings of the Royal Entomological Society of London* **1**:11–13.

Horne, P.D. (1978) The autopsy of John Paul Jones: The story of a lost corpse. In papers presented at the 5th Annual Meeting of the Paleopathology Association at Toronto, Ontario, Canada, April 15. *Paleopathology Newsletter* **22**:77.

Horne, P.D. (1979) Head lice from an Aleutian mummy. *Paleopathology Newsletter* **25**:7.

Horne, P. (1985) A review of the evidence of human endoparasitism in the Pre-Columbian New World through the study of coprolites. *Journal of Archaeological Sciences* **12**:299–310.

Horne, P. (1986a) Case no. 23: Angiokeratoma circumscriptum. (Department of Pathology, Medical College of Virginia, Richmond, VA) *Paleopathology Club Newsletter* **27**:1.

Horne, P.D. (1986b) Angiokeratoma in Inca mummy (poster). Presented at the 13th Annual Meeting of the Paleopathology Association held in Albuquerque, New Mexico, April 12–13. *Paleopathology Association Newsletter* **54** (June):16 (Abstract).

Horne, P. (1995) Aspergillosis and dracunculiasis in mummies from the tomb of Parennefer. *Paleopathology Newsletter* **92**:10–12.

Horne, P. D. (1996) The Prince of El Plomo: A frozen treasure. In *Human Mummies: A Global Survey of Their Status and the Techniques of Conservation,* ed. K. Spindler, H. Wilfing, E. Rastbichler-Zissernig, D. zur Nedden & H. Nothdurfter, pp. 3–8. New York: Springer-Verlag.

Horne, P.D. & Ireland, R.R. (1991) Moss and a Guanche mummy: an unusual utilization. *The Bryologist* **94**:407–408.

Horne, P.D. & Jarzen, D.M. (1993) Spores or red blood cells? A case from Leptiminus. *Paleopathology Newsletter* **84**:13–14.

Horne, P.D., Summerbell, R.C. & Redford, S. (1996) Aspergillus in the tomb of Parennefer: Minerva story of the week. *British Medical Journal* **313**:240.

Horton, D. (1948) The Mundurucû. In *Handbook of South American Indians*, ed. J.H. Steward, Bureau of American Ethnology, Bulletin **143**, Vol. 3, *The Forest Tribes*, pp. 271–283. Washington, DC: Smithsonian Institution Press.

Houghton, P. (1975) A renal calculus from proto-historic New Zealand. *OSSA* **2**:11–14.

Hourani, G.G. (2000) The mummies of the holy valley of Qannobine in Lebanon. *Chungará* **32**:103–109.

Howard, R.J., Uni, S., Aikawa, M., Aley, S., Leech, J.H., Lew, A.M., Wellems, T.E., Rener, J. & Taylor, D.W. (1986) Secretion of a malarial histidine-rich protein (PfHRII) from *Plasmodium falciparum*-infected erythrocytes. *Journal of Cell Biology* **103**:1269–1277.

Howey, M.O. (1972) *The Cults of the Dog.* Ashington, Essex: C.W. Daniel Co.

Hrdlička, A. (1935) Ear exostoses. *Smithsonian Miscellaneous Collections* **93**(6):1–98.

Hrdlička, A. (1941) Exploration of mummy caves in the Aleutian Islands. *Scientific Monthly* **52**:5–23.

Hudson, M. (1996a) Royal burials of Early Imperial China. In *Tombs, Graves and Mummies,* ed. P.G. Bahn, pp. 152–153. London: Weidenfeld & Nicolson.

Hudson, M. (1996b) The mummies of the northern Fujiwara. In *Tombs, Graves and Mummies,* ed. P.G. Bahn, pp. 198–199. London: Weidenfeld & Nicolson.

Huffman, K. (1976) The hidden world of Mbotgo't. *Australian Natural History* **18**:415–419.

Hughes, J. (1998) Kennewick Man poses religious, medical questions. *Bend* (Washington), June 15.

Hunan Medical University (1980) *Study of an Ancient Cadaver in Mawangtui Tomb No. 1 of the Han Dynasty in Changsha.* Beijing, China: Cultural Relics (Wen Wu) Publishing House.

Hunter, D. (1943) *Papermaking: The History and Technique of an Ancient Craft.* New York: Knopf.

Hutt, M.S.R. & Burkitt, D.P. (1986) *The Geography of Non-Infectious Disease.* Oxford: Oxford University Press.

Ikram, S. & Dodson, A. (1998) *The Mummy in Ancient Egypt.* London: Thames & Hudson.

Indriati, E. (1998) A dental anthropological approach to coca-leaf chewing in the Andes. Unpublished doctoral thesis on file in the Department of Anthropology at the University of Chicago, Chicago, Illinois.

Indriati, E. & Buikstra, J.E. (2001) Coca chewing in prehistoric coastal Peru: Dental evidence. *American Journal of Physical Anthropology* **114**:242–257.

Iserson, K. (1994) *Death to Dust*. Tucson, AZ: Galen Press.

Iskander, Z. (1973) Mummification in Ancient Egypt. In *Mummification in Ancient Egypt,* ed. G. Mokhtar, H. Riad & Z. Iskander, pp. 9–18. Cairo: Ministry of Culture, Egyptian Antiquities Organization, Cairo Museum.

Iskander, Z. (1980) Mummification in Ancient Egypt: Development, history and techniques. In *An X-ray Atlas of the Royal Mummies,* ed. J.E. Harris & E.F. Wente, pp. 1–52. Chicago: University of Chicago Press.

Jackson, G. (1909) Etiology of exostoses of the external auditory meatus. *British Medical Journal (Laryngology, Otology and Rhinology)* **2**:1137–1138.

James, E.O. (1957) *Prehistoric Religion*. New York: Frederick A. Praeger.

James, T.G.H. (1969) W.R. Dawson. *Journal of Egyptian Archaeology* **55**:211–214 and plate XXXIV.

James, T.G.H. (1974) History of Egypt. *The New Encyclopedia Britannica*, Vol. 3, pp. 460–471. London: Helen Hemingway Benton, Publisher.

Janssens, P.A. (1970) *Paleopathology: Diseases and Injuries of Prehistoric Man*. London: Pall Mall London.

Jantz, R.L. & Owsley, D.W. (1997) Pathology, taphonomy, and cranial morphometrics of the Spirit Cave mummy. *Nevada* **40**(1):62–84.

Jantzen, G. (1968) Der blinde Harfner auf dem Grabrelief Paätenemheb [The blind harpist on the tomb relief of Paätenemheb]. *Materia Medica Nordmark* **20**:689–694.

Jarcho, S. (ed.) (1966) *Human Palaeopathology*. New Haven, CT: Yale University Press.

Jerkic, S., Horne, P.D. & Aufderheide, A.C. (1995) The Beothuk Mummy from Newfoundland (abstract). *CAPA (Canadian Association of Physical Anthropologists) Newsletter,* Spring:13.

Jijon y Camano, J. (1949) Maranga. *Contribución al conocimiento de los aborigines del Valle del Rimac, Peru* [Contribution to knowledge about the indigines of the Rimac Valley, Peru]. Quito, Ecuador: La Prensa Catolica.

Jochelson, W. (1933) *History, Ethnology and Anthropology of the Aleut*. Washington, DC: Carnegie Institution.

Johansson, A. (1989) Final preservation of mummies by gamma irradiation. In *The Mummies From Qilakitsoq-Eskimos in the 15th Century*, Series: Man and Society, ed. J.P.H. Hansen & H.C. Gulov, Vol. 12, pp. 134–136. Copenhagen: Meddelelser om Gronland.

Johnson, E.C., Johnson, G.R. & Williams, M.J. (1990) The origin and history of embalming. In *Embalming: History, Theory and Practice,* ed. R.G. Mayer & G.S. Bigelow, pp. 419–478. Norwalk, CT: Appleton & Lange.

Johnson, P.H., Olson, C.B. & Goodman, M. (1981) Isolation and characterization of deoxyribonucleic acid from the tissue of a woolly mammoth. *Paleopathology Newsletter* **34**(Suppl.): D5 (Abstract).

Johnston, W.D. (1993) Tuberculosis. In *The Cambridge World History of Human Disease*, ed. K.F. Kiple, pp. 1059–1068. New York: Cambridge University Press.

Jolliffe, D.M. (1993) A history of the use of arsenicals in man. *Journal of the Royal Society of Medicine* **86**:287–289.

Josephus, F. (ca. AD80/1970) *Loeb Classical Library*. Cambridge, MA: Harvard University Press.

Jucker, H. (1984) Book review of *Ritratti di Mummie* [Mummy portraits] by Klaus Parlasca (book in Italian, review in German). *Gnomon (Kritische Zeitschrift fuer die Gesamte Klassische Altertumwissenschaft)* **56**:542–547.

Junker, H. (1914) The Austrian Excavation, 1914. Excavations of the Vienna Imperial Academy of Sciences at the Pyramids of Gizah. *Journal of Egyptian Archaeology* **1**:250–253.

Junker, H. (1944) Giza VII., Gragungen auf dem Friedhof des alten Reiches bei den Pyramides von Giza [Giza VII, Graves of the Old Kingdom cemetery at the Giza Pyramids]. *Akademie der Wissenschaften in Wien Denkschriften* [Vienna Academy of Sciences Memoranda] **72**:113–116 and plates 24 and 25. Vienna and Leipzig: Hoelder-Pichler-Ternsky.

Kamberi, D. (1994) The three-thousand-year-old Charchan Man preserved at Zaghunuq. *Silo-Platonic Papers* **44**:1–15.

Kaner, S. (1996) Danish Bronze Age log coffin burials. In *Tombs, Graves and Mummies,* ed. P.G. Bahn, pp. 96–99. London: Weidenfeld & Nicolson.

Karasch, M.C. (1993) Ophthalmia (conjunctivitis and trachoma). In *The Cambridge World History of Human Disease,* ed. K.F. Kiple, pp. 897–906. New York: Cambridge University Press.

Karpowicz, A. (1981) Ageing and deterioration of proteinaceous media. *Studies in Conservation* **26**:153–160.

Kaufmann, B. (1996a) Mummification in the Middle Ages.

In *Human Mummies: A Global Survey of Their Status and the Techniques of Conservation,* ed. K. Spindler, H. Wilfing, E. Rastbichler-Zissernig, D. zur Nedden & H. Nothdurfter, pp. 231–238. New York: Springer-Verlag.

Kaufmann, B. (1996b) The corpse from the Porchabella-glacier in the Grisons, Switzerland (Community of Berguen). In *Human Mummies: A Global Survey of Their Status and the Techniques of Conservation,* ed. K. Spindler, H. Wilfing, E. Rastbichler-Zissernig, D. zur Nedden & H. Nothdurfter, pp. 239–246. New York: Springer-Verlag.

Kauffmann-Doig, F. (1996) Los Andes Amazonicos y su pasada arqueológico [The Amazonian Andes and its archaeological history]. *Politica Internacional* **46**:113–143.

Kauffmann-Doig, F. (1997) Los mausoleos de la Laguna de las Momias [The mausolea of the Lagoon of the Mummies Arkinka]. **24**:93–112 .

Kauffmann-Doig, F. (1998a) Primera Expedición arqueológica a los mausoleos Chachapoyas de La Laguna de Las Momias (The first archaeological expedition to the mausolea of the Chachapoyas of the Laguna de los Momias). Manuscript read at the 3rd World Congress on Mummy Studies held in Arica, Chile, May 14–18.

Kauffmann-Doig, F. (1998b) *Ancestors of the Incas*, transl. E. Guzman. Memphis, TN: Wonders.

Kauffmann-Doig, F. (1998c) *Monograph of the Ligabue Research and Study Centre.* Lima: Instituto de Arqueología Amazonica.

Kaup, Y., Baumer, U., Koller, J., Hedges, R.E.M., Werner, H., Hartmann, H.J., Etspueler, H. & Weser, U. (1994) Zn_2Mg alkaline phosphatase in an early Ptolemaic mummy. *Zeitschrift fuer Naturforschung* **49c**:489–500.

Keilin, D. & Wang, Y.L. (1947) Stability of haemoglobin and of certain endoerythrocytic enzymes *in vitro. Biochemistry Journal* **41**:491–500.

Kellerman, G.D., Waterman, N.G. & Scharfenberger, L.F. (1976) Demonstration *in vitro* of postmortem bacterial transmigration. *American Journal of Clinical Pathology* **106**:911–915.

Kelley, M.A. & Smeenk, L.M. (1995) Cranial injuries found in Ancient Guanche remains of Tenerife. In *Proceedings of the First World Congress on Mummy Studies*, February, 1992, Vol. 1, pp. 201–208. Santa Cruz, Tenerife, Canary Islands: Archaeological and Ethnographical Museum of Tenerife.

Kelson, W. (1900) Cause of aural exostoses. *Journal of Laryngology, Rhinology and Otology* **15**:667–676.

Kemp, B. (1989) *Ancient Egypt: Anatomy of a Civilization.* New York: Rutledge.

Kennedy, G.E. (1986) The relationship between exostoses and cold water: A latitudinal analysis. *American Journal of Physical Anthropology* **71**:401–415.

Keuchenmeister, F. (1881) Ueber Lithopaedien (On lithopedion). *Archiv fuer Gynaekologie* (Berlin) **17**:153–252.

Kimberling, W.J., Kumar, S., Gabow, P.A., Kenyon, J.B., Connolly, C.J. & Somlo, S. (1993) Autosomal dominant polycystic kidney disease: Localization of the second gene to chromosome 4 a13-q23 *Genomics* **18**:467–472.

King, G. (1954) Advanced extrauterine pregnancy. *American Journal of Obstetrics and Gynecology* **67**:712–740.

Klaatsch, H. (1905) Mumie aus Australien [An Australian Mummy]. *Zeitschrift fuer Ethnologie* **5**:772–781.

Kleiss, E. (1967) Zum Problem der natuerlichen Mumifikation und Konservierung [About the problem of natural mummification and conservation]. *Zeitschrift fuer Morphologie und Anthropologie* **2**:204–213.

Kleiss, E. (1975) La momificación natural y artificial [Mummification, natural and artificial]. *Archivas de Anatomía e Antropología* **1**:37–56.

Kleiss, E. (1977) Some examples of natural mummies. *Paleopathology Newsletter* **20**:5–6.

Kleiss, E. (1995) Momias naturales en tumbas y catacumbas (Natural mummies in tombs and catacombs). In *Proceedings of the First World Congress on Mummy Studies*, February, 1992, Vol. 2, pp. 829–836. Santa Cruz, Tenerife, Canary Islands: Archaeological and Ethnographical Museum of Tenerife.

Kleiss, E. (1998) Shrunken heads. In *Mummies, Disease and Ancient Cultures*, ed. A. Cockburn, E. Cockburn & T. Reyman, pp. 210–215. Cambridge: Cambridge University Press.

Kleiss, E. & Simonsberger, P. (1984) *La Parafinización como Metodo Morfologico* [Paraffinization as a morphologic method]. Merida, Venezuela: Universidad de los Andes.

Klenck, J.D. (1995) Bedouin animal sacrifice practices: Case study in Israel. In *The Symbolic Role of Animals in Archaeology,* ed. K. Ryan & P.J. Crabtree, Vol. 5, pp. 57–72. Philadelphia: MASCA, University of Pennsylvania Museum of Archaeology and Anthropology.

Klohn, M., Susini, A., Baud, C.A., Sahni, M. & Simon, C. (1988) Taphonomy of an ancient Kerma (Sudan) burial: A biophysical and biochemical study. *Rivista di Antropologia* **66**(Suppl.):21–34.

Knudson, R.J. (1993) Emphysema. In *The Cambridge World History of Human Disease,* ed. K.F. Kiple, pp. 706–708. New York: Cambridge University Press.

Kobler, J. (1960) *The Reluctant Surgeon: A Biography of John Hunter*, pp. 325–327. Garden City, NY: Doubleday,

Koenig, R. (2001) Unwrapping a modern mummy mystery. *Science* **292**:2417.

Koenig, W. (1896) *14 photographien mit Roentgen-Strahlen.* Leipzig: Verlag von Johann Ambrosius Barth.

Kolata, A.L. & Ortloff, C.R. (1996) Tiwanaku raised-field agriculture in the Lake Titicaca basin of Bolivia. In *Tiwanaku and its Hiterland,* ed. A.L. Kolata, pp. 109–152. Washington, DC: Smithsonian Institution Press.

Koller, J., Baumer, U., Daup, Y., Etspuler, H. & Weser, U. (1998) Embalming was used in Old Kingdom. *Nature* **391**:343–344.

Kolman, C., Centurion-Lara, A., Lukehart, S.A., Owsley, D.W. & Tuross, N. (1999) Identification of *Treponema pallidum* subspecies *pallidum* in a 200-year-old skeletal specimen. *Journal of Infectious Diseases* **180**:2060–2063.

Kolta, K.S. (1995) On sense and nonsense of unwrapping mummies. In *Proceedings of the First World Congress on Mummy Studies*, February, 1992, Vol. 2, pp. 523–533. Santa Cruz, Tenerife, Canary Islands: Archaeological and Ethnographical Museum of Tenerife.

Kowal, W.A., Krahn, P.M. & Beattie, O.B. (1989) Lead levels in human tissues from the Franklin Forensic Project. *International Journal of Environmental and Analytical Chemistry* **35**:119–126.

Kowal, W.A., Beattie, O.B., Baadsgaard, H. & Krahn, P. (1991) Source identification of lead found in tissues of sailors from the Franklin Arctic expedition of 1845. *Journal of Archaeological Science* **18**:193–203.

Kramar, C., Baud, C.A. & Lagier, R. (1983) Presumed calcified leiomyoma of uterus: morphological and clinical studies of a calcified mass dating from the Neolithic period. *Archives of Pathology and Laboratory Medicine* **107**:91–103.

Kramer, J. (1996) Mammoth pathologies and megafauna predation: Evidence for opportunistic predation of the Hebior mammoth. (Abstract). Manuscript read at the 6th North American Paleontology Convention (NAPC VI) held in Washington, DC in June.

Krings, M., Stone, A., Schmitz, R.W., Krainitzki, H., Stoneking, M. & Pääbo, S. (1997) Neandertal DNA sequences and the origin of modern humans. *Cell* **90**:19–30.

Kroeber, A.L. (1926) Culture stratifications in Peru. *American Anthropologist* **28**:331–351.

Kroeber, A.L. & Collier, D.C. (1960) Untitled manuscript on file. Field Museum of Natural History, Chicago.

Krogman, W.M. (1940) The skeletal and dental pathology of an early Iranian site. *Bulletin of the History of Medicine* **8**:28–35.

Krogman, W.M. & İşcan, M.Y. (1986) *The Human Skeleton in Forensic Medicine.* Springfield, IL: Charles Thomas.

Kunerth, J. (1985) It takes magic to hit a scientific jackpot. *The Orlando Sentinel*, January 2, p. D1.

Kuo-Liang, Y., Bing-Sen, Q., You-E, Z., Zhen-Yuan, H., Ming-Yan, C., Gen-Sheng, W., Zu-Yi, L., Zeng-Shou, C., Guo-Fang, L. & Wen-Ying, C. (1982) Skin changes of 2100 year old Changsha female corpse. *Chinese Medical Journal* **93**:765–776.

Lamb, D. (1901) Mummification, especially of the brain. *American Anthropologist* **3**:294–307.

Lambert, L.T., Cox, K., Mitchell, R., Rosella-Mora, C. del Cueto, Dodge, D. & Cano, R. *et al.* (1998) Isolation and characterization of a novel organism, *Staphylococcus succinus* sp. from 25–35 million year old Dominican amber. *International Journal of Systematic Bacteriology* **48**:511–515.

Lamendin, H. (1974) Observation with SEM of rehydrated mummy teeth. *Journal of Human Evolution* **3**:271–274.

Langsjoen, O. (1996) Dental effects of diet and coca leaf-chewing in two prehistoric cultures in northern Chile. *American Journal of Physical Anthropology* **101**:475–489.

Lantis, M. (1947) Aleutian and Alaskan Eskimo Ceremonialism and Burial Rites. In *Monographs of the American Ethnological Society*, ed. M.W. Smith, pp. 1-20. Seattle: University of Washington Press.

Lasky, I.J. (1982) Autopsy of Admiral John Paul Jones 113 years postmortem. *New York State Journal of Medicine* **87**:1110–1115.

Laughlin, W.S. (1963) Eskimos and Aleuts: Their origins and evolution. *Science* **142**:633–645.

Laughlin, W.S. (1980a) Aleut mummies. *Alaska Geographic Society* **7**(3):90–91.

Laughlin, W.S. (1980b) *Aleuts: Survivors of the Bering Land Bridge.* New York: Holt, Rinehart & Winston.

Laughlin, W.S., Harper, A.B. & Thompson, D.D. (1979) New approaches to the pre-and post-contact history of arctic peoples. *American Journal of Physical Anthropology* **51**:579–588.

Lausarot, P.M., Ambrosino, C., Favro, F., Conti, A. & Rabino-Massa, E. (1972) Preservation and amino acid composition of Egyptian mummy structure proteins. *Journal of Human Evolution* **1**:489–499.

Laves, W. (1950) Postmortale Veraenderungen des menschlichen Koerpers [Postmortem changes in the human body]. In *Lehrbuch der Gerichtlichen Medizin*, ed. A. Ponsold, pp. 117–135. Stuttgart: Georg Thieme Verlag.

Laytner, R. (1975) Strange curse of the subway mummy. *The Toronto Star*, November 25.

Leek, F.F. (1969) The problem of brain removal during embalming by the ancient Egyptians. *Journal of Egyptian Archaeology* **55**:112–116.

Leek, F.F. (1986) Cheops' courtiers: their skeletal remains. In *Science and Egyptology*, ed. R.A. David, pp. 183–199. Manchester: Manchester University Press.

Leeson, T.S. (1959) Electron microscopy of mummified material. *Stain Technology* **34**:317–320.

Leonard, W. & Loat, S. (1914) Ibis cemetery at Abydos. *Journal of Egyptian Archaeology* **1**:40–42.

Lerche, P. (2000) Quest for the lost tombs of the Peruvian Cloud People. *National Geographic* **198**:64–81.

LeRoy, C. (1972) The Aegean world. In *Larousse Encyclopedia of Archaeology*, ed. G. Charles-Picard, transl. A. Ward, pp. 257–265. Secaucus, NJ: Chartwell Books.

Levy, J.E. (1995) Animals: Bronze Age Scandinavia and Ohio Hopewell. In *The Symbolic Role of Animals in Archaeology*, ed. K. Ryan & P.J. Crabtree, MASCA Research Papers in Science and Archaeology, Vol. 5, pp. 9–19. Philadelphia: University of Pennsylvania Museum of Archaeology and Anthropology.

Levy, M. (1965) *The Moons of Paradise.* New York: Citadel Press.

Levey, M. (transl.) (1966) *The Medical Formulary or Aqrabadhin of Al-Kindi.* Madison, WI: University of Wisconsin Press.

Lewin, P.K. (1968) The ultrastructure on mummified skin cells. *Canadian Medical Association Journal* **98**:1011–1012.

Lewin, P.K. & Hardwood-Nash, D.C. (1977) X-ray computed axial tomography of an ancient Egyptian brain. *IRCS Medical Science* **5**:78.

Lewis, M., Roberts, C.A. & Manchester, K. (1994) Maxillary sinusitis in paleopathology: Its character and prevalence in urban and rural British medieval populations. Manuscript read at the 10th Biennial European meeting of the Paleopathology Association held in Goettingen, Germany, 29 August–3 September.

Lhuillier, A.R. (1953) The mystery of the Temple of the Inscriptions. *Archaeology* **6**:3–11.

Li, H.C., Fujiyoshi, T., Lou, H., Yashiki, S., Sonoda, S., Cartier, L. *et al.* (1999) The presence of ancient human T-cell lymphotropic virus type I provirus DNA in an Andean mummy. *Nature Medicine* **5**:1428–1432.

Lichtenberg, R. (1994) La momification en Égypte à l'époque tardive [Mummification in Egypt during the Late Period]. In *Aufstieg und Niedergang der roemischen Welt: Bb Geschichte und Kultur Roms im Spiegel der neueren Forschung* [The Rise and Fall of the Roman World as Reflected in Recent Research], ed. H. Temporini & W. Haase, Vol. 37.3, Part 2, pp. 2741–2760. Berlin: Walther de Gruyter.

Lieberman, A. & Bradbury, R. (1978) *The Guanajuato Mummies.* New York: Henry Adams.

Lindahl, T. (1993) Instability and decay of the primary structure of DNA. *Nature* **362**:709–715.

Lindahl, T. (1997) Facts and artifacts of ancient DNA. *Cell* **90**:1–3.

Lindenfelser, L.A. (1967) Antimicrobial activity of propolis. *American Bee Journal* **107**:90–92 and 130–131.

Linne, S. (1929) Darien in the past: The archaeology of Eastern Panama and northwestern Colombia. In *Kungl Vetenskaps-och Vitterhets-Samhalles Handlingar* Ser. A, Band 1, no. 3, pp. 212–318. Goeteborg: Elanderes Boktryckeri Aktiebolag.

Lippert, A. & Spindler, K. (1995) The discovery of a Late Neolithic Glacier Mummy at the Hauslabjoch in the Oetstal Alps. In *Proceedings of the First World Congress on Mummy Studies*, February, 1992, pp. 691–699. Santa Cruz, Tenerife, Canary Islands: Archaeological and Ethnographical Museum of Tenerife.

Lithell, U. (1981) Breast-feeding habits and their relation to infant mortality and marital fertility. *Journal of Family History* **6**:182–194.

Llagostera, A., Torres, C.M. & Costa, M.A. (1988) El complejo psicotropico en Solcor-3 (San Pedro de Atacama)

[The psychotropic complex in Solcor-3 (San Pedro de Atacama]. *Estudios Atacamenos* **9**:61–98.

Locksley, R. M. (1991) Leishmaniasis. In *Harrison's Principles of Internal Medicine,* ed. J.D. Wilson, E. Braunwald, K.J. Isselbacher, R.G. Petersdorf, J.B. Martin, A.S. Fauci, *et al.*, 12th edn, pp. 789–791. New York: McGraw-Hill.

Logan, G.A., Collins, M.J. & Eglinton, G. (1991) Preservation of organic biomolecules. In *Taphonomy: Releasing the Data Locked in the Fossil Record*, ed. P.A. Allison & D.G.E. Briggs, pp. 1–18. Topics in Geobiology Series. Gainesville, FL: University of Florida.

Lombardi, G.P. (1995) Detección de *Mycobacterium tuberculosis* en una momia Nasca con mal de Pott [Detection of *Mycobacterium tuberculosis* in a Nasca mummy with Pott's disease]. Manuscript read at the 2nd World Congress on Mummy Studies in Cartagena, Colombia, February.

Lombardi, G.P. (1997) Evidence of Aleut (Unganan) Century mummy from the Aleutian Islands, at Tulane University. *Human Mosaic* **31**(2):1–13.

Lombardi, G.P. (1998) Anasazi mummification: A photographic approach. In 108 Papers on Paleopathology read at the 25th Annual Meeting of the Paleopathology Association. *Paleopathology Newsletter,* **108** (June) (Suppl.):8 (Abstract).

London Observer Service (1996) 137 sign up for mummification in Utah. *The Arizona Daily Star*. August 20.

Long, A.R. (1931) Cardiovascular renal disease: Report of a case 3000 years ago. *Archives of Pathology* **12**:92–96.

Long, E.R. (1965) *A History of Pathology.* New York: Dover Publications.

Loomis, C.C. (2000) *Weird and Tragic Shores*. New York: Alfred A. Knopf.

Lothrop, S. & Mahler, J. (1957) A Chancay-style grave at Zapallan, Peru. *Papers at the Peabody Museum of Archaeology and Ethnology*, **50**(1).Cambridge, MA: Harvard University Press.

Loud, L.L. & Harrington, M.R. (1929) *Lovelock Cave.* University of California Publications in American Archaeology and Ethnology, Vol. 25, pp. 29–32 and 167–169. Berkeley, CA: University of California Press.

Lowis, G.W. (1993) Epidemiology of puerperal fever: The contributions of Alexander Gordon. *Medical History* **37**:399–410.

Lubec, G., Weniger, M. & Anderson, R. (1994) Racemization and oxidation studies of hair protein in the *Homo tirolensis*. *FABEC Journal* **8**:1166–1169.

Lucas, A. (1908) A preliminary note on some preservative materials used by the ancient Egyptians in connection with embalming. *Cairo Scientific Journal* **2** April:133.

Lucas, A. (1926) *Ancient Egyptian Materials and Industries.* London: Edward Arnold Ltd.

Lucas, A. (1932) The use of natron in mummification. *Journal of Egyptian Archaeology* **18**:125–140.

Lucas, A. & Harris, J.R. (1962) *Ancient Egyptian Materials and Industries*, 4th edn. London: Edward Arnold Ltd.

Lucas, M.L., Aufderheide, A.C. & el-Najjar, M.Y. (2000) The identification and analysis of Jordan's first mummy. *American Journal of Physical Anthropology* Suppl. **30** (February) (Sixty-Ninth Annual Meeting Issue): 214 (Abstract).

Lukasewycz, A. (1992) The effect of lead, copper and arsenic on the postmortem histologic preservation of animal tissues. Unpublished manuscript on file in the Paleobiology Laboratory in the Department of Pathology at the University of Minnesota, Duluth School of Medicine in Duluth, MN, 55812.

Lumbreras, L.G. (1971) Towards a re-evaluation of Chavin. In *Dumbarton Oaks Conference on Chavín*, ed. E.Benson, pp. 1–28. Washington, DC: Dumbarton Oaks Research Library & Collection.

Lumbreras, L.G. (1974) The peoples and cultures of ancient Peru. In *The Moche Culture,* transl. B. Meggers, pp. 99–111. Washington, DC: Smithsonian Institution Press.

Luna, L. (ed.) (1968) *Manual of Histologic Staining Methods of the Armed Forces Institute of Pathology*, 3rd edn. New York: McGraw-Hill.

Lupton, C. (1986) Where are the mummies? *Lore* **36**(2):24–25.

Lupton, C. (1995) Contribution of the CT to the understanding of Egyptian mummification. In *Proceedings of the First World Congress*, February, 1992, Vol. 2, pp. 505–513. Santa Cruz, Tenerife, Canary Islands: Archaeological and Ethnographical Museum of Tenerife.

Lurker, M. (1974) *The Gods and Symbols of Ancient Egypt*. London: Thames & Hudson.

Lutz, R. & Lutz, G. (1995) *The Secret of the Desert*. Innsbruck: Golf Verlag.

Lyke, M.L. (1997) Of spirits, skeletons, science: Kennewick man in demand. *Seattle Post-Intelligencer*, August 28, p. 1.

Lyman, R.L. (1994) *Vertebrate Taphonomy*. Cambridge: Cambridge University Press.

Lyons, A.S. & Petrucelli, R.J. (1987) *Medicine: An Illustrated History*. New York: Harry N. Abrams, Abradale Press.

Maat, G.J.R. (1991) Ultrastructure of normal and pathological fossilized red blood cells compared with pseudopathological biological structures. *International Journal of Osteoarchaeology* **1**:209–214.

Maat, G.J.R. & Baig, M.S. (1990) Scanning electron microscopy of fossilized sickle cells. *International Journal of Anthropology* **5**:271–276.

Maat, G.J.R. & Baig, M.S. (1996) A survey of the analysis of human remains from Ikaros. *Medical Principles and Practice* **6**:142–151.

Macalister, A. (1894) Notes on Egyptian mummies. *Journal of the Anthropological Institute* **33**:115–126.

MacArthur, J.D. (2000) *The Pyramid and I*. On Website the Franklin Institute Online, 1 December 2000. Musing with Mac. http: //www.fi.edu/qa98/macindex.html.

MacCurdy, G.G. (1923) Human skeletal remains from the highlands of Peru. *American Journal of Physical Anthropology* **6**:217–294.

Mace, A.C. & Winlock, H.E. (1916) *The Tomb of Senebtisi at Lisht*. New York: Metropolitan Museum of Art.

Macfarlan, D. (1915) The oldest autopsy in history. *Hahneman Monthly* **50**:343–351.

Macias-López, M.M. (2000) Dismorfogenesis de hombro en una nina mommificada [Dysmorphogenesis of a man and a female child mummy]. *Chungará* **32**:89–96.

Macintosh, N.W.G. (1949) Crania in the Macleay Museum. *Proceedings of the Linnean Society of New South Wales* **74**:161–191.

Mackenzie, D.A. (1978) *Egyptian Myth and Legend*. New York: Bell Publishing Co.

Macleod, R.I., Wright, A.R., Mcdonald, J. & Eremin, K. (2000) Mummy 1911–2101. *Journal of the Royal College of Surgeons of Edinburgh* **45**:85–92.

Macon, K., Petrich, M., Roufs, T. & Aufderheide, A.C. (1983) The McLeod Museum mummy. Unpublished report on file in Paleobiology Laboratory, Department of Pathology, University of Minnesota Duluth, School of Medicine, Duluth, MN, 55812.

Madden, M., Salo, W.L., Streitz, J., Aufderheide, A.C., Fornaciari, G., Jaramillo, C. *et al.* (2001) Hybridization screening of very short PCR products for paleoepidemiological studies of Chagas' disease. *BioTechniques* **30**:102–109.

Maegraith, B.G., Crewe, W. & Jones, K. (1991) The ship and the eye. *Annals of Tropical Medicine and Parasitology* **85**:1–3.

Maekawa, S. & Valentin, N. (1996) Development of a prototype storage and display case for the royal mummies of the Egyptian Museum in Cairo. In *Human Mummies*, ed. K. Spindler, H.Wilfing, E. Rastbichler-Zissernig, D. zur Nedden & H. Nothdurfter, pp. 47–56. New York: Springer-Verlag.

Maekawa, S., Preusser, F. & Lambert, F. (1995) An hermetically sealed display and storage case for mummified objects in inert atmospheres. In *Proceedings of the First World Congress on Mummy Studies*, February, 1992, Vol. 1, pp. 213–220. Santa Cruz, Tenerife, Canary Islands: Archeological and Ethnographical Museum of Tenerife.

Mafart, B-Y (1994) Approche de la mortalité maternelle au Moyen Age en Provence [An approach to maternal mortality during Middle Ages in Provence]. *Actes des 6e Journales Anthropologiques, Dussier de Documentation Archéologique 17*. CNRS Editions, Paris. Cited and annotated in *Paleopathology Newsletter* **89** (January 1995):17.

Mainardis, G. (1976) *Venzone: Studi Geologi Sul Terreno*. Udine, Italy: Arti Grafiche Friulane.

Mair, V.H. (1995) Prehistoric caucasoid corpses of the Tarim Basin. *Journal of Indo-European Studies* **23**:281–307.

Mair, W. (1914) On the lipoids of ancient Egyptian brains. *Journal of Pathology and Bacteriology* **18**:179–184.

Majno, G. (1975) *The Healing Hand*. Cambridge, MA: Harvard University Press.

Malek, J. (1993) *The Cat in Ancient Egypt*. London: British Museum Press.

Malkin, H.H. (1993) *Out of the Mist*. Berkeley, CA: Vesalius Books.

Manchester, K. (1983) *The Archaeology of Disease*. Bradford: University of Bradford Press.

Manialawi, M., Meligy, R. & Bucaille, M. (1978) Endoscopic examination of Egyptian mummies. *Endoscopy* **10**:191–194.

Mann, R.W., Bass, W.M. & Meadows, L. (1990) Time since death and decomposition of the human body: Variables and observations in case and experimental field studies. *Journal of Forensic Sciences* **35**:103–111.

Mann, R.W., Farmer, B.B. & Verano, J.W. (1992) South American shrunken heads: Genuines and fakes. *Bioantropología* **2**:8–13.

Mant, A.K. & Furbank, R. (1957) Adipocere – a review. *Journal of Forensic Medicine* **4**:18–35.

Manzi, G., Sperduti, A. & Passarello, P. (1991) Behavior-induced auditory exostoses in imperial Roman society: Evidence from coeval, urban and rural communities near Rome. *American Journal of Physical Anthropology* **85**:253–260.

Maples, W.R., Gatliff, B.P., Ludena, H., Benfer, R. & Goza, W. (1989) The death and mortal remains of Francisco Pizarro. *Journal of Forensic Sciences* **34**:1021–1036.

Marchetti, A., Pelligrini, S., Bevilacqua, G. & Fornaciari, G. (1996) K-ras mutation of the tumour of Ferrante I of Aragon, King of Naples. *Lancet* **347**:1272.

Marchiafava, V., Bonucci, E. & Ascenzi, A. (1974) Fungal osteoclasia: A model of dead bone resorption. *Calcified Tissue Research* **14**:195–210.

Marennikova, S.S., Shellukhina, E.M., Zhukova, O.A., Yanova, N.N. & Loparev, V.N. (1990) Smallpox diagnosed 400 years later: Results of skin lesions examination of 16th century Italian mummy. *Journal of Hygiene, Epidemiology, Microbiology and Immunology* **34**:237–231.

Mariette, A. (1882) *Le Serapéum de Memphis* [The Serapium at Memphis]. Paris: Gide.

Marin, V.T.W. & Cappelletti, E. (1997) The embalming process of St. Gregorio Barbarigo and his survey in 1997]. *Paleopathology Newsletter* **100**:5-7.

Márquez Morfín, L. & González-Crespo, N. (1985) *Las Momias de la Iglesia de Santa Elena, Yucatán* [The Mummies of the Church of Santa Elena, Yucatán]. Córdoba, Mexico: Instituto Nacional de Antropología e Historia.

Marsadolov, L.S. (1993) *Program of Sino-Altai Archaeological Expedition of the State Hermitage Museum*. St Petersburg: Hermitage Museum.

Martínez-Cortizas, A., Pontevedra-Pombal, X., García-Rodeja, E., Novoa-Muñoz, J.C. & Shotyk, W. (1999) Mercury in a Spanish peat bog: Archive of climate change and atmospheric metal deposition. *Science* **284**:939–942.

Masson, P.J. (1929) Trichrome stainings and their preliminary technique. *Journal of Technical Methods* **12**:75–90.

Masters, P.M. (1984) Age determination of an Alaskan mummy by amino acid racemization. *Arctic Anthropology* **21**:64–67.

Masters, P.M. (1986) Amino acid racemization – a review. In *Dating and Age Determination of Biological Materials*, ed. M.R. Zimmerman & J.L. Angel, pp. 39–58. London: Croom Helm.

Masters, P.M. (1987) Preferential preservation of noncollagenous protein during bone diagenesis: Implications for chronometric and stable isotopic measurements. *Geochimica et Cosmochimica* **51**:3209–3214.

Masters, P.M. & Zimmerman, M.R. (1978) Age determination of an Alaskan mummy: Morphological and biochemical correlation. *Science* **210**:811–812.

Masters, P.M., Bada, J.L. & Zigler, J.S. Jr (1978) Aspartic acid racemization in heavy molecular weight crystallins and water-insoluble protein from normal human lenses and cataracts. *Proceedings of the National Academy of Sciences, USA* **75**:1204–1208.

Matossian, M.K. (1989) *Poisons of the Past*. New Haven, CT: Yale University Press.

Matranga, J.P. (transl.) (1983) *The Capuchin Catacombs*. Museum Catalogue. Palermo: Capuchin Monastery of Palermo.

May, A.G. (1951) Mummies from Alaska. *Natural History* **60**:114–119.

Mayer, B.X., Reiter, C. & Bereuter, T.L. (1997) Investigation of the triacylglycerol composition of Iceman's mummified tissue by high-temperature gas chromatography. *Journal of Chromotography B* **692**:1–6.

Mayes, S. (1961) *The Great Belzoni*. New York: Walker & Co.

Mayne-Correia, P.M. (1997) Fire modification of bone: A review of the literature. In *Forensic Taphonomy*, ed. W.D. Haglund & M.H. Sorg, pp. 275–293. New York: CRC Press.

McConnel, U. (1957) *Myths of the Munkan*. Melbourne: Melbourne University Press.

McCraken, H., Wedel, W., Edgar, R., Moss, J.H., Wright, H.E., Husted, W. & Mulloy, W. (1978) *The Mummy Cave Project in Southwestern Wyoming*. Cody, WY: Buffalo Bill Historical Center.

McCreery, J.H. (1935) The Mummy Collection of the University of Cuzco El Palacio **39**:118–120.

McDonald, K. (1998) Researchers battle for access to a 9,300 year old skeleton. *The Chronicle of Higher Education* May 22, pp. A18–A22.

McEwan, G. (1987) The Middle Horizon in the Valley of Cuzco, Peru: The impact of the Wari occupation of Pikillacta in the Lucre Basin. *BAR International Series* **372**, pp. 1-203. Oxford: British Archaeological Reports.

McGee, M.B. & Coe, J.I. (1981) Postmortem wound dehiscence: A medicolegal masquerade. *Journal of Forensic Sciences* **26**:216–219.

McGee, W.J. (1894) The Remains of Don Francisco Pizarro. *American Anthropologist* **7**:1–25.

McHargue, G. (1972) Islands and outbacks. In *Mummies*, pp. 105–117. New York: Lippincott.

McKinley, L.M. (1977) Bodies preserved from the days of the Crusades in St Micham's Church. *Dublin Journal of Pathology* **122**:27–28 and Plate XIV.

McLaren, D. (1979) Nature's contraceptive. Wet-nursing and prolonged lactation: The case of Chesham, Buckinghamshire. *Medical History* **23**:426–441.

McManamon, F.P. (1999) Kennewick Man: The initial scientific examination, description, and analysis of the Kennewick Man human remains. Department of the Interior and National Park Service website: http: //www.cr.nps.gov/aad/kennewick/mcmanamon.htm

Meier, D.K. (1997) Conserving mummies: A problem for museums. *South American Explorer* **50**:15.

Meighan, C.W. (1980) Archaeology at Guatacondo, Chile. In *Prehistoric Trails of the Atacama: Archaeology of Northern Chile,* ed. C.W. Meighan & D.L. True, pp. 99–126. Los Angeles: University of California.

Mejanelle, P., Bleton, J., Goursaud, S. & Tschapla, A. (1997) Identification of phenolic acids and inositols in balms and tissues from an Egyptian mummy. *Journal of Chromatography* **767**:177–186.

Mekota, A.M., Zink, A., Esterhazy, M.G. & Nerlich, A.G. (2000) Determination of rehydration and staining conditions for the histological and immunohistochemical analysis of mummified soft tissues (poster). In papers on paleopathology presented at the 13th Biennial European Members Meeting held in Chieti, Italy. *Paleopathology Newsletter* **112** (Suppl.):20 (Abstract).

Melbye, J. & Stuart-Macadam, P. (1995) Two mummies from the northwest Coast of Canada. In *Proceedings of the First World Congress on Mummy Studies*, February, 1992, Vol. 2, pp. 701–706. Santa Cruz, Tenerife, Canary Islands: Archaeological and Ethnographical Museum of Tenerife.

Meldgaard, M. & Hart Hansen, J.P. (1998) Ancient mummified animal remains in Greenland. Manuscript read at the 3rd World Congress on Mummy Studies held in Arica, Chile, May 1998.

Meloy, H. (1971) *Mummies of Mammoth Cave.* Shelbyville, IN: Micron Publishing Co.

Meloy, H. & Watson, P.J. (1969) Human remains: "Little Alice" of Salts Cave and other mummies. In *Patty Jo Watson: The Prehistory of Salts Cave, Kentucky*. Reports of Investigations, no. 16. Springfield, IL: printed by the Authority of the State of Illinois.

Melville, B. (1995) Examination and conservation considerations of an unwrapped mummy in the National Museums of Scotland. In *Conservation in Ancient Egyptian Collections,* ed. C.E. Brown, F. Macalister & M.M. Wright, pp. 77–84. London: Archetype Publications.

Melvin, J.R. Jr, Cronholm, L.S., Simson, L.R. & Isaacs, A.M. (1984) Bacterial transmigration as an indicator of time of death. *Journal of Forensic Sciences* **29**:412–417.

Mendelsohn, S. (1944a) Preservation of human remains through natural agencies. *Ciba Symposia* **6**:1782–1789.

Mendelsohn, S. (1944b) Embalming from the Medieval Period to the present time. *Ciba Symposia* **6**:1805–1812.

Mendelsohn, S. (1952) Historical aspects of mummy in pharmacy and medicine. *American Journal of Pharmacy and the Sciences* **124**:53–63.

Merriwether, D.A. (1999) Freezer anthropology: New uses for old blood. *Philosophical Transactions of the Royal Society of London* **354**:121–130.

Merriwether, D.A., Rothhammer, F. & Ferrell, R.E. (1994) Genetic variation in the New World: Ancient teeth, bone and tissue as sources of DNA. *Experientia* **50**:592–601.

Merriwether, D.A., Rothhammer, F. & Ferrell, R.E. (1995) Distribution of the four-founding lineage haplotypes in Native Americans suggests a single wave of migration for the New World. *American Journal of Physical Anthropology* **98**:411–430.

Meryon, C.L. (1815/1987) Embalming a patriarch,

November 1815: An unusual task for a British physician in the Lebanon. In *Eyewitness to History*, ed. J. Carey, pp. 293–294. New York: Avon Books.

Metraux, A. & Kirchhoff, P. (1948) The Northeastern Extension of Andean Culture. *US Bureau of American Ethnology Bulletin* **143**:349–368.

Meyer, A.J.P. (1995) *Oceanic Art*, Vol. 1. *The Iatmul People*, pp. 220–234. New York: Knickerbocker Press.

Meyer, J. (1904) Ueber die biologische Untersuchung von Mumien-Materiel vermittelst der Praezipitinreaktion [Concerning the biological investigation of mummy tissue using the precipitin reaction]. *Muenchner Medizinische Wochenschrift* **15**:663–664.

Micozzi, M.S. (1982) Skeletal tuberculosis, pelvic contraction and parturition. *American Journal of Physical Anthropology* **458**:441–445.

Micozzi, M.S. (1986) Experimental study of postmortem change under field conditions: Effects of freezing, thawing and mechanical injury. *Journal of Forensic Sciences* **31**:953–961.

Micozzi, M.S. (1991) *Postmortem Change in Human and Animal Remains*. Springfield, IL: C.C. Thomas.

Middendorf, E.W. (ed.) (1894) *Peru*, (Tomo II). Berlin, republished 1973 by Lima: Universidad Nacional Mayor de San Marcos.

Mikhailov, M.I., Kuznetsov, S.V. & Zhdanov, V.M. (1984) Electron microscopy of the intestinal content of a mammoth. *Lancet* **2**:111–112.

Miller, M.F. II & Wyckoff, R.W.G. (1968) Proteins in dinosaur bones. *Proceedings of the National Academy of Sciences, USA* **60**:176–178.

Miller, P.M. (1812) Preservation of human bodies in a cave in Tennessee. *Medical Repository and Review of American Publications on Medicine, Surgery, and the Related Branches of Philosophy* **12**:147–149.

Miller, R.L., de Jonge, N., Krijger, F.W. & Deelder, A.M. (1993) Predynastic schistosomiasis. In *Biological Anthropology and the Study of Ancient Egypt*, ed. W.V. Davies & R. Walker, pp. 54–60. London: British Museum Press.

Miller, R.L., Ikram, S., Armelagos, G.J., Walker, R., Harer, W.B., Shiff, C.X.J. *et al.* (1994) Diagnosis of *Plasmodium falciparum* infections in mummies using the rapid ParaSight-F test. *Transactions of the Royal Society of Tropical Medicine and Hygiene* **88**:31–32.

Miller, S. (1998) Glimpsing the Ice Age. In *Federal Archaeology Program*, ed. D. Haas, pp. 40–41. Secretary of the Interior's Report to Congress 1994–1995.Washington, DC: US Department of the Interior.

Minelli, L.L. (1993) The Incan Mummy from the National Museum of Anthropology and Ethnology in Florence, Italy. *Anthropology and Ethnology Archives* **123**:725–728.

Moers, M.E.C. (1989) Occurrence and fate of carbohydrates in recent and ancient sediments from different environments of deposition. PhD thesis, Delft Technical University.

Møller-Christensen, V. (1958) *Bogen om Abelholt Kloster* [Rickets in Abelholt Cloister]. Copenhagen: Munksgaard.

Moore, D.R. (1984) *The Torres Straits Collections of A. C. Haddon: A Descriptive Catalogue.* London: British Museum Publications, Inc.

Moore, P.D. & Bellamy, D.J. (1974) *Peatlands*. New York: Springer-Verlag.

Monro, T.K. (1933) *The Physician as a Man of Letters, Science and Action* Glasgow: Jackson Wylie.

Monsalve, M.V., Groot de Restrpo, H., Espinel, A., Correal, G. & Devine, D.V. (1994) Evidence of mitochondrial DNA diversity in South America. *Annals of Human Genetics* **58**:265–273.

Monsalve, M.V., Cárdenas-Arroyo, F., Guhl, F., Delaney, A.E. & Devine, D.V. (1996) Phylogenetic analysis of mtDNA lineages in south American mummies. *Annals of Human Genetics* **60**:293–303.

Moodie, R.L. (1923) *Paleopathology*. Urbana, IL: University of Illinois Press.

Moodie, R.L. (1931) Roentgenologic studies of Egyptian and Peruvian mummies. *Anthropological Series* 3. Chicago: Field Museum of Natural History.

Moorhouse, G. (1997) *Sun Dancing.* New York: Harcourt Brace.

Moraga, M.L., Rocco, P., Miquel, J.F., Nervi, F., Llop, E., Chakraborty, R., Rothhammer, F. & Carvallo, P. (2000) Mitochondrial DNA polymorphisms in Chilean aboriginal populations: Implications for the peopling of the southern cone of the continent. *American Journal of Physical Anthropology* **113**:19–29.

Morgagni, G.B. (1682–1771) *De Sedibus et Causis Morborum* [On the Seats and Causes of Diseases]. Padova. 5 vols.

Mori, P. & Ascenzi, A. (1959) La mummia infantile di Uan Muhuggiag: Osservazioni antropologiche [An infant mummy from Uan Muhuggiag: Anthropological observations]. *Rivista di Antropologia* **46**:128–148.

Morimoto, I. (1988) Mummies of Ibaloy in Luzon, Philippines. *Anthropological Reports* **46**:89–107.

Morimoto, I. (1993) Buddhist mummies in Japan. (In Japanese but with a three-page English summary.) *Acta Anatomica Nipponica* **68**:381–98.

Morimoto, I. & Hirata, K. (1993) A Bolivian Mummy Eviscerated, Dating to 100 B.C. *St. Marianna Medical Journal* **21**:285–297.

Morris, A.G. (1981) Copper discolouration of bone and the incidence of copper artefacts with human burials in South Africa. *South African Archaeological Bulletin* **36**:36–42.

Morris, A.G. & Rogers, A.L. (1989) A probable case of prehistoric stone disease from the Northern Cape province, South Africa. *American Journal of Physical Anthropology* **79**:521–527.

Morris, C. (1988) Pan-Andean empire and the uses of documentary evidence. In *Peruvian Prehistory,* ed. R.W. Keatinge, pp. 233–236. Cambridge: Cambridge University Press.

Morrison, R.T. & Boyd, R.N. (1973) *Organic Chemistry*, 3rd edn. Boston: Allyn & Bacon, Third Edition.

Morrison-Scott, T.C.S. (1952) The mummified cats of ancient Egypt. *Proceedings of the Zoological Society of London* **121**:861–867.

Morse, D. (1961) Prehistoric tuberculosis in America. *American Review of Infectious Diseases* **83**:489–504.

Morton, S.G. (1839) *Crania Americana* [American Crania]. Philadelphia: J. Dobson.

Morton, S.G. (1844) Observations on Egyptian ethnography, derived from anatomy, history and the monuments. *Transactions of the American Philosophical Society* **9**:93–159.

Mostny, G. (1957) La momia del Cerro el Plomo [The mummy from El Plomo mountain]. *Boletin del Museo Nacional de Historia Natural* **28**:1–142.

Motter, M.G. (1898) A contribution to the study of the fauna of the grave. A study of one hundred and fifty disinterments, with some additional experimental observations. *New York Entomological Society* **6**:201–232.

Mould, R.F. (1996a) Les reliques authentiques. In *Mould's Medical Anecdotes*, pp. 16-18. Bristol and Philadelphia: Institute of Physics Publishing Co.

Mould, R.F. (1996b) Oliver Cromwell's head. In *Mould's Medical Anecdotes*, pp. 16–18. Bristol & Philadelphia: Institute of Physics Publishing Co.

Moulton, C.W. (1905) *A Biographical Cyclopedia of Medical History*. New York, Akron & Chicago: The Saalfield Publishing Co.

Mujais, S. (1999) The future of the realm: Medicine and divination in ancient Syro-Mesopotamia. *American Journal of Nephrology* **19**:133–139.

Munck, W. (1956) Patologisk Anatomisk og Retsmedicinsk Undersøgelse af Moseliget fra Grauballe [Pathological anatomy and forensic study results on Graubelle Man]. *Arbog fo Jysk Ariceologisk Selskae* [Annual Publication of the Archeological Society of Jutland], pp. 131–137.

Munizaga, J., Allison, M.J. & Parades, C. (1978) Cholelithiasis and cholecystitis in pre-Colombian Chileans. *American Journal of Physical Anthropology* **48**:209–212.

Munizaga, J., Allison, M.J., Gerszten, E. & Klurfeld, D.M. (1975) Pneumoconiosis in Chilean miners of the sixteenth century. *Bulletin of the New York Academy of Medicine* **51**:1281–1293.

Muñoz-Ovalle, I., Arriaza-Torres, B. & Aufderheide, A.C. (1993) *Acha-2 y los Origenes Del Poblamiento Humano en Arica* [Acha-2 and Human Origins in Arica]. Arica, Chile: University of Tarapaca.

Murdock, G.P. (1934) *Our Primitive Contemporaries*. New York, Macmillan.

Murray, J.A.H., Bradley, H., Craigie, W.A., Onions, C.T. & Burchfield, R.W. (1989) *The Oxford English Dictionary*, 2nd edn. Oxford: Clarendon Press.

Murray, M.A. (1910) *The Tomb of Two Brothers*. Manchester: Sherratt & Hughes.

Naim, A. & Fahmy, K. (1965) Vaginal fistulae complicating salt atresia in Arabia. *Alexandria Medical Journal* **11**:218–226.

Napton, L.K. (1997) The Spirit Cave mummy. *Nevada* **40**(1):97–104.

Nedden, D. zur, Wicke, K., Knapp, R., Seidler, H., Wilfing, H., Weber, G. *et al.* (1994) New findings on the Tyrolean "Ice Man": Archaeological and CT-body analysis suggest personal disaster before death. *Journal of Archaeological Science* **21**:809–818.

Needham, J. (1974) *Science and Civilisation in China*, Vol. 5:

Chemistry and Chemical Technology. Cambridge: Cambridge University Press.

Neel, J.V. (1982) The thrifty gene revisited. In *The Genetics of Diabetes*, ed. J. Kobberling & Tattersall, pp. 283–293. New York: Academic Press.

Neilson, W.A., Knott, T.A. & Carharet, P.W. (1939) *Webster's New International Dictionary of the English Language*, 2nd edn. Unabridged. Springfield, Massachusetts: G. & C. Merriam Co.

Nerlich, A.G., Parsche, F., Driesch, A. von den & Lohrs, U. (1993) Osteopathological findings in mummified baboons from ancient Egypt. *International Journal of Osteoarchaeology* **3**:189–198.

Nerlich, A.G., Haas, C.J., Zink, A., Szeimes, U. & Hagedorn, H.G. (1997) Molecular evidence for tuberculosis in an ancient Egyptian mummy. *Lancet* **350**:1404.

Neira-Avedaño, M. & Coelho, V.P. (1972) Enterramientos de cabezas de la cultura Nasca [Burials of Nasca culture heads]. *Revista do Museu Paulista* n.s. **20**:109–142.

Neumann, G.K. (1938) The human remains from Mammoth Cave, Kentucky. *American Antiquity* **4**:339–353.

Nials, F.L., Deeds, E.E., Moseley, M.E., Pozorski, S.G., Pozorski, T.G. & Feldman, R. (1979) El Niño: The catastrophic flooding of coastal Peru. Part II. *Field Museum of Natural History Bulletin* **50**(8):4–10.

Nicholson, P.T. (1994) Preliminary report on work at the sacred animal necropolis, North Saqqara, 1992. *Journal of Egyptian Archaeology* **80**:1–10.

Nicholson, P.T. & Shaw, I. (eds.) (2000) *Ancient Egyptian Materials and Technology*. Cambridge: Cambridge University Press.

Nield, T. (1986) An Iron Age murder mystery. *The Sciences* 4:4–6.

Nielsen, H. & Thuesen, I. (1998) Paleogenetics. In *Mummies, Disease and Ancient Cultures*, 2nd edn, ed. A. Cockburn, E. Cockburn & T.A. Reyman, pp. 355–362. Cambridge: Cambridge University Press.

Nissenbaum, A. (1992) Molecular archaeology: Organic geochemistry of Egyptian mummies. *Archaeology* **19**:1–6.

Nissenbaum, A. (1999) Ancient and modern medicinal applications of Dead Sea asphalt (bitumen). *Israel Journal of Earth Sciences* **48**:301–308.

Noro M., Masuda, R., Dubrovo, I.A., Yoshida, M.C. & Kato M. (1998) Molecular phylogenetic inference of the woolly mammoth *Mammuthus primigenius*, based on complete sequences of mitochondrial cytochrome *b* and 12 S ribosomal RNA genes. *Journal of Molecular Evolution* **46**:314–326.

Notman, D.N.H. (1983) Use of nuclear magnetic resonance imaging of archeological specimens. *Paleopathology Newsletter* **43**:9–12.

Notman, D.N.H. (1986) Ancient scannings: Computed tomography of Egyptian mummies. In *Science in Egyptology*, ed. R.A. David, pp. 251–320. Manchester: University of Manchester Press.

Notman, D.N.H. (1995) Paleoradiology and human mummies: A basic guide to imaging the past. In *Proceedings of the First World Congress on Mummy Studies*, February, 1992, Vol. 2, pp. 587–592. Santa Cruz, Tenerife, Canary Islands: Archaeological and Ethnographical Museum of Tenerife.

Notman, D.N.H. & Aufderheide, A.C. (1995) Experimental mummification and computed imaging. In *Proceedings of the First World Congress on Mummy Studies*, February, 1992, Vol. 2, pp. 821–828. Santa Cruz, Tenerife, Canary Islands: Archaeological and Ethnographical Museum of Tenerife.

Notman, D.N.H. & Beattie, O. (1995) The paleoimaging and forensic anthropology of frozen sailors from the Franklin Arctic expedition mass disaster (1845–1848): A detailed presentation of two radiological surveys. In *Human Mummies: A Global Survey of Their Status and the Techniques of Conservation*, ed. K. Spindler, H. Wilfing, E. Rastbichler-Zissernig, D. zur Nedden & H. Nothdurfter, pp. 3–8. New York: Springer-Verlag.

Notman, D.N.H. & Lupton, C. (1995) Three-dimensional computed tomography and densitometry of human mummies and associated materials. In *Proceedings of the First World Congress on Mummy Studies*, February, 1992, Vol. 2, pp. 479–484. Santa Cruz, Tenerife, Canary Islands: Archaeological and Ethnographical Museum of Tenerife.

Notman, D.N.H., Tashijian, J., Aufderheide, A.C., Cass, O.W., Shane, O.C. III, Berquist, D.H. *et al.* (1986) Modern imaging and endoscopic biopsy techniques in Egyptian mummies. *American Journal of Roentgenology* **146**:93–96.

Notman, D.N.H., Anderson, L., Beattie, O.B. & Amy, R. (1987) Arctic paleoradiology: Portable radiographic examination of two frozen sailors from the Franklin

Expedition (1845–1848). *American Journal of Roentgenology* **149**:347–350.

Nriago, J.O., Pfeiffer, W.C., Malm, O., de Souza, C.M.M. & Mierle, G. (1992) Mercury pollution in Brazil. *Nature* **356**:389.

Nuorteva, P. (1977) Sarcosaprophagus insects as forensic indicators. In *Forensic Medicine*, ed. C.G. Tedeschi, Vol. II, W.G. Eckert & L.G. Tedeschi, pp. 1072–1095. Philadelphia: Saunders.

Oakley, K.P. (1960) Ancient preserved brains. *Man* (London) **60**(122):90–91.

O'Floinn, R. (1988) Irish bog bodies. *Archaeology Ireland* **2**(3):94–97.

O'Floinn, R. (1995) Ireland. In *Bog Bodies*, ed. R.C. Turner & R.G. Scaife, pp. 221–234. London: British Museum Press.

Oliveros, A. (1990) *Las Momias de Tlayapacan*. Mexico City: Instituto Nacional de Antropología y Historia.

Omar, S., McCord, M. & Daniels, V. (1989) The conservation of bog bodies by freeze-drying. *Studies in Conservation* **34**:101–109.

Orchiston, D.W. (1967) Preserved human heads of the New Zealand Maoris. *Journal of the Polynesian Society* **76**:297–329.

Orchiston, D.W. (1968) The practice of mummification among the New Zealand Maori. *Journal of the Polynesian Society* **77**:186–190.

Oriel, J.D. (1973a) Anal and genital warts in the ancient world. *Paleopathology Newsletter* **3**:5–7.

Oriel, J.D. (1973b) Gonorrhea in the ancient world. *Paleopathology Newsletter* **4**:7–9.

Ortiz de Montellano, B.R. (1990) *Aztec Medicine, Health and Nutrition*. New Brunswick: Rutgers University Press.

O'Shea, J.M. (1984) *Mortuary Variability*. New York: Academic Press.

Pääbo, S. (1985) Preservation of DNA in ancient Egyptian mummies. *Journal of Archaeological Science* **12**:411–417.

Pääbo, S. (1989) Ancient DNA: Extraction, characterization, molecular cloning and enzymatic amplification. *Proceedings of the National Academy of Sciences, USA* **86**:1939–1943.

Pääbo, S. & Wilson, A.C. (1991) Miocene DNA sequences – a dream come true? *Molecular Evolution* **1**:45–46.

Pääbo, S., Gifford, J.A. & Wilson, A.C. (1988) Mitochondrial DNA sequences from a 7000 year-old brain. *Nucleic Acid Research* **16**:9775–9787.

Pabst, M.A. & Hofer, F. (1998) Deposits of different origin in the lungs of the 5300-year-old Tyrolean Iceman. *American Journal of Physical Anthropology* **107**:1–12.

Paddock, F.K., Loomis, C.C. & Perkons, A.K. (1970) An inquest on the death of Charles Francis Hall. *New England Journal of Medicine* **282**:784–786.

Page, H. (1897) Post-mortem artificially contracted Indian Heads. *Journal of Anatomy and Physiology* **31**:252–261.

Pahl, W.M. (1986) Radiography of an Egyptian "cat mummy." An example of the decadence of the animal worship in the Late dynasties? *OSSA* **12**:133–140.

Pahl, W.M. & Parsche, F. (1991) Rawtselfafte Befunde and anthropologischem Untersuchungs-material aus Aegypten. Addenda zu Herodouts "Historien", Lib. II, 86–88 und zum aegyptischen "Sparagmos?" [Enigmatic findings during Egyptian anthropological research. Addendum to Herodotus Book II:86–88 and to the Egyptian "Sparagmos"]. *Anthropologie Anzeuger* **49**(1/2):39–48.

Pahl, W.M. & Undeutsch, W. (1988) Differential diagnosis of facial skin ulcerations in an Egyptian mummy. *Medezinhistorisches Journal* **33**:152–165.

Pahl, W.M., Parsche, F. & Ziegelmayer, G. (1988) Innovationen in der Computertomographie (CT): Software-Routinen und ihre Relevanz fuer Mumienforschung und Anthropologie [Innovations in computed tomography (CT): Software routines and their relevance for mummy research and anthropology]. *Anthropologie Anzeiger* **46**(1):17–25.

Pahor, A.L. (1992) Ear, nose and throat in Ancient Egypt. Part III. *Journal of Laryngology and Otology* **106**:863–873.

Pahor, A. & Kimura, A. (1990) History of removal of nasal polyps. *A Folha Medica* **102**:183–186.

Paine, C. (1992) Standards in the museum care of biological collections. *Series: Standards in the Museum Care of Collections no. 2.* London: Museums and Galleries Commission.

Painter, T.J. (1991) Lindow Man, Tollund Man and other peat-bog bodies: The preservative and antimicrobial action of sphagnan, a reactive glycuronoglycan with tanning and sequestering properties. *Carbohydrate Polymers* **15**:123–142.

Painter, T.J. (1994) Preservation in Peat. *News Warp: Newsletter of Wetland Archaeology Research Project* **15**:30–376.

Painter, T.J. (1995) Chemical and microbiological aspects of the preservation process in sphagnum peat. In *Bog Bodies*, ed. R.C. Turner & R.G. Scaife, pp. 88–99. London: British Museum Press.

Palmer, G. (1998) The Xinjiang mummy in Shanghai Museum. Food for tot. *Newsletter of the History of Infant Feeding Association* **10**(3):2–5, Spa Road, Gloucester, England, GL1 1UY.

Pankhurst, R. (1990) The use of thermal baths in the treatment of skin disease in old-time Ethiopia. *International Journal of Dermatology* **29**:451–456.

Parker, R.B. & Toots, H. (1970) Minor elements in fossil bone. *Geological Society of America Bulletin* **81**:925–932.

Parkinson, R.B. (1991) *Voices From Ancient Egypt*. Norman, OK: University of Oklahoma Press.

Paterakis, A.B. (1996) Conservation: Preservation versus analysis? In *Archaeological Conservation and Its Consequences*, ed. A. Roy & P. Smith, pp. 143–148. London: International Institute for Conservation of Historic and Artistic Works.

Paton, D. (1925) *Animals of Ancient Egypt*. Princeton, NJ: Princeton University Press.

Patten, R.C. (1981) Aflatoxins and disease. *American Journal of Tropical Medicine and Hygiene* **30**:422–425.

Paul, A. (1991) Paracas: An ancient cultural tradition on the South Coast of Peru. In *Paracas: Art and Architecture*, ed. A. Paul, pp. 1–34. Iowa City: University of Iowa Press.

Paulshock, B.Z. (1980) Tutankhamun and his brothers. *Journal of the American Medical Society* **244**:160–164.

Payne, J.A., King, E.W. & Beinhart, G. (1968) Arthropod succession and decomposition of buried pigs. *Nature* **219**:1180.

Peacock, G. (1855) *Life of Thomas Young*. London: John Murray.

Pearce, S. (1990) *Archaeological Curatorship*. Washington, DC: Smithsonian Institution Press.

Pearlstein, E. (1995) Conservation for the New Egyptian Galleries at the Brooklyn Museum. In *Conservation in Ancient Egyptian Collections*, ed. C.E. Brown, F. Macalister & M.M. Wright, pp. 93–102. London: Archetype Publications.

Peck, W. & Ross, J.G. (1978) *Egyptian Drawings*. New York: Dutton.

Peng, L.X. (1995) Study of an ancient cadaver excavated from a Han Dynasty (207 B.C.–A.D. 220) Tomb in Hunan Province. In *Proceedings of the First World Congress on Mummy Studies*, February, 1992, Vol. 2, pp. 853–856. Santa Cruz, Tenerife, Canary Islands: Archaeological and Ethnographical Museum of Tenerife.

Péréz-Martínez, C. (1960) Cranial deformations among the Guanes Indians of Colombia. *American Journal of Orthopedics* **46**:539–543.

Perotti, B., Barale, E., Costamagna, F., Fulcheri, E., Gerbore, R. & Masali, M. (1998) Environmental biological agents and ancient bone remains: Definition of preservation status (poster). In papers on paleopathology, presented at the Twelfth Biennial European Members Meeting of the Paleopathology Association held in Prague and Pilsen, Czech Republic, August 26–29. *Paleopathology Newsletter* **104** (December) (Suppl.): 6.

Persoon, C.H. (1822) *Mycologia Europaea*. Vol. 1, p. 65. Erlangae: Inpensibus I.I. Palmi.

Petrie, W.M.F. (1898) *Deshashesh 1897: Fifteenth Memoir of the Egypt Exploration Fund*. London: Egypt Exploration Fund.

Pettigrew, T.J. (1834) *A History of Egyptian Mummies*. London: Longman, Rees, Omre, Brown, Green & Longman.

Péwé, T.L., Rivard, N.R. & Llano, G.A. (1959) Mummified seal carcasses in the McMurdo Sound Region. *Antarctica Science* **130**:716.

Pezzia, A. (1968) *Ica y el Peru Precolombino* [Precolumbian Ica and Peru]. Tomo I. *Arqueologia de la Provincia de Ica* [Archaeology of the Ica Province]. Ica: Editora Ojeda, S.A.

Pezzia, A. (1969) *Guia del Mapa Arqueológico Pictográfico del Departamento de Ica* [Guide and Archaeological Pictographic Map of the Department of Ica]. Lima: Editorial Italperu.

Pfeiffer, S. (1991) Adipocere in a Canadian pioneer burial [Letter]. On file in the Paleobiology Laboratory, Department of Pathology, University of Minnesota Duluth, School of Medicine, Duluth, MN, 55812.

Pfeiffer, S. & Varney, T. (2000) Quantifying histological and chemical preservation in archaeological bone. In *Biochemical Approaches to Paleodietary Analysis*, ed. S. Ambrose & M.A. Katzenberg, pp. 141–158. New York: Kluwer Academic/Plenum Press.

Pfeiffer, S., Milme, S. & Stevensom, R.M. (1998) The natural decomposition of adipocere. *Journal of Forensic Sciences* **43**:368–370.

Piepenbrink, H., Frahm, J., Haase, A. & Matthaei, D. (1986) Nuclear magnetic resonance imaging of mummified corpses. *American Journal of Physical Anthropology* **70**:27–28.

Pinniger, D.B. (1994) *Insect Pests in Museums*. London: Archetype Publications.

Pinniger, D.B. & Harmon, J.D. (1999) Pest management, prevention and control. In *Care and Conservation of Natural History Collections*, ed. D. Carter & A.K. Walker, pp. 153–176. Oxford: Butterworth/Heinemann.

Pinto, E. (2000) Santificados pelo povo [Sanctified hair]. In Portuguese. Website: http: //www.terravista.pt/ bilene/3224/santos.html.

Piperno, M. (1976) Scoperta di una sepoltura doppia epigravettiana nella Grotta dell'Uzzo (Trapani). *Kokalos* **22–23**:720–751.

Place, R. (1995a) The Paraca mummies of Peru. In *Digging Up the Past, Bodies from the Past*, pp. 30–31. Hove, Sussex: Wayland.

Place, R. (1995b) *Bodies from the Past: The Plaster People of Pompeii*. Hove, Sussex: Wayland.

Poinar, H.N., Hoss, M., Bada, J. & Pääbo, S. (1996) Amino acid racemization and the preservation of ancient DNA. *Science* **272**:864–866.

Polack, J.S. (1838) *New Zealand: A Narrative of Travels and Adventure in that Country Between the Years 1831 and 1837*. London: Bentley.

Polack, J.S. (1840) *Manners and Customs of the New Zealanders*. London: Madden & Co.

Pollicard, A. & Collet, A. (1952) Deposition of siliceous dust in the lungs of the inhabitants of the Saharan regions. *Archives of Industrial Hygiene and Occupational Medicine* **5**:527–534.

Polosmak, N. (1994) A mummy unearthed from the pastures. *National Geographic* **186**(4):80–103.

Poma De Ayala, F.G. (1613/1956) *Nueva Corónica Y Buen Gobierno* [New Crown and Good Government], transl. F. Pease. Paris: Institute d'Éthnologie.

Ponce-Sanguines, C. (1966) Time and Tiwanaku. *MD* **10**:161–163.

Ponce-Sanguines, C. & Iturralde, L. (1966) *Comentario antropologico acerca de la determinación paleoseologica de grupos sanguineos en momias prehispanicas del altiplano boliviano* [Anthropological commentary on paleoserological tests on prehispanic mummies from the Bolivian highlands]. Publication no. 15. La Paz, Bolivia: Academia Nacional de Ciencias de Bolivia.

Poncet, F. (1895) A prehistoric truss. *Lancet* **2**:869.

Pond, A.W. (1937) Lost John of Mummy Ledge. *Natural History* **39**:176–184.

Pope, F. (1995) After the autopsy: the continuing conservation, research and interpretation of an Egyptian Mummy, nakht (ROM 1). In *Proceedings of the First World Congress on Mummy Studies*, February, 1992, Vol. 1, pp. 231–235. Santa Cruz, Tenerife, Canary Islands: Archaeological and Ethnographical Museum of Tenerife.

Porter, H. (1905) The recovery of the body of John Paul Jones. *Century Magazine* **70**:927–955.

Porter, R. (1996) *The Cambridge Illustrated History of Medicine*. Cambridge: Cambridge University Press.

Post, P. & Daniels, F., Jr (1969) Histological and histochemical examination of American Indian scalps, mummies and a shrunken head. *American Journal of Physical Anthropology* **30**:269–294.

Post, P. & Daniels, F., Jr. (1973) Ancient and mummified skin. *Cutis*, June:779–781.

Postek, M.T., Howard, K.S., Johnson, A.H. & McMichael, K.L. (1980) *Scanning Electron Microscopy: A Student's Handbook*. Baton Rouge, FL: Ladd Research Industries.

Powers, R. (1960) Ancient preserved brains: An additional note. *Man* (London) **60**(122):91.

Prager, E.M., Wilson, A.C., Lowenstein, J. & Sarich, V.M. (1980) Mammoth albumin. *Science* **209**:287–289.

Prescott, W.H. (1936) *History of the Conquest of Mexico and History of the Conquest of Peru*. New York: The Modern Library.

Preston, D.J. (1980) A mummy's travels. *Natural History* **89**:90–92.

Pretty, G.L. (1969) The Macleay Museum mummy from Torres Straits: A postscript to Elliot Smith and the diffusion controversy. *Man* **4**(1):24–43, with VII plates.

Price, T.D. (1989) *The Chemistry of Prehistoric Bone*. Cambridge: Cambridge University Press.

Pringle, H. (1999) Unlocking the secrets of the Iceman. *Canadian Geographic* **119**(7):19.

Prioreschi, P. (1990) *A History of Human Responses to Death*. Studies in Health and Human Services, Vol. 17. Lewiston, NY: Edwin Mellen Press.

Proefke, M.L., Rinehart, K.L., Raheel, M., Ambrose, S.H. &

Wisseman, S.U. (1992) Probing the mysteries of ancient Egypt. *Analytical Chemistry* **64**(2):5a-110a.

Prominska, E. (1986) Ancient Egyptian traditions of artificial mummification in the Christian period in Egypt. In *Science in Egyptology,* ed. R.A. David, pp. 113–121. Manchester: Manchester University Press.

Proulx, D.A. (1971) Headhunting in Ancient Peru. *Archaeology* **24**(1):16–21.

Putnam, E.K. (1914) The Davenport Collection of Nazca and other Peruvian pottery. *Proceedings of the Davenport Academy of Sciences* **13**:17–46 plus 13 plates.

Putnam, J. (1993) *Mummy.* New York: Knopf.

Quackenbush, L.S. (1909) Notes on Alaskan mammoth expeditions of 1907 and 1908. *Bulletin of the American Museum of Natural History* **26**:107–130.

Quekett, J. (1848) On the value of the microscope in the determination of minute structures of a doubtful nature, as exemplified in the identification of human skin attached many centuries ago to the doors of churches. *Transactions of the Microscopical Society of London* **2**(o.s.):151–158.

Quibell, J.E. (1908) The Tomb of Yuaa and Thuiu. In *Catalogue Générale du Musée du Caire,* pp. 68–73. Cairo: Cairo Museum.

Quigley, C. (1998) *Modern Mummies.* McFarland & Company: Jefferson, North Carolina.

Quinn-Judge, P. (1998) Final rites for the czar. *Time,* July 27, pp. 34–35.

Ra, C. (1998) *Summum Mummification.* Brochure from Summum, 707 Genesee Avenue, Salt Lake City, UT 84104.

Rabino-Massa, E. (1982) Postpartum inversion of the uterus in an Egyptian dynastic mummy. In papers on paleopathology presented at the 4th European Members Meeting of the Paleopathology Association in Antwerp, Netherlands on September 16–19. *Paleopathology Newsletter* **39** (September) (Suppl.):10 (Abstract).

Rabino-Massa, E., Cerutti, N., Marin, A. & Savoia, D. (2001) Malaria in ancient Egypt: Paleoimmunological investigation on predynastic mummified remains. *Chungará* **32**(1):7–9.

Radanov, S., Stoev, S., Davidov, M., Nachev, S., Stanchev, M. & Kirova, E. (1992) A unique case of naturally occurring mummification of human brain tissue. *International Journal of Osteoarchaeology* **105**:173–175.

Rae, A. (1996) Dry human and animal remains: Their treatment at the British Museum. In *Human Mummies: A Global Survey of Their Status and the Techniques of Conservation*, ed. K. Spindler, H. Wilfing, E. Rastbichler-Zissernig, D. zur Nedden & H. Nothdurfter, pp. 3–8. New York: Springer-Verlag.

Rae, A., Ivanovich, M., Green, H.S, Head, M.J. & Kimber, R.W.L. (1987) A comparative dating study of bones from Little Hoyle Cave, South Wales, U.K. *Journal of Archaeological Science* **14**:243–250.

Rafi, A., Spigelman, M., Stanford, J., Lemma, E., Donoghue, H. & Zias, J. (1994) DNA of *Mycobacterium leprae* by PCR in ancient bone. *International Journal of Osteoarchaeology* **4**:287–290.

Rampa, T. L. (1957) *The Third Eye: The Autobiography of a Tibetan Lama.* Garden City, NY: Doubleday & Co.

Ransford, O. (1978) *David Livingstone: The Dark Interior.* London: John Murray.

Ransome, H.M. (1937) *The Sacred Bee.* London: George Allen & Unwin.

Regöly-Mérei, R. (1961) Beitraege zur Geschichte der Krankheiten (Üeber einege interresante paelaopathologische Faelle) [Contributions to the history of diseases: concerning several interesting paleopathological cases]. *Therapia Hungarica* **9**:33–36.

Reinhard, J. (1992) Sacred Peaks of the Andes. *National Geographic*, March:88–111.

Reinhard, J. (1996a) *The Nazca Lines.* Lima: Editorial Los Piños E.I.R.L.

Reinhard, J. (1996b) Peru's Ice Maiden: Unwrapping the secrets. *National Geographic* **189**(6):62–81.

Reinhard, J. (1997) Sharp Eyes of Science Probe the Mummies of Peru. *National Geographic* **191**(1):36–43.

Reinhard, J. (1998) New Inca mummies. *National Geographic* **194**(1):128–135.

Reinhard, J. (1999) Frozen in time. *National Geographic* **196**(5):36–55.

Reinhard, K. (1990) Coprolite report on mummy from Ventana Cave, Arizona. Unpublished manuscript in the files of the Paleobiology Laboratory, Department of Pathology at the University of Minnesota, Duluth, MN 55812.

Reinhard, K. & Aufderheide, A.C. (1990) Diphyllobothriasis in pre-Columbian Chile and Peru: Adaptive radiation of a helminth species to Native American populations. In papers on paleopathology presented at the 8th European Members Meeting of the Paleopathology Association in Cambridge, England, September 19–22. *Paleopathology Newsletter* **72** (Suppl.):18–19 (Abstract).

Reischek, A. (1930) *Yesterday in Maoriland*, transl. H.E.L. Priday. London: Cape.

Reisner, G.A. (1910) The Archaeological Survey of Nubia. Report for 1907–1908. Vol. 1 *The Archeological Report*, pp. 1–348. Cairo: National Printing Department.

Reisner, G.A. (1928) The empty sarcophagus of the mother of Cheops. *Bulletin of the Museum of Fine Arts* **26**:75–89.

Reiss, W. & Stuebel, A. (1880–1887) *The Necropolis of Ancon in Peru: A Contribution to our Knowledge of the Culture and Industries of the Empire of the Incas*, transl. A. H. Keane, Vol. 1. Berlin: A.Asher & Co, agents in America; New York: Dodd, Mead & Co

Reuter News Agency (1996) Tut, tut! Price of a drink fit for a pharaoh. *The Toronto Star*, June 25, p. 131.

Reverte, J.M. (1986) The mummies in the School of Legal Medicine of Madrid: A preliminary report. In *Science in Egyptology*, ed. A.R. David, pp. 485–509. Manchester University Press.

Reyman, T.A. (1977) Schistosomal cirrhosis in an Egyptian mummy (circa 1200 B.C.). *Yearbook of Physical Anthropology* (1976) **20**:356–358.

Reyman, T.A. & Goodman, M. (1981) The histology of ancient mammoth tissue. *Paleopathology Newsletter* **33**:13–15.

Reyman, T.A. & Peck, W.H. (1980) Egyptian mummification with evisceration per ano. In *Mummies, Disease and Ancient Cultures*, ed. A. Cockburn & E. Cockburn, pp. 85–100. Cambridge: Cambridge University Press.

Richardson, D.H.S. (1981) *The Biology of Mosses.* Oxford: Blackwell Scientific Publications.

Richardson, J.B. (1994) People of the Andes. In *The Great Artisans*, pp. 101–113. Washington, DC: Smithsonian Books.

Richburg, K.B. (2001) Scientists trace iceman's death to arrow wound. *Duluth Herald*, 28 July, p. 7A.

Riddel, F. & Belan, A. (2000) *Informe del Proyecto de Rescate Arqueológico INC-CIPS en el Sitio de Tambo Viejo (PV-74–1). Valle de Acari, Departamento de Arequipa. [Report of the Archaeological Salvage project INC-SIPS at the Tambo Viejo site (PV-74-1) in the Acari Valley, Department of Arequipa].* Report to the Instituto Nacional de Cultura, Lima.

Riddick, E.B.Jr. (1995) The Ozark Bluff-dweller mummies. In *Proceedings of the First World Congress on Mummy Studies*, February, 1992, Vol. 2, pp. 837–839. Santa Cruz, Tenerife, Canary Islands: Archaeological and Ethnographical Museum of Tenerife.

Riddle, J. & Vreeland, J.M. Jr (1982) Identification of insects associated with Peruvian mummy bundles by using scanning electron microscopy. *Paleopathology Newsletter* **39**:5–9.

Ridgway, G.L., Powell, M. & Mirza, N. (1986) The microbiological monitoring of Lindow Man. In *Lindow Man: The Body In The Bog*, ed. I.M. Stead, H.B. Bourke, & D. Brothwell, 21. Ithaca, NY: Cornell University Press.

Riding, A. (1997) Under the probing gaze of the Egyptian dead. *New York Times*, April 26: *The Arts*:13.

Rising, Lieut. R.N. (1866) On the artificial eyes of certain Peruvian mummies. *Transactions of the Ethnological Society of London* **4** (n.s.):59–60.

Rivera, M.A. (1977) Prehistoric chronology of Northern Chile. Doctorate thesis from the University of Wisconsin-Madison, microfilmed by the University of Wisconsin-Madison Photographic Media Center.

Rivera, M.A. (1995–1996) La sequencia de Azapa del centro sur Andino: Una revision de las contribuciones de Percy Dauelsberg Hahmann [The Azapa sequence within the South Central Andes: A revision of Dauelsberg's contributions]. *Dialogo Andino* (Arica, Chile) **15/16**: 17–37.

Roark, R.P. (1965) *From* Monumental to proliferous in Nasca pottery. Nawpa Pacha (Berkeley, California) **3**:1–92.

Robertson, A.L. (1988) Autopsy findings in Pre-Columbian North Americans. (Department of Pathology, Medical College of Virginia, Richmond, VA) *Paleopathology Club Newsletter* **33**:4–5.

Roche, A. (1964) Aural exostoses in Australian aboriginal skulls. *Annals of Otology, Rhinology and Laryngology* **73**:82–91.

Rodríguez-Albarrán, M.S., Casas-Seanchez, J.D., García-Bartrual, M. & Arroyo-Pardo, E.(1999) Aspartic acid racemization: Its usefulness as an ageing measure and for aDNA preservation. *Paleopathology Newsletter* **106**:9–12.

Rodríguez-Martín, C. (1995) Una historia de las momias Guanches [A History of Guanche mummies]. In *Proceedings of the First World Congress on Mummy Studies*, February, 1992, Vol. 1, pp. 151–162. Santa

Cruz, Tenerife, Canary Islands: Archaeological and Ethnographical Museum of Tenerife.

Rodríguez-Martín, C. (1996) Guanche mummies of Tenerife (Canary Islands): Conservation and scientific studies in the CRONOS Project. In *Human Mummies: A Global Survey of Their Status and the Techniques of Conservation,* ed. K. Spindler, H. Wilfing, E. Rastbichler-Zissernig, D. zur Nedden & H. Nothdurfter, pp. 183–194. New York: Springer-Verlag.

Rodríguez, W.C. & Bass, W.M. (1983) Insect activity and its relationship to decay rates of human cadavers in East Tennessee. *Journal of Forensic Sciences* **28**:423–432.

Rodríguez, C. & Bass, W.M. (1985) Decomposition of buried bodies and methods that may aid in their location. *Journal of Forensic Sciences* **30**:835–852.

Roentgen, W.C. (1895) Eine neue arte von Strahlen [A new kind of rays]. Sitzungsberichte de Wuerzburger Physik-medizinische Gesellschaft (*Proceedings of the Wuerzburger Physico-medical Society*) **137** (December):132–141.

Rofes, J. (2000) Sacrificio de cuyes en el Yaral, comunidad prehispanica del extremo sur Peruano [Guinea pig sacrifices at Yaral, a pre-Hispanic community in extreme southern Peru]. *Bulletin de Instituto para Estudes Andines* **29**(1):1–12.

Rogers, L. (1949) Meningiomas in pharaoh's people. Hyperostosis in ancient Egyptian skulls. *British Journal of Surgery* **36**:423–424.

Rogers, J. & Waldron, T. (1986) Iatrogenic paleopathology. In *Sixth European Meeting of the Paleopathology Association,* ed. F. Gómez-Bellard & J.A. Sánchez, pp. 31–35. Madrid: Universidad Complutense de Madrid.

Rosenzweig, W. (1983) Disease in art: A case for carcinoma of the breast in Michelangelo's *La Notte*. *Paleopathology Newsletter* **41**:8–11.

Ross, A. (1967) *Pagan Celtic Britain*. London: Routledge & Kegan Paul; New York: Columbia University Press.

Ross, A. & Robins, D. (1989) *The Life and Death of a Druid Prince*, pp. 130–167. Summit Books.

Roth, W.E. (1907) North Queensland Ethnography, Bulletin No. 9: Australian aboriginal burial ceremonies and disposal of the dead. *Records of the Australian Museum* **6**:365–403.

Rothenberg, R.B. (1993) Gonorrhea. In *The Cambridge World History of Human Disease,* ed. K.F. Kiple, pp. 756–763. New York: Cambridge University Press.

Rothschild, H. & Chapman, C.F. (1981) *Biocultural Aspects of Disease.* New York: Academic Press.

Rothschild, B.M., Rothschild, C. & Bement, L.C. (1994) Lithopedion as an archaic occurrence. *International Journal of Osteoarchaeology* **4**:247–250.

Rowe, J.H. (1945) Absolute chronology in the Andean area. *American Antiquity* **10**:265–284.

Rowe, J.H. (1946) Inca culture at the time of the Spanish Conquest. In *Handbook of South American Indians Bulletin,* ed. J.H. Stewart II, pp. 183–330. Washington, DC: Bureau of American Ethnology.

Rowland, R. & Farnham, J. (1959) Microradiographic measurements of mineral density. *Radiation Research* **10**:234–242.

Rowling, J.T. (1961) Pathological changes in mummies. *Proceedings of the Royal Society of Medicine* **54**:17–22.

Rowling, J.T. (1967) Hernia in Egypt. In *Diseases in Antiquity*, ed. D. Brothwell & A.T. Sandison, pp. 444–446. Springfield, IL: Charles Thomas.

Royal, W. & Clark, E. (1960) Natural preservation of human brain at warm mineral springs, Florida. *American Antiquity* **26**:285–287.

Rubin, E. & Farber, J.L. (eds.) (1994) *Pathology*, 2nd edn. Philadelphia: Lippincott.

Rudberg, R.D. (1954) Acute otitis media: Comparative therapeutic results of sulphonamide and penicillin administered in various forms. *Acta Otolaryngologica* (Stockholm) **113** (Suppl.):9–79.

Rudenko, S.I. (1970) *Frozen Tombs of Siberia: The Pazyryk Burials of Iron Age Horsemen*, transl. and with a preface by M.W. Thompson. Berkley & Los Angeles: University of California Press.

Ruffer, M.A. (1910) Remarks on the histology and pathological anatomy of Egyptian mummies. *Cairo Scientific Journal* **10**:3–7.

Ruffer, M.A. (1911a) Megacolon in a child of the Roman period. *Mémoires de l'Institute d'Égypte* **6**(3):1–39.

Ruffer, M.A. (1911b) On arterial lesions found in Egyptian mummies. *Journal of Pathology and Bacteriology* **15**:453–462.

Ruffer, M.A. (1913) On pathological lesions found in Coptic bodies. *Journal of Pathology and Bacteriology* **18**:149–62.

Ruffer, M.A. (1920) Note on the presence of Bilharzia haematobia in Egyptian mummies of the twentieth dynasty (1250–1000 B.C.). *British Medical Journal* **1**:16.

Ruffer, M.A. (1921a) *Studies in the Paleopathology of Egypt by*

Sir Marc Armand Ruffer Kt, C.M.G., M.D., ed. R.L. Moodie. Chicago: University of Chicago Press.

Ruffer, M.A. (1921b) Pathological notes on the royal mummies of the Cairo museum. In *Studies in the Paleopathology of Egypt*, ed. R.L. Moodie, pp. 166–178. Chicago: University of Chicago Press.

Ruffer, M.A. & Ferguson, A.R. (1911) Note on an eruption resembling that of variola in the skin of a mummy of the twentieth dynasty (1200–1100 B.C.). *Journal of Pathology and Bacteriology* **15**:1–3.

Ruffer, Sir M.A. & Rietti, A. (1912) On osseous lesions in ancient Egyptians. *Journal of Pathology and Bacteriology* **16**:439–465.

Rullkoetter J. & Nissenbaum, A. (1988) Dead Sea asphalt in Egyptian mummies: Molecular evidence. *Naturwissenschaften* **75**:245–262.

Russel, H.K. Jr (1972) A modification of Movat's pentachrome stain. *Archives of Pathology* **94**:187–189.

Rutherford, P. (1999) Immunocytochemistry and the diagnosis of schistosomiasis: Ancient and modern. *Parasitology Today* **15**:390–391.

Saccardo, P.A. (1899) *Sylloge Fungorum.* Vol. 14. Pavia.

Sadovsky, O.J. (1985) Siberia's frozen mummy and the genesis of California's Indian culture. *The Californians* 3(6):8–20.

Saggs, H.W. (1962) *The Greatness That Was Babylon.* New York: Hawthorn Books.

Saiki, R.K., Scharf, S., Faloona, F., Mullis, K.B., Horn, G.T., Erlich, H.A. & Arnhem, N. (1985) Enzymatic amplification of beta-globin genomic sequences and restriction site analysis for diagnosis of sickle-cell anemia. *Science* **230**:1350–1354.

Saiki, R.K., Gelfand, D.H., Stoffel, S., Scharf, S.J., Higuchi, R., Horn, G.T. *et al.* (1988) Primer-directed enzymatic amplification of DNA with a thermostable DNA polymerase. *Science* **239**:487–491.

Sakurai, K., Ogata, T., Morimoto, I., Long-Xiang, P. & Zhong-Bi, W. (1998) Mummies from Japan and China. In *Mummies, Disease and Ancient Cultures,* ed. A. Cockburn, E. Cockburn & T.A. Reyman, pp. 308–335. Cambridge: Cambridge University Press.

Salo, W.L., Aufderheide, A.C., Buikstra, J. & Holcomb, T.A. (1994) Identification of *Mycobacterium tuberculosis* DNA in a pre-Columbian mummy. *Proceedings of the National Academy of Sciences, USA* **91**:2091–2094.

Sánchez Sánchez, J.A. & Gómez-Bellard, F. (1988) Renal and biliary calculi: A paleopathological analysis. In papers on paleopathology presented at the 8th European Members Meeting of the Paleopathology Association in Cambridge, England, September 19–22, 1990. *Paleopathology Newsletter* **72** (Suppl.):19–22 (Abstract).

Sandison, A.T. (1955a) Reconstitution of dried-up tissue specimens for histological examination. *Journal of Clinical Pathology* **19**:522–523.

Sandison, A.T. (1955b) The histological examination of mummified material. *Stain Technology* **30**:277–283.

Sandison, A.T. (1957) The eye in the Egyptian mummy. *Medical History* **1**:336–339.

Sandison, A.T. (1963) The study of mummified and dried human tissues. In *Science in Archaeology,* ed. D. Brothwell, E. Higgs & G. Clarke, pp. 413–425. New York: Basic Books.

Sandison, A.T. (1967) Diseases of the skin. In *Diseases in Antiquity,* ed. D. Brothwell & A.T. Sandison, pp. 449–463. Springfield, IL: Charles Thomas.

Sandison, A.D. & Wells, C. (1967) Diseases of the reproductive system. In *Diseases in Antiquity*, ed. D. Brothwell & A.T. Sandison, pp. 498–520. Springfield, IL: Charles Thomas.

Sandweiss, D.H. & Wing, E.S. (1997) Ritual rodents: The guinea pigs of Chincha, Peru. *Journal of Field Archaeology* **24**(1):47–58.

Sanford, M. (ed.) (1993) *Investigations of Ancient Human Tissue.* Langhorne, PA: Gordon & Breach Science Publishers.

Saskin, H. (1984) Case no. 20: Vasculitis. (Department of Pathology, Medical College of Virignia, Richmond, VA.) *Paleopathology Club Newsletter* 20:1.

Sawicki, V.A., Allison, M.J. & Dalton, H.P. (1976) Presence of *Salmonella* antigens in feces from a Peruvian mummy. *Bulletin of the New York Academy of Medicine* **52**:805–813.

Saxe, A.A. (1970) Social dimensions of mortuary practice. Doctoral dissertation on file in the Department of Anthropology, University of Michigan.

Scaife, R.G. (1986) Pollen in human palaeofaeces and a preliminary investigation of the stomach and gut contents of Lindow Man. In *Lindow Man: The Body In the Bog,* ed., I.M. Stead, J.B. Bourke & D. Brothwell, pp. 126–135. Ithaca, NY: Cornell University Press.

Schaedler, J.M., Krook, L., Wootton, J.A.M., Hover, B., Brodsky, B., Naresh, M.D. *et al.* (1992) Studies of collagen in bone and dentin matrix of a Columbian mammoth (Late Pleistocene) of central Utah. *Matrix* **12**:297–307.

Schaefer, J., Pirsig, W. & Parsche, F. (1995) Videoprinting: An excellent method for documentation during flexible endoscopy of mummies. In *Proceedings of the First World Congress of Mummy Studies*, February, 1992, Vol. 2, pp. 591–596. Santa Cruz, Tenerife, Canary Islands: Archaeological and Ethnographical Museum of Tenerife.

Schechter, D.C. (1962) Breast mutilation in the Amazons. *Surgery* **51**:554–560.

Schmidt, W.A. (1908) Chemical and biochemical examination of Egyptian mummies, including some observations on the chemistry of the embalming process of the ancient Egyptians. *Cairo Scientific Journal* **2** (April):147.

Schobinger, J. (1966) La "momia" Del Cerro El Toro [The "Mummy" of Cerro El Toro]. Anales de Arqueología y Etnología **31**(Suppl.):5–218 [English summary, pp. 215–218.].

Schobinger, J. (1991) Sacrifices of the High Andes. *Natural History* **4**:62–69.

Schottelius, J.W. (1946) *Archeología de la Mesa de los Santos* [*Archaeology of the Boletin de Arqueología*] (Bogotá) 3:213–225.

Schreiber, K. (1996a) The Paracas necropolis. In *Tombs, Graves and Mummies*, ed. P.G. Bahn, pp. 78–81. London: Weidenfeld & Nicolson.

Schreiber, K. (1996b) Inca mountain sacrifices. In *Tombs, Graves and Mummies,* ed. P.G. Bahn, pp. 160–163. London: Weidenfeld & Nicolson.

Schulting, R.J. (1995) Preservation of soft tissues by copper in the Interior Plateau of British Columbia, Canada. In *Proceedings of the First World Congress on Mummy Studies*, February, 1992, Vol. 2, pp. 771–780. Santa Cruz, Tenerife, Canary Islands: Archaeological and Ethnographical Museum of Tenerife.

Schultz, M. (1993) Initial stages of systemic bone disease. In *Histology of Ancient Human Bone,* ed. G. Grupe & A.N. Garland, pp. 185–203. New York: Springer Verlag.

Schwabe, C.W. (1978) *Cattle, Priests and Progress In Medicine.* Minneapolis: University of Minnesota Press.

Schwartz, P. & Schultz, M. (1994) Postmortem destruction of archaeological skeletal material caused by insects: A scanning electron microscopic study (Poster). In papers in paleopathology presented at the 10th European meeting of the Paleopathology Association, Goettingen. *Paleopathology Newsletter* **88** (Suppl.):25 (Abstract).

Scott, C.R. (1837) *Rambles in Egypt and Canada*, Vol. II. London: Henry Colburn.

Screven, C.G. (1984) Educational evaluation and research in museums and public exhibits: A bibliography. *Curator* **27**:147–165.

Scully, R.E., Mark, E.J., McNeely, W.F. & McNeely, B.U. (1995) Case records of the Massachusetts General Hospital: Xanthogranulomatous pyelonephritis. *New England Journal of Medicine* **332**:174–179.

Seipel, W. (1996) Mummies and ethics in the museum. In *Human Mummies: A Global Survey of Their Status and the Techniques of Conservation,* ed. K. Spindler, H. Wilfing, E. Rastbichler-Zissernig, D. zur Nedden & H. Nothdurfter, pp. 3–8. New York: Springer-Verlag.

Seligman, R. (1870) Ueber Exostosen an Peruanaschaedeln [Concerning exostoses in Peruvian skulls]. *Archiv fuer Anthropologie, Volkerforschung und Kolonialen Kulturinandel (Braunschweig)* **4**:147–148.

Sengstake, F. (1892) Die Leichenbestattung aus Darnley Island [Mortuary practices on Darnley Island]. *Globus* **61**(16):248–249.

Sensabaugh, G.E., Wilson, A.C. & Kirk, P.L. (1971) Protein stability in preserved biological remains. Survival of biologically active proteins in an 8-year-old sample of dried blood. *Journal of Biochemistry* **2**:545–557.

Service, E.R. (1958) *A Profile of Primitive Culture.* New York: Harper & Brothers, Publishers.

Shattock, S.G. (1909) A report upon the pathological condition of the aorta of King Menephtah, traditionally regarded as the pharaoh of the Exodus. *Proceedings of the Royal Society of Medicine* **2** (pt 3):122–127.

Shaw, A.B. (1981) Knapper's rot: Silicosis in East Anglian flint-knappers. *Medical History* **25**:151–168.

Shaw, A.F.B. (1938) A histologic study of the mummy of Har-Mose, the singer of the eighteenth dynasty. *Journal of Pathology and Bacteriology* **47**:115–123.

Sheard, H.L., Mountford, C.P. & Hackett, C.J. (1927) An unusual disposal of an aboriginal child's remains from the Lower Murray, South Australia. *Transactions and Proceedings of the Royal Society of South Australia* **5**:173–176.

Sheehy, J.L. (1958) Osteoma of the external auditory canal. *Laryngoscope* **68**:1667–1673.

Shepartz, I. & Subers, M.H. (1964) The glucose oxidase

of honey. I. Purification and some general properties of the enzyme. *Biochemica et Biophysica Acta* **85**:228–237.

Sherwin, R.P., Barman, M.L. & Abraham, J.L. (1979) Silicate pneumoconiosis in farm workers. *Laboratory Investigation* **40**:576–582.

Shipman, P., Foster, G. & Schoeninger, M. (1984) Burnt bones and teeth: An experimental study of color, morphology, crystal structure and shrinkage. *Journal of Archaeological Science* **11**:307–325.

Shufeldt, R.W. (1892) Notes on paleopathology. *Popular Science Monthly* 42:679–684.

Siegel, L. (1998) Anasazi may have known how to mummify. *Last Lake Tribune*, April 2, p. A6.

Sievers, M. & Fisher, J.R. (1981) Diseases of North American Indians. In *Biocultural Aspects of Disease,* ed. H.R. Rothschild, pp. 191–252. New York: Academic Press.

Sillen, A. (1989) Diagenesis of the inorganic phase of cortical bone. In *The Chemistry of Prehistoric Human Bone,* ed. T.D. Price, pp. 211–229. Cambridge: Cambridge University Press.

Silva Celís, E. (1945) Contribución al conocimiento de la civilización de los lache [Contribution to knowledge about the Lache civilization]. *Boletin de Arqueología* **2**:396–424.

Silverman, H. (1987) A Nasca 8 occupation at an early Nasca site: The room of the posts at Cahuachi. *Andean Past* **1**:5–55.

Silverman, H. (1988) Cahuachi: Non-urban cultural complexity on the south coast of Peru. *Journal of Field Archaeology* **15**:403–430.

Simandl, I. (1928) A contribution to the histology of the skin. *Anthropologie* (Prague) **6**:56–60.

Simpson, K. (ed.) (1965) Death and Post-mortem changes. In *Taylor's Principles and Practice of Medical Jurisprudence*, 12th edn, ed. K. Simpson, pp. 70–104. London: J. & A. Churchill.

Simpson, W.K. (ed.) (1972) *The Literature of Ancient Egypt.* New Haven, CT: Yale University Press.

Sinclair, D. (2000) *The Maori.* Christchurch, New Zealand: The Claxton Press.

Singer, N. (1993) *Burmah: A Photographic Journey 1855–1925.* Gartmore, Stirling, Scotland: Paul Strachan-Kiscadale.

Sjovold, T., Swedborg, I. & Diener, L. (1974) *A Pregnant Woman from the Middle Ages with Exostosis Multiplex.* Sonona, Sweden: Osteological Research Laboratory, University of Stockholm.

Skinner, H.A. (1961) *The Origin of Medical Terms.* Baltimore, MD: Williams & Wilkins.

Skottsberg, C. (1924) Notes on the old Indian necropolis of Arica. *Meddlanden Fran Geografiska Føreningen i Gøteberg* 3:27–78.

Slatkin, D.N., Friedman, L., Irsa, A.P. & Gaffney, J.S. (1978) The $^{13}C/^{12}C$ ratio in black pulmonary pigment: A mass spectrometry study. *Human Pathology* **9**:259–267.

Sledzik, P.S. & Bellantoni, N. (1994) Brief communication: Bioarchaeological and biocultural for the New England vampire folk belief. *American Journal of Physical Anthropology* **94**:269–274.

Sledzik, P.S. & Ousley, S. (1991) Analysis of six Vietnamese trophy skulls. *Journal of Forensic Sciences* **36**:520–530.

Sluglett, J. (1980) Mummification in ancient Egypt. *MASCA Journal* **1**(6):163–167.

Smith, G.E. (1902) On the natural preservation of the brain in the ancient Egyptians. *Journal of Anatomy and Physiology, Normal and Pathological, Human and Comparative* **36** (n.s. **16**):375–380.

Smith, G.E. (1907) Report on the unrolling of the mummies of the Kings Siptah, Seti II, Ramses IV, Ramses V and Ramses VI. *Bulletin de l'Institute Égyptien* (Cairo) **1**:45–67.

Smith, G.E. (1910) Mode of Burial and Treatment of the Body. *The Archaeological Survey of Nubia. Report for 1907–1908*, ed. G.A. Reisner, Vol. II *Report of the Human Remains*, G.E. Smith & F. Wood Jones, pp. 181–220. Cairo: National Printing Department.

Smith, G.E. (1911a) *The Ancient Egyptians.* London: Kegan Paul Ltd.

Smith, G.E. (1911b) "Heart and reins" in mummification. *Journal of the Manchester Oriental Society* **1**:41–48.

Smith, G.E. (1912a) The earliest evidence of attempts at mummification in Egypt. *Report of the British Association for the Advancement of Science* (Dundee Meeting) 1912, pp. 612–613.

Smith, G.E. (1912b) *The Royal Mummies.* Cairo: Catalogue Général des Antiquitees Égyptiennes du Musée du Caire.

Smith, G.E. (1915) *The Migrations of Early Culture.* Manchester: Manchester University Press, London: Longmans, Green & Co.

Smith, G.E. & Dawson, W.R. (1924) *Egyptian Mummies.*

London: George Allen & Unwin, Ltd (reprinted 1991 by Kegan Paul International).

Smith, G.E. & Wood-Jones, F. (1910) *Report of the Human Remains*. In *The Archaeological Survey of Nubia. Report for 1907–1908*, ed. G.A. Reisner, Vol. II, pp. 7–367. Cairo: National Printing Department.

Smith, G.S., Bradley, Z.A., Kreher, R.E. & Dickey, T.P. (1978) The Kialegak Site, St. Lawrence Island. *Alaska Anthropology and Historic Preservation Occasional Paper* no. 10, *Anthropology and Historic Preservation*, pp. 65–73. Fairbanks, Alaska: Cooperative Park Studies Unit, University of Alaska.

Smith, K.G.V. (1986) *A Manual of Forensic Entomology*. Ithaca, NY: Comstock Publishing Associates, Cornell University Press.

Smith, M.W. (1950) Archaeology of the Columbia-Fraser region. *Memoirs of the Society for American Archaeology* no. 6, published as supplement to *American Antiquity* **15**(4, part 2), 1–46. Menasha, WI: Society for American Archaeology and Columbia University.

Snow, C. & Reyman, T.A. (1977) The life and afterlife of Elmer J. McCurdy. *Paleopathology Association Newsletter* **19** (Suppl.): s1–s8.

Sobotta, J. & McMurrich, J.P. (1939) *Atlas of Human Anatomy*, 5th revised English edn, Vol. 1. New York: G.E.Stechert.

Sonoda, S., Li, H.C., Cartier, L., Nuñez, L. & Tafima, K. (2000) Ancient HTLV type 1 provirus DNA of Andean mummy. *AIDS Research Human Retroviruses* **124**:1614–1618.

Soto-Heim, P. (1987) Evolución de deformaciones intentionales, peinados, tocados y practicas funerarias en la prehistoria de Arica [Evolution of intentional deformation, hair dressings, coiffures and funerary practices in the prehistory of Arica]. *Chungará* (Arica, Chile) **19**:129–214.

Spencer, A.J. (1982) *Death in Ancient Egypt*. Harmond, Middlesex: Penguin.

Spielman, P.E. (1932) To what extent did the ancient Egyptians employ bitumen for embalming? *Journal of Egyptian Archaeology* **18**:177–180.

Spigelman, M. & Lemma, E. (1993) The use of the polymerase chain reaction to detect *Mycobacterium tuberculosis* in ancient skeletons. *International Journal of Osteoarchaeology* **3**:137–143.

Spigelman, M., Fricker, C.R. & Fricker, E.J. (1995) Addendum: Extracting DNA from Lindow Man's gut contents. Modern technology looking for answers from ancient tissues. In *Bog Bodies*, ed. R.C. Turner & R.G. Scaife, pp. 59–61. London: British Museum Press.

Spindler, K. (1992) *Der Mann Im Eis. Sandoz Bulletin* **99**:21–29.

Spindler, K. (1994) *The Man in the Ice. The Preserved Body of a Neolithic Man Reveals the Secrets of the Stone Age*. London: Weidenfeld & Nicolson.

Spirtos, N.M., Einsenkop, S.M. & Mishell, D.R., Jr. (1987) Lithokelyphos. *Journal of Reproductive Medicine* **32**:43–46.

Standen, V., Arriaza, B. & Santoro, C. (1995) Una hipótesis ambiental para un marcador óseo: La exsosis auditiva externa en las poblaciones humanas prehistóricas del desierto del norte de Chile [An environmental hypothesis for an osseous marker: The external auditory canal exostosis in prehistoric human populations of northern Chile]. *Chungará* (Arica, Chile) **27**:99–116.

Stanish, C. & Rice, D.S. (1989) The Osmore drainage, Peru: An introduction to the work of Prgrama Contisuyo. In *BAR International Series*, ed. D.S. Rice, C. Stanish & P.R. Scarr, Vol. 545, Ecology, Settlement and History in the Osmore Drainage, Peru, pp. 1-15. Oxford: Oxford University Press..

Stanley, A. (1994) Tattooed lady, 2000 years old, blooms again. *New York Times*, July 12.

Stead, I.M., Bourke, J.B. & Brothwell, D. (eds.) (1986) *Lindow Man: The Body in the Bog*. London: British Museum Publications Ltd.

Steel, L. (1996) The shaft graves of Mycenae. In *Tombs, Graves and Mummies*, ed. P.G. Bahn, pp. 92–95. London: Weidenfeld & Nicolson.

Steinbock, R.T. (1985) The history, epidemiology and paleopathology of kidney and urinary bladder stone disease. *Anthropological Research* Paper 34:177–200. Tempe, AZ: Arizona State University.

Steinbock, R.T. (1989) Studies in ancient calcified soft tissues and organic concretions. II. Urolithiasis (renal and urinary bladder stone disease). *Journal of Paleopathology* **3**:35–59.

Steinbock, R.T. (1990) Studies in ancient calcified soft tissues and organic concretions. III Gallstones. *Journal of Paleopathology* **3**(2):95–106.

Steinbock, R.T. (1993a) Urolithiasis (renal and urinary bladder stone disease). In *The Cambridge World History of Human Disease*, ed. K.F. Kiple, pp. 1088–1092. New York: Cambridge University Press.

Steinbock, R.T. (1993b) Gallstones (cholelithiasis). In *The Cambridge World History of Human Disease*, ed. K.F. Kiple, pp. 738–741. New York: Cambridge University Press.

Stenn, F., von Seggen, W. & Farnham, J.E. (1975) Microradiography for anthropologists: A survey of current research. *Paleopathology Newsletter* **11**:5–6.

Stephenson, F. (1985) The Windover Archaeological Project. *Florida State University Bulletin Research in Review* **79**(4):3–10.

Stevens, W.K. (1997) Disease is new suspect in ancient extinction. *New York Times*, April 29, p. C1.

Stiner, M.C., Kuhn, S., Weiner, S. & Bar-Yosef, O. (1995) Differential burning, recrystallization and fragmentation of archaeological bone. *Journal of Archaeological Science* **22**:233–237.

Stirling, M. W. (1938) *Historical and Ethnographical Material on the Jivaro Indians.* Smithsonian Institution US Bureau of American Ethnology, Bulletin no. 117. Washington, DC: US Government Printing Office.

Stone, I. (1975) *The Greek Treasure*. Garden City, NY: Doubleday & Co. Inc.

Stone, R. (1999) Siberian mammoth find raises hopes, questions. *Science* **286**:876–887.

Stora, N. (1971) *Burial Customs of the Skolt Lapps.* FF (Folklore Fellows) Communications no. 210, Helsinki: Suomalainen Tiedeakatemia Academia Scientiarum Fennica.

Stothert, K.E. (1979) Unwrapping an Inca mummy bundle. *Archaeology* **32**:8–17.

Strabo (ca. 10 AD/1977) *Loeb Classical Library*, Vol. 2, Cambridge, MA: Harvard University Press.

Stout, S.D. (1986) The use of bone histomorphometry in skeletal identification: The Case of Francisco Pizarro. *Journal of Forensic Sciences* **31**:296–300.

Strano, A.J. & Font, R.(1976) Trachoma and inclusion conjunctivitis. In *Pathology of Tropical and Extraordinary Diseases*, ed. C.H. Binford & D.J. Connor, Vol. 1, pp. 79–81. Washington D.C.: Armed Forces Institute of Pathology.

Streitz, J.M., Aufderheide, A.C., el-Najjar, M. & Ortner, D. (1981) A 1500-year-old bladder stone. *Journal of Urology* **126**:452–453.

Strong, W. D. (1957) Paracas, Nazca and Tiahuanacoid Cultural Relationships in South Coastal Peru. *Memoirs of the Society for American Archaeology* no. 13. Salt Lake City: Society for American Archaeology.

Strong, W.D. & Evans, C. Jr (1952) *Cultural Stratigraphy in the Viru Valley, Northern Peru.* Columbian Studies in Archaeology and Ethnology, no. 4.

Strouhal, E. (1978) Ancient Egyptian case of carcinoma (nasopharyngeal). *Bulletin of the New York Academy of Medicine* **54**:290–292.

Strouhal, E. (1986) Embalming excerebration in the Middle Kingdom. In *Science in Egyptology,* ed. A.R. David, pp. 141–154. Manchester: Manchester University Press.

Strouhal, E. (1991) Vertebral tuberculosis in Egypt and Nubia. In *Human Paleopathology. Current Synthesis and Future Options*, ed. D.J. Ortner & A.C. Auferheide, pp. 181–194. Washington, DC, and London: Smithsonian Institution Press.

Strouhal, E. (1992a) Whence no traveler returns. In *Eugen Strouhal: Life of the Ancient Egyptians*, pp. 253–266. Norman, OK: University of Oklahoma Press.

Strouhal, E.(1992b) *Life of the Ancient Egyptians.* Norman, OK: University of Oklahoma Press.

Strouhal, E. (1994) Malignant tumors in the Old World. *Paleopathology Newsletter* **85** (March) (Suppl.):1–6.

Strouhal, E. (1995) Secular changes of embalming methods in Ancient Egypt. In *Proceedings of the First World Congress on Mummy Studies,* February, 1992, Vol. 2, pp. 859–865. Santa Cruz, Tenerife, Canary Islands: Archaeological and Ethnographical Museum of Tenerife.

Strouhal, E. & Gaballah, F. (1993) King Djedkare and his daughters. In *Biological Anthropology and the Study of Ancient Egypt,* ed. W.V. Davies & R. Walker, pp. 104–118. London: British Museum Press.

Strouhal, E. & Jungwirth, J. (1977) Ein Verkalktes Myoma Uteri aus der Spaete Roemerzeit in Egyptisch-Nubien [A calcified uterine myoma from the Late Roman Period in Egyptian Nubia]. *Mitteilungen der Anthropologischen Gesellschaft in Wien* **107**:215–221.

Strouhal, E., Gaballah, M.F., Bonani, G. & Woelfli, W. (1995) Re-examination of the alleged remains of King Djoser and an unknown girl from the step pyramid at Saqqara. In papers read at the 22nd Annual Meeting of the Paleopathology Association held in Oakland, California, March 28–29 *Paleopathology Newsletter* **90** (June) (Suppl.):9 (Abstract).

Stuart-Macadam, P. & Kent, S. (eds.) (1992) *Diet, Demography and Disease.* New York: Aldine de Gruyter.

Subert, F., Dvorak, V., Spalek, E., Navratilova, D., Vachala, B. & Kucera, J. (1993) *Problems of Contemporary Mummy Conservation.* Prague: Charles University.

Sudhoff, K. (1911) Aegyptische Mumienmacher-Instrumente [Egyptian mummification instruments]. *Archive fuer Geschichte der Medizin* **5**:165–171.

Swales, J.D. (1973) Tutankhamun's breasts. *Lancet* **1**:201.

Tacitus (98 AD/1981) *The Agricola and the Germania*, transl. H. Mattingley & S.A. Handford. New York: Penguin Books.

Taitz, L.S. (1973) Tutankhamun's breasts. *Lancet* **1**:149.

Takatori, T. (1996) Investigations on the mechanism of adipocere formation and its relation to other biochemical reactions. *Forensic Science International* **80**:49–61.

Takatori, T. & Yamaoka, A. (1977) The mechanism of adipocere formation. *Forensic Science* **9**:63–73.

Takatori, T. & Yamaoka, A. (1979) Separation and identification of 9–chloro-10–methoxy (9–methoxy-10–chloro) hexadecanoic and octadecanoic acids in adipocere. *Forensic Science International* **14**:63–73.

Takatori, T., Ishiguro, N., Tarao, H. & Matsumiya, H. (1986) Microbial production of hydroxy and oxo fatty acids by several microorganisms as a model of adipocere formation. *Forensic Science International* **32**:5–11.

Takatori, T., Gotuda, H., Terezawa, K., Mizukami, K. & Nagao, M. (1987) The mechanism of experimental adipocere formation: Substrate specificity on microbial production of hydroxy and oxo fatty acids. *Forensic Science International* **35**:277–281.

Tapp, E. (1992) The histologic examination of mummified tissue. In *The Mummy's Tale,* ed. A.R. David & E. Tapp, pp. 121–131. New York: St Martin's Press.

Tapp, E., Curry, A. & Anfield, C. (1975) Sand pneumoconiosis in an Egyptian mummy. *Lancet* **2**:276.

Tapp, E., Stanworth, P. & Wildsmith, K. (1984) The endoscope in mummy research. In *Evidence Embalmed,* ed. R. David & E. Tapp, pp. 65–77. Manchester: Manchester University Press.

Taubenberger, J.K., Reid, A.H., Krafft, A.E., Bijwaard, K.E. & Fanning, T.G. (1997) Initial genetic characterization of the 1918 "Spanish" influenza virus. *Science* **275**:1793–1796.

Taylor, G.M., Crossey, M., Saldanna, J. & Waldron, T. (1996) DNA from *Mycobacterium tuberculosis* identified in medieval human skeletal remains using polymerase chain reaction. *Journal of Archaeological Science* **23**:789–798.

Taylor, R.E. (1983) Non-concordance of radiocarbon and amino acid racemization deduced age estimates on human bone. *Radiocarbon* **25**:647–654.

Taylor, R.E., Payen, L.A., Prior, C.A., Slota, P.J., Gillespie, R., Gowlett, J.A.J. *et al.* (1985) Major revisions in the Pleistocene Age assignments for North American human skeletons by C-14 mass spectrometry: None older than 11,000 C-14 years BP. *American Antiquity* **50**:136–140.

Taylor, R.E., Hare, P.E. & White, T.D. (1995) Geochemical criteria for thermal alteration of bone. *Journal of Archaeological Science* **22**:115–119.

Taylor, R.E., Kirner, D.L., Southon, J.R. & Chatter, J.C. (1998) Radiocarbon dates of Kennewick Man. *Science* **280**:1171–1172.

Teegen, W.R. & Schultz, M. (1994) Epidural hematoma in fetuses, newborns and infants from the early medieval settlements of Elisenhof and Starigard-Oldenburg (Germany). In papers on paleopathology presented at the 10th European Members Meeting of the Paleopathology Association in Goettingen, Germany August 29 to September 3. *Paleopathology Newsletter* **88** (Suppl.):26 (Abstract).

Tello, J.C. (1929) *Antiguo Peru (Ancient Peru).* Lima: Excelsior.

Tello, J.C. (1943) Discovery of the Chavín Culture in Peru. *American Antiquity* **9(1)**:135–160.

Tello, J.C. & Mejia-Xesspe, T. (1979) *Paracas. II Parte. Cavernas y Necropolis* [Paracas. Part II. Caverns and Necropolis]. Lima: Universidad Nacional de San Marcos, Dirección Universitaria de Biblioteca y Publicaciones.

Thieme, F.P. & Otten, C.M. (1957) The unreliability of typing aged bone. *American Journal of Physical Anthropology* **15**:387–398.

Thompson, D.D. & Cowen, K.S. (1984) Age at death and bone biology of the Barrow mummies. *Arctic Anthropology* **21**:83–88.

Thompson, D.L. (1976) *The Artists of the Mummy Portraits.* Los Angeles: J. Paul Getty Museum.

Thompson, J., Bassett, D.A., Davies, D.G., Lewis, G.D. & Prince, D.R. (eds.) (1984) *Manual of Curatorship: A Guide to Museum Practice.* Boston: Butterworths..

Thompson, L.G., Moesly-Thompson, E., Bolzan, J.J. & Koci, B.R. (1985) A 1500-year record of tropical precipitation in ice cores from the Quelccaya ice cap, Peru. *Science* **229**:971–973.

Thornton, F. (1988) Extraordinary human teeth. *Journal of Paleopathology* **2**:173–176.

Thorp, R.L. (1991) Mountain tombs and jade burial suits: Preparations for eternity in the Western Han. In *Ancient Mortuary Traditions of China,* ed. G. Kuwayama, pp. 26–39. Los Angeles: Far Eastern Art Council, Los Angeles County Museum of Art.

Thorvildsen, K. (1947) *Moseliget rea Borrenose i National museets Arbejsmark,* pp. 57–67.

Thorvildsen, K. (1951) *Moseliget fra Tollund Aarboger for Nordisk Oldkyndighed og Historie* 1950, pp. 302–330.

Thorvildsen, E. (1952) Menneskeofringer i Oldtiden. *KUML* 1952, pp. 32–48.

Thuesen, I., Engberg, J. & Nielson, H. (1995) Ancient DNA from a very cold and a very hot place. In *Proceedings of the First World Congress on Mummy Studies*, February, 1992, Vol. 2, pp. 561–569. Santa Cruz, Tenerife, Canary Islands: Archaeological and Ethnographical Museum of Tenerife.

Tieszen, L.L., Matzner, S. & Buseman, S.K. (1995a) Dietary reconstruction based on stable isotopes (13C, 15N) of the Guanche, pre-Hispanic Tenerife, Canary Islands. In *Proceedings of the First World Congress on Mummy Studies*, February, 1992, Vol. 1, pp. 41–57. Santa Cruz, Tenerife, Canary Islands: Archaeological and Ethnographical Museum of Tenerife.

Tieszen, L.L., Iversen, E. & Matzner, S. (1995b) Dietary reconstruction based on carbon, nitrogen, and sulfur stable isotopes in the Atacama Desert, Northern Chile. In *Proceedings of the First World Congress on Mummy Studies*, February, 1992, Vol. 1, pp. 427–441. Santa Cruz, Tenerife, Canary Islands: Archaeological and Ethnographical Museum of Tenerife.

Tindale, N.B. & Mountford, C.P. (1936) Results of the excavations of Kongarati Cave, near Second Valley, South Australia. *Records of the South Australian Museum* **5**:487–502.

Tkocz, I., Bytzer, P. & Bierring, F. (1979) Preserved brains in medieval skulls. *American Journal of Physical Anthropology* **51**:197–202.

Tomita, K. (1984) On the production of hydroxy fatty acids and fatty acid oligomers in the course of adipocere formation. *Japan Journal of Legal Medicine* **38**:257–272.

Tonkelaar, D., Henkes, H.E. & Van Leersum, G.K. (1991) Herman Snellen (1834–1908) and Mueller's "Reform-Auge". A short history of the artificial eye. *Documenta Ophthalmologica* **77**:349–354.

Torodash, M. (1989) Great Seal of the United States. In *Academic American Encyclopedia*, Vol. 9, p. 322. Danbury, CT: Grolier, Inc.

Torres, C.M., Repke, D., Chan, K., Mckenna, D., Llagostera, A. & Schultes, R.E. (1991) Snuff powders from pre-Hispanic San Pedro de Atacama: Chemical and contextual analysis. *Current Anthropology* **32**:640–649.

Toynbee, J. (1849) Osseous tumours growing from the walls of the meatus externus and on the enlargement of the walls themselves with cases. *Provincial Medical and Surgical Journal* **14**:533–537.

Trigger, B.G., Kemp, B.J., O'Connor, D. & Lloyd, A.B. (1983) *Ancient Egypt: A Social History.* Cambridge: Cambridge University Press.

Trinkaus, E. (1983) *The Shanidar Neanderthals.* New York: Academic Press.

Trinkaus, E. (1985) Pathology and posture of the La Chapelle aux Saints Neandertal. *American Journal of Physical Anthropology* **67**:19–41.

Troyer, H. & Bablich, E. (1981) A hematoxylin and eosin-like stain for glycol-embedded tissue sections. *Stain Technology* **56**(1):39–43.

Tryon, D. (1996) The peopling of Oceania: The linguistic evidence. In *Arts of Vanuatu,* ed. J. Bonnemaison, C. Kaufmann, K. Huffman & D. Tryon, pp. 54–61. Honolulu: University of Hawaii Press.

Tschentscher, F. (1999) Too mammoth an undertaking. *Science* **286**:2084.

Turner, P. J. & Holtom, D.B. (1981) The use of a fabric softener in the reconstitution of mummified tissue prior to paraffin wax sectioning for light microscopical examination. *Stain Technology* **56**:35–38.

Turner, R.C. (1986) Discovery and excavation of the Lindow Bodies. In *Lindow Man: The Body in the Bog,* ed. I.M. Stead, J.B. Bourke & D. Brothwell, pp. 10–13. Ithaca, NY: Cornell University Press.

Turner, R.C. (1995a) Recent research into British bog bodies. In *Bog Bodies,* ed. R.C. Turner & R.G. Scaife, pp. 108–122. London: British Museum Press.

Turner, R.C. (1995b) Gazetteer of Bog Bodies in the British Isles. In *Bog Bodies,* ed. R.C. Turner & R.G. Scaife, pp. 205–220. London: British Museum Press.

Turner, R.C. (1995c) Discoveries and excavations at Lindow Moss 1983–8. In *Bog Bodies,* ed. R.C. Turner & R.G. Scaife, pp. 10–18. London: British Museum Press.

Turner, R.C. & Briggs, C.S. (1986) The bog burials of Britain and Ireland. In *Lindow Man: The Body in the Bog,* ed.

I.M. Stead, J.B. Bourke & D. Brothwell, pp. 144–161. Ithaca, NY: Cornell University Press.

Turner, R.C. & Scaife, R.G. (1995) *Bog Bodies: New Discoveries and New Perspectives.* London: British Museum Press.

Tuross, N. (1991) Recovery of bone and serum proteins from human skeletal tissue: IgG, osteonectin and albumin. In *Human Paleopathology: Current Syntheses and Future Options*, ed. D.J. Ortner & A.C. Aufderheide, pp. 51–54. Washington, DC: Smithsonian Institution Press.

Tuross, N. & Stathopolos, L. (1993) Ancient proteins in fossil bones. *Methods in Enzymology* **224**:121–129.

Tuross, N., Behrensmeyeer, A.K. & Eanes, E.D. (1989) Strontium increases and crystallinity changes in taphonomic and archaeological bone. *Journal of Archaeological Science* **16**:661–672.

Twain, M. (1903) *The Innocents Abroad.* New York: Harper & Brothers.

Tyson, R.A. (1995) Mummies at the San Diego Museum of Man: Considerations for the future. In *Proceedings of the First World Congress on Mummy Studies*, February, 1992, Vol. 1, pp. 221–223. Santa Cruz, Tenerife, Canary Islands: Archaeological and Ethnographical Museum of Tenerife.

Tyson, R.A. & Elerick, D.V. (eds.) (1985) *Two Mummies from Chihuahua, Mexico: A Multidisciplinary Study. San Diego Museum Papers* no. 19. San Diego: Museum of Man.

Ubaldi, M., Luciani, S., Marota, I., Fornaciari, G., Cano, R.J. & Rollo, F. (1998) Sequence analysis of bacterial DNA in the colon of an Andean mummy. *American Journal of Physical Anthropology* **107**:285–295.

Ubelaker, D. (1989) *Human Skeletal Remains*, 2nd edn. Washington, DC: Taraxacum.

Uhle, M. (1910) Ueber die Fruehkulturen in der Umgebung von Lima [Concerning the early cultures in the region of Lima] In *Proceedings of the XVIth International Congress of the Americas*, Vienna, 1908, Vol. 2, pp. 347–370. Vienna and Leipzig: Zweite Haelfte.

Uhle, M. (1914) The Nazca pottery of Ancient Peru. *Proceedings of the Davenport Academy of Sciences* **13**:1–16.

Uhle, M. (1917) Los aborigines de Arica. *Publicaciones del Museo de Etnología y Antropología* **1**:151–176.

US Patent Office (1856) Patent No. 15,972: Method of Preserving Dead Bodies. J.A. Gaussardia. Washington, DC: Government Printing Office.

Vahey, T. & Brown, D. (1984) Comely Wenuhotep: Computed tomography of an Egyptian mummy. *Journal of Computer Assisted Tomography* **8**:992–997.

Vail, L.G. (1936) Disposal of the dead among the Buang. *Oceania* **7**:63–68.

Valentin, N., Parra, E., Dolores-Gayo, M. & García, M. (1995) Métodos de analisis para evaluar el estado de conservación de momias de origen Guanche [Analytical methods to evaluate the conservation of Guanche mummies]. In *Proceedings of the First World Congress on Mummy Studies*, February, Vol. 1, 1992, pp. 225–230. Santa Cruz, Tenerife, Canary Islands: Archaeological and Ethnographical Museum of Tenerife.

Vallentyne, J.R. (1963) Geochemistry of carbohydrates. In *Organic Geochemistry*, ed. I.A. Breger, pp. 456–502. New York: Macmillan.

van der Sanden, W.A.B. (1995) Bog bodies on the continent: Developments since 1965, with special reference to the Netherlands. In *Bog Bodies*, ed. R.C. Turner & R.G. Scaife, pp. 146–167. London: British Museum Press.

van der Sanden, W.A.B. (1996) *Through Nature to Eternity: The Bog Bodies of Northwest Europe.* Amsterdam: Batavian Lion International.

van Gerven, D.P. (1981) Nubia's last Christians: The cemeteries of Kulunarti. *Archaeology* **34**(3):22–30.

van Gilse, P.H.G. (1938) Des observations ulterieures sur la génèse des oxostoses du donduit externe par l'iritation de l'eau froid. [Recent observations on the origin of external auditory canal exostoses by cold water]. *Otolaryngology* (Stockholm) **26**:343–352.

van Tilburg, J. (1994) *Easter Island.* Washington, DC: Smithsonian Institution Press.

Vandenberg, P. (1975) *The Curse of the Pharaohs.* New York: Barnes & Noble.

Vasilievsky, R.S. (1975) Problems of the origin of ancient sea hunters' cultures in the northern Pacific. In *Maritime Adaptations of the Pacific*, ed. R.W. Casteel & G.I. Quimby, pp. 67–75. Presented at Ninth International Congress of Anthropological and Ethnological Sciences, Chicago, IL. Mouton Publishers: The Hague, Paris.

Veniaminov, I. (1840) *Notes on the islands on the Unalaska Division.* St Petersburg: Great Russia–America Company

Verano, J.W. (1995) Where do they rest? The treatment of human offerings and trophies in Ancient Peru. In *Tombs for the Living*, ed. T. Dillehay, pp. 1–45. Washington, DC: Dumbarton Oaks.

Verano, J.W. (1997) Human skeletal remains from Tomb 1, Sipan (Lambayeque river valley, Peru) and their social implications. *Antiquity* **71**:670–682.

Verano, J.W. (1998) Paleopathological analysis of sacrificial victims at the pyramid of the moon, Moche River valley. *Chungará* **32**:61.

Verano, J. (2001) Mummies of the north coast of Peru. In *Studies on Ancient Mummies and Burial Archaeology*, ed. F. Cárdenas & C. Rodríguez-Martín, pp. 57–65. Bogotá, Fundación Erigaie, Instituto Canario de Bioantropología, Departamento de Antropología de Universidad de Los Andes.

Vescia, F. (1980) On Chinese mummies. *Paleopathology Newsletter* **31**:9–10.

Volenec, F.J., Clark, G.M., Mani, M. & Humphrey, L.J. (1979) Burn wound biopsy bacterial quantitation: A statistical analysis. *American Journal of Surgery* **138**:695–697.

von Daniken, E. (1968) *Chariots of the Gods?* New York: G.P. Putnam's Sons.

von Endt, D.W. & Ortner, D.J. (1984) Experimental effects of bone size and temperature on bone diagenesis. *Journal of Archaeological Science* **11**:247–253.

von Hagen, A. & Guillén, S. (1998) Tombs with a view. *Archaeology* **51**(2):48–54.

von Luschan, F. (1896a) Die Trepanirte von Tenerife (Trephined skulls from Tenerife). *Verhandlungen der Berliner Gesellschaft fuer Anthropologie* **28**:63–65.

von Luschan, F. (1896b) *Über eine Schädelsammlung von den Canarischen Inseln* [The cranial collection of the Canary Islands]. *In Die Insel Tenerife* [The Island Tenerife], ed. H. Meyer, pp. 285–319.

Vreeland, J.M. Jr (1978) Fungal spores from a Peruvian mummy ca. 800 A.D. *Paleopathology Newsletter* **21**:11–12.

Vreeland, J.M. Jr (1998) Mummies of Peru. In *Mummies, Disease and Ancient Cultures,* ed. A. Cockburn, E. Cockburn & T.A. Reyman, pp. 154–189, 2nd edn. Cambridge: Cambridge University Press.

Wade, R.S. (1998) Medical mummies.The history of the Burns collection. *The Anatomical Record* **253**(16): 158–151.

Wade, R.S. & Lyons, W. (1995) The restoration of anatomical and archaeological specimens using the S-10 plastination method: With special reference to preserving the good heart of a good priest. In *Book of Abstracts for the 2nd World Congress on Mummy Studies* in Cartagena, Colombia, February, p. 9.3. Bogotá, Colombia: Department of Anthropology, Universidad de Los Andes.

Wainwright, G. (1910) The mastaba of Nefermaat. In *Meydum and Memphis (III),* ed. W.M. Flinders-Petrie, pp. 18–22. London: School of Archaeology in Egypt, University College.

Wakely, J., Manchester, K. & Charlotte R. (1989) Scanning electronic microscope study of normal vertebrae and ribs from early medieval human skeletons. *Journal of Archaeological Sciences* **16**:627–642.

Waldie, P. (2000) Mummy from oddity museum could be Rameses I. *National Post* (Toronto), June 14, p. A3.

Waldron, T. (1995) Some mummies from Theban tombs 253, 254 & 294. In *Proceedings of the First World Congress on Mummy Studies*, February, 1992, Vol. 2, pp. 847–852. Santa Cruz, Tenerife, Canary Islands: Archaeological and Ethnographical Museum of Tenerife.

Walker, R. & Guillén, S. (1995) Methods of mummification: malnutrition and changes in Chiribaya Burial Patterns (poster). *Paleopathology Newsletter* **89** (Suppl.):28–29 (Abstract).

Walker, A., Zimmerman, M.R. & Leakey, R.E.F. (1982) A possible case of hypervitaminosis A in *Homo erectus*. *Nature* **296**:248–250.

Walker, R., Parsche, F., Bierbrier, M. & McKerrow, J.H. (1987) Tissue identification and histologic study of six lung specimens from Egyptian mummies. *American Journal of Physical Anthropology* **72**:43–48.

Wallen, L.L., Benedict, R.G. & Jackson, R.W. (1962) The microbiological production of 10–hydroxystearic acid from oleic acid. *Archives of Biochemistry and Biophysics* **99**:249–253.

Wallgren, J.E., Caple, R. & Aufderheide, A.C. (1986) Contributions of nuclear magnetic resonance studies to the question of alkaptonuria (ochronosis) in an Egyptian mummy. In *Science in Egyptology,* ed. R.A. David, pp. 321–327. Manchester: Manchester University Press.

Walsche, J.M. (1973) Tutankhamun: Klinefelter's or Wilson's? *Lancet* **1**:109–110.

Walter, P. (1929) Fossilized human brains found. *El Palacio* **27**(13–18):203–204.

Wang, B.H. (1996) Excavation and preliminary studies of the ancient mummies of Xinjiang in China. In *Human Mummies: A Global Survey of Their Status and the*

Techniques of Conservation, ed. K. Spindler, H. Wilfing, E. Rastbichler-Zissernig, D. zur Nedden & H. Nothdurfter, pp. 3–8. New York: Springer-Verlag.

Warrick, L.P. (1998) Personal communication (letter). Copy of letter from Lucille P. Warrick, granddaughter of the embalmer (John Johnson of Pawhuska, Oklahoma) to Theodore A. Reyman, in the file of the Paleobiology Laboratory, Department of Pathology, University of Minnesota, Duluth School of Medicine, Duluth, MN, 55812.

Webb, S. (1990) Prehistoric eye disease (trachoma?) in Australian aborigines. *American Journal of Physical Anthropology* **81**:91–100.

Wei, O. (1973) Internal organs of a 2100–year old female corpse. *Lancet* **2**:1198.

Weil, T. (1992) *The Cemetery Book*. New York: Hippocrene Books.

Weisgram, J., Splechtna, H., Hilgers, H., Walzl, M., Leitner, W. & Seidler, H. (1996) Remarks on the anatomy of a cat regarding the extent of preservation. In *Human Mummies. A Global Survey of Their Status and the Techniques of Conservation*, ed. K. Spindler, H. Wilfing, E. Rastbichler-Zissernig, D. zur Nedden & H. Nothdurfter, pp. 289–294. New York: Springer-Verlag.

Weisman, A. (1965) *The Weisman Collection of Pre-Columbian Medical Sculpture.* Arlington, VA: American Anthropological Society.

Weller, M. (1972) Tutankhamun: An adrenal tumor? *Lancet* **2**:1312.

Wells, C. (1963) Ancient Egyptian pathology. *Journal of Laryngology and Otology* **77**:261–265.

Wells, P.S. (1980) Iron Age Central Europe. *Archaeology* **33**(5):7–11.

Wenkan, X. (1995) The discovery of the Xinjiang mummies and studies of the origin of the Tocharians. *Journal of Indo-European Studies* **23**:357–369.

Wente, E.F. (1974) History of Egypt, Dynasty 18 to 330 B.C. In *The New Encyclopedia Britannica*, 15th edn, ed. P. Goetz (editor-in-chief), pp. 471–481. Chicago: Helen Hemingway Benton.

Weser, U., Miesel, R., Hartmann, H.J. & Heizmann, W. (1989) Mummifed enzymes. *Nature* **341**:696.

Weser, U., Etspueler, H. & Kaup, Y. (1995) Enzymatic and immunological activity of 4000 years aged bone alkaline phosphatase. *FEBS Letters* **375**:280–282.

West, A. (1984) The Torres Strait mummy from HMS Rattlesnake, 1849. In *Merseyside County Museum, Liverpool Museum Ethnographic Group Newsletter* (University of Hull Center for South East Asian Studies) **15**:12–26.

Wetherill, C.M. (1860) On adipocere and its formation. *Transactions of the Philosophical Society II* (n.s.):1–7.

Wheeler, J.C., Russell, A.J.F. & Redden, H. (1998) Preconquest llama and alpaca breeding as reconstructed from the El Yaral mummies. Manuscript read at the 3rd World Congress on Mummy Studies held in Arica, Chile, May.

White, C.D. (1993) Isotopic determination of seasonality in diet and death from Nubian mummy hair. *Journal of Archaeological Science* **20**:657–666.

White, J.W. Jr (1963) The identification of inhibine, the antibacterial factor in honey, as hydrogen peroxide and its origin in a honey glucose–oxidase system. *Biochimica et Biophysica Acta* **73**:57–70.

Wickersheimer, J. (1879) Patent no. 220, 103. Washington, DC: US Patent Office, US Government Printing Office.

Wigand, P.E. (1997) Native American diet and environmental contexts of the holocene revealed in the pollen of human fecal material. *Nevada* **40**:105–116.

Wilder, H.H. (1904) The restoration of dried tissues, with especial reference to human remains. *American Anthropologist* **6**:1–17.

Willard, H.H., Merritt, L.L., Dean, J.A. & Settle, F.A. Jr (1981) *Instrumental Methods of Analysis*, 6th edn. Belmont, CA: Wadsworth Publishing Co.

Williams, H.U. (1927) Gross and microscopic anatomy of two Peruvian mummies. *Archives Pathology and Laboratory Medicine* **4**:26–33.

Williams, H.U. (1929a) The examination of the bodies of mummies by laboratory methods. *Journal of Technical Methods and the Bulletin of the International Association of Medical Museums* **12**:25–28.

Williams, H.U. (1929b) Human paleopathology. *Archives of Pathology* **7**:890–895.

Williams, M. (1983) The voice in the text: Poe's "Some Words With A Mummy". *Poe Studies* (Washington State University) **16**(1):1–4.

Williams, S.A., Buidstra, J.E., Clark, N.R., Lozada-Cerna, M.C. & Pino, E.T. (1989) Mortuary site excavations and skeletal biology in the Osmore Project. In *BAR International,* Series 545, ed. D.S. Rice, C. Stanish & P.R. Scarr. Ecology, Settlement and History in the

Osmore Drainage, Peru, Vol. 2, pp. 329–346. Oxford: Oxford University Press.

Willoughby, C.C. (1907) The Virginia Indians in the seventeenth century. *American Anthropologist* **9**:62, 71, 72.

Wilson, A.S., Dixon, R.A., Dodson, H.I., Janaway, R.C., Pollard, C., Pollard, M.A. *et al.* (2001) Yesterday's hair: Human hair in archaeology. *Biologist* **48**:213–218.

Wilson, D. (1865) *Researches into the Origins of Civilisation in the Old and the New World*. London: Macmillan & Company.

Wilson, J.A. (1964) *Signs and Wonders Upon Pharaoh*. Chicago: Chicago University Press.

Wilson, J.D. (1991) Endocrine disorders of the breast. In *Harrison's Principles of Internal Medicine*, 12th edn, ed. J. Wilson, E. Braunwald, K.J. Isselbacher, R.G. Petersdorf, J.B. Martin, A.S. Fauci & R.K. Root, pp. 1795–1798. New York: McGraw-Hill.

Wilson, R.R. (1972) Artificial eyes in ancient Egypt. *Survey of Ophthalmology* **16**:322–331.

Winlock, H.E. (1922) The Egyptian Expedition 1921–1922. Part II. *Bulletin of the Metropolitan Museum of Art* **17**:32–43.

Winlock, H.E. (1945) *The Slain Soldiers of Neb-Hepet-Re Mentu-Hotpe*. New York: Metropolitan Museum of Art.

Winterstein, E. (1895) Zur Kenntnissder in den Membranen einiger Cryptogrammen enthalten Bestandtheile [Recognition of components in the well walls of Cryptogrammae]. *Zeitschrift fuer Physiologischer Chemie* **21**:152–154.

Wise, K. (1991) Complexity and variation in mortuary practices during the preceramic period in the south-central Andes. Manuscript presented at the American Anthropological Association held in Chicago, Illinois, 1965.

Wong, P.A. (1981) Computed tomography in paleopathology: Technique and case study. *American Journal of Physical Anthropology* **55**:101–110.

Woo, D. (1995) Apoptosis and loss of renal tissue in polycystic kidney diseases. *New England Journal of Medicine* **333**:18–25.

Worcester, D.C. (1906) The Non-Christian Tribes of northern Luzon. *Philippine Journal of Science* **1**:791–863.

Workman, B.K. (1964) Deification of Septimus Severus by Herodian. In *They Saw It Happen In Classical Times*, ed. B.K. Workman, pp. 191–193. New York: Barnes & Noble.

Wright, I.M. & Kerfoot, O. (1966) The African baobab – object of awe. *Natural History* **75**:50–53.

Yamada, T.K., Kudou, T. & Takahashi-Iwanaga, H. (1990) Some 320-year-old soft tissue preserved by the presence of mercury. *Journal of Archaeological Science* **17**:383–392.

Yamada, T.K., Kudou, T., Takahashi-Iwanaga, H., Ozawa, T., Uchihi, R. & Katsumata, Y. (1996) Collagen in 300 year-old tissue and a short introduction to the mummies in Japan. In *Human Mummies: A Global Survey of Their Status and the Techniques of Conservation*, ed. K. Spindler, H. Wilfing, E. Rastbichler-Zissernig, D. zur Nedden & H. Nothdurfter, pp. 3–8. New York: Springer-Verlag.

Yarrow, H.C. (1881) Study of the mortuary customs of the North American Indians. *First Annual Report of the Bureau of American Ethnology*, pp. 130–137. Washington, DC: Smithsonian Institution Press.

Yesko, S.A. (1940) History of human dissections. *Medical Record* **51**:238–241.

Zaki, A. & Iskander, Z. (1942) Materials and method used for mummifying the body of Amentefnekht, Saqqara 1941. *Annales du Service des Antiquités de l'Égypte* **41**:223–250.

Zhiyi, C. & Yongqing, X. (eds.) (1982) *An Investigation of the Loulan Mummy*. Shanghai: Natural History Museum of Shanghai.

Zimmer, C. (1999) New date for the dawn of dream time. *Science* **284**:1243–1246.

Zimmerman, M.R. (1972) Histological examination of experimentally mummified tissues. *American Journal of Physical Anthropology* **37**:271–280.

Zimmerman, M. R. (1977) The mummies of the tomb of Nebwenenef: paleopathology and archaeology. *Journal of the American Research Center in Egypt* **14**:33–36.

Zimmerman, M.R. (1978) The mummified heart: A problem in medicolegal diagnosis. *Journal of Forensic Sciences* **23**:750–753.

Zimmerman, M.R. (1979) Pulmonary and osseous tuberculosis in an Egyptian mummy. *Bulletin of the New York Academy of Medicine* **55**:604–608.

Zimmerman, M.R. (1980) The paleopathology of the human remains of Nebwenenef's tomb: The 1979 season. *Paleopathology Newsletter* **31**:15–16.

Zimmerman, M.R. (1981) A possible histiocytoma in an Egyptian mummy. *Archives of Dermatology* **117**:364–365.

Zimmerman, M.R. (1985) Paleopathology in Alaskan mummies. *American Scientist* **73**:20–25.

Zimmerman, M. R. (1986a) Case no. 22: Anthracosis and fibrosis. (Department of Pathology, Medical College of Virginia, Richmond, VA.) *Paleopathology Club Newsletter* **26**:1.

Zimmerman, M.R. (1986b) Mummification styles as a guide to the dating of Egyptian mummies. In *Dating and Age Determination of Biological Materials*, ed. M.R. Zimmerman & J.L. Angel, pp. 166–169. London: Croom Helm.

Zimmerman, M.R. (1990) The paleopathology of the liver. *Annals of Clinical and Laboratory Science* **20**:301–306.

Zimmerman, M.R. (1993) The paleopathology of the cardiovascular system. *Texas Heart Institute Journal* **20**:252–257.

Zimmerman, M.R. (1995) Paleohistology of ancient soft tissue tumors. Paper presented at the 22nd Annual Meeting of the Paleopathology Association held in Oakland, CA on 28–29 March, 1995. *Paleopatholology Newsletter* **90** (June) (Suppl.):14–15 (Abstract).

Zimmerman, M.R. (2001) Rectal carcinoma in Dakhleh Egyptian mummy. Manuscript in the files of Paleobiology Laboratory in Department of Pathology at University of Minneosta Duluth, School of Medicine, Duluth, MN, 55812.

Zimmerman, M.R. & Aufderheide, A.C. (1984) The frozen family of Utqiagvik: The autopsy findings. *Arctic Anthropology* **21**:53–64.

Zimmerman, M.R. & Clark, W.H. (1976) A possible case of subcorneal pustular dermatosis in an Egyptian mummy. *Archives of Dermatology* **117**:364–365.

Zimmerman, M.R. & Kelley, M.A. (1982) *Atlas of Human Paleopathology*. New York: Praeger.

Zimmerman, M.R. & Smith, G.S. (1975) A probable case of accidental inhumation of 1600 years ago. *Bulletin of the New York Academy of Medicine* **51**:828–837.

Zimmerman, M.R. & Tedford R.H. (1976) Histologic structures preserved for 21,300 years. *Science* **194**:183–184.

Zimmerman, M.R., Yeatman, G.W., Sprinz, H. & Titterington, W.P. (1971) Examination of an Aleutian mummy. *Bulletin of the New York Academy of Medicine* **47**:80–103.

Zimmerman, M.R., Trinkaus, E., LeMay, M., Aufderheide, A.C., Reyman, T.A., Marrocco, G.R. *et al.* (1981a) The paleopathology of an Aleutian mummy. *Archives of Pathology and Laboratory Medicine* **105**:638–641.

Zimmerman, M.R., Trinkaus, E., LeMay, M., Aufderheide, A.C., Reyman, T.A., Marrocco, G.R. *et al.* (1981b) Trauma and Trephination in a Peruvian Mummy. *American Journal of Anthropology* **55**:497–501.

Zimmerman, M.R., Saul, F.P., Celenko, T. Jr & McLaughlin, G. (1993) Curios, African art, and paleopathology. *Paleopathology Newsletter* **84**:9–12.

Zlonis, M. (1995) Studies in Human Taphonomy. Visceral organ preservation in spontaneously mummified human bodies buried in the Atacama Desert. Unpublished manuscript in the files of the Paleobiology Laboratory of the Department of Pathology at the University of Minnesota, Duluth School of Medicine, Duluth, MN, 55812.

Zugibe, F.T. & Costello, J.T. (1985) A new method for softening mummified fingers. *Journal of Forensic Sciences* **31**:726–731.

Zuki, A. & Iskander, Z. (1943) Materials and method used for mummification of the body of Amentefnekht, Saqqara. *Annales du Service des Antiquités de l'Égypte* **42**:223–255.

Zumwalt, R.E. & Fierro, M.F. (1990) Postmortem changes. In *Handbook of Forensic Pathology*, ed. R.C. Froede, pp. 78-84. Northfield, IL: College of American Pathologists.

Index

Bold number refer to figures

Index